CONVERSION FACTORS

P9-DVG-195

MASS
1 kg = 2.204623 lb mass = 0.06852 slug; 1 t (metric ton) = 10^3 kg
1 lb = 0.45359237 kg = 0.03108 slug; 1 slug = 32.17 lb = 14.59 kg

LENGTH
1 m = 39.37 in. = 3.281 ft = 6.214 × 10^{-4} mi = 10^{10} Å = 10^{15} fermis
1 in. = 0.02540000 m; 1 ft = 0.3048 m; 1 mi = 1609 m
1 nautical mile = 1852 m = 1.1508 mi = 6076.10 ft
1 Å = 10^{-10} m; 1 mil = 10^{-3} in.; 1 rod = 16.5 ft; 1 fathom = 1.8288 m

AREA
1 m^2 = 10.76 ft^2 = 1550 in.^{2}; 1 hectare = 10^4 m^2 = 2.471 acres
1 ft^2 = 929 cm^2; 1 in.^{2} = 6.452 cm^2 = 1.273 × 10^6 circular mils; 1 acre = 4047 m^2

VOLUME
1 m^3 = 35.31 ft^3 = 6.102 × 10^4 in^3; 1 liter (L) = 1 × 10^{-3} m^3 = 61.02 in^3
1 ft^3 = 0.02832 m^3; 1 U.S. gallon = 3.7854 L; 1 British gallon = 4.546 L;
1 fluid ounce = 0.0284131 L

TIME AND FREQUENCY
1 year = 365.25 days = 8.766 × 10^3 h = 5.259 × 10^5 min = 3.156 × 10^7 s
1 sidereal day (period of earth's revolution) = 86,164 s
1 Hz = 1 cycle/s

SPEED
1 m/s = 3.281 ft/s = 3.6 km/h = 2.237 mi/h = 1.944 knots
1 km/h = 0.2778 m/s = 0.9113 ft/s = 0.6214 mi/h
1 mi/h = 1.467 ft/s = 1.609 km/h = 0.8689 knot

FORCE
1 N = 10^5 dyn = 0.1020 kg wt = 0.2248 lbf
1 lb = 4.448 N = 0.4536 kg wt = 32.17 poundals

PRESSURE
1 N/m^2 = 9.869 × 10^{-6} atm = 1.450 × 10^{-4} lb/in.^{2} = 0.02089 lb/ft^2 = 7.501 × 10^{-3} mm Hg = 10^{-5} bar = 1 Pascal
1 lb/in.^{2} = 144 lb/ft^2 = 6895 N/m^2 = 51.71 mm Hg = 27.68 in. H_2O
1 atm = 1.01325 × 10^5 N/m^2 = 760 mm Hg = 760 torr — 2116 lb/ft^2 = 14.70 lb/in.^{2}

DENSITY
1 g/cm^3 = 1000 kg/m^3 = 62.43 lb/ft^3 = 1.940 slug/ft^3
1 lb/ft^3 = 0.03108 slug/ft^3 = 16.02 kg/m^3 = 0.01602 g/cm^3

WORK, ENERGY, HEAT
1 J = 0.2389 cal = 9.481 × 10^{-4} Btu = 0.7376 ft-lb = 10^7 ergs = 6.242 × 10^{18} eV
1 kcal = 4186 J = 3.968 Btu = 3087 ft-lb
1 eV = 1.602 × 10^{-19} J; 1 unified amu = 931.50 MeV
1 kWh = 3.6 × 10^6 J = 3413 Btu = 860.1 kcal = 1.341 hp-h

POWER
1 hp = 2.545 Btu/h = 550 ft-lb/s = 745.7 W = 0.1782 kcal/s
1 W = 2.389 × 10^{-4} kcal/s = 1.341 × 10^{-3} hp = 0.7376 ft-lb/s

ELECTRIC CHARGE
1 C = 0.1 abcoulomb = 2.998 × 10^9 statcoulombs (esu)
1 faraday = 96,487 C; 1 electronic charge = 1.602 × 10^{-19} C

ELECTRIC CURRENT
1 A = 0.1 abampere (emu) = 2.998 × 10^9 statamperes (esu)

ELECTRIC POTENTIAL DIFFERENCE
1 abvolt (emu) = 10^{-8} V; 1 statvolt (esu) = 299.8 V

CAPACITANCE
1 statfarad (esu) = 1 cm of capacitance = 1.1127 × 10^{-12} F

MAGNETIC FLUX
1 Wb = 10^8 maxwells = 10^5 kilolines

MAGNETIC INTENSITY *B*
1 T = 1 N/A-m = 1 W/m^2 = 10,000 gauss (G) = 10^9 gamma

MAGNETIZING FORCE *H*
1 A(turn)/m = 0.01257 oersted = 0.01257 gilbert/cm

ELEMENTS OF
PHYSICS

ELEMENTS OF
PHYSICS

NINTH EDITION

Alpheus W. Smith
Late Professor of Physics, Ohio State University

John N. Cooper
Professor of Physics, U.S. Naval Postgraduate School

McGRAW-HILL BOOK COMPANY

New York   St. Louis   San Francisco   Auckland   Bogotá   Düsseldorf
Johannesburg   London   Madrid   Mexico   Montreal   New Delhi
Panama   Paris   São Paulo   Singapore   Sydney   Tokyo   Toronto

Library of Congress Cataloging in Publication Data

Smith, Alpheus Wilson, dates
 Elements of physics.

 Includes index.
 1. Physics. I. Cooper, John N., joint author.
II. Title.
QC23.S65 1979   530   78-17919
ISBN 0-07-058634-9

ELEMENTS OF PHYSICS

Copyright © 1979, 1972, 1964, 1957 by McGraw-Hill, Inc. All rights reserved.
Copyright 1948, 1938, 1932, 1927, 1923 by McGraw-Hill, Inc. All rights reserved.
Copyright renewed 1951, 1955, 1960, 1966 by Alpheus W. Smith. Copyright renewed
1976 by Theodore B. Smith and Robert B. Smith. Printed in the United States of
America. No part of this publication may be reproduced, stored in a retrieval system,
or transmitted, in any form or by any means, electronic, mechanical, photocopying,
recording, or otherwise, without the prior written permission of the publisher.

   567890   DODO   832

This book was set in Optima with Times Roman Italic by York Graphic Services, Inc.
The editors were C. Robert Zappa and Laura D. Warner;
the designer was Elliot Epstein; the production supervisor was Dennis J. Conroy.
New drawings were done by York Graphic Services, Inc.
R. R. Donnelley & Sons Company was printer and binder.

PREFACE

Every student who enters a college or university knows a substantial amount of physics and has a broad familiarity with many physical phenomena. Some knowledge of physics is a necessity for every person who leads a normal existence. For many the information is purely qualitative and often haphazard as well. In a physics text this knowledge is organized and made quantitative. Relationships between physical observables are stated in the form of mathematical equations. Since this particular book is written for students who have not studied calculus, the equations are algebraic. The treatment of physical laws and their applications has been handled as quantitatively and rigorously as the mathematical sophistication of the prospective student permits. While there are certain situations which can be treated better by use of calculus, there are others in which careful physical reasoning can give a more penetrating understanding of a resulting equation than does the routine application of calculus. A few formulas involve definitions from trigonometry, but there is no need for a prior course because all the required relationships are developed in the text.

The role of physics in modern life continues to expand. Since the publication of the eighth edition of *Elements of Physics* in 1972, hand-held calculators have become more numerous than slide rules in the classroom, computers have proliferated for process control and the handling of information, and the uses of laser beams have increased steadily. Innumerable other applications of physics have been made in communications, in transportation, and in industry. A knowledge of the principles of physics is increasingly imperative if one wishes to keep abreast of what is happening in the broad technological segment of human activity. It is the purpose of this book to introduce the student to the fundamental laws of physics and to give some feeling of their power and beauty. Toward this end, a wealth of applications and examples is introduced, with particular attention to the physics of everyday experience. It is our belief that an understanding of physical phenomena adds a new dimension to the beauty of the rainbow and an enriched appreciation of such scientific advances as those which permit us to view in full color events taking place thousands of kilometers away.

The classic organization of physics into parts on mechanics, heat, sound, light, electricity and magnetism, and modern physics has been retained in this edition, but considerable rearrangement has taken place within the major divisions. In particular, the statics of a rigid body has been placed after kinematics and dynamics; teachers who prefer to discuss statics early can schedule Chapter 10 immediately after Chapter 3 and thus obtain the topic order of previous editions. There is ample material in this text for three full semesters of study. The classical physics sections alone require a full academic year for thorough coverage. For a two-semester course which includes considerable modern physics, it is possible to omit many selected chapters without seriously neglecting the background knowledge required for most subsequent chapters. Among the topics which are sometimes omitted are rotational motion, fluids, elastic properties of matter, sound, atmospheric physics, spectra and color, chemical and thermal electromotive forces, generators and motors, and alternating currents.

The Système International d'Unités (SI) has primacy everywhere in the book, but both English and cgs units are introduced because they are still in wide use. Throughout its 55 years of existence *Elements of Physics* has emphasized the physics of everyday life, and that dictates the need to relate the units in the newspaper and older scientific literature to the new international units. Of course, our practical units in electricity are a part of the international system toward which the entire world is moving.

In 1978 the United States is the only large country which is not firmly committed to the international units for general use. By 1973, 136 countries had adopted the SI and another 33 were going metric, including Canada, the United Kingdom, Australia, and New Zealand. The United States became officially metric in 1893, when the meter and the kilogram were declared the nation's "fundamental standards" of length and mass. Ever since then the yard, the pound, and other customary units have been defined as fractions of the metric standards, but these old units continued to dominate in trade and commerce. In 1968 Congress passed an act "to authorize the Secretary of Commerce to make a study to determine advantages and disadvantages of increased use of the metric systems in the United States." In 1971 the U.S. Metric Study concluded that the country should change to the metric system through a coordinated national program. Metric units are widely used in the physical sciences, pharmacy, and medicine, but in many other areas progress toward metrication has been extremely slow.

Of the many people who have been helpful in offering constructive criticism of previous editions and suggestions for improving the present one, the author can acknowledge only a few by name, but he extends his thanks to all. He is indebted to Alan F. A. Harper, Executive Member of Australia's Metric Conversion Board, for sharing his expertise regarding the best use of the SI system. Over several years Professor G. N. Koehl, of George Washington University, has done much to improve the manuscript. Colleagues at the Naval Postgraduate School have found many errors and proposed numerous improvements; among those to whom the author is grateful are Professors E. C. Crittenden, Jr., A. R. Frey, H. E. Handler, D. E. Harrison, S. H. Kalmbach, R. L. Kelly, J. R. Neighbours, L. O. Olsen, J. D. Riggin, J. V. Sanders, and W. B. Zeleny. Finally, special thanks are due to the author's wife Elaine, who typed and helped prepare the manuscript; to Elizabeth P. Richardson, who edited the entire manuscript and made many improvements; and to Laura Warner, who supervised and coordinated the editorial process, including figure production.

John N. Cooper

ELEMENTS OF
PHYSICS

ONE

MECHANICS

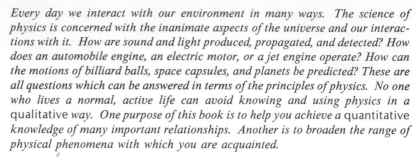

1 INTRODUCTION TO PHYSICS

Every day we interact with our environment in many ways. The science of physics is concerned with the inanimate aspects of the universe and our interactions with it. How are sound and light produced, propagated, and detected? How does an automobile engine, an electric motor, or a jet engine operate? How can the motions of billiard balls, space capsules, and planets be predicted? These are all questions which can be answered in terms of the principles of physics. No one who lives a normal, active life can avoid knowing and using physics in a qualitative *way. One purpose of this book is to help you achieve a* quantitative *knowledge of many important relationships. Another is to broaden the range of physical phenomena with which you are acquainted.*

Probably no aspect of physics is more familiar than the motions of such objects as balls or automobiles, and we therefore start our study with mechanics, the branch of physics dealing with the motions of bodies and the influence of forces on these motions. To obtain precise relationships, we require quantitative measurements in terms of carefully established standards. In this first chapter we define the fundamental unit of length (the meter*), of time (the* second*), and of mass (the* kilogram*). We can express any quantity we meet in mechanics in terms of these* three basic units.*

1.1 WHAT IS PHYSICS?

The word *physics* is derived from the Greek meaning *science of nature*. In a broad sense physics is the branch of knowledge which describes and explains the material world and its phenomena. In terms of this sweeping definition, other physical sciences, such as chemistry, geology, astronomy, and the engineering disciplines, are branches of the basic science *physics*. For centuries the body of knowledge we now call physics was known as *natural philosophy*.

At present, it is customary to use *physics* in a more restricted sense. Aspects of nature which are ordinarily regarded as clearly in the domain of physics are mechanics, sound, heat, electricity and magnetism, atomic and nuclear structure, and the properties of solids, liquids, gases, and plasmas. Broad areas of knowledge in which physics overlaps related sciences are described by names such as astrophysics, biophysics, chemical physics, and geophysics. Of course, these fields are not distinct; they merge into one another. Sharp distinctions between the sciences are neither necessary nor desirable.

Throughout the ages, intelligent people have endeavored to explain what went on in the world about them. There has been a never ending struggle to formulate a systematic set of concepts about the world we live in. Physics is as old as human curiosity about the environment and the wealth of processes it manifests. Early philosophers found many regularities and a great orderliness in the phenomena which impinge on our senses, suggesting that these phenomena are subject to quantitative laws. The philosophers sought to discover these laws of nature and to organize them into a logically elegant structure.

When we speak of a scientific law, we mean a *general* statement expressing some specific connection between observable quantities. For example, *Boyle's law* states how the pressure and the volume of a gas are related when the temperature is held constant; *Newton's law of universal gravitation* gives the force between two particles in terms of their masses and the distance between them. Usually a law can be written in the form of a mathematical equation. We believe that some laws are exact and are never violated; others are obeyed over a well-defined range but are not valid outside this range. As an example, *Hooke's law* states that the extension of a spring is directly proportional to the stretching force; however, if we continue to increase the force, eventually Hooke's law is no longer applicable.

Everyone is familiar with at least some of the regularities of nature. A small child soon learns that released objects fall to the floor; a more mature observer may be impressed by the regularity of the movement of the sun, moon, and stars. To find the relationship between these apparently unrelated processes required the genius of Newton. Yet many an important contribution to physics has been made by men of modest intellectual achievements. In the words of Aristotle:

> The search for Truth is in one way hard and in another easy.
> For it is evident that no one can master it fully nor miss it wholly.
> But each adds a little to our knowledge of Nature, and from all
> the facts assembled there arises a certain grandeur.

In the more than 2,000 years since Aristotle spoke, scientists have been adding steadily to our knowledge of nature, and today we experience the thrill of understanding many phenomena which were completely beyond the ken of the greatest scientists of earlier times. Albert Einstein, a great physicist of the early twentieth century, once said:

> The most incomprehensible fact of Nature is the fact that Nature
> is comprehensible.

To be sure, our knowledge is never complete, and there are always questions to which we seek answers. But many of nature's laws can be examined and understood by anyone who makes an effort to become familiar with them. The rewards are well worth the effort.

1.2 THE EVOLUTION OF PHYSICS

Physics today is a rapidly expanding science with many branches in widely varying stages of development. To appreciate the current state of the science fully it may be helpful to learn something of the historical background of physical science.

The writings of early scientists, such as Thales (640–547 B.C.) and Pythagoras (sixth century B.C.), are lost, but some of their contributions have been reported by later philosophers, of whom Aristotle (384–322 B.C.) is a prominent example. Typically the early philosophers observed naturally occurring phenomena and advanced *hypotheses* in terms of which a variety of observations could be understood. By a *hypothesis* we mean a proposal (often a pure guess) about how diverse observations in a given area can be logically related. Needless to say, many hypotheses turned out to be unacceptable. For example, about 450 B.C. Empedocles proposed the hypothesis that all material things were made from four "elements," earth, water, air, and fire. On the other hand, before 250 B.C. Aristarchus of Samos wrote that the earth revolves around the sun; this hypothesis, rejected for centuries, eventually won acceptance.

The emphasis of physical scientists before the seventeenth century was largely on qualitative aspects of physics: *How do things behave* in a given environment and *why* do they respond as they do? To be sure, early philosophers did know a great deal about the motions of the stars and the planets. Before 500 B.C. Pythagoras studied stringed musical instruments and found how the pitch depended on the length of the vibrating string (Chap. 21). About 250 B.C. Eratosthenes made a good measurement of the circumference of the earth, which he took to be a sphere (see Prob. 32). In spite of these and many other important quantitative contributions, the usual concerns of early philosophers were with describing and explaining phenomena rather than with performing experiments and making measurements. In their environment answers to quantitative questions were less important than they are in ours. Another reason is that they had few good measuring instruments. To measure time they had sundials, sand glasses, and water clocks, none of which were very good for measuring short time intervals. Nevertheless, they managed to understand many phenomena, and their contributions to modern knowledge are all too often forgotten.

As the understanding of natural phenomena expanded and the number of scientists grew, new facets of physical science became evident. Several aspects of physics, which were relatively neglected initially, came to play important roles in the evolution of modern science. These aspects are intimately related, but, interwoven as they are, it is convenient to introduce them separately.

The Experimental Method In much of early physics the scientist observed natural happenings and endeavored to explain his observations. Typically he did not cause the happenings and he did not control them. As one example, the Greek philosophers wrote about stones falling to the earth, but it was not until the time of Galileo (1564–1642) that this topic became the subject of well-controlled experiments. Galileo rolled spheres down inclined planes and made actual measurements on his system. He varied the angle of the inclined plane and studied how that affected the motion. In the laboratory environment he could repeat an experiment or refine it. Now there are laboratories all around the earth where carefully controlled experiments are carried out.

As the experimental approach grew, there was a great stimulus for new and better measuring instruments. They in turn made it possible to make better measurements and to gain further knowledge.

The Quantitative Aspect As better measuring instruments become available and better laboratory techniques develop, it becomes possible to accumulate many data. In order to formulate valid relationships between physical quantities, accurate measurements must be made. Thus, physics becomes a quantitative science. As Lord Kelvin said in 1883:

> I often say that when you can measure what you are speaking about, and express it in numbers, you know something about it; but when you cannot measure it, when you cannot express it in numbers, your knowledge is of a meager and unsatisfactory kind. It may be the beginning of knowledge, but you have scarcely, in your thoughts, advanced to the stage of science, whatever the matter may be.

The Role of Mathematics Paralleling the accelerated growth of physical knowledge which became evident in the seventeenth century was a corresponding growth in mathematical knowledge. Often the same man contributed to both; one of the most famous was Isaac Newton, but there were many others. Physics as a quantitative science rapidly became and has remained a mathematical science. The laws of physics are typically expressed in the form of mathematical expressions; as such, they are the same for physicists all over the world. Thus mathematics provides a universal language for physics and a means of rigorous, logical reasoning which is difficult to achieve in any other way.

With a growing literature of precise, reproducible measurements and new mathematical tools for manipulating ideas, it became possible to examine hypotheses more critically. If one assumed the truth of some hypothesis, one could in some cases calculate what the result of some particular experiment should be and compare this prediction with actual observations. Indeed, the treatment might suggest that some heretofore unobserved result would occur under some prescribed set of conditions. Under these circumstances the original hypothesis has become a *theory*. In general, a theory is much more than a hypothesis in that it involves not only a presumption about how nature behaves but also a mathematical formulation which gives correct quantitative values in agreement with previous experiments and correct predictions of what will be observed if some completely new experiment is performed.

An example of the transition from hypothesis to theory may help clarify the ideas. Several Greek philosophers, including Democritus, assumed that gases (and indeed all matter) consist of small particles separated by void and that these particles are in constant motion. This kinetic hypothesis was rejected by Aristotle and others. However, Daniel Bernoulli in 1738 calculated the pressure which would be produced by huge numbers of gas "molecules" colliding with the walls of the container. Other physicists made improvements in this model, and *kinetic theory* emerged. This theory correctly predicts the velocity distribution of the molecules of any gas as a function of the temperature as well as the thermal conductivity and the viscosity of the gas. Thus the *kinetic hypothesis*, an intuitive guess about the nature of a gas, has evolved into the *kinetic theory*, a powerful tool for quantitatively understanding many kinds of phenomena exhibited by gases.

Precise Language In many areas, e.g., politics, whenever people exchange ideas, it soon becomes evident that the same words are being used by different people to express different ideas. This familiar semantic problem also arose in the earlier days of physics. A classic example lies in a sometimes bitter argument in the late seventeenth and early eighteenth centuries between the followers of Descartes and those of Leibnitz as to whether "the quantity of motion" of a body was proportional to the velocity of the body or to the square of the velocity. The lengthy disagreement evaporated into nothingness when, in 1743, d'Alembert showed that to Descartes and his partisans "quantity of motion" meant what we call "momentum" (which is indeed proportional to velocity) while to Leibnitz and his protagonists "quantity of motion" referred to what we call "kinetic energy" (which is proportional to the square of velocity). The term "quantity of motion" is no longer used as a technical term in physics, and all physicists agree on what is meant by the terms *momentum* and *kinetic energy* of a body. In modern science it is the practice to define quantities in a completely unambiguous way so that everyone using the term in a technical context means the same thing by it. Actually, in physics we use some common words like *force, power,* and *work* which have a variety of meanings, but when they are used in physics, they have a definite and precise meaning.

In many situations it is possible to express physical relationships in terms of mathematical formulas, which in turn can be manipulated in accord with established mathematical procedures to yield new mathematical relationships and to obtain new physical insights. When physical relationships can be expressed in the language of mathematics, great advantages accrue. Mathematical language is an impersonal, unemotional, objective, and universal tool, which is richly rewarding to those who put forth the effort to master it.

The Creative Aspect of Physics As we have seen, early philosophers were concerned chiefly with observing natural occurrences and explaining them. In the seventeenth century physicists became able to make careful measurements in the laboratory on phenomena which they controlled. The primary objective was to satisfy the urge to understand nature. Later scientists became interested in learning how to apply the laws of physics to achieve certain desired ends. For example in the eighteenth century scientists made great progress in understanding heat. This knowledge led to the development of heat engines, and their use opened the way for steamships, railroads, automobiles, and aircraft in our transportation system and for tractors on our farms. In the nineteenth century physicists investigated and began to understand electrical phenomena. Creative synthesis of this knowledge brought about the widespread generation and distribution of electric energy for light, heat, electric motors, and radio and television. Investigation into the discharge of electricity through gases led to the discovery of x-rays, which not only provided a vital tool for the study of crystals and atomic structure but also paved the way for revolutionary advances in the diagnosis and treatment of diseases. In the twentieth century research in physics has expanded to unprecedented levels and has given us insight into the properties of solids and into the structure of atoms and nuclei. This knowledge in turn has been utilized to develop solid-state circuits, computers, and nuclear power sources.

Progress in physics brings with it progress in commerce and industry, in medicine, and in the related sciences. Much of the difference between the way we now live in the United States and the way the Indians lived 500 years ago is associated with our greater knowledge of physical phenomena and how they can be controlled.

1.3 THE UNIT OF TIME

To formulate valid relationships between physical quantities, accurate measurements must be made. If scientists in varying parts of the world are to communicate with each other and confirm each other's findings, they must make measurements in terms of some agreed units. Time is one of the most important physical variables and we first introduce the *second* as the unit of time because it is the one fundamental unit which is common to both the metric and the British systems of units. The second (s) had its origin in the rate of rotation of the earth. The time from the instant the center of the sun is on the meridian plane (the meridian plane of an observer is a plane determined by the axis of rotation of the earth and the point at which the observer is located on the earth's surface) until the sun is centered on the meridian the following day is called a *solar day.* For several reasons solar days are not all of identical duration; the average over 1 year is called the *mean solar day.* This period is divided into 24 h, each hour into 60 min, and each minute into 60 s. Thus, a solar day contains 86,400 s. For many years the second was defined as 1/86,400 of a mean solar day.

However, the earth's rotation is subject to small variations which are not thoroughly understood. As a result, the mean solar day is not an entirely satisfactory quantity for defining a standard of time. The 1960 International Conference on Weights and Measures redefined the second as 1/31,556,925.9747 of the solar year 1900. It is evident that this definition was not very useful for measuring small time intervals with precision, and in 1968 a new definition based on an "atomic clock" was adopted. It had earlier been discovered that it is possible to excite certain atoms in such a way that they emit radiation with a very well-defined *period* (time to complete one cycle). According to the 1968 definition, *The second is the duration of 9,192,631,770 periods of the radiation corresponding to the transition between the two hyperfine levels of the fundamental state of the atom of ce-sium 133.*

This standard for a second has evolved over the years. As higher precision is demanded, new standards may be created, usually in such a way that they are compatible with, but better than, the displaced stand-ard. It is not unlikely that the 1968 definition of the second may be replaced by some still more precise or useful one. For ordinary measure-ments the new unit would be indistinguishable from the old one; only in the most precise kind of measurements would any difference show up. In the next section we shall see that other units have also evolved as science progressed.

1.4 STANDARDS OF LENGTH AND MASS

In making measurements of physical quantities, we ordinarily perform a series of operations which involve the comparison of an unknown quan-tity with an accepted standard of the same kind. Thus when we say that a road is 10 meters (m) wide, we mean that a standard of length called a *meter* must be applied to it 10 times in succession in order to cover its entire width. Similarly, a body is said to have a mass of 25 kilograms (kg) if its mass is 25 times as great as the mass of a standard kilogram.

At one time the length of a king's foot or the span of his hand served as an adequate standard of length and a selected stone as a standard of mass. For example, the British yard was defined by Henry I (1068–1135) to be the distance from the point of his nose to the end of his thumb; much later it was established as the distance at 62°F between two lines ruled on a bronze bar preserved in London. The inch was defined in 1324 to be the length of three dry barleycorns laid end to end.

As commerce between countries increased, and as scientists began to make more and more accurate measurements, established standards of length and mass became of increasing importance. In 1791 the French National Assembly adopted the *metric system*, in which the standard of length was the *Meter† of the Archives*, a platinum bar fabricated to have a length one ten-millionth of the distance from the North Pole to the equator as then measured on the quadrant through Paris. As a standard mass, French scientists prepared a platinum cylinder, known as the *Kilogram of the Archives*, which had a mass as near as they could make it to the mass of 1,000 cubic centimeters (cm^3) of pure water at maximum density. Later measurements show that 1 kg of water at maximum density occupies 1,000.028 cm^3. A volume of exactly 1,000 cm^3 (10^{-3} m^3) is named the *liter* (L). The French system, slightly modified, is now used almost universally in scientific work and increasingly in world commerce.

Meantime in the United States a chaotic set of standards had evolved, in which a bushel basket in South Carolina held 68 in.^{3} more than a bushel basket in New York, while a pound of meat in Massachusetts was $\frac{1}{4}$ oz lighter than a pound of meat in Maine. These facts were pointed out to the Congress in 1821 by John Quincy Adams, then Secretary of State. In his "Report on Weights and Measures" Adams wrote:

> Weights and Measures may be ranked among the necessaries of life, to every individual of human society. They enter into the economical arrangements and daily concerns of every family. They are necessary to every occupation of human industry; to the distribution and security of every species of property; to every transaction of trade and commerce; to the labors of the husbandman; to the ingenuity of the artificer; to the studies of the philosopher; to the researches of the antiquarian; to the navigation of the mariner, and the marches of the soldier; to all the exchanges of peace, and all the operations of war. The knowledge of them, as in established use, is among the first elements of education, and is often learned by those who learn nothing else, not even to read and write. This knowledge is riveted in the memory by the habitual application of it to the employments of men throughout life.

In response Congress established a bureau which later became the National Bureau of Standards, an invaluable national asset.

In 1875, representatives from many civilized countries met in Paris to discuss international standards of measurement. It was generally agreed that the French units were most suitable and that copies should be made for nations which wished to adopt the *metric system* of units. Thirty bars were made of an alloy of platinum (90 percent) and iridium (10 percent). On these bars were marked fine transverse lines separated as nearly as possible by a distance equal to the length of the Meter of the Archives. These new standards were made in 1880, and in 1889 they were distributed to the governments participating in the convention of 1875. The United States obtained two of these standard "meters," of which Bar 27 was the primary standard of length for the United States until 1960. Bar 6 was most nearly equal to the Meter of the Archives, and it was declared

---

† Meter and liter are the American spellings. Many countries, e.g., Australia, use the French spellings metre and litre.

the *International Standard Meter*. It is kept at the International Bureau of Weights and Measures at Sèvres, near Paris.

At the same time, 40 standard kilograms, with masses as nearly as possible equal to that of the Kilogram of the Archives, were constructed from the platinum-iridium alloy. The new mass which agreed most closely with the Kilogram of the Archives is now the *International Mass Standard*. The primary mass standard of the United States is Kilogram 20 (Fig. 1.1), one of the two delivered to this country in 1889. With the kilogram the legal unit of mass, the pound mass is defined to be 0.45359237 kg.

By the middle of the twentieth century the prototype meter bar was no longer good enough as a standard of length. Secondary standards could be compared with it only to an accuracy of roughly 1 part in 10 million, while technological requirements for modern industry called for tolerances of the order of 10^{-7} m. In general, it is desirable that a master standard be at least 10 times more accurate than the scientific and industrial measuring systems which are derived from it. As a consequence, at the 1960 International Conference on Weights and Measures the meter was redefined in terms of the wavelength (Fig. 1.2) of the orange light emitted by the krypton isotope of mass 86. Krypton is found in the air, and a standard based on the light from krypton 86 has the advantage that it can never be destroyed, stolen, damaged, or lost. With this standard of length it is possible to make measurements with an accuracy of the order of 1 part in 100 million. Further, anyone with the appropriate equipment can reproduce the present standard of length, defined as follows:

The meter is a length 1,650,763.73 times the wavelength of the orange-red light given off by the pure krypton isotope of mass 86 when it is excited in an electric discharge.

The meter is the legal standard of length in the United States; however, most everyday measurements are made in British units, which are related to the meter by the definition:

One inch is 0.0254 m exactly.

1.5 THE INTERNATIONAL SYSTEM OF UNITS

The 1960 General Conference of the International Bureau of Weights and Measures established the Système International d'Unités (SI) by selecting the *meter* as the basic unit for *length*, the *kilogram* for *mass*, the *second* for *time*, the *ampere* (Sec. 36.8) for *current*, the *kelvin* (Sec. 15.3) for *temperature*, and the *candela* (Sec. 28.1) for *luminous intensity*. Larger and smaller units are related to the standard units in terms of powers of 10. Thus the kilometer is 1,000 m, while the millimeter is 0.001 m. Table 1.1 lists the prefixes for fractions and multiples together with their abbreviations. In most scientific work metric units are used, and the SI is widely accepted although the cgs system, based on the centimeter, the gram, and the

FIGURE 1.1
The mass standard of the United States is this standard kilogram, a platinum-iridium cylinder about 39 mm high and 39 mm in diameter. (*National Bureau of Standards.*)

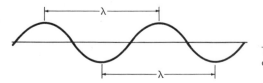

FIGURE 1.2
The wavelength λ is the distance from crest to adjacent crest (or trough to adjacent trough) of a wave.

TABLE 1.1
Fractions and Multiples (SI)

Factor	Prefix	Abbreviation
10^{-1}†	deci	d
10^{-2}†	centi	c
10^{-3}	milli	m
10^{-6}	micro	μ
10^{-9}	nano	n
10^{-12}	pico	p
10^{-15}	femto	f
10^{-18}	atto	a
10†	deka	da
10^2†	hecto	h
10^3	kilo	k
10^6	mega	M
10^9	giga	G
10^{12}	tera	T
10^{15}	peta	P
10^{18}	exa	E

† It is recommended that these factors be used only when there is a strongly felt need. Compound prefixes, for example, mμ, should be avoided.

second (together with other electrical units), retains an important following.

Many other systems have been devised. The *British absolute system* has as its fundamental units the *foot*, the *pound mass*, and the *second*. It is not necessary that the fundamental units of a system involve length, mass, and time. Indeed, as we shall see, the *British engineering system* is based on the *foot*, the *second*, and the *pound force*.

An overwhelming number of countries, and well over 90 percent of the world's population, now use the metric system, including virtually all the nations of Europe, Asia, and South America and most of those of Africa. For several years the United States has been in the process of converting to the SI.

1.6 DERIVED UNITS

When a physical observable is expressed in terms of a combination of the fundamental quantities, its units are said to be *derived*. For example, the units of area can be written as the square of units of length. Thus, we may express areas in square meters or square feet. Similarly, the dimensions of volume are lengths cubed—cubic meters, cubic feet, etc. Speed and density are examples of the many quantities in mechanics which are written in terms of derived units.

Speed The average speed of a body is defined as the ratio of the distance it travels to the time required to pass over that distance. Thus, speed is obtained by dividing a length by a time. Appropriate units of speed are meters per second, feet per second, miles per hour, etc. The speeds of ships and aircraft are often measured in *knots*. One knot is one nautical mile (1,852 m or 6,076 ft) per hour.

□ **Example** A track man runs a measured kilometer in 2 min 27 s. Find his average speed in meters per second.

$$\text{Average speed} = \frac{\text{distance}}{\text{time}} = \frac{1{,}000 \text{ m}}{147 \text{ s}}$$
$$= 6.80 \text{ m/s} \qquad \square$$

Density The denisty of a body is defined as the ratio of its mass to its volume. Thus

$$d = \frac{m}{V} \qquad (1.1)$$

where d is the density, m the mass, and V the volume. Among the familiar units for density are kilograms per cubic meter, slugs (Sec. 4.5) per cubic foot, and grams per cubic centimeter.

□ **Example** A quantity of mercury has a mass of 2.05 kg and occupies a volume of 151 cm³. What is its density?

$$d = \frac{m}{V} = \frac{2.05 \text{ kg}}{151 \text{ cm}^3} = 0.0136 \text{ kg/cm}^3$$
$$= 13.6 \text{ g/cm}^3 = 13{,}600 \text{ kg/m}^3 \qquad \square$$

□ **Example** Find the mass of air in a room which is 3.00 by 8.00 by 6.00 m. The density of the air is 1.29 kg/m³.

$$m = dV = 1.29 \text{ kg/m}^3 \times 144 \text{ m}^3 = 186 \text{ kg} \qquad \square$$

The term "density" is often used in conjunction with certain other words to suggest the amount of some quantity per given amount of some other quantity. For example, we use the term *population density* to describe number of people per unit area. In electricity the term *charge density* means the ratio of electric charge to volume, and *surface charge density* is the ratio of electric charge to surface area. The ratio of the weight of an object to its volume is called its *weight density*. The ratio of the mass of a rope to its length is sometimes called its *linear density*. Whenever the word density appears in this text in any sense other than that of mass per unit volume, it will be qualified by another word.

1.7 CONVERSION OF UNITS AND SIGNIFICANT FIGURES

In the measurement of any physical quantity, the choice of units is ordinarily dictated by convenience and by the habits of the observer. In reporting any measurement, it is important to list not only the number of times the standard unit was included but also to state what this standard unit is. For example, the height of a man might be 6.08 ft, 73.0 in., or 1.85 m. The number which describes the height is meaningless unless the accompanying units are known.

We live in a country where many everyday measurements are made in the British system, but most scientific work involves metric units. Since anyone who has to deal with scientific phenomena in the United States is essentially forced to use both British and metric units, it is important to be able to transform from one system to another. To facilitate this operation an extensive list of conversion factors is presented inside the front cover.

It is convenient to convert a quantity from one unit to another in the form of the examples below. Note that one can cancel units just as though they were algebraic quantities. Note further that the factors by which one multiplies are all unity.

□ **Example** Convert 60.0 mi/h to feet per second and meters per second.

$$60.0\frac{\text{mi}}{\text{h}} = 60.0\frac{\text{mi}}{\text{h}} \times \frac{1 \text{ h}}{3{,}600 \text{ s}} \times \frac{5{,}280 \text{ ft}}{1 \text{ mi}}$$

$$= \frac{60.0 \times 5{,}280 \text{ mi h ft}}{3{,}600 \text{ h s mi}} = 88.0\frac{\text{ft}}{\text{s}}$$

$$= 88.0\frac{\text{ft}}{\text{s}} \times \frac{0.3048 \text{ m}}{1 \text{ ft}} = 26.8\frac{\text{m}}{\text{s}} \qquad \square$$

□ **Example** The density of aluminum is 2,699 kg/m³. Convert this density to grams per cubic centimeter and to pounds per cubic foot.

$$2{,}699\frac{\text{kg}}{\text{m}^3} = 2{,}699\frac{\text{kg}}{\text{m}^3} \times \frac{1 \text{ m}^3}{(100 \text{ cm})^3} \times \frac{1{,}000 \text{ g}}{1 \text{ kg}}$$

$$= 2.699\frac{\text{g}}{\text{cm}^3}$$

$$2{,}699\,\frac{\text{kg}}{\text{m}^3} = 2{,}699\,\frac{\text{kg}}{\text{m}^3} \times \frac{1\text{ lb}}{0.4536\text{ kg}} \times \frac{(0.3048\text{ m})^3}{1\text{ ft}^3}$$

$$= 168.5\,\frac{\text{lb}}{\text{ft}^3} \qquad\qquad \square$$

In making conversions and in solving problems it is both improper and a waste of time to carry out calculations with more figures than are meaningful. For example, if we measure the thickness of a brick at six places and obtain the values 6.85, 6.88, 6.87, 6.86, 6.84, and 6.88 cm, the average of these values is 6.8633333 ··· cm, but obviously the numbers after 6.86 overstate the reliability of the average. Clearly, even the last 6 is not certain; all numbers following it have *no physical significance*. It is not only pointless to retain numbers which have no physical meaning, but it may be misleading as well. *All the figures in the numeric of a measurement are significant if they do not overstate the reliability.* In the first example above, the initial datum and the solutions are given to three significant figures, while in the second four significant figures are involved. Usually we shall treat all numbers as though we knew them to three significant figures and calculate answers accordingly.

With modern hand-held calculators it is common to have 8 or 10 numerals displayed, but it is seldom indeed that we know the data entering a calculation with anything like that precision. For example, if you measure the radius of a circle with a good ruler, you may find it to be 2.43 cm, where the last numeral is estimated. If we now use the relation area $= \pi r^2$ on a typical calculator, we may obtain 18.55079046 on our display, but we should record the area as 18.6 cm^2. Even now, the 6 is uncertain; it could easily be 5 or 7 . In general, when we write down a result, we should stop after the first numeral which is uncertain. To quote more figures implies improperly that we know the result more accurately than we actually do.

In tables and inside the front cover we often state more than three significant figures for one of two reasons: (1) Sometimes for some special reason you may need to know the value of the quantity to more significant figures. For example, the speed of light in vacuum, an important physical constant, is 3.00×10^8 m/s to three significant figures, but its value is known to be less than that by about 0.07 percent. (2) Whenever an answer depends on the *difference* of two nearly equal numbers, we must know the latter to several significant figures to get the difference to three significant figures. For example, $2{,}539.2 - 2{,}517.8 = 21.4$; in this case we need five significant figures in the original numbers to have three significant figures in the difference.

1.8 MEASUREMENT OF ANGLES

Degrees, minutes, and seconds are familiar units in which angles are measured. A degree is the angle subtended at the center of a circle by an arc of length $\frac{1}{360}$ of the circumference of the circle. A minute is $\frac{1}{60}$ degree, and a second is $\frac{1}{60}$ minute.

To measure the angle turned through by a rotating wheel, we are likely to use the *revolution*. The angle subtended by the full circumference is one revolution (r).

Still another unit which is often convenient for measuring angles is the radian. *A radian is the angle subtended at the center of a circle by an arc of length equal to the radius.* Thus, the angle *AOC* of Fig. 1.3 is 1 radian (rad), since the arc length *AC* is equal in length to radius *OA*.

The angle *AOB* of Fig. 1.3 is represented by θ (the Greek letter theta); in radians it is *s* (the length of arc *AB*) divided by the radius *R*:

$$\theta = \frac{s}{R} \qquad (1.2)$$

The angle θ is the ratio of two lengths. Since the units of the denominator cancel the units of the numerator, an angle has no dimensions in terms of length, mass, or time; we say it is *dimensionless*. Since the circumference of a circle is 2π times the radius, it follows that there are 2π rad in a revolution. Hence,

$$1\ r = 2\pi\ rad = 360°$$

Thus one radian is approximately 57.3°.

1.9 FORCE

One of the important concepts in physics is that of force. For the moment, we may think of a force as a push or a pull exerted upon some object. The most familiar forces are those which we experience through our own muscular activities. More generally, a force is an action, exerted by one body on another, that tends to change the state of motion of the body acted upon. Thus, someone throwing a baseball exerts a force upon it which changes its state of motion. We shall later define force explicitly in terms of effects on the motions of objects.

The basic force unit for the SI system is the *newton* (N), which is properly defined (Sec. 4.5) in terms of its effect on the motion of a one-kilogram mass in a manner which is *independent of the location of this mass.* For this reason it is called an *absolute* unit of force.

In addition to the newton there are two other force units which are often used. They are called *gravitational* units because each is defined in terms of *the pull of the earth on a particular standard mass at a specified location* (essentially mean sea level, 45° north latitude). The net pull of the earth, at the specified location, on a standard kilogram mass is known as the *kilogram force* (1 kgf = 9.807 N).† Similarly *the pound force (lbf) is the net gravitational pull of the earth on a standard one-pound mass (lb) at the specified location (1 lbf = 4.448 N).*

1.10 WEIGHT AND MASS

We know that every body near the surface of the earth is pulled toward the earth. This gravitational attraction arises from the fact that every particle in the universe attracts every other particle; according to *Newton's law of universal gravitation,* between any two particles in the universe there exists a gravitational attractive force which is proportional to the product of the masses of the particles and inversely proportional to the square of the distance between them. Consider a small stone held 1 m

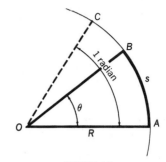

FIGURE 1.3
The angle θ in radians is the ratio of the arc length *AB* to the radius *OA*. The angle *AOC* is 1 rad (= 57.2958°).

---

† The kilogram force is also called the kilogram weight (kg-wt) or simply kilogram. The abbreviations kgf and lbf are used only where lb and kg might be ambiguous.

Spring balance

Weight, kg

FIGURE 1.4
The weight of a body is equal and opposite to the force required to hold the body at rest against the net gravitational pull on it.

above the earth's surface. We may regard the stone as one particle and the earth as being composed of a huge number of small particles, each of which attracts the stone according to Newton's law. The net gravitational force on the stone is the result of the forces exerted by all the particles constituting the earth. Newton's law of universal gravitation is treated in Chap. 8.

We now define the *weight* of a body in terms of a series of operations. Suppose that we have a standard kilogram near sea level at 45° north latitude along with a spring balance with an unmarked scale. For an ideal spring the extension is directly proportional to the force producing the stretching. If we indicate the spring location with no load applied by zero and the location of the pointer with the standard kilogram supported as 9.807 N, we can lay out a linear scale for the balance so that it reads force in newtons directly (Fig. 1.4), and we define the *weight* of the standard kilogram to be equal and opposite to the force exerted by the spring when the kilogram is held at rest by the balance. If we travel around with our scale and kilogram, we shall find that the kilogram weighs slightly less than 9.807 N at the earth's equator and slightly more at the North Pole. Similarly it would weigh less atop Pikes Peak than it would in New York City. These changes in weight are not large, but they are easy to detect with sensitive instruments. If an object weighing 12 N on the earth is taken to the moon, its weight there will be about 2 N; this 2 N is due to the pull of moon. The weight of a body would be almost zero if we took it far enough out in space (Chap. 8) as long as we did not go too near some other massive body. Thus the weight of a body is a variable which depends on where we measure it.

On the other hand, the mass of a body at rest is the same at all points in the universe. This is one reason that the concept of mass is so important in physics. It is sometimes said that the mass of a body is the "quantity of matter in the body," a property of the body which does not change as long as its speed is negligible compared with the speed of light. An object has mass whether or not it is near enough to the earth to have observable weight. The mass of a body reveals itself as *inertia*, the tendency of matter to resist any change in its state of motion. This important attribute of mass is discussed further in Chap. 4.

The larger the mass of a body, the harder the earth pulls on it. Indeed, the weight of a body at any place is directly proportional to its mass, and the mass of a body is often determined by comparing the pull of the earth on it with the pull on a standard mass at the same place. An unfortunate consequence is that the word "weight" is frequently used improperly in place of mass. For example, an analytical balance typically determines the mass of the unknown in terms of standard masses, yet we refer to this operation as "weighing" and call the result the "weight" of the unknown.

1. What is your weight in pounds and in newtons? Your mass in kilograms? Your height in inches and in meters?
2. What advantages does the new international standard of length have over the platinum-iridium bar it replaced? What advantages did the meter bar have over the new standard?
3. If you had a ruler with a smallest division of 1 mm, how accurately could you measure the length of this page? The thickness of this page alone? If you wished to know the thickness of the page more precisely, what could you do with just the ruler?
4. If someone told you that the dimensions of every object in the universe, including you, had decreased by one-fourth last night while you slept, how could you refute him?
5. How can length be measured along a curved line?
6. What repetitive phenomena could serve as reasonable time standards?
7. In what sense is it true that the earth makes one rotation each 24 h? In what sense is it false? The sidereal day, widely used by astronomers, is the time interval between two successive passages of a star across the meridian. How many sidereal days are there in 1 year? Show that the length of a sidereal day is 86,164 s.
8. A simple pendulum 1 m in length has a period very close to 2 s. What would be the disadvantages of defining the time unit as one-half the period of a pendulum 1 m in length?
9. The density of a cylinder is to be determined by measuring its mass, length, and diameter. How big an error would be introduced in the final result by a 1 percent error in (a) the mass, (b) the length, and (c) the diameter? Why does the 1 percent error in the diameter introduce twice as large an error in the density as the corresponding error in mass or in length?
10. What are some of the more important characteristics of a physical standard unit?
11. What are the dimensions of speed? Of density? List at least two kinds of units in which each may be measured. In general, what is the distinction between dimensions and units?
12. How could you determine the height of a tree or building from the level ground around it? What apparatus would you need, and how would you use it?
13. What advantages will accrue to the nation when the United States adopts the metric system for general commercial and government use? What are the disadvantages of abandoning the use of British units of length and force?

14. The time from one full moon to the next is one lunar month, while the time for the moon to return to the same position against the background of the stars is one sidereal month. Which is longer? Why?

PROBLEMS

Unless otherwise specified, the numerics stated in the problems should be regarded as valid to three significant figures; in these cases answers should be given to three significant figures. For conversion factors see inside the front cover.

1. The meter was selected originally as one ten-millionth of the distance from the North Pole to the equator along the quadrant of the earth's surface through Paris. Assuming that the earth is spherical, find the radius of the earth in kilometers and in miles.   *Ans:* 6,370 km; 3,960 mi
2. The highest peaks in South and North America are respectively Aconcagua at 6,960 m and McKinley at 6,194 m. Compute their altitudes in feet.
3. The 600 pages of a book have a total thickness of 30 mm. Find the average thickness of a single sheet in micrometers and in mils (1 mil = 0.001 in.).   *Ans.* 100 μm; 3.94 mils
4. Mount Logan is the highest mountain in Canada, with an elevation of 19,850 ft. Find its altitude in meters.
5. Mount Kilimanjaro in Tanzania has an altitude of 19,340 ft, greatest in Africa, while Mount Cook has an elevation of 12,349 ft, greatest in New Zealand. Find these heights in meters.
     Ans. 5,895 m; 3,764 m
6. The Washington monument is reported to be 555 ft 5⅛ in. high. To how many significant figures is the height given? How meaningful do you think the last significant figure is? Express the height in meters to three significant figures.
7. From San Francisco it is 2,393 mi to Honolulu and 7,416 mi to Sydney by air. Find these distances in kilometers.   *Ans.* 3,851 km; 11,934 km
8. Airline distances between New York and London and New York and Capetown are respectively 5,589 and 12,552 km. What are these distances in miles?
9. On a certain day the speed of sound in air is

350 m/s. Find the speed in kilometers per hour, in feet per second, and in miles per hour.
Ans. 1,260 km/h; 1,150 ft/s; 783 mi/h

10. A supersonic aircraft has a speed of 2,200 km/h. Find its speed in meters per second, in miles per hour, and in feet per second.

11. A jet airplane is moving at a speed of 615 mi/h. Find its speed in feet per second, in meters per second, and in kilometers per hour.
Ans. 902 ft/s; 275 m/s; 990 km/h

12. Find the area of a rectangular plot of ground 50 m long and 45 m wide in (*a*) hectares, (*b*) square feet, and (*c*) acres.

13. A rectangular farm is 2 km long and 1 km wide. Find its area in (*a*) hectares, (*b*) acres, and (*c*) square feet.
Ans. (*a*) 200 ha; (*b*) 494 acres; (*c*) 2.15×10^7 ft^2

14. The gasoline (petrol) tank of an automobile holds 50 liters (L). Find its volume in U.S. gallons (1 U.S. gal = 231 in.^{3}) and in British imperial gallons (277.4 in.^{3} = volume of 10 lb of water at 62°F).

15. Before 1977, bottles containing one-half and one-fifth of a gallon were widely used. They have been replaced by 1.75 L and 750 mL bottles, respectively. Find the ratio of the volume of each of the new sizes to the volume of the old size which it replaced. *Ans.* 0.925; 0.991

16. An aquarium has dimensions of 90 by 50 by 30 cm. What mass of water does the aquarium hold when full? What is the volume in cubic feet?

17. A lake has an area of 6×10^8 m^2. A layer of water 1.9 mm thick evaporated from this lake over a hot weekend. What mass of water was evaporated in this period? *Ans.* 1.14×10^9 kg

18. An irregular crystal of quartz has a mass of 170 g. When it is submerged in water in a test tube of radius 30 mm, the water level is observed to rise 22.6 mm. Find the density of the quartz crystal.

19. In going around a sharp curve, the driver of an automobile turned the steering wheel through an angle of 4 rad. Through how many degrees did it turn? Through how many revolutions? If the wheel has a radius of 170 mm, through what distance did a point on the outer edge of the wheel move? *Ans.* 229°; 0.637 r; 680 mm

20. A curve on a scenic highway is 350 m long and subtends an angle of 50° at the center of the circle of which it is a part. What is the radius of the curve?

21. Solid carbon dioxide (dry ice) has a density of 1,530 kg/m^3. What volume would be occupied by 24 cm^3 after evaporation to gas with a density of 1.9 kg/m^3. *Ans.* 19,300 cm^3

22. The angle subtended by the sun's diameter at the surface of earth is 0.5°. Find the approximate radius of the sun if it is at a distance of 1.5×10^8 km from the earth. The moon subtends essentially the same angle at the earth. What is the approximate radius of the moon if it is 3.8×10^5 km away?

23. The effective radius of the wheel of a small automobile is 0.24 m. Find the angle in revolutions and in radians through which the wheel rotates in going a distance of 1 km.
Ans. 663 r; 4,167 rad

24. The circumference of the circular magnet ring of the Brookhaven 33-GeV accelerator is approximately $\frac{1}{2}$ mi. Find the area circumscribed by the magnet ring in square feet and in square meters.

25. A mountain peak 15 km (measured horizontally) away from the observer is sighted from a telescope at an elevation of 800 m above sea level. If the line of sight makes an angle of 12° with the horizontal, find the height of the peak above sea level. *Ans.* 3,990 m

26. If 1 gal = 231 in.^{3}, find the volume in gallons of an aquarium with dimensions of 3 by 2 by 1.2 ft. If 1 ft^3 of water weighs 62.4 lb, what weight of water does the tank hold when full?

27. What mass of water falls on a rectangular field 500 m long and 350 m wide in 1 year at a place where the annual rainfall is 1.25 m?
Ans. 2.19×10^8 kg

28. The speed of an automobile is 96 km/h. Find the speed in meters per second, miles per hour, and feet per second.

29. To determine the width of a canyon to be spanned by a horizontal bridge, a distance AB of 140 m is marked off parallel to one edge of the canyon. Point C on the opposite side of the canyon is sighted from points A and B. If AC is perpendicular to AB and BC makes an angle of 35° with AB, what is the width of the canyon?
Ans. 98.0 m

30. In astronomy distances are so huge that it is convenient to have some very large units in which to measure them. Among them are the *astronomical unit* (the average distance from the earth to the sun, about 1.496×10^{11} m), the *light-year* (distance traversed by light in vacuum in a time of 1 year), and the *parsec* (the distance at which 1 astronomical unit subtends an angle of 1 second of arc). Show that 1 parsec is 206,265 astronomical units, or 3.26 light-years. (The speed of light in vacuum is given along with other important constants inside the front cover.)

31. In surveying for a straight road to traverse rough terrain between points A and B by a series of cuts and fills, a horizontal baseline AC 90 m long is laid out. A transit shows that AB makes an angle of 36° with AC while CB makes an angle of 74° with CA. Determine the distance AB.

Ans. 92.1 m

32. About 250 B.C. Eratosthenes measured the circumference and hence the radius of the earth. He observed that at noon on the first day of summer sunlight struck the bottom of a vertical well at Syene (near Aswan), Egypt. At Alexandria, 5,000 stadia to the north, at the corresponding time he observed that an obelisk (see figure) cast a horizontal shadow; from the length of the shadow and the height of the obelisk he calculated that the sun was one-fiftieth of a circle south of the zenith (straight up). Calculate the radius of the earth if his stadion was equal to 160 m (no one knows for sure).

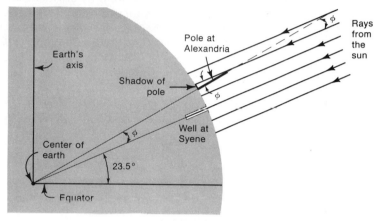

PROBLEM 32

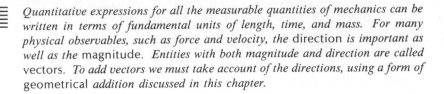

Quantitative expressions for all the measurable quantities of mechanics can be written in terms of fundamental units of length, time, and mass. For many physical observables, such as force and velocity, the direction is important as well as the magnitude. Entities with both magnitude and direction are called vectors. To add vectors we must take account of the directions, using a form of geometrical addition discussed in this chapter.

2
VECTOR QUANTITIES

2.1 DISPLACEMENT

When a body moves from one location to another, it is said to undergo a *displacement.* Suppose, for example, that a helicopter takes off, flies due east 12 kilometers, and lands. It has undergone a displacement of 12 km east. If the helicopter takes off again and flies 25 km in a straight line before landing, how far is it from its original starting point? A specific answer to this question cannot be given until we have one *additional* piece of information: the direction of this second displacement. From the information now available we can only be sure that the helicopter lies somewhere on a circle with center at B and a radius of 25 km (Fig. 2.1). *A displacement is completely specified only when we state both its magnitude and direction.* The magnitude consists of a numeral and a unit.

If the second displacement is in the same direction as the first, the helicopter is at point C and the total displacement is 37 km east. If the second displacement is in the direction opposite to the displacement AB, the helicopter is at D and the displacement is 13 km west. If the displacement is due north, the helicopter is at E and its distance from the starting point is $R = \sqrt{(25)^2 + (12)^2}$ km $= 27.7$ km.

To specify the final position of the helicopter we need not only its distance from the starting point but also its direction. For this we use a few ideas from trigonometry. (For this course it is not necessary to have studied trigonometry; all that is required is to become familiar with three definitions in Sec. A.1 in Appendix A. We make immediate use of them.) By definition we know that *in any right triangle the sine of an angle is the ratio of the side opposite the angle to the hypotenuse, the cosine is the ratio of the side adjacent to the angle to the hypotenuse, and the tangent of the angle is the ratio of the side opposite the angle to the side adjacent to the angle.* In the triangle ABE of Fig. 2.1 we have

$$\sin \theta = \frac{25}{27.7} = 0.902 \qquad \cos \theta = \frac{12}{27.7} = 0.433 \qquad \tan \theta = \frac{25}{12} = 2.08$$

To find the angle θ we look in the trigonometric tables (Appendix A) and we find that $\theta = 64.4°$ by all three relations above. In solving for θ, just one of the functions would have been sufficient, of course. By using two we have a check on part of our work.

Now suppose that the 25-km displacement is at an angle of 53.1°N of E (Fig. 2.2). In the figure each displacement is represented by a directed line segment (arrow) pointed in the direction of the displacement with length proportional to the displacement. One way of determining the distance from the starting point to the final position is by making a scaled drawing and measuring the distance with a ruler, but in most cases it is easier to solve a problem of this type analytically. We observe that the displacement BP of 25 km at an angle of 53.1°N of E takes the helicopter a distance BG to the east and a distance GP to the north. By definition

$BG/BP = \cos 53.1° = 0.600$; since $BP = 25.0$ km, BG must be 15.0 km. Similarly, $GP/BP = \sin 53.1° = 0.800$, from which $GP = 20.0$ km. Thus, a displacement of 25.0 km at an angle of 53.1°N of E is equivalent to a displacement of 15.0 km east plus a displacement of 20.0 km north. We can replace our 25-km displacement with the two mutually perpendicular displacements BG and GP. We observe from Fig. 2.2 that the helicopter is now 27.0 km east of its original starting point and 20.0 km north. To find the distance AP we make use of the *pythagorean theorem. The square of the length of the hypotenuse of any right triangle is equal to the sum of the squares of the lengths of the other two sides.* Therefore

$$AP = \sqrt{(27.0)^2 + (20.0)^2} = 33.6 \text{ km}$$

From the right triangle AGP we see that $\tan\theta = 20.0/27.0 = 0.741$, from which we find $\theta = 36.5°$ in Table A.1.

2.2 VECTORS AND SCALARS

In the preceding section we have examined the problem of adding two displacements which were *not* in the same direction. This is our first example of adding vector quantities; many more will follow. *Physical quantities which have both magnitude and direction and which add like displacements are called vectors.* A vector quantity can be represented graphically by a directed line segment (arrow) or line vector. A knowledge of both the magnitude and the direction of a vector is required for its complete description. Vectors in different directions cannot be added, subtracted, or multiplied by ordinary arithmetic; they are treated by geometrical techniques. In this chapter we deal with three types of classical| vector quantities: displacements, velocities, and forces. In later chapters, many more vector quantities such as acceleration, momentum, torque, angular velocity, and angular acceleration are introduced.

Other quantities in physics, such as mass, time, and volume, have only magnitude and can be completely specified by a numeral and a dimension. They do not involve any idea of direction. Such quantities are called *scalars.* They obey the ordinary laws of addition, subtraction, multiplication, and division. If 30 L of gasoline is added to a tank which originally contained 7 L, the tank contains 37 L. If a 10-g mass is removed from the pan of a balance which holds 80 g, the mass remaining on the pan is 70 g. These are illustrations of the addition and subtraction of *scalar quantities.*

Vectors are usually denoted in print by boldface type, such as **F** for force, **v** for velocity, etc. Whenever a boldface symbol is used in text or in a formula, it indicates that direction as well as magnitude is of importance. Italic type is used for scalar quantities and for the magnitudes of vectors.

2.3 THE GRAPHICAL ADDITION OF VECTORS

A vector quantity can be represented graphically by a directed line segment, or arrow, in the direction of the vector with length proportional to

†Many quantities which behave like vectors in everyday situations fail to satisfy the definition above when speeds comparable to the speed of light are involved. For example, very high velocities do not add like displacements even though they do have magnitudes and directions.

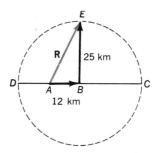

FIGURE 2.1
A displacement from point A of 12 km east followed by a 25-km displacement in an arbitrary direction would bring a helicopter to *any* point on the dashed circle DEC.

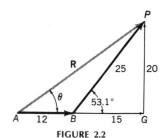

FIGURE 2.2
The vector **R** is the resultant of the displacements 12 km east and 25 km 53.1°N of E.

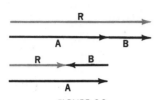

FIGURE 2.3
The resultant of two displacements in the same direction (above) and in opposite directions (below).

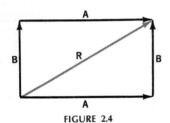

FIGURE 2.4
Resultant of two vectors at right angles to each other. Note that $R = A + B = B + A$.

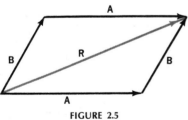

FIGURE 2.5
Resultant of two vectors; as always, $R = A + B = B + A$.

FIGURE 2.6
Subtraction of vector **B** from vector **A** can be accomplished (a) by adding −**B** to **A** or (b) by finding the vector which must be added to **B** to give **A**. Observe that the difference vector **D** has the same magnitude and the same direction in both (a) and (b).

the magnitude of the vector. To add two vectors **A** and **B**, we can represent each of them by an arrow of scaled length and find their sum (Figs. 2.3 to 2.5) by placing the tail of the second arrow at the point of the first. Alternatively, we can start with the second vector and add the first, as shown in Figs. 2.4 and 2.5. This gives rise to the *vector parallelogram*. The vector sum, known as the *resultant*, is given by the diagonal **R**. If the graphical representation of the vectors **A** and **B** is performed carefully, the resultant vector **R** can be measured with considerable accuracy.

To subtract one vector from another, we need only reverse the direction of the vector we wish to subtract and add this negative vector to the first. This is shown schematically in Fig. 2.6a, where $D = A - B$. An alternative way of subtracting **B** from **A** is to find that vector **D** which must be added to **B** to give **A** (Fig. 2.6b), since $B + D = A$ if $D = A - B$.

There is no limit to the number of vectors which may be combined graphically. If we wish to determine the resultant of four vectors, we can add the third vector to the resultant of the first two, and the fourth vector to the resultant of the first three. For example, suppose that an object has four forces acting upon it: 8 newtons east, 5 newtons 30°N of E, 7 newtons 37°W of N, and 3 newtons due south, as shown in Fig. 2.7. The resultant is given by the vector **R**. The order in which the vectors are added makes no difference in determining the resultant. We shall calculate the value of **R** in Sec. 2.7.

2.4 VELOCITY

Suppose you take an automobile ride. The *distance* you travel, as read by the odometer, is a *scalar* quantity. If at the end of 3 h you have traveled a distance of 240 km, your average speed has been 80 km/h. Suppose that at this point you are 75 km due east of your starting point (Fig. 2.8). Then your displacement is 75 km east. *Displacement* is a *vector* quantity. It is not necessarily related in any simple way to the total *distance* traversed. The *average velocity* during the trip is *the ratio of the displacement to the time*. In this case the average velocity is 25 km/h east. Note that *velocity is a vector quantity* while *speed is a scalar*. When you return to your starting point, the average *velocity* is zero since the *displacement* is zero; the average speed is very different from zero.

□ **Example** On a transcontinental trip a family begins to drive at 6 A.M. and stops at 5 P.M. In this period the car goes 495 mi and is 350 airline miles southwest of its starting point. Find the average speed and average velocity for the day.

$$\text{Average speed} = \frac{\text{distance}}{\text{time}} = \frac{495 \text{ mi}}{11.0 \text{ h}} = 45 \text{ mi/h}$$

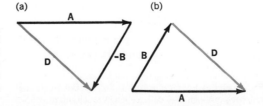

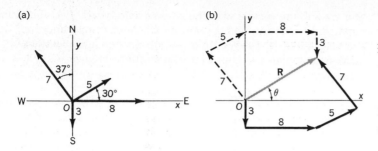

FIGURE 2.7
(a) Four forces acting on a body at O. (b) The resultant $\mathbf{R}$ of these four forces is independent of the order in which they are added.

$$\mathbf{v}_{av} = \frac{displacement}{time} = \frac{350 \text{ mi southwest}}{11.0 \text{ h}} = 31.8 \text{ mi/h southwest} \quad \square$$

In this chapter we shall add velocities, confining ourselves to cases in which the velocity is constant, or *uniform*, i.e., in which equal displacements are traversed in equal times. In later chapters we shall consider types of motion in which the velocity varies with time.

2.5 FRAME OF REFERENCE

Whenever we refer to the displacement or velocity of a body, we are always, by implication, referring these measurements to some other body or framework of lines which is regarded as fixed. Such a framework is called a *frame of reference*. Some problems can be greatly simplified by the choice of a suitable frame of reference.

With few exceptions we refer quantities to a system of axes fixed on the earth. When we say an object is at rest, we ordinarily mean it is at rest relative to the earth. Relative to the sun it is moving through space at a speed of 30 km/s by virtue of the orbital motion of the earth (and at a speed which depends on latitude because of the spin of the earth on its axis). A passenger seated in a train traveling 100 km/h is at rest relative to the car and the other passengers. If she walks forward in the car with a speed of 6 km/h, her speed relative to the earth is 106 km/h.

It is meaningless to speak of the "absolute" position of a body in space or to assign to the body an "absolute" velocity. Any position vector or velocity vector must be specified relative to some frame of reference.

In general, if the velocity of a body B relative to one frame of reference is $\mathbf{v}_{B1}$, its velocity relative to a second frame $\mathbf{v}_{B2}$ is given by the vector equation

$$\mathbf{v}_{B2} = \mathbf{v}_{B1} + \mathbf{v}_{12} \tag{2.1}$$

where $\mathbf{v}_{12}$ is the velocity of the first frame relative to the second. For example, if an airplane is flying with a velocity $\mathbf{v}_{PA}$ relative to the air and the air has a velocity $\mathbf{v}_{AG}$ relative to the ground, the velocity $\mathbf{v}_{PG}$ of the plane relative to the ground is given by

$$\mathbf{v}_{PG} = \mathbf{v}_{PA} + \mathbf{v}_{AG} \tag{2.1a}$$

If $\mathbf{v}_{PA}$ is 500 km/h east and $\mathbf{v}_{AG}$ is 80 km/h east, $\mathbf{v}_{PG} = 580$ km/h east. If $\mathbf{v}_{AG}$ is 80 km/h west, $\mathbf{v}_{PG}$ is 420 km/h east.

In general, if we have three bodies A, B, and C and we let $\mathbf{v}_{AB}$ represent the velocity of A relative to B, and so forth, then $\mathbf{v}_{AC} = \mathbf{v}_{AB} + \mathbf{v}_{BC}$. (Note

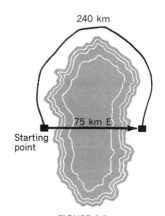

FIGURE 2.8
Displacement is a vector quantity; distance is a scalar quantity.

that we add the vectors with the two *B*'s adjoining. This is an example of what is sometimes called the *domino rule* of vector addition.) If we want the velocity of *C* relative to *B*, we observe that $\mathbf{v}_{CB} = \mathbf{v}_{CA} + \mathbf{v}_{AB}$. The same rule applies to other vector quantities.

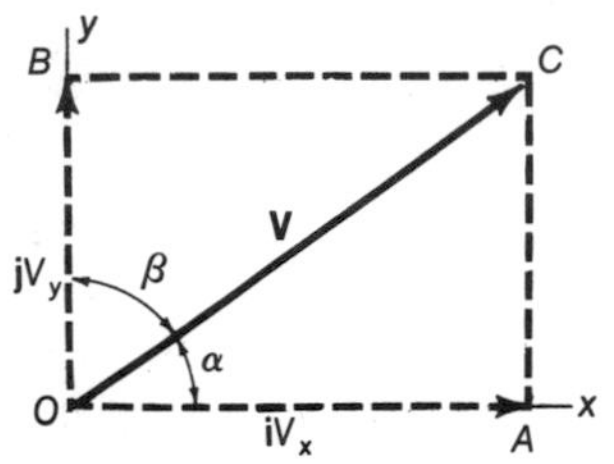

2.6 RECTANGULAR COMPONENTS OF A VECTOR

Consider a vector **V**, 5 units in length and directed at an angle of 36.9°N of E (Fig. 2.9). This displacement is equivalent to displacements of 4 units east and 3 units north; to put it in another way, it is the resultant of the two displacements 4 units east and 3 units north. These last two vectors, which are perpendicular to each other, are called the *rectangular components* of the 5-unit vector. In any problem involving vectors in a plane we can always replace any vector by its rectangular components and thus reduce the problem so that we are dealing with components which all lie either parallel or perpendicular to a chosen line.

To find the eastward and northward components of the 5-unit vector of Fig. 2.9 we proceed as follows. We place the origin of a coordinate system on the tail of the vector with its mutually perpendicular axes in the directions desired; in this case, the *x* axis runs eastward and the *y* axis northward. Next we drop perpendiculars from the tip of the vector to each of the coordinate axes. The *x* component is then *OA* in Fig. 2.9, and the *y* component is *OB*. Since *OAC* is a right triangle, we have, by definition, $OA/OC = \cos \alpha$ or $OA/5 = \cos 36.9° = 0.80$, so $OA = 4$. Similarly, $OB/OC = \cos \beta$ and $OB/5 = 0.60$, so $OB = 3$ (α is the Greek letter alpha, and β is the Greek letter beta; the Greek alphabet is given inside the front cover). We observe that *the magnitude of the rectangular component in any direction is given by the magnitude of the vector multiplied by the cosine of the angle between the vector and the axis along which we seek the component.* The vector **V** of Fig. 2.9 is the resultant of a vector in the *x* direction of magnitude $V_x = 4$ units and one in the *y* direction of magnitude $V_y = 3$ units. If we let **i** and **j** represent vectors of unit length in the *x* and *y* directions respectively, we can write

$$\mathbf{V} = V_x\mathbf{i} + V_y\mathbf{j} = 4\mathbf{i} + 3\mathbf{j}$$

□ **Example** A 50-N weight rests on an inclined plane (Fig. 2.10) making an angle of 30° with the horizontal. Find the components of the weight parallel and perpendicular to the plane.

We place the origin of a system of axes at the body, with the *x* axis parallel to the plane and the *y* axis perpendicular to the plane. Next we drop perpendiculars from the end of the vector **w** to find the lengths of *X* and *Y*:

$$X = w \cos \alpha = w \sin \theta = (50 \text{ N}) (\sin 30°) = 25 \text{ N}$$
$$Y = w \cos \theta = (50 \text{ N}) (\cos 30°) = 43.3 \text{ N} \qquad □$$

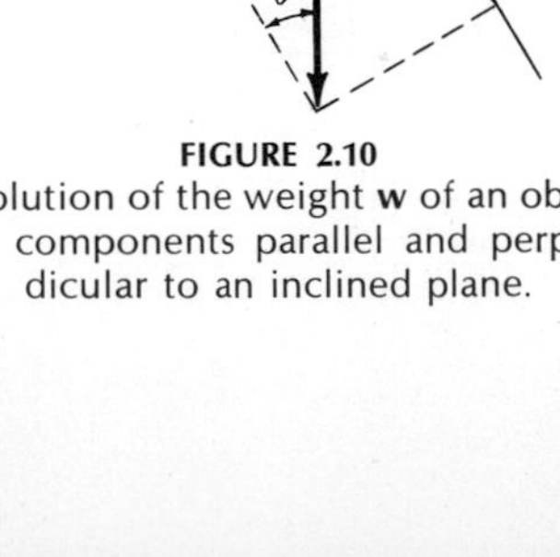

Although in this course we shall be concerned primarily with vectors in a plane, the ideas above can readily be expanded to handle a vector in space which typically has three components. If we place the origin of a cartesian coordinate system at the tail of some vector **A** with the mutually perpendicular *x*, *y*, and *z* axes oriented in any way we choose, we can find

FIGURE 2.9
Resolution of a vector **V** into its rectangular components V_x and V_y.

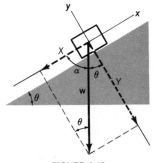

FIGURE 2.10
Resolution of the weight **w** of an object into components parallel and perpendicular to an inclined plane.

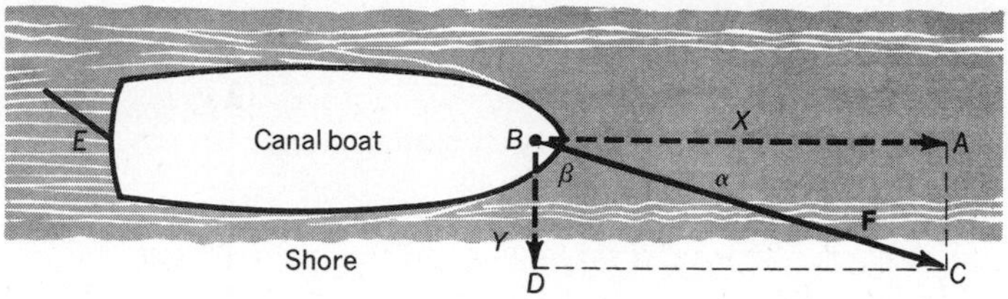

FIGURE 2.11
Components *BA* and *BD* of the force *BC* exerted on a canalboat by a cable.

the x component A_x geometrically by dropping a perpendicular from the tip of **A** to the x axis. Analytically A_x is given by $A \cos \alpha$, where α is the angle between the vector **A** and the x axis. Similarly, we can find the y component A_y and the z component A_z geometrically by dropping perpendiculars to the y and z axes, respectively; if β is the angle between **A** and the y axis, $A_y = A \cos \beta$, while $A_z = A \cos \gamma$, where γ is the angle between **A** and the z axis. If we let **i**, **j**, and **k** stand for unit vectors† in the x, y, and z direction, respectively, we can write

$$\mathbf{A} = A_x \mathbf{i} + A_y \mathbf{j} + A_z \mathbf{k} \qquad (2.2)$$

As another example of the resolution of a vector in a plane, consider a canalboat (Fig. 2.11) upon which a cable pulls with a force **F** in the direction BC. This force simultaneously moves the boat forward and pulls it toward the bank. Two separate forces X and Y could be applied with the same result. These two forces are the rectangular components of **F**; their magnitudes are given by $X = F \cos \alpha$ and $Y = F \cos \beta$.

Similarly, the velocity of the wind propelling a sailboat may be resolved into components perpendicular and parallel to the sail; the force perpendicular to the sail is then resolved into components perpendicular and parallel to the axis of the boat. The component perpendicular to the axis produces little effect because of the shape of the boat.

2.7 THE ANALYTICAL ADDITION OF VECTORS

In adding vectors analytically, it is desirable first to make a rough sketch of the vectors to be added. Then we choose a convenient set of mutually perpendicular axes. Usually, this involves a choice of the east-west and north-south axes or vertical and horizontal axes. However, in some cases it may be more convenient to choose a set of mutually perpendicular axes oriented in some special way, as in Fig. 2.10. Next, we find the components of all the vectors along the axes chosen. We can then ignore the vectors and work only with the components. We can readily add together all the vector components whose arrows are in the direction of each of the axes and then find their resultant by the use of the pythagorean theorem (Appendix A).

In general, if we are adding vectors **A**, **B**, and **C**, where $\mathbf{A} = \mathbf{i}A_x + \mathbf{j}A_y$, and so forth, the resultant **R** is given by

$$\begin{aligned}
\mathbf{R} &= \mathbf{A} + \mathbf{B} + \mathbf{C} \\
&= \mathbf{i}A_x + \mathbf{j}A_y + \mathbf{i}B_x + \mathbf{j}B_y + \mathbf{i}C_x + \mathbf{j}C_y \\
&= \mathbf{i}(A_x + B_x + C_x) + \mathbf{j}(A_y + B_y + C_y)
\end{aligned}$$

†A unit vector in any direction is a vector of unit length in the direction specified; thus **i** is a vector 1 unit long pointing in the x direction.

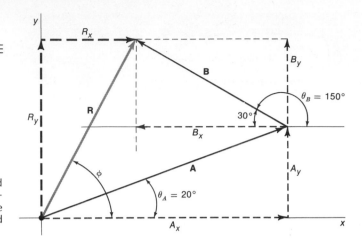

FIGURE 2.12
Analytical addition of two vectors **A** and **B** by resolving each into x and y components. The x and y components of the resultant **R** are $R_x = A_x + B_x$ and $R_y = A_y + B_y$.

Thus each component of **R** has magnitude equal to the sum of the corresponding components of the vectors added. In dealing with the components of vectors along any given axis, we select one direction as positive; a component in the opposite direction is then assigned a negative sign. Thus, if we have a north-south axis and elect to call north the positive direction, a component of 5 km southward is written as -5 km north.

□ **Example** Given two displacements; $A = 52.6$ m at an angle of 20° with the x axis and $B = 34.8$ m at an angle of 150° with the x axis (Fig. 2.12). (a) Find the x and y components of both vectors. (b) Find the magnitude and the direction of the resultant **R**.

(a) The x component of **A** is $A_x = A \cos \theta_A = 52.6 \cos 20° = 52.6 \times 0.940 = 49.4$ m. The y component of **A** is $A_y = A \sin \theta_A = 52.6 \sin 20° = 52.6 \times 0.342 = 18.0$ m. The x component of **B** is $B_x = B \cos \theta_B = -B \cos 30° = -34.8 \times 0.866 = -30.1$ m. The y component of **B** is $B_y = B \sin \theta_B = B \sin 30° = 34.8 \times 0.500 = 17.4$ m. Thus $A = 49.4i + 18.0j$, while $B = -30.1i + 17.4j$.

(b) The x component of the resultant is the sum of the x components of **A** and **B**, so $R_x = A_x + B_x = 49.4 - 30.1 = 19.3$ m; the y component of the resultant is the sum of the y components of A and B, so $R_y = A_y + B_y = 18.0 + 17.4 = 35.4$ m. By the pythagorean theorem, $R^2 = (19.3)^2 + (35.4)^2 = 1,626$, so $R = 40.3$ m. The tangent of ϕ, the angle between **R** and the x axis, is $35.4/19.3 = 1.834$, so the angle ϕ is 61.4°. Consequently we can write $R = (19.3i + 35.4j)$ m $= 40.3$ m at an angle of 61.4° with the x axis. □

□ **Example** Find the resultant of the four forces shown in Fig. 2.7 (repeated here as Fig. 2.13).

First we find the x (east) and y (north) components of each force. It is convenient to list them in a table as shown. For the 5-N force the x component is $(5 \text{ N})(\cos 30°) = 4.33$ N, and the y component in newtons is $5 \sin 30° = 2.50$. For the 7-N force the x (east) component is

(a)

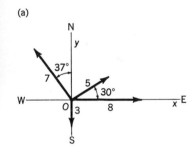

(b)

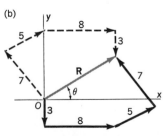

FIGURE 2.13
(Figure 2.7 repeated) The resultant of the four forces in (a) can be found graphically (b) or analytically, as in the text. The forces are given in newtons.

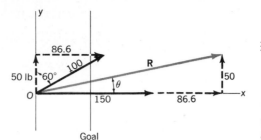

FIGURE 2.14
Finding the resultant **R** of two forces by adding their components and then applying the pythagorean theorem.

$(-7 \text{ N})(\sin 37°) = -4.20 \text{ N}$, and the y component is $(7 \text{ N})(\cos 37°) = 5.60 \text{ N}$. (In calculating these components we have tacitly assumed that the forces involved were 5.00 and 7.00 N, i.e., that they were known to three significant figures at least.) From the table we see that the components of the resultant are 8.13 N east and 5.10 N north. The resultant **R** is given by

$$R = \sqrt{(8.13)^2 + (5.10)^2} \text{ N} = \sqrt{92.1} \text{ N} = 9.60 \text{ N}$$

and the angle between **R** and the x axis is given by

$$\sin \theta = \frac{5.10}{9.60} = 0.531$$

so that $\theta = 32.1°$ north of east. □

| | | Components |
	Force, newtons	x (east)	y (north)
8 E		8.00	0
5 30°N of E		4.33	2.50
7 37°W of N		−4.20	5.60
3 S		0	−3.00
R		8.13	5.10

□ **Example** In a football game two linemen are assigned to open a hole by blocking a guard. One of them exerts a force of 150 lb straight toward the goal line (Fig. 2.14), and the other exerts a force of 100 lb at an angle of 60° with the goal line. Find the resultant force.

The x component of the 100-lb force is $(100 \text{ lb})(\sin 60°) = 86.6 \text{ lb}$. The y component is $(100 \text{ lb})(\cos 60°) = 50.0 \text{ lb}$. The net force toward the goal line is then 236.6 lb, while the new force parallel to the goal line is 50.0 lb. The resultant is

$$R = \sqrt{(237)^2 + (50.0)^2} \text{ lb} = \sqrt{58,700} \text{ lb} = 242 \text{ lb}$$

and the angle θ is given by $\sin \theta = 50.0/242 = 0.207$, so that $\theta = 12°$. □

□ **Example** An airplane cruises at a speed of 300 km/h relative to the surrounding air. If the wind is blowing from the southwest at 40 km/h, find the direction in which the pilot must point the airplane if he wishes to go due east. What is the velocity of the airplane relative to the ground?

The pilot must fly at an angle θ (Fig. 2.15) south of east such that the southward component of $\mathbf{V}_{PA}$ (velocity of plane relative to air) is exactly equal to the northward component of $\mathbf{V}_{AG}$ (velocity of air relative to ground).

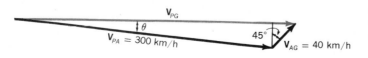

FIGURE 2.15
Flying due east relative to the ground when the wind is blowing from the southwest is accomplished by flying at an angle θ south of east relative to the air.

$$300 \sin \theta = 40 \cos 45° = 28.3$$

$$\sin \theta = \frac{28.3}{300} = 0.0943$$

$$\theta = 5.4°S \text{ of } E$$

The velocity $\mathbf{V}_{PG}$ of the airplane relative to the ground is

$$\mathbf{V}_{PG} = (300 \cos \theta + 40 \sin 45°) \text{ km/h east}$$
$$= (299 + 28.3) \text{ km/h east} = 327 \text{ km/h east} \qquad \square$$

QUESTIONS

1. What justification can you offer for the assertion that force is a vector quantity?
2. Can the resultant of two unequal forces be zero? Of three unequal forces?
3. If the resultant of three vectors is zero, is it necessary that all three lie in the same plane? If so, prove it; if not, propose an example.
4. An airplane has an airspeed of 400 km/h, and the wind is blowing with a speed of 50 km/h. What is the maximum possible speed of the plane relative to the ground? The minimum possible speed? Could the speed relative to the ground also be 400 km/h? If so, explain how.
5. Why does an airplane land and take off into the wind? When a plane is landing on an aircraft carrier, what is the most favorable direction of carrier and plane relative to the wind and to each other? Why?
6. When two vectors **A** and **B** are drawn from a common origin to form sides of a vector parallelogram (Fig. 2.5), one diagonal is the resultant. What is the other diagonal?
7. Can a 0.5-m steel tube be packed in a rectangular box not over 0.3 m long? Explain.
8. How should three vectors of equal magnitude be arranged if they are to have a resultant of zero?
9. Does the odometer of an automobile measure displacement? If not, what does it measure?
10. With the aid of a figure explain how an iceboat can sail faster than 20 km/h in a 20 km/h breeze. What limits the speed the boat can attain?
11. What can you conclude about two vectors **A** and **B** if:
    (a) $\mathbf{A} + \mathbf{B} = \mathbf{R}$ and $R^2 = A^2 + B^2$
    (b) $\mathbf{A} + \mathbf{B} = \mathbf{A} - \mathbf{B}$
    (c) $\mathbf{A} + \mathbf{B} = \mathbf{R}$ and $A + B = R$

PROBLEMS

1. A body is displaced 7 m east and 24 m north. What are the magnitude and direction of the resultant displacement? *Ans.* 25 m 73.7°N of E
2. Calculate the horizontal and vertical components of a vector which has a magnitude of 40 N and makes an angle of 35° with the horizontal.
3. An airplane flies 300 km in a direction 55°S of W. Find the components of its displacement toward the south and toward the west.
   Ans. 246 km, 172 km
4. Find the magnitude of the single displacement that is equivalent to successive displacements of 80 and 50 m, the direction of the second displacement being perpendicular to that of the first.
5. Find the magnitude and direction of the resultant of the three displacements 6 km east, 4 km south, and 5 km 65°N of W. *Ans.* 3.92 km 7.8°N of E
6. Find the resultant of the following four displacements: 7 m east, 8 m south, 3 m west, and 2 m south.
7. A helicopter carrying mail flies 5 km due north and lands. It then flies 15 km in a direction 30°N of E. Find (a) how far north the helicopter is from its original starting point and (b) the magnitude and direction of its resultant displacement.
   Ans. 12.5 km; 18.0 km; 43.9°N of E
8. A man pushes along the handle of a lawnmower which makes an angle of 40° with the horizontal ground. One component of the 90-N force he exerts drives the lawnmower along; another component pushes the lawnmower against the ground. Calculate the force components parallel and perpendicular to the ground.
9. A boy walks 4 km west, 2 km south, and then 1 km east in 1.2 h. Determine the distance traveled, the displacement, the average speed, and the average velocity.
   Ans. 7 km, 3.61 km 33.7°S of W, 5.83 km/h; 3.01 km/h 33.7°S of W
10. In 40 min an automobile is driven 30 km east, then 12 km south, and then 14 km west. Determine the distance traveled, the displacement, the average speed, and the average velocity.
11. Let **i**, **j**, and **k** be unit vectors of 1 cm length in the x, y, z directions. A cube with 2-cm sides is placed with one corner at the origin and three edges of the cube lying along the x, y, and z axes.

Write the position vector for each of the eight corners of the cube in terms of **i**, **j**, and **k**.

Partial ans. $\mathbf{r} = 0\mathbf{i} + 2\mathbf{j} + 2\mathbf{k}$

12. A helicopter undergoes three displacements, all in kilometers, in the xy plane. If the displacements are $\mathbf{r}_1 = 4\mathbf{i} - 5\mathbf{j}$, $\mathbf{r}_2 = 3\mathbf{i} + 10\mathbf{j}$, and $\mathbf{r}_3 = -\mathbf{i} + 3\mathbf{j}$, write the resultant displacement $\mathbf{r}$ in the **ij** notation. What is the magnitude and direction of the resultant displacement?

13. An automobile is driven 35 km east and then 70 km 50°N of E in going from one city to another. Write the displacement vector in the component form of Eq. (2.2). How far apart are the two cities?   *Ans.* $(80\mathbf{i} + 53.6\mathbf{j})$ km; 96.3 km

14. Find the resultant of the following three forces: a force of 50 N making an angle of 40° with the $+x$ axis, a force of 90 N making an angle of 135° with the $+x$ axis, and a force of 50 N making an angle of 210° with the $+x$ axis.

15. Two helicopters take off from the same landing area. One proceeds 10 km 50°S of W; the second flies 8 km 65°S of E. Find the displacement of the second helicopter relative to the first.

Ans. 9.82 km 2.4°N of E

16. The current in a river flows south at a rate of 2 km/h. A man rows a boat west at a rate of 3 km/h relative to the water. Find the speed of the boat relative to the land.

17. Three football players participating simultaneously in a tackle exert the following forces on the ball carrier: 400 newtons north, 480 newtons 20°N of E, and 320 newtons 35°W of N. Find the resultant of these forces.

Ans. 868 newtons at 72.1°N of E

18. A ferryboat goes straight across a river in which there is a current of 3 km/h. If the speed of the boat relative to the water is 8 km/h, find the direction in which it is pointed. What is its velocity relative to the earth?

19. A helicopter has a velocity relative to the air of 40 knots 25°N of W. The velocity of the wind is 15 knots from west to east. Find the velocity of the helicopter relative to the ground.

Ans. 27.2 knots 38.5°N of W

20. A ball is thrown from a moving car with a velocity relative to the car of 16 m/s perpendicular to the direction of the car's motion. If the speed of the car is 25 m/s, find the velocity of the ball relative to the ground.

21. An airplane with an airspeed of 200 km/h is to fly due west. The wind has a speed of 50 km/h from 30°S of W. In what direction should the aircraft be pointed? What is the groundspeed?

Ans. 7.2°S of W; 155 km/h

22. An airplane flies a compass course due south with speed relative to the air of 300 km/h. After 1 h the pilot finds himself over an airport 270 km

south and 40 km east of his starting point. Calculate (*a*) the speed of the wind and (*b*) the direction in which he should set his course to fly due south with this wind.

23. A boat which has a speed of 9 km/h crosses a river in which the current is 2 km/h. At what angle must the boat be headed to land at a point directly opposite its starting point? What is the speed of the boat relative to the ground?

Ans. 12.8° upstream; 8.77 km/h

24. An aircraft has a speed of 300 km/h relative to the air. On a day when the wind speed is 60 km/h, a pilot flies due east relative to the ground above an east-west railroad track. In 30 min the plane is 150 km from its starting point. Calculate the wind direction and the heading of the plane (angle between its axis and the railroad) if the heading is south of east.

25. Two trains traveling toward each other leave stations at different times. One train travels at a speed of 90 km/h, the other at a speed of 80 km/h. The stations are 180 km apart. How much later should the faster train start in order for the trains to meet halfway between the stations?

Ans. 7.5 min

26. The velocity of an aircraft relative to the air is 270 km/h south. The wind is blowing 75 km/h from the south and west with the wind velocity making an angle of 36.9° with east. Find the velocity of the aircraft relative to the ground.

27. The pilot of an airplane is trying to keep on a beam toward an airport due west. The airspeed of this plane is 900 km/h. If the wind is blowing from the southwest at 100 km/h, in what direction must he head the plane? Find the speed relative to the ground.

Ans. 4.5°S of W; 827 km/h

28. A motor launch (see figure) cruises at 15 knots due west while the wind has a velocity of 5 knots due south. In what direction will a flag point if it is hung from the mast of the boat?

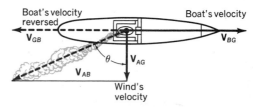

PROBLEMS 28 TO 30

29. A steamboat (see figure) is moving eastward with a speed of 40 km/h relative to the earth. The wind is blowing from north to south with a speed of 10 km/h. Find the velocity of the smoke relative to the boat.   *Ans.* 41.2 km/h 14°S of W

30. A steamboat (see figure) is moving eastward with a speed of 20 mi/h. The wind is blowing from the north with a speed of 8 mi/h. Find the velocity of the smoke relative to the boat.

31. The pilot of an aircraft maintains an airspeed of 600 knots in flying from city A to city B, which is 2,000 nautical miles due east. There is a 90-knot wind from 30°S of W. What course does the pilot maintain? Find his groundspeed and flying time.
    Ans. 4.3°S of E; 676 knots; 2.96 h

32. A boat moves east at 35 km/h. The wind is blowing from the northwest at 10 km/h. To an observer on the boat in what direction does the trail of smoke from the boat lie? What is the speed of the smoke relative to the boat?

33. A room is 10 m long, 6 m wide, and 3 m high. A fly starts on the floor at one corner, assumed to be the origin of a system of axes. The fly ends at the diagonally opposite corner. (*a*) Write a possible vector expression for its displacement. (*b*) Calculate the magnitude of this displacement. (*c*) If the fly crawled from one corner to the other, find the shortest possible distance.
    Ans. (*a*) $\mathbf{r} = (10\mathbf{i} + 6\mathbf{j} + 3\mathbf{k})$ m; (*b*) 12.0 m; (*c*) 13.5 m

34. Show that for any triangle with sides **A**, **B**, and **C**

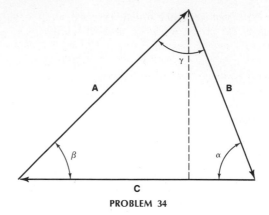

PROBLEM 34

($\mathbf{A} + \mathbf{B} + \mathbf{C} = 0$) and angles α, β, and γ (see figure) $(\sin \alpha)/A = (\sin \beta)/B$. *Hint:* Write the length of the dashed line perpendicular to **C**, first in terms of A and α and then in terms of B and β.

35. Ship A is steaming due north at 20 knots and observes ship B 14 nautical miles dead ahead. After 30 min B is 3 nautical miles due east of A. Both ships maintain constant velocity. (*a*) Write the velocity of B in the form $\mathbf{v}_B = a\mathbf{i} + b\mathbf{j}$, where $\mathbf{i}$ and $\mathbf{j}$ are unit vectors east and north, respectively. (*b*) Write an expression for the distance between the ships as a function of the time in hours.
    Ans. (*a*) $6\mathbf{i} - 8\mathbf{j}$ knots
       (*b*) $(196 - 784\,t + 820\,t^2)^{1/2}$ nmi

36. Show that if ϕ is the angle between two vectors **A** and **B** when they are placed tail to tail, the magnitude of their resultant is given by $R = \sqrt{A^2 + B^2 + 2AB \cos \phi}$. Show also that the angle θ between **A** and **R** is given by $\tan \theta = (B \sin \phi)/(A + B \cos \phi)$.

We have been discussing situations in which the velocity of some body does not change, but in the world of footballs, automobiles, baseballs, and airplanes velocities seldom remain constant. To treat problems in which the velocity varies we introduce the concept of acceleration *and concern ourselves with several kinds of accelerated motion.*

3

TRANSLATIONAL MOTION

3.1 TYPES OF MOTION

When a ball is dropped or an automobile chassis moves along a straight road or an elevator goes up and down its shaft, all points of the body move along parallel lines with the same velocity. Any object which moves in this way is said to undergo a motion of *pure translation*. On the other hand, a merry-go-round or the flywheel of a stationary engine turns about a fixed axis. Such a body is said to undergo a motion of *pure rotation*, or an angular motion. All points in the object describe concentric circles about the axis. When we deal with a rigid body, at any instant we can resolve its motion, no matter how complex, into an appropriate pure translation and a pure rotation. The motion of a spinning tennis ball can be treated as arising from a translation of the center of the ball plus a rotation about the center. Similarly, the motion of the wheel of a train or of a boomerang flying through the air can be regarded as a combination of translational and rotational motions.

In this chapter we develop some of the equations of *kinematics*, the science of motion, for translation (we defer detailed consideration of rotational motion to Chap. 11). The problems we deal with here are relatively simple in terms of the mathematics involved, but nevertheless they are important examples of linear motion which introduce the underlying physical ideas. More advanced problems in kinematics require *calculus*, a branch of mathematics developed by Newton, Leibnitz, and others to solve such problems.

3.2 INSTANTANEOUS VELOCITY

In dealing with the motion of an object, we are concerned with how its displacement and velocity change in time. Consider the motion of an automobile. Let **r** be the radius vector from some chosen origin fixed on the earth to the center of the steering wheel at any time t, which we measure with a stopwatch. Suppose that when we start the stopwatch (time $t = 0$), the radius vector is $\mathbf{r}_0$. (Here and elsewhere we shall use the subscript zero to indicate the value of a quantity at time $t = 0$.) As time advances, the position of the automobile changes as shown in Fig. 3.1, and at time t the radius vector from the origin is **r**. During the time interval t the automobile moves from $\mathbf{r}_0$ to **r**, and its displacement is $\mathbf{s} = \mathbf{r} - \mathbf{r}_0$. The average velocity (Sec. 2.4) is

$$\mathbf{v}_{av} = \frac{\mathbf{r} - \mathbf{r}_0}{t - 0} = \frac{\mathbf{s}}{t} \tag{3.1}$$

In general, the average velocity over any time interval Δt is

$$\mathbf{v}_{av} = \frac{\Delta \mathbf{s}}{\Delta t} \tag{3.2}$$

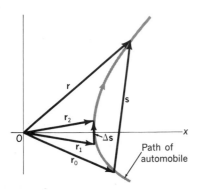

FIGURE 3.1
The average velocity over the time interval t is the ratio of the displacement $\mathbf{s} = \mathbf{r} - \mathbf{r}_0$ to t; the instantaneous velocity is the limit of the ratio $\Delta\mathbf{s}/\Delta t$ as Δt approaches zero.

where $\Delta\mathbf{s}$ is the displacement occurring in the time Δt (the Greek capital delta Δ is mathematical shorthand for "the change in"). Thus, in Fig. 3.1 if the car is located by $\mathbf{r}_1$ at time t_1 and by $\mathbf{r}_2$ at t_2, the displacement $\Delta\mathbf{s}$ is $\mathbf{r}_2 - \mathbf{r}_1$ and the corresponding change in time Δt is $t_2 - t_1$. Consequently the average velocity over the time interval t_1 to t_2 is $(\mathbf{r}_2 - \mathbf{r}_1)/(t_2 - t_1)$.

The velocity of the automobile may well be changing. If so, we may be interested in how fast (and in what direction) it is moving *at a given instant*—a quantity which we call the *instantaneous velocity*. If we want to know the velocity as the car passes a telephone pole, we proceed as follows. We measure the average velocity of the car in the block containing the telephone pole by dividing the length of the block by the time required to cover the block. This will probably be closer to the velocity at the instant of passing than the average velocity over some longer period would be. We might come closer to the instantaneous velocity by determining where the car was 1 s before it reached the pole and 1 s after it passed the pole, but this still gives us an average velocity over a 2-s interval. To get closer to the instantaneous velocity, we measure the displacement over shorter and shorter time intervals. As we make Δt smaller, this average velocity comes closer to the instantaneous velocity. We define the instantaneous velocity $\mathbf{v}$ as the limit approached by $\Delta\mathbf{s}/\Delta t$ as Δt decreases, approaching zero as a limit. We write this

$$\mathbf{v} = \lim_{\Delta t \to 0} \frac{\Delta\mathbf{s}}{\Delta t} \quad \left(= \frac{d\mathbf{s}}{dt} \text{ in calculus notation} \right) \qquad (3.2a)$$

The instantaneous speed is the magnitude of the instantaneous velocity. Ideally, the speedometer of an automobile reads instantaneous speed.

3.3 ACCELERATION

When an automobile is driven in city traffic, its velocity changes frequently, sometimes in magnitude, sometimes in direction, and sometimes in both. When it is going around a curve at constant speed, the velocity is changing only in direction. When the car is proceeding due south along a street with numerous traffic lights, its velocity as a function of time might be that shown in Fig. 3.2. Just before $t = 0$ the car is waiting for a red traffic light to change. When the light turns green, the speed (magnitude of the velocity) increases for a while and then falls to zero in response to another red light. After a few seconds the light

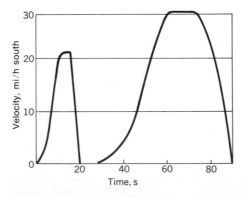

FIGURE 3.2
Velocity plotted as a function of time for an automobile in city traffic.

changes and the car picks up speed again, this time less rapidly than the first time. For the situation described by Fig. 3.2, the velocity is changing only in magnitude; the direction remains the same. Finally, when the car is slowing down as it turns a corner, the velocity is changing in both magnitude and direction.

In treating motion in which the velocity is changing, we define a new term: *acceleration. Acceleration is the time rate of change of velocity.* The average acceleration over any period of time t is the change in velocity divided by the corresponding change in time.

$$\mathbf{a}_{av} = \frac{\mathbf{v} - \mathbf{v}_0}{t} \qquad (3.3)$$

The acceleration vector need not be in the direction of either $\mathbf{v}$ or $\mathbf{v}_0$. For example, when a ball is thrown horizontally, $\mathbf{v}_0$ is horizontal and $\mathbf{a}$ is vertical; 0.1 s later $\mathbf{v}$ will be sloping downward. When $\mathbf{v}$ and $\mathbf{v}_0$ have the same direction all the time, $\mathbf{a}$ must be in the same or in the opposite direction; all three vectors lie along the same line.

In the rather complex motion shown in Fig. 3.2 the acceleration varies, and we define instantaneous acceleration $\mathbf{a}$ as

$$\mathbf{a} = \lim_{\Delta t \to 0} \frac{\Delta \mathbf{v}}{\Delta t} \qquad \left(= \frac{d\mathbf{v}}{dt} = \frac{d^2\mathbf{s}}{dt^2} \text{ in calculus notation} \right) \qquad (3.4)$$

From its definition *acceleration* has the *units of velocity divided by time;* it tells us how much the velocity changes per unit of time. Since velocity is expressed in units of length divided by time, acceleration has units of (length divided by time) divided by time. For example, acceleration may be expressed in meters per second per second (m/s^2). In discussing the motion of an automobile, we might measure the change in velocity in miles per hour and the change in time in seconds. Then the acceleration would be in miles per hour per second. If an automobile decelerates from 80 to 60 km/h in 10 s, the average acceleration is -20 km/h divided by 10 s, or -2 km/h-s. In metric units we shall usually measure acceleration in meters per second per second.

□ **Example** An automobile manufacturer advertises that his product can start from rest and reach 100 km/h (62 mi/h) in less than 10 s. What minimum average acceleration is required?

$$a_{av} = \frac{v - v_0}{t} = \frac{100 \text{ km/h}}{10 \text{ s}} = 10 \text{ km/h-s}$$
$$= 2.78 \text{ m/s}^2 = 9.11 \text{ ft/s}^2 \qquad \square$$

The mathematical treatment of motions in which the acceleration varies irregularly is beyond the scope of this text. We shall confine our discussion to several very important types of motion in which the acceleration is either constant or varies in some relatively simple way. We begin by considering the case in which the acceleration is constant and the motion takes place along a straight line.

3.4 UNIFORMLY ACCELERATED RECTILINEAR MOTION

If a ball is dropped and its velocity measured as a function of time, it is found that the velocity is 0.98 m/s 0.1 s after release, 1.96 after 0.2 s, 2.94

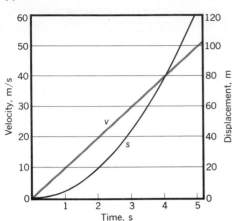

FIGURE 3.3
Velocity (left scale) and displacement (right scale) of a freely falling body as functions of time, (*a*) in SI units and (*b*) in British engineering units.

FIGURE 3.4
A feather and a steel ball have the same accelerations and equal velocities when they fall freely in an evacuated tube.

after 0.3 s, and so forth. The velocity and displacement vary with time after release as shown in Fig. 3.3. Each tenth of a second the velocity increases 0.98 m/s; each second it increases 9.8 m/s (or 32 ft/s).

The acceleration of a freely falling body at mean sea level and 45° latitude is 9.80665 m/s², or 32.17398 ft/s². By a freely falling body we mean one for which the air resistance is negligible. A feather does not fall through the air with this large an acceleration because of the substantial air friction, but if it is released in an evacuated tube, it has the same acceleration as a steel ball (Fig. 3.4). When a body falls through the air, the effect of air friction depends on the velocity, size, shape, density, and surface of the falling object. This friction naturally reduces the acceleration somewhat for a falling body, but if a ball is thrown upward, air friction increases the downward acceleration on the way up. We represent the acceleration due to gravity by g. It varies slightly from place to place for reasons discussed in Sec. 8.9. In our problems on falling bodies we shall take $g = 9.80$ m/s² or 32.0 ft/s², thereby making the arithmetic of the problems a little simpler.

Although freely falling bodies are the most common examples of uniformly accelerated motion, they are by no means the only ones. An automobile either speeding up or slowing down may have a constant linear acceleration over several seconds. Similarly an airplane may have a constant acceleration along the runway as it prepares to take off or a constant deceleration as it slows to a stop.

When the acceleration is constant, the average velocity during any time interval is given by

$$v_{\text{av}} = \frac{v + v_0}{2} \tag{3.5}$$

If a ball is released from rest and falls for 3 s, its initial velocity is 0 and its final velocity is 29.4 m/s. During these 3 s the average velocity is $(29.4 + 0)/2 = 14.7$ m/s. Similarly, if an automobile is *uniformly accelerated* from 20 to 60 km/h, the average velocity during the acceleration is 40 km/h. Equation (3.5) together with the equations defining average velocity and acceleration are the three fundamental equations for uniformly accelerated motion. When we measure displacement from the position of the body at $t = 0$,

$$v_{\text{av}} = \frac{s}{t} \tag{3.1a}$$

$$a = \frac{v - v_0}{t} \tag{3.3}$$

$$v_{\text{av}} = \frac{v + v_0}{2} \tag{3.5}$$

These basic equations can be combined in many ways, two of which are particularly useful:

$$s = v_0 t + \tfrac{1}{2} a t^2 \tag{3.6}$$

$$v^2 = v_0^2 + 2as \tag{3.7}$$

Equation (3.6) can be obtained by combining Eqs. (3.1a), (3.3), and (3.5) as follows:

$$s = v_{\text{av}} t \qquad \text{by Eq. (3.1a)}$$

$$= \frac{v + v_0}{2} t \qquad \text{by use of Eq. (3.5)}$$

$$= \frac{v_0 + at + v_0}{2} t \qquad \text{by use of Eq. (3.3)}$$

$$= v_0 t + \tfrac{1}{2} a t^2$$

Equation (3.7) can be derived by multiplying $s = [(v + v_0)/2]t$ by $a = (v - v_0)/t$, which gives $as = (v^2 - v_0^2)/2$ or $2as = v^2 - v_0^2$.

In the relations developed above, it is important to observe that s is the *displacement* of the body and not the total distance traversed. For example, if we throw a ball up into the air with a speed of 96 ft/s and ask what the displacement is at the end of 5 s, the relations above will provide the answer, which is 80 ft. Figure 3.5 shows the displacement measured upward and velocity as functions of time. During the 5 s the ball traverses a much greater distance. Indeed, at the end of 3 s it is at a height of 144 ft. Further, in the equations above the displacement s is measured *from the position at $t = 0$.* If we wish to measure the displacement from

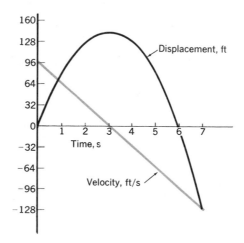

FIGURE 3.5
Velocity (gray) and displacement (black) as functions of time for a ball thrown upward with a speed of 96 ft/s at $t = 0$.

some other point, Eq. (3.6) must be replaced by

$$s = s_0 + v_0t + \tfrac{1}{2}at^2 \qquad (3.6a)$$

where s_0 is the displacement of the body at $t = 0$ from the new reference point.

A second caution has to do with the fact that displacement, velocity, and acceleration are *vector* quantities, and due regard must be given to direction. Which direction one elects to choose as positive is one's own free choice, but once it is decided that upward is positive, any vector quantity which is downward must be given a negative sign. If we throw a ball upward with a speed of 30 m/s, we may decide to choose upward as the positive direction. If we do, $v_0 = 30$ m/s, but $a = -9.8$ m/s². An alternative choice would be to call downward positive. Then v_0 becomes -30 m/s, and $a = +9.8$ m/s².

Equations (3.5) to (3.7) have been developed *only* for the case in which **s**, **v**, **v₀**, and **a** all lie along the same straight line. If these vectors have different directions but **a** is constant, the x components of the vectors satisfy these equations—and so do the y and z components. Indeed, Eq. (3.6) remains a valid equation when the quantities are written as vectors, but Eqs. (3.5) and (3.7) do not.

☐ **Example** Galileo is reputed to have shown that a heavy ball and a light one fall together by dropping them from the leaning tower of Pisa, which is 54 m high. How long does it take a ball to fall this distance? What is the speed of the ball just before it strikes the ground?

Let us take downward as positive. Then $a = 9.8$ m/s², $s = 54$ m, and $v_0 = 0$. By Eq. (3.6)

$$s = v_0t + \tfrac{1}{2}at^2$$
$$54 \text{ m} = 0 + \tfrac{1}{2}(9.8 \text{ m/s}^2)t^2$$
$$t^2 = 11.0 \text{ s}^2$$
$$t = 3.32 \text{ s}$$

By Eq. (3.7)

$$v^2 = v_0{}^2 + 2as$$
$$= 0 + 2(9.8 \text{ m/s}^2)(54 \text{ m})$$
$$= 1,060 \text{ m}^2/\text{s}^2$$
$$v = 32.6 \text{ m/s}$$

Check:

By Eq. (3.5)

$$v_{\text{av}} = \frac{v_0 + v}{2} = \frac{0 + 32.6}{2}$$
$$= 16.3 \text{ m/s}$$

By Eq. (3.1)

$$s = v_{\text{av}}t$$
$$= (16.3 \text{ m/s})(3.32 \text{ s})$$
$$= 54 \text{ m} \qquad\qquad ☐$$

☐ **Example** A ball is thrown upward from a bridge with a speed of 48 ft/s. It misses the bridge on the way down and lands in the water 160 ft below. Find how long the ball rises, how high it goes, how long it is in the air, and its velocity when it strikes the water.

Let us take upward as positive. Then $v_0 = 48$ ft/s and $a = -32$ ft/s². The ball rises until $v = 0$. By Eq. (3.3)

$$a = \frac{v - v_0}{t}$$

$$-32 \text{ ft/s}^2 = \frac{0 - 48 \text{ ft/s}}{t}$$

$$t = \frac{48 \text{ ft/s}}{32 \text{ ft/s}^2} = 1.5 \text{ s}$$

During this 1.5 s

$$v_{av} = \frac{v + v_0}{2} = \frac{0 + 48 \text{ ft/s}}{2} = 24 \text{ ft/s}$$

$$s = v_{av}t = (24 \text{ ft/s})(1.5 \text{ s}) = 36 \text{ ft up}$$

By Eq. (3.6)

$$s = v_0 t + \tfrac{1}{2}at^2$$

$$-160 = 48t + \tfrac{1}{2}(-32)t^2 \qquad \text{(Note that } s \text{ is negative.)}$$

$$t^2 - 3t - 10 = 0$$

$$t = +5 \text{ s (or } -2)$$

(Clearly we want the positive answer here. The negative answer tells us how much before we threw the ball we would have had to fire it upward from the water to have it pass the bridge going 48 ft/s at $t = 0$.) By Eq. (3.3)

$$a = \frac{v - v_0}{t}$$

$$-32 \text{ ft/s}^2 = \frac{v - 48 \text{ ft/s}}{5 \text{ s}}$$

$$v = -112 \text{ ft/s} \qquad \text{(The minus sign means } \textit{downward}.)$$

Check:

$$s = v_{av}t = \frac{v + v_0}{2}t$$

$$-160 \text{ ft} = \frac{-112 + 48}{2} \text{ ft/s} \times 5 \text{ s} = -160 \text{ ft} \qquad \square$$

3.5 PATH OF A PROJECTILE FIRED HORIZONTALLY

If a body is projected horizontally from the top of a tower of height h (Fig. 3.6) with a velocity v_x, it continues to move with the same horizontal velocity it had at the beginning of its path (any decrease caused by the resistance of the air is neglected). At the same time the body falls because of the attraction of the earth. Hence, at any instant the velocity of the projectile has two components: a horizontal component which remains constant and a downward component which increases with the time. The horizontal component of the velocity and the horizontal displacement are independent of whether or not the body is falling. Similarly, the vertical components of the displacement, velocity, and

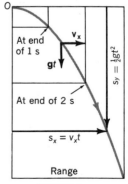

FIGURE 3.6
Path (grey) of a body projected horizontally with a velocity v_x. The displacement at time t has horizontal component $v_x t$ and vertical component $\tfrac{1}{2}gt^2$; the velocity has components v_x and gt.

acceleration are independent of whether or not the body is moving horizontally.

If a bullet is fired horizontally with a velocity of 700 m/s, it falls 4.9 m in 1 s, 19.6 m in 2 s, and so forth, exactly like a bullet which was simply dropped from the end of the gun. In this case the vertical velocity of the body after t seconds is given by $v_y = gt$. The horizontal component of the velocity remains constant, so that the resulting velocity, which is tangent to the path of the body, has a magnitude $v = \sqrt{v_x^2 + v_y^2} = \sqrt{v_x^2 + g^2t^2}$. The horizontal displacement s_x of the body in time t is equal to $v_x t$, while the distance the body falls in the same time is $s_y = \frac{1}{2}gt^2$.

When an aircraft flying straight and level releases a bomb, the bomb continues to move forward with constant horizontal velocity until it strikes the ground (if air friction is neglected). The downward velocity component, which was zero at release, increases linearly with time because of the acceleration of gravity. In order to hit the target, the bomb must be released some time before the aircraft is directly above the target.

□ **Example** A ball is thrown horizontally with a velocity of 20 m/s from a tower 40 m high. Find the time of flight, the horizontal range, and the speed of the ball just before it strikes the ground.

Let us choose downward as positive. For the vertical motion, $a_y = 9.8$ m/s^2, $v_{0y} = 0$, and $s_y = 40$ m. By Eq. (3.6)

$$s_y = v_{0y}t + \tfrac{1}{2}a_y t^2$$
$$40 \text{ m} = 0 + \tfrac{1}{2}(9.8 \text{ m/s}^2)t^2$$
$$t = \tfrac{20}{7} = 2.86 \text{ s}$$

The horizontal range s_x is the horizontal velocity (which is constant at 20 m/s) multiplied by the time of flight.

$$s_x = v_x t = (20 \text{ m/s})(2.86 \text{ s}) = 57.2 \text{ m}$$

Just before hitting, $v_x = 20$ m/s, and, by Eq. (3.3),

$$v_y = a_y t = (9.8 \text{ m/s}^2)(\tfrac{20}{7} \text{ s}) = 28 \text{ m/s}$$
$$v = \sqrt{v_x^2 + v_y^2} = \sqrt{(20)^2 + (28^2)} \text{ m/s} = 34.4 \text{ m/s} \qquad \square$$

3.6 PROJECTILE FIRED AT AN ANGLE WITH THE HORIZONTAL

If an object is given a velocity **V** at an angle θ with the horizontal, it follows a parabolic path (Fig. 3.7). If there were no gravity, this projectile would move along the line **V**t. In time t the effect of gravity is to bring the projectile a distance $\frac{1}{2}gt^2$ below this line. In treating projectile problems we shall assume that air friction is negligible, that the horizontal component of the velocity remains constant, and that the vertical motion has uniform acceleration **g**. These are reasonable approximations for our purposes, but they are not entirely justified in many practical situations.

The first step in analyzing the motion of a projectile fired at an angle with the horizontal is to resolve the velocity of projection into vertical and horizontal components. If the velocity of projection **V** makes an

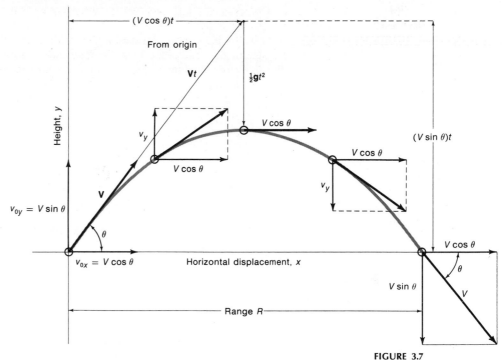

FIGURE 3.7
Path (grey) of a body projected with
velocity **V** at an angle θ above the horizontal axis; the displacement at any time
t is the resultant of the vectors **V**t and
$\frac{1}{2}\mathbf{g}t^2$. At five positions of the body the
instantaneous velocity vector is shown
along with its components.

angle θ with the horizontal, the horizontal component is $v_x = V \cos \theta$, the vertical component $v_y = V \sin \theta$. The horizontal velocity remains constant, but the vertical component changes. After time t the vertical component is $v_y = V \sin \theta - gt$. The projectile rises until the vertical component of the velocity becomes zero. At this instant the height is maximum, and the time of rise t_r is given by

$$0 = V \sin \theta - gt_r \quad \text{or} \quad t_r = \frac{V \sin \theta}{g}$$

The total time the projectile is in the air is the sum of the time of rise t_r and the time of fall t_f. Throughout the time of flight the horizontal component of the velocity remains constant at $V \cos \theta$. The horizontal range of the projectile is therefore $V \cos \theta \, (t_r + t_f)$. If the projectile falls to the same height from which it was projected, the time of fall is the same as the time of rise, and the time of flight is given by $t_r + t_f = (2 \, V \sin \theta)/g$. In this case maximum range is attained when $\theta = 45°$ (Fig. 3.8) if air friction is negligible. Note that under these conditions the range is the same for any two complementary angles.

☐ **Example** A punter kicks a football with a velocity of 80 ft/s at an angle of 36.9° with the horizon. Find the time during which the ball rises, how high it goes, and the length of the kick (from punter's toe to receiver's hands).

The vertical component of the initial velocity is $(80 \text{ ft/s})(\sin 36.9°) = 48$ ft/s, and the horizontal component $(80 \text{ ft/s})(\cos 36.9°) = 64$ ft/s. If we choose upward as positive for the vertical motion, $v_{0y} = 48$ ft/s and

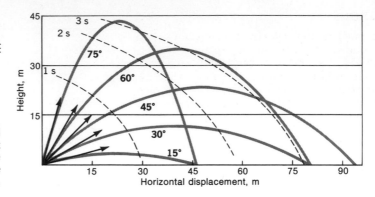

FIGURE 3.8
Approximate paths of balls projected with the same initial speed (30 m/s) at angles of 15, 30, 45, 60, and 75° above the horizontal. Dashed lines show the positions of the balls after 1, 2, and 3 s.

$a_y = -32$ ft/s². At the top of the path $v_y = 0$; thus, by Eq. (3.3),

$$-32 \text{ ft/s}^2 = \frac{0 - 48 \text{ ft/s}}{t}$$

$$t = 1.5 \text{ s}$$

On the way up, the average velocity is [by (Eq. 3.5)] 24 ft/s. Since $s_y = v_{av}t$, we have $s_y = (24 \text{ ft/s})(1.5 \text{ s}) = 36$ ft, which is how high the ball goes.

It takes as long for the ball to come down as it did to go up, so that the total time in the air is 3.0 s. In this time the horizontal velocity is constant at 64 ft/s, so that

$$s_x = (64 \text{ ft/s})(3.0 \text{ s}) = 192 \text{ ft or 64 yd} \qquad \square$$

□ **Example** A stone is thrown from the edge of a cliff with a velocity of 16 m/s at an angle of 30° above the horizontal, and 4 s later it strikes the ocean below. Find the height of the cliff above the ocean and the horizontal range of the stone.

The vertical component of the initial velocity is 16 sin 30° = 8 m/s upward. If we choose upward as positive, $v_{0y} = 8$ m/s and $a_y = -9.8$ m/s². The vertical distance down to the ocean is given by

$$
\begin{aligned}
s_y &= v_{0t}t + \tfrac{1}{2}at^2 \\
&= (8 \text{ m/s})(4 \text{ s}) + \tfrac{1}{2}[(-9.8 \text{ m/s}^2)(16 \text{ s}^2)] \\
&= 32 \text{ m} - 78.4 \text{ m} \\
&= -46.4 \text{ m}
\end{aligned}
$$

The horizontal range s_x is $s_x = v_x t$ since v_x is constant. Then

$$v_x = (16 \text{ m/s})(\cos 30°) = 13.9 \text{ m/s}$$
$$s_x = (13.9 \text{ m/s})(4 \text{ s}) = 55.6 \text{ m} \qquad \square$$

1. Can a body with a velocity directed northward have southward acceleration? Explain.
2. Under what circumstances can a body have (*a*) an acceleration but no velocity, (*b*) an acceleration but constant speed, (*c*) a velocity but no acceleration, and (*d*) an acceleration which is constantly changing?
3. Does a negative acceleration necessarily imply that a body is slowing down? What (or who) determines the sign of the acceleration in a particular problem?
4. Why are the sights on a rifle adjusted when the distance to the target is changed?
5. If a body has a uniform velocity, what is the relationship between its initial, final, and average velocities? If a body has a uniform acceleration, what is the relationship between its initial, final, and average velocities?
6. For a given initial speed a football kicked at 35° has the same range as one kicked at 55°. Why would the lower trajectory be better for a quick kick when the receiving safety man is close to the line of scrimmage? Why would the higher trajectory be preferred if the receiver is back to catch the ball?
7. If a body is thrown vertically upward and air resistance is not negligible, is the time of rise equal to the time of fall? Which is greater? Explain.
8. Why does an outfielder trying to cut off a run in a baseball game ordinarily try to get the ball to the catcher on the first bounce rather than on the fly?
9. Show that in the absence of air friction, the speed of a ball is the same at any two points on its trajectory which are at the same height. How do the velocities at these two points compare?
10. What is the effect of air friction on the vertical motion of a baseball? On the horizontal motion? Compare the path of the baseball assuming air friction with the ideal parabolic path by means of a sketch.
11. A man stands at the edge of a bridge high above a river. He throws one ball straight upward with speed v_0 and another identical ball straight downward with the same initial speed. If air friction is negligible, which ball will have the greater speed when it strikes the river?
12. Discuss why it is not possible in general to take account of the effects of air on the motion of a ball by simply reducing the value of g.
13. Draw the path of a stone thrown backward with a speed of 15 m/s from the rear of a train traveling 20 m/s (*a*) from the point of view of the thrower and (*b*) from the frame of an observer on the ground.

14. What are the effects of air friction and the curvature of the earth on the motion of a long-range ballistic missile? What other factors must be taken into account?
15. Suppose a man threw a rock at a monkey on the limb of a tree 20 m away. If the initial velocity vector pointed straight at the monkey's heart, would the rock strike the monkey if he remained on the limb? Suppose at the instant the rock was released, the monkey dropped straight down from the limb. Now what would happen? (Consider Fig. 3.7.) Does the initial speed of the rock or the initial height of the monkey make any difference?

PROBLEMS

Unless otherwise specified, assume that air resistance is negligible and that g is 9.8 m/s² or 32 ft/s².

1. A uniformly accelerated body is moving with a velocity of 6 m/s south; 4 s later it has a velocity of 2 m/s² north. What is the acceleration?
   Ans. 2 m/s² north
2. An airplane starting from rest has a uniform acceleration of 2 m/s² south. What is its velocity at the end of 25 s if this acceleration is maintained?
3. Calculate your average speed if you (*a*) walk 800 m at a speed of 1.6 m/s and then jog 800 m at a speed of 4 m/s; (*b*) walk for 500 s at a speed of 1.6 m/s and then jog for 500 s at a speed of 4 m/s.
   Ans. (*a*) 2.29 m/s; (*b*) 2.8 m/s
4. A ball is dropped from a bridge. If it requires 3 s for the ball to strike the river below, how high is the bridge in meters? How fast is the ball traveling just before it reaches the river?
5. A freight train starts from rest, is uniformly accelerated, and travels 150 m in 50 s. Find the acceleration, the average speed, and the final speed.
   Ans. 0.12 m/s²; 3 m/s; 6 m/s
6. A woman has a 180-mi trip to make, of which the first third is over congested highways while the remainder is over a good freeway. If she averages 40 mi/h over the first 60 mi, at what average speed must she drive the remaining 120 mi to average 45 mi/h for the entire trip?
7. A bullet has a speed of 350 m/s as it leaves a rifle. If it is fired horizontally from a cliff 6.4 m above a

lake, what is the horizontal range of the bullet?
Ans. 400 m

8. A train has an acceleration of 0.25 m/s² in a direction opposite that of its motion. How long will it take the train to stop if it is initially going 30 m/s (67 mi/h)? How far will it travel in this time?

9. An automobile traveling 30 m/s can be stopped on a dry highway in 75 m, while on a wet pavement it requires 120 m, and on an icy pavement it requires 600 m. (*a*) Calculate the average deceleration in each case. (*b*) Estimate the corresponding stopping distances for an initial speed of 15 m/s. (*c*) If the reaction time required for a typical driver to set his brakes is 0.75 s, calculate the total distance traversed after a driver traveling 30 m/s on a wet pavement sees something requiring an emergency stop.
Ans. (*a*) 6, 3.75, and 0.75 m/s²; (*b*) 18.8, 30, and 150 m; (*c*) 143 m

10. In March 1954 Col. J. P. Stapp rode a rocket-propelled sled down a track at 1,017 km/h (632 mi/h). He blacked out but suffered no serious injury when he and the sled were stopped in 1.4 s. Calculate the average acceleration in meters per second per second and in g's.

11. A naval aircraft is to be launched by a catapult from a carrier. If the catapult acts over a distance of 30 m, find the minimum average acceleration which will give the aircraft a speed of 45 m/s at launching. *Ans.* 33.8 m/s² = 3.44 g's

12. A particle moves in a straight line with an acceleration of 2 m/s². If it starts from rest, how far does it go in 5 s? What is its velocity after 5 s? How long is required for the particle to go 49 m? What is its speed as it passes the 49-m point?

13. An object passes the origin at $t = 0$ with a velocity of 18 m/s to the right. It has a constant acceleration, and 12 s later it is moving to the left with a speed of 6 m/s. Calculate (*a*) the acceleration, (*b*) the position of the particle at $t = 12$ s, and (*c*) the maximum displacement from the origin during the 12-s period.
Ans. (*a*) 2 m/s²; (*b*) 72 m to right; (*c*) 81 m

14. The takeoff speed of an airplane is 40 m/s. What uniform acceleration must this plane have if it is to become airborne after a 500-m run? How long does the takeoff require?

15. In a football game a passer throws the ball with a speed of 60 ft/s. If he produces speed by moving the ball over a distance of 5 ft along a straight line, find the average acceleration during the throwing operation. *Ans.* 360 ft/s²

16. The speed of a baseball pitched by Bob Feller was once determined to be 44.2 m/s. What average acceleration must Feller have given the ball if this speed was produced over a distance of 3 m?

17. With what initial velocity must a ball be thrown upward to reach a maximum height of 40 m? How long is the ball in the air, assuming it is caught at the same height at which it was released? *Ans.* 28 m/s; 5.71 s

18. A basketball player can reach to a height of 2.5 m when he is standing. If he is a good jumper, he can reach 3.5 m. With what vertical velocity must he leave the floor to reach to 3.5 m?

19. If a car going 60 km/h can be stopped in 30 m, how much distance is required to stop the car when it is traveling 120 km/h, assuming that the same constant acceleration is involved?
Ans. 120 m

20. The maximum deceleration which the tires of an automobile can produce on a certain wet pavement is 18 ft/s². Find the minimum distance in which a car traveling 100 ft/s (68 mi/h) can be stopped, measuring from the point at which the brakes are first applied. How long does it take?

21. When an aircraft lands on a carrier, it is arrested by a restraining cable, which is engaged by a tail hook as the aircraft passes over the cable. If the cable produces an average deceleration of 2.5 g's (24.5 m/s), how far does the aircraft move after it engages the cable at a speed of 40 m/s?
Ans. 32.7 m

22. A woman is reported to have survived a 144-ft fall from the seventeenth floor of a building when she landed on a metal ventilator box, which she crushed to a depth of 18 in. What was her average deceleration in feet per second per second and in g's? (No doubt air friction was important here, so her maximum speed was less than that of a freely falling body.)

23. A ball is batted straight up and returns to the level of the batter in 6.5 s. How high does the ball travel? What are its velocity and acceleration at the top of its flight? What are its initial and final velocities?
Ans. 51.8 m; 0 m/s, 9.8 m/s²; 31.9 m/s

24. If an aircraft lands at a speed of 120 mi/h (176 ft/s) and the maximum deceleration which the braking system can produce is 10 ft/s², find the minimum distance in which the plane can be stopped on level ground. How long does it take to bring the aircraft to rest?

25. A ball is thrown upward from the roof of a building. It has an initial velocity of 15 m/s. How high is the building if on its downward flight it just

misses the thrower and falls to the ground 5 s after it is thrown? With what speed does it strike the ground? *Ans.* 47.5 m; 34 m/s

26. A rifle fires a bullet with a muzzle speed of 300 m/s. If the rifle sights are aligned with the barrel, find how far above a bull's-eye 100 m away the rifle should be aimed to hit the center. (The usual practice is to adjust the sights to allow for the fall of the bullet for some prescribed target distance.)

27. An automobile starts from rest, maintains a constant acceleration of 2 m/s² until its speed is 20 m/s, and then continues at this speed. Find the time required for the car to go 800 m.
Ans. 45 s

28. At the kickoff a football is given a velocity of 80 ft/s at an angle of 53.1° with the horizontal. Find how long the ball is in the air, how high the ball goes, how far from the kickoff point the ball lands, and the magnitude of the velocity of the ball 3 s after it is kicked. Compare your results with those of the first example in Sec. 3.6.

29. A ski jumper takes off horizontally from a level ramp. He lands 8.28 m below the point of takeoff and 20 m from a vertical line through this point. Find the speed of the skier as he leaves the ramp.
Ans. 15.4 m/s

30. A baseball is batted, and 6 s later it is caught by the center fielder 420 ft from the batter. (Assume the ball was hit and caught 4 ft above the ground; neglect air friction.) Find (*a*) the horizontal component of the ball's velocity and (*b*) the maximum height *above the ground* reached by the ball.

31. A boy kicks a soccer ball off the ground, giving it a speed of 18 m/s at an angle of 53.1° with the horizontal. Find how long the ball is in the air, how high it goes, and the horizontal distance it traverses before it strikes the ground.
Ans. 2.94 s; 10.6 m; 31.7 m

32. A girl on a bridge throws a stone horizontally with a speed of 25 m/s, releasing the stone from a point 19.6 m above the surface of a river. How far from a point directly below the girl will the stone strike the water?

33. A golf ball is driven horizontally with a speed of 40 m/s from a cliff 25.6 m high, and it lands on a level plain below. How long is the ball in the air? How far from the base of the cliff does it strike the ground? *Ans.* 2.29 s; 91.4 m

34. A baseball is batted with a speed of 120 ft/s at an angle of 36.9° with the ground. Find how long the ball is in the air, how high the ball goes, how far from the plate it is caught, and the components of the velocity of the ball 3 s after it is hit.

35. A golf ball on a level fairway is given an initial velocity of 50 m/s at an angle of 35° with the horizontal. If air friction is neglected, find

(*a*) how long the ball is in the air, (*b*) how high it goes, and (*c*) how far away it lands.
Ans. (*a*) 5.85 s; (*b*) 42.0 m; (*c*) 240 m

36. A golfer chips a ball over a sand trap to an elevated green, and 3 s after she hits the ball it lands on the green 50 m east and 12 m above her initial lie. (*a*) Find the components of the initial speed of the ball as it left the club. (*b*) Find the speed of the ball when it strikes the green.

37. A boy throws a stone toward the ocean with an initial velocity of 24.5 m/s 36.9° above the horizontal. The stone just clears a fence at the edge of a cliff; the height of the fence is luckily the same as the height at which the stone is released. The stone lands in the ocean 50 m below the top of the fence. Find (*a*) the distance from the boy to the fence, (*b*) the horizontal distance from the fence to the point at which the stone hits the ocean, and (*c*) the speed of the stone the instant before it hits.
Ans. (*a*) 58.8 m; (*b*) 39.8 m; (*c*) 39.8 m/s

38. A golf ball is hit with a speed of 150 ft/s at an angle of 30° with the horizontal from an elevated tee 40 ft above the fairway. How long is the ball in the air? At what horizontal distance from the tee does the ball strike the ground?

39. A basketball player releases a ball 7 ft above the floor when he is 30 ft from the basket. The ball goes through the rim of the basket 10 ft above the floor 1.5 s later. Find the horizontal component of the initial velocity, the vertical component of the initial velocity, and the maximum height above the floor reached by the ball.
Ans. 20 ft/s; 26 ft/s; 17.6 ft

40. A golfer chips a ball high over a sand trap. If the ball leaves her club at an angle of 53.1° with the horizontal, what velocity must it be given to land 90 m distant? (Assume the green is at the same level as the initial lie of the ball.) How high does the ball go? How long is it in the air?

41. A baseball is given a velocity of 100 ft/s at an angle of 53.1° above the horizontal. It leaves the bat at a height of 3 ft above the ground. What are its height and distance from home plate (*a*) after 2 s and (*b*) after 4 s? (*c*) Will the ball clear a 15-ft fence 280 ft from the plate for a home run? By how much?
Ans. (*a*) 99 ft, 120 ft; (*b*) 67 ft, 240 ft; (*c*) yes, about 12 ft

42. Find the minimum speed with which a baseball

must be hit at an angle of 53.1° with the horizontal to leave the bat at a height of 4 ft and clear a 30-ft screen at a distance of 300 ft from home plate. Assume that air friction is negligible.

43. An automobile traveling 100 km/h in a 50-km/h zone passes a parked patrol car. The automobile maintains a constant speed. The patrol car has a constant acceleration of 3 m/s² until it reaches a speed of 120 km/h and thereafter holds its speed constant. How long will it take the patrol car to catch the speeding vehicle, assuming the former starts as the latter passes by?   *Ans.* 33.3 s

44. A boy riding a bicycle at a speed of 5 m/s passes a parked car; 20 s later the car starts and is uniformly accelerated for 12 s, at which time it passes the cyclist. What was the acceleration of the car?

45. A ball is projected upward from the bottom of a tower that is 88.2 m high, and at the same instant another ball is dropped from the top of the same tower. If the balls meet at a point halfway between the top and the bottom of the tower, with what initial velocity was the ball projected upward? How high will this ball rise?

Ans. 29.4 m/s; 44.1 m

46. A station wagon starts from rest and has a linear acceleration of 1.5 m/s². A police car parked 121 m behind starts at the same instant and accelerates at 2 m/s². Calculate (a) how long it takes before the two vehicles are abreast and (b) the distance traveled by the station wagon.

47. Balls rolling on an inclined plane have an acceleration of 0.5 m/s². One ball is released from rest 16 m up the plane, and at the same instant a second ball at the bottom is rolled upward along the plane. If the second ball returns to the bottom at the same moment as the first, find the initial speed of the second ball and how far it rolled up the plane.   *Ans.* 2 m/s; 4 m

48. A ball is dropped from a window and is caught by a boy 19.6 m below. A second ball is tossed upward by his sister at the instant the first was dropped, and it returns to her at the same instant the first ball is caught. How high did the second ball rise above the point from which it was tossed and caught?

49. An elevator moves upward at a speed of 3 m/s along the side of a building under construction. When the elevator is 7 m above the ground, a boy on the ground throws a ball vertically upward with a speed of 20 m/s, releasing the ball 2 m above the ground. Write expressions (numerical as far as possible) for the displacement, velocity, and acceleration as functions of time (a) for a reference frame with its origin on the ground and (b) a reference frame riding with the elevator.
Ans. (a) $y = 2 + 20t - 4.9t^2$ m;
$v = 20 - 9.8t$ m/s; $a = -9.8$ m/s²
(b) $y = -5 + 17t - 4.9t^2$ m;
$v = 17 - 9.8t$ m/s; $a = -9.8$ m/s²

50. A projectile fired with initial velocity v_0 at an angle θ above the horizontal has only a downward acceleration of magnitude g in the flat-earth, frictionless approximation. (a) Write the vector equation giving the displacement $\mathbf{r}$ as a function of time. (b) Show that the equation of the trajectory is

$$y = x \tan \theta - \frac{gx^2}{2v_0^2 \cos^2 \theta}$$

51. An object of mass m initially at rest at the top of a frictionless inclined plane of length L slides to the bottom in a time t. With what velocity must the body be started up the plane if it is to slide $0.81L$ before it starts back down?   *Ans.* $1.8L/t$

52. Show that in the flat-earth, frictionless-atmosphere approximation the horizontal range R of a mass projected with initial speed V at an angle θ above the horizontal is given by $R = (2 V^2 \sin \theta \cos \theta)/g$ and that the ratio of the maximum height reached to R is $(\tan \theta)/4$.

It is obvious from experience that the motion of a body can be controlled by forces, but it was not until the seventeenth century that Newton discovered precisely how force and motion are related. He formulated three celebrated laws of motion, which we now introduce and discuss.

4
NEWTON'S LAWS OF MOTION

4.1 NEWTON'S FIRST LAW OF MOTION

It is a familiar fact that the motion of a body is intimately related to the forces which act upon it. However, for centuries the discovery of how force and motion are related eluded the philosophers who speculated about it. Aristotle expressed the idea that a body in motion comes to rest unless it has a force acting upon it continuously. For centuries men wondered what it was that pushed the planets around in the sky and what made the moon go around the earth. Although brilliant minds worked on the problem, no consistent and acceptable relationship between force and motion was established until the seventeenth century.

Galileo performed a number of careful experiments on the motions of bodies; it is largely to his work that we owe the equations of kinematics in the preceding chapter. Galileo studied accelerated motion by dropping bodies and by rolling balls down inclined planes. He observed that when friction was very small, a ball would roll for a great distance on a horizontal plane without stopping. Eventually, he became convinced that a ball on a perfectly frictionless horizontal plane would persist forever in its motion at constant speed. This revolutionary idea proved most fruitful in developing an understanding of motion. Galileo discovered that bodies do not fall with constant speed but have a *constant acceleration.* This suggested that the pull of the earth produces not the motion itself but the change in the state of motion.

Isaac Newton (1642–1727) accepted Galileo's conclusions and formulated this idea as his first law of motion, which may be stated as follows:

Every body continues in its state of rest or of uniform velocity in a straight line unless it is compelled to change that state by the application of some resultant external force.

No actual body is ever completely free from external forces, but there are situations in which it is possible to make the resultant force approximately zero. In those cases we find that the body behaves in accordance with the first law of motion. Since we can never eliminate friction completely in our experiments, and since our efforts to compensate for it are imperfect, we must recognize that Newton's first law is an idealization. However, it is an idealization which provided the key for building a consistent and understandable theory of motion. There are, of course, many ways in which this great principle can be enunciated. Another statement is the following:

A body at rest remains at rest, and a body in motion remains in motion with constant velocity, unless acted upon by some resultant force.

When a very small force acts on a reasonably massive body for a very short time, the resulting change in the velocity of the body is very small, and the zero-force requirement in Newton's first law is *almost* satisfied. Under these conditions we can snap a card from under a coin (Fig. 4.1) or jerk a tablecloth from under a glass. Newton's first law can also be invoked to explain how we put the head of a hammer onto the handle (Fig. 4.2).

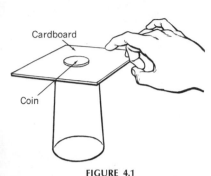

Cardboard

Coin

FIGURE 4.1
A piece of cardboard can be snapped from under a coin, which falls into the glass.

FIGURE 4.2
When the handle stops suddenly, the head of the hammer continues downward in accord with Newton's first law until external forces stop it.

4.2 INERTIA

In Newton's first law of motion there appears an important property of matter known as *inertia*—that property of matter by which it maintains a constant velocity in the absence of an unbalanced external force. When an automobile is suddenly stopped, the passengers obey Newton's first law and continue in motion with constant velocity until some external force changes their state of motion. Seat belts in an automobile can provide such an external force—one much preferred to that exerted by the windshield or dashboard. A person running on an icy sidewalk finds it difficult to stop suddenly because friction is inadequate to provide the necessary external force. When a baseball leaves the pitcher's hand, it continues to move with essentially constant velocity until it reaches the catcher's glove. No force is required to keep it moving. Of course, the ball is slowed down slightly by air resistance and is pulled toward the earth by its weight.

If a heavy ball is suspended from a string (Fig. 4.3) and an identical string is fastened below, the upper string breaks under a slowly increasing steady pull from below. But if a sudden jerk is applied to the lower string, it breaks. The inertia of the ball is so great that it isolates the upper string from the jerk.

All matter has inertia. The concept of mass was introduced by Newton as a measure of inertia. At any point on the earth the weight of any body is proportional to its mass, but if the body were in interplanetary space where it no longer had observable weight, it would still have inertia. It would still take the same force of toe on ball to give a football a specified acceleration. If the football were filled with mercury, its inertia would be greatly increased. To kick such a football would be just as painful in a rocket ship as on the earth, even though the mercury-filled ball might have little or no weight in interplanetary space.

4.3 NEWTON'S SECOND LAW

If no external unbalanced force acts upon a body, it maintains a constant velocity. What happens if there is an external unbalanced force? To answer this question quantitatively, let us consider a set of idealized experiments.

1. Suppose that we have a perfectly level and frictionless table along which we can accelerate a mass of several kilograms. (We could make the friction very small by using small wheels with roller bearings.) If we now take an accurately calibrated spring balance and exert a force **F** on the mass, a certain acceleration is produced. Let us measure this acceleration. Next, let us exert exactly twice as great a force and again measure the acceleration (Fig. 4.4). We find the acceleration to be exactly double the first acceleration. If we again double the resultant force, the acceleration doubles once more and is thus 4 times the first acceleration. By measuring the acceleration for a large number of different resultant forces, we find that (within experimental error) the *acceleration* of our chosen mass *is directly proportional to the resultant force* **F** and its direction is that of the *resultant* force.

2. Next suppose that we choose a resultant force **F** and measure the acceleration it produces on a mass of 1 kg. Then let us keep the force

constant but increase the mass accelerated to 2 kg (Fig. 4.5). We find the acceleration is half as great as it was the first time. If we increase the mass accelerated to 3 kg and measure the acceleration, we find it is one-third as great. If this experiment is performed for a number of different masses, we find that the data are consistent with the idea that *the acceleration* **a** *is inversely proportional to the mass accelerated.*

If we combine the results of these two series of experiments, we conclude that

$$\mathbf{a} = \frac{k\mathbf{F}}{m} \qquad (4.1)$$

This equation is a statement (in restricted form) of Newton's second law of motion. *The acceleration of a body is directly proportional to the resultant force acting upon it and is inversely proportional to the mass of the body.*

Newton's second law has proved to be a powerful tool for understanding natural phenomena. For any body it relates the resultant force acting on the body, its mass, and its acceleration. If any two are known, the third can be found; thus:

1. If we know the resultant of all the forces acting on a body as well as its mass, we can deduce its acceleration and predict its motion.

2. If we know the forces acting on a body and can measure its acceleration, we can calculate its mass; indeed, this is one of the most successful ways of determining the masses of atoms and subatomic particles.

3. If we know the mass of a body and can find its acceleration by observing its motion, we can determine what net force is acting on the body. Even if we do not know the mass, from knowledge of the motion we can infer the ratio of force to mass; it was in this way that Newton discovered the law of universal gravitation (Sec. 8.5).

Newton's second law is a cornerstone of mechanics; it is also of fundamental importance in acoustics, electricity, and other areas of classical and modern physics.

In applying Newton's second law to a real situation, the "body" involved may be a system of several objects. For example, it could be an airplane with people, luggage, and freight aboard. Always in using Eq. (4.1) to solve a problem, *m* is the total mass of the "body" and **F** is the resultant of all forces which act on that body. Newton's second law applies equally well to the airplane system and to a particular suitcase aboard.

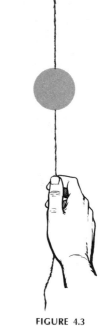

FIGURE 4.3
Inertia is an important aspect of mass. Here a slow continuous increase in the downward force breaks the string above the heavy ball, but the lower string breaks when the lower string is jerked downward.

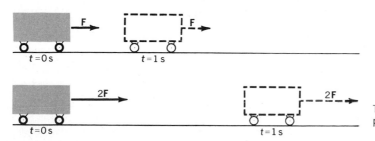

FIGURE 4.4
The acceleration of a body is directly proportional to the resultant force acting on the body.

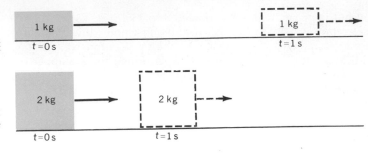

FIGURE 4.5
The acceleration due to a given resultant force is inversely proportional to the mass accelerated.

4.4 NEWTON'S THIRD LAW

Consider a cord of negligible weight from which a ball is suspended (Fig. 4.6*a*). If the cord is at rest, the upward pull of the ceiling is equal and opposite to the downward pull of the ball. The force exerted on the cord at either end is known as the *tension* in the cord. The cord, subjected to pulls at each end, is said to be "in tension."

We now direct our attention to the ball of Fig. 4.6*b*. When the ball hangs at rest, there are two forces acting *on it:* its weight **w**, which is the downward pull of the earth, and the pull **T** of the cord upward. By Newton's first law, the resultant of these two forces must be zero, so they must be equal in magnitude and opposite in direction.

Note that the cord pulls upward on the ball with the force **T** while the ball pulls downward on the cord with an equal and opposite force. This is an example of the application of Newton's *third law*, which states: *If body A exerts a force on body B, body B exerts an equal and opposite force on body A.* Forces in nature always occur in pairs which are equal and opposite. However, the two members of the pair always act on *different* bodies. In Fig. 4.6*a* the ceiling pulls up on the cord, which in turn pulls down on the ceiling with an equal and opposite force.

When you hold this book in your hand, you exert an upward force *on the book;* at the same time the book exerts an equal downward force *on your hand.* If a man pulls *on a rope* with a force of 40 N, the rope pulls back *on the man* with a force of 40 N.

In the examples above we have introduced Newton's third law of motion in treating a system of bodies at rest. This law is equally valid in dealing with bodies in motion, either uniform or accelerated. The wheels of an automobile in motion push backward on the road, but the road pushes forward on the wheels with an equal force. *Whenever one body exerts a force upon a second body, the second body exerts an equal and opposite force on the first.*

Whenever a force acts upon a body, there is always an equal and opposite force exerted by the body (Figs. 4.7 and 4.8). A train pulls the locomotive back with a force which is exactly as great as the forward force which the locomotive exerts on the train. A helicopter pushes downward on the air with a force equal to that which the air exerts upward on the helicopter. The sun pulls on the earth and the earth pulls on the sun with equal and opposite forces. Newton stated his third law in the form: *To every action there is an equal and contrary reaction.* Here the term "action" is used to imply "force."

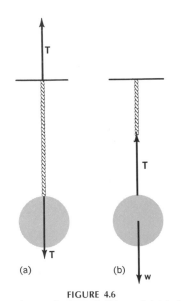

FIGURE 4.6
(*a*) The cord of negligible weight is in tension under the influence of the upward force **T** exerted by the support and the equal downward force due to the weight. (*b*) The upward force **T** exerted on the weight by the string is equal to the downward force **w** exerted on the string by the weight.

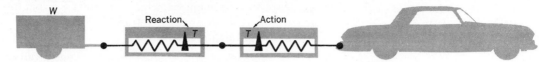

W
Reaction
Action
T
T

FIGURE 4.7
The force exerted on the trailer by the car (the action) is equal and opposite to the force exerted on the car by the trailer (the reaction).

4.5 UNITS FOR APPLYING NEWTON'S SECOND LAW

In Sec. 4.3 we wrote Newton's second law as an equation

$$\mathbf{a} = \frac{k\mathbf{F}}{m} \qquad (4.1)$$

This equation tells us that the acceleration $\mathbf{a}$ of a mass m is proportional to the resultant $\mathbf{F}$ of all forces which act upon this mass. Since it is a vector equation and both m and k are positive, the equation correctly predicts that the acceleration is in the same direction as the resultant force.

The proportionality constant k in Eq. (4.1) depends upon the units in which the force, the mass, and the acceleration are measured. By a suitable choice of these units one can make the constant k equal to unity. Such a choice results in convenience in handling the great structure of mechanics, which is built around Newton's second law of motion. Once we agree to measure force, mass, and acceleration in units for which $k = 1$, we can write Newton's second law as

$$\mathbf{F} = m\mathbf{a} \qquad (4.2)$$

Actually there is no limit to the number of different ways in which we could make $k = 1$. To see that this is true, note that we could choose any units for any two of the quantities $\mathbf{F}$, m, and $\mathbf{a}$. Then we could define a new unit for the third of such a size that Eq. (4.2) is satisfied.

Unfortunately, many different schemes for making $k = 1$ have been used. Each has advantages and disadvantages. In our work with Newton's second law and its consequences, we shall use two of the many possible choices, one for working in the SI and a very different one for the British system.

The Newton as a Unit of Force In the SI the fundamental units are the kilogram for mass, the meter for length, and the second for time. Acceleration then has the dimensions meters per second per second. With the units for both mass and acceleration chosen, we can use Eq. (4.2) provided we measure force in terms of an appropriate unit. The required unit is the newton. *One newton is that resultant force which produces an acceleration of one meter per second per second in a mass of one kilogram.*

The newton is the basic force unit in the SI, and $1 \text{ N} = 0.2248 \text{ lb}$. The newton is a so-called "absolute" unit of force because its definition is made without any reference to the earth or its pull. From $\mathbf{F} = m\mathbf{a}$, we see that 1 N is equivalent to 1 kg-m/s^2. The weight of 1 kg at sea level and 45° north latitude is 9.80665 N.

In dealing with tiny forces in the metric system of units, the *dyne* is sometimes used ($1 \text{ dyne} = 10^{-5} \text{ N}$). The dyne is the force required to produce an acceleration of one centimeter per second per second in a mass of one gram.

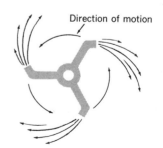

Direction of motion

FIGURE 4.8
The force of the water pushing on the nozzle is exactly equal and opposite to the force the nozzle exerts on water.

□ **Example** An automobile has a mass of 1,600 kg. What force is required to give this automobile an acceleration of 1.20 m/s² if there are frictional retarding forces totaling 200 N?

By Newton's second law, the unbalanced force is given by

$$F = ma$$
$$= (1{,}600 \text{ kg})(1.20 \text{ m/s}^2)$$
$$= 1{,}920 \text{ N}$$

This is the resultant, or unbalanced, force on the automobile. The total force is the sum of the force required to produce the acceleration (1,920 N) and the force required to overcome friction (200 N); thus the total force is 2,120 N. □

The Slug as a Unit of Mass For the *British engineering system* of units an entirely different approach is adopted. The constant k in Newton's second law of motion is made equal to unity by choosing not a new unit of force but a new unit of mass.† In this system the basic unit chosen for acceleration is the foot per second per second, and the basic force unit is the pound force, which is the gravitational attraction of the earth for a 1-lb mass at a place where the acceleration due to gravity has the value 32.17398 ft/s². This force unit is known as a *gravitational unit*, since it is defined in terms of the pull of gravity on a standard object. Once we have chosen such a force unit and an acceleration unit, we make k equal to unity by choosing a new unit of mass. This unit, called the *slug* (*slug*gishness = inertia), is defined as follows: *One slug is the mass of a body which experiences an acceleration of one foot per second per second when acted upon by an unbalanced external force of one pound.*

A force of 1 lb gives a 1-lb mass an acceleration of 32.17 ft/s² and gives a 1-slug mass an acceleration of 1 ft/s². Therefore, the slug has a mass slightly more than 32 times that of the standard pound mass. In the British engineering system, forces are measured in pounds, masses in slugs, and accelerations in feet per second per second.

When the unbalanced force acting on a body is its weight **w**, the resulting acceleration is **g**, the acceleration due to gravity. For this case **F** = *m***a** becomes

$$\mathbf{w} = m\mathbf{g} \tag{4.3}$$

If we divide Eq. (4.2) by Eq. (4.3), we obtain **F**/**w** = **a**/**g**, or

$$\mathbf{F} = \frac{w}{g}\mathbf{a} \tag{4.4}$$

a form in which it is often convenient to put Newton's second law, especially when using the British engineering system of units.

□ **Example** A 160-lb woman stands in an elevator; find the force exerted on her by the floor of the elevator when it has (*a*) a constant downward velocity of 4 ft/s, (*b*) an upward acceleration of 4.0 ft/s², and (*c*) a downward acceleration of 4.0 ft/s².

---

†An alternative giving rise to the British *absolute* system of units is to retain the pound as a unit of mass and to define a new unit of force called the *poundal*. One poundal is that force which produces an acceleration of one foot per second per second in a mass of one pound.

(a) If the velocity is constant, the acceleration is zero and the unbalanced force is zero. The upward force **E** exerted by the elevator is just equal in magnitude to the downward pull of gravity, which is the woman's weight **w**:

$$E = w = 160 \text{ lb}$$

(b) When the elevator has an upward acceleration of 4.0 ft/s², the mass accelerated is 5 slugs (160 lb ÷ 32 ft/s²) and the resultant force is given by

$$F = ma = \frac{w}{g}a$$

$$= \frac{160 \text{ lb}}{32 \text{ ft/s}^2} \times 4.0 \text{ ft/s}^2 = 20 \text{ lb}$$

This is the *unbalanced force* on the woman. If E is the force exerted by the elevator, $F = E - w$, $20 = E - 160$, or $E = 180$ lb upward.

(c) The unbalanced force is now 20 lb downward, so that E is 20 lb less than w:

$$E = 160 - 20 = 140 \text{ lb} \qquad \text{still upward} \qquad \square$$

4.6 MASS: ANOTHER POINT OF VIEW

Ordinarily, when we wish to determine the mass of a body, we do so by comparing its mass with that of standard masses on an equal-arm balance or by some equivalent method. Given a standard mass m_s, can we measure the mass of some other body by a method which is completely independent of the earth's gravitational pull? We can see that the answer is affirmative by considering an imaginary but fundamentally straightforward experiment.

Consider a spaceship located in a region in which there is no observable gravitational pull. In this ship imagine several bodies, one of which we arbitrarily agree is to be our standard mass, to which we assign a value m_s. To determine the mass m_1 of one of the other bodies, we allow it to interact with the standard body in the several different ways suggested by Fig. 4.9. In each case the interaction produces accelerations of magnitudes a_s and a_1 in opposite directions. We measure both these accelerations at the same instant and find that, for any interaction, the ratio a_1/a_s is always the same. We now define the ratio of the masses to be

$$\frac{m_s}{m_1} = \frac{a_1}{a_s} \tag{4.5}$$

so that $m_1 = (a_s/a_1)m_s$, or $m_1a_1 = m_sa_s$, which is just what Newton's second and third laws predict.

In a similar way the mass m_2 of a second body can be determined by having it interact with the standard mass m_s. In this case $m_2a_2 = m_sa_s$, where a_2 and a_s are the magnitudes of the accelerations of m_2 and m_s, respectively, when they are interacting only with each other. If m_2 and m_1 interact, we find $m_1a_1 = m_2a_2$. Thus, a consistent set of mass values can be assigned to any number of particles, even in a region in which none of the particles has any weight.

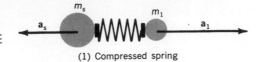

(1) Compressed spring

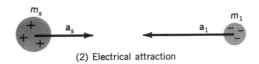

(2) Electrical attraction

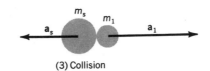

(3) Collision

FIGURE 4.9
Equal but opposite forces are exerted on m_1 and n_s by each other when they interact. In the absence of any other force on either body, $m_1\mathbf{a}_1 = -m_s\mathbf{a}_s$, regardless of the particular way in which the bodies interact.

4.7 MOMENTUM AND NEWTON'S SECOND LAW

In Sec. 4.3 we introduced Newton's second law in a somewhat restricted form. In order to express this law in a more general form, we introduce linear *momentum*. *The linear momentum of a body is the product of the mass and the linear velocity.* Momentum is a *vector* quantity which has the direction of the velocity.

☐ **Example** A fullback with a mass of 90 kg is running with a velocity of 8 m/s due north. Find his momentum.

$$\text{Momentum} = \text{mass} \times \text{velocity}$$
$$= (90 \text{ kg})(8 \text{ m/s north}) = 720 \text{ kg-m/s north} \qquad ☐$$

A force changes the momentum of a free mass upon which it acts. The rate at which the momentum of the body changes is directly proportional to the force. Indeed, Newton's second law may be written in the form: *The time rate of change of the momentum of a body is proportional to the net force acting upon the body and is in the direction of this resultant force.* We can write

$$\mathbf{F} = \frac{\Delta(m\mathbf{v})}{\Delta t} \quad \left[= \frac{d(m\mathbf{v})}{dt} \text{ in calculus notation} \right] \qquad (4.6)$$

When we deal with a single simple mass and a velocity small compared with that of light, the mass is constant and we can write Eq. (4.6) as $\mathbf{F} = m\,\Delta\mathbf{v}/\Delta t = m\mathbf{a}$. However, there are many situations in which we must use Newton's second law in the form of Eq. (4.6). For example, at a speed approaching the speed of light, the mass of a body increases as the speed increases (Sec. 42.8). Similarly, if a sailor pulls a rope along a deck from a large coil, the mass in motion constantly increases. In rocket propulsion (Sec. 7.9) the mass of the rocket decreases as the hot gases are ejected. In each of these cases the mass being accelerated is not constant, and to find the velocity and displacement as functions of time we use Newton's second law in the form $\mathbf{F} = \Delta(m\mathbf{v})/\Delta t$.

1. A book rests on a horizontal desk top. What two forces act on the book? What are the reactions to these forces, and on what body is each reaction exerted?
2. Describe and discuss the problems of a man standing in a bus (a) when the brakes are suddenly applied and (b) when the bus has a large forward acceleration.
3. What is the fallacy in the assertion that you cannot throw a baseball because (a) by Newton's third law the ball exerts as big a force on you as you can exert on it, (b) the sum of these forces is zero, and (c) therefore, by Newton's second law, there is no acceleration?
4. If the same retarding force is available to stop a car regardless of its speed, how much is the stopping distance increased by doubling the speed?
5. A serious neck injury is sometimes incurred by the driver of a stopped car which is struck from behind. Discuss the underlying physics.
6. In a tug of war do both teams pull equally hard on each other? If so, how can one team beat the other?
7. Aristotle argued that if one stone lies on another, the upper one pushes down on the lower; as a consequence two stones fall faster than one, and a heavy stone falls faster than a light one. What is wrong with this argument?
8. A woman stands on the platform of a large, sensitive scale. Discuss the reading as a function of time if the woman tosses a heavy ball upward and catches the ball on its way down. How would things differ if she tossed the ball horizontally to someone standing nearby?
9. Discuss the possibility of driving a sailboat by means of a blower which directs a stream of air against the sails. Which way would the boat be most likely to move? Why?
10. One 0.5-kg mass is suspended from a spring balance, and an identical 0.5-kg mass is balanced on one pan of an equal-arm balance. If both masses are in an elevator, what happens in each case when the elevator is given an upward acceleration?
11. Show that if two bodies A and B initially at rest are acted upon only by their mutual interaction, the distance moved by A is to the distance moved by B as m_B is to m_A.
12. What are the dimensions of mass in the British engineering system for which length, time, and force are the fundamental quantities?
13. A frictionless plane in an elevator makes an angle θ with the horizontal. What is the acceleration of a block on the plane when the elevator has

(a) constant velocity v upward, (b) constant velocity v downward, (c) constant acceleration a upward, (d) constant acceleration a (less than or equal to g) downward?
14. Newton's third law says that action and reaction are always equal and opposite. If this is true, why don't they always cancel each other and leave no unbalanced force acting on any body? (Is your answer to this question compatible with your answer to Question 3?)

PROBLEMS

1. A body has a mass of 8 kg. (a) Find its weight in newtons. (b) If a resultant force of 12 N acts on the body, determine its acceleration.
   Ans. (a) 78.4 N; (b) 1.5 m/s²
2. A net force of 10 N acting on a body produces an acceleration of 2 m/s². Find the mass of the body and its weight.
3. What resultant force is required to accelerate a 1,200-kg automobile uniformly from 4 to 20 m/s in 10 s? *Ans.* 1,920 N
4. Find the resultant force required to give a 1,000-kg automobile an acceleration of 2 m/s².
5. Find the resultant force required to accelerate a 2,400-lb car from rest to 30 mi/h (44 ft/s) in 6 s. What is the mass of the car in slugs?
   Ans. 550 lb; 75 slugs
6. The maximum braking force available for a 6,400-lb airplane is 800 lb. What deceleration does this force produce? If the plane lands at a speed of 120 ft/s, how long does it take to stop? How far does it go in that time?
7. A 2-g bullet is uniformly accelerated in a rifle barrel 1.2 m long. If the bullet leaves the barrel with a speed of 300 m/s, find the net force acting on the bullet in the barrel *Ans.* 75 N
8. A particular type of string breaks when the tension exceeds 11 lb. If a piece of this string is tied to an 8-lb weight, what is the maximum vertical acceleration which can be achieved by pulling upward on the string?
9. A 90-kg man stands in an elevator. Find the force which the floor of the elevator exerts on the man (a) when the elevator has an upward acceleration of 2 m/s²; (b) when the elevator is rising at constant speed; (c) when the elevator has a downward acceleration of 2 m/s².
   Ans. (a) 1,062 N; (b) 882 N; (c) 702 N

10. A man stands on a spring scale in an elevator. When the elevator is at rest, the scale reads 192 lb. What does it read when the elevator has an upward acceleration of 4 ft/s²? A downward acceleration of 4 ft/s²? What is the acceleration when the scale reads 202 lb?

11. A 160-lb woman stands on a spring balance in an elevator. What does the balance read when the elevator has (a) a constant upward speed of 8 ft/s, (b) a constant upward acceleration of 8 ft/s², and (c) a constant downward acceleration of 8 ft/s²? *Ans.* (a) 160 lb; (b) 200 lb; (c) 120 lb

12. A baseball has a mass of 0.145 kg. Find the average resultant force required to give this baseball a speed of 30 m/s over a distance of 2.5 m.

13. A football has a mass of 0.42 kg. What unbalanced force is required to give this football an acceleration of 300 m/s²? If this force acts over a distance of 1.5 m, with what speed does the ball leave the passer? *Ans.* 126 N; 30 m/s

14. A string that can sustain a tension of 60 N is fastened to a 5-kg mass lying on a table. What is the largest vertical acceleration that can be imparted to the mass without breaking the string?

15. A man weighing 160 lb slides down a rope that can sustain only 140 lb. What is the smallest acceleration the man can have without breaking the rope? *Ans.* 4 ft/s²

16. A porter carries a 15-kg bag into an elevator. What force must he exert on the bag in order to hold it when the car is started with an upward acceleration of 3 m/s²? When the car has a constant upward velocity of 3 m/s? When the car has a downward acceleration of 3 m/s²?

17. A Boeing 707 jet aircraft has a takeoff mass of 1.2×10^5 kg. Each of its four engines has a net thrust of 75 kN. Calculate the acceleration and the length of the ground run if the takeoff speed is 73 m/s. *Ans.* 2.5 m/s²; 1,070 m

18. A 70-kg woman slides down a rope that serves as a fire escape. The maximum force that can be applied to the rope without breaking it is 620 N. Find the least acceleration the woman can have without breaking the rope.

19. The brakes of a 1,000-kg automobile can exert a retarding force of 4,000 N. Find the distance the car will move before stopping if it is traveling at the rate of 25 m/s when the brakes are applied. *Ans.* 78.1 m

20. If the retarding forces on a 3,200-lb automobile add up to 200 lb, what would be the deceleration of the car coasting to a stop on a level road? How far would it coast from an initial speed of 50 ft/s?

21. A 4-kg mass interacts with a 5-kg mass through a light stretched spring. If the 5-kg mass has an acceleration at some instant of 1.6 m/s², find the net force on the 4-kg mass and its acceleration at that instant. *Ans.* 8 N; 2 m/s

22. A 12,000-lb aircraft lands on a level field at a speed of 100 ft/s. If the greatest acceleration which its brakes can produce is −5 ft/s², find the minimum distance in which the aircraft can stop, the time required to stop, and the retarding force available.

23. A 200,000-kg DC-10 aircraft uses a run of 3.6 km to achieve its takeoff speed of 320 km/h. If each of the three engines produces the same thrust and the acceleration can be assumed to be uniform, find the net thrust per engine.
Ans. 73.2 kN per engine

24. When the DC-10 of Prob. 23 reaches its destination, its mass is 150,000 kg (50,000 kg of fuel have been consumed). It lands at 270 km/h and decelerates uniformly to a stop in 2.4 km. What is the average net retarding force?

25. A toboggan slides down a steep hill which makes an angle of 25° with the horizontal. If friction between the snow and the toboggan is negligible, find the acceleration. *Hint:* See Fig. 2.10.
Ans. 4.14 m/s²

26. What is the acceleration of a skier coming down an incline which makes an angle of 20° with the horizontal? Assume negligible friction between the skis and the snow.

27. A catapult for launching aircraft from a carrier produces an acceleration of 3 g's (29.4 m/s²) on a 10,000-kg aircraft. Find the unbalanced force required to produce this acceleration. Through what distance must this unbalanced force act to give the plane a speed of 60 m/s?
Ans. 294 kN; 61.2 m

28. The car of a frictionless elevator and its contents exert a force of 2 tons on the cables when at rest. How great is the force when an upward acceleration of 7 ft/s² is being given to the elevator? When the acceleration is numerically the same but downward?

29. Two carts are in contact on a horizontal frictionless air track; cart *A* has a mass of 3 kg, and cart *B* a mass of 2 kg. (a) Find the force on *A* needed to give the two carts an acceleration of 0.8 m/s². What is then the force exerted on *B* by *A*? (b) If the carts are given an equal acceleration in the opposite direction by pushing on *B*, find the force exerted on *A* by *B*. Explain why the force exerted on *B* by *A* in part (a) is not equal to the force exerted on *A* by *B* in part (b).
Ans. (a) 4 N, 1.6 N; (b) 2.4 N

30. Each car of a toy train weighs 2 lb. In a train containing four cars and the engine, the engine pulls on the first car with a force of 3 lb. If the acceleration of the train is 5 ft/s^2, with what force does the first car pull on the engine? What is the unbalanced force on the first car? What is the total frictional force on the four cars?

31. The three bodies A, B, and C of the accompanying figure have masses of 6, 3, and 2 kg, respectively. If they are given a vertical acceleration of 2 m/s^2, find the tensions T_1, T_2, and T_3.

Ans. 129.8, 59, and 23.6 N

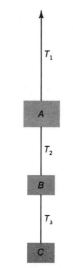

PROBLEMS 31 AND 32

32. If A, B, and C of the accompanying figure weigh 10, 5, and 3 lb, respectively, find the vertical acceleration, T_1, and T_3 if T_2 is 9 lb.

33. A 2-kg block rests on a horizontal frictionless table top 1.225 m high. A constant horizontal force F pushes the block from rest at a point 1.6 m from the edge of the table until it leaves the table with a speed of 4 m/s. Find (a) the force F and (b) the horizontal distance from the table edge to the point at which the block reaches the floor. *Ans.* (a) 10 N; (b) 2 m

34. A 16,000-lb airplane starts from rest and has a uniform acceleration of 5 ft/s^2. If it requires a speed of 90 ft/s to take off, find (a) the minimum run for a takeoff, (b) the minimum time from rest to takeoff, and (c) the net thrust from the engine available for acceleration.

35. A 2-kg mass is placed on one of two identical carts initially at rest. The carts are then driven apart by an ideal massless spring and receive velocities of 6 and -9 m/s. (a) Find the mass of the carts. (b) An unknown mass is added to the second cart, and the experiment is repeated; the observed velocities are now 4 and -2.5 m/s, respectively. What mass was added?

Ans. (a) 4 kg; (b) 5.6 kg

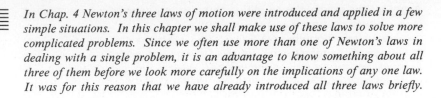

5

FORCES AND MOTION

In Chap. 4 Newton's three laws of motion were introduced and applied in a few simple situations. In this chapter we shall make use of these laws to solve more complicated problems. Since we often use more than one of Newton's laws in dealing with a single problem, it is an advantage to know something about all three of them before we look more carefully on the implications of any one law. It was for this reason that we have already introduced all three laws briefly.

5.1 STATICS AND DYNAMICS

Our everyday experience supports the idea that forces must be exerted on bodies whenever we wish to change their state of motion. It is now our objective to obtain a thorough understanding of the quantitative relationships between forces and motions, a subject called *dynamics*. But first we examine the conditions which must be satisfied in order that a body be at *rest*. This study is known as *statics*, and it is here that we begin our formal development.

According to Newton's first law, a body at rest remains at rest and a body in motion remains in motion with constant velocity so long as the resultant force acting on it is zero. If the resultant force is not zero, Newton's second law states that the body has an acceleration in the direction of the resultant force. Regardless of whether the body is at rest or in motion, it is the *resultant of all forces acting on the body* which must be determined.

It is important to distinguish between the forces exerted *on* a body and the forces exerted *by* that body. According to Newton's third law, if body *A* exerts a force on body *B*, body *B* exerts an *equal and opposite* force on body *A*. Thus, forces always come in equal and opposite pairs, but the members of each pair always act on *different* bodies. In calculating the resultant force *on* a body one must consider *all* forces which act *on the body* and *only* those forces. Often it is helpful to draw a sketch of the body, showing each force acting on the body by an appropriate vector. Such a sketch is called a *free-body diagram*.

5.2 CONCURRENT FORCES

When two or more forces act upon a body and the lines of action of these forces pass through a common point, the forces are said to be *concurrent*. By the "line of action" of a force we mean simply the line along which the force acts, extended indefinitely in both directions.

The two forces **w** and **T** acting on the ball of Fig. 5.1 are concurrent; both act along a line which passes through the center of the ball. Now, suppose that someone were to push on the ball with a force of 5 N to the right, and you wish to keep the ball at rest. What must you do? Clearly, you must push with a force of 5 N in the opposite direction (Fig. 5.1). In general, if we wish to keep the ball at rest, we must balance any force applied in any given direction with an equal force in the opposite direction. (Of course, we can replace any force with its rectangular components acting at the same point.)

A body that remains at rest or moves with constant velocity is said to be in a state of *equilibrium* (equal balance). For a body to be in equilibrium under the influence of any number of concurrent forces, *the vector sum of the forces acting on the body must be zero*. If we draw a vector diagram to represent the forces, the condition that the resultant be zero is equivalent to the condition that the force polygon close (Fig. 5.2).

In treating the equilibrium of any body remember that we must be sure to include *all* the forces which act *on that body* and *only* those forces. The first question to be resolved in working any equilibrium problem is: Of what body is the equilibrium being considered? Then *all* the forces exerted *on this body*, and no other forces, are taken into account. A figure showing the body in question with *all* forces which act upon it (a free-body diagram) is valuable in solving problems involving the equilibrium or acceleration of any object.

5.3 STATICS OF A PARTICLE

Clearly, if the resultant of all the forces acting on a body is zero, the sum of the rectangular components of these forces along any axis must be zero. Thus, when we resolve the forces into components in mutually perpendicular x and y (and z if necessary) directions, for a body in equilibrium the sum of the x components is zero, or the sum of the components in the $+x$ direction is equal to the sum of the components in the $-x$ direction. Similarly, the sum of the y (and z) components is zero.

Usually, but not always, it is desirable to choose vertical and horizontal axes. In this case:

1. *The sum of all upward force components is equal to the sum of all downward force components.*

2. *The sum of all force components to the right is equal to the sum of all force components to the left.*

In mathematical shorthand these conditions can be written as

$$\Sigma F_{up} = \Sigma F_{down} \tag{5.1}$$

$$\Sigma F_{right} = \Sigma F_{left} \tag{5.2}$$

where Σ (Greek capital sigma) is a standard abbreviation for "the sum of."

Consider next the case in which the ball of Fig. 5.1 is pulled to the left by a horizontal force **F**, as indicated in Fig. 5.3. The ball is acted upon by three concurrent forces **F**, **T**, and **w**, the vector sum of which must be zero.

In treating a problem of this kind it is convenient to resolve the tension **T** into a vertical component $T\cos\theta$ and a horizontal component $T\sin\theta$. If we apply Eqs. (5.1) and (5.2) above, we obtain immediately

$$T\cos\theta = w \qquad \text{and} \qquad T\sin\theta = F$$

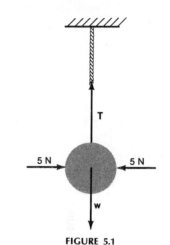

FIGURE 5.1
If the ball is to remain at rest when a force is exerted on it to the right, an equal force to the left is required. The ball is in equilibrium when the resultant of all forces acting on it is zero.

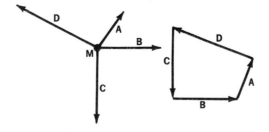

FIGURE 5.2
The point mass M is in equilibrium under the influence of four forces if the resultant force is zero; this is equivalent to requiring that the force polygon close.

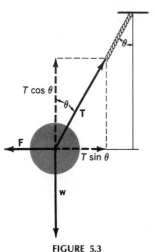

FIGURE 5.3
A ball in equilibrium under the influence of three concurrent forces.

□ **Example** If $w = 20.0$ N and $\theta = 35°$, find F and T.

$$T \cos \theta = w \qquad F = T \sin \theta$$
$$0.819T = 20.0 \text{ N} \qquad = (24.4 \text{ N})(0.574)$$
$$T = 24.4 \text{ N} \qquad = 14.0 \text{ N} \qquad \square$$

□ **Example** A weight of 40 N is supported as shown in Fig. 5.4. Find the tension in each rope.

The lower rope supports the 40-N weight, so the tension in it is 40.0 N. Now let us consider that the knot where the ropes meet is the body in equilibrium. It is acted upon by three forces w, T_1, and T_2.

The vertical component of T_1 is $T_1 \sin 53.1° = 0.800T_1$, and the horizontal component is $T_1 \cos 53.1° = 0.600T_1$. The vertical component of T_2 is $T_2 \sin 30° = 0.500T_2$, and the horizontal component is $T_2 \cos 30° = 0.866T_2$. Applying the force conditions (5.1) and (5.2) yields

$$0.800T_1 + 0.500T_2 = 40.0 \text{ N}$$
$$0.866T_2 = 0.600T_1$$

from which $T_1 = 34.9$ N and $T_2 = 24.2$ N. (Note that the larger force is exerted by the shorter rope in this case.) □

A force which is equal and opposite to the resultant of two or more forces is called the *equilibrant*. Figure 5.5 shows the resultant and equilibrant of two forces **P** and **Q** which are not mutually perpendicular.

5.4 FRICTION

Frictional forces play an important role in mechanics, both for moving bodies and for those at rest. Unlike the weight of a body, which is constant at any particular location, we find that the frictional force between a body and a surface depends on many factors. For example, when a heavy block of wood is pushed along the top of a table, the

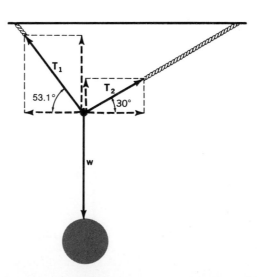

FIGURE 5.4
A knot in equilibrium under the influence of three concurrent forces T_1, T_2, and **w**.

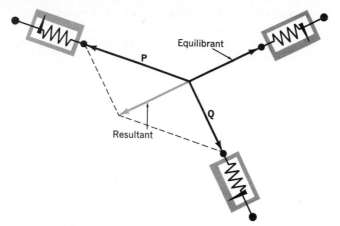

FIGURE 5.5
Resultant and equilibrant of two forces
P and **Q**.

frictional resistance depends on the surface of the table, the surface of the block, how clean the surfaces are, the speed of the block, and the force pressing the surfaces together. When a body moves on a surface, there is always such resistance to the motion. In many situations friction is undesirable, and elaborate means are employed to make it as small as possible.

Friction has advantages as well as disadvantages. If it were not for friction between their shoes and the floor, people would have great difficulty in moving about. When the pavement is covered with ice, friction is small and driving is difficult. Because of friction, belts cling to pulleys and drive machinery. Screws and nails stay in place in objects into which they are driven by means of friction.

5.5 KINETIC FRICTION

When an object is sliding over a surface, the direction of the frictional force is parallel to the surface and opposite to the direction of motion (Fig. 5.6). Let us denote by f the force which is just necessary to overcome the frictional force and keep a body moving with *constant speed* across the surface. It is found experimentally that the force necessary to overcome friction is proportional to the force pressing the surfaces against one another. This fact is expressed by the equation

$$f = \mu_k N \tag{5.3}$$

where N is the force pressing the surfaces together and μ_k is a constant called the *coefficient of kinetic* (or sliding) *friction. The coefficient of kinetic friction between two surfaces is the ratio of the force required to overcome friction to the normal force pressing the surfaces together when one surface is sliding over the other at constant speed.* (Here *normal* means *perpendicular;* the normal force is at right angles to the surfaces.)

For ordinary surfaces the frictional force does not depend on the gross area of the rubbing surfaces. For our rough calculations we assume that friction is independent of velocity, although this is only a crude approximation. At very low speeds μ_k sometimes increases with speed but not always; for moderate speeds (15 mm/s to a few meters per second) it is

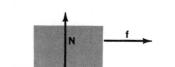

FIGURE 5.6
The coefficient of friction is the ratio f/N, where f is the force required to overcome friction and N is the force pressing the surfaces together.

(a)

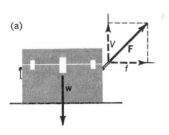

(b)

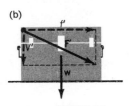

FIGURE 5.7
The force N pressing the surfaces together is (a) reduced by the upward component V of the applied force **F** so that $N = w - V$ and (b) increased by the downward component V' of the applied force **F'** so that $N = w + V'$.

often roughly independent of speed; at higher speeds μ_k typically becomes smaller as the speed goes up.

It should be noted that these statements about friction are worded rather guardedly. This is necessary because the rules which govern friction are only good working approximations. Friction is a very complex phenomenon. Even the smoothest surface is composed of hills and valleys from the microscopic point of view. The actual microscopic area of contact is generally only a tiny fraction of the macroscopic areas of the surfaces. A substantial body of evidence supports the idea that the frictional force is proportional to the area of contact. Ordinarily the frictional force is proportional to the load because the actual contact area is proportional to the load. Apparently, friction arises primarily from adhesion of molecules of the two surfaces which are in contact. Indeed, if we carefully polish and degas two steel surfaces, the friction is tremendously increased over that for rougher surfaces of the same material. Very thin contamination layers can result in large changes in observed friction. Some substances melt and flow under the influence of frictional forces, particularly when high speeds are involved. Such behavior, called *plastic flow*, often lubricates the surface.

It is important to observe that the force N pressing two surfaces together is not necessarily the weight of the sliding object. To move a heavy trunk across a rough floor, one possibility is to pull on the handle as in Fig. 5.7a. We resolve the force **F** into two components; the horizontal component f is useful in overcoming the friction, while the vertical component V reduces the force pressing the surfaces together. In this case, $N = w - V$, where w is the weight of the trunk. To move the trunk by pushing downward (Fig. 5.7b) requires a larger force **F'**, since the force N pressing the surfaces together becomes $w + V'$, where V' is the downward component of **F'**.

5.6 THE COEFFICIENT OF STATIC FRICTION

If a body is at rest on a surface, a larger force is required to overcome the friction and put the body in motion than is required to keep the body in motion at constant speed. If we let f_s represent the force necessary to overcome friction in starting the body, we find that this force is also proportional to the force N pressing the surfaces together. Thus,

$$f_s = \mu_s N \tag{5.4}$$

where μ_s is called the *coefficient of static friction. The coefficient of static friction between two surfaces is the ratio of the force required to overcome friction to the normal force pressing the surfaces together when the surfaces are at rest relative to one another.* Since the force necessary to overcome static friction is greater than the force necessary to overcome sliding friction, the coefficient of static friction is greater than the coefficient of kinetic friction. Table 5.1 lists the coefficients of kinetic and static friction for several types of surfaces.

Because static friction is always greater than kinetic friction, a good automobile driver is careful not to keep his wheels locked when he is braking to a stop. Once the wheels stop rotating, the tires slide over the pavement and the coefficient of kinetic friction is in effect. When the wheels are rolling without sliding, the point of contact between tire and

TABLE 5.1
Approximate Coefficients of Friction

Surface	μ_k	μ_s
Steel on:		
Steel, clean	0.50	0.75
Greasy	0.15	0.20
Ice	0.01	0.02
Oak on oak	0.4	0.5
Greased surfaces	0.05	0.06
Rubber tire on concrete:		
Dry, low speed	0.7	0.9
High speed	0.35	0.6
Wet, low speed	0.5	0.7

pavement is at rest and the coefficient of static friction is applicable.

If a body is at rest on a surface, the frictional force is just large enough to permit the conditions for equilibrium to be satisfied. If the body is resting on a horizontal surface and no horizontal forces act, the frictional force is zero. If the surface is tilted slightly, the frictional force becomes just great enough to prevent the body from sliding. Thus the force of friction can be computed by the relation $f_s = \mu_s N$ only when the body is just at the point of sliding. The product $\mu_s N$ gives the *maximum* value of the friction when the body is at rest. The frictional force may be substantially less than this.

5.7 FRICTION ON AN INCLINED PLANE

In Fig. 5.8 a block of weight w rests on a plane inclined at an angle θ with the horizontal. Let the angle θ be varied until the block just slides down the plane with uniform velocity when once started. Represent this angle by θ_k. Let X represent the component of the weight down the plane and Y the component of the weight normal to the plane. The force of friction between the block and plane is parallel to and up the plane. Since the block experiences no acceleration, the frictional force up the plane is equal to the component of the weight parallel to the plane; thus $X = f_k = w \sin \theta_k$. The force pressing the surfaces together is just the component of the weight perpendicular to the plane; $N = Y = w \cos \theta_k$. By definition, the coefficient of kinetic friction is

$$\mu_k = \frac{f_k}{N} = \frac{w \sin \theta_k}{w \cos \theta_k} = \tan \theta_k \qquad (5.5)$$

The tangent of the angle at which the block slides down the plane with uniform velocity is equal to the coefficient of kinetic friction.

If the angle θ is varied until it is the largest angle at which the block remains at rest on the plane, the component of the weight parallel to the plane ($w \sin \theta_s$) is equal to the force f_s necessary to overcome static friction, while the force pressing the surfaces together is equal to $w \cos \theta_s$, the component of the weight perpendicular to the plane. The angle θ_s at which we have satisfied these conditions is known as the *angle of repose*, and

$$\mu_s = \frac{f_s}{N} = \frac{w \sin \theta_s}{w \cos \theta_s} = \tan \theta_s \qquad (5.6)$$

5.8 REDUCING FRICTION

If a layer of liquid is introduced between two surfaces, the liquid flows over the surfaces and adheres to them. Instead of friction between two solids, there is friction between layers of the liquid. When oil is poured into the bearings of a machine, the sliding takes place primarily between layers of oil. Since friction between oil layers is much less than between metal surfaces, the frictional force is decreased. Oil is usually preferred to water as a lubricant, not because there is less friction between oil layers than between water layers but because oil films have the property of staying between the metal surfaces while water layers are readily squeezed out.

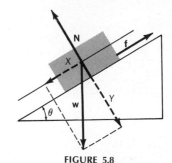

FIGURE 5.8
When the block slides down the inclined plane with constant velocity, the frictional force **f** is up the plane and equal in magnitude to X, the component of the weight **w** down the plane; the normal force **N** has the direction shown and magnitude equal to Y, the component of the weight perpendicular to the plane.

When frictional forces cannot be reduced sufficiently by lubrication, it is customary to substitute rolling friction for sliding friction. The friction of a solid rolling on a surface is far less than the friction of a solid sliding over the surface. For this reason, automobiles and railroad trains operate on wheels rather than runners and often use ball bearings or roller bearings at the axles. When a car wheel rolls on a level track, it makes a slight depression in the track and the wheel is somewhat flattened. As the wheel rolls, it is continually forced to climb out of the depression, which is one reason why even rolling wheels require a force to overcome friction. The amount of depression depends on the nature and area of the surfaces in contact.

5.9 APPLICATION OF NEWTON'S SECOND LAW

When a body is in motion, frictional forces are often of major importance. Now that we have a way to approximate the frictional force acting on a body, we are in a position to include this force in calculating the resultant force acting on a body if it is needed. We now return to Newton's second law and discuss it in more detail.

In a problem involving Newton's second law, the first question which we must ask ourselves is: To what body do we wish to apply $\mathbf{F} = m\mathbf{a}$? Once this question is answered, we must find *all the forces* which act on this body and determine their resultant $\mathbf{F}$. In many situations it is desirable to isolate the body and draw a diagram showing every force which acts on the body. Note that $\mathbf{F}$ is the resultant force while m is the mass accelerated by $\mathbf{F}$.

Newton's second law can be applied to a system of bodies as well as to individual bodies. For example, it applies equally well to a system composed of an elevator, several people, and their luggage as to any one individual in the elevator.

□ **Example** Consider two masses m_1 and m_2 $(m_2 > m_1)$ suspended by an inextensible string which passes over a frictionless pulley (Fig. 5.9). Such an arrangement is known as an *Atwood's machine*. We wish to find the tension T in the string and the acceleration a of the system.

Since the string is inextensible, the accelerations of both masses have the same magnitude. We now apply Newton's second law *to each mass separately*. For m_2 there is a downward force m_2g and an upward force T; the resultant force is $m_2g - T$. For m_1, which rises, the unbalanced force is $T - m_1g$. Since frictional forces are negligible, we have

For m_2: $\qquad\qquad\qquad m_2g - T = m_2a$

For m_1: $\qquad\qquad\qquad T - m_1g = m_1a$

Adding these equations gives

$$m_2g - m_1g = (m_1 + m_2)a$$

which is just what we obtain from applying Newton's second law to the system composed of m_1 plus m_2.

If $m_1 = 1.20$ kg and $m_2 = 1.80$ kg, we have

$$(1.80 \text{ kg})(9.80 \text{ m/s}^2) - T = (1.80 \text{ kg})(a)$$
$$T - (1.20 \text{ kg})(9.80 \text{ m/s}^2) = (1.20 \text{ kg})(a)$$

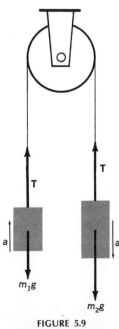

FIGURE 5.9
An Atwood's machine.

Adding these equations gives

$$17.64 - 11.76 \text{ kg-m/s}^2 = (3.00 \text{ kg})(a)$$
$$a = 1.96 \text{ m/s}^2$$

whence
$$T = 14.1 \text{ kg-m/s}^2 = 14.1 \text{ N} \qquad \square$$

☐ **Example** As a second illustration of the application of Newton's second law of motion, consider a mass M sliding along a smooth horizontal table (Fig. 5.10) with a second mass m fastened to it by means of a string that passes over a frictionless pulley. (The word *smooth* here means *frictionless.*)

Let T be the tension in the string and a the acceleration of the system. For M the resultant force is T, since the weight Mg is exactly balanced by the upward force exerted by the table. Applying Newton's second law of motion to each mass separately, we obtain

For M:
$$T = Ma$$
For m:
$$mg - T = ma$$

If M weighs 8.0 lb, and m 2.0 lb, these equations become

$$T = \frac{8.0}{32} \text{ slug} \times a$$

$$2.0 \text{ lb} - T = \frac{2.0}{32} \text{ slug} \times a$$

from which $a = 6.4 \text{ ft/s}^2$ and $T = 1.6 \text{ lb}$. ☐

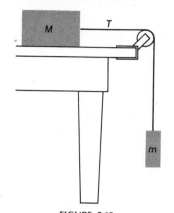

FIGURE 5.10
Newton's second law of motion applies to the system as a whole and to each of the masses independently.

☐ **Example** Next, let us suppose that the coefficient of kinetic friction between M and the table is 0.20. Then there is a frictional force (0.20×8.0) lb acting to the left, and the resultant force on M becomes $T - 1.6$ lb. Applying Newton's second law to the two masses now yields

For M:
$$T - 1.6 \text{ lb} = \frac{8.0}{32} \text{ slug} \times a$$

For m:
$$2.0 \text{ lb} - T = \frac{2.0}{32} \text{ slug} \times a$$

from which $a = 1.3 \text{ ft/s}^2$ and $T = 1.9 \text{ lb}$. ☐

☐ **Example** Find the tension T in the cord of Fig. 5.11 required to give a 5-kg block an acceleration of 2 m/s² up the inclined plane, which makes an angle of 30° with the horizontal. The coefficient of kinetic friction between the block and plane is 0.3.

The resultant force F required is given by Newton's second law

$$F = ma = (5 \text{ kg})(2 \text{ m/s}^2) = 10 \text{ N}$$

Since $Y = N$, the resultant force is $F = T - f_k - X$, where f_k is the frictional force opposing the motion which we must determine. By Eq. (5.5)

$$f_k = 0.3Y$$
$$X = w \sin 30° = mg \sin 30° = 5(9.8)(0.500) = 24.5 \text{ N}$$
$$N = Y = w \cos 30° = mg \cos 30° = 5(9.8)(0.866) = 42.4 \text{ N}$$

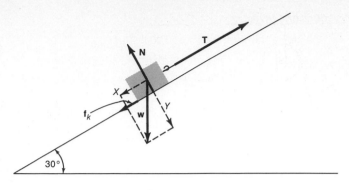

FIGURE 5.11
The tension **T** must be enough greater than the sum of X, the component of the weight down the plane, and the frictional force f_k to provide the mass with the specified acceleration up the plane.

$$f_k = \mu_k Y = 0.3(42.4) = 12.7 \text{ N}$$

Therefore

$$T = F + X + f_k$$
$$= 10 + 24.5 + 12.7 = 47.2 \text{ N} \qquad \square$$

QUESTIONS

1. A book rests on a board which makes an angle of 30° with the horizontal. What forces act on the book? By what bodies are these forces exerted? What are the reactions to these forces, and upon what bodies are they exerted?

2. Explain why a piece of paper can be pulled out from under a glass of water if the paper is jerked suddenly although the glass moves with it when the paper is pulled slowly?

3. If a woman and her parachute have a total mass M and she descends through the air at constant speed, what is the retarding force exerted by the air? How does the retarding force as the parachute opens compare with Mg?

4. A weight is hung from the middle of a rope. Will it ever be possible to get the rope to be horizontal by pulling on the ends? Discuss with a vector diagram.

5. Under what circumstances will the frictional force on a block on a rough inclined plane be directed down the plane?

6. In city driving an automobile starts from rest, has a uniform acceleration until its speed is 15 m/s, proceeds at constant speed for two blocks, and then decelerates uniformly to rest at a stop sign. Is the force required to stop the car equal in magnitude to the one which brought it up to speed? Justify your answer.

7. If frictional forces are independent of area, why is it a good idea to make automobile brake shoes of large area?

8. One end of a rod is held up by a string while the other end rests on frictionless ice (or on a floating cork). Why does the string take up a vertical line in the absence of any additional forces regardless of the angle made by the rod with the horizontal?

9. If two surfaces are very rough, polishing them may reduce the coefficient of static friction between them. Further polishing may lead to an increase in the friction. Why?

10. What is the smallest number of forces not all acting in the same plane which can put a point mass in equilibrium?

11. A man lies in a hammock stretched between two trees. Are the supporting ropes more likely to break if the hammock is strung tightly or loosely? Explain using a vector diagram.

12. Prove that if a body is in equilibrium under the action of a set of forces, any one of the forces is equal and opposite to the resultant of all the other forces.

13. Show that for a body at rest on a rough horizontal plane, no force making an angle with the vertical smaller than the angle of repose can slide the body along the plane.

14. If the pulleys in the accompanying figure are of negligible mass, why can't the system as shown in the figure be in equilibrium? What must happen before equilibrium can be achieved for this pulley system? If all the pulleys had the same mass, what would be the smallest value which would allow the system as shown to be in equilibrium?

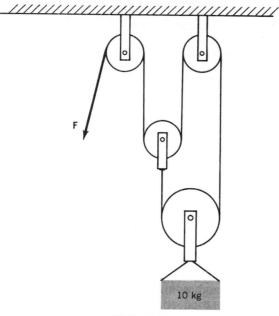

QUESTION 14

PROBLEMS

Draw an appropriate sketch for each problem.

1. The ball of Fig. 5.3 has a weight of 8 N. Find the horizontal force F required to hold the ball at rest with $\theta = 25°$. What is the tension in the cord? *Ans.* 3.73 N; 8.83 N

2. A frictionless cart standing on an inclined plane that makes an angle of 15° with the horizontal is kept from rolling downhill by a force of 10 N applied in a direction parallel to the plane. What is the weight of the cart? What is the normal force exerted on the cart by the plane?

3. A pendulum bob with a weight of 25 N hangs from a cord. A horizontal force sufficient to bring the cord to an angle of 30° with the vertical is applied to the bob. Find the horizontal force and the tension in the cord. What is the smallest force which will hold the bob in equilibrium? What is the tension in this case?

   Ans. 14.4 N; 28.9 N; 12.5 N; 21.7 N

4. A block weighing 40 N rests on a frictionless plane inclined 35° to the horizontal. It is connected to a weight w by a rope of negligible weight. The rope runs up parallel to the plane and passes over a pulley so that the weight w hangs vertical. What must be the magnitude of the weight if the system is in equilibrium? What is the normal force exerted on the block by the plane? Draw an appropriate vector diagram showing all forces acting on the block.

5. A carton weighing 80 N rests on a horizontal loading platform. A horizontal force of 32 N is required to start the carton in motion, but once it is moving, a force of 28 N is adequate to keep it in motion at constant speed. Find the coefficients of static and kinetic friction.

   Ans. 0.40; 0.35

6. A brake shoe is pressed against the rim of a moving wheel with a force of 50 lb. If the coefficient of kinetic friction between the surfaces is 0.45, how much frictional force is developed?

7. If the angle of repose for a 60-N block of metal on an incline is 40°, what force parallel to the incline is necessary to cause the body to begin to move up the incline? *Ans.* 77.1 N

8. What force acting at an angle of 30° above the horizontal is necessary to move a mass of 0.5 slug with uniform velocity along a horizontal surface, the coefficient of friction being 0.35?

9. A box weighing 80 N is on a horizontal floor. The coefficient of static friction between surfaces is 0.35, and that of sliding friction is 0.30. Find the frictional force under each of the following conditions: (*a*) no additional force is applied; (*b*) a force of 15 N is applied toward the east; (*c*) a force of 25 N is applied toward the west; (*d*) a force of 30 N to the south is applied steadily to the box.

   Ans. (*a*) 0; (*b*) 15 N west; (*c*) 25 N east if block at rest; 24 N east if in motion; (*d*) 24 N north

10. A 4-kg block pulled along a level surface by a horizontal force of 20 N has an acceleration of 2 m/s². Find the coefficient of friction. Draw a diagram showing all forces acting on the block.

11. (*a*) If blocks A and B of the accompanying figure have masses of 2 and 8 kg, respectively, find μ_k between B and the table if a force of 35 N is required to pull the blocks at constant speed. (*b*) What force would be required to pull B if A were held fixed by a string between the hooks? Take $\mu_k = 0.40$ for the surface between the two blocks. *Ans.* (*a*) 0.357; (*b*) 42.8 N

12. (*a*) If block A of the accompanying figure weighs 6 lb and block B weighs 24 lb, what force F is required to pull them along at constant speed if $\mu_k = 0.25$? (*b*) If a string is fastened between the two hooks so A does not move, what is F, assuming the coefficient of kinetic friction is 0.40 between the two blocks?

13. A book rests on a horizontal seat in an automobile traveling 20 m/s. If the coefficient of static

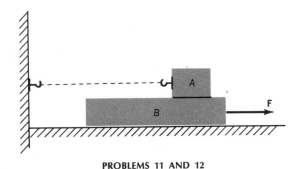

PROBLEMS 11 AND 12

friction is 0.25, calculate the shortest distance in which the car can be stopped without the book's sliding on the seat. *Ans.* 81.6 m

14. A block of mass m slides down a plane that is inclined at 36.9° with the horizontal. If the coefficient of friction between the block and the surface of the incline is 0.3, find the acceleration of the block.

15. A book tossed on a table slides off and lands 0.5 m from the edge of the table, which is 0.625 m high. (*a*) Find the horizontal velocity of the book as it left the table. (*b*) If the book had a speed of 2.4 m/s when it was sliding 0.8 m from the edge of the table, what was the coefficient of kinetic friction between book and table?
Ans. (*a*) 1.4 m/s; (*b*) 0.242

16. The coefficient of sliding friction between a block and the floor is 0.35. If a force of 30 N acting 36.9° above the horizontal gives the block a constant velocity, find the weight of the block.

17. A man is pushing a 40-kg box along a level floor with a uniform velocity of 1.2 m/s by exerting a force on the box at an angle of 30° downward from the horizontal. If the coefficient of friction between sliding surfaces is 0.40, what is the magnitude of the force? *Ans.* 235 N

18. A man exerts a constant force of 60 N for 5 s on a 25-kg sled initially at rest on a frozen pond. If the coefficient of sliding friction is 0.1, find the acceleration of the sled while the man is pushing on it. What is the acceleration after he stops pushing? What is the maximum speed of the sled? How far does the sled move?

19. A 2-kg block rests on top of a 6-kg block which is on a table. If $\mu_k = 0.20$ and $\mu_s = 0.25$ for all surfaces involved, find the largest horizontal force

which can be applied to the lower block without having the upper block slide. *Ans.* 35.3 N

20. What horizontal force is needed to push a 19.5-kg box up an inclined plane which rises 5 m in a distance of 13 m measured along the plane if the coefficient of friction is 0.35? What horizontal force is required to give this box an acceleration of 2 m/s² up the plane?

21. In the accompanying figure block A has a mass of 10 kg, and block B a mass of 15 kg. The coefficient of kinetic friction between the blocks and the surface is 0.40. Force F is applied to pull B up the plane at constant speed. (*a*) Draw free-body diagrams for A and B, showing *all* forces acting on each. (*b*) Find the tension in the cord between A and B. (*c*) Find the magnitude of F.
Ans. (*b*) 39.2 N; (*c*) 174 N

22. If A and B in the accompanying figure each weigh 10 lb and rest on surfaces such that the coefficient of friction is 0.35, what is the tension in the cord between A and B? Find the force F needed to pull B up the plane.

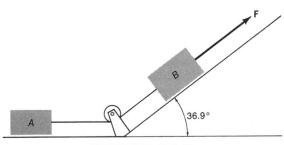

PROBLEMS 21 AND 22

23. A 50-N block rests on a plane that is inclined at an angle of 36.9° to the horizontal. The coefficient of kinetic friction between the block and the plane is 0.30. What force parallel to the incline is necessary to keep it sliding up the plane? What horizontal force would be required?
Ans. 42 N; 67.7 N

24. A 15-lb frictionless car rests on an inclined plane that makes an angle of 30° with the horizontal. What force must be exerted to keep it from sliding down the plane if the force is applied so that it makes an angle of 36.9° downward with the horizontal?

25. An Atwood's machine (Fig. 5.9) consists of masses of 2 and 2.5 kg suspended from a massless, frictionless pulley. When the larger mass is released from rest, how long does it take to descend 1.6 m? What is the tension in the cord during the descent? *Ans.* 1.71 s; 21.8 N

26. A block of mass 4.6 kg resting on a horizontal

frictionless surface is connected to a hanging 3.4-kg block by a cord passing over a frictionless pulley (Fig. 5.10). Find the tension in the cord and the acceleration of the blocks.

27. Two bodies are suspended by a flexible string that passes over a weightless pulley. If one body weighs 14 lb and the other 10 lb, what are the acceleration of the system and the tension in the string? *Ans.* 5.33 ft/s²; 11.7 lb

28. An elevator weighs 10,000 N. A counterweight of 10,000 N is attached to it, and the cable between the two passes over a pulley. A person weighing 800 N steps onto the elevator. If all other forces are neglected, what is the acceleration of the system? The tension in the cable?

29. A mass of 12 kg rests on a horizontal table and is attached to a mass of 5 kg by a flexible cord passing over a weightless pulley (Fig. 5.10). If the coefficient of friction between the mass and the table is 0.3, what is the acceleration of the system? The tension in the cord?
Ans. 0.807 m/s²; 45.0 N

30. A 5-kg block resting on a horizontal surface is connected to a hanging block of 4-kg mass by a cord passing over a light, frictionless pulley (Fig. 5.10). When the system is released, the blocks have an acceleration of 0.5 m/s². Find the unbalanced force on each mass, the tension in the cord, and the coefficient of friction between the 5-kg block and the horizontal surface.

31. A brick with a mass of 2 kg is given an initial velocity of 5 m/s up an inclined plane which makes an angle of 36.9° with the horizontal. If the coefficient of sliding friction is 0.40, find (a) the frictional retarding force, (b) the deceleration of the brick, (c) how far the brick moves up the plane, and (d) the acceleration of the brick as it slides back down the plane.
Ans. (a) 6.27 N; (b) 9.02 m/s²; (c) 1.39 m; (d) 2.74 m/s²

32. Find the acceleration and tension in both cords of the accompanying figure if m_1 is 0.3 slug, m_2 is 0.6 slug, M is 0.7 slug, and the coefficient of friction between M and the table is 0.25.

33. (a) If $m_1 = 1$ kg, $M = 6$ kg, and $m_2 = 3$ kg, find the acceleration of the system and the tension in both cords of the accompanying figure, assuming that there is no friction and the pulleys are massless. (b) If the coefficient of friction between the table and mass M is 0.15, find the acceleration and the tension in both cords.
Ans. (a) 1.96 m/s², 11.8 N, 23.5 N; (b) 1.08 m/s², 10.9 N, 26.2 N

34. A 10-lb block (A in the accompanying figure) on a 36.9° inclined plane is connected by a massless string which passes over a frictionless pulley to a 2-lb block B. If the 2-lb block has an upward acceleration of 4 ft/s², find (a) the tension in the cord, (b) the frictional force on the 10-lb block, and (c) the coefficient of sliding friction.

35. If A has a mass of 4 kg in the accompanying figure and B a mass of 3 kg, θ is 25°, and the acceleration of the system is 0.5 m/s², find the tension in the cord and the coefficient of sliding friction. Assume the pulley is frictionless.
Ans. 27.9 N; 0.263

36. The angle of the rough plane in the accompanying figure can be varied from 0 to 90°. (a) If A has mass m_1, B mass m_2, the pulley is frictionless, and μ_k is the coefficient of kinetic friction between A and the plane, write an equation for the acceleration in terms of m_1, m_2, μ_k, g, and θ. (b) If $m_1 = 3$ kg, $m_2 = 2$ kg, and $\mu_k = 0.2$, find the acceleration for $\theta = 0, 20, 40, 60,$ and 80°. (c) Over

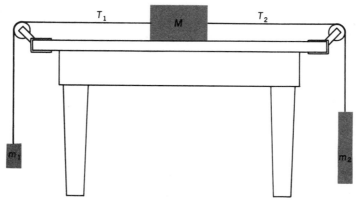

PROBLEMS 32 AND 33

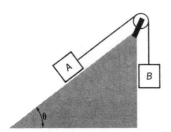

PROBLEMS 34 TO 37

what range of angles does friction either prevent sliding or make the speed decrease?

37. A 5-kg block A on a rough 30° plane is connected by a massless string which passes over a frictionless pulley to a 4-kg block B, as shown in the accompanying figure. If $\mu_k = 0.25$, find the acceleration and the tension in the string.

Ans. 0.455 m/s²; 37.4 N

38. Body A of the accompanying figure weighs 5 lb, and the coefficient of kinetic friction for it is 0.4. Body B weighs 8 lb, and for it $\mu_k = 0.125$. Find the tension in the connecting cord.

39. Body A of mass 3 kg and body B of mass 2 kg are connected by a light string (see figure). If the coefficients of friction for A and B are, respectively, 0.4 and 0.2, find the acceleration of the system and the tension in the string.

Ans. 3.37 m/s²; 1.88 N

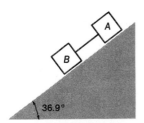

PROBLEMS 38 AND 39

40. Find the acceleration of the blocks and the tension in the cord of the accompanying figure if A has a mass of 4 kg, B has a mass of 5 kg, θ is 36.9°, ϕ is 30°, the planes are frictionless, and the pulley is massless.

41. The planes of the accompanying figure are frictionless, and the pulley is massless. Find the acceleration of the blocks if A weighs 10 lb, B

weighs 6 lb, θ is 36.9°, and ϕ is 30°. What is the tension in the cord? *Ans.* 6 ft/s²; 4.13 lb

42. A 30-kg traffic light is supported by two wires; one which makes an angle of 20° with the horizontal, and the second makes an angle of 10°. Find the tension in each.

43. A tow rope 10 m long is stretched taut between a tree and an automobile. A force of 300 N at the center of the rope moves it 0.5 m perpendicular to the original line of the rope. Find the force exerted on the tree and on the automobile.

Ans. 1,508 N

44. Find the tension in each rope of the accompanying figure if the weight w is 100 lb.

45. Find the tension in each rope of the figure if the weight w is 70 N.

Ans. (a) 50 N, 56.6 N; (b) 87.5 N, 52.5 N; (c) 43.75 N, 58.3 N, 20.4 N

46. A force of 250 N exerted at an angle of 30° above the horizontal is needed to keep a 800-N block moving along a cement floor. Find the coefficient of kinetic friction.

47. Two cylindrical lengths of pipe, each weighing 320 N, lie against each other on a smooth sidewalk. What horizontal force on each pipe is needed to keep them together when an identical length is placed in the groove between them?

Ans. 92.4 N

48. A body is in equilibrium under the influence of three forces, two of which are equal in magnitude, the third having a magnitude half as great. Find the angle between the two equal forces.

49. The end of a horizontal electric transmission line in which there is a tension of 3 kN is fastened to the top of a vertical pole and to a guy wire which can sustain a tension of 8 kN. What is the smallest angle the guy wire can make with the pole if there is to be no resultant lateral force on the pole? *Ans.* 22.0°

50. A picture is supported by two wires fastened to the ends of the upper edge of the picture frame, which is horizontal. Each of the wires makes an angle of 65° with the vertical. What is the tension in each wire if the picture weighs 9 lb?

51. A bird weighing 6 N alights on a weightless telephone wire midway between two poles 50 m apart. Assume that the wire between the bird and each of the poles forms a straight line. If the center of the wire is 0.15 m below its level at the poles, and if the weight of the wire is neglected,

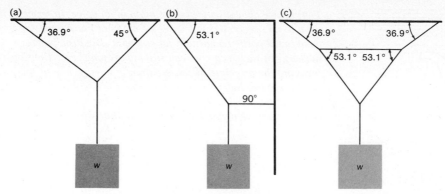

PROBLEMS 44 AND 45

calculate the tension in the wire due to the weight of the bird. *Ans.* 500 N

52. Find the compression in the strut and the tension in the cable of the accompanying figure if the strut is weightless and w is 200 lb. The strut is free to rotate about a pin at the wall and therefore exerts a force only along its length.

53. Find the compression in the strut and the tension in the cable of the accompanying figure if the strut is weightless and w is 500 N. The strut is free to rotate about a pin at the wall and therefore exerts a force only along its length.
Ans. (*a*) 883 N, 667 N; (*b*) 866 N; 1,000 N; (*c*) 1,430 N; 1,070 N

54. A block slides with constant acceleration a down a plane inclined at an angle θ with the horizontal. Find the coefficient of sliding friction μ_k in terms of a, θ, and g.

55. Show that for the frictionless, massless Atwood's machine of Fig. 5.9 the acceleration of each mass is $g(m_2 - m_1)/(m_2 + m_1)$ and the tension T is $2m_1m_2g/(m_1 + m_2)$.

56. A uniform flexible cable of weight w is supported by hooks at the same height on neighboring telephone poles. If the cable makes an angle θ with the horizontal at the hooks, find (*a*) the force exerted on the cable by the hook on the right and (*b*) the tension in the cable at its lowest point.
Ans. (*a*) $w/(2 \sin \theta)$ at angle θ with horizontal; (*b*) $(w \cos \theta)/(2 \sin \theta)$

57. A tightrope walker of weight w stands one-third of the way along a horizontal tightrope of length L, which sags a maximum distance s below the horizontal line. If s is very small compared with L and the rope has negligible weight, show that the tension T is given by $2wL/9s$.

58. A block of mass m rests on a rough plane inclined at an angle θ with the horizontal. Show that if a gradually increasing horizontal force F is exerted on the block, the block begins to slide up the plane when

$$F = \frac{mg\,(\sin \theta + \mu_s \cos \theta)}{\cos \theta - \mu_s \sin \theta}$$

where μ_s is the coefficient of static friction.

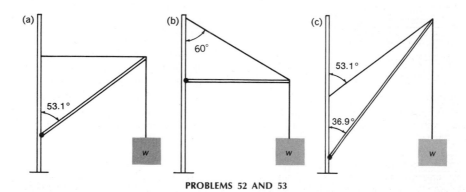

PROBLEMS 52 AND 53

6

WORK, POWER, AND ENERGY

Thus far we have discussed the motion of an object in terms of the forces acting upon it. However, in some cases we may not know what the forces are at every point in the path of a particle, for example, a neutron passing close to a uranium nucleus. In such a case we can still make many predictions about the motion by invoking the principle of conservation of energy, one of the great conservation laws of nature. Before we can apply the conservation of energy in any problem, we must learn what energy is and in what forms it appears. This in turn requires that we define and become familiar with work *as this word is used in physics.*

6.1 WORK

The word *work* is a familiar one, used in many everyday senses. In physics, however, it is used in a restricted and carefully defined manner. When a force **F** acts upon a body to produce a displacement **s**, the *work done by the force is defined*† *as the product of the displacement and the component of the force in the direction of the displacement*. If the angle between the displacement vector and the force vector is θ (Fig. 6.1), the component of the force in the direction of the displacement is $F \cos \theta$, and the work W is, by definition,

$$W = Fs \cos \theta \qquad (6.1)$$

If the force does not produce a displacement, no work is done in the sense in which work is used in physics. A woman holding a 50-N weight at rest does no work on it. A desk or cement post could hold the weight indefinitely without any difficulty. Although the woman holding the weight does *no work on it*, the muscles in her arm do stretch and contract, and thus work is done internally, which may result in fatigue. Similarly, someone walking at a constant horizontal velocity carrying a suitcase does no work *on the suitcase*, since the force he exerts on the suitcase and the displacement are mutually perpendicular.

6.2 UNITS OF WORK

Since work is measured by the product of force and displacement, its units involve a unit of force multiplied by a unit of length. In the metric system the basic units of force and distance are the newton and the meter. Work is measured in newton-meters; this unit has been named the *joule* (J) in honor of James Prescott Joule, a distinguished British physicist, whose work on the relation between heat and mechanical work was of great importance. *One joule is the work done when a force of one newton acts through a distance of one meter.*

There are many other units in which work can be measured. In the British engineering system force is measured in pounds and the displacement in feet; the resulting unit of work is the *foot-pound. One foot-pound is the work done when a force of one pound acts through a distance of one foot.* Large amounts of work are sometimes reported in ton-miles. Small amounts in the metric system are sometimes measured in ergs; 1 erg is equal to 10^{-7} J.

† In vector notation $W = \mathbf{F} \cdot \mathbf{s} = Fs \cos \theta$, where $\mathbf{F} \cdot \mathbf{s}$ is called the *scalar* or *dot product* of **F** and **s**.

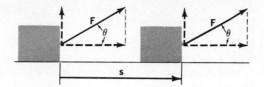

FIGURE 6.1
The work W done by a force F acting through a displacement s is $Fs \cos \theta$, where θ is the angle between F and s.

☐ **Example** How many joules of work are done by a force in lifting a mass of 2 kg upward a distance of 3 m?

Work in joules = force in newtons × distance in meters

$$F = mg = (2 \text{ kg})(9.8 \text{ m/s}^2) = 19.6 \text{ N}$$
$$W = (19.6 \text{ N})(3 \text{ m}) = 58.8 \text{ J} \qquad \square$$

☐ **Example** A force of 10 lb is used to move a box across a horizontal floor a distance of 5 ft. If the force makes an angle of 30° with the floor, how much work is done?

$$\text{Work} = Fs \cos \theta$$
$$W = (10 \text{ lb})(5 \text{ ft})(0.866)$$
$$= 43.3 \text{ ft-lb} \qquad \square$$

6.3 POWER

In physics, *power is the time rate of doing work;* the average power $\mathcal{P}_{av}$ is equal to work performed divided by the time required:

$$\mathcal{P}_{av} = \frac{\text{work}}{\text{time}} = \frac{W}{t} \qquad (6.2)$$

An engine of high power can do work rapidly.

In the SI system power is measured in joules per second or *watts* (W) (in honor of James Watt, developer of the steam engine). One watt of power is expended when one joule of work is done each second. In the British engineering system power may be expressed in foot-pounds per second, but more often it is given in *horsepower;* 1 hp = 550 ft-lb/s = 745.7 W.

The origin of the horsepower is of interest. When James Watt tried to sell steam engines for British coal mines, he was asked how many horses one of these new engines would replace. Watt found that, on the average, the horses were doing about 550 ft-lb of work per second; he called this unit the horsepower. He measured the rate at which his steam engines could work and thus rated them in "horsepower."

☐ **Example** A building crane lifts a 1,500-lb steel beam to a height of 44 ft in 10 s. Find the average power developed.

$$\mathcal{P}_{av} = \frac{\text{work}}{t} = \frac{(1,500 \text{ lb})(44 \text{ ft})}{10 \text{ s}} = 6,600 \text{ ft-lb/s}$$

$$= 6,600 \frac{\text{ft-lb}}{\text{s}} \times \frac{1 \text{ hp}}{550 \text{ ft-lb/s}} = 12 \text{ hp} \qquad \square$$

☐ **Example** An electric motor exerts a force of 400 N on a cable and pulls it a distance of 30 m in 1 min. Find the power supplied by the motor.

$$\mathcal{P}_{av} = \frac{work}{t} = \frac{(400 \text{ N})(30 \text{ m})}{60 \text{ s}} = 200 \text{ J/s}$$

$$= 200 \text{ W} \qquad \qquad \square$$

When a constant **F** force does work in accelerating a body, the instantaneous power supplied increases as the speed increases. If Δ**s** is the displacement during a short time interval Δt, the instantaneous power delivered is, by Eq. (6.1),

$$\mathcal{P} = \lim_{\Delta t \to 0} \frac{F \Delta s \cos \theta}{\Delta t} = Fv \cos \theta \qquad (6.3)$$

The instantaneous power delivered by a force **F** which acts in the direction of the motion is

$$\mathcal{P} = \lim_{\Delta t \to 0} \frac{F \Delta s}{\Delta t} = Fv \qquad (6.3a)$$

When the force and velocity are constant, the instantaneous power is equal to the average power. Equation (6.3a) is useful in resolving such practical questions as how powerful a motor must be provided to lift an elevator at a specified speed or how many diesel units are required to pull a railroad train at a given speed. If one wishes to make the speed of an airplane or a ship greater, Eq. (6.3a) shows that there are basically two things one can do: increase the power and decrease the retarding forces.

$\square$ **Example**   A Boeing 747 aircraft has four engines, each capable of exerting a thrust of 200,000 N (45,000 lb). When the aircraft is cruising at 1,000 km/h (621 mi/h), each engine is delivering 3×10^7 W (40,300 hp). Find the drag. What fraction of its maximum thrust is each engine delivering?

Metric (SI):

$$\mathcal{P} = Fv$$

$$4(3 \times 10^7 \text{ W}) = F_{drag} \left(1,000 \times \frac{1,000}{3,600} \right) \text{m/s}$$

$$F_{drag} = 4.32 \times 10^5 \text{ N}$$

$$\frac{4.32 \times 10^5}{4 \times 2 \times 10^5} = 0.54 = 54\%$$

British engineering:

$$4(40,300 \text{ hp}) \times \frac{550 \text{ ft-lb/s}}{1 \text{ hp}} = F \times 621 \text{ mi/h} \times \frac{1.467 \text{ ft/s}}{1 \text{ mi/h}}$$

$$F = 9.76 \times 10^4 \text{ lb}$$

$$\frac{9.76 \times 10^4}{4 \times 45,000} = 0.54 = 54\% \qquad \qquad \square$$

6.4   ENERGY AND ITS CONSERVATION

Energy is defined as the ability or capacity to do work. It occurs in many forms. A swinging hammer can do work by virtue of its motion; energy associated with motion is known as *kinetic energy*. A raised pile driver can do work by virtue of its elevated position; it has what we call *potential energy*. When we buy gasoline and food, we buy chemical energy. We purchase electric energy to obtain work from electric motors as well as for heat and light. The earth's primary source of energy lies in the electromagnetic radiation it receives from the sun. Much of physics involves the relationships between the many forms of energy and the transformations from one form to another.

One of the most important principles of nature and one of the most powerful generalizations of science is the *law of conservation of energy:*

Energy cannot be created or destroyed; it can be transformed from one form into another, but the total amount of energy never changes.

As an example of the transformations which energy may undergo, consider radiation coming to the earth from the sun. Some of this energy falls on plants, where it is transformed into chemical energy through photosynthesis. The energy stored in the plant may be converted eventually into coal or oil, or the plant may be eaten by some animal which utilizes the energy to carry on its existence. Part of the radiant energy from the sun goes into evaporating water from the ocean. The water vapor, lifted high above the earth, eventually returns in the form of rain, which may be trapped behind a dam. The water thus stored has potential energy, which can be converted by use of a giant turbine into energy of motion. This kinetic energy in turn can be converted into electric energy in a generator. Electric energy is distributed to houses, where it is converted into heat or light or used to perform work through an electric motor. Throughout all these transformations, the total energy remains constant.

Early in the twentieth century it was found that mass itself could be converted into energy; in this sense mass is one of the forms or manifestations of energy. The sun's mass is decreasing because mass energy is converted into radiant energy in the sun. In nuclear reactors mass is converted into energy. Einstein showed that whenever mass is transformed into another form of energy (or vice versa), the equivalence can be expressed by the equation

$$E = mc^2 \tag{6.4}$$

where E is the amount of some other form of energy appearing or disappearing, m is the mass disappearing or appearing, and c is the speed of light.

Mass is converted into energy in nuclear reactors and nuclear weapons as well as in the sun and the other stars. For reactors using uranium 235 as fuel, about one-thousandth of the mass of each fissioning atom is converted into other forms of energy. Although the fraction of our energy needs supplied by nuclear reactors on earth is relatively small, it is increasing rapidly as energy from oil and natural gas becomes less plentiful and more expensive.

□ **Example** What is the total energy in a gram of matter at rest?

$$m = 10^{-3} \text{ kg}$$

The speed of light c is 3×10^8 m/s, and $c^2 = 9 \times 10^{16}$ m²/s²

$$E = mc^2 = 9 \times 10^{13} \text{ kg-m}^2/\text{s}^2 = 9 \times 10^{13} \text{ J}$$

This energy is sufficient to lift a mass of 1 million kilograms through a distance of 9,000 km against the action of gravity! Only a *very* small fraction of this energy is available under the best of conditions.   □

6.5 ENERGY DEMANDS IN A MODERN SOCIETY

Energy is an important commodity in economics and commerce. It is bought and sold in many forms, as coal, petroleum products, food, electricity, and many others. Energy is used to operate automobiles, refrigerators, stoves, washing machines, and television sets. In the past century the amount of energy consumed per person has undergone a remarkable increase. Primitive people used energy only for food and warmth; today we have hundreds of appliances which transform energies for entertainment, comfort, and convenience. In recent decades there has been a major increase in the annual per capita energy use, particularly in the more developed countries. This, plus a burgeoning world population, has resulted in the widespread realization that our energy supplies are not unlimited and that continual prodigal expenditure of our hydrocarbon and other energy resources is sure to lead to future deficiencies.

In the United States on the average each person obtains about 12 MJ (12×10^6 J) per day from food. (One food calorie corresponds to 1 kcal, or 4,186 J.) There are large differences between individuals, of course, but there seems to be general agreement that the collective health would be better if less were eaten. The oxidation of carbohydrates, fats, and proteins provides energy for performing external work as well as for maintaining body temperature and for carrying on internal vital functions. Pumping blood through the veins and arteries by the heart muscle is one example of internal work performed in the body. For a person at rest the heart beats about 70 times per minute and develops a power of about 1.3 W.

In the Middle Ages in Europe people needed some energy from fires to cook and to keep warm plus additional energy for their animals; it is estimated that 50 MJ per person was about the average daily energy consumption. By 1880 one person in the industrial society used about 314 MJ daily; for that person only one-twentieth of the energy consumption was for food. In 1975 the average daily energy consumption per capita in the United States was 1.2 GJ; this corresponds to about 60 barrels of oil or 11 tons of coal per year. Also in 1975, the worldwide average was about 170 MJ per day per person and in some of the less developed countries the average was 50 MJ.

Accompanying the dramatic growth in energy use has been a great shift in energy sources for the United States. In 1850 wood supplied 90 percent and coal 10 percent of the fuel requirements. By 1910 the share supplied by wood had fallen to 10 percent and coal had risen to 80 percent, with oil, natural gas, and hydropower sources providing the remaining 10 percent. In 1973 the share of coal had dropped to 18 percent, oil supplied 46 percent, natural gas 31 percent, and hydropower sources and nuclear reactors most of the remaining 5 percent.

6.6 POTENTIAL AND KINETIC ENERGY

In mechanics we are concerned primarily with two forms of energy, *potential energy* (PE) and *kinetic energy* (KE).

Kinetic energy is the energy a body possesses by virtue of its motion. Any body in motion can set other bodies in motion by colliding with them. The moving head of an axe does work in splitting a log. The bullet leaving the muzzle of a gun has kinetic energy and can do work in penetrating a board.

To find the kinetic energy a body possesses, we consider the work which must be done on the body in order to give it its speed. When the body is stopped, it gives up this amount of energy. By definition, this is its kinetic energy. Consider a mass m, initially at rest, upon which a constant force F is applied through a displacement s in the direction of the force. The work done on the body is Fs. By Newton's second law, $F = ma$; thus the work is equal to mas. Since the acceleration of the body is constant, Eq. (3.7) is applicable and $v^2 = 2as$, since $v_0 = 0$. If we replace as with $v^2/2$ and recall that the work done appears as kinetic energy, the kinetic energy is

$$KE = \tfrac{1}{2}mv^2 \qquad (6.5)$$

In metric units we express the kinetic energy in joules (or the equivalent kg-m^2/s^2), while in the British system the kinetic energy is given in foot-pounds (or the equivalent slug-ft^2/s^2).

☐ **Example** If a 1,000-kg automobile is moving with a speed of 20 m/s, what is its kinetic energy in joules?

$$KE = \tfrac{1}{2}mv^2 = \tfrac{1}{2}(1{,}000 \text{ kg})(20 \text{ m/s})^2$$
$$= 2 \times 10^5 \text{ kg-m}^2/\text{s}^2 = 2 \times 10^5 \text{ J} \qquad ☐$$

☐ **Example** What force is required to stop a bullet that has a mass of 15 g and a velocity of 400 m/s in a distance of 20 cm?

$$\text{Force} \times \text{distance} = \text{change in KE}$$
$$(0.20 \text{ m})(F) = \tfrac{1}{2}(0.015 \text{ kg})(400 \text{ m/s})^2$$
$$= 1{,}200 \text{ kg-m}^2/\text{s}^2 \text{ (or joules)}$$
$$F = \frac{1{,}200 \text{ kg-m}^2/\text{s}^2}{0.2 \text{ m}}$$
$$= 6{,}000 \text{ kg-m/s}^2 = 6{,}000 \text{ N} \qquad ☐$$

The energy that a body (or system of bodies) has by virtue of its position or configuration is called potential energy (Fig. 6.2). When a mass is lifted above the surface of the earth, it has energy because of its position. When a spring is compressed or a bow bent, potential energy is stored up. Other

FIGURE 6.2
The drawn bow and the elevated weight both have potential energy; both have the ability to do external work by virtue either of position or of configuration.

examples of potential energy are the mainspring of a watch and a stretched rubber band. In these cases the material of a body is in a state of strain; because of this strain, the body possesses potential energy.

The measure of the potential energy which a body has because of its elevated position is the work done against gravity in lifting the body from some level chosen as the zero for potential energy. The upward force required is equal to the weight of the body w, and the work done in lifting the body through a height h is given by the product wh; therefore the potential energy is

$$PE = wh = mgh \tag{6.6}$$

☐ **Example**   A block weighing 3 lb is lifted 6 ft against gravity. What potential energy is stored?

$$PE = wh = (3 \text{ lb})(6 \text{ ft})$$
$$= 18 \text{ ft-lb} \qquad \square$$

☐ **Example**   Find the potential energy given to the 50-kg hammer of a pile driver when it is raised 4 m.

$$PE = mgh = (50 \text{ kg})(9.8 \text{ m/s}^2)(4 \text{ m})$$
$$= 1,960 \text{ J} \qquad \square$$

A second important kind of potential energy is that associated with a stretched or compressed spring. Consider the spring of Fig. 6.3. Let its length when there is no load applied be l_0. If we add a load w_1, we find a new length l_1 for the spring. The change in length $l_1 - l_0$ we denote by y_1; in general, we shall represent the extension of the spring by y. If we now double the load on the spring, we find that y is twice as great. The fact that the change in length is proportional to the applied force was first reported by the English physicist Robert Hooke in 1678 in the words "Ut tensio, sic vis" (as the extension, so the force). In general, the force we apply F_a is proportional to the extension y; thus

$$F_a = ky \tag{6.7}$$

where k is known as the *force constant* of the spring. For many springs Eq. (6.7) is applicable for compression as well as for extension of the spring; y for such a *linear* spring is proportional to, and in the direction of, the applied force.

The potential energy of a stretched (or compressed) ideal spring is the amount of work the spring can do, and this is equal to the work required to distort the spring. To produce a displacement y the applied force F_a is zero at first and increases linearly to ky. The work W done by F_a is the product of the displacement y and the average force $(0 + ky)/2$. The potential energy P stored is equal to this work, and so

$$P = \tfrac{1}{2}ky^2 \tag{6.8}$$

The potential energy of a stretched or compressed ideal spring is proportional to the square of the displacement y *relative to the configuration in which $y = 0$.* Note that y^2 is positive regardless of whether y is positive or negative. Thus energy is stored either by stretching or by compressing a spring.

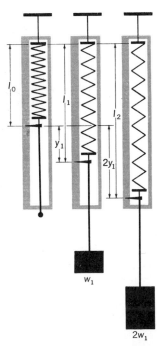

FIGURE 6.3
The elongation of a linear spring is proportional to the stretching force; the potential energy stored is proportional to the square of the elongation.

It is extremely important to remember that potential energy is always measured relative to some configuration chosen as the zero. For a mass suspended from a spring, the equilibrium position (Fig. 6.4) is the usual choice for this zero configuration. For the gravitational potential energy mgh, the height h is measured relative to some chosen zero level. When we deal with energy of motion, the kinetic energy depends on the reference frame relative to which we measure velocities. A passenger seated in an aircraft traveling 1,000 m/s has no kinetic energy relative to a set of axes tied to the plane but much kinetic energy relative to axes fixed on the earth.

6.7 ENERGY TRANSFORMATION

The potential energy of a body due to its position depends upon its height h, and this in turn depends on what one wishes to call the zero for height. A book on a table has no potential energy *relative to the tabletop*, but it does have potential energy *relative to the floor* and *relative to mean sea level*. If the book weighs 6 N and the tabletop is 1 m above the floor, the potential energy of the book is 6 J *relative to the floor*. Potential energy is always determined *relative* to some level or configuration which is assigned the value of zero. The choice of this level is up to the person working the problem.

In many situations we have a simple transformation of potential into kinetic energy with no other forms of energy involved. For example, when a pile driver is lifted into position, it is given potential energy. This potential energy is transformed into kinetic energy before the hammer strikes the pile upon which it is to do work. Similarly, consider water going over a dam (Fig. 6.5). Just above the dam the water has potential energy and (if its velocity is negligible) no significant kinetic energy. As the water falls, its potential energy decreases and its kinetic energy increases. When the water strikes bottom, part of this kinetic energy is transformed into heat.

The swinging of a pendulum illustrates how potential energy can be transformed into kinetic energy and then back to potential energy (Fig. 6.6). When a pendulum is pulled to one side, the mass has no velocity and no kinetic energy, but it does have potential energy. When the mass is released, it swings downward, acquiring kinetic energy and losing potential energy. At the bottom of the swing the pendulum bob has maximum kinetic energy. As it swings upward, this kinetic energy is retransformed into potential energy. A very similar transformation of energy can be traced for a mass on the end of a spring (Fig. 6.4).

☐ **Example** A pendulum bob is pulled to one side until its center of gravity has been raised 10 cm above its equilibrium position. Find the speed of the bob as it swings through the equilibrium position.

$$\text{PE at top} = \text{KE at bottom}$$
$$mgh = \tfrac{1}{2}mv^2$$

whence
$$v^2 = 2gh$$
$$= 2(9.8 \text{ m/s}^2)(0.1 \text{ m}) = 1.96 \text{ m}^2/\text{s}^2$$
$$v = 1.4 \text{ m/s}$$ ☐

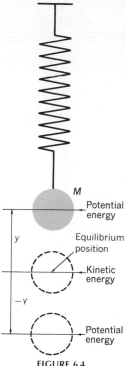

FIGURE 6.4
Relative to the equilibrium position as the configuration of zero potential energy, the system has potential energy $\tfrac{1}{2}ky^2$ when the mass on the end of the spring is displaced a distance y from the equilibrium position. If the mass is released from the position shown, the potential energy is converted into kinetic energy and then back into potential energy at the low point of the motion.

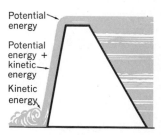

FIGURE 6.5
The potential energy of the water at the top of the dam is converted into kinetic energy at the bottom.

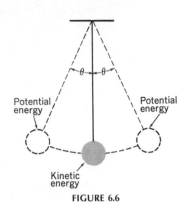

FIGURE 6.6
In a pendulum potential energy is transformed into kinetic energy as the mass moves toward the equilibrium position and then back into potential energy as the mass moves away from the equilibrium position.

Consider a girl on a sled at the top of a hill. Let P_i and K_i be the initial potential and kinetic energies, respectively. Suppose that the girl's father pushes on the sled to speed it up, doing an amount of work W_{in}. As the sled goes down the hill, its potential energy decreases, its kinetic energy increases (at least for a while), and the sled does some work W_{out} against friction. The law of conservation of energy requires that the sum of the initial potential energy, the initial kinetic energy, and the *work done on the sled* (the work *input*) be equal at any time to the sum of the potential energy P_f, the kinetic energy K_f, and the *work done by the sled* (the work *output*); thus

$$W_{in} + P_i + K_i = P_f + K_f + W_{out} \tag{6.9}$$

The law of conservation of energy as expressed in Eq. (6.9) is applicable not only to the sled and to other individual bodies but also to systems of bodies. In the latter case K_i is the sum of the initial kinetic energies, P_i the initial potential energy for the system, W_{in} the *total external work done on the system*, and so forth. In most of our problems one or more of the quantities of Eq. (6.9) will be zero; indeed, we frequently deal with situations in which only one quantity on each side of the equation is nonzero.

6.8 SIMPLE MACHINES

To help us do work we frequently make use of simple machines. A machine is a device for overcoming a resisting force at one point by the application of a force at some other point. Consider a specific example: suppose a man wishes to move a boulder which is too large to shove. He might find a strong stick and use it to move the boulder (Fig. 6.7). If he pushes down on one end of the stick with a force F_{in}, a very much larger force F_{out} is exerted on the boulder. When the stick is in equilibrium, the torque condition gives the relation $F_{in}a = F_{out}b$. If a is 20 times b, F_{out} is 20 times F_{in}. By using this lever, our man can exchange an input force F_{in} for an output force F_{out} 20 times as great.

At first glance this may seem to violate the principle of conservation of energy, but further consideration shows that this is not the case. In order to move the boulder 10 mm, the input force must be exerted for a distance a/b mm. The work done by the man on the stick is equal to the work done on the boulder by the stick. By the use of a simple machine, *it is possible to exchange a small force acting through a large distance for a large force acting through a small distance*, or vice versa.

There are many kinds of simple machines. Among them are levers, inclined planes, pulleys, gear wheels, screws, wheels and axles, the

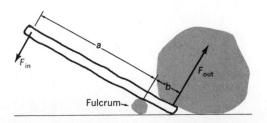

FIGURE 6.7
The lever was one of the first simple machines; here $F_{out}/F_{in} = a/b$.

wedge, and the differential pulley. Complicated machines, such as the automobile or the mechanical cotton picker, are composed of large numbers of interconnected simple machines.

6.9 MECHANICAL ADVANTAGE AND EFFICIENCY

As we have seen, a simple machine is a device for providing some output force in return for an input force. *The ratio of the output force F_{out} to the input force F_{in} is the actual mechanical advantage (AMA) of the machine:*

$$AMA = \frac{F_{out}}{F_{in}} \tag{6.10}$$

In defining the actual mechanical advantage of the machine, we concern ourselves only with the magnitudes of the forces and not with their directions.

In our example of a man moving a boulder, we used a lever with a large mechanical advantage, but useful machines do not always have a large mechanical advantage. The simple pulley has a mechanical advantage of essentially unity and serves only to change the direction of the force, but if we are trying to lift some mortar to the top of a building, it is convenient to be able to stand on the ground and pull down on the rope to lift the mortar. Some of our most useful machines have mechanical advantages of less than 1. In the human forearm (Fig. 6.8) the output force exerted by the hand is very much smaller than the input force exerted by the muscles, so that the mechanical advantage is much less than unity. In such a machine we have a smaller output force moving through a greater distance than the input force.

In real machines friction opposes the motion, and work must be done to overcome this friction. In addition, it may be necessary to raise some part of the machine which gives no useful work output. For example, in Fig. 6.9 the lower pulley block must be raised when we lift the load w. The work done against friction and in lifting parts of the machine does not appear as useful work output. Therefore, in any real machine the useful work output is less than the work input. *The efficiency (Eff) of a machine is the ratio of the useful work output to the work input.*

$$Eff = \frac{work_{out}}{work_{in}} = \frac{W_{out}}{W_{in}} \tag{6.11}$$

For an ideal machine, with no friction and massless moving parts, the efficiency would be 100 percent. The mechanical advantage of an ideal machine is called the *ideal mechanical advantage* (IMA). Let d_{out} be the distance moved by the output force and d_{in} the distance moved by the input force. For an ideal machine the efficiency is 100 percent, and $W_{out} = W_{in}$; hence, $F_{out}d_{out} = F_{in}d_{in}$. Therefore, *the mechanical advantage F_{out}/F_{in} for an ideal machine is d_{in}/d_{out}. This ratio is the ideal mechanical advantage.*

$$IMA = \frac{d_{in}}{d_{out}} \tag{6.12}$$

For any real machine the ideal mechanical advantage is greater than the actual mechanical advantage. Indeed, the efficiency can be seen to be

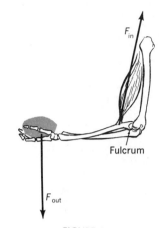

FIGURE 6.8
The human forearm is a lever with mechanical advantage much less than unity.

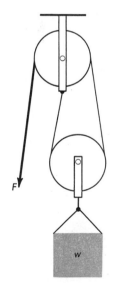

FIGURE 6.9
In raising the load work must be done to lift the lower pulley and to overcome friction; consequently, the useful work output is less than the total work input, and the efficiency is less than 100 percent.

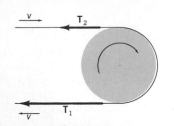

FIGURE 6.10
When the belt drives the pulley, as shown here, T_1 exceeds T_2; when the pulley drives the belt, T_2 is larger than T_1.

equal to the ratio of the actual mechanical advantage to the ideal mechanical advantage by observing that

$$\text{Eff} = \frac{W_{\text{out}}}{W_{\text{in}}} = \frac{F_{\text{out}}}{F_{\text{in}}} \frac{d_{\text{out}}}{d_{\text{in}}} = \frac{\text{AMA}}{\text{IMA}}$$

The concepts of actual mechanical advantage, ideal mechanical advantage, and efficiency may be applied to all types of simple machines.

6.10 ROTATING SYSTEMS

When a pulley is being driven by a belt (Fig. 6.10), the tensions in the straight parts of the belt are not equal. They differ by the friction that is exerted between the belt and the pulley. Let T_2 be the tension in that part of the belt which is moving toward the pulley, T_1 the tension in that part which is moving away from the pulley, and v the velocity with which the belt is moving. The net frictional force between the belt and the pulley is $T_1 - T_2$. The power delivered to the pulley is $(T_1 - T_2)v$, provided the belt does not slip over the pulley.

□ **Example** The tension on one side of a belt is 350 N, and that on the other side is 150 N. The belt is moving at 4 m/s. Find the power delivered to the pulley in watts.

$$\text{Power} = (350 - 150 \text{ N})(4 \text{ m/s})$$
$$= 800 \text{ N-m/s} = 800 \text{ W} \qquad \square$$

Engines and motors are rated in terms of the brake horsepower they develop. It can be measured by using the engine to drive a shaft similar to that of Fig. 6.10 while the belt is held stationary by spring balances which read the tensions in the two sides. Such an arrangement is known as a *Prony brake*. Under these conditions, when the shaft is rotated in the direction shown, T_2 exceeds T_1. The power output, dissipated in heat by friction, is determined by measuring the net frictional force $F = T_2 - T_1$ and the linear speed v of a point on the surface of the shaft and is given by the product Fv.

QUESTIONS

1. Give an example of a situation in which a force acts for an extended time but still does no work. Under what condition may a force move and yet perform no work?
2. A stone is thrown from a cliff with an initial speed v_0. How does its speed as it hits a body below depend on the angle at which it is thrown? Explain.
3. What happens to the kinetic energy given to the blood by pumping action of the heart?
4. An automobile travels with constant velocity along a level road. Is there any resultant force on the car? Is any work being done on the car?
5. Why does a railroad track often wind up a long hill although a road over the same hill may go directly up and over?
6. A boy walks up a downward-moving escalator. If he walks so that his elevation and potential energy stay constant, does he do external work on himself? Does he do external work at all? If so, on what?
7. One boy tosses a ball to another boy on a moving airplane. Does the kinetic energy of the ball depend on the velocity of the airplane? Explain.
8. Why is a high jumper less likely to be injured if he falls in sawdust or sand than on hard clay?
9. Why does it hurt your bare hand more to catch a baseball if you hold your hand in a fixed position than if you move your hand back as you catch the ball?

10. A man is swimming against the current in a river. If he remains stationary relative to the ground, does he do any work? Explain.
11. An object at rest is dropped on a sidewalk and bounces above its original height. How can this occur?
12. Why do scissors for cutting paper have long blades and short handles while metal-cutting shears have short blades and long handles?
13. If two springs of different stiffnesses but the same length are stretched by the same force, is more work done on the stiffer or the softer spring? Explain. What is the situation if they are stretched through the same distance?
14. Why is it that for a simple pulley system with two blocks, one fixed and one movable, as in Fig. 6.9, the ideal mechanical advantage is the number of strands directly supporting the load?
15. A barrel is rolled up an inclined plane which makes an angle of 30° with the horizontal. If no work is done against friction, what is the mechanical advantage if the applied force is parallel to the plane and acts through the center (and center of gravity) of the barrel? What is the mechanical advantage if the force acts tangentially to the circumference of the barrel? Explain the difference.

PROBLEMS

1. How much work is required to hoist an elevator and its contents with total mass of 2,400 kg to the top of a building 150 m high? What average power is required if this work is done in 40 s?
Ans. 3.53 MJ; 88.2 kW
2. A 70-kg man is lifted by an elevator through a distance of 60 m in 20 s. What is the increase in his potential energy? What average power is expended in raising him?
3. A girl pushes a lawn mower 15 m in 10 s by exerting a force of 120 N at an angle of 36.9° with the horizon. Find the work done and the average power expended. *Ans.* 1,440 J; 144 W
4. In how many seconds must a 192-lb man climb 55 ft to the fourth floor of a building if he wishes to expend power at the rate of 1 hp?
5. The locomotive of a freight train must exert a force of 180 kN on the train as it pulls it along on a level track at a speed of 100 km/h. Find the power required from the engine and the work done on the train in a distance of 1 km.
Ans. 5 MW; 180 MJ
6. If the net retarding force on a ship is 20,000 lb when it is traveling 33 ft/s (22.5 mi/h), find the work required to overcome fluid friction in going

1 mi and the power which must be delivered by the screws to overcome this friction.
7. What is the retarding force on a small plane if 60 hp is required to overcome the drag when the plane is flying level at a constant speed of 120 mi/h? *Ans.* 188 lb
8. A cylindrical standpipe 10 m high has an internal diameter of 4 m. How much work would be required to fill the standpipe with water (*a*) if the water were pumped in at the bottom and (*b*) if it were pumped in at the top?
9. An escalator is designed to lift 100 people of 70-kg average mass from one floor of a department store to another floor 5 m higher in 1 min. What power is required, assuming that 80 percent of the power goes into lifting people?
Ans. 7.15 kW
10. A 2,400-lb automobile requires 50 hp to overcome retarding forces on a level road when traveling at a speed of 110 ft/s. Find the net retarding force on the automobile. If the power and retarding force remain the same, find the speed of the car on a 5 percent grade, i.e., one which rises 5 ft in 100 ft of roadway.
11. A Boeing 727 aircraft cruising at 720 km/h at 10.6 km altitude has a thrust of 18 kN from each of its three turbofan jet engines. Find the power expended in both megawatts and in horsepower.
Ans. 10.8 MW; 14,500 hp
12. A force of 50 N stretches a spring 250 mm. What is the force constant of the spring? How much force is required to stretch the spring an additional 150 mm? What is the potential energy of the spring when it is stretched 300 mm?
13. A spring balance reads forces in newtons. The scale is 20 cm long and reads from 0 to 60 N. Find the potential energy of the spring (*a*) when it reads 40 N, (*b*) when it is stretched 20 cm, and (*c*) when a mass of 4 kg is suspended from the spring. *Ans.* (*a*) 2.67 J; (*b*) 6 J; (*c*) 2.56 J
14. The spring constant for a spring gun is 32 N/m. If the spring is compressed 100 mm by a 20-g ball in cocking the gun, find the speed of the ball when the trigger is pulled.
15. A mass of 0.5 kg suspended from a vertical spring produces an elongation of 100 mm. Find the force constant of the spring. What elongation would a force of 3 N produce? What is the potential energy of the spring when the elongation is 55 mm? *Ans.* 49 N/m; 61.2 mm; 74.1 mJ
16. A 50-kg mass rests on an inclined plane that

makes an angle of 36.9° with the horizontal. If the coefficient of friction between the mass and the plane is 0.2, how much work is done in moving the mass up the plane a distance of 4 m? If the mass slides back down, what will its speed be when it reaches its original position?

17. A 0.8-kg pendulum bob on a 2-m cord is pulled sideways until the cord makes an angle of 36.9° with the vertical. Find the work done on the bob and the speed of the bob as it passes through the equilibrium position after being released at rest.
   Ans. 3.136 J; 2.8 m/s

18. A 5-lb block slides down the quarter-circular surface of the accompanying figure. If the radius is 1.2 ft and the block has a speed of 6 ft/s at the bottom, find the work done by the block against friction.

19. A block of mass 3 kg starts from rest and slides down a surface which corresponds to a quarter circle of 1.6 m radius (see figure). (*a*) If the curved surface is smooth, find the speed at the bottom. (*b*) If the speed at the bottom is 4 m/s, find the energy dissipated by friction in the descent. (*c*) After the block reaches the level region at 4 m/s, it slides to a stop in 3 m. Find the frictional force.
   Ans. (*a*) 5.6 m/s; (*b*) 23 J; (*c*) 8 N

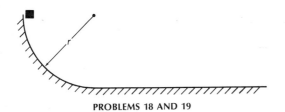

PROBLEMS 18 AND 19

20. Two men using a block and tackle are lifting a safe weighing 3,300 lb to a height of 20 ft. Each man develops $\frac{1}{2}$ hp. How long will it take to raise the safe if only half the work done by each man is useful?

21. A 1,200-kg car starts from rest and coasts down a uniform grade, at the bottom of which it has a speed of 20 m/s. If the car has traveled 800 m and has descended 35 m, how much energy was dissipated on the way down? To what average frictional retarding force does this energy dissipation correspond? *Ans.* 172 kJ; 215 N

22. A 2,000-lb car starts from rest and rolls down a hill 60 ft high and 1,000 ft long as measured along the road. At the bottom of the hill the car's speed is 40 ft/s. How much energy has been dissipated? To what average frictional retarding force does this energy dissipation correspond?

23. A boy throws a 0.15-kg stone from the top of a 20-m cliff with a speed of 15 m/s. Find its kinetic energy and speed when it lands in a river below.
   Ans. 46.3 J; 24.8 m/s

24. A 64-lb girl on a swing is pulled backward and upward until her center of gravity has been raised 4 ft. What is her potential energy? If the swing is released, what will her maximum speed be?

25. A 0.6-kg mass dropped through 1.5 m so that all its potential energy is changed into kinetic energy. Find its energy in joules. Calculate its final velocity by applying the law of conservation of energy. Verify the result by calculating the velocity of a body that falls freely through 1.5 m.
   Ans. 8.82 J; 5.42 m/s

26. A 0.25-lb ball is given an upward velocity of 50 ft/s. If 10 percent of its energy is lost because of air resistance during its rise, how high does the ball go?

27. Starting from rest, a boy slides down a hill 49 ft high on a sled. If friction is negligible, find the speed of the sled at the bottom. If 25 percent of the energy is dissipated in friction, what will his speed at the bottom be? *Ans.* 56 ft/s; 48.5 ft/s

28. A ski jumper glides down a 30° slope for 80 ft before taking off from a negligibly short horizontal takeoff. If her speed at takeoff is 45 ft/s, find the coefficient of kinetic friction on the slide.

29. A 30,000-kg airplane takes off at a speed of 50 m/s, and 5 min later it is at an elevation of 3,000 m and has a speed of 100 m/s. What average power is required during this 5 min if 40 percent of the power is used in overcoming dissipative forces? *Ans.* 5.52 MW

30. A pendulum bob has a mass of 0.5 kg. It is suspended by a cord 2 m long which is pulled back through an angle of 36.9°. Find (*a*) its maximum potential energy relative to its lowest position, (*b*) its potential energy when the cord makes an angle of 10° with the vertical, (*c*) its maximum speed, and (*d*) its speed when the cord makes an angle of 10° with the vertical, assuming the bob is released at 36.9°.

31. A block falls from a table 0.6 m high. It lands on an ideal, massless, vertical spring with a force constant of 2,400 N/m. The spring is initially 25 cm high, but it is compressed to a minimum height of 10 cm before the block is stopped. Find the mass of the block. *Ans.* 5.51 kg

32. A 5-lb block is dropped from a height of 3 ft above a tabletop onto an ideal massless spring.

The spring is vertical, and its top was initially 1 ft above the table. If the spring constant is 300 lb/ft, what is the length of the spring when the block reaches its lowest point?

33. If dissipative resisting forces total 600 N, find the power necessary to give a 1,200-kg automobile an acceleration of 1.5 m/s² when it has a velocity of 48 km/h. *Ans.* 32 kW

34. A motor drives a hoist which lifts a 3-ton load a distance of 40 ft in 50 s. The efficiency of the motor is 85 percent, and that of the hoist is 45 percent. What power is supplied to the load? To the hoist? To the motor?

35. A water bucket weighing 400 N is raised by a crank-and-axle arrangement. The axle has a radius of 75 mm, and the crank has a radius of 225 mm. If a force of 160 N is required on the crank, find the actual mechanical advantage, the ideal mechanical advantage, and the efficiency. *Ans.* 2.5; 3; 0.83

36. In the pulley system shown in Fig. 6.9 a force of 40 lb is required to lift a 60-lb weight. Find the ideal mechanical advantage, the actual mechanical advantage, and the efficiency.

37. An inclined plane (see figure) 12.5 m long is used as a simple machine to pull 4-kN boxes up to a height of 3.5 m. A force of 1.5 kN up the plane is required to pull the boxes up. If it takes 50 s to bring a box to the top, find the actual mechanical advantage of the inclined plane, the ideal mechanical advantage, the efficiency, and the power required. *Ans.* 2.67; 3.57; 0.75; 375 W

38. An inclined plane 13 ft long (see accompanying figure) is used to slide a 390-lb box up to a loading platform 5 ft above ground level. A force of 200 lb is required. Find the efficiency of the inclined plane for this job. What force is required to overcome friction? What is the coefficient of friction between the plane and the box?

39. The upper end of an inclined ramp 7.5 m long is 2.1 m above the lower end. A force of 460 N is required to push a 900-N box up the incline. Find the coefficient of friction and the work done against friction. If the box is released at the top of the ramp, will it slide down? If so, what will its speed at the bottom be?
Ans. 0.24; 1.56 kJ; yes; 2.68 m/s

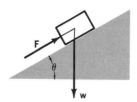

PROBLEMS 37 TO 39

40. The jackscrew in the accompanying figure has a pitch of 3 mm and a handle 600 mm long. A force of 60 N must be applied when a load of 25,000 N is being lifted. Calculate the ideal mechanical advantage, the actual mechanical advantage, and the efficiency.

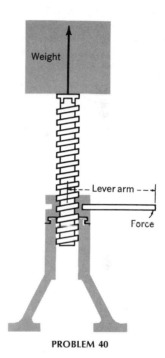

PROBLEM 40

41. A force of 1,200 N is required on the bar of a capstan to obtain an output force of 10 kN for lifting an anchor. The axle of the capstan has a radius of 55 mm, and the input force is applied 0.88 m from the axis. Find the work done by the input force in lifting the anchor 15 m in a time of 2 min, the actual mechanical advantage, the ideal mechanical advantage, the efficiency of the capstan, and the power provided by the input force.
Ans. 288 kJ; 8.33; 16; 0.52; 2.4 kW

42. Show that the ideal mechanical advantage of a differential pulley (see accompanying figure) is given by $2R/(R - r)$, where R and r are the radii of the larger and smaller sheaves, respectively. Find the ideal mechanical advantage of a differential pulley in which the radius of the larger

pulley is 115 mm and that of the smaller pulley is 105 mm.

43. A spring of neutral length 0.4 m and force constant 200 N/m is suspended from a second spring of neutral length 0.3 m and spring constant 500 N/m, which hangs from a hook. If the bottom of the first spring is pulled down until the combined length of the two springs is 1.05 m, find the individual lengths of the two springs and the stretching force required.

Ans. 0.65 m; 0.40 m; 50 N

44. Show that the efficiency of a jackscrew must be less than 50 percent if the jack remains set when no force is applied at the handle.

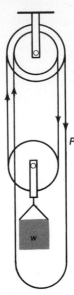

PROBLEM 42

In Chap. 6 we learned the principle of conservation of energy and saw how useful such a principle can be. Next we shall make use of Newton's second and third laws of motion (Chap. 4) to derive a second great conservation principle, that of the conservation of momentum. With the aid of conservation of momentum we can treat some problems of motion in which we do not know the precise forces acting on an object.

7

MOMENTUM

7.1 MOMENTUM AND IMPULSE

The *momentum* of a body is a vector defined as the product of its mass m and its velocity $\mathbf{v}$. Consider a constant unbalanced force $\mathbf{F}$ acting on a body of mass m and initial velocity $\mathbf{v}_0$ for a time Δt. By Newton's second law [Eq. (4.6)],

$$\mathbf{F} = \frac{\Delta(m\mathbf{v})}{\Delta t} = \frac{m(\mathbf{v} - \mathbf{v}_0)}{t}$$

from which
$$\mathbf{F}t = m\mathbf{v} - m\mathbf{v}_0 \tag{7.1}$$

The right side of this equation is just the momentum at time t minus the initial momentum, or the change in momentum of the body. The quantity on the left side of the equation, *the product of the force and the time during which the force acts, is called the impulse.* From Eq. (7.1) we see that *the impulse is equal to the change in the momentum.*

In many interactions the net force acting on a body varies during the reaction. For example, when a football is kicked off, the toe of the kicker exerts zero force on the ball until it comes in contact, and the force increases rapidly as the ball is distorted (Fig. 7.1). As the ball returns to its initial shape, the force diminishes. A graph of the force as a function of time might be that of Fig. 7.2. In this case the impulse is given by the area under the force-as-a-function-of-time curve. Alternatively, we can express the impulse as the product of the average force and the time during which the force acts. The average force is indicated by the dashed line of the figure.

In many situations the force varies so rapidly that it is not practical to know what the force is at any particular instant, but the change in the momentum of the body and the total impulse can be measured. If we know how long the impulsive force acts, we can calculate the average value of the force even though we may not know the instantaneous value, i.e., the exact shape of the force curve. We have impulsive forces in play when a pitched ball is hit by a bat or when two billiard balls collide.

☐ **Example**   A baseball of mass 0.145 kg is thrown by the pitcher at 30 m/s. The bat is in contact with the ball for 0.01 s and gives it a speed of 40 m/s in a direction straight toward the pitcher. Find the impulse and the average value of the force.

Impulse = change in momentum
= $(0.145 \text{ kg})[30 - (-40) \text{ m}\}\text{s}]$
= 10.1 kg-m/s
= 10.1 N-s

FIGURE 7.1
High-speed x-ray picture of a football being kicked. (*Courtesy of C. M. Slack, Westinghouse Corp.*)

$$F_{av} = \frac{impulse}{time}$$

$$= \frac{10.1 \text{ N-s}}{0.01 \text{ s}}$$

$$= 1{,}010 \text{ N} \qquad \square$$

7.2 THE CONSERVATION OF MOMENTUM

Consider a freely suspended rifle, cocked and ready to fire (Fig. 7.3). When the trigger is pulled, a force $\mathbf{F}_b$ is exerted on the bullet. By Newton's third law, if the gun exerts a force $\mathbf{F}_b$ on the bullet, the bullet exerts an equal and opposite force $\mathbf{F}_g$ on the gun, so $\mathbf{F}_g = -\mathbf{F}_b$. Applying Newton's second law yields

$$\mathbf{F}_b = \frac{\Delta(m_b \mathbf{v}_b)}{\Delta t} = \frac{-\Delta(M_g \mathbf{V}_g)}{\Delta t} = -\mathbf{F}_g$$

Since the forces $\mathbf{F}_b$ and $\mathbf{F}_g$ are not only equal and opposite but also act for exactly the same time, the impulses for the gun and the bullet are equal in magnitude and opposite in direction. Therefore,

$$m_b \mathbf{v}_b = -M_g \mathbf{V}_g \qquad (7.2)$$

The momentum gained by the bullet is equal in magnitude and opposite in direction to that received by the gun. Since momentum is a vector quantity, the resultant of the momenta is zero, just as it was before the trigger was pulled. The momentum of the system composed of the gun and the bullet has not changed.

It follows directly from Newton's second and third laws of motion that *if two or more bodies interact, the momentum after the interaction is equal to the momentum before the interaction.*

This important principle is known as the *law of conservation of momentum.* It may be stated in the following form: *The total momentum of any system of bodies is unchanged by any interactions between the different members of the system.*

When a battleship fires a broadside, the shells are given a large momentum in one direction. The battleship recoils with an equal momentum in the opposite direction. When two billiard balls collide, momentum is transferred from one ball to the other but the resultant momentum after the collision is equal to the momentum before the collision. When two football players collide in midair, the law of conservation of momentum determines which way they fall.

□ **Example** An 80-kg halfback dives over the line of scrimmage at a velocity of 7 m/s. He is met in midair by a 100-kg linebacker going

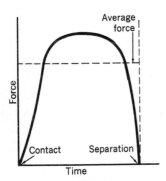

FIGURE 7.2
Force on the football of Fig. 7.1 as a function of time.

FIGURE 7.3
The momentum of the system comprising the bullet and the rifle is zero both before and after the bullet is fired. The momentum given the rifle is equal and opposite to that given the bullet.

5.4 m/s in the opposite direction. If they fall together, find the horizontal component of their velocity.

Momentum before = momentum after

$$(80 \text{ kg} \times 7 \text{ m/s}) - (100 \text{ kg} \times 5.4 \text{ m/s}) = 180 \text{ kg} \times v$$

$$20 \text{ kg-m/s} = 180 \text{ kg} \times v$$

$$v = \tfrac{20}{180} = 0.11 \text{ m/s in direction of halfback's initial velocity} \qquad \square$$

Consider one aircraft chasing another of about the same maximum speed. If the pursuing craft opens fire, the bullets are given forward momentum and the plane loses an equal momentum. On the other hand, when the pursued craft opens fire, the bullets have momentum toward the rear and this aircraft gains speed.

7.3 COLLISION PHENOMENA

When two bodies collide, the laws of conservation of momentum and conservation of energy are both always applicable, but in some collisions part of the kinetic energy of the bodies is transformed into some non-mechanical form of energy such as heat or sound. In this case the application of the law of conservation of energy to the problem becomes exceedingly difficult because many kinds of energy may be involved, some of which are difficult to measure. If a steel ball is dropped on a steel plate, the ball and plate are distorted during the action of the impulsive forces, but they return to their original shapes and the ball springs away from the plate. In this case the bodies are only temporarily deformed, and they regain their original shapes immediately after collision. Bodies which return to their original shapes after collision are said to be *elastic*.

In a perfectly elastic collision, kinetic energy, as well as momentum, is conserved. Collisions between atomic nuclei, atoms, molecules, and electrons are often elastic. Such an ideal situation is never achieved with large-scale objects; we have no perfectly elastic collisions of macroscopic bodies.

Any collision for which the final kinetic energy is less than the initial kinetic energy is said to be *inelastic*. When a bullet is fired into a block of wood, the bullet sticks in the block and the two remain together indefinitely. Such a collision is a *perfectly inelastic* one. The bodies are permanently deformed and never separate. Collisions between billiard balls or between a basketball and floor are not perfectly inelastic since the colliding objects spring apart. Such *semielastic* collisions are very common in nature.

In every collision momentum is conserved; kinetic energy is conserved only in perfectly elastic collisions. In the next three sections we consider head-on collisions between two bodies of masses m and M (Fig. 7.4) which have initial velocities $\mathbf{v}_i$ and $\mathbf{V}_i$, respectively, before collision, and final velocities $\mathbf{v}_f$ and $\mathbf{V}_f$ after collision. Since velocities are vector quantities, we must consider directions as well as magnitudes. In the equations which follow we adopt the convention that velocities to the right are positive, and velocities to the left negative.

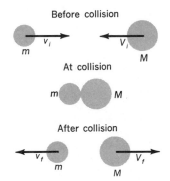

FIGURE 7.4
Head-on collision between two moving spheres.

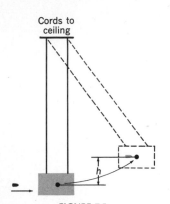

Cords to
ceiling

FIGURE 7.5
The speed of a bullet can be determined by firing it into the middle of the bob of a ballistic pendulum and measuring how much the center of the bob is raised at the end of the swing.

7.4 PERFECTLY INELASTIC COLLISIONS

In a completely inelastic collision the two bodies stick together after they have made contact. As a consequence, both have the same final velocity. Conservation of momentum alone is sufficient to permit us to compute the final velocity if the masses and initial velocities are known:

$$m\mathbf{v}_i + M\mathbf{V}_i = (m + M)\mathbf{V}_f \qquad (7.3)$$

□ **Example** A 2-g bullet is fired into a 2.398-kg block (Fig. 7.5) of wood suspended from a long cord. The bullet is embedded in the block, and the two start off together with a speed of 0.700 m/s. Find the velocity of the bullet before collision.

$$(0.002 \text{ kg} \times v_i) + (2.398 \text{ kg} \times 0) = (2.400 \text{ kg})(0.700 \text{ m/s})$$
$$v_i = 840 \text{ m/s} \qquad \square$$

An arrangement like that in Fig. 7.5 by which the speed of a moving body is measured by capturing it in a more massive pendulum bob is known as a *ballistic pendulum*. In Fig. 7.5 the moving body is a bullet and the pendulum bob is a uniform rectangular block of wood initially at rest. The bullet is fired horizontally toward the center of the block, so that after impact the bob moves off smoothly. (If the bullet hits off center, the block wobbles and we cannot treat the collision as a simple one between two particles.) Immediately after the collision the bullet-block system has kinetic energy; this energy is converted into potential energy as the pendulum bob swings upward. Because the block is uniform and the bullet is stopped near the center, we can treat the system as though its entire weight were concentrated at the center of the block. (For any body the point at which we can consider the entire weight to be concentrated is called the *center of gravity*, as we shall see in Sec. 10.2; in this particular case the center of gravity coincides with the center of the block.)

To calculate the speed V of the system immediately after the collision, we measure the height h of the highest point reached by the center of gravity relative to its initial position. By the law of conservation of energy, the kinetic energy of the bullet-bob system as it leaves the original position is equal to the potential energy at the end of the swing; $\frac{1}{2}(m + M)V^2 = (m + M)gh$, where m and M are the masses of the bullet and block respectively. Finally, to find the original speed of the bullet, we apply Eq. (7.3), since the collision of bullet and block is completely inelastic.

7.5 ELASTIC COLLISIONS

In a perfectly elastic collision both momentum and kinetic energy are conserved. If two bodies collide along the line connecting their centers, we write

$$m v_i + M V_i = m v_f + M V_f \qquad \text{conservation of momentum} \quad (7.4)$$

and

$$\tfrac{1}{2}m v_i{}^2 + \tfrac{1}{2}M V_i{}^2 = \tfrac{1}{2}m v_f{}^2 + \tfrac{1}{2}M V_f^2 \qquad \text{conservation of energy} \quad (7.5)$$

These equations can be rewritten as

$$m(v_i^2 - v_f^2) = M(V_f^2 - V_i^2) \quad \text{and} \quad m(v_i - v_f) = M(V_f - V_i)$$

If we divide the first by the second, we obtain $v_i + v_f = V_f + V_i$ or

$$v_i - V_i = V_f - v_f = -(v_f - V_f) \tag{7.6}$$

Note that $v_i - V_i$ is the *velocity of approach*, or the velocity of the smaller mass relative to the larger one before the collision, while $v_f - V_f$ is the *velocity of separation*, or the velocity of the smaller mass relative to the larger after the collision. In a perfectly elastic collision the velocity of approach is equal in magnitude to the velocity of separation but opposite in direction.

☐ **Example** A 40-g ball traveling east with a speed of 5.0 m/s has a head-on collision with a 60-g ball traveling 3.0 m/s west (Fig. 7.4). If the collision is elastic, find the velocities after the collision.

Let us choose east as the positive direction, so velocities to the west are negative. By Eq. (7.4)

$$(0.040 \text{ kg})(5.0 \text{ m/s}) + (0.060 \text{ kg})(-3.0 \text{ m/s})$$
$$= (0.040 \text{ kg})(v_f) + (0.060 \text{ kg})(V_f)$$
$$v_f + 1.5V_f = 0.50 \text{ m/s} \tag{A}$$

By Eq. (7.6),

$$5.0 \text{ m/s} - (-3.0 \text{ m/s}) = V_f - v_f$$
$$-v_f + V_f = 8.0 \text{ m/s} \tag{B}$$

Adding Eqs. (A) and (B) yields

$$2.5V_f = +8.5 \text{ m/s}$$
$$V_f = +3.4 \text{ m/s}$$
$$v_f = -4.6 \text{ m/s}$$

The minus sign means the 40-g ball is moving westward. ☐

7.6 SEMIELASTIC COLLISIONS

In a perfectly elastic collision kinetic energy is conserved, and the velocity of approach is equal to the velocity of separation in magnitude but opposite in direction. In a perfectly inelastic collision the velocity of separation is zero, since the two bodies remain together. Most ordinary collisions of macroscopic bodies lie between these two extremes. The bodies spring apart but with kinetic energy reduced. Newton studied collisions between many types of spheres and concluded that in general the velocity of separation is proportional to, but less than, the velocity of approach:

$$-(v_f - V_f) = e(v_i - V_i) \tag{7.7}$$

where, for a given pair of materials, e is a constant which lies between 0 and 1. This constant, called the *coefficient of restitution*, is 0 for perfectly inelastic collisions in which the bodies remain together. It is 1 for perfectly elastic collisions. For semielastic collisions it lies between 0 and 1,

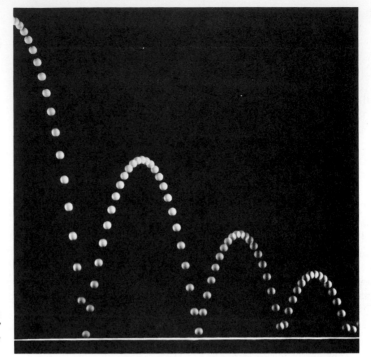

FIGURE 7.6
Semielastic collisions of a steel ball with a steel plate. (*From PSSC "Physics," D. C. Heath and Company, 1965. Courtesy Education Development Center.*)

being near 1 for a golf club striking a golf ball and near 0 for the same club striking a knitted practice ball. If the coefficient of restitution is known, the use of Eqs. (7.4) and (7.7) is sufficient to find the final velocities of the two bodies. Figure 7.6 shows imperfectly elastic collisions of a steel ball and a hardened steel plate. Since the plate is fastened to the earth, M is essentially infinity and $V_i = V_f = 0$. Hence, $e = -v_f/v_i$.

□ **Example** If the ball of Fig. 7.6 is dropped from a height of 40.0 cm, and if the coefficient of restitution is 0.80, find the height attained on the first bounce.

By conservation of energy, $mgh = \frac{1}{2}mv_i^2$, or $v_i^2 = 2gh$. Therefore, the velocity v_i for the first collision is $\sqrt{2(9.8)(0.400)} = 2.80$ m/s. By Eq. (7.7), $-v_f = ev_i$, since V_f and V_i are essentially zero.

$$-v_f = 0.80v_i = 2.24 \text{ m/s}$$

This initial upward speed will carry the ball to a height h_1 such that $v_f^2 = 2gh_1$ or

$$h_1 = \frac{(2.24)^2 \text{ m}}{2(9.8)} = 0.256 \text{ m} = 25.6 \text{ cm} \qquad \square$$

7.7 COLLISIONS IN SPACE

Thus far we have considered only collisions in one dimension. Typical collisions between billiard balls involve two dimensions, while collisions between air molecules occur in three dimensions. In every case momen-

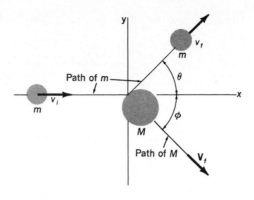

FIGURE 7.7
Glancing elastic collision of a moving
sphere with a second sphere initially at
rest.

tum is conserved. Since momentum is a vector quantity, its conservation requires that the components of the resultant momentum be conserved also. Thus, in a three-dimensional collision each component of the resultant momentum remains the same, and we obtain three equations by application of the conservation of momentum.

Consider a body of mass m, moving with a velocity $\mathbf{v}_i$, which has an elastic collision in two dimensions with a body of mass M initially at rest. After the collision the first body moves off at an angle θ to its original path (Fig. 7.7). Let the x axis be determined by $\mathbf{v}_i$. Applying the conservation of momentum yields

x component: $\quad mv_i + M(0) = mv_f \cos \theta + MV_f \cos \phi$

y component: $\quad\quad 0 + 0 = mv_f \sin \theta - MV_f \sin \phi$

Conservation of kinetic energy provides the equation

$$\tfrac{1}{2}mv_i{}^2 = \tfrac{1}{2}mv_f{}^2 + \tfrac{1}{2}MV_f^2$$

Thus, if m, M, v_i, and θ are given, we have three equations in three unknowns (v_f, V_f, and ϕ).

7.8 CENTER OF MASS

When a shell explodes in midair (Fig. 7.8), fragments are ejected in all directions. However, the resultant momentum of all fragments just after the explosion is the same as the momentum of the shell just before the explosion. Indeed, a unique point associated with the system, known as the *center of mass*, continues to move along the same path whether the shell explodes or not. The x coordinate of the center of mass of a system

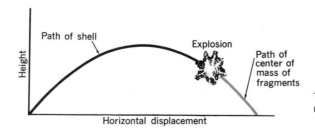

FIGURE 7.8
The center of mass of the shell fragments moves along the same trajectory
before and after the shell explodes.

of particles with masses $m_1, m_2, \ldots, m_N$ is defined at any instant by the relation

$$x_c = \frac{m_1 x_1 + m_2 x_2 + \cdots + m_N x_N}{m_1 + m_2 + \cdots + m_N} \tag{7.8}$$

where x_1 is the x coordinate of m_1, etc. Similar relations locate the y and z coordinates of the center of mass. The sum of the individual masses in the denominator is simply the total mass M of the system of particles.

Consider once again a shell which explodes into N fragments at time t. Just before the explosion let x_c be the x coordinate of the center of mass, m_1 and x_1 the mass and x coordinate of what becomes fragment 1, and so forth. A time Δt later the explosion has occurred and fragment 1 has x coordinate $x_1 + \Delta x_1$, fragment N is at $x_N + \Delta x_N$, and the x coordinate of the center of mass is at $x_c + \Delta x_c$. From

$$\frac{x_c + \Delta x_c}{\Delta t}$$

$$= \frac{m_1(x_1 + \Delta x_1)/\Delta t + m_2(x_2 + \Delta x_2)/\Delta t + \cdots + m_N(x_N + \Delta x_N)/\Delta t}{M}$$

we have for the x component of the velocity of the center of mass

$$v_{xc} = \frac{\Delta x_c}{\Delta t} = \frac{m_1 v_{x1} + m_2 v_{x2} + \cdots + m_N v_{xN}}{M} \tag{7.9}$$

Similar equations give the y and z components of this velocity, and we see that *the velocity of the center of mass of a system of particles is equal to the resultant momentum of the system divided by the mass of the system.*

The forces of explosion for the shell are forces exerted by one part of the system on other parts of the system. Such *internal* forces do not change the resultant momentum of the system. Even though the explosion produces drastic changes in the velocities $\mathbf{v}_1, \mathbf{v}_2, \ldots, \mathbf{v}_N$ of the fragments, it does not affect the velocity of the center of mass. However, the weight of the shell (and of the fragments) is an *external* force which does change the momentum of the system. *The resultant momentum $M\mathbf{v}_c$ of a system remains constant when no external forces act on the system; if external forces act, the center of mass has an acceleration proportional to the resultant external force and inversely proportional to the mass of the system.*

Figure 7.9 is a multiple-flash photograph of a rotating wrench moving freely in the absence of any external force. The center of mass, indicated by the black cross, moves with constant velocity; every other mass ele-

FIGURE 7.9
Succession of pictures taken at $\frac{1}{30}$-s intervals of a rotating wrench moving freely in the absence of any external force. The velocity of the center of mass (marked by a black cross) remains constant as the wrench undergoes simultaneous translation and rotation. (*From PSSC "Physics," D. C. Heath and Company, 1965. Courtesy Education Development Center.*)

ment of the wrench has a more complicated velocity, which is the resultant of the translational velocity of the center of mass and velocity due to
the rotation (to be discussed in Chap. 11). Any net external force applied
to the wrench would produce an acceleration of the center of mass;
whether it would affect the rotational velocity depends on the line along
which the resultant force acts. In general, when the applied force on a
body acts along a line which passes through the center of mass, there is
no change in its rotational motion; conversely, if the force does not act
through the center of mass, the rotational motion is changed.

7.9 ROCKET PROPULSION

Rockets are used for propelling long-range missiles and for placing satellites in orbit. A jet of hot gases is ejected from the combustion chamber
of a rocket (Fig. 7.10). Since the rocket exerts a rearward force on the
gases, the gases exert an equal and opposite forward force on the rocket.
Let **u** be the velocity of the ejected gases relative to the rocket and Δm
the mass of gas ejected at a constant rate in time Δt. Then **u** $\Delta m/\Delta t$ is the
time rate of change of momentum of the exhaust gases, which, by New-
ton's second law of motion, is the force exerted on the gases by the
rocket. This, by Newton's third law, is equal and opposite to the reaction
force $\mathbf{F}_R$ exerted on the rocket by the gases; $\mathbf{F}_R$, called the *thrust* of the
rocket, has the magnitude

$$F_R = u\frac{\Delta m}{\Delta t} \tag{7.10}$$

and a direction opposite to **u**.

Ordinarily the rocket has other forces, such as its weight, acting upon
it. If the resultant of these external forces is represented by $\mathbf{F}_E$, the net
force acting on the rocket is $\mathbf{F}_E + \mathbf{F}_R$. By Newton's second law,

$$\mathbf{F}_E + \mathbf{F}_R = M\mathbf{a} \tag{7.11}$$

where **a** is the acceleration of the rocket relative to some fixed coordinate
system and M is the mass of the rocket. Note that M varies during the
flight as fuel is burned or as some portion of the rocket system is jetti-
soned.

☐ **Example** The Thor ballistic missile had a gross weight of roughly
425,000 N at the beginning of flight. Its rocket engine produced 600,000 N
of thrust. If the exhaust velocity of the gases was 2,060 m/s, find the rate
at which mass was ejected, the acceleration at launch, and the acceleration after 100 s of operation.

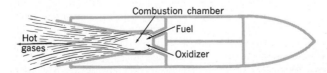

Combustion chamber
Fuel
Hot gases
Oxidizer

FIGURE 7.10
Schematic diagram of a rocket engine.

By Eq. (7.10)

$$F_R = u \frac{\Delta m}{\Delta t}$$

$$600{,}000 \text{ N} = 2{,}060 \text{ m/s} \times \frac{\Delta m}{\Delta t}$$

$$\frac{\Delta m}{\Delta t} = 292 \text{ kg/s}$$

At launch the unbalanced force is the difference between the thrust and the weight. By Newton's second law,

$$600{,}000 \text{ N} - 425{,}000 \text{ N} = \frac{425{,}000}{9.8} \text{ kg} \times a$$

$$a = 4.03 \text{ m/s}^2 = 0.41g$$

After 100 s of operation 29,200 kg has been ejected, and the mass of the rocket is 14,200 kg. If we assume that air friction is negligible but that the rocket is not so high that g is appreciably different from 9.8 m/s^2, the weight of the rocket is 139,000 N and the unbalanced force is 600,000 N − 139,000 N, or 461,000 N. Then

$$461{,}000 \text{ N} = 14{,}200 \text{ kg} \times a$$

$$a = 32.4 \text{ m/s}^2 = 3.3g \qquad \square$$

QUESTIONS

1. A shell explodes in midair. Is its total momentum greater than before the explosion? Is its kinetic energy changed? Explain.
2. Only a resultant *external* force can change the momentum of the center of mass of a body or system of bodies. If this is true, how can the internal forces of the brakes stop a moving automobile?
3. Does the center of mass of a solid body necessarily lie within the body? If not, what are some familiar bodies for which the center of mass lies outside the body?
4. The speed of an airplane flying at a constant altitude is increased 10 percent. What happens to its kinetic energy? Is energy conserved? Explain. What happens to the momentum of the airplane? Is momentum conserved? Explain.
5. Can a single body have momentum without having kinetic energy? Without having any energy? Can a body have energy without having momentum? Contrive a simple system of two or three bodies which can have kinetic energy without having momentum.
6. Identical bullets strike a wooden block and a steel block of identical mass. If the bullet sticks in the wood but rebounds from the steel, which block receives the greater impulse? Why?
7. Discuss what happens when a novice marksman fires a rifle while holding the butt about a centimeter from his shoulder.
8. How can a rocket engine accelerate a spacecraft even when there is no air for it to push against?
9. The blades of a turbine are curved rather than plane; consequently the fluid makes a U-turn instead of meeting the blade head on. Explain why the curved blade is advantageous.
10. Discuss qualitatively what happens when a light sphere collides head on and elastically with a more massive sphere initially at rest. How does this compare with what happens if the more massive sphere is in motion and strikes the light sphere head on? What happens if the two spheres have the same mass?
11. Initially an hourglass sits at rest on the pan of a sensitive spring balance. Then the hourglass is turned over and sand streams downward from the upper volume. Does the balance read the same as before? If not, does it read more or less? Explain.
12. Prove that if a ball makes an elastic collision with a smooth wall, the angle of incidence (between the incident path and a perpendicular to the surface) and the angle of reflection are equal.
13. In a certain guided missile 80 percent of the weight is fuel. Assuming that the thrust exerted by the rocket motor is constant, discuss the factors which influence the acceleration of the rocket as it rises vertically to a height of 300 km.

14. A bee is temporarily imprisoned in a light can which is completely closed. If the container is placed on a sensitive balance, is the weight the same when the bee is flying about and when it is at rest? Is the weight affected by anything the bee does?

PROBLEMS

1. A 3-g bullet is fired from a 2.4-kg rifle with a velocity of 360 m/s north. Find the momentum of the bullet and the recoil velocity of the rifle, assuming that no other bodies are involved.
   Ans. 1.08 kg-m/s north; 0.45 m/s south
2. A bullet of mass 5 g is projected from a gun of mass 4 kg with a speed of 320 m/s. With what speed does the gun recoil?
3. (*a*) An 80-kg man driving at 72 km/h falls asleep, and his automobile strikes a bridge abutment. If the driver comes to a stop in 0.06 s, what is the impulse and average stopping force acting on him? (*b*) If an airbag were activated and the stopping time were 0.2 s, what would the average force be?
   Ans. (*a*) 1,600 N-s, 26,700 N; (*b*) 8,000 N
4. A pitcher exerts an average resultant force of 7 lb for 0.12 s on a baseball weighing 0.32 lb. Find the average acceleration, the impulse, the speed of the ball, and the magnitude of the momentum.
5. A 0.144-kg baseball approaches a batter with a speed of 30 m/s. The batter lines the ball directly back to the pitcher with a speed of 40 m/s. Find the change in momentum and the impulse. If bat and ball were in contact for 0.012 s, find the average force exerted on the ball during this period.   *Ans.* 10.1 kg-m/s; 10.1 N-s; 840 N
6. A 180-lb swimmer dives from a 200-lb rowboat initially at rest. If the swimmer leaves the boat with a horizontal speed of 10 ft/s, find the recoil speed of the boat, neglecting the transfer of momentum to the water. What was the impulse on the boat?
7. A football of mass 0.42 kg is passed with a velocity of 25 m/s due south. A defending player lunges at the ball and deflects it so that the new velocity is 20 m/s 36.9°W of S. Find the magnitude of the impulse. If the player is in contact with the ball for 0.05 s, what is the magnitude of the average force he exerts? *Ans.* 6.3 N-s; 126 N
8. An aircraft with a mass of 12,000 kg catapulted from a carrier has a uniform acceleration of 25 m/s^2 over a distance of 40.5 m. Find the resultant force required, the time the force acts, the momentum of the aircraft as it leaves the catapult, and the impulse provided to the aircraft.

9. A 0.005-kg bullet going 300 m/s strikes a 1.995-kg wooden block which is the bob of a ballistic pendulum. Find the speed at which block and bullet leave the equilibrium position and the height which the center of gravity of the bullet-block system reaches above the initial position of the center of gravity.   *Ans.* 0.75 m/s; 28.7 mm
10. A 4-g bullet is fired into a 2.996-kg block of wood which is the bob of a ballistic pendulum. If the bob leaves its equilibrium position with a speed of 0.5 m/s, find the speed of the bullet and the height above the equilibrium position reached by the center of gravity of the block.
11. A 5-g bullet traveling 250 m/s strikes and embeds itself in a 2.495-kg block held on a frictionless table by a spring with constant $k = 40$ N/m. Find the speed of block and bullet immediately after the collision and the distance the spring is compressed.   *Ans.* 0.5 m/s; 125 mm
12. A freight car weighing 25 tons runs into another freight car of the same weight. One car was stationary, and the other was moving at 6 mi/h (8.8 ft/s). If the cars move off together after collision, with what speed do they move?
13. A 3-g bullet is fired into a 2.997-kg block suspended by a long cord. The bullet remains in the block, which swings until its center of gravity is raised by 4 mm. Find the speed of the block and bullet as they leave the equilibrium position of the block. What was the initial speed of the bullet?   *Ans.* 0.28 m/s; 280 m/s
14. A 4-g bullet is fired horizontally with a speed of 300 m/s into a 0.8-kg block of wood at rest on a table. If the coefficient of friction between the block and the table is 0.3, how far will the block slide? What fraction of the bullet's energy is dissipated in the collision itself?
15. A 3-g bullet is fired horizontally into a 597-g block of wood suspended from a cord 2 m long. The block swings because of the impact, deflecting the cord to a position 15° from the vertical. Find the initial speed of the bullet.
   Ans. 231 m/s
16. A ball of 0.4 kg mass and a speed of 3 m/s has a completely elastic collision with a 0.6-kg mass initially at rest. Find the speed of both bodies after the collision.
17. A proton of mass 1.66×10^{-27} kg collides head on with a helium atom at rest. The helium atom has a mass of 6.64×10^{-27} kg and recoils with a speed of 5×10^5 m/s. If the collision is elastic, find the

initial and final speeds of the proton and the fraction of its initial energy transferred to the helium atom.

Ans. $v_i = 1.25 \times 10^6$ m/s; $v_f = -7.5 \times 10^5$ m/s; 0.64

18. Two perfectly elastic balls, one weighing 2 lb and the other 3 lb, are moving in opposite directions with speeds of 8 and 6 ft/s, respectively. Find their velocities after head-on impact.

19. In the accompanying figure the smaller ball has a mass of 0.3 kg and the larger one a mass of 0.5 kg. If the smaller one is pulled back and released so that it has a speed of 4 m/s just before collision, find the velocities of both balls immediately after they have an elastic collision.

Ans. −1 and 3 m/s

20. If the small steel ball in the accompanying figure has a mass of 0.2 kg and the large one a mass of 0.4 kg, find the recoil velocity of each ball if the larger one is pulled out and released in such a way that it has an elastic collision when it is moving to the left with a speed of 0.6 m/s.

21. Both balls of the accompanying figure are pulled back and released together so that they collide elastically at the equilibrium position. The smaller ball has a mass of 0.15 kg and is moving to the right with a speed of 4 m/s, while the larger ball has a mass of 0.25 kg and is moving to the left at 3 m/s. Find the velocity of each after the head-on collision. *Ans.* −4.75 m/s, 2.25 m/s

22. A golf ball is dropped on a sidewalk from a height of 2 m and rebounds to a height of 1.50 m. What was the impulse? Assuming the ball was in

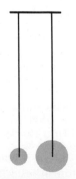

PROBLEMS 19 TO 21

contact with the concrete for 0.007 s, find the average acceleration. What was the coefficient of restitution? The mass of the ball is 45.8 g.

23. A mass A of 0.8 kg moving to the right with a speed of 5 m/s collides head on with a mass B of 1.2 kg moving in the opposite direction with a speed of 4 m/s. After the collision A is moving to the left with a speed of 4 m/s. Find the velocity of B after collision and the coefficient of restitution. *Ans.* 2 m/s to the right; 0.667

24. A tennis ball bounces down a flight of stairs, striking each step in turn and rebounding to the height of the step above. If the height of each step is 0.25 m, find the coefficient of restitution.

25. If the coefficient of restitution for a golf ball dropped on a sidewalk is 0.9, find the height to which the ball returns if it is dropped from a height of 2 m. *Ans.* 1.62 m

26. A 0.3-kg ball traveling 5 m/s collides head on with a 0.15-kg ball initially at rest. If the latter proceeds in the original direction with a speed of 4 m/s, find the speed of the 0.3-kg ball. What was the coefficient of restitution?

27. A 0.3-kg block slides down a frictionless hemispherical bowl with a radius of 0.1 m. At the bottom it collides perfectly inelastically with a 0.4-kg mass initially at rest. What impulse is given the 0.4-kg mass? Find the maximum angle α which the radius vector to the two blocks will make with the vertical after the collision.

Ans. 0.24 N-s; 35.3°

28. A ball dropped from a height of 4 ft rebounds from a flat surface to a height of 2.56 ft. Find the speed of the ball just before it strikes and just after it leaves the floor. What is the coefficient of restitution?

29. A 3-g bullet with a speed of 300 m/s passes right through a 400-g block suspended on a long cord. The impulse gives the block a speed of 1.5 m/s. Find (*a*) the speed of the bullet after it has passed through the block, (*b*) the distance which the center of mass rises as the block swings upward after the bullet passes through, (*c*) the work done by the bullet in passing through the block, and (*d*) the mechanical energy converted into heat. *Ans.* (*a*) 100 m/s; (*b*) 0.115 m; (*c*) 120 J; (*d*) 119.55 J

30. A system consisting of two masses connected by a massless rod lies along the x axis. A 0.4-kg mass is at $x = 2$ m while a 0.6-kg mass is at $x = 7$ m. Find the x coordinate of the center of mass.

31. Three spheres are located with their centers along a straight line as follows: a 4-kg mass is 0.3 m to the right of a 5-kg mass, and a 6-kg mass is 0.7 m to the right of the 4-kg mass. Find the center of mass of the system.

Ans. 0.18 m to the right of 4-kg mass

32. Find the distance from the center of the earth to the center of mass of the earth-moon system if the earth-moon separation is 3.8×10^5 km and the mass of the earth is 81.3 times the mass of the moon. To what fraction of the earth's radius of 6,370 km does this distance correspond?

33. A rigid body consists of a 3-kg mass connected to a 2-kg mass by a massless rod. The 3-kg mass is located at $r = 2i + 5j$ m, and the 2-kg mass has $r = 4i + 2j$ m. Find the length of the rod and the coordinates of the center of mass.

Ans. 3.61 m; $2.8i + 3.8j$ m

34. A machine gun fires six bullets per second into a target. The mass of each bullet is 3 g, and the speed 500 m/s. Find the average force required to hold the gun in position. What is the power delivered to the bullets?

35. A machine gun fires 50 bullets per minute with a speed of 800 m/s. If each bullet has a mass of 15 g, what average force is required to hold the gun against the recoil? What power is delivered to the bullets by the gun? *Ans.* 10 N; 4 kW

36. A machine gun fires 200 bullets each minute with a speed of 2,000 ft/s. If the weight of each bullet is 0.016 lb, what horsepower is developed by the gun? What average force is required to hold the gun in position?

37. A ballistic missile has a mass of 5,000 kg on the launch pad. It is fired off vertically, propelled by a rocket engine with a thrust of 54 kN, which ejects gases at the rate of 27 kg/s. Find (a) the speed of the exhaust gases relative to the rocket engine, (b) the initial acceleration of the missile, and (c) the acceleration 40 s after launch.

Ans. (a) 2,000 m/s; (b) 1.0 m/s²; (c) 3.98 m/s²

38. The Redstone ballistic missile was used by the United States for sending its first astronauts on ballistic trajectories. The Redstone had a rocket engine of 75,000 lb thrust. It weighed 40,000 lb at launch, and its jet exhaust had a speed of 6,000 ft/s. If the Redstone rose vertically, find its initial acceleration, the rate at which mass was ejected, the acceleration after 15 s of operation, and the time at which the acceleration became 2 g's.

39. A small vernier rocket ejects 1.2 kg of gases with a speed of 2,000 m/s in order to make a small corrective increase in the speed of a space capsule of 1,600 kg mass. Find (a) the momentum of the exhaust gases relative to the capsule and (b) the change in speed of the space capsule.

Ans. (a) 2,400 kg-m/s; (b) 1.5 m/s

40. A rocket has a takeoff mass of 6×10^4 kg, 80 percent of which is fuel. The exhaust speed is 2 km/s. (a) Find the minimum burning rate required for lift-off from the launching pad. (b) If

the burning rate is 400 kg/s, find the acceleration at the lift-off and at burnout, assuming that the rocket rises vertically and that g remains constant over the flight.

41. Find the net thrust of a jet engine traveling 300 m/s which takes in 16 kg of air each second and burns fuel at the rate of 1.5 kg/s, exhausting the 17.5 kg of combustion products at a speed of 800 m/s relative to the plane. (To get the net thrust one must take into account the fact that the air burned was not initially moving with the plane.) *Ans.* 9.2 kN

42. Two cars on a linear air track have an elastic collision. They approach each other from opposite directions, each with speed v_0. After the collision the car at the right, which has mass m, is at rest. Find the mass of the car on the left and its speed just after the collision in terms of m and v_0.

43. A manned space capsule and its propulsion system have a gross weight at launch of 980 kN and a total thrust from three engines of 1,078 kN; its engines exhaust 550 kg of gases each second. Assuming that g remains constant at 9.8 m/s², that aerodynamic forces are negligible, and that the capsule goes straight up, find (a) the velocity of the gases relative to the rocket, (b) the expression giving the acceleration of the system as a function of time, and (c) the acceleration 20 s after launch.

Ans. (a) 1,960 m/s; (b) $9.8(200 + 11t)/(2,000 - 11t)$; (c) 2.31 m/s²

44. A body of mass m and speed v_i has an elastic head-on collision with a mass M at rest. What must M be (in terms of m) if m and M have equal energies after the collision?

45. A spherical mass of 2 kg moving along the x axis at 5 m/s collides elastically with a spherical mass of 6 kg initially at rest. After the collision the 2-kg mass moves off at an angle of 36.9° with the x axis. Find the speed of both masses after the collision and the angle which the path of the 6-kg sphere makes with the x axis.

Ans. 4.67 m/s; 1.03 m/s; 65.8°

46. Show that when a moving particle of mass m collides head on and elastically with a particle of mass M initially at rest, it transfers to the stationary particle the fraction $4mM/(m + M)^2$ of its initial kinetic energy.

47. Sphere A of mass m and velocity v_i has a perfectly elastic collision with an identical sphere B initially at rest. (a) Show that for a head-on collision sphere A is left at rest while sphere B is given a velocity equal to v_i. (b) If the collision is not head on, show that the angle between the velocities of the spheres after collision is 90°.

48. A ball is dropped from a height h_0 onto a large flat surface (Fig. 7.6). If the coefficient of restitution is e, show that the time T from the first impact until the ball stops bouncing is given by

$$T = \frac{e}{1 - e} \sqrt{\frac{8h_0}{g}}$$

Hint: $1/(1 - e) = 1 + e + e^2 + \cdots$.

In Chap. 3 we studied trajectories of short-range projectiles, assuming that the acceleration g is constant both in magnitude and direction. Next we discuss another special case of curvilinear motion, that in which a body describes a circular path with uniform speed and an acceleration constant in magnitude but directed always toward the center of the circle. For a space capsule or a planet a typical path is an ellipse, with the acceleration changing in magnitude as well as in direction. From the accumulated knowledge of the motions of the planets Newton deduced the law of force which predicts the observed accelerations of the planets and in terms of which we can understand the acceleration g of freely falling bodies. We follow some of the reasoning which led to the formulation of Newton's law of universal gravitation and apply this law to such problems as those involving space satellites.

UNIFORM CIRCULAR MOTION AND GRAVITATION

8.1 UNIFORM CIRCULAR MOTION

When a body moves in a circular path with constant speed, it is said to describe *uniform circular motion*. While the magnitude of the velocity is constant, the direction is always changing. A body describing uniform circular motion has an acceleration $\mathbf{a} = \Delta\mathbf{v}/\Delta t$. In Fig. 8.1 the vector difference $\Delta\mathbf{v} = \mathbf{v} - \mathbf{v}_0 = \mathbf{a}\,\Delta t$ is shown for a sizable Δt. If we draw a vector diagram similar to that of Fig. 8.1 for Δt (and hence θ) approaching zero, the acceleration is directed toward the center of the circle. The acceleration is constant in magnitude throughout the circular path, but it varies in direction.

To calculate this acceleration, we proceed as follows. Suppose that the particle passes over the arc $AB = v\,\Delta t$ with constant speed v in time Δt. If θ is small, the chord† AB is essentially equal in length to the arc AB. In the velocity vector triangle of Fig. 8.1, the angle between $\mathbf{v}$ and $\mathbf{v}_0$ is also θ (since the velocity of uniform circular motion is always perpendicular to the radius). Therefore, the gray triangles are similar, from which we conclude that $v\,\Delta t/r = a\,\Delta t/v$ and

$$a = \frac{v^2}{r} \tag{8.1}$$

The acceleration of a body describing uniform circular motion is directed toward the center of the circle and has the magnitude v^2/r. This relationship was derived by Christiaan Huygens; it was one of many in his great book *Horologium oscillatorium* published in 1673.

9.2 CENTRIPETAL FORCE

If a body is accelerated toward the center of a circle, there is a net force in that direction, called the *centripetal force*. To find its magnitude, we recall that $\mathbf{F} = m\mathbf{a}$ and that $a_c = v^2/r$, whence

$$F_c = ma_c = m\frac{v^2}{r} \tag{8.2}$$

Consider a model airplane constrained to move in a horizontal circle by a control wire. The wire pulls on the airplane toward the center of the

---

†This chord AB is the straight line connecting points A and B, while the arc AB is the shortest path along the circumference of the circle from A to B.

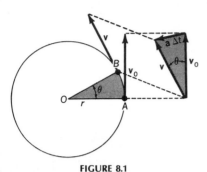

FIGURE 8.1
In uniform circular motion the magnitude of the velocity remains constant while the direction changes continuously. The acceleration is always directed toward the center of the circle.

circle. By Newton's third law, the airplane must exert an equal and opposite force on the wire. This force, *away* from the center of the circle and *acting on the wire*, is known as the *centrifugal force*.

Note carefully that the force *on the airplane is toward the center* of the circle. If this force were suddenly removed and no unbalanced force acted on the airplane, it would move along a tangent to the circular path. No force acts upon the airplane outward along the radius. The airplane does exert a force outward on the restraint, which in this case is the control wire. This outward force on the restraint is the centrifugal force; it is the reaction to the centripetal force which acts inward on the body describing circular motion.

When an automobile goes around a curve (Fig. 8.2) on a perfectly level road, friction between the tires and roadway is required to provide the necessary centripetal force mv^2/r. If the road is coated with ice, this frictional force may be too small and the car may slide off the road. *There is no force pulling the car outward* along the radius. By Newton's first law, the car would continue to move indefinitely with constant velocity in the absence of any force. If there is a force toward the center of the circle which is inadequate to provide the full centripetal force, the car is accelerated toward the center but not enough to keep it on the road.

If the road is suitably banked, the car may go around the curve without requiring any frictional force. When friction between tires and roadway is absent, the road pushes on the car only in the direction perpendicular to the surface. If the automobile goes around the curve without skidding, the vertical component of this normal force (Fig. 8.3) is exactly the weight of the car and the horizontal component is just mv^2/r, the necessary centripetal force. To satisfy these conditions the angle θ must be such that

$$\tan \theta = \frac{mv^2/r}{mg} = \frac{v^2}{gr} \tag{8.3}$$

Thus the road can be ideally banked for only a single speed. The angle at which the road should be banked depends on the speed of the car but not on its mass.

8.3 OTHER APPLICATIONS OF CENTRIPETAL FORCE

When an aircraft makes a turn in a horizontal plane, it banks in such a way that the vertical component of the forces on the wings is equal to the weight of the aircraft, while the horizontal component provides the necessary centripetal force. Very-high-speed aircraft cannot make exceedingly sharp turns because it is not feasible to provide the necessary centripetal force.

When an aircraft pulls out of a steep dive, large forces must be exerted by the wings. More than one pilot has stripped the wings off his aircraft when he has tried to pull out of a dive too quickly. Consider an aircraft which is descending at the rate of 280 m/s and tries to pull out of the dive in a circle of radius 800 m (Fig. 8.4). The centripetal acceleration becomes

$$a_c = \frac{v^2}{r} = \frac{(280 \text{ m/s})(280 \text{ m/s})}{800 \text{ m}} = 98 \text{ m/s}^2$$

or 10 times the acceleration due to gravity. Thus, the *unbalanced upward*

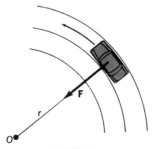

FIGURE 8.2
When an automobile goes around a curve, the centripetal force must be provided either by friction of the tires, or by banking the road, or by some combination of the two.

force required on the aircraft is 10 times the weight. At the bottom of the circle the wings must not only provide this centripetal acceleration but overcome the pull of the earth as well. They must provide $10 \times 9.8 \text{ m/s}^2 = 10 \, g$'s centripetal acceleration plus 1 g to take care of the pull of the earth on the aircraft, a total of 11 g's. The wings of most aircraft are not designed for such heavy loading. A typical fighter aircraft is designed to take accelerations of the order of 8 or 9 g's, while passenger aircraft, bombers, and transports are designed for lower g loadings. At the top of the circular path (Fig. 8.4) the pull of gravity helps to provide the centripetal acceleration, and the net force exerted by the wings is $mv^2/r - mg$ downward.

Not only does the aircraft itself have to stand these accelerations, but so do all the objects in the aircraft, including the organs of the pilot's body. Consider a gram of blood in a pilot's brain when the aircraft is performing a 6-g turn. The total force on this gram of blood must be 6 times the ordinary force exerted if the blood is to remain in place. In the absence of this force the blood may rush out of the pilot's brain, causing him to "black out."

The centrifuge, which is widely used in the separation of liquids of unequal densities, depends for its operation on centripetal force. One type of centrifuge (Fig. 8.5) consists of a wheel that rotates in a horizontal plane. To this wheel are attached buckets that are vertical when the wheel is at rest. When the wheel is revolving rapidly, the buckets assume a position such that their axes are almost horizontal. If a mixture of liquids of unequal densities is introduced into the buckets and the wheel is rotated rapidly, the liquids separate, the heavy liquids farther from the axis of rotation and the lighter liquids nearer to it. This means that the heavier liquids are at the bottom of the buckets when the centrifuge is stopped. The cream separator works on the same principle.

8.4 THE PTOLEMAIC AND COPERNICAN SYSTEMS

When Newton formulated his laws of motion, he was particularly interested in explaining the movements of the moon, the planets, and other heavenly bodies, which had fascinated people since the beginning of history. In Greek mythology the daily journey of the sun was attributed to the god Apollo driving a flaming chariot across the sky. Most of the early Greek philosophers, including Pythagoras, Aristotle, and Hipparchus, assumed that the earth was at the center of the universe, but about 260 B.C. Aristarchus of Samos proposed that the earth and the other planets revolved about the sun. The idea of a heliocentric (*helios* means "sun") system as opposed to the geocentric (*geos* means "earth") system received relatively little support. In the second century Ptolemy produced a great synthesis of geocentric astronomical ideas, most of which had originated with his predecessors. He accepted the postulate that the stars were mounted (Fig. 8.6) on a great transparent sphere which rotated at a constant rate around the earth at its center. The sun and the moon were mounted on similar but smaller spheres which rotated about the earth at rates different from that of the stars. To explain the motions of the planets (*planet* means "wanderer"), which move most of the time in the same direction as the stars but occasionally retrogress, the Ptolemaic theory assumed that each planet moved in a circular path (*epicycle*) about

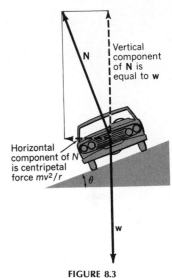

FIGURE 8.3
A curve is perfectly banked for a given speed v when the force exerted by the road perpendicular to its surface has a horizontal component equal to the required centripetal force and a vertical component equal and opposite to the weight mg of the vehicle.

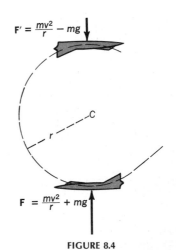

FIGURE 8.4
Force exerted by the wings and other surfaces of an airplane at the upper and lower extremities of a circular loop.

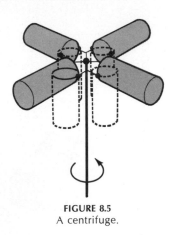

FIGURE 8.5
A centrifuge.

a center, which in turn followed a much larger circular path (*deferent*) about the earth.

In 1543, Nicholas Copernicus, a Polish monk, published the results of some 35 years of patient study in a celebrated book which showed that the need for epicyclic paths of planets and many of the other complications of the Ptolemaic system disappeared if one took the sun as the center of the universe (Fig. 8.7). His work was a great stimulus to the field of astrophysics. Supporting evidence for the Copernican theory grew over the next century.

Galileo constructed a telescope shortly after a Dutch spectacle maker Lippershey discovered in 1608 that he could use two spectacle lenses to form an enlarged image of a distant object. With his telescope Galileo discovered four of Jupiter's moons, mountains on the surface of the moon, Saturn's rings, and dark spots moving across the disk of the sun, which gave evidence that the sun was in rotation. Some of these discoveries were readily explained on the basis of the Copernican theory but not by the Ptolemaic theory. Galileo became one of the major supporters of Copernicus, but he soon suffered persecution and imprisonment because his views differed from those held by certain high officials of the church.

The Danish astronomer Tycho Brahe (1546–1609) devoted most of his professional life to making accurate measurements of the positions of stars and planets. His data were analyzed by the mathematical physicist Johan Kepler (1571–1630). Kepler found that the motions of the planets did not always agree exactly with those predicted by the Copernican theory of circular paths. After many years of work he was able to show that the motions of the planets could be predicted by use of three generalizations, which we know as *Kepler's laws*.

1. *The orbit of each planet is an ellipse with the sun at one focus.*

2. *The speed of the planet varies in such a way that the line joining the planet and the sun sweeps out equal areas in equal times* (Fig. 8.8).

3. *The cubes of the semimajor axes of the elliptical orbits are proportional to the squares of the times for the planets to make a complete revolution about the sun.*

Although the general motions of the planets about the sun occur in elliptic orbits, the motions of the earth and of several of the other planets around the sun can be reasonably well described by circular orbits, as can the motion of the moon around the earth. Since the geometrical com-

FIGURE 8.6
In the Ptolemaic system the earth is at
the center of the universe.

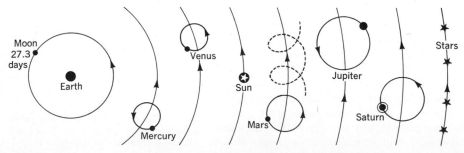

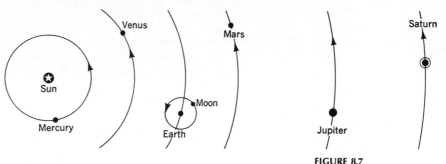

FIGURE 8.7
In the Copernican scheme the sun is at the center of the solar system, and the earth is one of the planets.

plexity of circular motion is substantially less than that of general elliptic motion, and since the physical ideas involved in both types are equivalent, we shall focus our attention on circular motion and leave elliptic motion to more advanced treatments. We observe that the circle may be regarded as a limiting case of the ellipse, namely, that in which the two foci merge into one.

8.5   NEWTON'S LAW OF UNIVERSAL GRAVITATION

Let us now apply these ideas and equations to the motion of the earth about the sun. Imagine for the moment that the sun itself is fixed in space. If we assume that the path of the earth is circular, the earth has an acceleration toward the sun (the center of the circle) given by $a_c = v^2/r$. To produce this acceleration there must be an unbalanced centripetal force $F_c = mv^2/r$.

The magnitude of the earth's velocity is $2\pi r/T$, where T is the period (1 year = time for one complete revolution). Therefore, the centripetal force is

$$F_c = m\frac{4\pi^2 r^2}{T^2}\frac{1}{r} = \frac{4\pi^2 mr}{T^2} \tag{8.4}$$

For a circular orbit the semimajor axis is equal to the radius r. By Kepler's third law, T^2 is proportional to r^3, or $T^2 = cr^3$, where c is a constant. Substituting this in Eq. (8.4) yields

$$F_c = \frac{4\pi^2 m}{c}\frac{1}{r^2} \tag{8.4a}$$

so the centripetal force F_c is inversely proportional to r^2.

Newton properly inferred that this centripetal force is provided by the gravitational attraction between the sun and the earth (actually Newton worked first on the earth-moon system rather than the sun-earth system, but the transition from Kepler's laws is smoother for the latter). He reasoned that the inverse square relationship expressed by Eq. (8.4a) probably applied not only to the planets and the sun in our solar system but also to any two particles anywhere in the universe. In other words, he advanced the hypothesis that between these two particles there is a gravitational attractive force along the line connecting the particles given by

$$F_G = \frac{4\pi}{c}\frac{m_1}{r^2} \tag{8.4b}$$

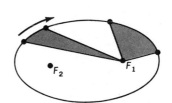

FIGURE 8.8
In the Keplerian scheme each planet describes an elliptic path with the sun at one of the foci. The line joining the planet and the sun sweeps out equal areas in equal time intervals.

where m_1 refers to one of two masses and r is the distance between their centers. The contribution from the second particle of mass m_2 to the gravitational attraction must be expressed somehow in the $4\pi^2/c$ factor in the equation. If the force exerted on m_1 by m_2 is proportional to the mass m_1, the force exerted on m_2 by m_1 should be proportional to m_2, so we assume that $4\pi^2/c = Gm_2$, where G is a constant. Substituting this in Eq. (8.4b) leads to

$$F_G = G\frac{m_1 m_2}{r^2} \tag{8.5}$$

This is *Newton's law of universal gravitation* in its mathematical formulation; in words this law states:

Between every two particles in the universe there is a force of gravitational attraction which is proportional to the product of the masses of the two particles and inversely proportional to the square of the distance between them.

The constant G, called the *universal gravitational constant*, has a value of 6.67×10^{-11} N-m²/kg². In the next section we describe two methods by which G has been measured.

When we deal with the gravitational attraction between large objects whose separation is very great compared with the dimensions of either body, we need not calculate the forces between every particle of one body and every particle of the other. If m_1 and m_2 are the masses of the bodies and r the distance between their centers of mass, Eq. (8.5) is directly applicable. Further, Newton was able to show that for any uniform distribution of mass over a spherical shell we may take the entire mass as concentrated at the center and apply Eq. (8.5) directly *for any point outside the sphere.* Thus, even if two uniform spheres touch each other, Eq. (8.5) is valid if r is taken to be the distance between their centers.

The force of attraction between two identical lead spheres of mass 1 kg with centers 1 m apart is only 6.67×10^{-11} N. The gravitational forces between two ordinary objects the size of people are detectable only under the most favorable conditions. If the interacting bodies are stars or planets, the masses involved are so great that the forces become exceedingly large. Evidence for the validity of Newton's law of gravitation is obtained from astronomical observations. By application of the law, it is possible to predict with great accuracy the motions of the planets and their satellites many years in advance. Such long-term predictions are a severe test of the law.

8.6 DETERMINATION OF THE GRAVITATIONAL CONSTANT AND THE MASS OF THE EARTH

The gravitational constant G was first measured by Cavendish in 1797. He used a torsion balance (Fig. 8.9) in which two spheres of mass m are fixed to the ends of a light rod suspended by a long, fine wire to form a torsion pendulum. When two heavy lead balls are placed at the positions marked M, the rod carrying the two small masses turns to a new position. If the lead balls are placed at M_1, the light rod with the small masses turns in the opposite direction. By measuring the elastic constant of the suspending wire it is possible to determine the attractive force between the movable masses and the heavy balls.

A method of measuring G which is simple to understand was devel-

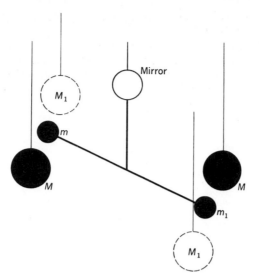

FIGURE 8.9
Schematic diagram of the apparatus with which Cavendish measured the universal gravitational constant. A light beam reflected from the mirror to a scale measured the amount of twisting when the large masses are moved from the positions marked M to the positions marked M_1.

oped by Jolly. A spherical vessel containing 5 kg of mercury was attached to one arm of a sensitive balance, and the balance put in equilibrium (Fig. 8.10). Then a lead sphere of mass 5,775 kg was rolled beneath the mercury flask with its center about 0.57 m below the center of mass of the mercury. The gravitational attraction between the lead and the mercury was found to be such that a mass of 0.589 mg on the right pan was able to restore equilibrium. All the quantities in Eq. (8.5) were thus known except for G, the value of which was calculated and reported in 1881.

Once the value of the gravitational constant is known, it is possible to calculate the mass of the earth m_E. For the moment, consider the earth to be a perfect sphere of radius $r = 6.4 \times 10^6$ m. The force which the earth exerts on a mass of 1 kg at its surface is 1 kgf $= 9.8$ N. Therefore, from $F_G = Gmm_E/r^2$,

$$9.8\ \mathrm{N} = 6.67 \times 10^{-11} \frac{\mathrm{N\text{-}m^2}}{\mathrm{kg^2}} \frac{(1\ \mathrm{kg})(m_E)}{(6.4 \times 10^6\ \mathrm{m})^2}$$

from which $m_E = 6 \times 10^{24}$ kg.

We can also find the mass of the sun M_s. The earth revolves around the sun in an elliptical path which is fairly close to a circle of radius 1.5×10^{11} m (93 million miles). The centripetal force for the earth is provided by the gravitational attraction between the earth and the sun. Thus,

$$F_G = G \frac{M_s m_E}{r^2} = m_E \frac{v_E^2}{r} \qquad (8.5a)$$

and

$$M_s = \frac{v_E^2 r}{G} = \frac{(2\pi r)^2 r}{T^2 G}$$

$$= \frac{(2\pi)^2 (1.5 \times 10^{11})^3}{(365.26 \times 86{,}400)^2 (6.67 \times 10^{-11})} = 2 \times 10^{30}\ \mathrm{kg}$$

Data on the solar system are listed in Table 8.1.

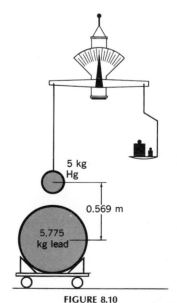

FIGURE 8.10
Jolly's method of measuring the universal gravitational constant.

TABLE 8.1
Data on the Solar System

Body	Mass, kg	Average radius,† km	g at surface, m/s²	Sidereal period, days	Radius of orbit,‡ km
Moon	7.35×10^{22}	1,738	1.62	27.3	3.8×10^5
Sun	1.97×10^{30}	695,000	274		
Mercury	3.28×10^{23}	2,570	3.9	88	5.8×10^7
Venus	4.82×10^{24}	6,310	8.9	245	1.08×10^8
Earth	5.98×10^{24}	6,370	9.80	365.26	1.50×10^8
Mars	6.37×10^{23}	3,430	3.8	687	2.28×10^8
Jupiter	1.88×10^{27}	71,800	26	4,333	7.78×10^8
Saturn	5.62×10^{26}	60,300	11	1.08×10^4	1.43×10^9
Uranus	8.62×10^{25}	26,700	10	3.07×10^4	2.87×10^9
Neptune	1.0×10^{26}	24,900	14	6.02×10^4	4.5×10^9
Pluto				9.09×10^4	5.9×10^9

†Bodies are not exactly spheres.
‡Orbits are actually elliptical; the values quoted are mean distances.

8.7 ARTIFICIAL SATELLITES

During the 1970s the dream of space stations circling the earth at a height of a few hundred kilometers became a reality. Now weather satellites in appropriate orbits transmit to earth periodic "pictures" of the cloud cover over all parts of the globe, while communications satellites relay messages and television programs from one continent to another. Let us find the speed required for a satellite to remain in circular orbit 630 km above the earth's surface (7,000 km from the center of the earth). If m_s is the mass of the satellite,

$$G\frac{m_s m_E}{r^2} = m_s \frac{v^2}{r}$$

or $\quad 6.67 \times 10^{-11} \dfrac{\text{N-m}^2}{\text{kg}^2} \dfrac{6 \times 10^{24} \text{ kg}}{(7 \times 10^6 \text{ m})^2} = \dfrac{v^2}{7 \times 10^6 \text{ m}}$

$$v = 7{,}600 \text{ m/s} \approx 17{,}000 \text{ mi/h}$$

At this speed the satellite would require a time to encircle the earth of $(2\pi \times 7 \times 10^6 \text{ m})/(7{,}600 \text{ m/s}) = 5{,}800 \text{ s} = 1.6 \text{ h}$.

If the only force on the satellite were gravitational, the satellite would encircle the earth indefinitely. Actually air friction and other small forces gradually reduce the speed, so that eventually the satellite approaches the earth and dissipates its energy in the atmosphere.

8.8 EXTENSIONS OF NEWTON'S THEORY

Newton not only proposed his law of universal gravitation but also applied it to explain previously unexplained observations. By combining it with Huygens' concept of centripetal force he was able to account for the equatorial bulge of the earth in terms of the earth's rotation and predict how the acceleration due to gravity should vary over the earth's surface (Sec. 8.9).

Proceeding further, he showed that gravitational action of the sun and moon on this equatorial bulge of the earth should give rise to the precession of the equinoxes, a phenomenon which had been observed but never explained. It had also been observed that several of the planets are not exactly spherical; Newton explained this by assuming that they too rotate about their axes, and he calculated their rotational periods from their measured deviations from sphericity.

Still another successful use of Newton's theory came when he recognized that comets are acted upon by gravitational forces and that they too are part of the solar system. In 1682 the astronomer Halley made observations on a brilliant comet (now called *Halley's comet*). Two years later Halley asked Newton what sort of orbit would be described by a body subject to inverse square gravitational attraction. Newton replied that the orbit would be an ellipse. Halley used his observations and Newton's theory to predict that the comet would return in about 75 years. The orbit of Halley's comet is a very eccentric ellipse about the sun, taking the comet from inside the earth's orbit to well outside Neptune's orbit and back in that period. The comet is visible to the eye only when it is relatively near the earth. Historical records contain numerous accounts of comets; we are confident that the one reported in 20 B.C., the one that appeared in 1066 when William of Normandy invaded England, and many others were earlier appearances of Halley's comet. Its most recent visit was in 1910, and its return is expected about 1986.

Another familiar and important phenomenon which Newton was able to explain in terms of the gravitational interactions is that of the tides. The attractive force exerted by the moon raises a tide (Fig. 8.11) on side A nearest to it and simultaneously on side B away from it. Consider the moon's gravitational pull acting on three identical mass elements, one at A, one at the center of the earth, and one at B. Clearly, A is nearest to the moon, and the force is greatest there; B is farthest from the moon, and the force is less than that either at A or at the center of the earth. A second factor making the net pull per unit mass less at B is associated with the fact that the moon does not revolve about a fixed earth; instead both bodies perform essentially uniform circular motion about the center of mass O of the earth-moon system, which lies almost three-fourths the earth's radius from the center of the earth. Since point B is much farther from O than point A, the centripetal force required for a kilogram of water at B is greater than that at A and in the opposite direction. The combination of a slightly smaller gravitational pull and the need for a slightly bigger centripetal force at B leads to a tide at B.

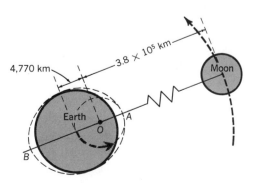

FIGURE 8.11
The center of mass of the earth-moon system is at O; both the center of the earth and the center of the moon revolve around O. Identical mass elements of the earth are at different distances from the moon and therefore experience different forces; those at A are most strongly attracted and those at B least strongly attracted. This fact, plus the effects due to the revolving of the earth about O, leads to a distortion of the earth's shape with tides raised at A and at B.

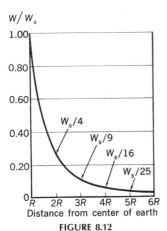

FIGURE 8.12
The weight of an object decreases as it moves away from the earth. If W_s is the weight at the earth's surface, the weight falls to $W_s/4$ at a distance of two earth radii, and so forth.

TABLE 8.2
Acceleration Due to Gravity at Sea Level†

Latitude	g, m/s^2
0° (equator)	9.78039
10°	9.78195
20°	9.78641
30°	9.79329
35°	9.79737
40°	9.80171
45°	9.80621
50°	9.81071
55°	9.81507
60°	9.81918
70°	9.82608
90° (pole)	9.83217

†For each meter above sea level g decreases by 3.086 μm/s^2.

If lunar attraction were the only factor, tides would have their greatest heights twice in a period of 24.85 h, as commonly observed in the Atlantic basin. Actually the inertia of the water, the earth's rotation, frictional forces, the interactions with land masses, and other causes complicate the picture. As a consequence, some places have only one high tide a day. Although the sun is much farther from the earth than the moon, the solar mass is so great that there is a solar contribution to the tides which is about four-tenths the lunar contribution. The resulting tide is a combination of the lunar and solar contributions. When the earth, moon, and sun are in line, unusually high tides result.

8.9 VARIATION OF WEIGHT WITH POSITION

The weight of an object of mass m is mg, where g is the observed acceleration due to gravity. If the earth were isolated, at rest, and a perfect sphere of radius r, the weight of an object would be the same at all points on its surface. Actually, the earth is somewhat flattened at the poles. Consequently, a body at one of the poles is closer to the center of the earth than an identical body at the equator. This leads to a somewhat greater weight at the poles than at the equator. Another factor which makes the weight of an object less at the equator is the rotation of the earth. At the equator a portion of the gravitational pull is used to supply the centripetal force to keep the body moving with uniform circular motion. Therefore, the force which must be applied to the body to prevent it from falling to the earth is smaller at the equator. The total variation from equator to pole is approximately 0.5 percent.

The gravitational pull is just adequate to produce the required centripetal acceleration for a body in a space capsule in orbit around the earth. The body is "weightless" in the sense that no force is required to hold it in its position in the capsule. An astronaut can place his camera in front of him, and it will "float" in that position. Here we have a special case in which the entire gravitational attraction of the earth is required to produce centripetal acceleration, leaving no *net* force on the camera as viewed from the capsule.

The weight of a body varies not only with latitude but also with altitude, being somewhat less atop Pikes Peak than in Death Valley. As we go above the earth, the weight of a body decreases gradually as r increases. Figure 8.12 shows how the weight varies with distance from the center of the earth (neglecting effects arising from rotation or revolution). Table 8.2 shows how g differs from place to place at the earth's surface.

In prospecting for oil and for heavy masses of ore below the surface of the earth, careful measurements of g are commonly made at selected points over a fairly large area. Modern gravimeters are sufficiently sensitive for a change in g of a few parts in 100 million to be detectable. The variation of g over a region is plotted on a map, and experts can frequently locate places where drilling for oil or ore is most likely to be successful.

The variations in g at different places on the earth's surface are small enough to be neglected in ordinary activities. However, at the surface of the moon the weight of an astronaut (force with which the moon holds him to its surface) is only one-sixth of that on the earth. Indeed, the gravitational pull on the moon is so small that gas molecules escape readily, and therefore the moon has essentially no atmosphere. On the planet Jupiter a man would be pulled down by a force 2.6 times his weight on earth. Under these circumstances it would be difficult for him to walk about.

8.10 GRAVITATIONAL POTENTIAL ENERGY

The potential energy of a body of mass m raised a distance h above the surface of the earth is, by Eq. (6.6), mgh. This relation is valid, however, only if the weight mg does not vary significantly over the range of heights involved. Let us now consider the potential energy when the mass is lifted or projected to a height h of hundreds of kilometers above the surface of the earth. Let R be the radius of the earth, M its mass, and r the distance from the center of the earth to a point a distance h above the earth's surface (Fig. 8.13). The potential energy of the mass relative to the earth's surface is the work required to move it from the surface of the earth to $h = r - R$. To calculate this work, we divide the displacement from R to r into many small intervals and calculate the work performed in each of these intervals. From R to r_1 the force varies from GMm/R^2 to GMm/r_1^2. If R and r_1 are close to each other, the average force in the interval from R to r_1 is GMm/Rr_1 and the work done in this interval is, by Eq. (6.1),

$$W_1 = \frac{GMm(r_1 - R)}{Rr_1} = GMm\left(\frac{1}{R} - \frac{1}{r_1}\right)$$

Similarly the work W_2 required to move the mass from r_1 to r_2 is

$$W_2 = GMm\left(\frac{1}{r_1} - \frac{1}{r_2}\right)$$

and so forth. The total work done in carrying m from R to r is

$$W = W_1 + W_2 + \cdots = GMm\left(\frac{1}{R} - \frac{1}{r_1} + \frac{1}{r_1} - \frac{1}{r_2} \cdots + \frac{1}{r_n} - \frac{1}{r}\right)$$

$$= GMm\left(\frac{1}{R} - \frac{1}{r}\right) = \text{PE at } r \text{ relative to 0 at } R \qquad (8.6)$$

In a frictionless atmosphere a projectile fired straight up with speed v rises until its kinetic energy is transformed into potential energy, or

$$\tfrac{1}{2}mv^2 = GMm\left(\frac{1}{R} - \frac{1}{r_{max}}\right)$$

where r_{max} is the maximum radius attained by the projectile.

□ **Example** A projectile is fired upward with a speed of 5,000 m/s. How high will it rise above the earth's surface if air friction is ignored?

KE at launch = PE at maximum height

$$\tfrac{1}{2}m(2{,}500 \times 10^4 \text{ m}^2/\text{s}^2) = 6.67 \times 10^{-11} \text{ N-m}^2/\text{kg}^2$$
$$\times m(5.98 \times 10^{24} \text{ kg})\left(\frac{1}{6.37 \times 10^6 \text{ m}} - \frac{1}{r_{max}}\right)$$

from which $r_{max} = 8.0 \times 10^6$ m, corresponding to a height above the earth's surface of about 1.6×10^6 m. □

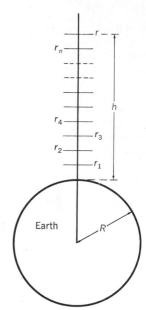

FIGURE 8.13
The potential energy of a mass at a great height above the earth's surface can be determined by dividing the displacement from R to r into many small intervals, calculating the change in potential energy for each interval, and adding the results. (Students familiar with integral calculus can readily find the potential energy by integration.)

QUESTIONS

1. Does the centripetal force do any work on a body in uniform circular motion? Why?
2. A coin is placed on the turntable of a phonograph. As the turntable gains angular speed, the coin moves outward and falls off. Explain.
3. An airplane makes a complete loop in a vertical plane. Under what condition does the pilot press upward on the seat at the top of the loop?
4. Can you imagine any situation in which a body can follow a curved path without being accelerated?
5. Explain carefully why an automobile may overturn if it goes around a curve too fast. Which wheels leave the ground first?
6. The moon falls about 0.28 ft toward the earth each minute, yet it always stays about the same distance away. Explain how this can be.
7. How would the weight of a radio transmitter vary on a rocket trip to the moon? Would its mass also vary?
8. If a hole were drilled to the center of the earth and a 1-kg mass were placed there, what would its weight be? Why?
9. A ball has a mass of 0.1 kg. If it were taken to the moon, what would its mass be? Would Newton's second law be valid? Would $m = w/g$ be valid? Would the weight of the ball be 0.98 N?
10. An astronaut in a stable orbit two earth radii from the center of the earth is weightless in his space capsule. How can this statement be reconciled with Fig. 8.12, which shows a weight one-fourth that at the earth's surface?
11. When a small continuous chain is set into rapid rotation around the axis of an electric motor and then released, it will roll along a lecture table and bounce over obstacles as though it were a solid hoop. What is the origin of this rigidity? Discuss the behavior of the cowboy's lariat in the same context.
12. Does a plumb bob at rest necessarily point to the center of the earth? Why not? Why is the vector acceleration due to gravity **g** not ordinarily directed toward the center of the earth? Where on the earth's surface might **g** be expected to point to the center of the earth?
13. If a spacecraft in a circular orbit turns on its rocket engines to produce an acceleration in the same direction as its velocity, its average speed in its new orbit is reduced. How could this come about?
14. Earth satellites which revolve with practically the same angular velocity as the earth have been launched as aids to navigation and as relay stations for television and other communication systems. Such a satellite appears stationary to observers on the earth. Could such a satellite remain on the zenith above Chicago for an extended period? Why not? Above what points can such a satellite remain virtually stationary?
15. The earth and the other planets travel about the sun in the same direction and almost in the same plane. Most of the time a more distant planet observed from the earth moves toward the east through the sky, but part of the time it appears to drift westward. Draw a diagram and explain this *retrograde* motion.
16. Recall that Galileo discovered four moons of Jupiter. How could you determine the mass of Jupiter from observations of these moons and their orbits?
17. The gravitational attraction between the earth and the sun is about 150 times that between earth and moon. Why then does the moon raise larger tides?

PROBLEMS

1. A 1,000-kg car begins to skid when traveling 108 km/h around a level curve of 150 m radius. Find the centripetal acceleration, the centripetal force, and the coefficient of friction between the tires and the road.   *Ans.* 6 m/s²; 6,000 N; 0.61
2. A 1,200-kg car is going around a curve of 100 m radius at a speed of 25 m/s. Find the centripetal acceleration and the centripetal force.
3. A 160-lb pilot in an aircraft moving at a constant speed of 500 ft/s pulls out of a vertical dive along an arc of a circle of 4,000 ft radius. For the pilot find the centripetal acceleration, the centripetal force, and the total force which the aircraft exerts on him at the bottom of the dive.
   Ans. 62.5 ft/s²; 313 lb; 473 lb
4. People's responses to high accelerations are checked in a test chamber at one end of a 7-m horizontal beam rotated about a vertical axis at its other end. How many revolutions per minute must the beam make to provide an acceleration of 9 g's (88.2 m/s²)?
5. What is the highest speed an automobile can travel around a 80-m-radius curve on a level road if the available coefficient of friction between the road and the tires is 0.49?   *Ans.* 19.6 m/s
6. What is the smallest speed an airplane can have making a vertical loop with a radius of 720 m if objects in the plane are not to start dropping at the top of the loop?

7. A curve of 220 m radius is banked at 18°. At what speed must an automobile go around this curve if *no* frictional forces are to be used to keep the automobile on its circular path?   *Ans.* 26.5 m/s

8. What is the angle at which a speedway must be banked for cars running at 50 m/s if the radius of curvature is 350 m and no frictional force is to be involved?

9. A 0.45-kg mass is suspended from a cord 1.5 m long to form a simple pendulum. If the mass is pulled to one side until it is raised 100 mm and then released, find the velocity of the mass and the tension in the cord at the bottom of the swing.   *Ans.* 1.4 m/s; 5.0 N

10. A 8,000-kg aircraft with a constant speed of 540 km/h pulls out of a vertical dive along an arc of a circle of radius 500 m. Find the centripetal acceleration and the total lift required at the bottom of the dive.

11. A 0.2-kg ball is whirled in a vertical circle on the end of a string 0.6 m long. At what speed will the tension in the string be zero at the top of the circle? What will the tension at the bottom be if the tension drops to zero at the very top? *Hint:* The ball gains speed as it descends.
    Ans. 2.42 m/s; 11.8 N

12. At what angle should a curve of 160 ft radius be banked if no frictional forces are to be required for a speed of 80 ft/s? What total force would the road exert on a 3,200-lb car?

13. A small cart on the circular loop-the-loop track of the accompanying figure can be approximated as a mass *m* sliding on a frictionless track. (*a*) Find the smallest height *h* in terms of the radius *r* from which the mass can be released and still remain in contact with the track throughout the path. (*b*) If the cart starts from a height of 3*r*, what is its acceleration at the top of the loop? With what force does the track press down on the cart at the top? When the cart starts at *h* = 3*r*, find the horizontal and vertical components of its acceleration at point *A* on the end of a horizontal diameter.   *Ans.* (*a*) 2.5*r*; (*b*) 2*g*; *mg*; 4*g*, *g*

14. The track in the accompanying figure is frictionless and has a radius of 1.5 ft. If a 2-lb block is released at *A*, find its speed at the bottom and the total force exerted on the block by the track at the bottom.

15. If the cart of Prob. 13 starts from a height of 5*r*, find its kinetic energy and the force it exerts on the track (*a*) at the bottom, (*b*) at the top of the circle, and (*c*) at point *A*.
    Ans. (*a*) 5*mgr*; 11*mg*; (*b*) 3*mgr*; 5*mg*; (*c*) 4*mgr*; 8*mg*

16. If the track in the accompanying figure has a radius of 0.3 m and is frictionless, find the speed and the force exerted on the track for a 40-g cube released from a height of 0.9 m (*a*) at the bottom, (*b*) at the top, and (*c*) at point *A*.

17. A 0.6-kg pendulum bob hangs from the roof of a moving van. If the van is traveling 35 m/s around a curve of 245 m radius, find the angle which the cord makes with the vertical when the bob is at rest relative to the van. What is the tension in the cord?   *Ans.* 27°; 6.6 N

18. A boy weighing 64 lb is swinging in such a way that he describes an arc of 12 ft radius. If his horizontal speed at the lowest point of the swing is 6 ft/s, what is the total force which the ropes of the swing must sustain at that instant?

19. Find the centripetal acceleration of an object at the equator due to the rotation of the earth. By how much does the rotation of the earth reduce the weight of a 70-kg man?
    Ans. 0.0339 m/s²; 2.37 N

20. How many revolutions per day would the earth have to make for the weight of a body at the equator to become zero?

21. With what speed must a 2-kg pendulum bob swing in the circular path of the accompanying

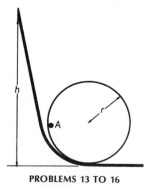

PROBLEMS 13 TO 16

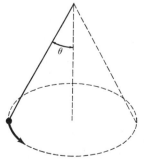

PROBLEMS 21 AND 22

figure if the supporting cord is 1.5 m long and θ is 30°? Find the tension in the cord.

Ans. 2.06 m/s; 22.6 N

22. Show that the period T (time for one complete revolution) of a conical pendulum of length l (see accompanying figure) is given by $T = 2\pi\sqrt{(l\cos\theta)/g}$, where θ is the angle between the cord and the vertical.

23. Find the gravitational attraction between two identical 40-kg lead spheres with centers 0.6 m apart. Compare this force with the earth's gravitational force on one of the spheres.

Ans. 2.96×10^{-7} N; 7.55×10^{-10}

24. Calculate the attractive force of the earth for the moon. How many times greater is the attraction of the earth for the sun?

25. When a lead sphere is placed 0.2 m from another lead sphere of mass 5 kg, the attraction of one for the other is found to be 7×10^{-8} N. What is the mass of the first sphere? *Ans.* 8.4 kg

26. From the masses and radii given in Table 8.1 estimate the acceleration due to gravity on the surface of (*a*) the moon, (*b*) Venus, (*c*) Jupiter, and (*d*) Mars. What would be the weight of a 2-kg mass at the surface of each?

27. It is proposed to put a space station in a circular orbit at a distance of one-third the earth's radius above the earth's surface. What will be the acceleration due to gravity at this elevation? Find the speed of the station if it is to go around the center of the earth in a circular orbit. How long will it take to make one complete revolution?

Ans. 5.51 m/s²; 6,840 m/s; 7,800 s

28. If a planet had a circular orbit about our sun with a radius 5 times that of the earth's orbit, what would its period of revolution and orbital speed be?

29. Find the radius of the orbit of a satellite which revolves about the earth in one sidereal day (86,164 s). What is its orbital velocity?

Ans. 4.22×10^{7} m; 3.08 km/s

30. Find the speed of a spacecraft circling the moon in a parking orbit at a radius of 1,800 km (60 km above the lunar surface). What is the period (time for one complete orbit) for this spacecraft?

31. Find the minimum speed with which a bullet must leave the gun of an astronaut on the moon if the bullet is to escape from the moon. If the moon had an atmosphere, what effect would that have on the required escape velocity? How high

would the bullet go if it had three-fourths the escape velocity? *Ans.* 2,380 m/s; 2,230 km

32. A satellite with a mass of 1,500 kg is to be placed in a circular orbit of radius 8,000 km about the earth's center. Find (*a*) the speed of the satellite in orbit and (*b*) the energy which must be provided to the satellite to place it in this orbit.

33. A rocket is 1,911 km (0.3 earth radius) above the surface of the earth. (*a*) What is g at this height? (*b*) Find the weight of a 5-kg mass at this point. (*c*) What is the potential energy of 5 kg at this position relative to the earth's surface as the zero? *Ans.* (*a*) 5.80 m/s²; (*b*) 29 N; (*c*) 7.2×10^{7} J

34. Sputnik, the first artificial earth satellite launched by the USSR on Oct. 4, 1957, took 96 min to traverse its almost circular orbit. Calculate its approximate height above the earth's surface and the acceleration due to gravity at that height.

35. Halley's comet has been observed on every passage near the sun at intervals of from 74 to 79 years since 240 B.C. (The variation in the period is due to perturbations produced by the major planets.) It is due about 1986, having appeared in 1910. Assuming that it describes an elliptical path about the sun with a 76-year period, calculate the approximate semimajor axis of its orbit.

Ans. 2.7×10^{9} km

36. Show that if a body describes a circular path under the influence of the gravitational inverse square force law, the square of the period is proportional to the cube of the path radius (Kepler's third law for a circular orbit) and find the proportionality constant in terms of the central mass M and universal constants.

37. How fast must an aircraft be able to fly due west to keep the sun always on the meridian (*a*) at the equator and (*b*) at 40° latitude? What is the centripetal acceleration in each case?

Ans. (*a*) 1,670 km/h, 0.034 m/s²; (*b*) 1,280 km/h, 0.026 m/s²

38. Imagine a planet having 3 times the average density of the earth and twice the radius. Calculate the acceleration due to gravity at the surface of this planet.

39. A 100-g mass (see accompanying figure) slides down a smooth (frictionless) hemisphere of

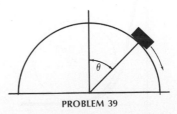

PROBLEM 39

0.25 m radius, starting at the top with a very tiny speed. Find the force exerted on the mass by the hemisphere as a function of the angle θ and the angle at which the mass leaves the surface of the hemisphere. *Ans.* 0.98(3 cos θ − 2) N; 48.2°

40. Show that for any homogeneous spherical planet of radius R the period of a satellite orbit with radius r only slightly greater than R depends only on the density of the planet and is given by $T = (198 \text{ min})(\rho_w/\rho)^{1/2}$, where ρ_w and ρ are the densities of water and the planet, respectively.

41. If a rocket ship moves toward the sun on a path such that it is always on the line connecting earth and sun, show that the net gravitational force will be zero (neglecting the fields of other planets) at a distance d from the earth given by $d = r\sqrt{m_E/M_S}$, where r is the distance from the sun to earth. *Hint:* $M_S \gg m_E$.

42. (*a*) A planet has a satellite which can be observed by a telescope. If the satellite has a period T and a radius (or semimajor axis) a, find the mass of the planet. (*b*) From the period of the moon and the radius of the moon's orbit, calculate the mass of the earth and compare your value with that quoted in Table 8.1. (*c*) Mars has satellites Phobos and Deimos, which have radii (semimajor axes) of 9.4×10^3 and 2.34×10^4 km and periods of 0.32 and 1.26 days, respectively. Estimate the mass of Mars from these data.

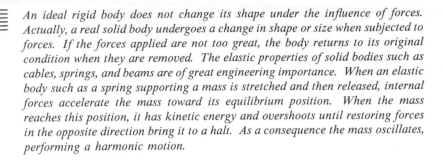

9

ELASTICITY
AND HARMONIC
MOTION

An ideal rigid body does not change its shape under the influence of forces. Actually, a real solid body undergoes a change in shape or size when subjected to forces. If the forces applied are not too great, the body returns to its original condition when they are removed. The elastic properties of solid bodies such as cables, springs, and beams are of great engineering importance. When an elastic body such as a spring supporting a mass is stretched and then released, internal forces accelerate the mass toward its equilibrium position. When the mass reaches this position, it has kinetic energy and overshoots until restoring forces in the opposite direction bring it to a halt. As a consequence the mass oscillates, performing a harmonic motion.

9.1 ELASTIC DEFORMATION AND HOOKE'S LAW

When a solid object is distorted by external forces, the changes in size or shape depend on the physical properties of the material it is made of. These changes reveal significant and useful information about the internal forces between the constituent atoms.

When we hang a weight from a spring, the spring stretches. If we increase the weight, the spring stretches still more. If we remove the weight, the spring returns to its original length. When a diver stands at the end of a diving board, the board is distorted, but when he leaves the board, it returns to its original position. When an archer prepares to shoot an arrow, he bends the bow, which springs back to its original form as the arrow is released. The spring, the diving board, and the bow are examples of elastic objects. *Elasticity is that property of a body by which it experiences a change in shape or volume when a deforming force acts upon it and by which it returns to its original size or shape when the deforming force is removed* (Fig. 9.1).

Materials which do not resume their original shape after being distorted are said to be *inelastic.* Mud, putty, and pastry dough are examples. Lead and solder are relatively inelastic, since it is easy to distort them permanently.

For an ideal spring the extension is directly proportional to the applied force (Sec. 6.6). For other sorts of elastic distortion we find that the magnitude of the deformation is proportional to the force (or torque) producing it. We can generalize Hooke's law and state that *the deformation of an elastic body is directly proportional to the applied force.* Whenever an elastic body is distorted by a force applied by some agent, the body exerts an equal and opposite force on the agent. For example, when we stretch a spring with an applied force F_a, the spring pulls on us with an equal and opposite restoring force F_r. By Hooke's law

$$F_a = ky \quad \text{and} \quad F_r = -ky \tag{9.1}$$

The act of stretching (or compressing) a spring gives it potential energy. It can do work through the action of the restoring force. The potential energy stored for a displacement y is given by Eq. (6.8):

$$P = \tfrac{1}{2}ky^2 \tag{9.2}$$

9.2 VIBRATIONS

When a mass on the end of a spring is pulled down and then released, it moves back and forth with a *vibratory* or *harmonic* motion (Fig. 9.2). Such periodic motions are very common and occur in all sorts of mechanical structures (automobiles, bridges, and buildings), in sound sources, in waves, and in electric oscillations. When a clock pendulum is displaced and then released, a vibratory motion is established. A further illustration of harmonic motion is afforded by clamping one end of a strip of steel (Fig. 9.3) in a vise and displacing the other end. When released, the strip vibrates back and forth with a period which depends on the characteristics of the strip of steel. Other examples of harmonic motion are those of a piston in a steam engine or an automobile and the vibration of a tree limb in a breeze.

Elastic forces lead to these cyclic motions. For a mass on the end of a spring, the restoring force is upward when the mass is displaced downward. The mass is pulled toward its equilibrium position, but when it reaches this position, it has a velocity and continues upward by virtue of Newton's first law. Thus, it overshoots the equilibrium position until there is a displacement in the opposite direction. The mass on the end of the spring moves back and forth until its energy is gradually dissipated in friction of various kinds.

FIGURE 9.1
Elastic deformation of a golf ball. (*Courtesy of Edgerton, Germeshausen, and Grier, Massachusetts Institute of Technology.*)

9.3 SIMPLE HARMONIC MOTION

There are many kinds of vibratory motion. We shall consider quantitatively only *simple harmonic motion*, which is by far the most important and universal. The basic ideas appear and reappear in many areas of physics, including vibrating mechanical systems, sound waves, alternating-current circuits, and electromagnetic waves. *Simple harmonic motion is motion in which the acceleration is proportional to the displacement from the equilibrium position and in the opposite direction.*

In discussing simple harmonic motion, several terms must be defined:

1. The *displacement* is the distance of a body from its equilibrium position.

2. The *amplitude* is the maximum value of the displacement, or the distance from the equilibrium position to the end of the vibrating path.

3. The *period T* is the time necessary for one *complete* vibration or cycle.

4. The *frequency ν* is the number of complete vibrations made in unit time. The basic unit of one cycle per second has been named the *hertz* (Hz) in honor of Heinrich Hertz, who produced the first radio waves in 1888. Clearly, the frequency is the reciprocal of the period; $ν = 1/T$.

To develop equations relating the displacement, the amplitude, the frequency, and the acceleration in simple harmonic motion, we make use of the fact that *the shadow of a body which is performing uniform circular motion describes simple harmonic motion.* Consider an object moving around a circle (Fig. 9.4) with uniform speed. Imagine the shadow of this body being projected on the wall at the right by a beam of light coming from the left. The shadow moves up and down as the particle moves around the circle with constant speed. Let R be the radius of the circle, which we call the *circle of reference*, and $θ$ the angle between the radius vector to the particle and the line OX. The displacement of the shadow from the "equilibrium position" is

$$y = R \sin θ \qquad (9.3)$$

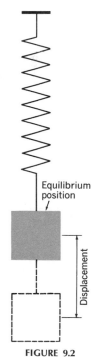

FIGURE 9.2
If a mass suspended from a spring is displaced from its equilibrium position and then released, the mass oscillates describing simple harmonic motion.

Let A represent the acceleration of the particle P toward the center of the circle and a the acceleration of the shadow on the wall. By Eq. (8.1), $A = V^2/R$, where V is the speed of the particle; a is the vertical component of A, which is $A \sin \theta$. Since a is directed downward when y is upward, and vice versa,

$$a = -A\frac{y}{R} = -\frac{V^2}{R^2}y \qquad (9.4)$$

Since neither V nor R changes in magnitude during the motion, V^2/R^2 is a constant. We observe that the *acceleration* of the shadow *is proportional to the displacement and in the opposite direction*. This establishes the fact that *the shadow describes simple harmonic motion.*

9.4 THE PERIOD OF A SIMPLE HARMONIC MOTION

Of all the characteristics of a harmonic motion, the period is one of the easiest to measure; therefore, let us rewrite our expression for the acceleration in terms of this quantity. The period T is the time for the body to make one revolution on the circle of reference; $T = 2\pi R/V$, and $V^2/R^2 = 4\pi^2/T^2$. Therefore,

$$a = -\frac{4\pi^2}{T^2}y \qquad (9.4a)$$

If ν represents the frequency, we can rewrite Eq. (9.4a) as

$$a = -4\pi^2\nu^2 y \qquad (9.4b)$$

When Eq. (9.4a) is solved for T, it yields

$$T = 2\pi\sqrt{\frac{-y}{a}} = 2\pi\sqrt{-\frac{\text{displacement}}{\text{acceleration}}} \qquad (9.4c)$$

The quantity under the square-root sign is positive since the displacement and acceleration are of opposite sign. *The period of a simple harmonic motion is independent of the amplitude.* If the amplitude is small, the body does not move far but it moves slowly, so that the time required to execute one complete vibration is the same as when the amplitude is larger. This interesting fact about simple harmonic motion was discovered by Galileo, who timed the swinging of a great chandelier in the cathedral at Pisa, using his own pulse as a clock.

9.5 VELOCITY IN SIMPLE HARMONIC MOTION

The displacement, the acceleration, and the velocity of a body describing simple harmonic motion are all changing constantly. As we have seen,

Displacement

FIGURE 9.3
Harmonic vibrations of a strip of steel.

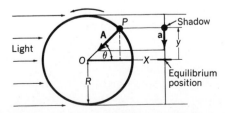

FIGURE 9.4
The shadow of a body moving with uniform circular motion executes simple harmonic motion with amplitude equal to the radius of the circle of reference.

the shadow of uniform circular motion on a wall (or its projection on a diameter of the circle) describes simple harmonic motion. From Fig. 9.5 we see that the velocity v of a body describing simple harmonic motion is given by

$$v = V \cos \theta = V \frac{\sqrt{R^2 - y^2}}{R} \tag{9.5}$$

or

$$v = \frac{2\pi}{T} \sqrt{R^2 - y^2} = 2\pi v \sqrt{R^2 - y^2} \tag{9.5a}$$

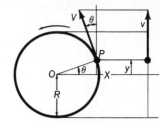

If we start a clock at the instant the body of Figs. 9.4 and 9.5 passes the line OX on its upward path, the angle θ is given by $\theta = 2\pi t/T = 2\pi v t$. Hence, by Eqs. (9.3), (9.5), and (9.4b),

$$y = R \sin 2\pi v t \tag{9.6}$$

$$v = V \cos \theta = 2\pi v R \cos 2\pi v t = 2\pi v R \sin \left(2\pi v t + \frac{\pi}{2}\right) \tag{9.7}$$

$$a = -4\pi^2 v^2 R \sin 2\pi v t = 4\pi^2 v^2 R \sin (2\pi v t + \pi) \tag{9.8}$$

FIGURE 9.5
The speed v of the shadow of a body describing uniform circular motion is $V \cos \theta$, where $V = 2\pi R/T$ is the speed of the body on the circle of reference.

We see (Fig. 9.6) that the velocity reaches its maximum 90° ($\pi/2$ rad) ahead of the displacement. Similarly, it passes through its zeros and reaches its minimum value one-fourth of a period ahead of the displacement. We say that the velocity *leads* the displacement by 90° or $\frac{1}{4}$ period. Similarly, the acceleration leads the velocity by 90° and the displacement by 180°. The displacement, velocity, and acceleration can all be represented by sine (or cosine) waves of the same frequency but differing in *phase*. Two variables described by sine waves are said to be *in phase* if they reach their maxima, zeros, and minima at the same instant.

☐ **Example** A body describes simple harmonic motion with an amplitude of 50 mm and a period of 0.25 s. Find the acceleration and speed of the body when the displacement is 50, 30, and 0 mm.

By Eq. (9.4a)

$$a = -\frac{4\pi^2}{T^2} y$$

and by Eq. (9.5a)

$$v = \frac{2\pi}{T} \sqrt{R^2 - y^2}$$

The amplitude $R = 50$ mm. When $y = 50$ mm,

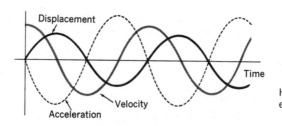

FIGURE 9.6
How displacement, velocity, and acceleration vary in time in simple harmonic motion.

$$a = -\frac{4\pi^2 \times 0.05 \text{ m}}{(0.25 \text{ s})^2} = -31.6 \text{ m/s}^2$$

$$v = \frac{2\pi}{0.25 \text{ s}} \sqrt{(0.05 \text{ m})^2 - (0.05 \text{ m})^2} = 0$$

When $y = 30$ mm,

$$a = -\frac{(4\pi^2)(0.03 \text{ m})}{(0.25 \text{ s})^2} = -18.9 \text{ m/s}^2$$

$$v = \frac{2\pi}{0.25 \text{ s}} \sqrt{(0.05 \text{ m})^2 - (0.03 \text{ m})^2} = \pm 1.01 \text{ m/s}$$

When $y = 0$ mm, $a = 0$ and

$$v = \frac{2\pi}{0.25 \text{ s}} (0.05 \text{ m}) = \pm 1.26 \text{ m/s} \qquad \square$$

9.6 FORCE AND ENERGY RELATIONS

When a shadow describes simple harmonic motion, of course no force need act on the shadow, but if a mass like the bob of a pendulum describes simple harmonic motion, there must be an unbalanced force F_r to produce the acceleration. By Newton's second law

$$F_r = ma = -4\pi^2 v^2 my \qquad (9.9)$$

where y is the displacement.

For a mass on the end of a spring, a displacement y brings into play a restoring force $F_r = -ky$, precisely that required to produce simple harmonic motion when the mass is released. If we replace F_r with $-ky$ in Eq. (9.9) and solve for $1/v$, we obtain

$$\frac{1}{v} = T = 2\pi \sqrt{\frac{m}{k}} \qquad (9.10)$$

When the mass has a displacement y, its potential energy is $\frac{1}{2}ky^2$, the work (Eq. 9.2) required to produce this displacement from the equilibrium position. The kinetic energy of the mass is $\frac{1}{2}mv^2$, or $2\pi^2 v^2 (R^2 - y^2)m$ if we make use of the fact that $v = 2\pi v \sqrt{R^2 - y^2}$ (Eq. 9.5a). The total energy, the sum of the kinetic and potential energies, remains constant in simple harmonic motion. At the ends of the motion the kinetic energy is zero and the total energy is $\frac{1}{2}kR^2$, while at the equilibrium position the potential energy is zero and the kinetic energy is $2\pi^2 v^2 R^2 m$. Thus

$$\text{Total energy} = \tfrac{1}{2}kR^2 = 2\pi^2 v^2 R^2 m = \tfrac{1}{2}ky^2 + 2\pi^2 v^2 (R^2 - y^2)m \qquad (9.11)$$

The total energy of a body performing simple harmonic motion is proportional to the square of the frequency and to the square of the amplitude.

□ **Example** A mass of 0.500 kg stretches a spring 200 mm. Find the force constant of the spring. If the mass is pulled down an additional 50 mm and then released, find the initial restoring force, the period, and the total energy of the resulting simple harmonic motion.

In this case $F_a = mg = (0.500 \text{ kg})(9.80 \text{ N/kg}) = 4.90 \text{ N}$, and $y = 0.200$ m. By Eq. (9.1)

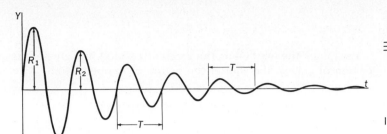

FIGURE 9.7
In a damped vibration, the amplitude decreases with time.

$$F_a = ky = -F_r$$
$$4.90 \text{ N} = k(0.200 \text{ m})$$
$$k = 24.5 \text{ N/m}$$

When m is displaced 5 cm from its new equilibrium position and released, the initial restoring force is upward and has a magnitude $ky_0 = 24.5 \times 0.05 = 1.23$ N. By Eq. (9.10)

$$T = 2\pi\sqrt{\frac{m}{k}} = 2\pi\sqrt{\frac{0.500 \text{ kg}}{24.5 \text{ N/m}}} = 0.898 \text{ s}$$

The amplitude $R = 0.0500$ m, since that is the maximum displacement. By Eq. (9.11)

$$\text{Total energy} = \tfrac{1}{2}kR^2$$
$$= \tfrac{1}{2}(24.5 \text{ N/m})(0.0500 \text{ m})^2$$
$$= 0.0306 \text{ joule} \qquad \square$$

If a particle is oscillating in a viscous medium, it does work on the medium and the energy of vibration is correspondingly reduced. For this reason the amplitude diminishes with time (Fig. 9.7). Such vibrations are said to be *damped.*

9.7 THE SIMPLE PENDULUM

A simple pendulum affords an illustration of simple harmonic motion. When the pendulum is displaced from its position of equilibrium (Fig. 9.8), the restoring force is $F_r = -mg \sin \theta$. Since $F_r = ma$, the acceleration which this force produces is $a = -g \sin \theta$. The displacement of the pendulum bob from its position of equilibrium N is $S = l\theta$. Hence,

$$\frac{\text{Displacement}}{\text{Acceleration}} = \frac{S}{a} = \frac{l\theta}{-g \sin \theta}$$

For small angles, $\sin \theta = \theta$ when the angle is measured in radians, and for small displacements of the pendulum,

$$\frac{\text{Displacement}}{\text{Acceleration}} = -\frac{l}{g} = \text{constant}$$

Consequently, the pendulum moves with nearly simple harmonic motion for small angles of swing; it deviates from simple harmonic motion for large angular amplitudes. Its period, by Eq. (9.4c), is

FIGURE 9.8
A simple pendulum moves with approximately simple harmonic motion (unless θ becomes large).

$$T = 2\pi \sqrt{\frac{l}{g}}$$

(9.12)

The longer the pendulum, the greater its period; the greater the acceleration of gravity, the shorter the period of the pendulum.

9.8 STRESS AND STRAIN: YOUNG'S MODULUS

To treat problems in elasticity in a convenient and consistent way, we introduce two new terms, *stress* and *strain*. When we apply a force to distort a body, internal forces within the body resist the distortion. For example, if we try to stretch a light bar (Fig. 9.9), the material at the left of some plane PP' exerts a force **F** to the left on the material at the right of PP'. Since the bar is in equilibrium, the material at the right of PP' exerts an equal and opposite force on the material at the left. The material is said to be under stress. For our bar *the stress is the ratio of the internal force to the area over which the force is distributed.* (When the stress varies from point to point in a material, we must take the limit of this ratio as the area chosen approaches zero to obtain the stress at a point.) We shall consider only situations in which the stress is uniform over the area in question. In this case

$$\text{Stress} = \frac{\text{force}}{\text{area}}$$

As a result of the applied stress, some change in the shape or size of the elastic body is produced. The term *strain* is applied to the relative change occurring in the dimensions or shape of a body when it is subjected to a stress. We shall define three particular types of strain later in this chapter. For all these types of strain Hooke's law states that *the stress applied is directly proportional to the strain produced.*

When a steel cable supports an elevator or a suspension bridge, it is stretched by the load. How much its length changes depends on the load, the cross-sectional area of the cable, its original length, and the material. The bigger the load, the larger the stretch expected; the greater the cross-sectional area of the cable, the smaller the stretch. The stress in the cable is the force applied divided by this area. *The stretching strain is the ratio of the change in length to the original length.*

FIGURE 9.9
The stretching stress in a bar is the ratio of the internal tensile force to the cross-sectional area over which the force is distributed.

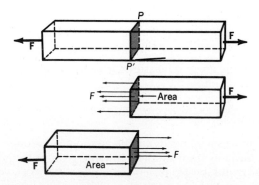

$$\text{Stretching (or longitudinal) strain} = \frac{\text{change in length}}{\text{original length}} = \frac{\Delta l}{l}$$

For an elastic material the stretching stress is directly proportional to the longitudinal strain. *The ratio of the stretching stress to the longitudinal strain is known as Young's modulus* (or sometimes as the *longitudinal modulus* or *stretch modulus*) of elasticity. We denote it by Y.

$$\frac{F}{A} = Y\frac{\Delta l}{l} \qquad (9.13)$$

□ **Example** A wire 3.00 m long with a cross section of 0.08 cm² hangs vertically. When a load of 2,000 N is applied to the wire, it stretches 3.9 mm. Find Young's modulus of elasticity.

$$\text{Young's modulus} = \frac{\text{stress}}{\text{strain}} = \frac{F/A}{\Delta l/l} = \frac{(2,000\text{ N})/(8 \times 10^{-6}\text{ m}^2)}{(3.9 \times 10^{-3}\text{ m})/(3\text{ m})}$$

$$= 1.9 \times 10^{11}\text{ N/m}^2 \qquad \qquad \square$$

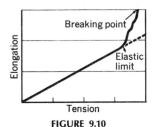

FIGURE 9.10
The elongation of a wire is approximately proportional to the applied tension until the elastic limit is reached.

9.9 LIMIT OF ELASTICITY

A body that has been deformed and then released returns to its original size and shape provided the stress has not exceeded the *elastic limit*, defined as the maximum stress from which the substance completely recovers its original size and shape. When the elastic limit is exceeded, the body acquires a *permanent set*. The elastic limit differs widely for different materials, being high for steel and low for lead; for any given substance it depends on the temperature. Figure 9.10 shows a typical relation between tension and elongation for a wire. The greatest stress which can be applied to the wire is the *ultimate tensile stress*. Table 9.1 lists some typical elastic constants.

When a metal rod is stretched beyond its elastic limit and the stress is increased still further, a stage is reached at which the rod begins to stretch rapidly, even though the stress is somewhat decreased. The stress at which this begins is called the *yield point*. Even though the metal is cold, it behaves as if it were in a semifluid state. Once a metal is strained beyond the elastic limit, the strain becomes a function of the time elapsed after application of the load. Under such conditions there is a viscous flow which depends on both temperature and the applied load.

TABLE 9.1
Typical Elastic Constants

Material	Young's modulus		Bulk modulus, GN/m²	Shear modulus, GN/m²	Elastic limit, GN/m²	Ultimate tensile stress, GN/m²
	10⁶ lb/in.²	GN/m²				
Aluminum	10	69	69	26	0.13	0.14
Brass	14	98	58	35	0.38	0.45
Copper	17	120	120	41	0.15	0.34
Medium steel	30	210	160	84	0.25	0.48

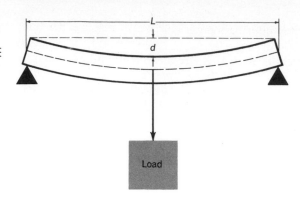

FIGURE 9.11
The bending of a beam under load.

9.10 STIFFNESS AND STRENGTH OF BEAMS

It is important in many engineering applications to be able to calculate the amount of bending that will be produced in a beam by a given load (Fig. 9.11). The deflection depends on the length of the beam, its breadth, and its depth.

$$\text{Deflection} = \frac{\text{load} \times \text{length}^3}{4 \times \text{Young's modulus} \times \text{breadth} \times \text{depth}^3}$$

The stiffer of two beams is not necessarily the stronger. The mere fact that the one beam bends more than the other does not mean that it will hold less. The strength depends on the same dimensions as the stiffness, but it depends on them in a different way: the strength is proportional to the breadth and to the square of the depth and inversely proportional to the length.

In the bending of a beam the top layer is shortened since it resists compression, while the lower layer is lengthened and must resist tension. A central layer of the beam remains the same length. Therefore, to make a beam as strong or as stiff as possible for a given amount of material, most of the material is put in the upper and lower layers, and relatively little in the middle. For this reason steel beams are usually manufactured in shapes similar to that of the letter I. They are called I beams and have great strength and stiffness for a given amount of steel.

9.11 VOLUME ELASTICITY

When a solid body such as a piece of iron is immersed in a liquid so that a uniform pressure is applied to its surface, a change in the volume of the solid occurs. By definition,

$$\text{Volume (or bulk) strain} = \frac{\text{change in volume}}{\text{original volume}} = \frac{\Delta V}{V}$$

The corresponding stress (force/area) is taken as $-\Delta p$, with the minus sign appearing because an increase in pressure produces a decrease in volume. The ratio of volume stress to volume strain is called the *bulk modulus* (or *volume modulus*). We denote it by B.

$$B = \frac{\text{volume stress}}{\text{volume strain}} = -\frac{\Delta p}{\Delta V/V} \tag{9.14}$$

Since an increase in pressure produces a decrease in volume, the bulk modulus B is inherently positive.

□ **Example** A sphere of copper having a volume of 100 cm³ is subjected to a pressure of 7×10^6 N/m². Find the change in volume that takes place.

$$-\Delta p = B\frac{\Delta V}{V} \tag{9.14a}$$

$$-7 \times 10^6 \text{ N/m}^2 = 12 \times 10^{10} \text{ N/m}^2 \frac{\Delta V}{100 \times 10^{-6} \text{ m}^3}$$

$$\Delta V = -5.8 \times 10^{-9} \text{ m}^3 = -5.8 \text{ mm}^3 \qquad □$$

Not only solids but also liquids and gases exhibit volume elasticity and have bulk moduli defined by Eq. (9.14). The reciprocal of the bulk modulus is called the *compressibility* k. Thus $k = 1/B$. It is customary to list compressibilities rather than bulk moduli for liquids (Table 9.2). The volume modulus of gases is discussed further in Sec. 14.2.

TABLE 9.2
Compressibility of Liquids

Substance	Tempera-ture, °C	Compressibility per atmosphere
Ethyl alcohol	14	0.0000987
Ethyl ether	0	0.000143
Kerosene	20	0.0000543
Mercury	20	0.0000039
Turpentine	20	0.000075
Water	20	0.000048

9.12 ELASTICITY OF SHAPE: SHEAR MODULUS

If one cover of a thick book is held firmly on the table, and a force parallel to the top of the table is applied to the other cover, the shape of the book is changed while its thickness and volume remain the same. Such a deformation in shape is called a *shear*. The *shearing stress* is defined as the tangential force per unit area of the surface, and the shearing strain is the angle θ of Fig. 9.12 in radians ($\theta = CC'/BC$). The ratio of the shearing stress to shearing strain is called the *shear modulus* S of the material. (It is also known as the *torsion modulus* or *modulus of rigidity*.)

$$\frac{F}{A} = S\theta \tag{9.15}$$

The shear modulus is the elastic constant involved when one twists a rod or a tube as in the drive shaft of an automobile. Clearly, this elasticity of shape is characteristic of solids only. It is not exhibited by liquids or gases, although liquids do evidence short-range forces between molecules which give rise to interesting surface phenomena.

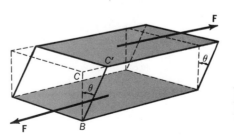

FIGURE 9.12
The shearing stress is the ratio of the force F to the area of one of the shaded faces; the shearing strain is the angle θ in radians ($\theta = CC'/BC$).

QUESTIONS

1. A uniform spring with force constant k is cut in half. What is the spring constant for one of the halves?

2. How does the period of a simple pendulum with a 200-g steel bob compare with that of a pendulum of the same length but with a 25-g wood bob? What differences in actual behavior would you expect?

3. Qualitatively, what is the effect on the period of a pendulum of taking it (a) to the North Pole, (b) to the equator, and (c) to the top of Mount Whitney?

4. An unstretched spring has a force constant k. A mass m is hung from it, and equilibrium is achieved. Is the spring constant changed, or is k the same for this new equilibrium situation?

5. Why is the motion of a simple pendulum not simple harmonic when the amplitude is large? Is the restoring force too big or too small? Does the period increase or decrease as the amplitude becomes larger?

6. Why are precisely simple harmonic motions relatively rare while approximately simple harmonic motions are common?

7. If the collisions between the ball and the plate of Fig. 7.6 were elastic, would the resulting motion be simple harmonic? Would the displacement follow Eq. (9.6) for a half-period?

8. For a uniform spring one should add one-third the mass of the spring to m in Eq. (9.10) if one wishes a more accurate relation. Why should the mass of the spring be involved in the period?

9. How does doubling the amplitude of a simple harmonic motion affect (a) the total energy of the oscillator? (b) the maximum velocity? (c) the frequency?

10. Two bodies of identical mass m are suspended from separate springs with force constants k_1 and k_2, where $k_1 > k_2$. If the bodies are given equal energies, for which spring is the amplitude greater? How do the speeds and equilibrium positions compare?

11. If one timing device is based on a mass oscillating on the end of a spring and a second one is based on a pendulum, would they keep the same time on the moon as on earth? Explain.

12. While an elevator has a downward acceleration of $0.2\,g$, what is the effect on the period of a simple pendulum suspended from the top of the elevator? What is the effect on the period of a mass on the end of a spring?

13. If a hole could be bored through the earth and a body dropped into the hole, it would describe simple harmonic motion with an amplitude equal to the earth's radius. At the center of the earth would the body have any (a) weight, (b) mass, (c) acceleration, or (d) momentum?

14. The cables supporting the roadway of a suspension bridge are identical in material and cross-sectional area, but they differ in length. If one cable is two-thirds the length of a second cable, how does the elongation of the second compare with that of the first for the same load? Will one cable be able to support a greater load without exceeding its elastic limit? Explain your answer.

15. If a body such as a bow is distorted until Hooke's law is no longer applicable, does it necessarily follow that the elastic limit has been exceeded? Explain.

16. Should steel reinforcing rods in a concrete beam be embedded near the upper or the lower side of the beam? Explain why.

PROBLEMS

1. A stone is swinging in a horizontal circle 0.8 m in diameter, making 30 r/min. A distant light causes a shadow of the stone to be formed on a nearby wall. What is the amplitude of the motion of the shadow? What is the frequency? What is the period?   *Ans.* 0.4 m; 0.5 Hz; 2 s

2. Find the length (in meters) of a pendulum which has a period of 2.4 s.

3. When a mass of 0.25 kg is suspended from a spring, the spring stretches 392 mm. Find the spring constant. What is the period of oscillation of the mass when it is given a small displacement? What is the frequency?
Ans. 6.25 N/m; 1.26 s; 0.796 Hz

4. A mass m suspended from a spring of constant k has a period T. If a mass M is added, the period becomes $3T$. Find M in terms of m.

5. A ball moves in a circular path of 0.15 m diameter with a constant angular speed of 20 r/min. Its shadow performs simple harmonic motion on the wall behind it. Find the acceleration and speed of the shadow (a) at the end of the motion, (b) at the equilibrium position, and (c) at a point 6 cm from the equilibrium position.
Ans. (a) 0.329 m/s²; 0; (b) 0; 0.157 m/s; (c) 0.263 m/s²; 0.0942 m/s

6. The mass on the end of a spring describes simple

harmonic motion with a period of 2 s. Find the acceleration when the displacement is 75 mm.

7. A 0.5-kg body performs simple harmonic motion with a frequency of 2 Hz and an amplitude of 8 mm. Find the maximum velocity of the body, its maximum acceleration, and the maximum restoring force to which the body is subjected.

Ans. 0.101 m/s; 1.26 m/s²; 0.632 N

8. A 200-g mass on the end of a spring describes simple harmonic motion with an amplitude of 50 mm and a period of 2.5 s. Find the maximum acceleration and maximum velocity of the mass and the force constant of the spring.

9. A body describing simple harmonic motion has a maximum acceleration of 8π m/s² and a maximum speed of 1.6 m/s. Find the period T and the amplitude R. *Ans.* 0.4 s; 102 mm

10. A 0.6-kg mass on the end of a spring has a period of 2 s and an amplitude of 0.12 m as it describes simple harmonic motion. Find the frequency of the motion, the acceleration of the mass at an endpoint of its motion, and the maximum velocity attained by the mass. What is the force constant of the spring?

11. A simple harmonic motion has a frequency of 1.6 Hz and an amplitude of 4 mm. What is the acceleration at maximum displacement? What is the acceleration 0.4 s later?

Ans. −0.404 m/s²; +0.258 m/s²

12. A pendulum executes 40 complete vibrations in 80 s. Find the length of the pendulum.

13. A mass of 0.5 kg is suspended from a spring. A force of 0.5 N pulls the mass downward 0.25 m. If the mass is then released from rest, find (a) the energy associated with the simple harmonic motion, (b) the frequency of the motion, (c) the speed and acceleration of the mass as it passes the equilibrium position, and (d) the speed and acceleration of the mass when its displacement is 0.20 m.

Ans. (a) 0.0625 J; (b) 0.318 Hz; (c) 0.5 m/s; 0; (d) 0.3 m/s; −0.8 m/s²

14. A 0.3-kg mass is suspended from a spring. If the spring is pulled downward with a force of 1 N, it stretches an additional 50 mm. If the spring is released, the mass describes simple harmonic motion with an amplitude of 50 mm. Find the period of the motion and the velocity and acceleration of the mass when the displacement is 40 mm. Find the maximum values of the potential energy and kinetic energy of the vibrating mass, taking the equilibrium position as the zero of potential energy.

15. A 2.5-kg mass hangs in equilibrium from a spring of force constant 40 N/m. How far must the spring be stretched to give the mass an accelera-

tion of 8 m/s² upward when released? What is the period of the motion? Find the speed of the mass 0.1 s after it passes the equilibrium position.

Ans. 0.5 m; 1.57 s; 1.84 m/s

16. Identical 0.4-kg masses are suspended from a spring and from a cord. The length of the cord is adjusted until the simple pendulum and the mass on the spring have identical periods of 1 s on the earth, where $g = 9.8$ m/s². If both were taken to the moon, where $g = 1.62$ m/s², what would be the period of each?

17. A particle describing simple harmonic motion has a period of 3 s. If the speed of the particle is 0.6 m/s when it has a displacement equal to half the amplitude, how long will it take the particle to reach the equilibrium position if it is moving (a) toward the equilibrium position and (b) away from the equilibrium position? What is the maximum speed of the particle?

Ans. (a) 0.25 s; (b) 1.25 s; 0.69 m/s

18. To be acceptable, a 5-kg airplane radio receiver must pass a shake test in which it is forced to describe simple harmonic oscillations at 120 Hz with a maximum acceleration of 10 g's ($g = 9.8$ m/s²). Find the maximum displacement, unbalanced force, and velocity.

19. A particle moves with simple harmonic motion along a line between two points 0.16 m apart. If the period is 2 s, what is the acceleration of the particle at maximum displacement? What is the speed of the particle as it passes the midpoint? If the displacement is maximum when $t = 0$, write the equations for the displacement and velocity as functions of time.

Ans. 0.79 m/s²; 0.251 m/s; $y = 0.08 \cos \pi t$ m; $v = -0.08\pi \sin \pi t$ m/s

20. In a gasoline engine the motion of a piston is approximately simple harmonic. A piston has a mass of 0.9 kg and a stroke (twice the amplitude) of 0.12 m. Find the maximum acceleration and the maximum unbalanced force on the piston if it is making 45 complete vibrations each minute.

21. A 0.2-kg mass suspended from a spring performs simple harmonic motion with a frequency of 1.5 Hz and an amplitude of 25 mm. Find the maximum acceleration of the mass, the maximum velocity, the velocity at the instant the displacement is 15 mm, and the force needed to stretch the spring 25 mm.

Ans. 2.22 m/s²; 0.236 m/s; 0.188 m/s; 0.444 N

22. A mass of 40 g is moving with simple harmonic motion with 45 mm amplitude and a period of 2 s. What is the acceleration of the mass one-twelfth of a period after it has passed the mid-point? What is the force acting on the mass? What is the speed?

23. A mass of 15 kg is supported by an aluminum wire 2 mm in diameter and 5 m long. Find the stress, the strain, and the elongation of the wire.
Ans. 46.8 MN/m²; 6.78×10^{-4}; 3.39 mm

24. An elevator weighing 25,000 lb is supported by two steel cables, each 1 in.² in cross section. Find the elongation produced in the cables if the elevator is 80 ft from the supporting cylinder.

25. Find how many cables of 400 mm² cross section should be used to support an elevator which weighs 60 kN if the elevator will have a maximum acceleration of 3 m/s² and the stress is not to exceed one-fifth of the elastic limit of the cable steel, which is 300 MN/m². *Ans.* 4

26. A wire 2.5 m long with a diameter of 1.2 mm is elongated 0.7 mm when a 6-kg mass is hung on it. What is Young's modulus for the material?

27. How much force is required to punch a hole 8 mm in diameter in a metal sheet 1.5 mm thick if the minimum shearing stress needed to rupture the metal is 250 MN/m²? *Ans.* 12.6 kN

28. The shearing stress on an aluminum rivet is not to exceed one-tenth of the elastic limit of 140 MN/m². How many rivets, each 120 mm² in cross-sectional area, are needed to hold two parts of an airplane wing together if each rivet carries the same fraction of the load and the total force on the rivets is 80 kN?

29. A wire made of medium steel is 6 m long and has an area of 8 mm². What is the maximum load which this wire can support? What is the greatest load which can be supported without exceeding the elastic limit? If the wire is fastened at its upper end, how far can it be stretched without exceeding the elastic limit?
Ans. 3,840 N; 2,000 N; 7.1 mm

30. A vertical steel column carries a load of 80,000 kg. The area of its cross section is 40 cm², and its length is 9 m. Find the decrease in length produced by this load.

31. Find the increase in pressure required to increase the density of medium steel by 0.05 percent.
Ans. 80 MN/m²

32. A hydraulic press contains 0.3 m³ of oil of compressibility 2×10^{-5} per atmosphere. Calculate the decrease in volume of the oil if it is subjected to a pressure of 500 atm. Find the bulk modulus of the oil in newtons per square meter.

33. Find the depth in a lake at which the density of the water is 0.1 percent greater than at the surface. Ignore temperature differences; 1 atm corresponds to a depth of 10.33 m. *Ans.* 215 m

34. Compute the compressibilities of copper and medium steel in atm⁻¹. How do they compare with that of water?

35. Two springs each 15 cm in unstressed length are fastened to pins on the frictionless table 50 cm apart. The force constant k_1 for the spring on the left is 3 N/m, while k_2 for the spring on the right is 6 N/m. The springs are stretched and connected to hooks on opposite ends of a 1-kg mass. It is 5 cm between hooks, so the sum of the extensions of the two springs is 15 cm. (a) Find the equilibrium position for the mass. (b) What is the effective spring constant for the arrangement? (c) Find the period of small oscillations about the equilibrium position.
Ans. (a) 5 cm to right of center; (b) 9 N/m; (c) 2.09 s

36. A spring of constant k_1 is suspended from a hook. A second spring of constant k_2 hangs from the bottom of the first spring and supports a mass m from its lower extremity. Assume that both springs have negligible mass. Show that when m is displaced from its equilibrium position, it describes simple harmonic motion with a frequency given by

$$\nu = \frac{1}{2\pi} \sqrt{\frac{k_1 k_2}{(k_1 + k_2)m}}$$

37. Show that l meters of liquid in a uniform U-tube will, if disturbed, perform simple harmonic motion with a period $T = \pi \sqrt{2l/g}$, provided friction is negligible.

38. Prove that if a particle is simultaneously subjected to simple harmonic motion in both the x and y directions, the resultant motion of the particle is a circle, provided the amplitudes and frequencies of the two harmonic motions are the same and they are 90° out of phase; i.e., the displacement in one direction is maximum when the other displacement passes through zero.

In Chap. 5 we learned that if the resultant force on a body is zero, the body maintains a constant velocity (which is zero if the body is initially at rest). However, the body may still have a rotational acceleration about its center of mass. We next consider what condition must be satisfied if there is to be no rotational acceleration, and we develop the general conditions for equilibrium which must be satisfied if a body is to remain at rest or move with constant linear and angular velocities.

10.1 TORQUE

If two boys who weigh 250 and 300 N, respectively, sit on opposite ends of a horizontal seesaw 4.8 m long (Fig. 10.1), we know that the seesaw does not remain horizontal; the 300-N boy moves downward and the 250-N boy upward. This occurs even though the upward force exerted by the support is equal to 550 N (plus the weight of the seesaw, which we assume for the moment is negligible). A glance at the figure reveals that the forces are not concurrent and that the conditions required for a body to be at rest under the influence of concurrent forces are not enough to keep the seesaw at rest. The weight of the 300-N boy acts to produce a clockwise rotation about the fulcrum, while the weight of the 250-N boy acts to produce a rotation in the opposite direction. When both boys are at the ends of the seesaw, the two rotational tendencies do not balance each other and there is a resultant clockwise rotation. We all know from experience that the two boys can sit on the seesaw and keep it at rest provided the heavier boy sits closer to the fulcrum. This suggests that the tendency to produce rotation depends not only on the force acting but also on the distance.

A simple experiment shows that if the 250-N boy sits 2.4 m from the fulcrum and the 300-N boy 2 m from the fulcrum, the seesaw is balanced. We observe that 2.4 m × 250 N = 2.0 m × 300 N. This suggests that the tendency to produce rotation depends on the product of force and distance, but we must inquire more closely into how this distance factor is to be determined. Suppose that our seesaw is elevated and that the 250-N boy is suspended from a rope 3.2 m long hanging at the end of the seesaw (Fig. 10.2). His distance to the fulcrum is then 4.0 m. Will the seesaw be balanced? Yes, indeed. An experiment would show that moving the 250-N boy to this position has no effect on the balance. Thus it is clear that it is *not* the distance from the fulcrum to the point at which the force is applied which is important but the *perpendicular distance from the fulcrum to the line along which the force acts*. This distance is known as the *lever arm* (or *moment arm*). *The lever arm for any axis is the perpendicular distance from the axis to the line along which the force acts.*

The product of the lever arm and the force (2.4 m and 250 N for the 250-N boy on the seesaw) is called the *torque* (or the *moment* of the force). If the seesaw is to be in equilibrium, the clockwise torque about the fulcrum must be balanced by the counterclockwise torque. Not only is the clockwise torque equal to the counterclockwise torque about the fulcrum, but the net counterclockwise torque *about any axis* is equal to the net clockwise torque *about the same axis*, provided the resultant of the forces is zero. For example, if the weight of the seesaw itself is negligible, Eq. (5.1) requires that the fulcrum exert an upward force of 550 N. If we choose an axis through the left end of the seesaw, we have a counter-

10
TORQUE AND EQUILIBRIUM

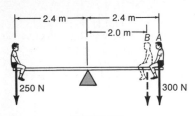

FIGURE 10.1
Rotational equilibrium of the seesaw is established when the 300-N boy at *A* moves to *B*.

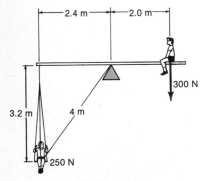

FIGURE 10.2
The torque about the fulcrum depends on the product of the force and the distance from the fulcrum to the line along which the force acts.

FIGURE 10.3
Torque is a vector quantity. By convention its direction is that in which a right-handed screw would advance if it were turned by the torque. Alternatively, the direction is pointed by the outstretched right thumb when the fingers point in the direction in which the screw would rotate.

clockwise torque of 2.4 m × 550 N = 1,320 m-N and a clockwise torque of 4.4 m × 300 N = 1,320 m-N.

The torque τ exerted by a force about a fixed axis is the product of the lever arm and the force, where the lever arm is the distance from the axis to the line along which the force acts. (τ is the greek letter *tau*.) Torque is a vector quantity with direction chosen by convention to be the direction in which a right-handed screw would advance if it were turned by the torque (Fig. 10.3). Alternatively, the torque has the direction in which the right thumb points when the right fingers curve to indicate the path along which points of the body would move under the influence of the torque. For the case shown in Fig. 10.4 the direction of the torque† is out of the page, and its magnitude is

$$\tau = rF \sin \theta \qquad (10.1)$$

where θ is the angle between **r** and **F**. Clearly, $F \sin \theta$ is the component of **F** perpendicular to the line joining the axis to the point of application of the force, so the magnitude of the torque about any axis is given by the product of the distance r from the axis to the point at which the force is applied and the component $F \sin \theta$ of the force perpendicular to **r**. Alternatively, since $r(F \sin \theta) = (r \sin \theta)F$, the magnitude of the torque is the product of the lever arm $r \sin \theta$ and the force F.

The units of torque are a length times a force. In the SI the basic units for torque are meter-newtons, in the British engineering system torques are usually given in foot-pounds or inch-pounds.

It is interesting that although both work and torque have the same units (length multiplied by a force), they are of an entirely different nature. In *work* we have the *scalar* product equal to the product of the magnitudes of the displacement and the force times the *cosine* of the angle between them. In *torque* we have the *vector* product given by the product of the magnitude of the radius vector from the axis to the point of application of the force, the magnitude of the force, and the sine of the angle between them.

We shall be concerned primarily with forces confined to a single plane. In this case the torques are all perpendicular to the plane containing the forces. In describing the torques, we shall refer to them as clockwise if they act to produce a rotation in the direction in which the hands of a

---

† In general, the torque τ exerted about any point is given by $\tau = \mathbf{r} \times \mathbf{F}$, where **r** is the vector from the point in question to the point where the force **F** is applied. The product $\mathbf{r} \times \mathbf{F}$ is known as the *vector* or *cross product* of **r** and **F**. It has magnitude $rF \sin \theta$ and the direction described in Fig. 10.3.

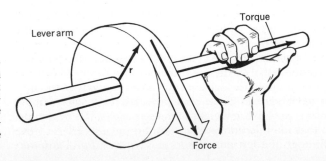

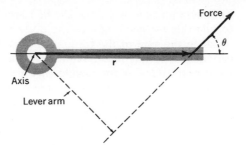

FIGURE 10.4
The torque about any axis is the product of the force and the *lever arm*, which is defined as the distance from the axis to the line along which the force acts.

clock move and counterclockwise if they act in the opposite direction. The conventional direction for a clockwise torque is into the plane and for a counterclockwise torque out of the plane.

As an illustration of the importance of the lever arm in computing the torque, consider the operation of a bicycle. If a boy pushes down on the pedal with his full weight (Fig. 10.5), the turning moment produced depends on the location of the pedal. If it is straight down, no rotational effect is produced. The torque in this case is zero, because the lever arm is zero. The maximum torque occurs when the pedal is in the position marked A; in this position the lever arm is R. When the pedal is in position B, the lever arm is reduced from R to X.

10.2 CENTER OF GRAVITY

Every particle of an extended body possesses weight, so that the pull of the earth on the body is composed of a large number of forces directed toward the center of the earth. For a body of ordinary size these forces are essentially parallel to each other. The body can, however, be supported in equilibrium by a single upward force, provided its line of action passes through a point called the *center of gravity* of the body. This point can be located for a rigid body as follows.

If we suspend the body from a single point, it comes to rest with a definite orientation. Let us determine a line through the body directed vertically downward from the point of suspension (Fig. 10.6). Next, let us suspend the body from several other points on its surface and, in each case, determine the position of the line through the point of support and directed vertically downward. We find that all the lines determined in this way intersect at a point which is the *center of gravity* of the body.

The center of gravity of a uniform sphere, cube, or rod is at its geometrical center. The center of gravity of an axe or hammer is nearer the head than the handle. The center of gravity of a telephone pole or a baseball bat lies on the axis but is closer to the thicker end. The center of gravity of a ring or of a tire is at the geometrical center, which, of course, does not lie in the material of the object. Although the center of gravity is a point fixed relative to the body, it does not necessarily lie within the body.

Since an extended body can be at rest when supported by a single upward force whose line of action passes through the center of gravity, the sum of the clockwise torques due to the individual forces on the particles of the body must be equal to the sum of the counterclockwise

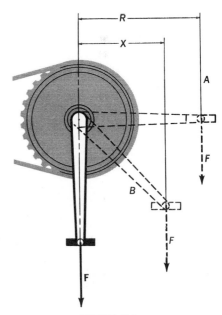

FIGURE 10.5
The vertical force produces no torque when the pedal is straight down; maximum torque is exerted when the pedal is at A.

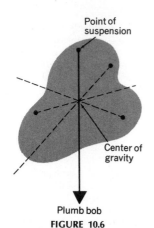

Point of suspension

Center of gravity

Plumb bob

FIGURE 10.6
If a body is suspended from a single point, the center of gravity lies directly below the point of suspension when equilibrium is achieved. In the figure the center of gravity must lie somewhere along the line determined by the plumb bob. Suspension from a second point gives a second line, and the center of gravity is at the intersection.

torques about an axis through the center of gravity. For any axis the torque due to the weight of the body acting at the center of gravity is equal to the sum of the torques due to the forces of gravity on all the various parts of the body. When we are dealing with the extended body, we can consider all the weight of the body to be concentrated at the center of gravity, and we can thus replace the individual forces acting on the many particles of the body by a single force equal to the weight. This is true not only for an axis determined by the center of gravity but for any axis. *For the purpose of computing torques the entire weight of an object can be considered to act on the center of gravity.*

The center of gravity coincides with the center of mass (Sec. 7.3) for a body of ordinary size. If the body is so large that the acceleration due to gravity is different at the various points occupied by the body, the two points do not coincide exactly.

□ **Example** Find the center of gravity of two spheres A and B of weights 0.1 and 0.4 N, respectively, connected by a light (negligible mass) rod 200 mm long (Fig. 10.7).

If the center of gravity is at a point C, the clockwise torque due to the larger weight w_2 about an axis through C is just equal to the counterclockwise torque due to the smaller weight w_1. Then

$$x_1 w_1 = x_2 w_2$$

Since $x_1 + x_2 = 200$ mm $= 0.200$ m,

$$(0.1 \text{ N})x_1 = (0.4 \text{ N})(0.200 \text{ m} - x_1)$$
$$x_1 = 0.160 \text{ m} = 160 \text{ mm} \qquad \square$$

10.3 EQUILIBRIUM

Thus far we have been considering the conditions which must be satisfied for a body to remain *at rest*. Exactly the same conditions apply to a body which is in motion with *constant velocity*. When we ride in an automobile over a smooth road at a constant velocity of 40 km/h south, the resultant of the forces which act on us is the same as when we are at rest. As we saw in Chap. 4, if the resultant force acting on a body is not zero, the body is accelerated.

The condition that the vector sum of the torques on a body be zero in order that there be no rotation for a body at rest can be extended to include bodies rotating about a fixed axis as follows: *If the vector sum of the torques acting on a rigid body is zero, the rotational velocity about a fixed axis is constant.* When a resultant torque acts on the body, there is a change in its angular velocity, as we shall see in Chap. 11. *A body which has a constant linear velocity and a constant rotational velocity is in a state of equilibrium.* Statics involves a special case of equilibrium, namely, the case in which both the linear velocity and the rotational velocity have the constant value *zero.*

10.4 THE CONDITIONS FOR EQUILIBRIUM

For a rigid body to be in equilibrium, two conditions must be satisfied:

1. *The vector sum (resultant) of all forces acting on the body must be zero;* $\Sigma \mathbf{F} = 0$.

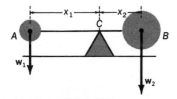

x_1 x_2
C
A B
w_1 w_2

FIGURE 10.7
The center of gravity of a system composed of two spheres connected by a light rod lies just above the fulcrum when the system is in equilibrium.

2. *The vector sum (resultant) of all torques* about any axis *must be zero;* $\Sigma \tau = 0.$

The first condition for equilibrium quarantees that the center of gravity (more accurately the center of mass) of the body moves with constant velocity. The second condition guarantees that the rotational velocity of the body remains constant. In applying the second condition there is no restriction on the axis chosen. It may be through the center of gravity of the body, through one end of the body, or at any convenient place.

In practice, it is usually convenient to use the conditions for equilibrium in a slightly different form. If we replace forces with their vertical and horizontal components, we can write the conditions† for equilibrium in the form:

1. *The sum of all upward forces is equal to the sum of all downward forces.*

2. *The sum of all forces to the right is equal to the sum of all forces to the left.*

3. *The sum of all clockwise torques* about any axis *is equal to the sum of all counterclockwise torques* about the same axis.

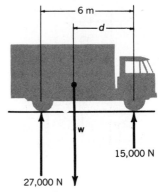

$$\Sigma F_{up} = \Sigma F_{down} \qquad (10.2)$$

$$\Sigma F_{right} = \Sigma F_{left} \qquad (10.3)$$

$$\Sigma \text{ torques}_{clockwise} = \Sigma \text{ torques}_{counterclockwise} \qquad (10.4)$$

FIGURE 10.8
Center of gravity of a truck.

In working equilibrium problems it is imperative that *all* forces acting on the body in question be considered and *only* those forces. The first thing one must do is decide what body is to be dealt with. Then it is usually helpful to make a rough sketch of this body, representing all forces acting on it by suitably placed arrows. Next, one resolves all those forces which are neither horizontal nor vertical and replaces them with their components. At this point one is ready to apply the conditions for equilibrium. It is not necessary to choose vertical and horizontal axes for applying the conditions for equilibrium, but in most cases they are the most convenient choice.

□ **Example** It is found by weighing that the front wheels of a truck support 15,000 N and the rear wheels 27,000 N. If the wheelbase (distance between axles) is 6 m, find the weight of the truck and the location of the center of gravity (Fig. 10.8).

By Eq. (10.2)

$$15,000 \text{ N} + 27,000 \text{ N} = w$$

$$w = 42,000 \text{ N}$$

Let d be the distance from the front wheels to the center of gravity, and find torques about the front axle as axis. By Eq. (10.4)

$$wd = (27,000 \text{ N})(6 \text{ m})$$

$$(42,000 \text{ N})d = (27,000 \text{ N})(6 \text{ m})$$

$$d = 3.86 \text{ m} \qquad \qquad □$$

†These conditions are written for the case of forces acting in a plane. The extension to three-dimensional systems is not difficult; all we need do is (1) add to the first condition for equilibrium that the sum of all components perpendicular to the plane determined by the first two axes equals zero and (2) apply the torque condition to three noncoplanar axes.

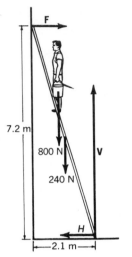

7.2 m

800 N

240 N

V

F

H

2.1 m

FIGURE 10.9
Equilibrium of a ladder leaning against a smooth wall.

□ **Example** A uniform ladder weighs 240 N and is 7.5 m long. It leans against a smooth (frictionless) wall at a point 7.2 m above a cement driveway with its base 2.1 m from the wall. An 800-N man stands six-tenths of the way up the ladder. (*a*) Find the horizontal and vertical components of the force exerted on the ladder by the driveway. (*b*) If the coefficient of static friction between ladder and driveway is 0.2, what fraction of the way to the top can the man climb before the ladder slips?

The object in equilibrium is the ladder, and the forces exerted on it are shown in Fig. 10.9. H and V are the horizontal and vertical components of the force exerted on the ladder by the driveway.

(*a*) By Eq. (10.2)

$$V = 800 \text{ N} + 240 \text{ N} = 1{,}040 \text{ N}$$

By Eq. (10.3)

$$H = F$$

If we choose the base of the ladder as axis, we obtain, by Eq. (10.4),

$$(7.2 \text{ m})F = (1.26 \text{ m})(800 \text{ N}) + (1.05 \text{ m})(240 \text{ N})$$

$$F = 175 \text{ N} \quad \text{and} \quad H = 175 \text{ N}$$

Note that we can use any axis for applying Eq. (10.4). It may be instructive to try other axes (such as one at the point where the ladder touches the wall or one through the center of the ladder) to assure yourself that we can indeed choose any axis.

(*b*) As the man climbs the ladder, F and H increase, but since H is provided by friction, it cannot exceed $0.2V = 208$ N. Let q represent the fraction of the length of the ladder which the man has climbed. When $H = 208$ N, F is also 208 N and applying Eq. (10.4) about the bottom of the ladder yields

$$(7.2 \text{ m})(208 \text{ N}) = (1.05 \text{ m})(240 \text{ N}) + (2.1q)(800) \text{ N-m}$$

from which $q = 0.74$. The ladder slips just before the man has climbed three-fourths of the way to the top. □

□ **Example** A derrick (Fig. 10.10) has a uniform boom 30 ft long which weighs 1 ton. A load of 3 tons is suspended from the end. The boom is hinged to a vertical mast and held up by a cable which makes an angle of 60° with the mast and which is fastened 10 ft from the end of the boom. If the boom makes an angle of 36.9° with the mast, find the tension in the cable and the force exerted on the boom by the hinge.

First, we resolve the tension into its horizontal and vertical components, which are $T \sin 60° = 0.866T$ and $T \cos 60° = 0.5T$, respectively. Next let H and V represent the components of the force exerted on the boom by the hinge. By Eq. (10.2)

$$V = 3 + 1 + 0.5T$$

By Eq. (10.3)

$$H = 0.866T$$

If we choose the hinge as the axis, Eq. (10.4) yields

$$(9 \times 1) + (0.5T \times 12) + (3 \times 18) = (0.866T \times 16)$$

$$T = 8.02 \text{ tons}$$

whence $V = 8.01$ tons and $H = 6.95$ tons.

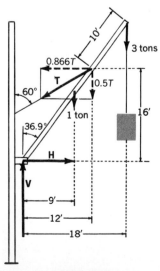

FIGURE 10.10
Equilibrium of a derrick boom.

10′

3 tons

0.866T

T

0.5T

60°

1 ton

16′

36.9°

H

V

9′

12′

18′

Note that we have replaced the force T with its components and then computed the torque due to each component. The resultant of the torques due to the components is equal to the torque due to the original force. □

10.5 TYPES OF EQUILIBRIUM

The equilibrium of a body may be *stable, unstable,* or *neutral* (Fig. 10.11). When a body returns to its original position after being slightly disturbed, the equilibrium is said to be *stable*. A cone standing on its base is an illustration of this type of equilibrium. When this cone is tilted slightly and released, it returns to its original position.

If the cone rests on its vertex, it can be in equilibrium only when its center of gravity lies directly above the vertex. If it is slightly displaced, the cone falls over; it is in *unstable* equilibrium. A body in unstable equilibrium does not return to the original equilibrium position when slightly displaced but moves farther away. Lastly, a billiard ball resting on a horizontal table is in *neutral* equilibrium. When it is slightly displaced, it neither returns to its former position nor moves farther away from the initial position. It remains in any position in which it finds itself. A cylinder or a cone lying on its side on a horizontal surface is also in neutral equilibrium.

The position of the center of gravity is of paramount importance in determining the stability of a body. The lower the center of gravity, the greater the stability of the body and the more difficult it is to overturn it. The body becomes unstable as soon as the vertical line through its center of gravity falls outside its base. The leaning tower of Pisa remains in stable equilibrium because, in spite of its leaning, the line of action of the weight falls inside the base. When an effort is made to turn a body over by pushing on it, the center of gravity rises for a body in stable equilibrium, descends for a body in unstable equilibrium, and remains at the same level for a body in neutral equilibrium.

10.6 EQUILIBRIUM OF THE HUMAN BODY

When a person is standing on level ground with feet together, the earth exerts a net upward supporting force F on the feet equal to the weight mg. In the absence of other forces this upward force must lie along a line passing through the center of gravity if the person is in equilibrium. In general, any body balanced on a plane surface which exerts a supporting force F can be in equilibrium if and only if its center of gravity is directly over the area of support. For a typical person standing erect, the center

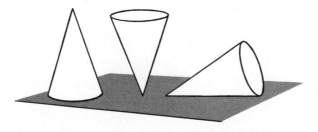

FIGURE 10.11
Three kinds of equilibrium are *stable* (left), *unstable* (center), and *neutral* (right).

of gravity is located near the top of the hipbone, and a vertical line through the center of gravity lies about 30 mm in front of the ankle joint. A complex neuromuscular control system must constantly act to keep the center of gravity properly positioned.

If you stand on your toes, your center of gravity must be moved forward to above the toes. The area of support is now much smaller, and keeping the center of gravity directly above it is more difficult. You can readily convince yourself that the center of gravity moves forward when you rise on your toes by performing a simple experiment. Place one foot on each side of a half-opened door and touch your nose to the edge of the door. Try to rise on your toes. You can't because the door does not allow you to move your center of gravity forward over the supporting toes.

As you walk, your center of gravity must be constantly shifted to a position over the supporting foot. This is accomplished by moving the entire body sideways, and it leads to a rocking motion which is readily observed. The rocking motion is much more prominent in the walk of birds. They typically have the center of gravity of the head, wings, and body below the hips, so they hang stably from their hips. For people the center of gravity of the head, arms, and body is well above the hips so that human equilibrium is precariously unstable.

FIGURE 10.12
In order to keep the center of gravity directly over the area of support, the legs and buttocks move backward as a person bends over. (*From A. H. Cromer, "Physics for the Life Sciences," copyright McGraw-Hill, Inc.; used with permission of McGraw-Hill Book Company.*)

Of course, a person is not a rigid body, and the location of the center of gravity depends on the body position. For example, if one keeps the knees stiff and bends over to touch the floor, the legs and buttocks move backward to balance the head, arms, and chest, which have moved forward. In this way the center of gravity remains over the area of support; it also moves downward and is ordinarily no longer within the body (Fig. 10.12).

10.7 THE STABILITY OF AIRCRAFT

An airplane flying horizontally at constant speed is in equilibrium under the influence of four generalized forces: the lift, the weight, the drag, and the thrust (Fig. 10.13). The weight of the airplane may, of course, be regarded as being concentrated at the center of gravity. The lift forces, which arise from pressure differences on the wing, fuselage, and tail sections of the aircraft, may be replaced in a similar way by a resultant lift vector. The friction resulting from the passage of air over the surface of the aircraft gives rise to a net retarding force known as the *drag*, while the aircraft engine, whether propeller or jet, provides *thrust* to overcome this drag.

For the aircraft to be longitudinally stable, the resultant of the torques about a transverse axis through the center of gravity must be zero. The airplane must be stable both longitudinally and laterally. Note that the conditions for equilibrium must be satisfied over a wide range of speeds and loadings. This can be accomplished by moving aileron, wing, or elevator surfaces in such a way as to keep the lift equal to the weight and the center of lift appropriately placed relative to the center of gravity. When the speed of an aircraft approaches the speed of sound, there may be radical changes in the way air flows over the aircraft surfaces. Such changes introduce sudden shifts in the position of the center of lift and thereby introduce large and sudden torques on the aircraft. For this and other reasons, it is dangerous to fly any type of aircraft above a maximum speed which depends upon the design.

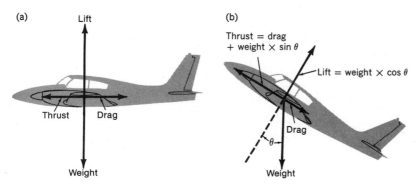

(a) Lift — Thrust — Drag — Weight

(b) Thrust = drag + weight × sin θ — Lift = weight × cos θ — Drag — θ — Weight

FIGURE 10.13
An aircraft in stable equilibrium; the velocity must be constant.

QUESTIONS

1. A Marine officer is standing at attention. Is he in stable, neutral, or unstable equilibrium? Explain.
2. Where are the positions of stable, neutral, and unstable equilibrium for a right circular cylinder?
3. How does pushing downward on the pedal of a bicycle result in the production of a horizontal force to push the bicycle forward?
4. Why does a person lean forward when pushing a heavy cart?
5. Discuss the stable, neutral, and unstable equilibria of a body in terms of potential energy.
6. Can a body be in equilibrium under the influence of three forces which are neither concurrent nor parallel?
7. How could a body estimated to weigh 10 lb be weighed using a meterstick and a spring balance calibrated to read only up to 4 lb?
8. Why is a wagon loaded with hay more likely to turn over on a hillside than one loaded with an equal weight of gravel?
9. Under what conditions does the center of mass of a body not coincide with its center of gravity? Suppose that the moon were to move slowly toward the earth until it came to rest on the earth's surface. Would the center of mass of the moon coincide with its center of gravity? Why not? Which would be closer to the earth?
10. Many cathedrals have large outside buttresses built against their masonry walls. What is the purpose of the buttresses? Why don't ordinary brick houses need buttresses?
11. A hollow sphere filled with sand serves as a pendulum bob. If there is a small hole in the bottom of the sphere so that sand runs out as the bob swings, it is observed that the period of the pendulum increases at first and then decreases. Why? *Hint:* Where is the center of gravity of the sand-sphere pendulum bob?
12. Is a ladder resting on a sidewalk and leaning against a wall any more likely to slip as a person climbs toward the top? Explain.
13. In the prototypes of the Concorde supersonic aircraft provision was made to pump 12,000 lb of fuel to a rear fuselage tank to give the desired handling and control characteristics. What does this fuel transfer do to the center of gravity? What does this provision imply about the net lift vector at supersonic speed compared with the same vector at subsonic speed?
14. Under what circumstances is it *not* necessary to consider torque in an equilibrium problem? If one does need to calculate torques, what considerations govern the choice of the axis about which torques are to be calculated?
15. Consider the compressive forces in the backbone and the tension in the muscles of the back when one bends over to lift a heavy load. Why is a back injury much more likely to occur if you bend your back to pick up a heavy load than if you bend your knees, keeping your back nearly vertical?
16. A brick rests on a horizontal sidewalk. An identical brick is placed on top of the first with three-tenths of its length overhanging. Can a third brick be added with three-tenths of its length overhanging the second? What is the limiting number which can be assembled in equilibrium in this way?
17. One end of a meterstick (or rod) rests on a pencil held by one hand and the other end on a horizontal eraser held by the other hand. If the two hands are slowly moved closer together until they touch, will the meterstick eventually roll off? Explain why it will not, even though the coefficients of friction are different?

PROBLEMS

Draw an appropriate sketch for each problem.

1. When only the front wheels of an automobile are run onto a platform scale, the scale balances at 8 kN; when only the rear wheels are run onto the scale, it balances at 6 kN. What is the weight of the automobile and how far is its center of gravity behind the front axle? The distance between axles is 2.8 m. *Ans.* 14 kN; 1.2 m
2. Weights of 8, 5, 3, and 10 N are located, respectively, at 25-, 50-, 75-, and 100-cm marks on a meterstick whose weight is negligible. What are the magnitude and location of the single upward force which will balance the system?
3. A tapered pole, which is 5 m long and weighs 400 N, can be balanced at a point 2 m from the thicker end. If it were to be supported at its ends, how much force would be needed at each end? *Ans.* 240 N, 160 N
4. A telegraph pole is placed on a two-wheeled dolly located 4 m from the thicker end, and an upward force of 1,500 N at the thinner end is required to keep it horizontal. The pole is 15 m long and weighs 9,000 N. Where is the center of gravity?
5. A 60-lb girl and an 80-lb boy sit at the ends of a

15-ft seesaw. At what point should a 50-lb boy sit for the seesaw to be in equilibrium?

Ans. 3 ft from fulcrum

6. A bar of uniform cross section is carried by a boy and his younger sister, one at either end of the bar. If the bar weighs 30 lb and is 10 ft long, where must a load of 50 lb be hung from the bar so that the boy will carry twice as much as his sister?

7. A 600-N bricklayer is 1.5 m from one end of a uniform 800-N scaffold which is 7 m long. A pile of bricks weighing 500 N is 3 m from the same end. If the scaffold is supported at the two ends, calculate the force on each support.

Ans. 1,157 and 743 N

8. A tapered pole 10 ft long weighs 40 lb. It balances at its midpoint when a 9-lb weight hangs from the slimmer end. Find the position of the center of gravity of the pole.

9. A 90-kg man stands on the balls of both feet. His shinbones exert downward forces at the ankle joints roughly 10 cm back of the balls of the feet, while the Achilles tendons pull (almost) upward 4 cm farther back. Using the crude approximation that the forces involved are all vertical, find the force exerted by each Achilles tendon and that exerted by each shinbone.

Ans. 1,544 N; 1,103 N

10. A uniform beam 4 m long weighs 300 N and is supported at its ends by two walls. Find the reaction of the walls against the beam when a man weighing 750 N stands on the beam at a distance of 1.2 m from one end.

11. A 60-lb uniform beam 12 ft long is supported at its ends by two walls. Find the forces exerted on the supports when a 180-lb man stands on the beam 4 ft from one end.   *Ans.* 150 lb; 90 lb

12. A uniform bar 2 yd long has a weight of 9 lb fastened to it at one end and a weight of 8 lb at the other end. The bar itself weighs 7 lb. Where could a single force be applied to balance the system, and how great would the force have to be?

13. The head, arms, and upper torso supported by the fifth lumbar vertebra have a mass of 60 kg [see (*a*) of the accompanying figure]. The center of gravity of this mass lies 80 mm in front of this vertebra. The resulting torque is balanced by tension in the back muscles; they have a lever arm of 50 mm about the vertebra, which serves as fulcrum. Find the compressional force *C* on the vertebra and the tension *M* in the muscles of the back.   *Ans.* 1,529 N; 941 N

14. In an obese man [see (*b*) of accompanying figure] the mass supported by the fifth lumbar vertebra is 100 kg, and its center of gravity is 150 mm in front of the vertebra. The lever arm for the back

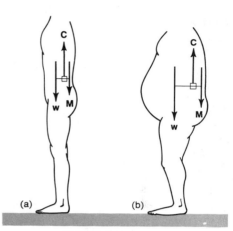

PROBLEMS 13 AND 14

muscles is 60 mm. Find the compressional force on the vertebra and the tension in the back muscles. Compare your answers with those of Prob. 13.

15. A uniform horizontal bar 2 m long weighs 250 N. Upward forces of 150 and 100 N are exerted at the ends. Is the bar in equilibrium? Calculate the torque about each end, about the center, and about a point 1.5 m from one end of the bar.

Ans. No; 50 m-N in each case

16. A uniform rod 120 cm long has a mass of 1.2 kg. It has a mass of 5 kg attached to it at one end, a mass of 4 kg at the other end, and a mass of 2 kg in the middle. Find the position of the center of gravity.

17. A wheel of 0.25 m diameter has an axle of 3 cm radius. If a force of 300 N is exerted along the rim of the wheel, what is the smallest force exerted on the outside of the axle which will result in zero net torque?   *Ans.* 1,250 N

18. A square is acted upon by the forces of 4, 9, 7, and 10 N, respectively, along the four sides. The forces all act to produce rotation in the same direction. If the length of a side of the square is 0.5 m, what is the resultant torque acting to rotate the square about an axis through its center?

19. A cubical box 1.5 m on a side weighs 800 N and rests on a rough floor for which the coefficient of static friction is 0.80. If the center of gravity of the box is at its center, what is the greatest height at which a horizontal force can be applied to slide the box without tipping it? What is the

lowest height at which a horizontal 500-N force can be applied to tip the box?

Ans. 0.938 m; 1.2 m

20. If weights A, B, C, D, and E are, respectively, located at distances a, b, c, d, and e from one end of a light bar, use the two conditions for equilibrium to show that the center of gravity is located a distance y from that end of the bar, where

$$y = \frac{Aa + Bb + Cc + Dd + Ee}{A + B + C + D + E}$$

21. A uniform ladder 25 ft long weighs 48 lb. It leans against a vertical frictionless wall with its lower end 7 ft from the wall and its upper end 24 ft from the ground. Find the force exerted on the ladder by the wall and the horizontal and vertical components of the force exerted on the ladder by the ground. *Ans.* 7, 7, and 48 lb

22. Two vehicles are crossing a bridge 20 m long. A passenger car weighing 12,000 N is 6 m from one end. A truck weighing 40,000 N is 7 m from the opposite end. If the bridge is symmetrical with respect to the center and weighs 1.2 MN, what are the forces on the two supports at the ends of the bridge?

23. The horizontal bar AB in the accompanying figure is supported by the cord BC at one end and by a pin at the wall AC at the other end. The bar is of uniform cross section, weighs 120 N, and is 2 m long. If θ is 40°, what are the tension in the cord and the horizontal and vertical components of the force exerted on the bar by the pin at A?

Ans. 93.3 N; 71.5 N; 60 N

24. In the accompanying figure AB represents a uniform sign 12 ft long and weighing 500 lb. If a

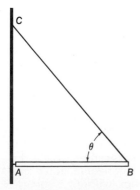

PROBLEMS 23 TO 26

180-lb man stands 4 ft from a pin at A and θ is 36.9°, find the tension in the cable and the vertical and horizontal components of the force exerted on the sign by the pin.

25. The 500-N horizontal bar AB in the accompanying figure is nonuniform, being 4 m long with its center of gravity 1.6 m from the wall. A 600-N weight is suspended on the bar 2.5 m from the wall. The angle θ is 53.1°. What is the tension in the cord BC? Find the horizontal and vertical components of the force exerted on the bar by the pin at A. *Ans.* 719 N; 431 N; 525 N

26. The horizontal bar AB in the accompanying figure is nonuniform, being 5 m long with its center of gravity 2 m from the wall. The bar weighs 300 N. A 900-N weight is suspended at the end B of the bar. What is the tension in the supporting cord if θ is 30°? What are the horizontal and vertical components of the force exerted on the bar by the pin at A?

27. A ladder is 8 m long, has its center of gravity 3 m from the bottom, and weighs 240 N. A girl weighing 400 N stands halfway up the ladder, which makes an angle of 20° with the vertical. Find the force exerted on the ladder by the smooth wall and the horizontal and vertical components of the force exerted on the ladder by the ground. *Ans.* 106 N; 106 N; 640 N

28. A uniform ladder 5 m long weighs 200 N. It leans against a smooth wall, making an angle of 60° with the horizontal ground. How far up the ladder can a man weighing 700 N climb before the ladder slips if the maximum frictional force the ground can provide is 400 N?

29. A nonuniform ladder is 10 m long. It weighs 200 N, and its center of gravity is 4.5 m from the bottom. The ladder leans against a smooth wall with its base on a sidewalk 6 m from the base of the wall. Find the force exerted on the ladder by the wall and the horizontal and vertical components of the force exerted on the ladder by the sidewalk. If the coefficient of static friction between ladder and sidewalk is 0.4, at what angle would the ladder be on the verge of slipping?

Ans. 67.5 N; 67.5 N; 200 N; 48.4°

30. A uniform 20-ft ladder weighing 40 lb rests against the smooth side of a house so that it makes an angle of 70° with the ground. A 180-lb man stands three-quarters of the way to the top. What is the horizontal force exerted on the ladder by the house? What are the horizontal and vertical components of the force exerted on the ladder by the ground?

31. A uniform 26-ft ladder weighing 52 lb leans against a smooth wall with its base 10 ft from the wall. How far up the ladder can a 156-lb woman climb if the maximum horizontal force available at the base is 60 lb? *Ans.* 19.7 ft

32. A horizontal uniform steel beam weighs 30,000 N and is 6 m long. It is supported at a vertical wall by a pin at one end and by a cable making an angle of 53.1° with the horizontal, attached 1 m from the other end and running to a clamping point on the wall. A load of 80,000 N is suspended from the end of the beam away from the wall. Find the tension in the cable and the horizontal and vertical components of the force exerted on the bar by the pin. What is the direction of the vertical component?

33. A light triangular frame ABC held together by pins (see accompanying figure) is equilateral with 1.6-m sides. If w is 250 N, find the tension in member AC and the compressional force exerted by member AB.　　　　*Ans.* 72 N; 144 N

34. A simple light triangular frame ABC (see accompanying figure) is held together by pins. If sides AB and BC are both 4 ft long and AC is 6.4 ft, find the horizontal and vertical components of the force exerted on the pin at A by member BC when w is 60 lb. What is the tension in member AC?

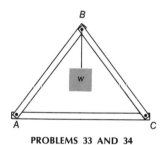

PROBLEMS 33 AND 34

35. If the load hanging from the end of the boom of Fig. 10.10 were increased from 3 to 5 tons, what would the new magnitudes of H, V, and T be?
　　　　Ans. 10.9, 12.3, and 12.6 tons

36. To a thin circular disk with radius 15 cm there is attached another circular disk of the same material with radius 5 cm. Find the center of gravity of the combination if the smaller disk has its center 6 cm from the center of the larger disk.

37. From a circular disk with a radius of 250 mm a circular hole with a radius of 75 mm is cut. Find the center of gravity of the remainder of the disk if the hole has one point of its circumference at the center of the disk.
　　　　Ans. 7.42 mm from center

38. From a circular disk of 0.5 m diameter a circle of 150 mm radius is cut out. Find the center of gravity of the remainder of the disk if the circumference of the hole passes through the center of the disk.

39. A 240-N door 2.5 m high and 1.2 m wide is hung by two hinges, each of which supports half the

weight. If the hinges are 0.3 m from the top and 0.3 m from the bottom of the door, find the horizontal components of the forces exerted on the door by the hinges. Assume that the center of gravity of the door is at the geometrical center.
　　　　Ans. 75.8 N for both hinges

40. A uniform 200-N gate is 1.6 m high and 1.2 m wide. It is supported by two hinges, the upper of which is 0.1 m from the top and bears three-fifths the weight. The lower hinge is 0.1 m from the bottom. Find the horizontal and vertical components of the force exerted on the gate by the lower hinge.

41. A derrick similar to that of part (c) of the figure for Probs. 52 and 53 in Chap. 5 has a uniform boom 10 m long which weighs 9,000 N. A load w of 20,000 N is suspended from one end. Find the tension in the cable and the horizontal and vertical components of the force exerted on the boom by the pin at the base.
　　　　Ans. 52,500 N; 42,000 N; 60,500 N

42. A derrick similar to that of part (a) in the figure for Probs. 52 and 53 in Chap. 5 has a uniform boom 20 ft long which weighs 2,500 lb. A load w of 6,000 lb is suspended from the end. Find the tension in the cable and the horizontal and vertical components of the force exerted on the boom by the pin at the base.

43. A stepladder is 2 m high. The step side is 2.5 m long, uniform, and weighs 60 N. The other side is also uniform, is 2.5 m long, and weighs 20 N. The sides are hinged at the top, and there is a massless tie rod 1.5 m long which connects the midpoints of the sides. A 640-N man stands three-fourths the way to the top. Find the forces exerted on the two sides of the ladder by the floor and the tension in the tie rod. *Hint:* Assume the floor is smooth initially; for the tension apply the equilibrium conditions to one or both sides.
　　　　Ans. 450 N; 270 N; 390 N

44. Prove that a uniform cube of edge L will be in stable equilibrium if it is placed on top of a rough cylindrical pipe of radius R provided the diameter of the pipe is greater than L.

45. A homogeneous cylinder of radius R and mass m slides along a table without rotating when acted upon by a horizontal force **F** applied at an appropriate height y. (a) If μ_k is the coefficient of sliding friction, find y. (b) Prove that the cylinder is not in equilibrium unless $y = 0$ and then only if $F = \mu_k mg$.　　　　*Ans.* (a) $y = R(1 - \mu_k mg/F)$

11

ROTATIONAL MOTION

In Chap. 10 we found that for a body at rest both the resultant of all forces and the resultant of all torques acting on the body are zero. In Chaps. 4 and 5 we considered the translational motion of a body subject to a resultant force and found that the body undergoes an acceleration proportional to the force and inversely proportional to the mass. Now we turn to a discussion of rotational motion and develop relations closely analogous to those of translational motion.

11.1 ANGULAR VELOCITY

Rotational motion is common in our everyday life. The flywheel of a stationary engine, the armature of an electric motor, and the spinning top all perform rotational motion. The wheel of a moving automobile describes rotational and translational motions simultaneously. In general, the motion of any rigid moving body can be regarded as a translation of the center of mass and a simultaneous rotation about the center of mass.

Consider a flywheel (Fig. 11.1) rotating about the axis O. A straight line at OA on the flywheel rotates from OA to OB in time t. The rate at which the line rotates, called its *angular velocity*, is usually expressed in radians per second, although it can also be measured in other units involving an angle divided by a time, such as revolutions per minute (r/min) or degrees per second. If the rate of rotation is constant, the angular velocity is equal to the angle turned through divided by the corresponding time. If we represent the angular velocity by ω (the Greek letter omega), the angular displacement by θ, and the time by t, we have

$$\omega = \frac{\theta}{t} \tag{11.1}$$

The angular velocity ω is a vector quantity with its direction along the axis of rotation in the sense indicated by the outstretched thumb of the right hand when the fingers point in the direction of rotation.

If we consider the linear velocity of the point A on the flywheel, we observe that θ, the angle swept through in radians, is equal to the length of arc AB divided by the radius r. The distance s traveled by the point is just the arc length AB, and

$$s = \theta r \tag{11.2}$$

The linear velocity v of the point is given by

$$v = \frac{s}{t} = \frac{\theta r}{t} = \omega r \tag{11.3}$$

For a rigid body the linear velocity of a point increases as the distance from the axis of rotation increases. Note that Eqs. (11.2) and (11.3) are valid *only* if angles are measured in radians.

□ **Example** The spoke of a wheel makes 60 r/min. It is 500 mm long. Find the linear velocities of a point on its outer end and of a point halfway between the axis and the outer end.

$$v_e = \omega r_e = (2\pi \text{ rad/s})(500 \text{ mm}) = 3{,}140 \text{ mm/s}$$
$$v_m = \omega r_m = (2\pi)(250) = 1{,}570 \text{ mm/s} \qquad \square$$

11.2 ANGULAR ACCELERATION

In situations in which rotating bodies are speeding up or slowing down, the angular velocity is no longer constant. The *instantaneous angular velocity* ω is then defined by the relation

$$\omega = \lim_{\Delta t \to 0} \frac{\Delta \theta}{\Delta t} \qquad (11.1a)$$

The instantaneous angular acceleration α (Greek letter alpha) of a rotating object is defined as the rate of change of angular velocity:

$$\alpha = \lim_{\Delta t \to 0} \frac{\Delta \omega}{\Delta t} \qquad (11.4)$$

When the angular acceleration is constant, its instantaneous value is equal to its average value, given by

$$\alpha_{av} = \frac{\Delta \omega}{\Delta t} = \frac{\omega - \omega_0}{t} \qquad (11.4a)$$

where ω and ω_0 are the angular velocities of the body at the times t and 0.

Any point in a rotationally accelerated body has a linear acceleration a_t tangential to its path and given by

$$a_t = \alpha r$$

if we measure α in radians per second per second. This equation gives only the tangential component of the total linear acceleration of the point. The point is moving in a circle and has also a centripetal acceleration a_c equal to v^2/r.

☐ **Example** At a certain instant the angular velocity of a wheel was 10 rad/s. In 20 s the angular velocity has become 50 rad/s. What is the average angular acceleration?

$$\text{Angular acceleration} = \frac{\text{change in angular velocity}}{\text{time}} = \frac{\Delta \omega}{\Delta t}$$

$$\alpha_{av} = \frac{50 \text{ rad/s} - 10 \text{ rad/s}}{20 \text{ s}}$$

$$= 2 \text{ rad/s}^2 \qquad \square$$

☐ **Example** What are the tangential, centripetal, and total linear accelerations of a particle that is 0.5 m from the axis of a wheel (Fig. 11.2) when the angular acceleration of the wheel is 3 rad/s² and ω is 2 rad/s?

$$\text{Tangential linear acceleration} = \alpha r$$

$$a_t = (3 \text{ rad/s}^2)(0.5 \text{ m}) = 1.5 \text{ m/s}^2$$

$$\text{Centripetal linear acceleration} = \frac{v^2}{r} = \omega^2 r$$

$$a_c = (2 \text{ rad/s})^2(0.5 \text{ m}) = 2.0 \text{ m/s}^2$$

$$\text{Total linear acceleration} = \sqrt{a_t^2 + a_c^2} = 2.5 \text{ m/s}^2 \qquad \square$$

11.3 THE EQUATIONS OF ANGULAR MOTION

The equations of angular motion have the same form as the corresponding equations of translational motion because the angular quantities are

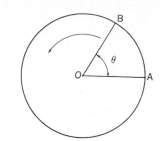

FIGURE 11.1
A flywheel rotating through an angle θ about an axis through O carries line OA to OB.

$\omega = 2$ rad/s; $\alpha = 3$ rad/s²

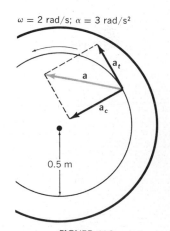

FIGURE 11.2
The linear acceleration **a** is the resultant of the tangential acceleration $\mathbf{a}_t$ and the centripetal acceleration $\mathbf{a}_c$.

defined analogously to the corresponding linear quantities. In the following paragraphs we treat several of the more common special cases.

Case 1 If a body rotates with uniform velocity ω, the angle θ through which it turns in t seconds is

$$\theta = \omega t \tag{11.1b}$$

☐ **Example** An electric motor rotates at a constant angular velocity of 1,800 r/min. Find the angle turned through in 10 s.

$$\omega = 1{,}800 \text{ r/min} \times \frac{1 \text{ min}}{60 \text{ s}} = 30 \text{ r/s}$$

$$= (30 \text{ r/s})(2\pi \text{ rad/r}) = 60\pi \text{ rad/s}$$

$$\theta = \omega t = 600\pi = 1{,}885 \text{ rad} \qquad\qquad ☐$$

Case 2 If a body begins to rotate from rest with a uniform angular acceleration α, the angular velocity ω at the end of t seconds is αt by Eq. (11.4a). During the acceleration the average angular velocity is given by

$$\omega_{av} = \frac{\omega_0 + \omega}{2} = \frac{0 + \alpha t}{2} = \tfrac{1}{2}\alpha t$$

The angle θ swept out in t seconds is the average angular velocity times the time, so that

$$\theta = \tfrac{1}{2}\alpha t^2$$

☐ **Example** A flywheel starts from rest and has a uniform angular acceleration of 4 rad/s² for a time of 10 s. Find the final angular velocity, the average angular velocity during the 10-s acceleration, and the angle turned through in the 10 s.

$$\omega = \alpha t = (4 \text{ rad/s}^2)(10 \text{ s})$$
$$= 40 \text{ rad/s}$$

$$\omega_{av} = \frac{\omega_0 + \omega}{2} = \frac{0 + 40}{2} = 20 \text{ rad/s}$$

$$\theta = \tfrac{1}{2}\alpha t^2 = \tfrac{1}{2}(4 \text{ rad/s}^2)(100 \text{ s}^2)$$
$$= 200 \text{ rad} \qquad\qquad ☐$$

Case 3 If a body has an initial angular velocity ω_0 and a uniform angular acceleration α, its angular velocity at the end of t seconds is given by

$$\omega = \omega_0 + \alpha t \tag{11.4b}$$

If the acceleration is uniform, the average angular velocity is

$$\omega_{av} = \frac{\omega_0 + \omega}{2} = \omega_0 + \frac{\alpha t}{2}$$

The angle θ swept out is given by

$$\theta = \omega_{av}t = \omega_0 t + \tfrac{1}{2}\alpha t^2 \tag{11.5}$$

If we eliminate t between Eqs. (11.4) and (11.5), we obtain

$$\omega^2 = \omega_0^2 + 2\alpha\theta \tag{11.6}$$

11.4 KINETIC ENERGY OF ROTATION

A rotating body has kinetic energy by virtue of the motion of its parts. To find the expression for kinetic energy of rotation, consider Fig. 11.3, which represents a rigid body rotating with an angular velocity ω about an axis through O perpendicular to the plane of the paper. For the particle at A having a mass of m_1 and linear velocity v_1, the kinetic energy is $m_1 v_1^2/2$. Since $v_1 = \omega r_1$, the kinetic energy of the particle at A becomes

$$\frac{m_1 r_1^2 \omega^2}{2} = m_1 r_1^2 \frac{\omega^2}{2}$$

For other particles m_2, m_3, m_4, . . . moving with velocities v_2, v_3, v_4, . . . similar expressions are found. The total kinetic energy is

$$\frac{m_1 r_1^2 \omega^2}{2} + \frac{m_2 r_2^2 \omega^2}{2} + \cdots = \frac{\omega^2}{2} (m_1 r_1^2 + m_2 r_2^2 + \cdots)$$

$$= \tfrac{1}{2}\omega^2 \sum_i m_i r_i^2$$

The quantity $\sum m_i r_i^2$ is called the *moment of inertia I* of the body about the axis through O. For a rotating body,

$$\text{Kinetic energy} = \tfrac{1}{2}I\omega^2 \qquad (11.7)$$

This equation is the rotational analog of $\tfrac{1}{2}mv^2$ for the kinetic energy of a mass m moving with a translational velocity v. Note that ω must be expressed in radians per second.

11.5 MOMENT OF INERTIA

To obtain the moment of inertia of a rigid body about any axis, we may regard the body as composed of a large number of small mass elements. If we multiply the mass of each element by the square of its distance from the axis and add the contributions of all the elements, we obtain the moment of inertia I:

$$I = m_1 r_1^2 + m_2 r_2^2 + m_3 r_3^2 + \cdots = \sum_i m_i r_i^2 \qquad (11.8)$$

Such a summing process for bodies of most shapes involves detailed numerical calculation or the application of integral calculus. However, we can compute the moment of inertia of a simple ring of mass m and radius r about its central axis by direct application of Eq. (11.8). Since all parts of the ring are essentially at the same distance R from the axis, the moment of inertia is $I = MR^2$. In Fig. 11.4 moments of inertia are given for bodies of several simple shapes—in every case for an axis through the center of mass as indicated. The moment of inertia may be expressed in kg-m², slug-ft², or in terms of the product of any mass unit multiplied by the square of a unit of length.

If we know the moment of inertia about an axis through the center of mass, the *parallel-axis theorem* states that the moment of inertia about any axis parallel to the one through the center of mass is given by

$$I = I_0 + Md^2 \qquad (11.9)$$

where M is the mass, I_0 the moment of inertia about an axis through the

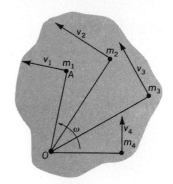

FIGURE 11.3
The kinetic energy of rotation of a body is the sum of the kinetic energies of all the particles (such as m_1, m_2, . . .) of which the body is composed; this sum is $\tfrac{1}{2}I\omega^2$.

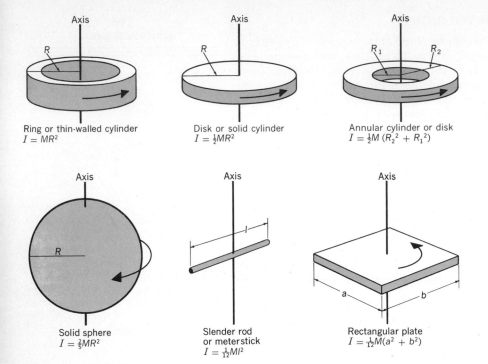

Ring or thin-walled cylinder
$I = MR^2$

Disk or solid cylinder
$I = \frac{1}{2}MR^2$

Annular cylinder or disk
$I = \frac{1}{2}M\,(R_2{}^2 + R_1{}^2)$

Solid sphere
$I = \frac{2}{5}MR^2$

Slender rod
or meterstick
$I = \frac{1}{12}Ml^2$

Rectangular plate
$I = \frac{1}{12}M(a^2 + b^2)$

FIGURE 11.4
Moments of inertia of common shapes about the indicated axes through the centers of mass.

center of mass, and d the distance from this axis to the *parallel* one about which the moment of inertia is I.

It is sometimes convenient to write the moment of inertia of an extended rigid object in the form $I = Mk^2$, where k is a constant called the *radius of gyration*. Clearly, k corresponds to the radius of a thin circular ring which has the same mass and the same rotational inertia as the rigid body in question.

□ **Example** Find the moment of inertia of a thin rod of mass 2 kg and length 1.6 m about an axis through its center. Also find the moment of inertia about an axis at one end.

For an axis through the center of mass,

$$I_0 = \frac{Ml^2}{12} = \frac{(2 \text{ kg})(1.6 \text{ m})^2}{12}$$

$$= 0.427 \text{ kg-m}^2$$

For an axis at one end,

$$I = I_0 + Md^2 = 0.427 \text{ kg-m}^2 + (2 \text{ kg})(0.8 \text{ m})^2 = 1.71 \text{ kg-m}^2 \qquad □$$

□ **Example** Find the moment of inertia of a 10-kg mass and a 15-kg mass about an axis of rotation which is 0.3 m from the 10-kg mass and 0.4 m from the 15-kg mass. Find the radius of gyration for this system.

$$I = m_1 r_1{}^2 + m_2 r_2{}^2$$
$$= (10 \text{ kg})(0.3 \text{ m})^2 + (15 \text{ kg})(0.4 \text{ m})^2$$
$$= 3.3 \text{ kg-m}^2$$

$$I = mk^2$$
$$3.3 \text{ kg-m}^2 = (25 \text{ kg})(k^2)$$
$$k = 0.363 \text{ m}$$

□

11.6 COMBINATION OF ENERGY OF TRANSLATION AND ENERGY OF ROTATION

Consider a solid cylinder and a ring which have identical radii R and masses M. When both are placed at the top of an inclined plane (Fig. 11.5), they have identical potential energies Mgh. If the cylinder and ring are allowed to roll down the plane starting from rest, which will reach the bottom first?

To answer this question, we apply the law of conservation of energy. At the bottom of the inclined plane both the cylinder and the ring have kinetic energy equal to the original potential energy. For each, the kinetic energy is associated partly with the translational velocity of the center of mass and partly with rotation about the center of mass. Let subscript c represent the cylinder, and subscript r the ring. Then

$$Mgh = \tfrac{1}{2}Mv_c{}^2 + \tfrac{1}{2}I_c\omega_c{}^2 \quad \text{and} \quad Mgh = \tfrac{1}{2}Mv_r{}^2 + \tfrac{1}{2}I_r\omega_r{}^2$$

For a cylinder of mass M and radius R, $I = MR^2/2$, while for the ring $I = MR^2$. Further, $\omega^2 = v^2/R^2$. Therefore, for the cylinder

$$Mgh = \frac{Mv_c{}^2}{2} + \frac{Mv_c{}^2}{4}$$

or
$$v_c{}^2 = \frac{4gh}{3} \tag{A}$$

For the ring

$$Mgh = \frac{Mv_r{}^2}{2} + \frac{Mv_r{}^2}{2}$$

or
$$v_r{}^2 = gh \tag{B}$$

The object with the smaller moment of inertia has the greater linear velocity and a smaller fraction of its energy associated with its rotational motion. Further, neither the mass nor the radius appears in Eqs. (A) and (B). Therefore the restrictions that both cylinder and ring have the same radius and mass were quite unnecessary; indeed, Eq. (A) is valid for any homogeneous disk or cylinder and Eq. (B) for any thin ring.

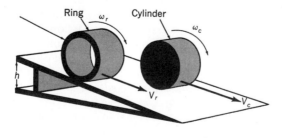

FIGURE 11.5
When a solid cylinder and a ring are released together at the top of an inclined plane, the solid cylinder reaches the bottom first, whether or not they have the same mass and outside diameter.

11.7 NEWTON'S LAWS FOR ROTATIONAL MOTION

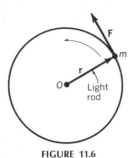

FIGURE 11.6
The angular acceleration is proportional to the unbalanced torque and inversely proportional to the moment of inertia.

Experiments show that the opposition of a body to being set in translation is proportional to the mass of the body and does not depend on the distribution of this mass. We now show for one simple situation that the opposition which a rigid body offers to being set in rotational motion about any fixed axis depends on the moment of inertia about that axis. Consider a small mass m fastened by a light rod to point O, about which the mass m is to be set in rotation (Fig. 11.6). Let us exert on this mass a force F tangential to the circle. The torque τ exerted by force F about the axis through O is Fr, and, since $F = ma$, $\tau = mar$ and

$$\tau = mr^2\alpha = I\alpha$$

In general, to change the angular velocity of a rotating rigid body, we must apply an unbalanced torque, just as we must apply an unbalanced force to change the linear velocity of a body. The angular acceleration is directly proportional to the torque applied:

$$\tau = I\alpha \tag{11.10}$$

where τ is the unbalanced torque acting, α the angular acceleration, and I the *moment of inertia* (or the *rotary inertia*) of the body. The moment of inertia, a measure of the opposition of the body to being set into rotation, is analogous to the mass of a body, which is a measure of the opposition of the body to being set into translational motion.

Equation (11.10) expresses Newton's second law for rotary motion, which we have developed from $F = ma$. *If an unbalanced torque acts about an axis fixed in a rigid body, an angular acceleration is produced which is proportional to the unbalanced torque and inversely proportional to the moment of inertia of the body about that axis.* The angular acceleration is in the direction of the unbalanced torque. As we saw in Chap. 10, the direction of a torque is that in which a right-handed screw moves when acted upon by a torque about its axis. Angular acceleration and angular velocity (but not angular displacement) are vector quantities with directions determined just as for torque (Fig. 11.7).

☐ **Example** A flywheel has a moment of inertia of 300 kg-m² about its fixed axis. Find the torque necessary to produce an angular acceleration of 3 rad/s².

Torque = moment of inertia × angular acceleration

$$\tau = (300 \text{ kg-m}^2)(3 \text{ rad/s}^2)$$
$$= 900 \text{ N-m} \qquad ☐$$

FIGURE 11.7
Angular velocity and angular acceleration are vector quantities and may be represented by arrows along the axis of rotation. The direction can be found by using the right-hand rule.

If we raise the front wheel of a bicycle off the ground and set it spinning, its angular velocity gradually decreases until the wheel comes to rest. Torques due to bearing friction and air friction produce the deceleration. If we reduce the friction, the wheel spins longer; if we could eliminate friction, the wheel would spin indefinitely with constant angular velocity. This is a direct analog of Newton's first law for translational motion and is known as Newton's first law of rotational motion. *A rigid body rotating about a fixed axis continues with constant angular velocity unless acted upon by some unbalanced external torque.* The rotation of the earth on its axis is a reasonable example, although the earth is not quite a rigid body and does not have an exactly fixed axis, and there are some unbalanced frictional torques introduced by tides and other considerations.

Newton's third law of rotational motion states: *For every torque acting on one body there is an equal and opposite torque about the same axis acting upon some other body (or bodies).*

11.8 ANGULAR HARMONIC MOTION

When a heavy disk is suspended by a wire (Fig. 11.8) and rotated through an angle θ from its equilibrium position, there is a restoring torque τ_r which is proportional to the angular displacement:

$$\tau_r = -K\theta \tag{11.11}$$

where K is the *torsion constant* of the wire. By Eq. (11.10), $\tau = I\alpha$. Consequently, for this *torsion pendulum,*

$$I\alpha = -K\theta \tag{11.12}$$

the angular acceleration is proportional to the angular displacement, and angular simple harmonic motion is produced. Equations (10.4a) to (10.11) are applicable to angular harmonic motion if angular displacement, angular amplitude, angular velocity, and angular acceleration are substituted for the corresponding linear quantities and moment of inertia, torque, and torsion constant replace mass, force, and spring constant, respectively. Thus by the analog of Eq. (10.10) for angular harmonic motion

$$\frac{1}{\nu} = T = 2\pi \sqrt{\frac{I}{K}} \tag{11.13}$$

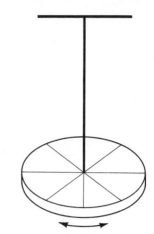

FIGURE 11.8
A torsion pendulum describes angular harmonic motion.

We have seen that for many of the equations which govern linear motion there are analogous equations in rotary motion. *These equations are valid when the angles are expressed in radians.* It is unnecessary to memorize many of the equations of angular motion if use is made of the analogy between linear and rotational motions. Table 11.1 lists corresponding quantities in the two kinds of motion. Knowing a relationship between linear quantities, one can often write the appropriate relationship between angular quantities by simply replacing each linear quantity with the corresponding angular one. For example, in linear motion, work = force × displacement. The corresponding expression for angular motion is work = torque × angular displacement. The angular analog of $P = Fv$ is $P = \tau\omega$.

TABLE 11.1
Linear Quantities and Their Angular Analogs

Linear		Angular	
Displacement	s	Displacement	θ
Velocity	v	Angular velocity	ω
Acceleration	a	Angular acceleration	α
Force	F	Torque	τ
Mass	m	Moment of inertia	I

11.9 ANGULAR MOMENTUM

When a mass m like that of Fig. 11.6 (or a planet) revolves about some point O, the *angular momentum* **A** of the mass about O is defined as the vector product of the radius vector **r** from O to the mass and the momentum $m\mathbf{v}$. The angular momentum has the magnitude

$$A = rmv \sin \theta \qquad (11.14)$$

where θ is the angle between **r** and **v**; the direction of **A** is given by the right-hand rule (Fig. 11.7). Newton's second law for rotational motion may be stated in a form more general than that of Sec. 11.7. *The unbalanced torque acting on a body is equal to the time rate of change of angular momentum.* If there is no unbalanced torque on the body, its angular momentum remains constant.

Not only is angular momentum conserved, but evidence from atomic and molecular physics shows that it is quantized (comes in small, well-defined chunks) as well. Changes in the angular momentum of an atomic system come in units of size $h/2\pi$, where h, Planck's constant, has the magnitude 6.6×10^{-34} kg-m^2/s (or J-s). This is a minuscule angular momentum, utterly insignificant for macroscopic bodies but of immense importance in systems of atomic size.

For a rigid body rotating about a fixed axis the angular momentum is the resultant of the angular momenta of all the N atoms of which the body is composed. Since all have the same angular velocity ω,

$$A = \sum_{i=1}^{N} r_i m_i \omega_i r_i = \omega \sum_{i=1}^{N} m_i r_i^2 = I\omega$$

Thus *the angular momentum is the product of the moment of inertia I and the angular velocity ω* for a body rotating about a fixed axis.

For a pair of bodies which interact only with each other, Newton's third law for rotational motion tells us that if body A exerts a torque about some axis on body B, body B exerts an equal and opposite torque on body A about *the same axis.* Therefore the change in angular momentum of body A is equal to and opposite that of body B. Thus *the angular momentum of the system remains constant in the absence of external torques.* The *conservation of angular momentum* is one of the key conservation laws of nature, which include conservation of energy and conservation of linear momentum.

In Fig. 11.9 a man stands on a platform mounted on ball bearings. In his hands he holds heavy weights. His moment of inertia is greater when his arms are outstretched than when they are folded. If this man is set in rotation with arms outstretched, he and the platform rotate with constant angular velocity as long as he does not change his moment of inertia. If he folds his arms as indicated in the figure, the moment of inertia I of the system (man plus weights and platform) is decreased. Since there is no external torque about the vertical axis, the angular momentum $I\omega$ remains constant; since I has decreased, the angular velocity ω must increase. If the man stretches his arms again, his angular velocity is reduced. When a diver performs a somersault from the high board, she doubles up to make her moment of inertia small while she is making her turn, and then she decreases her rotational velocity by increasing her moment of inertia through stretching out to her full length.

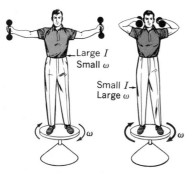

Large I
Small ω

Small I
Large ω

ω ω

FIGURE 11.9
The conservation of angular momentum about the vertical axis requires that the angular velocity increase when the moment of inertia is decreased.

11.10 GYROSCOPES

There are many kinds of gyroscopes, but all include a body spinning rapidly about an axis. In the gyroscopic compass this gyro element is mounted in such a way that no significant torques are applied. Under these circumstances the vector angular momentum remains constant. The axis of the gyro element always points in the same direction, regardless of the maneuvers of the aircraft or ship in which the gyro is mounted. Thus, it can be made the key element of an automatic pilot.

A gyroscope like that of Fig. 11.10, which is supported at point O only, does not fall to the ground when the gyro wheel is in rapid rotation; instead the entire frame rotates about a vertical axis through O at a constant speed. This motion is called *precession*.

The behavior of the gyroscope can be analyzed in terms of angular momentum. Let the angular momentum of the rotating gyro wheel at $t = 0$ be represented by A_0 (Fig. 11.10). The torque due to the weight of the rotor and frame produces an angular acceleration about the axis OY and in a brief time Δt gives the gyroscope a small angular momentum ΔA about OY. The resultant angular momentum A is the vector sum $A_0 + \Delta A$. Thus the gyroscope precesses about the axis OZ.

There is an alternative way of looking at the precession of a gyroscope. Suppose that the rotating element of the gyroscope consists of a metal disk in which balls are mounted in slots (Fig. 11.11). When this disk is put into rapid rotation about an axis perpendicular to the plane of the paper, the balls move in the directions indicated in the figure. Now if a torque is suddenly exerted about a vertical axis, the upper balls continue to move to the left and the lower ones to the right. Not only does the inertia of each ball keep it going in the same direction, but so also does the inertia

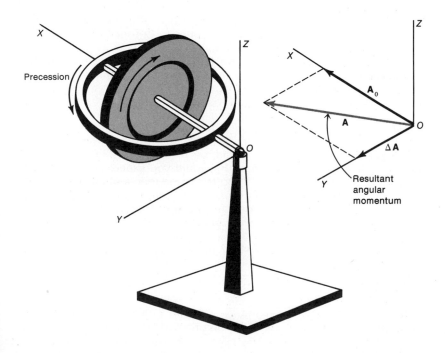

FIGURE 11.10
Precession of a gyroscope.

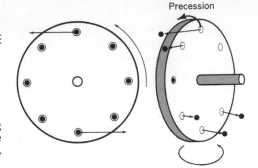

FIGURE 11.11
Precession occurs when the rotating wheel is turned about a vertical axis; the steel balls fly out of their retaining slots, and the disk tilts.

of each particle constituting the disk. Hence, the entire wheel tilts, and the axle precesses upward.

An aircraft moving eastward with a single propeller rotating clockwise tends to nose down when it turns south and to nose up when it turns north, since the rotating propeller acts like a gyroscope. If the airplane is traveling eastward and the propeller is rotating clockwise as viewed by the pilot, at any given instant the top blade of the propeller is traveling southward, and the lower blade northward. If at this instant the airplane is turned toward the north, the upper blade continues to move southward, and the lower one northward. As a result, the airplane noses up.

QUESTIONS

1. Would the kinematic equations (11.1) to (11.6) be valid if the angular quantities were expressed in terms of degrees rather than radians? Revolutions rather than radians? Which equations of Chap. 11 are written in such a form that they are correct only if angular quantities are expressed in radians?

2. If a solid sphere of mass M and radius R raced the ring and cylinder of Fig. 11.5 down the plane, would it reach the bottom first or last? What would its speed at the bottom be?

3. A flywheel rotating with constant angular velocity about a fixed axis is in equilibrium. Is every particle in the flywheel in equilibrium? Explain.

4. What fraction of the kinetic energy of a rolling billiard ball is rotational? Of a rolling solid cylinder? Of a rolling hoop?

5. How can someone in a swing pump the swing from a small amplitude to a large one? Discuss in terms of angular momentum and the work done against gravity.

6. Why does an automobile nose downward if it is traveling at a high speed when the brakes are applied? Would this still happen if there were no brakes on the front wheels?

7. Why does a typical helicopter have a small propeller exerting a torque on the craft about a vertical axis? Discuss the torques involved in helicopter flight.

8. A rifle barrel is rifled to impart a spin to the bullet. What is the advantage of spin? How does rifling affect the way the rifle recoils?

9. Discuss the conservation of angular momentum for a diver performing a $2\frac{1}{2}$-turn forward somersault from a high board.

10. Explain how to distinguish between a hard-boiled egg and a raw one by spinning them on a flat table top.

11. Why would you expect the period of a meterstick oscillating as a pendulum about an axis at one end to be longer than the period of a simple pendulum 50 cm long? After all, the center of gravity would be the same distance from the support. Why would you expect the meterstick to have a shorter period than a simple pendulum 1 m long?

12. Why does an airplane tend to yaw (tilt sideways) when pulling out of a dive? Explain with the aid of a vector diagram.

13. The flywheel of an automobile rotates counterclockwise when viewed from the rear. Will the rear wheels skid more readily during a right or left turn? Why?

14. A 150-g meterstick rests on its flatter side, supported by your fingers under each end. A 100-g mass M lies on one end. If both fingers are sud-

denly removed, M and the stick fall together; but if only the finger nearer M is removed, the stick falls away from M even though the latter has acceleration g. Why?

15. Tidal friction is calculated to dissipate the earth's rotational energy at the rate of 15×10^{11} W and is primarily responsible for a decrease in the length of the day by 0.0016 s per century. Eventually both the day and the month should be about 47 of our current days. Why should the month lengthen? Why will the friction due to lunar tides cease when the day and month are of equal duration?

PROBLEMS

1. A grinding wheel starts from rest and has a constant angular acceleration of 5 rad/s². At $t = 6$ s find the centripetal and tangential accelerations of a particle 75 mm from the axis, the angular velocity, and the angle turned through.
   Ans. 67.5 m/s²; 0.375 m/s²; 30 rad/s; 90 rad

2. A small motor starts from rest and attains its rated speed of 600 r/min in 7 s. Calculate the angular acceleration, assuming it to be uniform. Through how many radians does the motor turn in achieving its rated speed?

3. Compute the angular speed of an automobile wheel 356 mm in radius when the car is moving 72 km/h. What angular acceleration is needed to bring the wheel from rest to this speed in 11 s?
   Ans. 56.2 rad/s; 5.11 rad/s²

4. One method of determining the speed of a bullet is to fire it through two cardboard disks mounted on a long axle and rotated at high angular speed. If the disks are 1.5 m apart and are rotated at 1,800 r/min, find the speed of a bullet fired parallel to the axis if it makes holes which are 12° apart.

5. An electric motor operates at 1,800 r/min. Find its angular velocity in radians per second. What is the speed of a point 55 mm from the axis of rotation? What is its centripetal acceleration?
   Ans. 188 rad/s; 10.4 m/s; 1,950 m/s²

6. Construction of large space stations has been proposed. To provide a pseudo-gravitational field of 0.4 g at a radius of 10 m, what angular speed would the space station require?

7. A 1,000-kg automobile starts from rest and proceeds around a flat, circular track of 500 m radius with a constant tangential acceleration of 2 m/s². How long will it take for the centripetal acceleration to reach 4 m/s²? At that instant what is the resultant acceleration?
   Ans. 22.4 s; 4.47 m/s² at 63.4° with tangent

8. A bicycle wheel initially at rest is accelerated uniformly until it attains an angular speed of 20 rad/s in 5 s. Find the angular acceleration and the angle turned through in this time.

9. The coefficient of static friction between a wooden block and the floor of a merry-go-round is 0.30. If the merry-go-round has a uniform angular acceleration of 0.2 rad/s², how long will it be before the block begins to slide if it is initially 3 m from the axis? *Ans.* 4.95 s

10. A racing car starts from rest and accelerates uniformly along the track with an acceleration of 5 ft/s². If the track is circular with a radius of 900 ft, at what speed is the centripetal acceleration equal to the tangential acceleration? How long does it take to achieve this speed?

11. The turbine and associated rotating parts of a jet engine have a total moment of inertia of 25 kg-m². It is accelerated uniformly from rest to an angular speed of 150 rad/s in a time of 25 s. Find (a) the angular acceleration, (b) the net torque required, (c) the angle turned through during the acceleration, (d) the work done by the net torque, and (e) the kinetic energy of the turbine at its final velocity.
   Ans. (a) 6 rad/s²; (b) 150 N-m; (c) 1,875 rad; (d) 2.81 × 10⁵ J; (e) 2.81 × 10⁵ J

12. The moment of inertia of a body about its axis of rotation is 8 kg-m². A torque of 12 N-m is applied to it. If the body starts from rest, find the angular acceleration. What are the angular velocity and the kinetic energy at the end of 25 s?

13. A flywheel has a mass of 75 kg, all concentrated in the rim, and a radius of 0.6 m. It is rotating with an angular velocity of 30 rad/s. How much work was necessary to give the flywheel this angular velocity? If the wheel is stopped in 1 min by a friction brake, what torque is applied?
   Ans. 12,150 J; 13.5 N-m

14. If a flywheel has a moment of inertia of 5 kg-m², what angular speed does the wheel attain if 50 kJ of work is done in producing rotational kinetic energy? What torque is required to bring this flywheel to rest in 20 s?

15. The turbine of a jet engine has a moment of inertia of 80 kg-m². It is to be accelerated uniformly from rest to an angular velocity of 200 rad/s in 50 s. Find the angular acceleration, the unbalanced torque required, the angle turned through during the acceleration, and the

kinetic energy of the turbine at its final angular velocity.

Ans. 4 rad/s²; 320 N-m; 5,000 rad; 1.6 MJ

16. Calculate the moment of inertia of a drive shaft that has a kinetic energy of 15 kJ when it is making 75 rad/s. Find the radius of gyration if the mass of the wheel is 200 kg.

17. The tip of the propeller of an airplane is 1.5 m from the propeller shaft. The moment of inertia of the rotating propeller and associated shaft is 75 kg-m². If the propeller is making 1,200 r/min and is given an angular acceleration of 2 rad/s², how long is required to increase the angular speed to 1,500 r/min? What unbalanced torque acts on the propeller shaft during the acceleration? At what angular velocity will the tip of the propeller have a linear speed equal to the local speed of sound, which is 320 m/s at the position of the airplane? *Ans.* 15.7 s; 150 N-m; 213 rad/s

18. The radius of the earth is approximately 6,370 km, and it rotates about its axis once each sidereal day of 86,164 s. Find the angular velocity of the earth in radians per second, the linear speed due to the earth's rotation of a point near the equator, and the speed of a point on the earth's surface at 40° latitude.

19. Find the moment of inertia about an axis through the center of mass of a system consisting of two masses of 3 and 5 kg connected by a rod of negligible mass 0.8 m long. Calculate the kinetic energy of the system if the center of mass has a speed of 15 m/s and the angular speed of the system is 5 rad/s. *Ans.* 1.2 kg-m²; 915 J

20. Find the torque developed by an airplane engine which rotates a propeller at 1,800 r/min when supplying 900 hp.

21. A body consists of a solid horizontal cylindrical shaft with a mass of 60 kg and a radius of 45 mm and a solid disk having a mass of 120 kg and a radius of 0.4 m. Find the moment of inertia if the cylinder and the disk are mounted so that they have the same axis. What torque is necessary to give the body an angular speed of 20 rad/s at the end of 5 s, starting from rest?

Ans. 9.66 kg-m²; 38.6 N-m

22. A hollow cylinder has a mass of 25 kg and a diameter of 0.6 m. Its mass is concentrated in the rim. It is rolling with a linear speed of 1.6 m/s. What is its kinetic energy of rotation? Its total kinetic energy?

23. A turbogenerator with a moment of inertia of 100,000 slug-ft² rotates at the rate of 90 rad/s. The power and load are shut off simultaneously, and friction stops the rotation in 3,000 s. Find the initial kinetic energy of rotation, the initial angular momentum, and the torque due to friction.

Ans. 4.05 × 10⁸ ft-lb; 9 × 10⁶ slug-ft²/s; 3,000 ft-lb.

24. A circular hoop has a mass of 0.3 kg and a radius of 0.6 m. Find its moment of inertia (*a*) about an axis through the center of the hoop and perpendicular to its plane and (*b*) about an axis through the circumference of the hoop and parallel to the axis through the center.

25. A solid sphere with a radius of 75 mm rolls down an inclined plane that is 5 m long. What is the angular velocity of the sphere at the bottom of the plane if it requires 10 s for it to reach the bottom? What is the angle of the plane?

Ans. 13.3 rad/s; 0.819°

26. What linear velocity is acquired by a solid steel disk that has a radius of 65 mm and a mass of 3 kg when it rolls down an inclined plane which is 2 m long and makes an angle of 10° with the horizontal? What is the acceleration of the disk?

27. A 25-kg boy stands 2 m from the center of a frictionless playground merry-go-round which has a moment of inertia of 200 kg-m². If the boy begins to run in a circular path with a speed of 0.6 m/s *relative to the ground*, calculate (*a*) the angular velocity of the merry-go-round and (*b*) the speed of the boy relative to the merry-go-round. *Ans.* (*a*) 0.15 rad/s; (*b*) 0.9 m/s

28. A hoop weighing 8 lb and having a radius of 0.6 ft rolls down an inclined plane that makes an angle of 20° with the horizontal. Assuming that there is no slipping, what are the linear acceleration of the hoop, the angular acceleration, and the speed of the center of mass after the hoop has rolled a distance of 25 ft?

29. A meterstick has a mass of 0.2 kg. A small hole is bored in it at the 10-cm mark so the meterstick can be hung from a horizontal nail. (*a*) Find the moment of inertia of the meterstick around an axis at the 10-cm mark. (*b*) If the meterstick is given a small angular displacement while hung from the nail, find the period of the angular harmonic motion. *Hint:* For small angles $\sin \theta = \theta$ in radians.

Ans. (*a*) 0.0487 kg-m²; (*b*) 1.57 s

30. If the meterstick of Prob. 29 is rotated about the nail until it makes an angle of 30° with the vertical and is then released, find its angular velocity as it passes through its equilibrium position, assuming that friction is negligible.

31. A 1,000-kg automobile has a 2.4-m wheelbase. Its center of mass is 1.1 m behind the front axle and 0.8 m above the ground. Find the vertical force

on the front wheels when the car is at rest and when it is stopping with a deceleration of 4 m/s². Assume the center of mass does not move relative to the wheels. *Hint:* Recall that the retarding forces at the road produce a torque which is balanced by a redistribution of the force on the front and rear wheels.

Ans. 5,310 N; 6,640 N

32. A 3,200-lb automobile has a 9-ft wheelbase, and its center of mass is 4 ft behind the front axle and 2.25 ft above the road. Assuming that the center of mass does not move relative to the wheels, find the vertical force on the front wheels when the car has a uniform deceleration of 16 ft/s². *Hint:* See Prob. 31.

33. Find the moment of inertia of a thin spherical shell of mass M and radius R. *Hint:* Take the difference between the moments of inertia for spheres of radii $R + \Delta R$ and R; then assume that $\Delta R \ll R$. *Ans.* $\frac{2}{3} MR^2$

34. A 4-lb ring of radius 6 in. rolls down an inclined plane, one end of which is 0.9 ft above the other. If the ring starts from rest at the top, show that its kinetic energy of rotation is equal to its kinetic energy of translation at the bottom. Find the kinetic energy of rotation at the bottom and the speed of the center of mass at the bottom.

35. A pulley has mass M, radius R, and moment of inertia about its center of mass of $0.8 MR^2$. It has a cord wrapped around it with one end of the cord fastened to the ceiling. The pulley wheel is released from rest and falls a distance h. (*a*) Consider the motion to be a translation of the center of mass and a rotation about the center of mass. Apply Newton's second law to the translational and rotational motions and calculate the angular acceleration of the wheel and the tension in the cord. (*b*) Find the speed v of the wheel after it has fallen a distance h.

Ans. (*a*) 0.556 g/R; 0.444 Mg; (*b*) $\sqrt{10\,gh}/3$

36. What angular acceleration will be imparted to a solid disk with a fixed axis that has a radius of 1.5 ft and a weight of 48 lb by a 10-lb weight hanging from a cord wound around the disk? Find the tension in the cord.

37. A Yo-Yo has a moment of inertia about its center of mass of 4×10^{-5} kg-m² and a mass of 0.1 kg. If the string is wrapped around a center post of 5-mm radius and the Yo-Yo is released from rest, find the angular acceleration and the tension in the string. *Ans.* 115 rad/s²; 0.922 N

38. A solid uniform disk with a 75-mm radius starts from rest and rolls down a plane 2 m long and inclined at 36.9° to the horizontal. What is its linear speed at the bottom? Its angular velocity? How long does it take to reach the bottom?

39. A mass of 0.4 kg hangs by a cord wrapped around the axle of a frictionless flywheel. The axle has a radius of 15 mm. The tension in the cord as the mass falls is 3.72 N. Find (*a*) the acceleration of the mass, (*b*) the angular acceleration of the flywheel, and (*c*) the moment of inertia of the flywheel.

Ans. (*a*) 0.5 m/s²; (*b*) 33.3 rad/s²; (*c*) 0.00167 kg-m²

40. Show that if a solid cylinder is to roll down an inclined plane without slipping, the coefficient of static friction between cylinder and plane must be at least $(\tan \theta)/3$, where θ is the angle between the plane and the horizontal.

41. A bowling ball is released with speed v_0 and no angular velocity just above the alley. It skids until its angular speed reaches a value such that $\omega_f R = v_f$, where R is the radius of the ball and v_f is the speed at which it rolls without skidding. Show that $v_f = \frac{5}{7} v_0$.

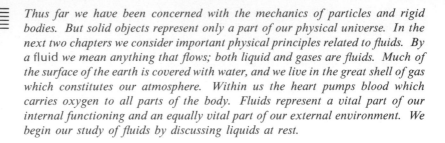

12

LIQUIDS
AT REST

Thus far we have been concerned with the mechanics of particles and rigid bodies. But solid objects represent only a part of our physical universe. In the next two chapters we consider important physical principles related to fluids. By a fluid we mean anything that flows; both liquid and gases are fluids. Much of the surface of the earth is covered with water, and we live in the great shell of gas which constitutes our atmosphere. Within us the heart pumps blood which carries oxygen to all parts of the body. Fluids represent a vital part of our internal functioning and an equally vital part of our external environment. We begin our study of fluids by discussing liquids at rest.

12.1 PRESSURE IN FLUIDS

A liquid occupies a well-defined volume. It has no shape of its own but takes the shape of the vessel containing it. On the other hand, a gas completely occupies any enclosed volume in which it is placed. Liquids yield to a continued application of force that tends to deform them or change their shape, but they manifest wide differences in their readiness to yield to distorting forces. Water, alcohol, and ether are very mobile liquids; glycerin is less mobile and tar still less so. Although a liquid readily changes shape in response to forces, it typically offers large resistance to efforts to change its volume. Most liquids are almost incompressible. On the other hand, a gas is readily compressible, and it is relatively easy to reduce the volume of a gas to half its original volume or less.

Any fluid exerts forces against the walls of the containing vessel. In order to discuss the interaction between the fluid and the walls, it is convenient to introduce the concept of *pressure*. Consider some area A of surface and suppose that a force F is exerted against this area in a direction perpendicular to the surface. The *average pressure* on the area A is defined as *the ratio of the force F to the area A:*

$$p_{av} = \frac{F}{A} \tag{12.1}$$

Pressure may be expressed in newtons per square meter, pounds per square inch, etc., in general, as any unit of force divided by a unit of area. The SI unit of pressure, newton per square meter, has been named the *pascal* (Pa). In meteorology pressures are commonly reported in millibars, where one bar, i.e., barometric unit, is 10^5 Pa.

In a fluid at rest the force acting on any area element is always perpendicular to the surface, but for a large area the pressure may be greater at some points than at others. In this case we consider an infinitesimal area ΔA and determine the normal force ΔF acting on it. Then the *pressure at any point* is the limit as $\Delta A \to 0$ of the ratio $\Delta F/\Delta A$, or

$$p = \lim_{\Delta A \to 0} \frac{\Delta F}{\Delta A} \tag{12.1a}$$

12.2 PRESSURE IN A LIQUID

If we explore a liquid confined in a vessel with a device for measuring pressure, we find that *at a given level in the liquid the magnitude of the force acting on a test area is the same regardless of how the area is oriented.* The force is always perpendicular to the area. Further, the deeper we go, the greater the pressure becomes. If p_0 represents the pressure at the surface of the liquid and p the pressure at a depth y, we find that the increase in pressure due to the liquid is directly proportional to the depth y and to the density d of the liquid. Let p_L represent the increase in pressure due to the liquid. Then

$$p_L = p - p_0 = dgy \qquad (12.2)$$

where g is the acceleration due to gravity. That this should be true is easy to understand if we consider the force on an area A at the bottom of a container (Fig. 12.1). This area supports the weight of the column of fluid above it. The density d is, by definition, the mass per unit volume of the liquid, while the volume of the column is Ay. So dAy is the mass of the column and, by Eq. (4.3), the weight of the column is $dgAy$. Then by Eq. (12.1) the pressure due to the liquid is dgy.

In some problems it is convenient to use *weight density* d_w (weight per unit volume) rather than *density* (mass per unit volume). The weight of a given volume is equal to $d_w V$, and the pressure due to the liquid is yd_w. In the British engineering system we ordinarily use weight density (pounds per cubic foot) rather than mass density (slugs per cubic foot). Hence the relation

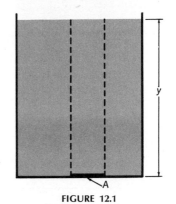

FIGURE 12.1
The pressure in a liquid increases in direct proportion to the depth and to the density of the fluid.

$$p_L = p - p_0 = d_w y \qquad (12.2a)$$

is sometimes more convenient than $p_L = dgy$. (Note that, since $w = mg$, $d_w = w/V = mg/V = dg$.)

Although we have derived Eqs. (12.2) and (12.2a) only for the bottom of our container, they are applicable at any point in the liquid at rest. Observe that y is the *vertical* height from the point to the free surface.

☐ **Example** The hatch of a submarine is 50 m under the surface of the ocean. If the density of seawater is 1,025 kg/m³, find the pressure at the hatch due to the water and the net force on the hatch if it is a rectangle 0.6 m wide and 1.2 m long. The pressure inside the submarine is the same as that at the surface.

$$
\begin{aligned}
p_L = p - p_0 &= dgy \\
&= (1{,}025 \text{ kg/m}^3)(9.8 \text{ N/kg})(50 \text{ m}) \\
&= 5.02 \times 10^5 \text{ N/m}^2 = 5.02 \times 10^5 \text{ Pa}
\end{aligned}
$$

The net force on the hatch is the difference between the inward force pA and the outward force p_0A. Therefore,

$$
\begin{aligned}
F = pA - p_0A &= (p_L + p_0)A - p_0A = p_LA \\
&= (5.02 \times 10^5 \text{ N/m}^2)(0.72 \text{ m}^2) \\
&= 3.61 \times 10^5 \text{ N}
\end{aligned}
$$

Many types of pressure gauges read the excess of pressure over atmospheric pressure; this is the so-called *gauge pressure*. When a typical gauge reads 1.65×10^5 N/m² (24 lb/in.²) and atmospheric pressure is 1.013×10^5 N/m² (14.7 lb/in.²), the absolute pressure p is 2.66×10^5 N/m² (38.7 lb/in.²). ☐

☐ **Example** The gauge pressure of water in the water mains is 2.5×10^5 Pa. How much work is required to pump 30,000 m³ of water at atmospheric pressure into the mains?

Work = net force × distance = $\dfrac{\text{force}}{\text{area}}$ × (distance × area)

= net pressure difference × change in volume

= $(2.5 \times 10^5 \, \text{N/m}^2)(3 \times 10^4 \, \text{m}^3)$

= 7.5×10^9 J ☐

12.3 LIQUIDS IN COMMUNICATING VESSELS

Liquids seek their own level in interconnected vessels. When tubes of various sizes and shapes are connected, liquid poured into one of them (Fig. 12.2) comes to the same level in all the tubes (provided surface-tension effects may be neglected). This result is to be expected from the fact that the pressure in the liquid depends on the depth below the free surface. *At all points at the same level within a liquid at rest, the pressure is the same.* If it were not, the liquid would flow from one point to the other until the pressure became equalized.

In Fig. 12.2 the areas of the openings at which vessels *A*, *B*, *C*, and *D* join the bottom reservoir are the same. The following question may be raised: If the liquid pressure is due to the weight of the fluid above, why is the pressure not greater in vessel *D*, in which the sides slope outward, than in vessel *B*, in which the sides slant inward? Certainly vessel *D* contains a greater weight of water. The answer is that in *D* the slanting sides exert forces on the liquid which have an upward component. Thus, a portion of the weight of the fluid is actually held up by the sloping sides. On the other hand, in *B* the force exerted by the slanting sides on the liquid is downward. When the force due to the slanting walls is added to the weight of the fluid, the pressure at the bottom is the same in all four vessels.

Let two liquids that do not dissolve in each other, e.g., water and mercury, be placed in a bent tube (Fig. 12.3*a*). When the liquids are at rest, the pressure exerted by the column of lighter liquid is balanced by the pressure due to the column of heavier liquid above the junction level of the liquids. Then, provided surface-tension effects are negligible, $d_1 g y_1 = d_2 g y_2$, from which $y_1/y_2 = d_2/d_1$. *The heights of two liquids above their surface of separation are inversely proportional to their densities.*

If the liquids are mutually soluble, e.g. water and alcohol, the bent tube can be inverted and the ends placed in cups containing the liquids (Fig. 12.3*b*). The air from the upper part of the bent tube is partly removed, and the stopcock closed. The pressure above both liquids is the same, and the atmospheric pressure on the liquids in the open vessels is the

FIGURE 12.2
The pressure in a liquid at rest is the same at all points at the given level, regardless of the shape of the containing vessel. In the figure the areas of all four arms at depth *y* are exactly the same; even though there is much less fluid in arm *B* than in arm *D*, the pressure at depth *y* is the same in all four arms.

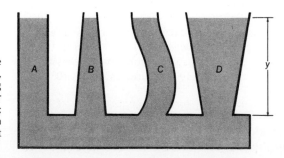

same. The difference between the pressure inside the tube and atmospheric pressure is balanced in each case by the rise of the liquid in the tube. These differences in pressure are the same, and again $d_1gy_1 = d_2gy_2$.

☐ **Example** If one of the beakers in Fig. 12.3b contains sulfuric acid and the other contains water, and if the height of the column of water is 40 cm when the height of the column of acid is 30 cm, find the density of the sulfuric acid.

$$y_a d_a g = y_w d_w g \qquad d_w = 1,000 \text{ kg/m}^3$$

$$\frac{d_a}{d_w} = \frac{y_w}{y_a} = \frac{0.40 \text{ m}}{0.30 \text{ m}}$$

$$d_a = 1,330 \text{ kg/m}^3 \qquad\qquad ☐$$

12.4 PASCAL'S PRINCIPLE

If we increase the pressure at any one point in a fluid which is completely enclosed (Fig. 12.4), the pressure increases by an equal amount at all other points. For example, if we add a mass of 1 kg to piston A of Fig. 12.4, an equal mass must be added to identical pistons B and C in order to keep them from moving. This is a special case of a general law known as *Pascal's principle. If the pressure at any point in an enclosed fluid at rest is changed, the pressure changes by an equal amount at all points in the fluid.* Note that Pascal's principle does *not* say that the pressure is the same everywhere within the enclosed fluid. The pressure continues to be greater at greater depth. Pascal's principle deals with the change in pressure, not with the absolute pressure.

Consider two cylinders which are connected together and filled with water (Fig. 12.5). If each cylinder is fitted with a piston which moves without friction, and if the two pistons are at the same level, the pressure at each will be the same. If A is the area of the larger piston and a the area of the smaller one, the force f on the smaller piston is given by pa, while the force F on the larger piston is given by pA. By applying a force f on the small piston, it is possible to produce a force $F = Af/a$ on the larger piston. When the larger piston is raised a distance S, the smaller moves through a greater distance s, where $s = SA/a$.

If the larger and smaller pistons are not at the same level, the pressures on the two pistons are not the same; they differ by dgy, where y is the difference between the heights of the two pistons. If the pressure is increased 10 kN/m² on the smaller piston, it also increases 10 kN/m² at the larger piston, but the pressures are not equal. A direct application of these ideas is found in the hydraulic press.

An important application of Pascal's principle occurs in the hydraulic system of an automobile (Fig. 12.6). When the pressure is increased in the master cylinder by pressing on the brake pedal, the pressure increases by an equal amount at every point in the hydraulic system. If all the pistons at the brake shoes have the same cross-sectional area, an equal force is applied at all the brake shoes.

Pascal's principle applies to all fluids, gases as well as liquids. A typical application of Pascal's principle to gases is the automobile lift pump. Compressed air exerts pressure on the oil in a reservoir. The oil in turn transmits the pressure to a cylinder which lifts the automobile.

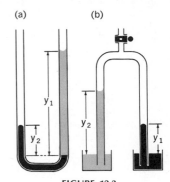

FIGURE 12.3
The pressure produced by the column of liquid 1 of height y_1 is equal to that due to the column of liquid 2 of height y_2 in both (a) and (b).

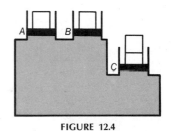

FIGURE 12.4
A pressure change at any point in an enclosed fluid at rest results in an equal pressure change at all other points; this fact is known as *Pascal's principle.*

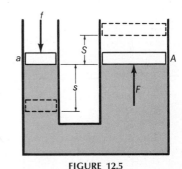

FIGURE 12.5
Multiplication of force by transmitted pressure in a hydraulic press. The ideal mechanical advantage of this simple machine is A/a, where A is the area of the output piston and a the area of the input piston.

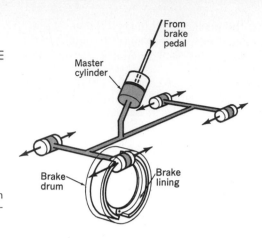

FIGURE 12.6
Pascal's principle as it is applied in the hydraulic brake system of an automobile.

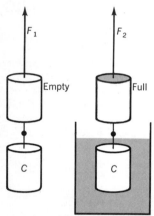

FIGURE 12.7
The solid cylinder C is buoyed up by a force equal to the weight of the fluid it displaces; F_2 is equal to F_1 when the upper container, which has the same volume as C, is filled with liquid.

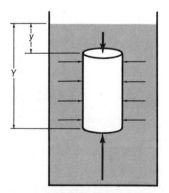

FIGURE 12.8
The net upward force on the cylinder is equal to the weight of the fluid displaced by the cylinder.

□ **Example** Find the minimum gauge pressure which must be supplied to an automobile lift pump with a piston of area 800 cm² to lift an automobile and piston with combined mass of 1,500 kg.

$$\text{Net force} = \text{net gauge pressure} \times \text{area}$$
$$1{,}500 \text{ kg} \times 9.8 \text{ N/kg} = p_g (800)(10^{-4} \text{ m}^2)$$
$$p_g = 1.84 \times 10^5 \text{ N/m}^2 = 184 \text{ kPa} \qquad □$$

12.5 ARCHIMEDES' PRINCIPLE

It is a matter of common experience that bodies are apparently lighter under water than in air. A fresh egg sinks in pure water but floats in salty water. A piece of iron sinks in water but floats in mercury. A diver who picks up a stone under water and brings it to the surface finds that the stone is much heavier above the surface. The principle which explains these observations, discovered by the distinguished Greek mathematician and physicist Archimedes, states;

A body immersed in a fluid is buoyed up by a force equal to the weight of the fluid displaced.

An experimental verification of Archimedes' principle can be obtained by using the equipment of Fig. 12.7. A hollow cylindrical cup and a solid brass cylinder C turned so that it will just fill the cavity inside the cup are suspended from one arm of a balance, and the weights necessary to restore equilibrium are added to the other pan. When a vessel of water is brought up in such a way that the cylinder C is completely submerged, the side of the balance carrying the cylinder rises, showing that the water is pushing upward on the cylinder. If water is now poured into the cup until it is full, the original equilibrium of the balance is restored. The cylinder is buoyed up by a force equal to the weight of the water displaced.

Archimedes' principle follows directly from the laws of fluid pressure. If a cylindrical block (Fig. 12.8) is immersed in a vessel filled with liquid, the resultant of the forces on the vertical sides is zero. Upon the upper face of the cylinder there is a downward force $ydgA$ equal to the pressure at the upper surface multiplied by the cross-sectional area A of the cylinder. On the lower face there is an upward force $YdgA$ equal to the pressure at the bottom multiplied by the area. The net upward force is $YdgA - ydgA = (Y - y)Adg = Vdg$, where V is the volume of the cylinder. *The net buoyant force is equal to the weight of the fluid displaced.* The same result holds for a body of any shape in any liquid. Therefore, a body immersed in any fluid is lighter by the weight of fluid which it displaces.

☐ **Example** An aluminum casting weighs 5.40 lb in air and 3.40 lb submerged in water. Find the volume of the casting and the weight density of aluminum.

$$\text{Loss of weight} = 5.40\ \text{lb} - 3.40\ \text{lb} = 2.00\ \text{lb}$$

the weight of water displaced. Since 1 ft³ of water weighs 62.5 lb, 2 lb occupies $2/62.5 = 0.032$ ft³.

$$d_w = \frac{w}{V} = \frac{5.40\ \text{lb}}{0.032\ \text{ft}^3} = 169\ \text{lb/ft}^3 \qquad ☐$$

Fish are capable of moving toward the surface or into deep water by regulating the quantity of water they displace and therefore the buoyant force. By distending the air bags in their bodies, they can change their volumes and thus change the buoyancy of the water on them. By contracting its air sacs, a fish diminishes its volume and sinks. Similarly, a submarine can submerge by taking water into tanks, thus making the submarine heavier than an equal volume of water. It rises from below the surface by blowing or pumping this water out of the tanks.

12.6 FLOATING BODIES

When a body floats on a liquid, the buoyant force is equal to its weight. The body sinks until it displaces its own weight (Fig. 12.9) and then remains in equilibrium. This explains why a ship rides lower in the water when loaded than when empty and why it rides higher in salt water than in fresh.

☐ **Example** A barge 10 m long and 5 m wide has vertical sides. When two automobiles are driven on board, the barge sinks 5 cm farther into the water. How much do the automobiles weigh?

$$\text{Volume of displaced water} = (10\ \text{m})(5\ \text{m})(0.05\ \text{m})$$
$$= 2.5\ \text{m}^3$$
$$\text{Weight of water displaced} = (2.5\ \text{m}^3)(1,000\ \text{kg/m}^3)(9.8\ \text{N/kg})$$
$$= 24,500\ \text{N}$$
$$\text{Weight of automobiles} = \text{weight of displaced water}$$
$$= 24,500\ \text{N} \qquad ☐$$

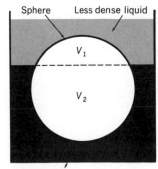

FIGURE 12.9
A solid sphere floating at the interface between two immiscible liquids of different densities. The sphere sinks until the sum of the weight of the first liquid displaced and the weight of the second liquid displaced is equal to the weight of the sphere.

The buoyant effect of all the displaced fluid can be replaced by a single force acting at the center of mass of the displaced fluid, known as *the center of buoyancy.* For a floating body such as a ship to be in stable equilibrium, there must be a torque which restores the body to its stable position whenever it undergoes an angular displacement. This torque is

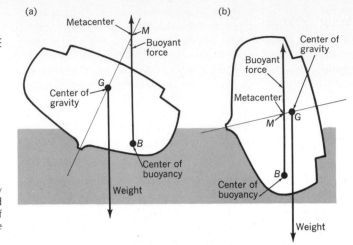

FIGURE 12.10
(*a*) When the center of gravity is below the metacenter *M*, the boat is stable and rights itself. (*b*) When the center of gravity is above the metacenter, the boat is unstable and rolls over.

TABLE 12.1
Densities of Solids and Liquids at 20°C

Substance	Density, kg/m³	Weight density, lb/ft³
Alcohol	789	49.3
Aluminum	2,650	164
Balsa wood	160	10
Brass	8,600	535
Brick	2,100	131
Copper	8,930	555
Cork	240	15
Diamond	3,520	220
Gasoline	790	49.4
Glass, crown	2,500	156
Flint	3,700	230
Glycerin	1,260	78.7
Gold	19,320	1,200
Ice (0°C)	917	57.2
Iron, cast	7,200	450
Wrought	7,800	480
Kerosene	820	51.2
Lead	11,370	710
Mercury	13,600	840
Oak	800	50
Pine	500	31.2
Silver	10,500	655
Tin	7,290	455
Turpentine	870	54.3
Water, fresh	1,000	62.5
Sea	1,030	64.4
Zinc	7,150	446.2

provided by the combined action of the weight *w* of the body acting downward at the center of gravity *G* (Fig. 12.10*a*) and the buoyant force acting upward through the center of buoyancy. The condition for stability will be realized if the metacenter *M* lies above the center of gravity of the body. The position of the metacenter is determined by the intersection of two lines, one drawn vertically through the center of buoyancy *B* and the other drawn vertically through the center of gravity *G* of the body *before displacement*. If the metacenter lies above the center of gravity, there is a restoring torque and the body is in stable equilibrium, but if the metacenter lies below the center of gravity, the torque that comes into play when the body is displaced from its normal position increases the displacement further (Fig. 12.10*b*). The body is in unstable equilibrium and turns over.

12.7 DENSITY AND SPECIFIC GRAVITY

Density is the ratio of *mass* to *volume*, while *weight density* is the ratio of *weight* to *volume*. Table 12.1 lists the densities of a number of common liquids and solids. When a body has an irregular shape, the volume can be determined by application of Archimedes' principle.

When a body is heavier than an equal volume of water (and is insoluble in water), its volume can be determined by finding the loss of weight in water. From the the mass and volume the density can be obtained.

□ **Example** A piece of iron has a mass of 0.0780 kg. Submerged in water, it has an apparent weight of 0.666 N. What is the volume of the iron? What is its density?

$$\text{Weight of iron } mg = (0.078 \text{ kg})(9.8 \text{ m/s}^2)$$
$$= 0.764 \text{ N}$$

$$\text{Weight of water displaced} = 0.764 \text{ N} - 0.666 \text{ N}$$
$$= 0.098 \text{ N}$$

Mass of water displaced $\dfrac{w}{g} = \dfrac{0.098 \text{ N}}{9.8 \text{ m/s}^2}$

$$= 0.010 \text{ kg} = 10 \text{ g}$$

Volume of water displaced $\dfrac{m}{d} = \dfrac{0.010 \text{ kg}}{1{,}000 \text{ kg/m}^3}$

$$= 10^{-5} \text{ m}^3 = 10 \text{ cm}^3$$

Volume of iron $= 10^{-5} \text{ m}^3 = 10 \text{ cm}^3$

Density of iron $\dfrac{m}{V} = \dfrac{0.078 \text{ kg}}{10^{-5} \text{ m}^3} = 7{,}800 \text{ kg/m}^3$ □

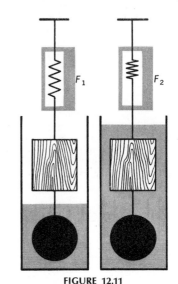

If the object is lighter than water (and insoluble), its volume can still be determined by this method by using a sinker large enough to pull it below the surface of the water. In this case (Fig. 12.11) the combined weight of the body and the sinker is first determined when the sinker is immersed in water and the body is above the surface of the water. The body is then also submerged, and the combined weight redetermined. The reduction in weight is equal to the weight of water displaced by the body.

□ **Example** A piece of cork has a mass of 0.025 kg. When it is fastened to a sinker and the sinker alone immersed in water, the combined apparent weight of sinker and cork is 1.960 N. When both sinker and cork are immersed, they weigh 0.735 N. What is the density of the cork?

FIGURE 12.11
The buoyant force on the block of wood, which would float if there were no sinker, is the difference between F_1 and F_2.

Loss of weight due to submerging body $= 1.960 - 0.735 \text{ N}$

$$= 1.225 \text{ N}$$

Mass of water displaced by cork $= \dfrac{1.225 \text{ N}}{9.8 \text{ m/s}^2}$

$$= 0.125 \text{ kg}$$

Volume of water displaced $= \dfrac{0.125 \text{ kg}}{1{,}000 \text{ kg/m}^3}$

$$= 1.25 \times 10^{-4} \text{ m}^3$$

Density of cork $= \dfrac{\text{mass of cork}}{\text{volume}}$

$$= \dfrac{0.025 \text{ kg}}{1.25 \times 10^{-4} \text{ m}^3}$$

$$= 200 \text{ kg/m}^3 \qquad □$$

The specific gravity of a body is the ratio of its density to the density of water. If the specific gravity of a body is equal to 5, the body weighs 5 times as much as an equal volume of water.

12.8 COHESION AND ADHESION

In liquids the molecules move freely with respect to each other but are held together by attractive forces. The molecules of a liquid cling not only to each other but also to the molecules of other substances, as can be seen when a piece of glass is dipped into a vessel of water. The molecules of water adhere to the glass and form a thin film over its surface. *The attraction of like molecules for one another is called cohesion; the attraction of unlike molecules for one another is called adhesion.* It is *cohesive* forces which hold the molecules of iron, copper, and other solid substances together so firmly.

If the molecules of a liquid have less attraction for each other than for the molecules of the solid with which they are in contact, the liquid adheres to the solid and wets it. Here *adhesive forces are greater than cohesive.* When the cohesive forces are greater than the adhesive, the solid is not wet by the liquid. Such is the case when mercury is in contact with glass. If a drop of mercury and a drop of distilled water are placed on a clean glass surface, the water spreads over the glass in a thin layer, while the mercury forms a distorted ball.

If molecules are further apart than a few millionths of a centimeter, the attractive forces are not appreciably great. Intermolecular forces have short range. It takes a moderate force to break a piece of chalk, but after it is broken, the pieces do not adhere if they are pressed back together. Because of the irregular surfaces, not enough molecules are able to exert appreciable attractive forces.

Molecules of water adhere to glass, forming a thin layer over its surface. When we place a little water between two sheets of glass and take care to see that no air is trapped between the plates, it requires a significantly greater force to separate the glass plates than it would if they were dry. The water is acting as an adhesive. Glues are much superior adhesives. Once a glue has set, it may be stronger than the materials it is binding together. If a thin gold foil is placed over a layer of metal and carefully pressed into place, the molecular attraction between the gold and the base metal is strong enough to keep the gold in place without a cement.

12.9 SURFACE ENERGY AND SURFACE TENSION

A greased needle gently placed on the surface of water floats although the density of the needle is greater than that of the water. Some insects can walk on the surface of a lake or stream, which behaves as though it were covered with a thin elastic film. It requires a force to break this film, the amount of force depending on the molecular attraction characteristic of the liquid.

The molecules of a liquid move under the influence of strong forces exerted by nearest neighbors. These forces are repulsive when molecules approach too closely and attractive when the separation is greater than normal. For a molecule within the liquid the time average of these internal forces is zero. However, if a molecule at a surface begins to leave the liquid, the resultant of these forces pulls it back. A molecule must have considerable energy to escape from the surface.

Consider Fig. 12.12, in which *CEHD* is a bent wire and *AB* is a straight

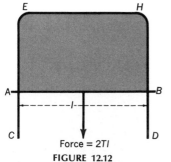

Force = 2Tl

FIGURE 12.12
A force 2Tl is required to hold the movable wire *AB* in equilibrium against surface tension.

wire which can be moved in the direction of the arrow. If we dip this system in a soap solution, we may have the area *AEHB* covered by a soap film. In order to prevent the soap film from contracting and pulling the side *AB* toward *EH*, a force *F* must be applied. This force is given by $F = 2Tl$, where l is the length of the wire *AB* and T is the force per unit length. The factor 2 is introduced because the film has two sides. *The surface force per unit length which must be applied to overcome molecular forces is the surface tension.*

If the wire *AB* is drawn down a distance s, the work done is

$$W = Fs = T(2ls) = T \, \Delta A \qquad (12.3)$$

where $\Delta A = 2ls$, the increase in area counting *both* sides of the film. Thus, T is the ratio of the work required to increase the area against the surface molecular forces to the change in area.

The surface tension of a liquid depends on the temperature (Fig. 12.13). The values of the surface tension for several liquids are listed in Table 12.2.

One way of measuring surface tension is to determine the force required to pull a platinum ring through the surface of a liquid. The total length of liquid surface which must be broken is twice the circumference of the ring. Thus the force required to break the surface is $T \times 4\pi r$, where r is the radius of the ring.

The splashes formed (Fig. 12.14) when drops of milk fall on a surface of milk illustrate the effects produced by surface tension. When a drop of oil is placed in a mixture of water and alcohol of the same density, the oil droplet does not rise or fall. Under the action of molecular forces it assumes a spherical form. Lead pellets for a shotgun can be produced by allowing liquid lead to fall from an opening of suitable size. Because of surface forces the freely falling lead drops assume a spherical shape, in which they solidify before they strike anything.

Surface tension plays a vital role in the formation of bubbles. The

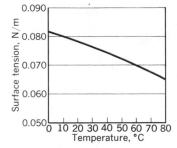

FIGURE 12.13
The surface tension of water decreases as the temperature rises.

TABLE 12.2
Surface Tension

Substance	Temperature, °C	Surface tension, N/m
Acetic acid	20	0.0234
Alcohol (ethyl)	20	0.0216
Ether (ethyl)	20	0.0169
Glycerin	18	0.0632
Mercury	18	0.545
Petroleum	20	0.0259
Turpentine	20	0.0289
Water	20	0.0728

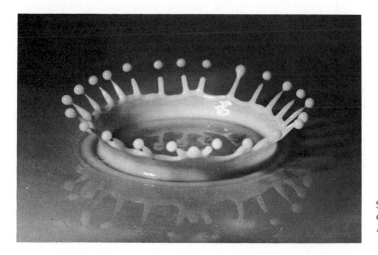

FIGURE 12.14
Surface tension creates splash patterns on the surface of milk. (*Courtesy of Edgerton, Germeshausen, and Grier, Massachusetts Institute of Technology.*)

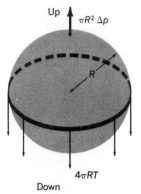

FIGURE 12.15

Excess pressure Δp inside a soap bubble produces an upward force $\pi R^2 \Delta p$ on the top hemisphere of the bubble; this force is balanced by the downward force $2(2\pi R)T$ due to the surface tension.

pressure p inside a bubble in equilibrium is greater than the external pressure p_0 because of surface tension. The forces across the equatorial plane of a stable soap bubble (Fig. 12.15) are balanced; hence, $\pi R^2(p - p_0) = 4\pi RT$. The excess pressure in the soap bubble is then

$$\Delta p = p - p_0 = \frac{4T}{R} \tag{12.4}$$

The surface force is $4\pi RT$ for the soap bubble because there are two surfaces to the bubble. A bubble in a carbonated liquid or a bubble of steam in boiling water has only a single surface. In these cases the excess pressure is only half that given by Eq. (12.4), or $2T/R$.

12.10 CAPILLARITY

If a piece of glass tubing of very small bore is thoroughly cleaned and then dipped into water (Fig. 12.16a), the water wets the inside of the tube and rises in it. If the liquid does not wet the tube (Fig. 12.16b) as in the case of mercury in a glass tube, the liquid is depressed. The smaller the bore of the tube, the greater the height to which the liquid rises or the greater the amount which it is depressed. This rise or depression of liquids in tubes of small bore, known as *capillarity*, is caused by the molecular forces responsible for surface energy. If the cohesive forces between molecules of the liquid are greater than the adhesive forces between liquid and wall, the liquid pulls away from the tube and is depressed. If the adhesive forces are greater, the liquid wets the capillary tube and rises.

When a glass tube is thrust into water, the molecules in the surface of the wall just above the water pull up on the molecules of water lying nearest to them and raise them above the level of the water in the vessel. This carries upward a column of water, which is supported by the surface forces. The net upward force available is the vertical component of the surface-tension forces if we are dealing with a vertical cylindrical tube. The angle of contact θ (Fig. 12.16) between the surface of the liquid and the tube depends on the liquid, the gas above, and the kind of tube involved. The upward force per unit length is $T \cos \theta$, and the total upward force is $T \cos \theta$ multiplied by the length of liquid in contact with the tube, which is the circumference. Thus, the net upward force is $2\pi r T \cos \theta$. The liquid rises until the weight of liquid supported, $\pi r^2 h d g$, is equal to this force. Hence, for equilibrium

$$2\pi r T \cos \theta = \pi r^2 h d g \tag{12.5}$$

or

$$h = \frac{2T \cos \theta}{rdg}$$

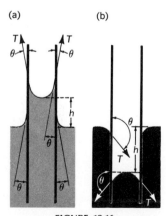

FIGURE 12.16

(a) Water wets glass, and the angle of contact θ approaches zero. (b) Mercury does not wet glass; the angle of contact θ is 140°.

For water in clean glass the angle θ is zero, and $\cos \theta$ becomes unity. Whenever a liquid wets a tube thoroughly, θ is zero. When cohesive forces exceed adhesive, θ is greater than 90°, the liquid is depressed in the capillary, and h is negative. For water on paraffin θ is 107°. Water does not wet paraffin. For mercury on glass the angle of contact is approximately 140°. Figure 12.17 shows how mercury and water behave in glass tubes of unequal radii.

□ **Example** The liquid in a capillary tube rises to the height of 7 cm. The radius of the tube is 0.1 mm, and the density of the liquid 800 kg/m³. If the angle between the liquid and the surface of the tube is zero, find the surface tension of the liquid.

$$2\pi rT \cos \theta = \pi r^2 h dg$$
$$2T(1) = (10^{-4} \text{ m})(0.07 \text{ m})(800 \text{ kg/m}^3)(9.8 \text{ N/kg})$$
$$T = 0.0274 \text{ N/m} \qquad □$$

In recent years the chemical industry has developed a number of *waterproofing* agents which prevent water from wetting a fabric. For these compounds θ is greater than 90°. Similarly, a number of *wetting agents*, or *detergents*, have been developed. These agents change θ to a smaller angle. The addition of a suitable detergent can make water wet paraffin.

There are many illustrations of capillary action in nature. The oil in a lamp rises in the wick by capillary action. Ink spreads in a blotter and water in a lump of sugar by this same action. If one end of a towel dips into a bucket of water and the other end hangs over the bucket, the whole towel soon becomes wet.

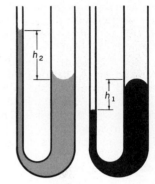

FIGURE 12.17
Unequal heights of water and mercury in connecting glass tubes of different radii.

QUESTIONS

1. Does the force exerted on a dam by an artificial lake depend on how far the water has been backed up? Upon what factors does the force depend?
2. An iron ball floats in a beaker half filled with mercury. If water is poured over the mercury until the beaker is filled, will the level at which the ball floats change? Explain.
3. A beaker partly filled with water rests on the pan of a spring balance. A small boat filled with lead balls floats on the water. If the boat is tipped over so that it and its contents sink to the bottom, does the water level in the beaker change? Does the balance reading change?
4. Explain why pressure is a scalar quantity even though it is defined in terms of the vector quantity *force*.
5. Is a sailboat equally stable in fresh and in salt water? If not, explain why.
6. A razor blade can be made to float on water. What forces act on this blade? Is Archimedes' principle applicable?

7. Water rises to a height of 5 cm in a certain capillary tube. If the tube is broken off 2 cm above the level of the liquid in the main vessel, will water then run out at the top of the tube? Explain.
8. Why does the pressure inside a soap bubble decrease as it is blown up? How does a rubber balloon behave in this respect? Explain.
9. Why does flame polishing a piece of glass by putting it in a bunsen flame round off sharp edges?
10. A large soap bubble is formed at one end of a capillary tube, and a small one at the other end. Which will grow at the expense of the other? Why?
11. Why is the radius of the stem of a sensitive hydrometer (Prob. 21) much smaller than the radius of the submerged bulb?
12. A smooth wooden block rests on the equally smooth bottom of an aquarium tank, and the contact is so good that no water gets under the block when the tank is filled. Is there a buoyant force on the block? Will it rise to the surface? Explain.

PROBLEMS

1. What is the average pressure due to the weight of a 60-kg person if the effective area of each foot is 95 cm² and the person is standing on both feet? On just one foot?

    Ans. 3.09 N/cm² = 3.09 × 10⁴ Pa; 6.19 × 10⁴ Pa

2. A 50-kg girl is skating on ice. Find the pressure exerted by one skate when the other is not touching the ice if the length of the blade in contact with the ice is 0.15 m and the runners are 0.0045 m wide.

3. The gauge pressure in an automobile tire is 1.8 × 10⁵ Pa. If the wheel on which the tire is mounted supports 3.6 kN, what area of the tire is in contact with the ground? (Neglect the mechanical strength of the casing.)   *Ans.* 0.02 m²

4. Find the force on the glass side of an aquarium containing salt water with a density of 1,030 kg/m³ if the glass is 0.6 m wide and the water behind it is 0.4 m deep.

5. A submarine is designed to withstand a maximum water pressure of 1.6 MPa. If the density of seawater is 1,028 kg/m³, find the maximum depth to which the submarine can be taken without exceeding the design pressure. Find the net force on a rectangular hatch of dimensions 1.1 by 0.7 m at this depth.   *Ans.* 159 m; 1.23 MN

6. The water level in a reservoir standpipe is 80 m above the lowest part of town. What is the maximum water pressure available due to this head of water?

7. The apparatus of Fig. 12.3*b* contains alcohol in one arm and distilled water in the other. When the level of the water is at 283 mm, the alcohol level is at 359 mm. Find the density of alcohol.

    Ans. 788 kg/m³

8. Water and oil are standing in opposite legs of a U-tube open at both ends (Fig. 12.3*a*). Water fills the bottom and stands 322 mm above the oil-water interface. How high does the oil stand above the interface if its density is 810 kg/m³?

9. A lock gate is 7 m wide and 8 m high. The height of water on one side is 7.5 m and on the other side is 3 m. What is the net horizontal force on the gate due to water pressure?   *Ans.* 1.62 MN

10. A submarine is at a depth of 80 ft in seawater which has a density of 2 slugs/ft³. Find the pressure due to the water and the resulting force on a hatch of dimensions 3 by 2 ft.

11. One end of a closed hydraulic system is 4 m above the other end. The system is filled with oil of density 800 kg/m³. At the higher end, a force of 200 N is applied on an area of 10 cm². What is the net force on a piston with an area of 20 cm² at the lower end?   *Ans.* 463 N

12. A liquid-filled reservoir is connected to two cylinders with closely fitting pistons which are at the same level. One cylinder has a diameter of 15 mm, while the other has a diameter of 450 mm. If a force of 24 N is exerted on the smaller piston, find the increase in pressure on the larger piston. How large a force is required on the larger piston to keep it at rest if frictional forces are negligible?

13. Assuming an average weight density of 995 kg/m³ for the human body, find the volume occupied by a 50-kg swimmer. What pressure due to the water does the swimmer experience at a depth of 3.2 m in fresh water?

    Ans. 0.0503 m³; 31.4 kPa = 31.4 kN/m²

14. A piece of copper is "weighed" with an equal-arm laboratory balance. When the copper is in air, the balancing mass is 0.224 kg. When it is immersed in water and in kerosene, the balancing masses are 0.199 and 0.204 kg, respectively. Find the densities of copper and kerosene.

15. A barge with vertical sides is used to transport automobiles down the Mississippi River. If 15 cars, each with a mass of 1,200 kg, are driven onto the barge, what additional volume of water must the barge displace? How much deeper does the barge ride in the water if it is 16 m long and 8 m wide?   *Ans.* 18 m³; 0.141 m

16. A truck loaded to a total weight of 10 tons drives onto a ferryboat, causing it to sink $\frac{3}{4}$ in. deeper into the water. What is the area of the horizontal section of the boat at the waterline?

17. Find the volume and the mass of a specimen of copper that has an apparent weight of 4.5 N under water.   *Ans.* 5.79 × 10⁻⁵ m³; 0.517 kg

18. A body was "weighed" in water, in oil, and in alcohol. Its loss of weight in water was 2.25 N, in oil 1.47 N, and in alcohol 1.80 N. What is the specific gravity of the oil? Of the alcohol?

19. A piece of glass with a mass of 0.644 kg is weighed immersed in various liquids. It weighs 4,575 N in water and 4,167 N in glycerin. Find the density of the glass and the specific gravity of the glycerin.   *Ans.* 3,635 kg/m³; 1.23

20. A piece of cork weighs 5.4 N. A 18-N sinker is attached to it. With only the sinker in water, the cork and sinker weigh 20.7 N. With the sinker and cork both in water, they weigh 0.9 N. What is the density of the cork? What is its specific gravity? What is the density of the sinker?

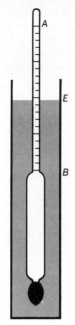

PROBLEM 21

21. A hydrometer (see accompanying figure) has a mass of 110 g. The graduated stem BA is 30 cm long and has a cross-sectional area of 0.5 cm², while the bulb below B has volume of 107.5 cm³. What fraction of the stem BA will be submerged when the hydrometer floats in water? What is the lowest specific gravity the hydrometer can read? What is the specific gravity of a liquid in which the hydrometer floats with the liquid level at B? *Ans.* $\frac{1}{6}$; 0.90; 1.023

22. A piece of wood with a weight of 3.72 N is immersed in water by use of a sinker that weighs 1.91 N in water. The combined weight of the wood and sinker when both are immersed is 1.26 N. Find the density of the wood.

23. A capillary tube with an inside radius of 0.3 mm is dipped in water. What is the weight of the water raised by capillary action above the normal water level? To what height is the water raised?
Ans. 1.37×10^{-4} N; 49.5 mm

24. A solid glass rod is immersed in water so that its axis is perpendicular to the surface of the water. The rod has a circular cross section and a diameter of 5.8 mm. Find the downward force exerted on the rod by surface tension.

25. A thin sheet of metal is bent in the form of a hollow square, and the open square end is dipped into a soap solution whose surface ten-

sion is 0.025 N/m. The length of each side of the square is 32 mm. With what force does the surface tension pull the metal into the solution?
Ans. 6.4 mN

26. What is the excess pressure inside a soap bubble that is 36 mm in diameter, assuming 0.026 N/m is the surface tension of the soap solution?

27. Find the work which must be done to increase the outside surface area of a soap bubble by 300 cm² if the surface tension of soap film is 0.028 N/m. *Ans.* 1.68 mJ

28. What is the diameter of a soap bubble when the excess pressure inside is 8.84 N/m² and the surface tension of the soap solution is 0.027 N/m?

29. One limb of a U-tube (Fig. 12.17) is 4 mm in diameter, and the other is 1 mm in diameter. What will be the difference in surface levels in the two tubes when mercury is poured into them? Take the angle of contact as 140° for mercury. *Ans.* 9.4 mm

30. A tube is bent in the form of a U (Fig. 12.17). The diameter of the larger arm is 6 mm, and that of the smaller arm is 0.8 mm. Find the difference in levels when the U-tube contains water.

31. A capillary tube whose inside diameter is 0.88 mm is dipped in glycerin. The glycerin rises 23.3 mm in the tube. If the density of glycerin is 1,260 kg/m³, what is its surface tension? Assume the angle of contact is zero. *Ans.* 0.0633 N/m

32. Find the volume of cork that must be employed in a life preserver if it is designed to support one-fifth of a man's body out of fresh water, assuming the man weighs 720 N and has a specific gravity of 0.995 and that all the cork is submerged.

33. A diver with his suit and apparatus weighs 1,400 N. Blocks of lead with a volume totaling 820 cm³ attached to his shoes just cause him to sink. What volume of water is displaced by the suit with the diver inside if the lead blocks are outside? *Ans.* 0.151 m³

34. A cylindrical block of wood has a cross-sectional area of 120 cm² and a height of 40 cm. It floats vertically in water because 500 g of lead is fastened to the lower edge. How much of the wood projects out of the water? (Take the density of wood to be 550 kg/m³.)

35. A piece of wood with a volume of 800 cm³ floats with seven-tenths of its volume under water. Find the smallest mass of iron of specific gravity

7.8 which will cause the wood to sink to the bottom.   *Ans.* 0.275 kg

36. A 0.4-kg plastic sphere has a density of 900 kg/m³. It floats at a water-kerosene interface (Fig. 12.9). What volume V_2 lies beneath the interface?

37. A 3-kg mass of wrought iron floats in a beaker of mercury. What fraction of the iron is submerged? If sufficient water is poured into the beaker to cover the iron completely, what fraction of the iron is submerged in the mercury?
                                    Ans. 0.57; 0.54

38. A cylindrical buoy of mass m and radius R floats in water of density d with two-thirds of its height h submerged. If it is depressed a distance $y \ll h$ and released at rest, show that the resulting motion is simple harmonic. Show that the frequency is $(R/2)\sqrt{gd/\pi m}$.

In liquids the pressure increases linearly with depth because the density does not change much as the pressure rises. In gases, however, the density is proportional to the pressure if the temperature doesn't change. In this chapter we consider first how the pressure varies with height in the atmosphere, and then we turn to fluids in motion. Whenever there is relative motion between a fluid and bodies in contact with the fluid, there are additional phenomena in which we shall be interested.

13

ATMOSPHERIC PRESSURE AND FLUIDS IN MOTION

13.1 PRESSURE OF THE ATMOSPHERE

Just as water is the most widely distributed and most important of liquids, so air is the most important and intimate of gases. It consists for the most part of nitrogen and oxygen. In spite of the fact that there is no chemical union between the two, the composition of the air is extraordinarily constant. Up to a height of 11 km it contains about 21 parts of oxygen to 78 parts of nitrogen by volume. Besides oxygen and nitrogen, the air contains small amounts of other gases, among which are water vapor, argon, and carbon dioxide. A cubic meter of air weighs over 12.7 N at sea level, although to an ordinary observer the air seems to have no weight and to offer little resistance to bodies moving through it.

Since it has weight, a column of air exerts pressure, just as a column of liquid does. However, Eq. (12.2) is not directly applicable, because the density of a gas, unlike that of a liquid, depends on the pressure (Sec. 14.2). Though a liter of air weighs little, even at ground level, the height of the atmosphere is large, and the weight of all this air pressing on the earth is great.

About 1644 Torricelli, a pupil of Galileo, measured atmospheric pressure in the following way. A glass tube closed at one end was filled with mercury. A finger was then placed over the open end of the tube, the tube inverted in a basin of mercury, and the finger removed from the open end of the tube under the mercury. The mercury in the tube sank until its level was about 760 mm (Fig. 13.1) higher than the level in the basin. The pressure of the atmosphere on the mercury in the basin supported the mercury in the tube. Since the density of mercury is 13,600 kg/m^3, the pressure exerted by the atmosphere as given by Eq. (12.2) is $p = dgy = 13,600$ kg/m$^3 \times 9.80$ N/kg $\times 0.760$ m $= 1.013 \times 10^5$ N/m^2. In meteorology pressures are often reported in millibars; one standard atmosphere = 1,013 millibars. Another unit in which pressures are measured is the *torr*, named in honor of Torricelli; since one torr is the pressure due to one millimeter of mercury, one atmosphere is 760 torr. One atmosphere can support a column of water 10.3 m (33.8 ft) high; in British units one atmosphere pressure corresponds to 14.7 lb/in.^{2}

Our atmosphere is a great ocean of air which becomes steadily less dense as we go upward. In 1646 the sickly French physicist Pascal sent his strong brother-in-law up a mountain with two Torricelli tubes to test the hypothesis that the mercury level drops as the altitude increases. He found that the mercury column fell 15 cm for an increase in altitude of 1.6 km. If we go to an altitude of 5.5 km, the pressure is close to half that at sea level. The approximate variation of pressure with altitude is shown in Fig. 13.2. Less than one-quarter of the atmosphere remains above a height of 11 km.

Breathing involves an application of atmospheric pressure. A reduction

FIGURE 13.1

Torricelli showed that the atmosphere supports a column of mercury approximately 760 mm high in a closed tube with vacuum above the mercury. The vertical height of the mercury is independent of the shape and length of the tube.

of pressure is caused by a movement of the diaphragm. The greater pressure of the outside air causes a fresh supply to flow into the lungs. Then air is forced out when the internal pressure is made greater than atmospheric. Sucking and drinking animals take advantage of atmospheric pressure to aid them in these operations. They reduce the pressure in the mouth and allow water to be forced into the mouth by the outside atmospheric pressure.

13.2 BUOYANT EFFECT OF THE AIR

Archimedes' principle applies to all fluids. In general, the densities of gases are much smaller than those of liquids, and the buoyant effects are correspondingly reduced. Nevertheless there are many examples of situations in which the buoyant effects of gases—in particular of air—are of considerable importance. The density of air is 1.293 kg/m³ at 0°C and one standard atmosphere pressure.

A simple demonstration of the buoyant effect of air can be made by suspending a lead ball (Fig. 13.3) from one side of a small balance and a large hollow brass sphere from the other side. The hollow sphere is just heavy enough to balance the lead ball when both are in air. If the balance together with the suspended spheres is placed under a bell jar and nearly all the air removed by means of a pump, the lead ball no longer balances the hollow sphere. This is because the buoyancy of the air on the hollow brass sphere is greater than on the lead ball; when this lift has been removed, the true weights of the spheres become evident, and the hollow sphere weighs more than the lead sphere.

The gross lifting capacity of a balloon is equal to the weight of the air it displaces. The pressure and density of the air become less at higher elevations. However, as the pressure outside the balloon decreases, the lighter gas inside the balloon expands, thus displacing a larger volume of the less dense air. Ballast may be carried along and thrown overboard to make the balloon rise higher. By allowing some of the gas in the balloon to escape the balloon can be made to descend.

☐ **Example** A balloon has a volume equal to that of a sphere 15 m in radius. What is the gross mass which it will lift when the density of the air is 1.2 kg/m³?

$$\text{Volume of balloon} = \tfrac{4}{3}\pi (15)^3 = 14{,}000 \text{ m}^3$$

$$
\begin{aligned}
\text{Mass of air displaced by balloon} &= \text{volume} \times \text{density} \\
&= (14{,}000 \text{ m}^3)(1.2 \text{ kg/m}^3) \\
&= 16{,}800 \text{ kg} \qquad \square
\end{aligned}
$$

13.3 FLUID FLOW

In hydrodynamics and aerodynamics we are concerned with the motion of a body through a fluid or the equivalent movement of fluid past a body. The forces involved depend only on the *relative motion* of body and fluid. The forces on an airplane wing at rest in a wind tunnel are the same as those on the same wing moving through air at rest, provided the relative velocities are the same in both cases and the air has the same composition, pressure, and temperature.

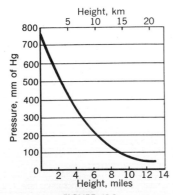

FIGURE 13.2

Variation of atmospheric pressure with height above sea level.

For the moment suppose that we have the fluid in motion. Imagine many small elements of the fluid which are individually colored so one can see the motions of these "fluid particles" during flow. The trajectory of a single particle of the fluid is called a *pathline* or a *line of flow*. In general, the velocity of each such particle changes along a line of flow, but let us suppose that every particle passing a given point follows the same pathline and further that the velocity of each particle which passes a given point is the same as that of every other particle as it passes that point. In such a case the flow is described as *steady* or *stationary*. (In nonsteady flow the velocity of particles at a given point is a function of time, and in *turbulent* flow the velocities change erratically from point to point as well as from one time to another.)

In any flow a *streamline* is a curve whose tangent at any point is in the direction of the particle velocity at that point. For steady flow the streamlines coincide with the pathlines. If we consider all the streamlines passing through the boundary of an element of area A (Fig. 13.4), they enclose a region called a *tube of flow*. In steady flow no particles cross the boundaries of a flow tube; every fluid particle which enters at one end leaves at the other.

Imagine a tiny paddle wheel carried along in the tube of flow of Fig. 13.4. In some situations the paddle wheel would travel without rotation; then the flow at the position of the paddle wheel would be *irrotational*. In other flow situations the paddle wheel would rotate, and the flow would be *rotational*. Swirling eddies are an evidence of rotational flow for which angular momentum is an important property. The rotationality is conserved.

13.4 PRESSURE IN A MOVING FLUID; BERNOULLI'S THEOREM

As in many branches of physics, it is convenient to begin by considering a highly idealized situation. In this section we discuss the *steady, irrotational* flow of an *incompressible, frictionless* (nonviscous) fluid. Such an analysis was characterized by von Neumann as the study of "dry water," and indeed there is no fluid which behaves exactly like our ideal fluid under all conditions. Nevertheless, the results derived for the ideal fluid apply *in some situations* to many fluids, including air.

Following Daniel Bernoulli (1700–1782), we consider an ideal liquid flowing through a pipe of varying cross section (Fig. 13.5). Because the cross section of the pipe is smaller at b than it is at a, the velocity of the liquid must be greater at b than at a so that the flow of the liquid through each cross section of the pipe can be the same. Consequently, the momentum of the liquid per unit volume increases in going from a to b. According to Newton's second law, a force is necessary to produce a change of momentum. In this case, the force results from a difference in pressure between a and b. The pressure at a is greater than the pressure at b, thereby producing a positive acceleration of the liquid and an increase in its momentum per unit volume. *The pressure in the fluid is smaller where the velocity is greater.*

The change in pressure can be measured by observing the heights of the liquid in transverse manometers at a, b, and c. Since the liquid is *incompressible*, the amount passing through every cross section of the tube per second is the same. Let v_a, v_b, and v_c represent the speeds and

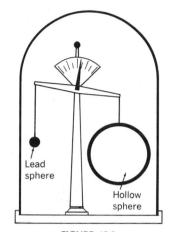

FIGURE 13.3
A body immersed in air is buoyed up by a force equal to the weight of the air it displaces. The two spheres balance each other at normal air pressure, but when the jar is evacuated, the larger sphere sinks.

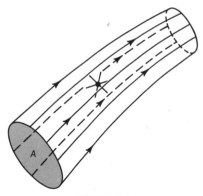

FIGURE 13.4
A tube of flow bounded by streamlines. A little paddle wheel in the fluid remains at rest anywhere the flow is irrotational but rotates anywhere the flow is rotational.

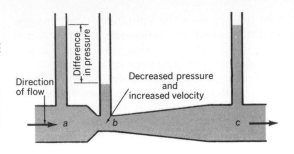

FIGURE 13.5
In a horizontal tube of varying cross section the flow velocity is greater in regions of reduced cross-sectional area; where the flow velocity is greater, the pressure is lower.

A_a, A_b, and A_c represent the cross-sectional areas at a, b, and c, respectively. Then

$$v_a A_a = v_b A_b = v_c A_c \qquad (13.1)$$

For steady flow the speed is inversely proportional to the cross section of the tube.

☐ **Example**   The cross section of the tube at a is 10 cm² and at b is 2 cm². If the velocity of the stream at a is 1.2 m/s, what is it at b?

$$\text{Velocity at } b = \text{velocity at } a \times \frac{\text{area at } a}{\text{area at } b}$$

$$= 1.2 \text{ m/s} \times \frac{10 \text{ cm}^2}{2 \text{ cm}^2} = 6 \text{ m/s} \qquad ☐$$

Consider the case in which all parts of a tube of varying cross section (Fig. 13.6) are not at the same height. Since (1) the liquid is incompressible, (2) the flow is streamlined, and (3) there is no fluid friction (no viscosity), we can use the conservation of energy to derive a relationship between the pressure, velocity, and height at any one point and pressure, velocity, and height at some other point. Let volume V of liquid be forced from a to c. The work done *on* the system in forcing the liquid past a is $F_a S_a = p_a A_a(V/A_a) = p_a V$, where the subscript a represents the value at point a. The work done *by* the system when volume V is forced past point c is $F_c S_c = p_c A_c(V/A_c) = p_c V$. The *net* work done *on* the system is $(p_a - p_c)V$. Since there are no energy losses, we have

Work done on system = energy of volume V at c − energy of V at a

$$(p_a - p_c)V = [\tfrac{1}{2}(dV) v_c^2 + dVgy_c] - [\tfrac{1}{2}(dV) v_a^2 + dVgy_a] \qquad (A)$$

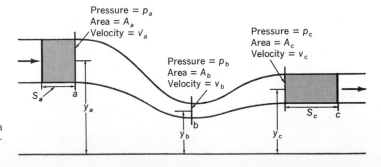

FIGURE 13.6
Streamlined flow of a liquid through a tube of variable cross section at different heights.

where d is the density of the liquid and dV the mass of the volume V. If we divide both sides of Eq. (A) by V and rearrange the terms, we obtain *Bernoulli's theorem*,

$$p_a + \tfrac{1}{2}dv_a{}^2 + dgy_a = p_c + \tfrac{1}{2}dv_c{}^2 + dgy_c = \text{constant} \qquad (13.2)$$

The sum of the pressure, the kinetic energy per unit volume, and the potential energy per unit volume is the same at all points along the same streamline in the case of streamlined flow for a nonviscous, incompressible fluid.

In the flow of a real fluid frictional forces act, and as we move in the direction of the fluid velocity, the sum of the three terms of Eq. (13.2) decreases (Sec. 13.6). Although we have derived Bernoulli's theorem for an incompressible fluid, Eq. (13.2) is approximately true for a gas so long as the speed is well below the speed of sound (about 340 m/s or 740 mi/h for air at 20°C).

□ **Example** Water of density 1,000 kg/m³ is passing through the tube of Fig. 13.6. The cross-sectional area at a is twice that at b. At a, $v = 6$ m/s, $y_a = 10$ m, and $p_a = 140{,}000$ N/m². If $y_b = 8$ m, find the velocity and pressure at b.

Since the area at a is twice that at b, the speed at b must be twice that at a by Eq. (13.1); hence $v_b = 12$ m/s. By Eq. (13.2)

$$140{,}000 \, \text{N/m}^2 + \tfrac{1}{2}(1{,}000\,\text{kg/m}^3)(36\,\text{m}^2/\text{s}^2) + (1{,}000\,\text{kg/m}^3)(9.8\,\text{m/s}^2)(10\,\text{m})$$
$$= p_b + \tfrac{1}{2}(1{,}000\,\text{kg/m}^3)(144\,\text{m}^2/\text{s}^2) + (1{,}000\,\text{kg/m}^3)(9.8\,\text{m/s}^2)(8\,\text{m})$$
$$256{,}000 \, \text{N/m}^2 = p_b + 150{,}400 \, \text{N/m}^2$$
$$p_b = 105{,}600 \, \text{N/m}^2 = 105.6 \, \text{kPa} \qquad \Box$$

13.5 APPLICATIONS OF BERNOULLI'S THEOREM

Several common phenomena involve Bernoulli's theorem.

A Ball in an Air Jet A ping-pong ball can be balanced on an air jet with the ball supported by the rate of transfer of momentum from the airstream. When the ball moves slightly to one side, the airspeed is lower and the pressure correspondingly greater on that side than on the other. Consequently there is a net force moving the ball back toward the center of the stream. Similarly, the ball may be held in an inverted funnel (Fig. 13.7) by a downward stream of air. The air speeds up as it passes the constricted region between the ball and the funnel. As a result the pressure there is reduced below atmospheric, and this pressure difference in turn gives rise to a net force holding the ball in the funnel.

The Sprayer The forward stroke of the piston (Fig. 13.8) produces a stream of air past the end of the tube D. The other end of this tube is

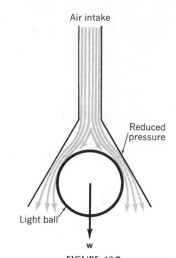

FIGURE 13.7
A light ball can be supported in an inverted funnel by the pressure reduction associated with the Bernoulli theorem.

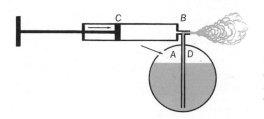

FIGURE 13.8
In a sprayer the high velocity of the air at B produces a reduced pressure there. As a result liquid rises in tube D and is atomized as it enters the air streaming toward the opening.

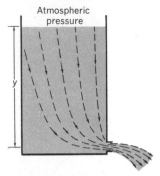

Atmospheric
pressure

y

FIGURE 13.9
Efflux of fluid from an orifice.

immersed in the liquid to be sprayed. The stream of air flowing past the open end of the tube reduces the pressure on the liquid in the tube. Atmospheric pressure acting on the surface of the liquid in A raises liquid in the tube D, from which it is carried away by the stream of air. A spray results from this mixture of air with the fine particles of liquid.

Flow of Liquid through an Orifice In Fig. 13.9 liquid is flowing out of an orifice. Both at the top of the liquid and at the orifice the pressure is atmospheric. At the top of the container the liquid has potential energy and no kinetic energy. If we pick the opening as our position of zero height, the escaping liquid has kinetic energy but no potential energy. Applying Bernoulli's theorem to points at the top of the fluid and at the orifice yields

$$p_{atm} + 0 + dgy = p_{atm} + \tfrac{1}{2}dv^2 + 0$$

where p_{atm} represents atmospheric pressure, d the density of the fluid, and v the escape velocity.

$$dgy = \tfrac{1}{2}dv^2 \qquad \text{or} \qquad v = \sqrt{2gy} \qquad (13.2a)$$

Figure 13.10 shows streamlines associated with flow through a sharp-edged orifice. In this case the fluid is still being accelerated as it passes the sharp edges at the opening. As the fluid gains speed, the area of flow decreases in accordance with Eq. (13.1). The flow lines converge to give a *vena contracta*, with the effective area a little less than two-thirds the actual area for a sharp circular orifice. A well-designed nozzle gives a flow which utilizes the actual opening far more effectively.

Two familiar and very important applications of the Bernoulli effect are the lift on the wing of an aircraft and the curving of a rapidly moving, spinning ball. Since both depend on the fact that there is some fluid friction, we introduce the concept of viscosity before we discuss them.

13.6 VISCOSITY

When real fluids are in motion and the shape as well as the position of a given fluid element may be changing, there is an internal friction between adjacent layers moving with different speeds; this internal friction

FIGURE 13.10
Converging of streamlines to form a vena contracta at a sharp-edged orifice. (*Courtesy of F. N. M. Brown.*)

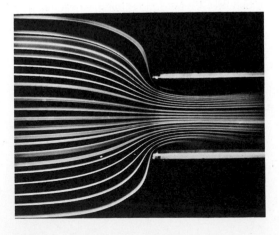

is known as *viscosity*. When two beakers, one containing heavy oil and the other water, are tilted from side to side, the mobility of the water is clearly greater than that of the oil; oil is more viscous than water. Gases have lower viscosity than liquids.

Consider the flow of a fluid over the surface SS' (Fig. 13.11). Whenever a fluid is in contact with a solid surface, there is an infinitesimal layer which does not move relative to the surface regardless of relative motion between fluid and solid. Thus the fluid in contact with SS' is at rest because of adhesion, but each successive layer moves with respect to the layer directly below it with the speed increasing with the distance from SS'. Each horizontal layer is acted upon by the layer above with a tangential force in the direction of the motion and by the layer below with a retarding force. These forces arise from momentum transfer between the molecules of the fluid. The dashed figures *abcd* and *efgh* (Fig. 13.11) show the distortion of the fluid as it moves. The upper layer *fg* travels faster than the lower layer *eh*. Consider a tiny imaginary cube represented by *abcd* which becomes *efgh* a short time later. Let F be the tangential force on the plane which cuts the page in *bc* (and later in *fg*). Let A be the area of this surface of the cube, Δl the distance *ab*, and Δv the speed of *bc* relative to *ad*. When Δv is not too great, F/A is proportional to $\Delta v/\Delta l$:

$$\frac{F}{A} = \eta \frac{\Delta v}{\Delta l} \qquad (13.3)$$

where η is a constant called the *coefficient of viscosity* (Table 13.1). It is equal to the force per unit area necessary to maintain unit difference of velocity between two parallel planes when the planes are unit distance apart. The coefficient of viscosity of a typical liquid changes with both temperature (Fig. 13.12a) and pressure (Fig. 13.12b). For gases the viscosity behaves differently; it is proportional to the square root of the temperature and independent of pressure over a broad range. From Eq. (13.3) it is clear that η may have the dimensions newton-seconds per square meter. In many handbooks viscosities are given in poises; 1 poise = 0.1 N-s/m^2 = 1 dyn-s/cm^2.

In writing Eq. (13.3) we have assumed that the fluid flows in thin sheets (lamina). Such flow is said to be *laminar*. It is observed only when the velocity is relatively small; at higher velocities the flow becomes turbulent. Since in turbulent flow the velocities of the fluid particles change erratically, the instantaneous velocity at a point is not the same as the time-average velocity. Then, if average velocities are used in Eq. (13.3), the proportionality constant between F/A and $\Delta v/\Delta l$ is much greater than η because small patches of turbulent flow (eddies) also carry transverse momentum from one layer to another.

13.7 LIFT AND DRAG

If a smooth sphere were to move at constant velocity through an ideal nonviscous fluid, theory predicts that there would be no resistance at all! This paradox, attributed to d'Alembert or Euler, arises because in the absence of viscosity the velocity of the fluid and hence the pressure at every point such as P' on the downstream side would be exactly equal to that at the corresponding point P on the upstream side (Fig. 13.13a). The

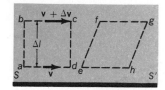

FIGURE 13.11
Displacement of layers of a moving fluid relative to a surface at rest.

TABLE 13.1
Viscosities of Fluids
(All at 20°C except as noted)

	η, N-s/m^2
Liquids:	
Ethyl alcohol	0.0012
Glycerin	1.48
Mercury	0.0015
Oil:	
SAE 10 at 55°C	0.16–0.22
SAE 20 at 55°C	0.23–0.30
Water	0.001
Gases:	
Air	18×10^{-6}
Argon	22×10^{-6}
Hydrogen	9×10^{-6}
Oxygen	20×10^{-6}

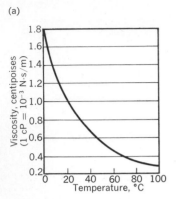

(a)

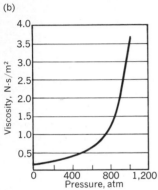

(b)

FIGURE 13.12
Viscosities of liquids vary with temperature and pressure. (*a*) The viscosity of water decreases as the temperature rises. (*b*) The viscosity of a typical oil becomes greater as the pressure increases.

FIGURE 13.13
(*a*) When the ball is not rotating, the flow of air past a moving ball is symmetric about the velocity vector. If there were no air friction, the pressure at P' on the downstream side would be equal to the pressure at the corresponding point P on the upstream side. With air friction the pressure is lower at P' than at P, and there is a net drag on the ball. (*b*) When the ball is rotating, the spin leads to reduced air velocity and increased pressure at A and to increased velocity and decreased pressure at B.

total fluid force on the back side would push the sphere forward just as hard as the fluid force on the front side holds it back. However, in any real fluid viscosity immediately introduces two retarding forces: (1) The fluid molecules in contact with the sphere remain attached, and shear stresses [Eq. (13.3)] cause a tangential decelerating force at the surface; and (2) the pressure at each point P' falls below that at the corresponding point P. The component of the resultant fluid force parallel to the relative velocity between the fluid and any object is called the *drag*.

For real fluids the fluid velocity relative to a solid surface is zero at the boundary. The *velocity gradient* $\Delta v/\Delta l$ and the shear stress have their maximum values at the surface and decrease into the fluid. At the extreme of high velocity and low viscosity the only significant viscous shear is found in a thin zone called the *boundary layer*. Outside this region the velocity changes only slowly with distance, and so the viscous stress is small. For an airplane wing the boundary layer may be less than a millimeter thick.

The Curving Ball A ball follows a curved path if it is hit or thrown in such a way that it has both a high speed and a large spin about an axis perpendicular to the velocity. This is readily explained in terms of Bernoulli's theorem and viscosity. When we consider the center of the ball as the origin of a coordinate system (Fig. 13.13), air is rushing past the ball from right to left. As the ball spins, it drags a thin neighboring layer of air with it since the molecules right next to the ball move with the surface. The combination of the spinning motion with a large translational speed results in a concentration of streamlines at side B. Near B the spin and translation carry molecules in the same direction, and the resultant air velocity there is large. On side A the spin of the ball carries air to the right and the translation to the left, so that the resultant airspeed near A is less than that near B. By Bernoulli's theorem the pressure is less where the velocity is greater; here the pressure is less on side B. Consequently there is a force on the ball directed from A toward B, and the ball curves.

(a)

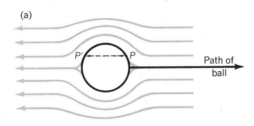

Motion of air relative to ball

(b)

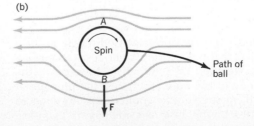

Such a force component perpendicular to the relative translational velocity is called *dynamic lift*. No lift results from symmetrical streamlines like those of Fig. 13.13a. To obtain dynamic lift on an object, an asymmetric flow pattern is required, corresponding to higher flow speeds on one side than on the other.

Lift on an Airfoil In a well-designed aircraft wing (or other airfoil) the velocity of the air above the wing relative to the aircraft is greater than the velocity below the wing (Fig. 13.14a). As a consequence, the pressure is less above the wing than it is below; this pressure difference leads to a *net lifting force*. Although the flow in Fig. 13.14a appears to be irrotational everywhere, this is not the case in the boundary layer. If the flow were perfectly irrotational, there would be no lift. In Fig. 13.14b the angle between the airfoil and the flow has been increased to the point where a large turbulent wake is produced, reducing the lift and increasing the drag.

When a stream of air strikes a surface inclined to the flow lines, such as the airfoil of Fig. 13.14b, there is a net force on the foil which arises from the rate of change of momentum of the air. The upward component contributes to the lift, the horizontal component to the drag. The lift due to the Bernoulli effect is enough to support many aircraft in normal flight, so that no lift by impact is needed and the lower wing surfaces can be horizontal. When a plane is taking off, part of the lift is usually contributed by the impact of air on the lower wing surfaces. To enhance these forces, the angle of attack is increased and the wing flaps are lowered. This increases drag as well as lift. The lift due to the Bernoulli effect can be simply demonstrated by blowing over a sheet of paper held at adjacent corners; lift due to impact can be shown by blowing under the sheet.

13.8 THE REYNOLDS NUMBER AND DYNAMIC SIMILARITY

For speeds considerably less than the speed of sound the nature of the flow about a sphere is determined essentially by the Reynolds number $\Re$, a combination of the density d and viscosity η of the fluid, the diameter D of the sphere, and the relative velocity v.

$$\Re = \frac{dvD}{\eta} \qquad (13.4)$$

FIGURE 13.14
(a) In airflow past a tapered airfoil the flow velocity is increased above the airfoil and the pressure correspondingly decreased; the result is net lift on the airfoil. (b) A stalled airfoil in which a large turbulent wake has reduced lift and increased drag. (*Courtesy of F. N. M. Brown.*)

(a)

(b)

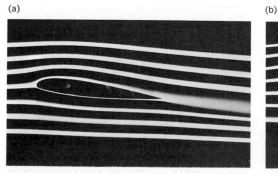
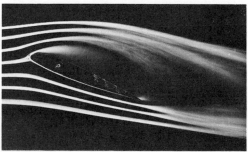

The Reynolds number is a measure of the ratio of the inertial force to the viscous force on a small element of the fluid; it relates the ratio of the rate of transfer of fluid momentum per unit area to the viscous drag per unit area. When $\mathfrak{R}$ is small, viscous forces are dominant and inertial effects subordinate. For large $\mathfrak{R}$ inertial reactions outweigh viscous forces, and the flow pattern is quite different. The Reynolds number is dimensionless, having the same value in any consistent system of units.

Not only does the Reynolds number determine the nature of the flow about a sphere, but *flow patterns are the same for any two objects of similar shape at the same Reynolds number* (provided certain conditions are satisfied). This important fact, known as the *principle of dynamic similitude*, permits us to build a model airplane, test it in a small wind tunnel, and make good predictions for a full-sized airplane in the same Reynolds number range. The concept of dynamic similarity is of paramount importance in several branches of engineering.

For low Reynolds number ($\mathfrak{R} < 5$) flow is laminar. The streamlines are symmetrical about a sphere or cylinder (Fig. 13.13a). Consider the motions of three spheres: a baseball falling in water, a dust particle, and a helium-filled balloon in air. Each is acted upon by (1) an external force mg (its weight) and (2) the net force exerted by the surrounding fluid. This latter force is the resultant of the buoyant force $\mathbf{F}_{buoyancy}$ associated with Archimedes' principle and the additional force $\mathbf{F}_{fluid\ dynamic}$ which comes into play when there is relative motion of sphere and fluid and which always opposes the motion. By Newton's second law, for each sphere

$$\mathbf{F}_{external} + \mathbf{F}_{buoyancy} + \mathbf{F}_{fluid\ dynamic} = m\mathbf{a} \tag{13.5}$$

[For the helium-filled balloon $\mathbf{F}_{buoyancy}$ is greater than the weight $\mathbf{F}_{external}$, so the balloon "falls" upward but Eq. (13.5) is still applicable.]

In 1845 Stokes showed that the viscous force on a sphere at low Reynolds number is given by

$$F_{viscous} = 3\pi\eta vD \qquad \text{for a sphere at low } \mathfrak{R} \tag{13.6}$$

Since the viscous force is proportional to D while the weight of sphere is proportional to D^3, it is clear why small dust particles have a small terminal velocity, achieved when the resultant force is zero in Eq. (13.5). In the low-Reynolds-number regime, drag is less on a smooth surface than on a rough one.

13.9 DRAG AT HIGHER REYNOLDS NUMBER

As the relative velocity v increases, there comes a point at which the flow lines no longer follow the surface of a sphere or cylinder all the way back to the center, but instead a pair of small vortices appear just behind the cylinder ($\theta = 180°$). For a cylinder this develops for $5 < \mathfrak{R} < 50$ [D in Eq. (13.4) is now the diameter of the cylinder]. The pressure at the downstream side is less than that at the upstream side; the streamlines begin to separate from the cylinder near $\theta = 180°$. As $\mathfrak{R}$ increases, the separation occurs earlier and earlier (Fig. 13.15), until, when $\mathfrak{R} > 5,000$, the separation begins very near $\theta = 90°$. The entire wake is essentially turbulent.

In these flow regimes the drag can be greatly decreased by streamlining; compare the flow of Fig. 13.15 with that in Fig. 13.14a. It is convenient to write the drag force in the form

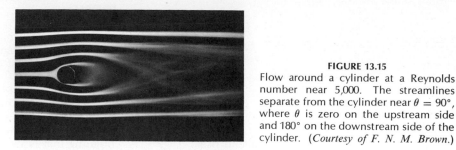

FIGURE 13.15
Flow around a cylinder at a Reynolds number near 5,000. The streamlines separate from the cylinder near $\theta = 90°$, where θ is zero on the upstream side and 180° on the downstream side of the cylinder. (*Courtesy of F. N. M. Brown.*)

$$F_{\text{drag}} = C_D S \left(\tfrac{1}{2} dv^2\right) \qquad (13.7)$$

where S is the cross-sectional area of the object normal to the flow, d the density of the fluid, v the relative velocity, and C_D is a factor called the *drag coefficient*. Figure 13.16 shows the drag coefficient for a cylinder as a function of $\Re$. In the low-Reynolds-number regime C_D decreases with v; recall that for low $\Re$ the drag force is roughly proportional to v, so C_D varies as $1/v$.

In the Reynolds number range $5,000 < \Re < 2 \times 10^5$, C_D remains almost constant at 1.2, but a sharp drop occurs in the vicinity of $\Re = 5 \times 10^5$, where the boundary layer on the upstream side becomes turbulent. As a consequence of the increased kinetic energy and momentum near the boundary, the boundary layer maintains attachment to the cylinder to $\theta \approx 140°$ to 150°. Increasing the roughness of the cylinder decreases the value of $\Re$ for which the drop in C_D occurs. Over a considerable range of $\Re$ values the drag is less on a blunt body if it is rough than if it is smooth. This is one reason why golf balls are dimpled.

Our discussion pertains to gases as well as liquids as long as the compressibility can be neglected, but that in turn requires that the ratio of v to the local speed of sound be less than 1. This ratio is called the *Mach number*. As v approaches the speed of sound, the drag coefficient increases abruptly. Once the flow is supersonic, C_D decreases again with v. For speeds near or above sound velocity, flows are the same for similar shapes only if both the Mach and Reynolds numbers are the same.

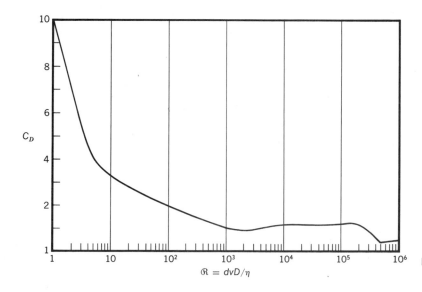

$\Re = dvD/\eta$

FIGURE 13.16
Drag coefficient C_D for a circular cylinder with axis normal to the flow.

QUESTIONS

1. What determines how high a balloon rises?
2. Explain why a balloon can remain in stable equilibrium at almost any given height in the atmosphere while a propulsionless submarine cannot ordinarily remain in stable equilibrium at a given depth in the ocean. Under what special circumstance might the submarine be in stable equilibrium at some particular depth?
3. When an object is dropped from a balloon at high altitude, its velocity may increase for a while and then decrease as it nears the earth. Why does this occur?
4. Why does a balloon descend when it releases gas, even though its weight is thereby decreased?
5. What physics underlies the act of filling a fountain pen?
6. An athlete stands on the platform of a sensitive scale. When she takes a deep breath, will her weight as read by the scale change?
7. When there is a steady, streamlined flow of liquid through a pipe of varying cross section, does a small volume element of the fluid remain in equilibrium as it moves through the pipe?
8. The surface of a particular river forms a long inclined plane with the speed of the water constant. If a cork is dropped on the surface, will it move down the plane faster than the water? What experiment could you perform to compare the water speed with that of the cork?
9. Why does the speed of a liquid increase and the pressure decrease when the liquid passes a constriction in a pipe?
10. A tennis ball travels southward, rotating counterclockwise as viewed from above. Which way does it curve? Explain carefully.
11. Discuss the operation of a siphon and a paint sprayer.
12. Bernoulli's theorem [Eq. (13.2)] was derived in the text for an incompressible fluid with no viscosity. How would a finite viscosity affect the relationship qualitatively?
13. By applying Bernoulli's theorem to the siphon (Prob. 16) show that the velocity of efflux is given by $v = \sqrt{2gy}$. What limits the height h over which the siphon can operate? If h is less than this limit, will v depend at all on h? If so, why doesn't h appear in the formula for v?

PROBLEMS

1. Find the mass of air in a small classroom 8 by 5 by 3 m on a day when the density is 1.29 kg/m³. What is the weight of this air?
   Ans. 155 kg; 1,517 N
2. What is the weight of the air in a living room 30 by 12 by 8 ft? The weight density of air is 0.081 lb/ft³.
3. What is the reading of a barometer on top of a building that is 150 m high when the barometer at the base of the building reads 745 mmHg if the average density of air between these heights is 1.24 kg/m³? *Ans.* 731 mmHg
4. A man carries a barometer from the bottom to the top of a building that is 150 m high. At the bottom of the building the barometer reads 751 mmHg, and at the top it reads 737 mmHg. Find the average density of the air between these heights.
5. A balloon with a volume of 708 m³ is filled with helium of density 0.176 kg/m³. The density of air is 1.29 kg/m³. What is the net buoyant force on the balloon? *Ans.* 7,730 N
6. The volume of the gas bags of an airship is 120,000 m³. The density of air at 0°C and at atmospheric pressure is 1.29 kg/m³, and that of the hydrogen with which the gas bags are filled is 0.092 kg/m³. If the mass of the airship and its necessary equipment is 70,000 kg, what is the maximum net lift available for a useful load?
7. Find the velocity of efflux of water from a hole in the side of a tank if the water surface is 1.6 m above the hole. What mass of water is discharged each second if the area of the hole is 30 mm²? (Assume frictionless water and neglect the contraction of the streamlines as they emerge from the hole.) *Ans.* 5.6 m/s; 168 g/s
8. Find the velocity of efflux of water from a hole in the side of a tank when the water in the tank is 3.6 m above the hole. If the area of the hole is 500 mm², how many kilograms will be discharged each minute? (Neglect the contraction of the streamlines as they emerge from the hole and assume frictionless water.)
9. Water flows at the rate of 5 L/min from a hole at the bottom of a tank in which the water is 0.9 m deep. Find the effective area of the hole. Find the rate at which the water would escape if an additional pressure of 10⁵ N/m² were applied to the surface of the water.
   Ans. 19.8 mm²; 17.6 L/min
10. With what speed will water discharge from the bottom of a tank 25 ft deep through an orifice? If the orifice has a diameter of 0.6 in., what weight of water will be discharged each second?
11. A stream of water projected horizontally from a

hose 3.6 m above the ground strikes the ground 7.2 m away. Find the velocity of efflux and the fluid pressure in the nozzle.

Ans. 8.4 m/s; 3.53×10^4 N/m²

12. A reservoir is filled with mercury. In its side there is an opening 0.8 mm in diameter. If the opening is 0.9 m below the surface, how many grams of mercury will escape per second?

13. In a paper machine, paper is formed on a Fourdrinier wire screen, which moves 5 m/s, by ejecting a mixture of paper fibers and water from a slit orifice onto the screen. What gauge pressure is required at the orifice to match the horizontal speed of efflux to the speed of the wire screen moving just below the slit orifice? Take the density of the mixture (known as *stuff*) to be 1,240 kg/m³. *Ans.* 15,500 Pa

14. Find the velocity with which water in an enclosed tank will be forced through an orifice in the side of the tank if the orifice is 1.6 m below the surface of the water and there is a gauge pressure of 10 atm on the surface of the water in the tank.

15. Water flowing from an orifice in the side of a tank situated at a height of 4.9 m from the ground strikes the ground 8 m from the foot of the tank. Compute the escape velocity of the water and the gauge pressure in the tank at the level from which the water is squirting.

Ans. 8 m/s; 3.2×10^4 N/m²

16. A siphon (see accompanying figure) has a tube with a cross-sectional area of 80 mm². Find the velocity of efflux and the volume of water transferred per second if $y = 125$ mm and it is assumed there is no fluid friction. In the real world such fluid friction is not negligible. How would this affect the rate of efflux from the siphon?

17. Find the velocity of efflux and the volume of gasoline transferred per minute by the siphon of the accompanying figure if $y = 0.4$ m, the cross-sectional area of the tube is 50 mm², and fluid friction is assumed to be negligible.

Ans. 2.8 m/s; 8.4 L/min

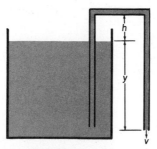

PROBLEMS 16 AND 17

18. The level of the water in the manometer at a (Fig. 13.5) reads 25 cm, and the cross section at a is 16 cm². The cross section of the constriction at b is 2 cm². Find the height of the water column at the constriction b, where the velocity is 1.2 m/s.

19. Water is flowing through the pipe (Fig. 13.5), which has an area of 0.075 m² at a and an area of 0.05 m² at b. The pressure at cross section a is 140 kPa, and that at b is 105 kPa. Find the speed in each of the cross sections. (Assume "dry water.") *Ans.* $v_a = 7.48$ m/s; $v_b = 11.2$ m/s

20. Water is flowing through a horizontal pipe of varying cross section. The pressure is 4 atm where the cross-sectional area is 2,500 mm² and 1.5 atm where the area is 1,250 mm². What is the rate of flow of the water?

21. Water in streamline flow is moving with a speed of 5 m/s through a pipe with a cross-sectional area of 480 mm². If the water gradually descends 10 m and the pipe increases in area to 960 mm², find the speed of flow and the pressure at the lower level if the pressure is 180 kPa at the higher level. *Ans.* 2.5 m/s; 207 kPa

22. A viscometer consists of two concentric cylinders of large diameter with a thin annular space between them. The cylinders have a mean radius of 5 cm and a length of 12 cm. The thickness of the annular ring is 2 mm. If the outer cylinder is rotated at 50 r/min while the inner one is held fixed, find the torque on the inner cylinder when the viscometer contains oil of viscosity 0.22 N-s/m².

23. If the viscometer of Prob. 22 is filled with a different kind of oil and a torque of 0.06 N-m is required to rotate the outer cylinder at 40 r/min when the inner one is fixed, find the viscosity of the oil. *Ans.* 0.30 N-s/m²

24. A pitot head (see accompanying figure), much used to measure the airspeeds of planes, consists of a *pitot tube* open in the direction of flight and a static tube with holes drilled in the sides to permit its pressure to be the same as that of the surrounding air. When the pitot head moves through the air, the pressure in the pitot line builds up to a value above that of the static line. Show that if Δp is the difference in pressure between the pitot and the static tubes, the speed is given by $v = \sqrt{2\Delta p/d}$, where d is the local density of the air.

25. A plane is flying at a speed of 100 m/s at an

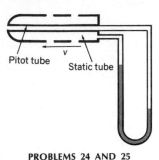

Pitot tube   Static tube

PROBLEMS 24 AND 25

altitude at which the density of the air is 0.8 kg/m³. Find the difference in pressure Δp between the pitot tube and the static tube (see accompanying figure). To how many millimeters of mercury does this correspond?

Ans. 4,000 Pa; 30 mmHg

26. Show that the terminal velocity v_T of a small solid sphere of diameter D and density d falling in a fluid of density d_f and viscosity η is given by

$$v_T = \frac{D^2 g(d - d_f)}{18\eta}$$

in the low-Reynolds-number regime where Stokes' law applies.

27. Air is streaming past a horizontal airplane wing in such a way that the speed is 120 m/s over the upper surface and 90 m/s at the lower surface. If the air density is 0.5 kg/m³, find the difference in pressure between the top and bottom of the wing. If the wing is 10 m long and has an average width of 2 m, calculate the gross lift of the wing.

Ans. 1,575 Pa; 31,500 N

28. A constriction in a pipeline similar to that of Fig. 13.5, with the associated pressure gauges, can be used to measure the rate of flow through the line. When so used, it is called a *venturi meter*. Show that the volume Q of fluid flowing through the pipe each second is given by

$$Q = A_a A_b \sqrt{\frac{2(p_a - p_b)}{d(A_a^2 - A_b^2)}}$$

TWO

KINETIC THEORY, HEAT, AND THERMODYNAMICS

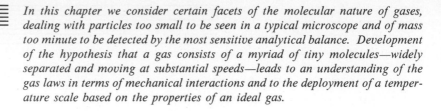

14

KINETIC THEORY
AND
TEMPERATURE

In this chapter we consider certain facets of the molecular nature of gases, dealing with particles too small to be seen in a typical microscope and of mass too minute to be detected by the most sensitive analytical balance. Development of the hypothesis that a gas consists of a myriad of tiny molecules—widely separated and moving at substantial speeds—leads to an understanding of the gas laws in terms of mechanical interactions and to the deployment of a temperature scale based on the properties of an ideal gas.

14.1 MOLECULAR COMPOSITION OF MATTER

In the fifth century B.C. Greek philosophers, among them Democritus, advanced the theory that all matter was composed of tiny particles called atoms (*atom* means "uncut" or "indivisible"). About 1800, the English chemist Dalton introduced experimental evidence for the existence of atoms and laid the groundwork for modern chemistry. Although we now know that the atom is composed of still smaller particles, we need not consider this fact in discussing the properties of interest here, and we defer discussion of the structure of atoms to later chapters.

One or more atoms may be bound together to form a chemical compound. The smallest unit into which a substance can be divided without chemical decomposition is known as a *molecule*. In gases such as helium and neon the molecule consists of a single atom. In other gases, such as hydrogen and oxygen, two atoms form a molecule, while molecules of carbon dioxide and of water vapor contain three atoms. In the gaseous state the molecules are usually separated by distances large compared with molecular dimensions. At room temperature molecules move with high velocities and have frequent collisions. A gas entirely fills the space of the containing vessel and exerts on it a pressure which results from the change in momentum of the molecules colliding with the walls.

14.2 COMPRESSIBILITY OF GASES: BOYLE'S LAW

When one attempts to decrease the volume of a liquid or a solid, an enormous pressure is required to achieve an appreciable change in volume. The behavior of gases in this respect is quite different. It is relatively easy to compress a gas so that it occupies only one-third of its original volume.

The relationship between the volume of any mass of gas and the pressure was investigated by Robert Boyle in 1660. Boyle's law states: *The product of the pressure and the volume of a given mass of gas is constant if the temperature is not changed.* Thus, if V_0 and p_0 denote the original volume and pressure while V and p are the final volume and pressure,

$$p_0 V_0 = pV = \text{constant} \qquad (14.1)$$

The mass m of gas, the product of the volume V and the density d, does not change as the pressure is varied, so Eq. (14.1) can be written

$$\frac{p_0 m}{d_0} = \frac{p_1 m}{d_1} \qquad \text{or} \qquad \frac{p_0}{d_0} = \frac{p_1}{d_1} \qquad (14.2)$$

Thus, *when the temperature is constant, the ratio of pressure to density remains constant.* In other words the density is proportional to the pressure. In applying Boyle's law and its corollary Eq. (14.2), it is necessary to use *absolute* pressures; gauge† pressures will not do.

Consider a mass of gas initially occupying a volume V at pressure p. Let us keep the temperature constant and change the pressure by a very small amount Δp, giving us a new volume $V + \Delta V$, where ΔV is also very small. Application of Boyle's law yields

$$pV = (p + \Delta p)(V + \Delta V) = pV + p\,\Delta V + V\,\Delta p + \Delta p\,\Delta V$$

Since Δp and ΔV are both small, we may neglect their product in comparison with $p\,\Delta V$ and $V\,\Delta p$. Therefore $p\,\Delta V = -V\,\Delta p$. Comparison with Eq. (9.14) reveals that $p = B$. Thus *when the change in pressure is small*, the *isothermal* (constant temperature) *bulk modulus of a gas is the initial pressure.*

14.3 KINETIC THEORY OF GAS PRESSURE

The molecular nature of matter is well established by chemical and physical evidence. We shall attribute the pressure exerted by a gas to the transfer of momentum to the wall by molecules which are bouncing off. Although we have previously advanced the idea that molecules are in constant motion, we have not discussed experimental evidence for this point of view. Perhaps the simplest and the most direct evidence for the motion of molecules was noted by a Scottish botanist, Robert Brown, in 1827. With a microscope he observed that very fine particles held in suspension in water were constantly in motion. The smaller the particles, the more freely they moved. The motion is caused by the incessant bombardment of the particles by the molecules of the liquid in which they are suspended. Although, on the average, the net momentum transferred to a particle over a period of time is zero, there are small statistical fluctuations. For particles of very small mass the instantaneous momentum transfers drive the particles hither and thither. For more massive particles the momentum transfers are far less obvious.

Although Brown had no explanation for what we call *brownian motion,* Daniel Bernoulli had laid the foundations for the explanation in the preceding century. Bernoulli developed the *kinetic theory of gases;* his ideas were expanded and refined by many others. To explain the physical properties of gases we accept the following basic assumptions of Bernoulli and his successors.

1. A gas consists of a very large number of rapidly moving molecules which are so small that the volume of the molecules themselves is negligible compared with the volume of the gas. One justification for this assumption is the fact that the gas can be compressed so that it occupies a small fraction of its normal volume at atmospheric pressure. When 1 cm³ of water at 100°C is converted into steam at 100°C, it occupies a volume of about 1,670 cm³. The distance between molecules in the steam is $\sqrt[3]{1,670}$ (roughly 12) times the distance between molecules in the liquid.

---

† Pressure gauges read the difference between the pressure in a container and the ambient pressure, which is typically atmospheric pressure. A tire gauge reads how much greater the pressure is than atmospheric pressure. (See Sec. 12.2.)

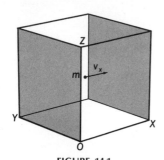

FIGURE 14.1

A single molecule of mass m moving with x component of velocity v_x bounces back and forth between the shaded walls.

2. When the rapidly moving molecules collide with each other or with the walls of the container, the collisions are elastic. (For polyatomic gases this assumption is too restrictive; it is only necessary to assume that on the average the kinetic energy of translation of the molecules remains constant; see Sec. 15.8.)

3. For an *ideal* gas the attractive forces between molecules are negligible.

4. Under identical conditions of temperature and pressure equal volumes of all gases contain equal numbers of molecules. This hypothesis was advanced in 1811 by Avogadro to explain the results of certain chemical experiments. For any gas at 0°C and one standard atmosphere pressure a mass in grams equal to the molecular weight M occupies a volume of 22.4 L. This quantity is known as *one mole* (mol). Similarly, the volume occupied under standard conditions by 1 kmol (M kg) is 22.4 m³. Avogadro's hypothesis implies that M kg of any gas contains the same unique number of molecules, which we call *Avogadro's number*. We designate it by N_A; $N_A = 6.022 \times 10^{26}$ molecules per kilomole.

Let us now calculate the pressure exerted by a gas on the walls of a container as the molecules strike against them. We begin with a single molecule of mass m moving with velocity v_x parallel to the x axis in a cubical box of length L (Fig. 14.1) and compute the average force which this molecule exerts against one of the faces parallel to the YZ plane. As the molecule approaches the wall, the x component of its momentum is mv_x. After the collision the x component of the velocity is reversed, since the collision is elastic. The change of momentum is $2mv_x$. Some time later, the molecule strikes the opposite wall and eventually returns to the first wall. Between successive collisions at this wall the molecule travels a distance $2L$. Therefore, the time between collisions at the same place is $2L/v_x$. The number of collisions per second is $v_x/2L$. Since at each collision the change of momentum is $2mv_x$, the change in momentum per second at one wall for one molecule is

$$2mv_x \frac{v_x}{2L} = \frac{mv_x^2}{L}$$

Let there be n kmol of identical molecules in the box. The number of molecules in the box is then $6.022 \times 10^{26}\, n$ or $N_A n$, since each kilomole contains Avogadro's number N_A molecules. To find the total change in momentum per second at the wall in question, we assume tentatively that all the molecules act independently and add the contributions of all the molecules to obtain

$$\text{Change in momentum per second} = \frac{m}{L}\left(v_{x1}^2 + v_{x2}^2 + \cdots + v_{xNn}^2\right)$$

$$= \frac{nN_A m}{L} \frac{\left(v_{x1}^2 + v_{x2}^2 + \cdots + v_{xNn}^2\right)}{N_A n}$$

$$= \frac{nN_A m}{L}\,\overline{v_x^2}$$

where $\overline{v_x^2}$ represents the average value of the square of v_x for all the molecules in the box.

By Newton's second law of motion the change in momentum per second is equal to the average force. Therefore,

$$F = \frac{nN_A m\overline{v_x^2}}{L} = pL^2$$

since the pressure p is the ratio of the force to the area. Hence,

$$pL^3 = nN_A m\overline{v_x^2}$$

Now L^3 is just the volume V of the box, so

$$pV = nN_A m\overline{v_x^2}$$

For a wall of the box parallel to the XZ plane, $pV = nN_A m\overline{v_y^2}$, and for a wall parallel to the XY plane, $pV = nN_A m\overline{v_z^2}$. We know from experience that the pressure due to the molecular bombardment is the same on all walls (we assume pressure due to the weight of the gas is negligible); thus we expect that $\overline{v_x^2} = \overline{v_y^2} = \overline{v_z^2}$. By the extension of the pythagorean theorem to three dimensions, $\overline{v_x^2} + \overline{v_y^2} + \overline{v_z^2} = \overline{v^2}$, so that $\overline{v_x^2} = \overline{v^2}/3$, and we may write

$$pV = \frac{nN_A m\overline{v^2}}{3} = \frac{2nN_A}{3}(\tfrac{1}{2}m\overline{v^2}) \tag{14.3}$$

The product of the pressure and the volume is given by two-thirds of the number of molecules multiplied by the average kinetic energy of translation of the molecules.

Before we relate the average kinetic energy of the molecules to the temperature, we must discuss temperature in some detail.

14.4 TEMPERATURE

Our first ideas about temperature come from our senses. By touching a body we determine whether it is hot or cold. For some purposes our senses give us an adequate description of temperature, but often sensory impressions are unreliable. For example, a room may feel hot to a person who has been outdoors in snow, while it may feel cold to a person entering it from a steam bath. Indeed, sensory impressions of temperature depend greatly on the environment during the recent past.

There is a second situation in which the senses give an unreliable comparison of temperature. If one removes a cardboard container and a metal ice tray from the freezing compartment of a refrigerator, both objects are at the same temperature. Nevertheless, the tray feels much colder to the hand than the cardboard container does, because the metal is a better conductor of heat. In view of the uncertainties associated with our sensations of temperature it is not surprising that scientists have developed objective and reproducible methods for measuring the relative "hotness" of bodies under various conditions.

Temperature is a quantity which cannot be defined in terms of mass, length, and time; it is a fourth fundamental quantity. To measure the temperature of a substance we immerse a *thermometer* in the substance, or at least put the thermometer into intimate contact in such a way that energy can be exchanged freely between the substance and the thermometer. A thermometer is a device with some observable property which changes with temperature; for example, in a mercury thermometer the variable is the length of a mercury column, while in a resistance thermometer the variable is the electric resistance of a conductor. As the thermometer exchanges energy with the substance, the observable property changes until eventually the "reading" becomes constant. Then *thermal equilibrium* has been achieved. By definition, *two systems in thermal equilibrium are at the same temperature*. When the thermometer is in thermal equilibrium with the substance, the substance is at the temperature of the thermometer, and vice versa.

Suppose that a thermometer reads the same when immersed in a beaker of water and when inserted in a hole in a copper sphere. If the sphere is then dropped into the water, it will be found that the temperature of both remains the same. Thus the water and the copper are in thermal equilibrium with each other. *Two systems in thermal equilibrium with a third system are in thermal equilibrium with each other.* This experimental fact has been called *the zeroth law of thermodynamics*. If the copper had had a higher temperature than the water, experiment would have shown the temperature of the copper to fall and that of the water to rise when the copper was dropped into the water.

14.5   THE CELSIUS AND FAHRENHEIT TEMPERATURE SCALES

The first recorded effort to make an instrument for measuring temperature was that of Galileo about 1593. He took a glass bulb with a long stem and submerged the end of the stem in water (Fig. 14.2). By heating the bulb, he drove some of the air out; as the bulb cooled, water rose in the stem. A change in the temperature of the bulb gave rise to a change in the water level in the stem. Such thermometers were used for many years by physicians and others. The galilean thermometer has several serious drawbacks, the most serious of which is that changes in atmospheric pressure also affect the height of the water in the stem.

In the seventeenth century thermometers using water or alcohol sealed in glass tubes were developed. Alcohol is still widely used in inexpensive thermometers. Early in the eighteenth century Fahrenheit introduced thermometers which used mercury as the thermometric substance. These thermometers rapidly won wide acceptance among scientific workers because they were consistent with each other over the whole length of scale and they were convenient, reliable, and reasonably cheap. Fahrenheit elected to call zero on his thermometer "the most intense cold obtained artificially in a mixture of water, of ice, and of sal ammoniac." The temperature of the human armpit he called 96°. The choice of these two "fixed points" established the Fahrenheit temperature scale.

The *Celsius*, or *centigrade*, temperature scale, proposed by the Swedish astronomer Celsius about 1742, took as its zero the temperature of a mixture of ice and water under standard pressure. The temperature at which water boils under standard atmospheric pressure was taken to be 100°C.

It was soon found that the fixed points of the Celsius scale could be reproduced easily and with far greater accuracy than the original fixed points of the Fahrenheit scale. The melting point of ice was approximately 32° on the Fahrenheit scale, and the boiling point of water 212°F. Eventually it became standard practice to use the ice point and the boiling point as the fixed points for the Fahrenheit scale.

Although the Celsius scale has replaced the Fahrenheit scale in most

FIGURE 14.2
Galilean thermometer using air inside the bulb as the working substance.

scientific work and for general use in most of the countries of the world, the Fahrenheit scale is still widely used in the United States. Therefore, it is frequently desirable to convert a Celsius temperature to Fahrenheit, and vice versa. This can be done readily if one recalls the fixed points of the two temperature scales. Let F be the temperature on the Fahrenheit scale and C the temperature on the Celsius scale (Fig. 14.3). The number of Fahrenheit degrees above the freezing point is related to the total temperature difference between the boiling point and the freezing point on the Fahrenheit scale (180°) as the number of Celsius degrees above the freezing point is to 100°C. This proportion is written

$$\frac{F - 32}{180} = \frac{C}{100} \quad \text{or} \quad \frac{F - 32}{9} = \frac{C}{5} \quad (14.4)$$

Since 100 Celsius degrees corresponds to 180 Fahrenheit degrees, clearly

$$1 \text{ C degree} = \tfrac{9}{5} \text{ F degree} \quad (14.5)$$

(Note that 1°C is 33.8°F.) Clearly 1 Celsius degree (C°) is not a temperature, but a temperature interval. Equation (14.4) relates a temperature on the Celsius scale to the corresponding temperature on the Fahrenheit scale, while Eq. (14.5) relates temperature differences. We shall write temperatures as °C or °F, while temperature intervals will be expressed in C° or F°.

14.6 GASES AND THE ABSOLUTE TEMPERATURE SCALE

Gases expand when their temperature is raised and contract when the temperature is lowered. If we keep the pressure on the gas constant, we find that a plot of the volume as a function of temperature gives a straight line (Fig. 14.4) described by

$$V = V_0(1 + \beta_0 t) \quad (14.6)$$

where V_0 is the volume of the gas at 0°C, V is the volume at some temperature t°C, and β_0 is a constant called the *coefficient of volume expansion at 0°C*. For any substance and any initial temperature *the coefficient of volume expansion is the change in volume divided by the product of the initial volume V_i and the change in temperature.* Thus

$$\beta = \frac{\Delta V}{V_i \, \Delta t} \quad (14.6a)$$

and $\beta_0 = (V - V_0)/V_0 t = \Delta V/V_0 t$.

The relation expressed by Eq. (14.6) is known as *Charles' law.* By careful experiments, Charles found that the coefficient of volume expansion of all noncondensing gases at constant pressure and 0°C is equal to 0.00367, or $\tfrac{1}{273}$. *The coefficient of volume expansion is the same at a given temperature for all gases.* Thus, if we extrapolate (Fig. 14.4) a plot of the volume as a function of temperature of any gas to the x axis, the intercept is −273°C, regardless of the original volume of gas taken, the gas chosen, or the pressure. The fact that this intercept of −273°C is the same for different gases, different original volumes, and different pressures suggests this temperature has a very special significance. It was the first indication of *absolute zero* for temperature. Of course, it should not be concluded that

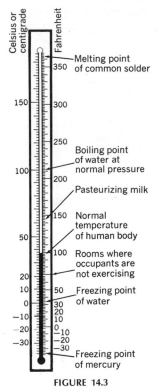

FIGURE 14.3
Comparison of the Celsius (centigrade) and Fahrenheit temperature scales.

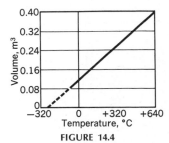

FIGURE 14.4
The volume of a gas as a function of temperature when the pressure is held constant.

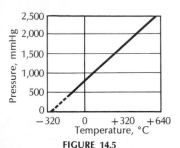

FIGURE 14.5

Pressure as a function of temperature for a gas at constant volume.

if the temperature of the gas were actually lowered to $-273°C$, it would follow the extrapolated line of Fig. 14.4 and occupy zero volume. Before any such low temperature is reached, the gas liquefies.

If we define a tentative temperature scale by $T = t(°C) + 273$, it turns out to be almost the same as the *absolute* temperature scale devised about 1848 by Lord Kelvin. The Kelvin scale, based on thermodynamics (Chap. 18), is independent of the properties of any particular substance. For more than a century temperatures on the absolute scale were measured in "degrees Kelvin," but the International Union of Pure and Applied Physics now recommends that the word "degrees" be omitted. Thus a temperature of $20°C$ corresponds to an absolute temperature of 293 K, read "293 kelvins." When Eq. (14.6) is rewritten in terms of absolute temperature, it takes the simpler form

$$V = V_0\left(1 + \frac{t}{273}\right) = V_0\frac{273 + t}{273} = V_0\frac{T}{273} \tag{14.6b}$$

The volume of a gas at constant pressure is proportional to the absolute temperature T. In general $V_1/V_2 = T_1/T_2$, where V_1 and V_2 are the volumes occupied by the gas at temperatures T_1 and T_2, respectively, when the pressure is kept constant.

If the temperature of a gas is raised and its volume is kept constant, the pressure varies, increasing $\frac{1}{273}$ of its value at $0°C$ for every rise of 1 Celsius degree. If p_0 represents the pressure at $0°C$ and p_t the pressure at $t°C$,

$$p_t = p_0\left(1 + \frac{t}{273}\right) = p_0\frac{T}{273} \tag{14.7}$$

Figure 14.5 shows the relation between the pressure and temperature of a gas at constant volume. Note that once again the intercept on the temperature axis is $-273°C$ (or more exactly $-273.15°C$).

14.7 THE GENERAL GAS LAW

From Boyle's law we know that the product of the pressure and the volume of a gas remains constant if the temperature is held fixed. From Charles' law we know that the volume of a gas is proportional to the absolute temperature if the pressure is constant. If the volume is kept constant, the pressure is proportional to the absolute temperature. If we combine these relations, we may write, for any given mass of gas,

$$pV = bT \tag{14.8}$$

where b is a constant for the particular mass of gas in question. If subscript 1 represents one particular set of conditions for the gas and subscript 2 another set of conditions, we may write

$$\frac{p_1V_1}{T_1} = \frac{p_2V_2}{T_2} \tag{14.9}$$

In both chemistry and physics it is often convenient to measure quantities of substances in moles. As we saw in Sec. 14.3, *one mole of any gas is that amount of the gas which has a mass in grams equal to the molecular weight.* For example, the molecular weight of gaseous oxygen is 32.0, and hence 480 g of oxygen represent 15.0 mol. A kilomole is 1,000 mol, or the amount of the gas which has a mass in kilograms equal to the molecular weight. Thus, since the molecular weight of hydrogen is 2.016, 10 kmol of hydrogen has a mass of 20.16 kg.

At 0°C and one standard atmosphere (760 mmHg) pressure, 1 kmol of any gas occupies a volume of 22.414 m³; thus the constant in Eq. (14.8) is the same for 1 kmol of any gas. For n kmol of any gas,

$$pV = nRT \qquad (14.10)$$

where R is the *universal gas constant*. It has the value

$$R = \frac{1.0133 \times 10^5 \text{ N/m}^2}{273.15 \text{ K}} \frac{22.414 \text{ m}^3}{1 \text{ kmol}} = 8{,}314 \text{ J/(kmol)(K)}$$

The relation $pV = nRT$ is the *equation of state* for an ideal gas. It is an excellent approximation for real gases so long as the temperature is far above the boiling point and the pressure is not extremely high.

When we compare Eqs. (14.10) and (14.3), we find that $pV = nRT = \frac{2}{3}nN_A(\frac{1}{2}m\overline{v^2})$, from which

$$\tfrac{1}{2}m\overline{v^2} = \frac{3}{2}\frac{RT}{N_A} = \tfrac{3}{2}kT \qquad (14.11)$$

where N_A is Avogadro's number. The ratio $R/N_A = k$ is known as *Boltzmann's constant*. The expression on the left of Eq. (14.11) is the average kinetic energy of translation for the gas molecules, and it is directly proportional to the absolute temperature. Thus, *the absolute temperature is a measure of the average translational kinetic energy of the molecules in a gas.* When we double the absolute temperature, we double this kinetic energy. To double the average velocity of the molecules in a gas sample, we must increase the absolute temperature by a factor of 4.

The average kinetic energy of a molecule of any gas at temperature T is $\frac{3}{2}kT$, where $k = R/N_A = 1.38 \times 10^{-23}$ J/K. At 27°C (300 K), the average kinetic energy of a gas molecule is $\frac{3}{2} \times (1.38 \times 10^{-23}$ J/K$) \times 300$ K $= 6.21 \times 10^{-21}$ J.

Let us now calculate the effective velocity of nitrogen molecules at 27°C. By the "effective velocity" we mean the velocity of a molecule which has the average energy, so $v_{\text{eff}}^2 = \overline{v^2}$. One kilomole of nitrogen has a mass of 28.0 kg, and one molecule of mass of $(28.0/N_A)$ kg $= 28.0$ kg$/6.022 \times 10^{26} = 4.65 \times 10^{-26}$ kg. Hence, by Eq. (14.11),

$$\tfrac{1}{2}m\overline{v^2} = \tfrac{1}{2}mv_{\text{eff}}^2 = \tfrac{3}{2}kT$$
$$\tfrac{1}{2}(4.65 \times 10^{-26} \text{ kg}) v_{\text{eff}}^2 = \tfrac{3}{2}(1.38 \times 10^{-23} \text{ J/K})(300 \text{ K})$$
$$v_{\text{eff}}^2 = 26.7 \times 10^4 \text{ m}^2/\text{s}^2$$
$$v_{\text{eff}} = 517 \text{ m/s}$$

14.8 DALTON'S LAW OF PARTIAL PRESSURES

We have derived the equations for the pressure exerted by a gas for the case in which the gas contains only molecules of one kind. If the gas contains several different kinds of molecules, each kind rebounds from the walls and contributes to the pressure. The total pressure is the sum of the pressures which each of the various kinds would exert if it occupied the volume alone. We may write

$$p_t = p_1 + p_2 + p_3 + \cdots \qquad (14.12)$$

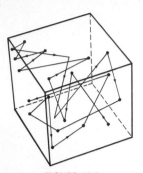

FIGURE 14.6
Schematic path of a molecule colliding with other molecules and with the walls of the containers.

where p_t is the total pressure exerted by all the molecules and p_1, p_2, p_3, ... are the pressures which molecules of type 1, 2, 3, ... would exert if each filled the volume alone. This is known as *Dalton's law of partial pressures.*

Air is a mixture of gases, with nitrogen, oxygen, water vapor, carbon dioxide, argon, hydrogen, and many other kinds of molecules present. The atmospheric pressure is the pressure exerted by the nitrogen plus that due to oxygen, plus that due to each of the other atmospheric constituents. Equation (14.11) shows that the average translational kinetic energy is the same for each kind of molecule. If we apply this fact to compare the effective velocities of hydrogen and oxygen molecules at a given temperature, we have

$$\frac{3}{2} \frac{R}{N_A} T = \tfrac{1}{2} m_h v_h{}^2 = \tfrac{1}{2} m_o v_o{}^2$$

where m_h and v_h are the mass and effective velocity of hydrogen molecules and m_o and v_o the corresponding quantities for oxygen molecules. Since $m_o = 16 m_h$, $v_h = 4 m_o$. Thus, the effective speed of hydrogen molecules in the atmosphere is four times that of oxygen molecules. The effective speed is also called the *root-mean-square* (rms) *speed*, because it is the square *root* of the *mean* (average) *square* of the speed.

14.9   DISTRIBUTION OF MOLECULAR VELOCITIES IN GASES

In writing the expression for the pressure exerted by the ideal gas, we have assumed that a given molecule travels back and forth striking the walls, without taking into account the effects of collisions with other molecules. Although molecules collide frequently with other molecules (Fig. 14.6), and although the velocity of any individual molecule undergoes sharp changes, the average kinetic energy of the molecules is the same at a given temperature. Some molecules are moving rapidly, and some slowly. Figure 14.7 shows the distribution of speeds for gas molecules. While the effective *speed* v_{eff} for air molecules at room temperature is of the order of hundreds of meters per second, the average *velocity* is *zero;* this is true because, on the average, for every molecule moving to the left there is some other molecule with a comparable velocity to the right. The most probable speed is $\sqrt{\tfrac{2}{3}}$ (or 0.82) times v_{eff}, while the average speed is $\sqrt{8/3\pi}$ (or 0.92) times v_{eff}.

FIGURE 14.7
Maxwellian distribution of speeds for gas molecules. Plotted on the ordinate is the relative number of molecules in each 1 m/s speed range and on the abscissa the ratio of the speed v to the effective (or rms) speed v_{eff}.

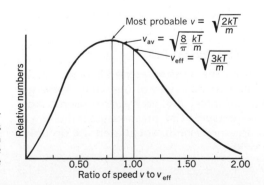

TABLE 14.1
191
KINETIC THEORY AND
TEMPERATURE

TABLE 14.1
Molecular Quantities at 0°C and 760 mmHg

Quantity	Hydrogen	Oxygen
Number of molecules per cubic meter	2.69×10^{25}	2.69×10^{25}
Diameter of each molecule, m	2.4×10^{-10}	3.2×10^{-10}
Mass of each molecule, kg	3.34×10^{-27}	5.3×10^{-26}
Mean free path, m	1.8×10^{-7}	1.0×10^{-7}
Number of collisions per second	1.00×10^{10}	4.6×10^{9}
Effective speed, m/s	1,840	461
Average speed, m/s	1,700	425
Most probable speed, m/s	1,500	376
Density, kg/m³	0.0899	1.43
Volume per kilogram, m³	11.1	0.699
Number of molecules in 1 kmol	6.022×10^{26}	6.022×10^{26}

The distance a given molecule travels between collisions differs considerably from collision to collision, but its average value is again well defined. The average length of the path over a large number of collisions is called the *mean free path*. For air molecules it is tens to hundreds of nanometers. Table 14.1 lists some of the important molecular quantities for oxygen and hydrogen molecules.

14.10 DEVIATIONS FROM THE GENERAL GAS LAW

In our discussions of gases we have made two important assumptions which are not always justified. We have assumed (1) that the volume of the molecules themselves is negligible in comparison with the total space occupied by the gas and (2) that the attractive forces between gas molecules are negligible. For experiments with gases such as oxygen and hydrogen at room temperature and ordinary pressures, these assumptions are justified. However, if they were true under all circumstances, we would never have molecules sticking together to form liquids or solids. Indeed, substances composed of molecules in which the attractive forces are relatively great are already liquid or solid at room temperature.

Let us keep the temperature of a given mass of a *real* gas constant and study the behavior of the pressure and volume. If Boyle's law held rigorously, the product pV would be constant for all pressures and volumes so long as the temperature was kept constant. But accurate observations show that the product is not constant when the pressure is varied over a large range. At exceedingly high pressures, the molecules are close together, and the space occupied by the molecules themselves is no longer negligible. Therefore an increase in the pressure of the gas results in too small a decrease in volume, so that pV increases with p instead of remaining constant according to Boyle's law. At exceedingly low pressure the product pV decreases somewhat as p is increased.

Van der Waals showed that the intermolecular attraction and the space occupied by the molecules themselves can be simply taken into account by a modification of the general gas law. According to van der Waals, for 1 kmol of gas,

$$\left(p + \frac{a}{V^2}\right)(V - b) = RT \qquad (14.13)$$

where a and b are constants characteristic of a given gas but independent of temperature, pressure, and volume. The constant b is a correction to take account of the fact that the molecules themselves occupy a finite amount of space. A detailed analysis shows that b is about 4 times the actual volume of the molecules. The term a/V^2 takes account of the attractive forces between the molecules, which have the effect of reducing the volume just as the pressure does.

14.11 DIFFUSION

Evidence supporting the idea that molecules are in constant motion is found in the way the odor penetrates to all parts of a room when a bottle of ammonia is opened. The spread of the odor can be explained on the assumption that the ammonia molecules escape from the bottle and wander about among the air molecules. They are buffeted back and forth, gradually spreading throughout the room. The process by which molecules of one kind penetrate and intermix with molecules of another kind is called *diffusion*.

If a porous jar (Fig. 14.8) is surrounded by another vessel into which hydrogen is introduced, hydrogen diffuses through the porous jar and increases the gas pressure inside it, causing gas to bubble up from the water into which the end of the tube leading from the porous jar is dipped. Molecules diffuse out of the porous walls as well as in, but the heavier air molecules diffuse outward far more slowly than the hydrogen molecules diffuse inward. The reason is the greater speeds of the hydrogen molecules; as we saw in Sec. 14.7, the effective speed of a gas molecule is inversely proportional to the square root of its mass. It is diffusion which keeps the air in such a state of uniform mixture and which accounts for the rapid disappearance of the fumes when a bottle of ether is broken out of doors.

In liquids as well as in gases, the molecules are free to move about. If a little sulfuric acid is placed in a bottle of water, it diffuses to form a uniform concentration. When a lump of sugar is placed in a cup of coffee, the contents of the whole cup are sweetened by the distribution of the sugar molecules throughout the coffee; stirring promotes the rapid mixing of the molecules.

If stirring and convection currents are avoided, diffusion in liquids occurs very slowly. This can be demonstrated by using a tall glass jar filled with water into which, by means of a thistle tube extending to the bottom of the jar, a solution of copper sulfate is carefully poured until the bottom of the jar is filled to a height of a centimeter or two. If the jar is allowed to stand for some days without being disturbed, the upward diffusion of the copper sulfate can be observed. Gravity tends to keep the copper sulfate on the bottom, but diffusion occurs, and the copper sulfate gradually rises in the jar until a uniform distribution is achieved.

When two metals like copper and nickel are placed in contact, atoms of each gradually diffuse into the other. If the specimens are cut into sections after they have been in contact for some time, the concentrations of the two metals at different distances from the contact plane can be determined. The rate of diffusion depends on the temperature, but in any case it is very slow for solids.

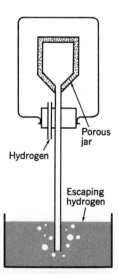

Hydrogen

Porous jar

Escaping hydrogen

FIGURE 14.8
The pressure within the porous jar increases because hydrogen diffuses in through its walls faster than the heavier air molecules diffuse out.

14.12 OSMOSIS

When the pores of a membrane are the right size to pass small molecules but not larger ones, the membrane is said to be *semipermeable*. The process of preferential diffusion through a semipermeable membrane is called *osmosis*. For example, if a carrot is hollowed out, and a thick sugar syrup placed in the cavity, there is a net flow of water into the carrot when it is placed in water. If a glass tube is sealed to the cavity, the water may rise to a height of several feet. The carrot serves as a semipermeable membrane (Fig. 14.9) having a great number of small holes through which water molecules can pass in or out. Sugar molecules, on the other hand, are too large to pass through. The rising water level in the tube means that more water flows in than flows out. The large sugar molecules act rather like valves for many of the openings. A water molecule moving inward may slightly displace the large sugar molecule at *A* in Fig. 14.9 and get into the idealized carrot, but a water molecule coming from the other side can only push the sugar molecule more tightly against the hole. There are many holes through which water molecules pass in either direction, but there are more holes through which water molecules enter than there are through which they leave. As water rises in the tube, the pressure inside increases, eventually becoming great enough so that there is no longer a net inflow. The pressure that just prevents further flow of the solvent is called the *osmotic pressure* of the solution. In dilute solutions in which the molecules are not dissociated, the osmotic pressure is proportional to the concentration of the dissolved substance. In a solution of given concentration, the osmotic pressure increases as the temperature is increased, the rate of increase being the same for all solutions because the increase in average kinetic energy is the same for all kinds of molecules.

When dried fruits like prunes and raisins are cooked, they swell and burst if the pressure inside becomes sufficiently large. The swelling is due to the fact that the vegetable sacs surrounding the fruit are semipermeable membranes through which there is a net diffusion of water from outside to inside. If marine animals such as oysters are transferred from salt water to fresh water, more water flows into than out of the animal through the membrane that serves as its covering beneath the shell. Dilation and death of the animal result.

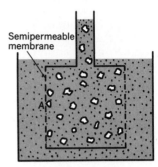

Semipermeable membrane

FIGURE 14.9
Sugar molecules prevent water molecules from leaving through many of the holes in a semipermeable membrane, but they do not prevent water molecules from entering the solution.

QUESTIONS

1. The temperature of a gas kept at constant volume is increased. Does the number of collisions per second of gas molecules with the walls change? Does the average momentum change per collision increase or decrease?

2. Explain why the pressure decreases as the volume of a gas is increased at constant temperature.

3. If the speed of every molecule of gas were doubled, what would happen to the temperature? Explain.

4. What volume is occupied by 1 kmol of any gas at 0°C under 1 atm (760 mmHg) pressure? These are known as the standard temperature and pressure, respectively, abbreviated STP.

5. On the basis of what happens to individual molecules, explain why the temperature of a gas rises when the gas is compressed.

6. What reasons can you advance in support of the molecular hypothesis?

7. What types of thermometers do you know about? What kind could be used to measure 4 K? 1400 K?

8. What are the advantages and disadvantages of mercury as a thermometric material?

9. What justification do we have for believing that 50°C should be halfway between 0°C and 100°C on a good mercury thermometer?

10. What is the temperature (98.6°F) of the normal human body on the Celsius scale?

11. Does the concept of temperature apply to a single molecule? Is temperature a microscopic or a macroscopic quantity?

12. At very low pressure and density, why does pV decrease slightly as p is increased at constant temperature?

13. Why does the product pV increase slightly as the pressure is raised when pressure and density are very high, even though the temperature remains constant?

14. Do more than half the molecules in a gas have a speed less than v_{eff}? Than v_{av}? Than the most probable speed? Which of these speeds corresponds to a molecule with average kinetic energy?

15. Although typical gas molecules have speeds of the order of hundreds of meters per second at room temperature, it takes several seconds for ammonia molecules to diffuse 5 m through a room. Why does it take so long?

16. Why may there be greater danger associated with the failure of a storage tank for gas at 30 MPa than with the failure of a water tank at the same pressure?

17. In collisions between macroscopic objects some mechanical energy is always converted into internal energy, resulting in an increase in temperature. Is there always a loss of mechanical energy in inelastic collisions between molecules?

PROBLEMS

1. A thermometer shaded from the sun on a hot day reads 90°F. What is the temperature on the Celsius scale? On the Kelvin scale?
   Ans. 32.2°C; 305 K

2. The temperature at which mercury boils is 675°F, and that at which it freezes is −40°F. Find the corresponding temperatures on the Celsius scale.

3. Solid carbon dioxide (dry ice) turns to the gaseous phase at −80°C. Find the corresponding temperature on the Fahrenheit and Kelvin scales.
   Ans. −112°F; 193 K

4. What is the temperature on the Celsius scale when a Fahrenheit thermometer indicates −10°? 50°? 70°?

5. Copper melts at 1084°C and iron at 1503°C. Find the corresponding Fahrenheit temperatures.
   Ans. 1983°F; 2737°F

6. The boiling point of pure methanol (methyl alcohol) is 64.65°C. Find the corresponding temperature on the Fahrenheit and Kelvin scales.

7. Aluminum melts at 1220°F and titanium at 3272°F. Find the corresponding Celsius temperatures.
   Ans. 660°C; 1800°C

8. On the Reaumur temperature scale the ice point is 0°Re and the steam point 80°Re. Extend the relationship

$$\frac{C}{100} = \frac{F - 32}{180}$$

to include a term for the Reaumur scale. Find the Celsius and Fahrenheit temperatures corresponding to 30°Re.

9. The Scottish engineer Rankine devised an absolute temperature scale based on Fahrenheit degrees. To what Fahrenheit temperature does zero on the Rankine scale correspond? What is a temperature of 20°C on the Rankine scale?
   Ans. −460°F; 528°R

10. A thermostat is set to maintain the temperature at 85°C. What is the corresponding temperature on the Fahrenheit scale? On the Kelvin scale?

11. A flask with negligible expansion coefficient contains air at a temperature of 22°C and a pressure of 755 mmHg. Find the pressure in the flask after it is sealed and cooled to 0°C.
   Ans. 699 mmHg

12. Illuminating gas is stored in a tank designed so that the volume can change but the pressure remains constant. If the tank contains 4,000 m³ under standard pressure, how much does the volume change when the temperature rises from 7 to 27°C?

13. In the early morning, when the temperature is 7°C, the gauge pressure in an automobile tire is 165 kPa (24 lb/in.²). In the afternoon, after hard driving, the temperature of the tire is 47°C. If atmospheric pressure has remained constant at 100 kPa (14.5 lb/in.²), find the new gauge pressure.
   Ans. 203 kPa

14. A sample of gas occupies 500 cm³ at 0°C and 750 mmHg pressure. What would be its volume if it were heated to 60°C and the pressure increased to 880 mmHg?

15. Find the density of nitrogen gas at 17°C and 8 atm pressure. The molecular weight of N_2 is 28.
   Ans. 9.41 kg/m³ = 9.41 g/L

16. A stratosphere balloon has a partially filled gas bag containing 5,000 m³ of helium when the barometer reads 745 mmHg and the temperature is 27°C. Find its volume when the balloon has risen to such a height that the atmospheric pressure is 149 mmHg and the temperature is −23°C.

17. A McLeod gauge indicates pressure as low as

10^{-5} torr (or 10^{-5} mmHg). How many molecules per cubic centimeter would be required to produce this pressure at 27°C?   *Ans.* 3.2×10^{11}

18. How many molecules of air remain in each cubic centimeter of a radio tube at 17°C in which the pressure has been reduced to 10^{-9} torr (10^{-9} mmHg)?

19. Pure oxygen at 0°C is enclosed in a cylinder having a volume of 9 L. If the pressure in the cylinder is 3 MPa, what is the mass of the enclosed gas? How much pressure would it exert if the temperature were increased to 77°C?
    Ans. 0.380 kg; 3.85 MPa

20. An air bubble has a volume of 1.8 cm^3 at the surface of a lake, where the temperature is 27°C. What was its volume at a depth of 80 m, where the temperature was 7°C? Atmospheric pressure is 10^5 Pa.

21. An air bubble released at the bottom of a lake expands to 3 times its original volume by the time it reaches the surface. How deep is the lake if atmospheric pressure is 100 kPa, the surface is at 22°C, and the bottom of the lake is at 17°C?
    Ans. 19.9 m

22. The bag of a partially inflated balloon contains 120 m^3 of hydrogen at 756 mmHg pressure and 27°C. The balloon rises to a height of 8 km, where the temperature is -23°C and the pressure is 275 mmHg. Find the volume of the hydrogen under these circumstances.

23. A cylindrical diving bell, open at the bottom, has a volume of 8 m^3 and a cross-sectional area of 1.8 m^2. How high will the water rise in the bell when the bottom is lowered in fresh water to a depth of 15 m if atmospheric pressure is 100 kPa and no air leaves the bell? Assume that the temperature remains constant.   *Ans.* 2.73 m

24. A vessel having a capacity of 2.5 L contains nitrogen under a pressure of 3 atm. It is connected to a vessel of 1.5 L capacity containing oxygen at a pressure of 4 atm. If the temperature remains constant, what will the pressure be when the gases have mixed?

25. Find the number of mercury molecules, each with a mass of 3.3×10^{-25} kg and an effective speed of 200 m/s, that would maintain normal atmospheric pressure against the walls of a containing vessel with a volume of 1 L. What is the temperature of the gas?   *Ans.* 2.3×10^{22}; 319 K

26. If a single hydrogen molecule occupies a cubical box 0.05 m on a side and the walls are perfectly smooth, calculate the average force exerted on each wall of the box if the molecule has velocity components $v_x = 900$ m/s, $v_y = 1,100$ m/s, and $v_z = 1,000$ m/s.

27. A helium tank has a volume of 20 L, is at 27°C, and is filled to a pressure of 1.5×10^7 Pa (148 atm). A leak develops in the valve, and after a week the pressure is 9×10^6 Pa when the temperature is 17°C. Find the mass of helium that was in the tank when it was full and the fraction of the original mass which has leaked out.
    Ans. 481 g; 0.379

28. A thoroughly evacuated vessel has a volume of 3.8 L. It develops a small leak through which 750 million molecules of air enter each second. How long will it take for the pressure of the air in the vessel to reach 1 N/m^2 if the temperature remains constant at 17°C? Assume the number of molecules in the vessel at the start is negligible.

29. Calculate the effective (rms) speed of molecules of carbon dioxide 37°C. The molecular weight of CO_2 is 44. What is the average speed? The most probable speed?
    Ans. 419 m/s; 386 m/s; 342 m/s

30. Compute the effective speed of oxygen molecules under standard pressure at 67°C, making use of the fact that one gram molecular weight (32 g) occupies 22.4 L under standard conditions (1 atm and 0°C).

31. If the effective (rms) speed of oxygen molecules at 27°C and 1 atm pressure is 484 m/s, find (*a*) the effective speed of hydrogen molecules under the same conditions, (*b*) the effective speed of oxygen molecules at 27°C and 500 mmHg pressure, and (*c*) the temperature at which the effective speed of oxygen molecules is 968 m/s.
    Ans. (*a*) 1,936 m/s; (*b*) 484 m/s; (*c*) 927°C

32. Find the temperature at which the effective (rms) speed of oxygen molecules is equal to the effective (rms) speed of nitrogen molecules at 27°C.

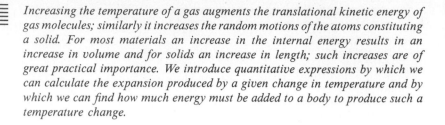

Increasing the temperature of a gas augments the translational kinetic energy of gas molecules; similarly it increases the random motions of the atoms constituting a solid. For most materials an increase in the internal energy results in an increase in volume and for solids an increase in length; such increases are of great practical importance. We introduce quantitative expressions by which we can calculate the expansion produced by a given change in temperature and by which we can find how much energy must be added to a body to produce such a temperature change.

15

HEAT CAPACITY AND THERMAL EXPANSION

15.1 LIQUID AND SOLID PHASES

The kinetic theory of matter leads to an understanding of many properties of liquids and solids, as well as of gases. The forces between molecules in an ideal gas are zero unless the molecules are "touching"; in a real gas there are small attractive forces between molecules. As we reduce the temperature of a gas, we reduce the average kinetic energy of translation of the molecules. As the molecules move more and more slowly, attractive forces between molecules, which played a negligible role at high temperatures, become important. Eventually, groups of molecules stick together. The gas begins to liquefy. In the liquid, molecules cling together but are free to move with respect to each other. Liquids do not resist forces tending to change their shape, but they strongly resist forces tending to change their volume. Because the molecules of the liquid can easily be displaced with respect to each other, layers flow relatively freely over each other and the liquid assumes the shape of the vessel in which it is placed. In the liquid two or more molecules may stick together and move about as a unit.

When the temperature of the liquid is reduced, the energy of the molecules decreases. Eventually, the translational kinetic energy becomes so small that the molecule is trapped by the attractive forces exerted by its neighbors. Once the molecule is able only to vibrate about some equilibrium position, the material loses its fluid properties and retains its shape. We then speak of it as a *solid.* If each atom of the material is located in a particular place in a regular array, we have a *crystalline solid* (Fig. 15.1a) of which diamond, quartz, and calcite are examples. In some cases each atom is confined to a small region about an equilibrium point by the attractive forces of neighboring particles, but no perfectly repetitive structure is formed. This is the case when glass cools or butter solidifies. The structure is not crystalline but *amorphous* (Fig. 15.1b). Solids resist an effort to change their shape or volume. The interatomic forces between atoms in a solid are reasonably idealized in terms of relatively stiff springs connecting nearest neighbors and weaker springs connecting next nearest neighbors. The forces are such that in a solid it is no longer meaningful to say that a particular atom belongs to a particular molecule. In the solid the molecule is no longer a separate entity; instead there is an array of atoms, and any one atom is bound to many others.

Under suitable conditions it is possible to change a given material from the solid to the liquid phase and from the liquid to the gaseous phase. Water can be frozen to form ice and evaporated to form steam. The change from one phase to another is discussed in Chap. 17.

The basic particles of all matter are in constant motion. Atoms in solids vibrate back and forth in complex motions about their equilibrium positions. Molecules in a liquid wander around among the other molecules, having frequent collisions with them and thus exchanging energy. In gases molecules travel about at high speeds and have frequent elastic collisions with their neighbors. *The sum of the kinetic and potential energies associated with the random motion of the atoms of a substance is the internal energy of the substance.*

Consider a bullet flying through the air with velocity v. This bullet has a kinetic energy $\frac{1}{2}mv^2$ associated with the speed of its center of mass (Sec. 6.6). In addition to the kinetic energy due to the organized, collective motion of the atoms of the bullet there is also energy associated with the *random* motion of the atoms relative to the center of mass. This latter energy is the *internal energy;* it depends on the temperature of the bullet and is independent of the external kinetic energy $\frac{1}{2}mv^2$. The internal energy is there even when the bullet is at rest. If the moving bullet strikes a block of wood, much of the external energy $\frac{1}{2}mv^2$ is converted into internal energy and the temperature of the bullet rises substantially.

When we heat a material, its internal energy is increased. The nature of the heating process was speculated upon and argued over for centuries before the answer began to evolve at the end of the eighteenth century. One idea which held wide support for many years was that a fluid called *caloric* entered a body when it was heated and leaked away as the body cooled. When measurements of the mass of a body showed no increase when the body was heated, the proponents of the caloric theory argued that the fluid was massless. As a rival of the caloric theory, there gradually evolved the modern point of view that heat is a form of energy. More specifically: *Heat is energy which is transferred between a substance and its surroundings or between one part of the substance and another as a result of temperature differences only.*

One of the decisive experiments supporting the theory that heat is a form of energy was performed by Count Rumford in 1798, where he observed that boring cannon resulted in a large increase in temperature although there was no flame or other source of caloric. When the drill was dull, the rise in temperature was exceedingly great and was related to the amount of mechanical work done in the drilling. Thus, adding energy to the cannon by doing work against friction led to the same temperature response as heating it in a furnace.

When we pound a nail with a hammer, the nail becomes hot. As the hammer is stopped, the atoms in the nail are given energy which shows up in the form of increased internal energy. After the hammer has struck, the atoms vibrate with greater amplitudes about their equilibrium positions. When a chisel is ground on an emery wheel, its thermal energy and its temperature are increased. When a moving automobile is stopped, its kinetic energy is transferred to internal energy in the brake drums, tires, and road. In all such cases work is done against friction with a resulting increase in the internal energy of the bodies involved.

We can increase the internal energy of a substance either by adding heat to it or by performing work on it in such a way as to increase the *random* motion of the atoms. We can raise the temperature of the air in a bicycle

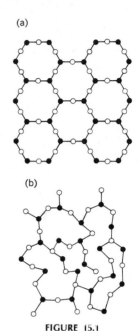

(a)

(b)

FIGURE 15.1

(*a*) In a crystalline solid the atoms are arranged in a regular pattern which is repeated over and over, so there is *long-range order*. (*b*) In an amorphous solid the atoms are arranged in various ways; there is *short-range* order but not long-range order.

pump either by heating the barrel of the pump in a flame or by doing work on the gas by vigorous pumping.

When we put a thermometer in a beaker of water, the random motions of water molecules produce collisions with the molecules of the glass in the thermometer. In turn, the glass molecules exchange energy with the mercury molecules until eventually the "level" of this random motion is the same for the water molecules, the glass molecules, and the mercury molecules. When this occurs, the position of the mercury in the thermometer becomes stationary, and we say the thermometer has come to the same temperature as the water. *Temperature is a measure of the level of internal energy.*

If we put a cool thermometer in hot water, heat flows from the water to the thermometer until the temperatures are equal. In general, whenever two bodies are placed in contact, heat is transferred from the one at higher temperature to the one at lower temperature, just as when two bodies of water are connected, water flows through the connection from the higher surface to the lower one. When two bodies are placed in contact and neither gains heat from the other, the two bodies are at the same temperature by definition, and the levels of internal energy are the same for both. When this condition is satisfied, the bodies are in *thermal equilibrium.*

15.3 THE KELVIN ABSOLUTE TEMPERATURE SCALE

If the temperature of a body is a measure of the level of internal energy for the body, there must be a lower limit, or *absolute zero,* of temperature. If we remove from the atoms and molecules all their available energy of random motion, we can properly assign them a temperature of absolute zero, since temperature measures the level of the energy of random motion. As we saw in Sec. 14.6, the gas laws had also suggested the existence of such an absolute zero of temperature. As a solid is cooled, the energy associated with random motions of the particles is reduced. At a temperature of 0 K, the motions of all atoms are reduced to their lowest possible value. At one time it was believed that this would correspond to a state of absolutely no motion, but modern quantum mechanics has shown there is necessarily some zero-point energy.

Lord Kelvin established an absolute temperature scale based on thermodynamic reasoning (Chap. 18) which uses degrees of the same size as the Celsius degree. In addition to absolute zero, this scale uses as a fixed point the *triple point of water* (Sec. 17.9), which was taken by international agreement in 1954 to be exactly 273.16 K. The triple point (the temperature at which ice, water, and water vapor are all in equilibrium) is chosen because it is highly reproducible. On the Kelvin absolute thermodynamic scale the ice point is 273.15 K, and the steam point 373.15 K. We use T to indicate absolute temperature, t to represent Celsius or Fahrenheit temperature, and the approximation

$$T(\text{K}) = t(^\circ\text{C}) + 273 \qquad (15.1)$$

The Kelvin scale is essentially equivalent to the ideal-gas scale introduced tentatively in Sec. 14.6.

15.4 HEAT UNITS

Heat is a form of energy and can be measured in the units of mechanics, e.g., joules. Long before this was known, however, heat units based on temperature increases in a known mass of water became well established. In the SI *the kilocalorie is the quantity of heat required to raise the temperature of one kilogram of water from 14.5 to 15.5°C.* In terms of mechanical-energy units 1 kcal = 4,185.8 J.

The heat required to raise the temperature of 1 kg of water 1 Celsius degree depends only slightly on the temperature (Fig. 15.2). It takes almost exactly 100 kcal to raise the temperature of 1 kg of water from 0 to 100°C, and in our problems we shall assume that 1 kcal will raise the temperature of 1 kg of water 1 Celsius degree at any temperature in this range.

As used by physical scientists, the *calorie* is the heat required to raise the temperature of one gram of water one Celsius degree. In the biological sciences, the kilocalorie is referred to as the *Calorie* (sometimes not capitalized). Thus, when a book on diet reports that an apple has 70 "calories" of food value and a large cooked egg has 80 "calories," the energies quoted are in the physicists' *kilocalories.*

In the English system heat is measured in British thermal units (Btu). *One British thermal unit is the heat required to raise the temperature of a one-pound mass of water from 58.5 to 59.5°F.* Since 1 lb is equal to 0.4536 kg and 1 Fahrenheit degree corresponds to $\frac{5}{9}$ Celsius degree, it follows that 1 Btu is equal to 0.252 kcal. In mechanical units 1 Btu = 778 ft-lb.

The number of units of mechanical energy necessary to produce the same increase in internal energy as one unit of heat energy is called the *mechanical equivalent of heat.* It is represented by J in honor of Joule, who was a pioneer in studying the relationship between work and heat.

$$J = 4,186 \text{ J/kcal} = 778 \text{ ft-lb/Btu}$$

One of several experiments devised by Joule to determine J utilizes a well-insulated vessel containing a known mass of water. Inside the vessel a series of paddle wheels, driven by weights hung over pulleys, are rotated to churn the water. When the weights descend, the paddle wheels do work on the water in the vessel (calorimeter), causing the water to increase in temperature. From the weights and the distance through which they descend, the work done on the water can be calculated. From the mass of the water, the mass of the container, and the increase in temperature, the heat corresponding to this amount of mechanical energy can be computed.

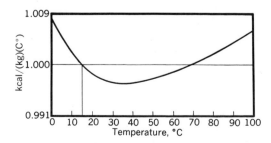

FIGURE 15.2
The specific heat of water as a function of temperature.

15.5 HEAT CAPACITY AND CALORIMETRY

If we add equal amounts of heat to 1 kg of water and to 1 kg of copper, the temperature of the copper goes up far more than the temperature of the water. It takes 0.093 kcal to raise the temperature of 1 kg of copper 1 Celsius degree. *The heat required to change the temperature of a unit mass of a substance one degree is the specific heat of the substance.* Specific heat has the dimensions *energy per unit mass per degree change in temperature;* some examples are kcal/(kg)(C°), Btu/(lb)(F°), J/(kg)(K). The specific heats of a number of solids and liquids are listed in Appendix Tables B.1 and B.2.

Let Q denote the quantity of heat added to a mass m, and let t_1 and t_2 be the initial and final temperatures, respectively. If c is the specific heat of the material at constant volume,

$$Q = mc(t_2 - t_1) = mc\, \Delta t \qquad (15.2)$$

The quantity $mc\, \Delta t$ of Eq. (15.2) is the increase in the internal energy of the mass.

□ **Example** Find the number of kilocalories required to raise the temperature of 0.100 kg of brass from 25 to 75°C. The specific heat of brass is 0.09 kcal/(kg)(C°).

$$\begin{aligned}
\text{Heat} &= \text{mass} \times \text{sp ht} \times \text{change in temperature} \\
&= m \times c \times (t_2 - t_1) \\
&= 0.100\ \text{kg} \times 0.09\ \text{kcal/(kg)(C°)} \times (75 - 25)\text{C°} \\
&= 0.45\ \text{kcal} \qquad\qquad\qquad\qquad\qquad\qquad □
\end{aligned}$$

One of the most familiar methods of measuring a quantity of heat is by imparting this heat to a known mass of water and observing the change it produces in the temperature of the water. Experiments of this kind are carried on in a vessel known as a *calorimeter* (Fig. 15.3), which is carefully insulated so that it neither gains heat from nor loses heat to its surroundings during the experiment. The basic principle underlying all calorimetry experiments is the *law of conservation of energy*, which, for calorimetry, takes the form: *The heat gained by those bodies which gain heat is equal to the heat lost by those bodies which lose heat.*

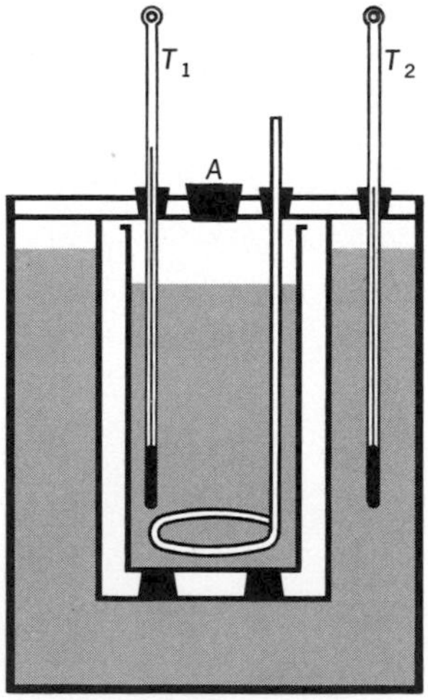

□ **Example** A 0.450-kg cylinder of lead is heated to 100°C and then dropped into a 50-g copper calorimeter containing 0.100 kg of water at 10°C. The water is stirred until equilibrium is established, at which time the temperature of the whole system is 21.1°C. Find the specific heat of lead.

$$\text{Heat gained} = \text{heat lost}$$
$$0.100[1(21.1 - 10)] + 0.050[0.093(21.1 - 10)] = 0.450c(100 - 21.1)$$
$$1.16 = 35.5c$$
$$c = 0.033\ \text{kcal/(kg)(C°)} \quad □$$

When something like a calorimeter is used over and over again in heat experiments, it is often convenient to compute its heat capacity. *The heat capacity of a body is defined as the heat required to raise the temperature of the entire body one degree.* To find the heat capacity, it is necessary only to multiply the mass m of the body by the specific heat c.

FIGURE 15.3
Calorimeter for measuring the heat transferred to the water in the inner container from a heated mass introduced through the opening closed by the plug A. The water in the outer container maintains an almost constant-temperature environment for the inner one.

$$\text{Heat capacity of body} = mc \qquad (15.3)$$

In the example above, the heat capacity of the calorimeter is $0.050 \text{ kg} \times 0.093 \text{ kcal}/(\text{kg})(\text{C}°) = 0.0047 \text{ kcal}/\text{C}°$. The heat capacity of a calorimeter is sometimes specified by giving its *water equivalent,* which is the mass of water that requires the same heat to raise its temperature one degree. Thus it takes the same heat to raise the temperature of the body a given number of degrees as it would to raise the temperature of a mass of water equal to the water equivalent by the same amount.

From Eq. (15.3) it is clear that the specific heat c is equal to the heat capacity of a body divided by its mass. Thus, heat capacity per unit mass is another name for specific heat; the two terms means exactly the same thing.

15.6 HEAT OF COMBUSTION

The heat of combustion is the heat liberated by burning a unit mass or unit volume of a fuel such as coal or gas. One method of finding it for a solid or liquid fuel is to place a measured quantity in a crucible inside a bell jar which is closed so that the products of combustion can escape only through the openings at the base of the jar. The bell jar is placed inside a calorimeter, the mass of which is known, and this vessel is then filled with a known weight of water. The temperature of the water is determined, and then the fuel is oxidized. A supply of oxygen is admitted through the opening at the top of the bell jar until all the fuel has been burned. The products of combustion bubble up through the water. When the combustion is complete, the temperature of the water is again determined. The heat energy gained by the water and the calorimeter is equal to the chemical energy released in the oxidation.

☐ **Example** A 6.8-g sample of coal was burned in a calorimeter like that described above. The mass of the water was 4.54 kg, and its temperature at the beginning of the experiment was 15°C; the final temperature was 26°C. Find the heat of combustion of the coal. The vessel forming the calorimeter has a water equivalent of 260 g (or a heat capacity of 0.260 kcal/C°).

$$\text{Energy released} = \text{heat gained}$$
$$\text{Heat of combustion} \times 0.0068 = (4.54 \times 11 \times 1) + (0.26 \times 11 \times 1)$$
$$= 52.8 \text{ kcal}$$

$$\text{Heat of combustion} = \frac{52.8 \text{ kcal}}{0.0068 \text{ kg}} = 7800 \text{ kcal/kg} \qquad ☐$$

15.7 HEAT CAPACITIES OF SOLID ELEMENTS

The heat capacities of six elements are listed in Table 15.1. When we multiply the specific heat of each element by its atomic weight, we obtain the *heat capacity per kilomole;* in Chap. 14 we defined the kilomole of an element as the number of kilograms of that element equal to the atomic weight. Since both the kilocalorie and the joule are units of energy with 1 kcal = 4.186 J, we can express heat capacities in terms of either energy unit. The joule is ordinarily used for measuring work and

TABLE 15.1
Heat Capacities of Solid Elements

Element	Atomic weight	Heat capacity per kilogram, or specific heat		Heat capacity per kilomole = specific heat × atomic weight	
		kcal/(kg)(C°)	J/(kg)(C°)	kcal/(kmol)(C°)	J/(kmol)(C°)
Aluminum	27	0.22	920	5.9	25,000
Titanium	47.9	0.14	590	6.7	28,000
Iron	55.8	0.11	460	6.1	26,000
Copper	63.5	0.093	390	5.9	25,000
Tin	118.7	0.054	230	6.4	27,000
Lead	207.2	0.031	130	6.4	27,000

mechanical energy, while the kilocalorie is often used with thermal energy.

In 1819 Dulong and Petit showed that *for many solid elements the heat capacity per kilomole is approximately 6 kcal/(kmol)(C°).* This value is very nearly 25 kJ/(kmol) (C°) $\approx 3R$, where R is the general gas constant, a fact on which further comment appears in Sec. 15.8. The physical meaning of the *law of Dulong and Petit* is readily seen when we recall that 1 kmol of any element contains $N_A = 6.02 \times 10^{26}$ atoms; thus, *for most solid elements, it takes the same amount of energy to raise the temperature of equal numbers of atoms 1 Celsius degree;* on the average we must add about 10^{-26} kcal per atom.

At room temperature there are a few solid elements which are notable exceptions to the law of Dulong and Petit. For example, carbon in the diamond form has a specific heat of only 1.46 kcal/(kmol)(C°) at room temperature, but the value approaches 6 as the temperature is raised (Fig. 15.4). For all elements the specific heat is small at low temperature, approaching zero as the temperature approaches 0 K. Modern quantum theory gives us an understanding of the reasons for the discrepancies between observed heat capacities and the law of Dulong and Petit.

15.8 THE HEAT CAPACITY OF GASES

For gases the specific heat depends on how the heating is carried out. If the volume is kept constant as heat is applied, all the heat goes into increasing the internal energy of the molecules. When the pressure is held constant as heat is added, the internal energy increases as before, but

FIGURE 15.4
The atomic heat capacity (product of the specific heat and the atomic weight) in kilocalories per kilogram atomic weight per Celsius degree as a function of temperature for three solid elements. For larger atomic weights the curves rise more sharply than the curve for aluminum.

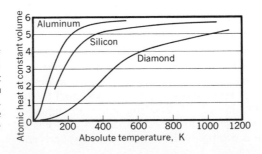

external work is done by the gas as it expands. Consequently it takes more heat to raise the temperature of a unit mass of gas 1 degree at constant pressure than it does at constant volume.

Heat Capacity at Constant Volume From Eq. (14.11) we know that $m\overline{v^2}/2 = 3RT/2N_A$. The total kinetic energy of translation of all the molecules in 1 kmol is $N_A m\overline{v^2}/2 = 3RT/2$. To increase the temperature by 1 K, we must increase the total kinetic energy of translation to $3R(T + 1)/2$, or by $3R/2$. Hence, for any gas in which all the thermal energy is associated with translational motion, the heat capacity per kilomole is $3R/2 = 2.98$ kcal/(kmol)(C°), since $R = 1.987$ kcal/(kmol)(K). This is just half the value given by the law of Dulong and Petit for solid elements. Actually the heat which goes to translational kinetic energy is the same for solid elements and gaseous molecules, but the atoms in solids have potential energy equal on the average to the kinetic energy.

For a monatomic gas (or any gas in which all the internal energy of the molecules is associated with translational motion) the specific heat is $3R/2M$, where $M (= N_A m)$ is the mass in kilograms of 1 kmol (numerically M is simply the molecular weight). Thus

$$c_V = \frac{3R}{2M} \qquad (15.4)$$

For helium, $M = 4$ kg/kmol, and $c_V = 0.745$ kcal/(kg)(K); for argon, $M = 39.95$ kg/kmol, and $c_V = 0.075$ kcal/(kg)(K).

For diatomic and triatomic gases at room temperature the specific heat is greater than the value $3R/2M$ because the molecule has kinetic energy of rotation as well as of translation. We may think of a diatomic molecule as a dumbbell which is rotating as it moves about (Fig. 15.5). When we raise the temperature, the kinetic energy of rotation increases, as well as the kinetic energy of translation. This is why the specific heat at constant volume for diatomic gases is greater than that predicted by Eq. (15.4). At room temperature the specific heat at constant volume for most diatomic gases is given by $5R/2M$. At higher temperatures it becomes still greater; the additional energy goes into exciting vibrations of the two atoms along the line connecting their centers of mass. For triatomic gases near room temperature, the specific heat is $3R/M$ or more, depending on how much energy is associated with vibration.

Heat Capacity at Constant Pressure When a gas is kept at constant pressure, the heat required to increase the internal energy of the molecules is the same as when the gas is kept at constant volume. In addition, however, it is necessary to supply heat energy *to do the external work required to expand the gas*. If a gas expands in a cylinder having at one end a moving piston (Fig. 15.6) by means of which the pressure is held constant, the work W done against the piston in moving it from a to b is $p \, \Delta V$, where p is the pressure and ΔV the change in volume. This can be seen as follows:

$$W = Fs = pAs$$

where F is the force, s the distance through which the piston moves, and A the area of the piston. The product $As = V_2 - V_1$ is the increase in volume ΔV of the gas during expansion. Therefore,

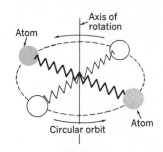

FIGURE 15.5
A diatomic molecule may have kinetic energy of rotation and of vibration as well as kinetic energy of translation.

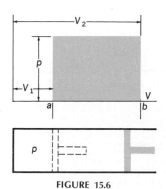

FIGURE 15.6
The external work done by a gas expanding at constant pressure is given by the product of the pressure and the change in volume.

$$W = p\,\Delta V \qquad (15.5)$$

This work is represented by the shaded area on the pV diagram of Fig. 15.6.

Consider the work done when 1 kmol of gas is heated 1 K under constant pressure. By the general gas law, $pV = RT$. If we raise the temperature 1° at constant pressure, we have $p(V + \Delta V) = R(T + 1)$. Therefore, $p\,\Delta V = R$. Thus, the external work done by 1 kmol of gas when its temperature is raised 1 K is numerically equal to R. The *heat capacity per kilomole at constant pressure is equal to that at constant volume plus the universal gas constant R.*

The energy which must be added per unit mass of gas to do external work is R/M, and the heat capacity of the gas at constant pressure is R/M greater than the heat capacity at constant volume:

$$c_p = c_V + \frac{R}{M} \qquad (15.6)$$

The ratio of the heat capacity of the gas at constant pressure to the heat capacity at constant volume is ordinarily represented by γ. For a monatomic gas c_V is $3R/2M$, and $c_p = 5R/2M$. Therefore,

$$\gamma = \frac{c_p}{c_V} = \frac{5R/2M}{3R/2M} = \frac{5}{3}$$

for a *monatomic* gas.

For typical diatomic gases at room temperature,

$$c_V = \frac{5R}{2M} \qquad \text{and} \qquad \gamma = \frac{7R/2M}{5R/2M} = 1.4$$

15.9 THERMAL EXPANSION OF SOLIDS

Most substances expand when heated (Fig. 15.7). This is not surprising if we remember that heating a solid increases the amplitude of vibration of the atoms and has the effect of moving the atoms somewhat farther apart. If the vibrations were simple harmonic, the average position of each atom would not change and there would be no expansion. However, the forces which restore an atom to its equilibrium position are greater when the atom approaches too close to a neighbor than when the atom moves an equal distance too far away; the forces are *anharmonic*. As the amplitudes of the vibrations increase, the average distance between atoms becomes slightly larger. The change of length of a typical solid is small—only a few parts in 10,000 for an ordinary daily range of outside temperature—but the forces associated with such expansion can be tremendous and must be allowed for in many structures. The amount of expansion depends on the nature of the substance and on the change in temperature. On railroads small distances are usually left between sections of rail to permit expansion without having the tracks bow. In concrete roads and sidewalks, expansion joints filled with tar are provided. Care is taken in selecting compatible types of glass and metal when one wishes to bring conductors through glass walls into vacuum tubes and light bulbs.

When we heat a solid, the change in length ΔL depends on the material, on the original length L, and on the change in temperature Δt. For most materials we can write the approximate equality

$$\Delta L = \alpha L\,\Delta t \qquad (15.7)$$

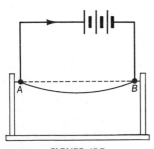

FIGURE 15.7

The length of a metal wire increases when its temperature is raised by passing an electric current through the wire.

where α is known as the *coefficient of linear expansion* of the material. From Eq. (15.7) the *coefficient of linear expansion* α *is the change in length per unit length for a one-degree rise in temperature.* Clearly, α is independent of the units in which the length is measured, but it does depend on the temperature unit. A change of 1 Fahrenheit degree corresponds to a change of $\frac{5}{9}$ Celsius degree. Therefore, α per Fahrenheit degree is only $\frac{5}{9}$ of α per Celsius degree. The coefficient of linear expansion varies with the temperature, usually slowly near room temperature but rapidly at very low temperature. The room-temperature coefficients of linear expansion of many solids are listed in Appendix Table B.1. In the construction of a bridge, or indeed in almost any structure, provision must be made for expansion and contraction. One possible arrangement for a small bridge is to mount one end on a roller. As the length of the bridge increases or decreases, this end of the bridge moves forward or back without injuring the piers.

□ **Example**   The main cable of a suspension bridge is 1,500 m long at 0°C. If the cable is made of steel, find its length on a hot summer day when the temperature is 35°C.

$$\Delta L = \alpha L\, \Delta t$$
$$= (1.2 \times 10^{-5})(1,500)(35)$$
$$= 0.63 \text{ m}$$

Thus the length is 1,500.63 m at 35°C.   □

15.10   DIFFERENTIAL EXPANSION

Consider a strip of brass and a strip of iron welded together (Fig. 15.8) to form a composite bar. When this bar is heated, it bends because the brass expands more rapidly than the iron; therefore the brass is on the outside of the curve when it is heated. Devices of this kind, called *bimetallic strips*, have found many uses. One of the most familiar is in the thermostat, which can be made to open and close electric circuits as the movable end of the bimetallic strip changes its position (Fig. 15.9). Bimetallic strips are used in a common type of thermometer.

In a watch, the balance wheel (Fig. 15.10) is made of bimetallic strips. As the temperature increases, ends A and B are carried inward, making the moment of inertia of the wheel less; at the same time the expansion of the radius carries the rim of the wheel farther from the center, thereby producing an increase in the moment of inertia. Meanwhile, the elasticity of the hairspring is reduced by the higher temperature. This would make the balance wheel move more slowly if the moment of inertia of the wheel were not reduced to compensate. By proper design the balance wheel can be made to change its moment of inertia just enough for the period of vibration to remain the same over a reasonable temperature range.

15.11   EXPANSION IN AREA

The area of a square of material with sides of length L is L^2. When we raise the temperature by Δt, the length of each side becomes $L + \Delta L$ and the area becomes

$$(L + \Delta L)^2 = L^2 + 2L\,\Delta L + (\Delta L)^2$$

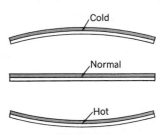

FIGURE 15.8
A bimetallic strip is formed by sealing a strip of a material with a lower coefficient of linear expansion (gray) to a strip of material with a higher coefficient of linear expansion. Unequal expansions or contractions lead to bending of the strip.

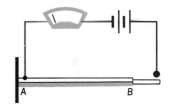

FIGURE 15.9
The bimetallic strip AB closes an electric circuit when it bends upward with a decrease in temperature. Again the gray strip has the lower coefficient of linear expansion.

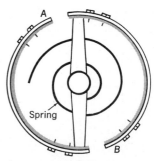

FIGURE 15.10
Balance wheel of a watch.

Since ΔL is very small compared with L, we may neglect $(\Delta L)^2$ in comparison with the other terms and write the new area as $L^2 + 2L \Delta L$. The change in area ΔA is then given by

$$\Delta A = 2L \Delta L = 2\alpha L^2 \Delta t = 2\alpha A \Delta t \tag{15.8}$$

From Eq. (15.8) we see that *the coefficient of area expansion is twice the coefficient of linear expansion.*

The area of a hole in a piece of material expands exactly as though it were filled with the material in question, regardless of the shape of the hole or the piece. The pioneers made use of this fact in putting iron tires on the wooden wheels of their wagons. The iron tire was made slightly smaller than the wooden wheel upon which it was to be placed. Then the tire was heated red hot, and, thanks to its expansion, it could be slipped over the wood. As the iron cooled, it shrank around the wood. In modern manufacturing, when one cylinder is to fit inside and be fastened to another cylinder, the two are often "sweated" together. In this process the inner cylinder is made slightly larger than the hole in the outer cylinder. Then the outer cylinder is heated, and the inner one cooled. Because of expansion of the outer cylinder and contraction of the inner one, the outer cylinder can be slipped over the inner. When the two come to the same temperature, they are held together by strong frictional forces.

15.12   VOLUME EXPANSION

Most materials, fluids as well as solids, expand when heated. The change in volume ΔV depends on the substance, the change in temperature Δt, and the original volume V according to the relation

$$\Delta V = \beta V \Delta t \tag{15.9}$$

where β is *the coefficient of volume expansion* of the substance. *The coefficient of volume expansion is the change in volume divided by the product of the original volume and the change in temperature.*

For most substances β varies somewhat with temperature. *For solids the coefficient of volume expansion β is three times the coefficient of linear expansion α*, as the following reasoning indicates. Consider a solid cube of length L on each side. When the temperature is raised an amount Δt, the length of each side becomes $L + \Delta L$. The volume is then $(L + \Delta L)^3 = L^3 + 3L^2 \Delta L + 3L(\Delta L)^2 + (\Delta L)^3$. Since ΔL is very small compared with L, we may neglect the last two terms in comparison with the first two. The initial volume was L^3, and the change in volume is then

$$\Delta V = 3L^2 \Delta L = 3\alpha L^3 \Delta t = 3\alpha V \Delta t \tag{15.9a}$$

Liquids such as alcohol and kerosene expand when heated. Coefficients of volume expansion for several liquids are listed in Appendix Table B.2. However, the most familiar liquid, water, does not behave in so simple a fashion. When warmed from 0 to 4°C, it contracts. As the temperature is raised above 4°, the water expands (Fig. 15.11). Thus a given mass of water has its minimum volume and its maximum density at 4°C under 1 atm pressure. The temperature for maximum density decreases about 0.02 Celsius degree per atmosphere as the pressure is increased. Although the temperature at the bottom of a deep northern lake may be less than 4°C in the winter, it still does not reach the freezing point, which also

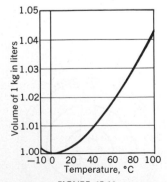

FIGURE 15.11
The volume of 1 kg of water in liters as a function of temperature. The volume per kilogram at its minimum (at 3.98°C) is 1.000028 L.

decreases with increased pressure. As a result the surface may freeze over in winter, but there is still water below. In the summer the upper layers of the lake are warmed, but very deep layers may not increase appreciably in temperature.

207

HEAT CAPACITY AND THERMAL
EXPANSION

QUESTIONS

1. What is the distinction between internal energy and heat? How are they related to temperature?
2. When a mercury-in-glass thermometer is thrust into boiling water, the mercury first descends and then rises. Why?
3. What is unique about the thermal expansion of water? What are some of the important consequences? Why wouldn't water in glass make a good thermometer?
4. Why are specific heats numerically the same in both British and metric units?
5. Why is the molar heat capacity of a diatomic gas ordinarily greater than the corresponding heat capacity of a monatomic gas?
6. How would you expect the molar heat capacity of hydrogen at constant volume to compare with that of oxygen? Why?
7. Nitrogen is sealed in a thin glass container which is placed under a bell jar. If the bell jar is evacuated and the container broken, what happens to the temperature of the gas? Explain. What would have happened if the bell jar had not been evacuated? Why is there a difference?
8. An iron rod lies along the diameter of a circular iron hoop and connects opposite sides of the hoop. If the temperature of the whole system is increased, will the hoop remain circular?
9. Does the coefficient of linear expansion depend on what thermometer scale is used? On what units of length are used? Explain.
10. How is the thermal expansion allowed for in (a) concrete pavements, (b) steel bridges, and (c) ordinary light bulbs?
11. A stretched rubber hose contracts when its temperature is increased. What does this mean about the coefficient of linear expansion? Can you explain why rubber and some other materials contract with increasing temperature?
12. Some pendulum clocks have a metal rod supporting a pendulum bob consisting of a hollow metal cylinder partially filled with mercury. How does this serve to make the period of the clock almost independent of temperature?

PROBLEMS

1. If a 120-g piece of copper at 250°C is dropped into 300 g of water at 12°C contained in a 100-g copper calorimeter, find the resulting equilibrium temperature. *Ans.* 20.3°C
2. A piece of lead heated to 100°C is dropped into 200 cm³ of water at 4.6°C. The final temperature is 9.8°C. Compute the mass of the lead.
3. A piece of aluminum with a mass of 80 g is heated in a steam jacket to a temperature of 99.5°C and then plunged into 160 g of water at 13°C, causing the temperature to rise to 17.5°C. What is the specific heat of aluminum, assuming that no heat was lost or gained during the experiment? *Ans.* 0.219 kcal/(kg)(C°)
4. To find the temperature of a furnace, a 180-g piece of platinum was placed inside until temperature equilibrium was reached. The platinum was then dropped into a 150-g copper calorimeter containing 400 g of water at 6°C. If the final temperature of the water was 14°C, find the temperature of the furnace.
5. A lead bullet that has a velocity of 320 m/s strikes a target. If 80 percent of the kinetic energy of the bullet is absorbed by the target and the other 20 percent is retained by the bullet as heat, what is the increase in temperature of the bullet? *Ans.* 78.9 C°
6. How much heat is generated in stopping a flywheel that is making 20 r/s about an axis through its center? The flywheel has a moment of inertia of 1.8 kg-m².
7. One kilogram of water falls from the top to the bottom of Niagara Falls, a distance of 49 m. If all the potential energy lost in the fall is transformed into heat, how much is the temperature of the water raised? *Ans.* 0.115 C°
8. A truck and its load have a mass of 4,500 kg. The brakes are used to bring it to rest from a speed of 100 km/h. How much heat is developed?
9. A soft drink at 0°C has a mass of 450 g. If 1 g of sugar yields 4 Cal (= 4 kcal), what mass of sugar would be required to provide the body with the energy to raise the temperature of the liquid to body temperature of 37°C? *Ans.* 4.16 g
10. Upper Yosemite Falls is 1,430 ft high. If all the potential energy lost in this fall is converted into heat, how much is the temperature of the water raised?
11. A hiker with his equipment weighs 800 N. If he climbs 650 m up a mountain, and if 20 percent of the energy of the food he consumes goes into potential energy, how many eggs at 80 kcal each

would he have to eat to supply himself with food energy? *Ans.* 8

12. A 70-kg man daily consumes food with a fuel value of about 2600 kcal. If this energy were used to heat 70 kg of water, by how many degrees would the temperature be raised?

13. Use the law of Dulong and Petit to estimate the specific heats of lanthanum, gold, and nickel. (Atomic weights are listed inside the back cover; the answers given are experimental values at 18°C from a handbook.)
    Ans. 0.0448, 0.0312, and 0.105 kcal/(kg)(C°)

14. Estimate by use of the law of Dulong and Petit the room-temperature specific heats of selenium, silver, and uranium. (Atomic weights are listed inside the back cover.)

15. Natural gas from the mains, with a heat combustion of 56 MJ/m³, is used to heat water. Assuming that half the available heat is wasted, how much gas will be required to heat 200 kg of water from 5 to 95°C? *Ans.* 2.69 m³

16. A specimen of coal with a mass of 3.96 g was burned in a copper calorimeter having a mass of 1.1 kg. The mass of the water in the calorimeter was 2.6 kg, and the initial temperature was 14.6°C. The final temperature of the water was 23.9°C. Find the heat of combustion.

17. A sample of methyl alcohol of 15 g mass was burned in a fuel calorimeter containing 6.8 kg of water. The water equivalent of the calorimeter was 535 g. The initial temperature was 11.2°C, and the final temperature was 22.1°C. Calculate the heat of combustion of the methyl alcohol.
    Ans. 5330 kcal/kg

18. The heat of combustion of natural gas is 1500 Btu/ft³. How many cubic feet of gas are needed to heat 20 ft³ of water from 40 to 180°F in a hot-water heater, assuming that 80 percent of the energy released is absorbed by the water?

19. Into a 100-g copper calorimeter containing 180 g of water at 10°C there are dropped simultaneously 70 g of silver at 100°C, 60 g of iron at 85°C, and 20 g of platinum at 90°C. What is the resulting equilibrium temperature? *Ans.* 14.5°C

20. A sample of coal weighing 0.8 oz with a heat of combustion of 12,000 Btu/lb was burned in a crucible in a calorimeter. The calorimeter contained 10 lb of water, and it had an initial temperature of 45°F. The parts of the calorimeter weighed 6.2 lb and had an average specific heat of 0.18 Btu/(lb)(F°). To what temperature was the water raised?

21. Suppose that 0.2 kmol of air at 0°C and two standard atmospheres pressure occupies 2.24 m³. With the pressure kept constant, heat is added until the temperature reaches 30°C. Find the volume of the gas at 30°C, the external work done by the gas, and the heat supplied to the gas.
    Ans. 2.49 m³; 49.8 kJ; 175 kJ = 41.7 kcal

22. Find the external work done and the increase in internal energy if 2 mol (0.056 kg) of nitrogen at 17°C and 1×10^5 N/m² pressure is expanded at constant pressure until the volume increases 40 percent.

23. Krypton is a monatomic gas with an atomic weight of 83.8. Find the specific heat of krypton at constant volume and at constant pressure. How much external work is done when 3 kg of krypton is heated from 0 to 5°C at constant pressure? How much is the internal energy increased?
    Ans. 0.0356 kcal/(kg)(C°); 0.0593 kcal/(kg)(C°); 1,488 J; 0.534 kcal

24. The specific heats of air at constant pressure and at constant volume are 0.24 and 0.17 kcal/(kg)(C°), respectively. If 5 kg of air is heated from 0 to 30°C at constant pressure, find (*a*) how much external work is done and (*b*) how much the kinetic energy of the air molecules is increased.

25. Three moles (12 g) of helium at 0°C occupies a volume of 33.9 L at a pressure of 2×10^5 N/m². If 500 cal of heat is added to the helium at constant pressure, find the final temperature of the gas, the external work done by the gas, and the increase in the internal energy of the gas.
    Ans. 33.6°C; 837 J; 1,256 J

26. When 2 kg of argon is heated from 17 to 127°C, how many kilocalories are required if the heating is done at constant volume? At constant pressure?

27. Five gram molecular weights (140 g) of nitrogen at 27°C is supplied with 8,000 J of heat. It expands, doing 2 kJ of work. What is the increase in internal energy? What is the increase in temperature? *Ans.* 6 kJ; 57.7 C°

28. A cylinder contains 600 cm³ of neon at 0°C and 760 mmHg pressure. What is the mass of this quantity of gas? If the temperature of this gas is raised to 100°C and the pressure is reduced to 350 mmHg, find the new volume of the gas and the increase in internal energy of the gas.

29. The steel supporting cables of a suspension bridge have a total length of 350 m at 0°C. Find the increase in the length at 30°C. *Ans.* 0.13 m

30. Steel rails in 20-m lengths are laid in winter at 0°C. How much space must be allowed between

consecutive rails to permit expansion in summer at a temperature of 35°C?

31. A bar of copper at 20°C is 1.2 m long. At what temperature will it be 1 mm longer?
    Ans. 69.9°C

32. The channel span of the steel Ohio River bridge in Cincinnati is 354 m long. Calculate the maximum change in its length if it is subject to extreme temperatures of −15 and 35°C.

33. A steam locomotive had steel driving wheels 1.782 m in diameter before the tires were shrunk on. If the diameter of the tire was 1.780 m at 20°C, find the temperature to which it had to be heated to make its diameter 1.783 m so it could be slipped over the wheel. *Ans.* 160°C

34. A steel ring which is 0.4500 m in diameter at room temperature (20°C) is to be shrunk onto a pulley which is 0.4509 m in diameter. Find the temperature to which the ring must be raised so that it will just slip over the pulley.

35. Find the coefficient of areal expansion per Fahrenheit degree for aluminum. An aluminum panel for an aircraft wing has an area of exactly 6 ft^2 at 50°F. What is the increase in area when the temperature is raised to 120°F?
    Ans. 2.83×10^{-5} per F°; 1.71 in.^{2}

36. A brass cap is to be sweated on a brass cylinder. The inner diameter of the cap is machined to 22.00 mm, while the cylinder has an outer diameter of 22.04 mm. What must the temperature of the cap be before it can be slipped over the cylinder if the cylinder is at 15°C?

37. A glass flask holds exactly 0.6 L of mercury at 15°C. If it is heated to 75°C, how many grams of mercury will run out? *Ans.* 77 g

38. A sheet of brass has an area of 8,000 mm^2 at 5°C. How much greater will its area be at a temperature of 95°C?

39. A rod of steel and one of brass have exactly the same length, 1 m at 0°C. The rods are heated together until they differ in length by 0.5 mm. What is the temperature of the rods?
    Ans. 68.5°C

40. A glass flask holds exactly 0.5 L. It is filled with ethyl alcohol at 0°C and then heated to 40°C. How much ethyl alcohol runs out?

41. A clock with a steel pendulum beats seconds correctly when the temperature of the room is 15°C. How many seconds per day will it gain or lose when the temperature of the room is 20°C.
    Ans. Loses 2.6 s/day

42. A long U-tube is filled with ethyl alcohol. One arm of the tube is kept at 5°C and the other at 30°C. Find the length of the column in the tube at the higher temperature if the length of the column in the tube at the lower temperature is 755 mm.

43. A steel rod 115 mm long has a cross-sectional area of 145 mm^2. What force would be required to extend the bar the same amount as the expansion produced by heating it 25°C?
    Ans. 9.14 kN

44. A brass pendulum is adjusted to beat seconds, i.e., has a period of 2 s at 20°C. What will the gain or loss per day be if the temperature of the clock drops to 10°C?

45. A bimetallic strip is made by soldering a strip of brass to a strip of steel. The strips are each 0.2 mm thick and 125 mm long. If one end is held fast, find how far the other end moves if the strip is heated 100 C°. (Assume the strips were initially straight and that after the temperature rise the brass and steel form arcs of concentric circles 0.2 mm apart.) *Ans.* 28 mm

46. Show that the stress in a heated rod which is not free to expand (length held constant) is given by $F/A = Y\alpha\,\Delta t$, where Y is Young's modulus, α the coefficient of linear expansion, and Δt the temperature change.

47. Compute the hydrostatic pressure required to prevent an aluminum block from expanding when its temperature is increased from 10 to 50°C. *Ans.* 2.1×10^8 N/m^2

48. Show that the coefficient of volume expansion of a gas is given by $1/T$, where T is the absolute temperature.

49. A 750-mm copper rod and a 250-mm aluminum rod are laid end to end between rigid supports which remain exactly 1 m apart. The rods have identical radii and do not bend. Calculate the stress in each rod and the length of each if the temperature is raised from 10 to 90°C.
    Ans. 1.53×10^8 N/m^2; 750.045 and 249.955 mm

16

HEAT TRANSFER

To warm water we increase the internal energy of the system by adding heat. In a calorimetry experiment we can find the specific heat of material by dropping a known mass of the material at high temperature into a known mass of water and determining the equilibrium temperature. In both these cases heat is conveyed from one body to another or from one part of a system to another. In this chapter we study the processes by which these heat transfers are accomplished.

16.1 HEAT-TRANSFER PROCESSES

As we saw in Sec. 15.2, heat is energy in transit from one body to another, or from one part of a substance to another, *by virtue of temperature differences.* There are three natural processes by which heat is transported from one point to another.

1. *Conduction is the process by which heat energy is transferred from particle to particle by collisions or direct interactions.*

    When the bowl of a spoon is placed in a hot liquid, heat is conveyed by conduction from the bowl to the handle. Conduction occurs in solids, liquids, and gases. It is the only method by which heat is transferred through opaque solids. The atoms in the hotter part of the material vibrate with greater amplitude than those in the colder part. Atoms with greater thermal energy transfer part of this energy to their neighbors, which in turn transmit energy to their own neighbors. In many materials this exchange occurs through interatomic interactions, but in metals free electrons are the primary agency through which heat energy is transported. These same electrons are responsible for electric conductivity, and it is for this reason that the best thermal conductors are also the best electric conductors.

2. *Convection is heat transfer by the actual movement of the heated material itself.*

    Steam produced in a boiler can be transported to radiators throughout a building. The transfer of heat by movement of the hot steam is a convective process. Convection is ordinarily the most important heat-transfer process for liquids and gases. The distribution of heat in practically all houses and buildings is achieved by means of convection currents of heated air, hot water, or steam.

3. *Radiation consists of electromagnetic waves which transmit energy from a source to an absorber.*

    All bodies radiate and absorb energy in the form of oscillating electric and magnetic fields. The human eye is sensitive to some of these electromagnetic waves but not all. Those we can see constitute *visible light;* they have wavelengths between 380 and 780 mm. Radiation with wavelengths too long to be seen is *infrared,* while that with too short wavelengths is called *ultraviolet.* Bodies at a higher temperature than their surroundings radiate more energy than they receive and are thereby cooled, while those at a lower temperature than their surroundings absorb more than they radiate and are thus warmed. The earth's *primary source of energy* is *radiation from the sun.* Radiation traverses a vacuum as well as transparent media. Indeed, radiation is of primary importance for bodies at high temperature and is the only means of transmission across regions in which there is no material medium. The atoms and molecules of a body at a higher temperature emit energy as radiation, which passes to a cooler body, where it is absorbed and reconverted to thermal energy. For temperatures only 100 C° or so above or below room temperature, radiation is relatively unimportant compared with conduction and convection for most material mediums.

If a steam radiator provides heat in a room, all three processes play a role in transferring the heat. Conduction brings the heat from the steam to the outside of the radiator. From here the primary heat transfer is by convection, although radiation does play a role in distributing the heat. The first evidence that convection is of principal importance is the fact that a person several feet from the radiator feels equally warm on both sides. It is characteristic of radiation transfer that the energy moves in straight lines from the heat source to the point where it is absorbed. The familiar "roast one side, freeze the other" situation which arises at an outdoor bonfire on a cold night is typical of radiative transfer.

16.2 CONDUCTION

If heat is to be conducted through a material, there must be a difference in temperature between two regions. If a constant temperature difference is maintained between the two faces of a slab of material (Fig. 16.1), the heat Q conducted through this slab is proportional to the area A of the slab, the temperature difference $t_h - t_c$ (h for hotter, c for cooler) between the two faces, and the time the temperature difference is maintained. It is inversely proportional to the thickness d of the slab.

$$Q = \frac{kA(t_h - t_c) \times \text{time}}{d} \tag{16.1}$$

The proportionality constant k for a given material is called the *coefficient of thermal conductivity*. Coefficients for a number of materials are listed in Table 16.1. In the SI k is the quantity of heat energy in joules which passes through one square meter of area in one second when a temperature difference of one Celsius degree is maintained across a thickness of one meter. Older tables usually list k in cal-cm/(cm²)(s)(C°). In the British system of units, k is numerically equal to *the number of Btu conducted per hour through an area of one square foot when a temperature difference of one Fahrenheit degree is maintained across a thickness of one inch.*

In problems of thermal conductivity it is important to use the actual temperatures of the slab faces in Eq. (16.1). For example, if one calculates the heat conducted through a glass window when the indoor temperature is 22°C and the outdoor temperature is −10°C, one obtains a heat loss which is far too great if one puts 22°C as t_h and −10°C as t_c. Actually, the temperature of the inner surface of the glass under conditions of this sort might well be 12 C° lower than the 22°C of the air in the room, while the outer face of the glass may similarly be some 12 C° or more warmer than the outside air. There are major temperature drops in the thin air layers adjacent to each side of the glass. Equation (16.1) gives acceptable values of Q only when the actual surface temperatures are used.

An apparatus for measuring the coefficient of thermal conductivity of a good conductor such as copper is represented in Fig. 16.2. It consists mainly of a rod, the coefficient of thermal conductivity of which is to be

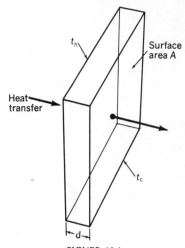

FIGURE 16.1
The heat transferred per second through a slab is directly proportional to the area and to the temperature difference between the faces and inversely proportional to the thickness of the slab.

TABLE 16.1
Thermal Conductivities at 20°C
$1 \text{ J-m}/(s)(m^2)(C°) = 1 \text{ W}/(m)(C°)$

Substance	Thermal conductivity k, W/(m)(C°)†
Air	0.023
Aluminum	230
Brass	100
Concrete	1.7
Copper	400
Cork	0.046
Glass	0.63
Ice	1.7
Rock or glass wool	0.038
Sand (white dry)	0.39
Silver	418
Soil (dry)	0.14
Steel	46
Water	0.60
Wood (across grain)	0.13
Zinc	110

† To obtain k in Btu-in./(ft²)(h)(F°), multiply by 6.9; in cal/(cm)(s)(C°) divide by 419.

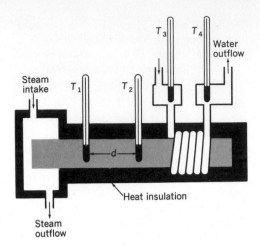

FIGURE 16.2

Apparatus for measuring the thermal conductivity of the material of a solid rod.

determined. This rod is typically about 30 cm long and has a diameter of about 3 cm. On one end is soldered a copper box through which steam can be passed to make the temperature of that end of the rod about 100°C. Around the other end is soldered a coil of copper tubing through which cold water is circulated. The temperature of this water as it enters is found by means of the thermometer T_3, and its temperature as it leaves is read on the thermometer T_4. The difference in temperature between incoming and outgoing water multiplied by the mass of water gives numerically the quantity of heat that has flowed down the rod. The coil and the rod are carefully insulated with felt to prevent heat losses. Small holes drilled at right angles to the axis of the rod contain thermometers T_1 and T_2 by which the fall of temperature across the thickness d is measured.

☐ **Example** In measuring the thermal conductivity of a rod (Fig. 16.2), the following data were obtained. Temperature of incoming water = 20.0°C; temperature of outgoing water = 30.0°C; temperature t_h read by thermometer T_1 = 80.0°C; t_c = 60.0°C; distance between thermometers = 10 cm; area of rod = 20 cm²; mass of water flowing through box = 650 g in a time of 180 s. Find the coefficient of thermal conductivity.

Heat gained by water = mass of water × temperature change × sp ht
$$= 0.650 \times 10 \times 1 = 6.50 \text{ kcal} = 27,200 \text{ J}$$

$$Q = \frac{kA(t_h - t_c)}{d} \times \text{time}$$

$$27,200 = k \times 20 \times 10^{-4} \times \frac{20}{0.10} \times 180 = 72k$$

$$k = \frac{27,200}{72}$$

$$= 380 \text{ J/(m)(s)(C°)} = 380 \text{ W/(m)(C°)} \qquad \square$$

The ratio of the temperature change Δt to the thickness over which it occurs is called the *temperature gradient.* In the example above it is given by $\Delta t/d = (t_h - t_c)/d = (80 - 60)\text{C°}/(0.10 \text{ m}) = 200 \text{ C°/m}$.

The thermal conductivity of carbon increases with the temperature. On the other hand, the thermal conductivities of metals ordinarily decrease

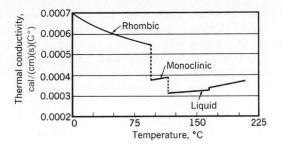

FIGURE 16.3
The thermal conductivity of sulfur changes when the crystalline form changes and when the solid sulfur liquefies.

with an increase of temperature. The thermal conductivity changes abruptly when a substance such as sulfur (Fig. 16.3) passes from one crystalline state to another.

The walls of a house or a refrigerator often consist of two or three layers of different materials. To compute the heat conducted through such a system of layers, we make use of the fact that the only heat which can be conducted through the second layer is that which passes through the first layer, and so forth. When equilibrium is established, the heat conducted per unit area per second through each of the layers is the same.

□ **Example** The wall of a shed in which ice is stored consists of an outer layer of wood 2 cm thick and an inner layer of rock wool 3 cm thick. Find the heat conducted through 50 m² of the wall in 1 h when the outer wood surface is at 20°C and the inner rock-wool surface is at 5°C. Also find the temperature of the wood–rock-wool interface.

Let t be the temperature of the interface. The heat conducted through both layers is the same, so that

$$Q = \left[\frac{kA\,\Delta t(\text{time})}{d} \right]_{\text{wood}} = \left[\frac{kA\,\Delta t(\text{time})}{d} \right]_{\text{rock wool}}$$

$$= \frac{0.13 \times 50(20 - t) \times 3{,}600}{0.02} \text{ J}$$

$$= \frac{0.038 \times 50(t - 5) \times 3{,}600}{0.03} \text{ J}$$

Solving for t yields $t = 17.55°C$, the temperature of the interface. If we put this value into the equation for Q, we obtain

$$Q = \left[\frac{0.13 \times 50(20 - 17.55) \times 3{,}600}{0.02} \right]_{\text{wood}} = 2.9 \times 10^6 \text{ J}$$

It may be of interest to check this answer by showing that the same heat is conducted through the rock wool. □

16.3 CONVECTION

The transfer of heat from one place to another by the motion of the heated substance is known as convection. The convection is said to be *forced* if the heated material is moved by means of fans or pumps, as is the case in a forced-air heating system in a home where fans blow the hot air from the furnace throughout the rooms. *Natural* convection arises from the change in density that takes place when a fluid is heated.

For example, when a gas or a liquid is heated, it expands and becomes lighter than the cold fluid. When water is heated on a stove, the liquid at the bottom is warmed, expands, and rises to the top. Meanwhile the cooler and denser water sinks. The flows thus set up in the liquid are known as *convection currents*. There is no simple formula which permits the calculation of heat transfer by convection in the general case. The heat transferred to or from a surface at one temperature in contact with a fluid at a different temperature depends on many factors, including the geometry and orientation of the surface as well as properties of the fluid.

16.4 RADIATION

Radiation from the sun travels through many gigameters of empty space until eventually it falls upon absorbing matter, where it is transformed into the mechanical energy of random motion. Sunlight falling on an object warms it above the temperature of the surrounding air. In like manner, radiation from a roaring fire warms objects on which it falls.

All bodies, whether cold or hot, radiate. If two bodies are exactly alike in every way except that one is at a higher temperature, the hotter body radiates more energy than the colder one. When a body is at a higher temperature than its surroundings, it radiates more heat than it receives; when it is at a temperature lower than that of its surroundings, it receives more energy than it radiates. The energy radiated by a body increases rapidly as its temperature is raised; it is proportional to the fourth power of the absolute temperature.

The rate at which heat is radiated depends not only on the temperature of the body but also on the character and area of the radiating surfaces. Some surfaces are good radiators, others poor. Generally, polished surfaces are poor radiators; rough, blackened ones are good radiators.

There is an intimate relationship between the rate at which a surface radiates energy and the rate at which the same surface under the same conditions absorbs heat. Good radiators are also good absorbers of heat. That this must be true can be seen easily. Consider a polished reflecting sphere and a rough black sphere in an insulated enclosure. Eventually they reach the same temperature as the enclosure (Fig. 16.4). Radiation from the enclosure falls upon each at the same rate, but the polished surface reflects most of the incident energy, while the blackened sphere absorbs the energy striking its surface. Therefore, the blackened surface is receiving energy at a more rapid rate than the polished one. However, once the bodies and the enclosure are all at the same temperature, they remain at the same temperature. Therefore, the blackened body, which absorbs energy at a greater rate, also radiates at a greater rate. Under equilibrium conditions each body is absorbing energy at the same rate at which it is radiating.

The arguments of the preceding paragraph apply to the situation in which equilibrium has been reached and all bodies are at the same temperature. If the blackened sphere and the polished one are placed in sunlight on a bright day, the blackened sphere attains a substantially higher temperature than the polished one. In this case the bodies and their surroundings (which must include the sun) never attain identical temperatures. However, it is still true that *a good absorber is a good radiator and a poor absorber a poor radiator.*

FIGURE 16.4
Once thermal equilibrium is attained, a black sphere and a reflecting one remain at the enclosure temperature. Although the black sphere absorbs more energy per unit area, it also radiates more.

16.5 BLACKBODY RADIATION

The best emitter of radiation is necessarily the best absorber. A perfect absorber of radiation is known as *an ideal blackbody. A blackbody is one which absorbs all the radiation incident upon it.* Although no body with a perfect absorbing surface exists, a surface covered with a thick coating of lampblack absorbs about 99 percent of the heat radiation incident upon it. An almost perfect blackbody can be produced by making a small opening in the wall of a rough cavity (Fig. 16.5); the surface area of the hole acts as the blackbody. Radiation incident on this small hole enters the cavity, in which it is reflected from wall to wall until completely absorbed. Meanwhile, each point of the surface of the cavity emits radiation which falls on other points of the cavity, where part is reflected and part absorbed. Radiation of all wavelengths is reflected back and forth in this cavity until there is a uniform density of radiation throughout it. If the temperature of the wall is increased, the radiation level is correspondingly increased.

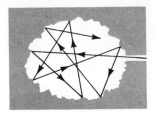

FIGURE 16.5
An ideal blackbody can be formed by producing a cavity of rough internal surface with a small opening in any solid material. The surface of the opening acts as the blackbody.

If radiation emerging from the opening of the cavity is examined with suitable instruments, a wide range of wavelengths is found to be present. The energy associated with each wavelength can be measured. The distribution of the energy in this spectrum changes with temperature. The total radiation emitted from such a cavity, called *blackbody radiation,* increases rapidly with temperature. In 1879 Stefan analyzed experimental data obtained by Tyndall and concluded that if R stands for the energy radiated per unit area per second by a blackbody.

$$R = \sigma T^4 \qquad (16.2)$$

where σ is a constant equal to 5.67×10^{-8} W/(m^2)(K^4) and T is the absolute temperature. This is known as *Stefan's law* (or the Stefan-Boltzmann law). If the temperature of a blackbody is doubled, the rate at which energy is radiated is increased by a factor of 16.

If the radiator is not a blackbody, it radiates less than a blackbody of identical size and temperature. If the *absorption factor* α represents the fraction of incident radiation absorbed by the surface of the radiator, the energy radiated per unit area per second is

$$R = \alpha \sigma T^4 \qquad (16.2a)$$

Thus a body which absorbs only half the radiation incident upon it has $\alpha = 0.5$, and it emits only half as rapidly as a blackbody of the same size and shape. For copper α is roughly 0.3; thus a copper radiator radiates only 0.3 as much power per unit area as a blackbody at the same temperature.

For a blackbody at 998 K most of the radiation is in the wavelength region between 1 and 10 μm, as shown by the lowest curve of Fig. 16.6. The curves of the figure show the energy emitted per second in a small (1-nm) wavelength interval centered at the wavelength shown on the abscissa. As the temperature of the blackbody increases, the power emitted at every wavelength increases but not in the same proportion; the increase is relatively much greater at the shorter wavelengths. At each temperature there is a wavelength at which the curve has a maximum. As the temperature rises, the wavelength at which maximum intensity is radiated moves toward smaller values. If λ_m is the *wavelength at which the radiation is maximum* for the absolute temperature T,

$$\lambda_m T = 2.898 \times 10^{-3} \text{ m-K} \qquad (16.3)$$

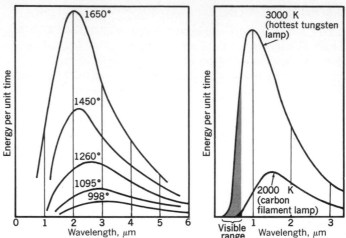

FIGURE 16.6
Blackbody radiation at several temperatures, showing the energy radiated per
unit time in a small fixed-wavelength
range as a function of wavelength.

This relation is known as *Wien's displacement law*. For temperatures up to 600°C or about 900 K, essentially all the radiation is in wavelengths too long for the human eye to see. However, by the time the temperature reaches 1000 K, the blackbody emits visible radiation (Fig. 16.6) and the body appears dull red. As the temperature is increased, more energy is emitted in the blue and violet regions of the visible spectrum and the color of the body changes from red to orange to yellow and eventually to white.

16.6 THE SURFACE TEMPERATURE OF THE SUN

If we assume that the sun is roughly a blackbody, we can estimate its surface temperature by applying either Stefan's radiation law or the Wien displacement law. Let us consider first the use of Stefan's law. Measurements at the earth reveal that a surface perpendicular to the sun's rays when the sun is directly overhead receives roughly 1,400 W/m². The energy received per unit area per unit time on a surface perpendicular to the sun's rays is called the *solar constant*. The solar constant of 1,400 W/m² is equivalent to about 2 cal/cm²-min. If we assume that the sun's energy is emitted equally in all directions, the total power in watts emitted is given by 1,400 times the area in square meters of a sphere of radius equal to the distance from the sun to the earth. The power radiated by the sun is therefore

$$P = (1{,}400 \text{ W/m}^2)(4\pi)(1.5 \times 10^{11} \text{ m})^2$$
$$= 4 \times 10^{26} \text{ W}$$

When we divide this by the area of the sun's surface, we obtain the energy emitted per square meter per second

$$R = \frac{4 \times 10^{26} \text{ W}}{4\pi(7 \times 10^8 \text{ m})^2} = 6.5 \times 10^7 \text{ W/m}^2$$

If we insert this in Stefan's law, assuming the sun is a blackbody, we obtain

$$R = \sigma T^4 \quad \text{or} \quad 6.5 \times 10^7 = 5.67 \times 10^{-8} T^4$$
$$T^4 = 1.14 \times 10^{15}$$
$$T = 5800 \text{ K}$$

If the spectral distribution of the radiation from the sun is studied and intensity is plotted as a function of wavelength, it is found that maximum energy is radiated at approximately 0.49 μm. When we put this value in Wien's displacement law, we obtain

$$0.49 \times 10^{-6}T = 2.898 \times 10^{-3}$$

$$T = 5900 \text{ K}$$

Both these values have been obtained on the assumption that the sun is a blackbody. Actually, this is not the case. The curves of the sun's radiation fall substantially below the blackbody curves for 5900 K in the short-wavelength region and are substantially above in the far-infrared region.

Our best methods for measuring very high temperatures involve the use of the radiation laws. Instruments called *optical pyrometers* are used to compare the radiation from the body whose temperature is to be measured with the radiation from a calibrated source.

16.7 NEWTON'S LAW OF COOLING

Whenever a body is at a somewhat higher temperature than its surroundings, the rates at which energy is lost due to conduction, convection, and radiation are all roughly proportional to Δt, the temperature difference between the body and its surroundings. Therefore, *the rate at which the body loses temperature is proportional to the difference between its temperature and that of its surroundings.* This is known as *Newton's law of cooling.* Thus, a cup of coffee loses temperature twice as rapidly when it is 80 C° above its surroundings as when it is 40 C° above.

If the body is somewhat cooler than its surroundings, Newton's law is still applicable. In this case both the "rate of cooling" and Δt are negative, since the body gets warmer and its temperature is lower than that of the surroundings.

□ **Example** A pan filled with hot food cools from 92 to 88°C in 2 min when the room is at 20°C. How long will it take to cool from 71 to 69°C?

The average of 92 and 88°C is 90°C, which is 70 C° above room temperature. Under these conditions the pan cools 4 C° in 2 min or 2 C°/min.

$$\frac{\text{Change of temperature}}{\text{Time}} = k \, \Delta t$$

$$\frac{4 \text{ C}°}{2 \text{ min}} = k (70 \text{ C}°) \tag{A}$$

The average of 69 amd 71°C is 70°C, which is 50 C° above room temperature. k is the same for this situation as for the original.

$$\frac{2 \text{ C}°}{\text{Time}} = k (50 \text{ C}°) \tag{B}$$

If we divide (A) by (B), we have

$$\frac{4 \text{ C}°/2 \text{ min}}{2 \text{ C}°/\text{time}} = \frac{k (70 \text{ C}°)}{k (50 \text{ C}°)}$$

$$\text{Time} = 1.4 \text{ min} \qquad \qquad □$$

QUESTIONS

1. Why wear woolen mittens to keep your hands warm when the thermal conductivity of wool is far greater than that of air?
2. A piece of paper tightly wrapped around a copper rod can be held briefly in a bunsen flame without catching fire, while the same paper on a wooden rod would ignite quickly. Why does this happen?
3. A wooden door and a metal doorknob are in thermal equilibrium at 5°C. Why does the metal doorknob feel much colder than the wood? Why would the metal feel warmer if both were at 50°C?
4. If all bodies radiate electromagnetic energy, why don't they always get colder? Under what circumstances do they become warmer?
5. Why are the contents of an automobile radiator less likely to freeze on a cold, clear night if the automobile is under the roof of a carport, even though it is completely exposed to the cold air?
6. Why does an automobile with closed windows get so hot inside when it is parked in the sun?
7. Why does a small hole leading to a rough cavity in any material act as a blackbody?
8. Steam radiators are seldom painted black, although a black radiator should radiate better than one painted a light color. Do you think blackening radiators would increase their effectiveness substantially? Why?
9. Why are pipes carrying hot fluids often covered with some thermal insulating material which is in turn covered with aluminum paint?
10. When a white plate, a piece of blackened steel, and a piece of reddish copper are heated in a furnace, why do they all glow red as they first begin to be incandescent?
11. If a white china plate with black decorative markings is heated to incandescence, which will be brighter, the markings or the white china? Why?
12. Why are stainless steel pans often coated on the bottom with a thin layer of copper?

PROBLEMS

1. The heat exchanger for a reactor is designed to transfer 0.8 MW through a steel conductor 1.5 mm thick. The temperature on one side of the steel is 140°C and on the other side 138°C. What area is required? *Ans.* 13.0 m²
2. A glass window has an area of 2 m² and a thickness of 3.5 mm. The outer surface is at 10.5°C, and the inner surface is at 11.9°C. How much heat flows through the window each hour?
3. Find the coefficient of thermal conductivity for asbestos paper if 8 cal flows each second through a slab 2 mm thick and 4,000 mm² in area when the temperature of one face is maintained at 96°C and that of the other at 34°C.

    Ans. 0.27 W/(m)(C°)
4. A brass rod 0.25 m long and 180 mm² in cross-sectional area is thermally insulated. One end is kept at 100°C in a steam bath, and the other end is cooled by circulating water, as shown in Fig. 16.2. In 5 min, 300 g of water enters at 15°C. Two thermometers 0.15 m apart along the rod read 85 and 30°C, respectively. Find the heat conducted down the rod each minute and the temperature of the water as it leaves.
5. A glass window is 4.5 mm thick, 1.2 m wide, and 1 m high. The temperature of the inner surface is 14.1°C, and that of the outer surface is 12.9°C. How much heat is lost per hour through the window? *Ans.* 726 kJ
6. How many calories will be lost each hour through a glass window 1.1 by 0.9 by 0.004 m thick if the inner surface is at 0°C and the outer surface is at −2.5°C?
7. How much water would be evaporated per hour per square meter by the heat which flows through a steel boiler plate 2.5 mm thick when there is a difference in temperature of 3.5 C° between the faces of the plate and the water enters at 20°C? Assume the pressure remains at 1 atm and that it takes 540 kcal to evaporate 1 kg of water at 100°C. *Ans.* 89.3 kg
8. If the bottom of an aluminum pan is 2 mm thick and the temperature of the inner surface is 100°C while that of the outer surface is 102°C, find the heat conducted through an area of 0.012 m² in 40 s.
9. A brass bar has a length of 0.8 m and a cross section of 500 mm². One end is kept in a mixture of water and ice at 0°C, consisting initially of 10 g of ice and 30 g of water. The other end is maintained at a uniform higher temperature until all the ice is melted, and then the temperature is raised to maintain the same difference in temperature between the two ends of the bar. Neglect losses of heat to other bodies. How many calories are needed to melt the ice? What is the difference in temperature between the two ends of the bar if the ice melts in 15 min? How soon after the ice melts will the water temperature be 20°C? It takes 80 cal to melt 1 g of ice at 0°C. *Ans.* 800 cal; 59.5°C; 15 min

10. A large meat-storage refrigerator has a window with an area of 0.2 m². If the window is 12 mm thick and the temperature of the inner glass surface is 10°C, how many calories per day pass through the glass when the outer surface has a temperature of 14°C?

11. Superheated steam is flowing through a steel pipe that has a length of 12 m. The internal diameter of the pipe is 75 mm, and its external diameter is 80 mm. If the outside surface of the pipe has a temperature of 106°C and the inner surface a temperature of 110°C, how much heat escapes in 10 min? (Assume that the wall of the pipe is approximately a thin slab 2.5 mm thick and of width equal to the average circumference of the pipe.) *Ans.* approx. 30,800 kcal

12. A brass bar has a length of 250 mm and a cross-sectional area of 225 mm². One end is kept in steam at 100°C, and the other in ice at 0°C. Neglecting losses due to radiation, find the number of grams of ice melted in 15 min if it takes 80 cal to melt 1 g.

13. The surface of a radiation pyrometer receives 0.8 cal/s from a furnace whose temperature is 527°C. How many calories per second will it receive when the temperature of the furnace is raised to 927°C? *Ans.* 4.05 cal/s

14. A blackbody radiates 15 W when its temperature is 600 K. Find the power radiated if the temperature is raised to 1200 K, and find the wavelength at which the radiation rate is greatest at 1200 K.

15. If the fireball of a nuclear weapon can be approximated at some instant as a spherical blackbody of 0.5 m radius with a surface temperature of 10^6 K, find the power radiated and the wavelength at which maximum energy is radiated. *Ans.* 1.8×10^{17} W; 2.9 nm

16. If the surface temperature of a star is 6000 K, what is the wavelength of its radiation maximum? What would the surface temperature of a star have to be if it were to have its radiation maximum in the blue at a wavelength of 380 nm?

17. The filament in a tungsten lamp radiates energy at the rate of 150 W when operating at a temperature of 2400 K. A second lamp operating at a temperature of 3000 K radiates 6 times as much energy. What is the ratio of the area of the second filament to that of the first? At what wavelength is maximum power radiated at each temperature? *Ans.* 2.46; 1.21 μm; 966 nm

18. To what temperature must a blackbody be raised to triple its total radiation if its original temperature is 827°C?

19. Calculate the power radiated by a hot body at a temperature of 1127°C if the radiating area is 0.02 m² and the absorption factor α is 0.8. *Ans.* 3.49 kW

20. A blackbody initially at 27°C is heated to 227°C. How many times as much power is radiated at the higher temperature? At what wavelength is the maximum power radiated in each case?

21. Find the heat conducted in 10 min through an area of 2 m² of a layer consisting of 25 mm of cork and 75 mm of wood. The outer surface of the wood is at 25°C, and the opposite cork surface is at 0°C. What is the temperature of the cork-wood interface? *Ans.* 26.8 kJ; 12.1°C

22. A copper-clad, stainless steel pan has a conducting area of 0.02 m² and a 0.1-mm layer of copper on 1 mm of stainless steel. The coefficient of thermal conductivity of stainless steel is 21 W/(m)(C°). (a) Find the temperature at the copper–stainless steel interface when the outer surface is at 112°C and the inner surface is at 110°C. (b) How many calories are conducted through the pan each second?

23. A glass of 20 g water equivalent contains 270 g of water and 10 g of ice. If it takes 20 min for the ice to melt, approximately how much longer will it take for the temperature to rise to 2°C if the surroundings remain at 20°C? Roughly how long will it take for the temperature to rise from 9.5 to 10.5°C? It takes 80 cal to melt 1 g of ice at 0°C. *Ans.* 15.8 min; 15 min

24. If it takes 10 min for a glass of water to warm from 4.5 to 5.5°C in a room at 20°C, approximately how long will it take for its temperature to rise from 9.5 to 10.5°C?

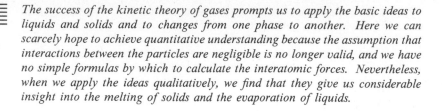

The success of the kinetic theory of gases prompts us to apply the basic ideas to liquids and solids and to changes from one phase to another. Here we can scarcely hope to achieve quantitative understanding because the assumption that interactions between the particles are negligible is no longer valid, and we have no simple formulas by which to calculate the interatomic forces. Nevertheless, when we apply the ideas qualitatively, we find that they give us considerable insight into the melting of solids and the evaporation of liquids.

17 CHANGE OF PHASE

17.1 MELTING OR FUSION

Removing energy from a gas lowers its temperature. For many gases liquefaction occurs, and if still more energy is taken away, the liquid solidifies. Such changes in phase are typically reversible, so that adding energy to a solid may lead to its fusion and eventually to its evaporation.

When ice is heated, the temperature rises until it reaches 0°C and then remains stationary until all the ice is melted. Once all the ice is melted, the temperature of the water begins to rise. *The temperature at which the solid changes into liquid is called the melting point.* At the melting point more heat serves to hasten the melting process without any change of temperature.

The temperature at which the liquid changes into the solid state is its *freezing point.* For crystalline substances such as ice or copper, the freezing point and the melting point coincide and are sharply defined. Substances that are not crystalline, such as wax or glass, gradually soften and do not have definite melting points. For certain fats, the melting point is not the same as the freezing point. For example, butter melts between 28 and 32°C and solidifies between 20 and 23°C.

If a pure liquid is carefully protected from mechanical disturbances, it can be cooled below the temperature at which it normally solidifies. This process is known as *supercooling.* Water has been cooled to about −40°C without becoming ice. The liquid at such a temperature is in a state of unstable equilibrium and immediately solidifies if it is disturbed or if a crystal of the solid is dropped into it.

To melt a solid such as ice, it is necessary to supply a given quantity of thermal energy to each unit mass. This is true even though the temperature of the ice at the beginning is the same as the temperature of the water at the end of the process. The ice has a crystalline structure, and the energy is necessary to tear down this structure. Water, in changing back to ice (in general, a liquid in changing back to a solid), gives up the heat that it absorbed in melting. *The heat of fusion of a substance is the heat necessary to convert a unit mass of solid into liquid at the same temperature and pressure.* To change 1 kg of ice to water at 0°C requires 80 kcal; to convert 1 lb of ice to water at 32°F requires 144 Btu. Heats of fusion of several solids are listed in Appendix Table B.1.

□ **Example** In an experiment designed to measure the heat of fusion, 25 g of ice at 0°C was dropped into 195 g of water at 30°C contained in a copper calorimeter of mass 100 g. The final temperature was 18°C. Find the heat of fusion of ice.

$$\text{Heat gained} = \text{heat lost}$$
$$0.025L + 0.025(1)(18 - 0) = 0.195(1)(30 - 18) + 0.100(0.093)(30 - 18)$$

where L is the latent heat of fusion and $0.025L$ is the heat needed to melt the ice.

$$0.025L + 0.450 = 2.45$$
$$L = 80 \text{ kcal/kg} \qquad \square$$

17.2 CHANGE OF VOLUME DURING FREEZING

It is a familiar fact that ice floats and that pipes or bottles filled with water burst when frozen; 1 L of water makes about 1.09 L of ice. Cast iron and type metal behave like water in this respect, expanding when they solidify. For this reason they are suitable for making castings in which it is desired to reproduce the detail of a mold. Most metals and other substances contract on solidification. This is one reason why many coins, as well as gold and silver medals, are stamped rather than cast.

That the forces exerted by freezing water are large can be seen from the fact that they are sufficient to burst strong water pipes on a cold night. These forces are of importance in the formation of soils from rocks. Water penetrates into the crevices in the rocks and freezes. The expansion breaks off fragments, which may eventually become soil. Alternate freezing and thawing of soils tends to pulverize them. Ice forming in the interstices of the soil serves to loosen compact land and give it better tilth.

The fact that water expands when it freezes is responsible for the fact that ice forms on top of a lake or stream rather than the bottom. A layer of ice is an excellent insulation which protects all but shallow water from freezing to the bottom. Fish and other water life can carry on in the water below the ice layer.

Since an increase of pressure causes a body to contract, such an increase favors the liquid phase for any substance which contracts on melting and results in a lower melting point. Ice contracts when it melts; therefore the melting point is lowered by application of pressure. Careful experiments show that this lowering is 0.0075°C for an increase of 1 atm of pressure (Fig. 17.1). If a substance expands upon melting, its melting point is raised by the application of pressure.

The effect of pressure on the melting point of ice can be shown by supporting a loop of wire on a piece of ice and hanging a heavy weight from the wire. The pressure of the wire lowers the melting point so that the ice is in a condition to melt as soon as the necessary heat is supplied. This heat is taken from the water above the wire, causing it to freeze again. This process, called *regelation*, continues until the wire cuts its way through the ice, leaving the block as solid as it was at the beginning.

17.3 FREEZING POINT OF A SOLUTION

When any material is dissolved in water, the freezing point of the solution is lower than the freezing point of water. This is also true for other liquids. The freezing point of a solution is lower than that of the pure solvent. This fact is used to prevent the water in the cooling system of an automobile from freezing. For solutions that are not too concentrated, the lowering of the freezing point is proportional to the number of dissolved particles per unit volume. A given number of salt (NaCl) mole-

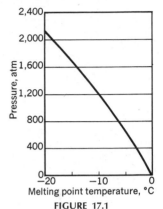

FIGURE 17.1
Ice melts at a lower temperature when it is subjected to additional pressure.

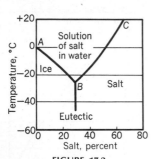

FIGURE 17.2

The freezing point of a salt solution as a function of salt concentration by weight.

cules lowers the freezing point more than an equal number of sugar molecules because the salt ionizes, thus producing two particles per dissolved molecule.

When a dilute solution begins to freeze, only solvent freezes out. This makes the remainder of the solution more concentrated and lowers its freezing point still further, until the solution becomes saturated. Then dissolved substance and solvent freeze out in such a way that the concentration of the solution remains unchanged. Figure 17.2 represents the relation between the freezing point and the percentage of salt in a solution. Curve *AB* shows the relation between temperature and concentration of the salt. The temperature at which the solvent and the dissolved substance crystallize out as a mixture is called the *eutectic temperature*. If the temperature of the solution is higher than the eutectic temperature and the solution is saturated, the dissolved substance crystallizes out as the temperature is lowered (curve *BC*).

17.4 SATURATED VAPOR

The molecules of a liquid move with a wide range of instantaneous velocities. When the liquid is heated, this range and the average speed increase. Some of the molecules near the surface attain sufficient kinetic energy to escape the forces of attraction which confine the less energetic molecules to the liquid (Fig. 17.3). When the space above the surface of the liquid is enclosed, some vapor molecules return to the surface of the liquid and are captured. As more molecules escape, the number returning to the surface also increases. When the number of molecules returning to the surface is equal to the number escaping, the space above the liquid is said to be *saturated*, and above the liquid we have a *saturated vapor*. The pressure exerted by a saturated vapor depends on the nature of the liquid (Fig. 17.4) and on the temperature. If the temperature is held constant and an attempt is made to increase the pressure on a saturated vapor, part of the vapor condenses. *The saturated vapor pressure of any given liquid depends only on the temperature.*

The saturated vapor pressures of several liquids as functions of temperature are plotted in Fig. 17.5. As the temperature increases, the pressure of the saturated vapor rises. A vapor-pressure curve can be obtained by measuring the difference in level between the surfaces of mercury in the apparatus of Fig. 17.6 as the temperature is increased.

17.5 EVAPORATION AND BOILING

Even at relatively low temperature the most energetic molecules evaporate from the surface of a liquid. *Evaporation is the escape of molecules from*

FIGURE 17.3

When a liquid evaporates in an enclosed region, eventually the number of molecules escaping the liquid per unit time becomes equal to the number returning to the liquid per unit time. When this situation applies, the vapor is said to be *saturated*.

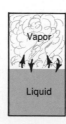

the surface of a liquid. As the temperature of a liquid is raised, evaporation proceeds more rapidly. Eventually bubbles of vapor form in the liquid and rise to the surface. That temperature at which vapor bubbles form in the volume of the liquid and rise to the surface is the *boiling point* or *boiling temperature.*

We shall now examine the boiling point from a somewhat different point of view. Consider a bubble of vapor within the liquid. This bubble will collapse unless the pressure of the saturated vapor within it is as great as the pressure in the adjacent liquid. The pressure at the bubble is the applied pressure at the surface of the liquid plus the hydrostatic pressure due to the liquid above the bubble plus the pressure due to surface tension. Even for a relatively small bubble near the surface, the last two contributions to the total pressure are usually negligible in comparison with the applied pressure. Thus, *the boiling point is the temperature at which the pressure of the saturated vapor of the liquid is equal to the applied pressure.*

The distinction between *evaporation* and *boiling* should now be clear. Both involve the change of phase from liquid to gas, which is called *vaporization.* Evaporation is vaporization from the surface alone whereas boiling is vaporization within the volume of the liquid. Since the boiling point of a liquid is the temperature at which the vapor pressure is the same as the applied pressure, it follows at once that when the applied pressure is changed, the boiling point also changes. At 1 atm (760 mmHg) pressure, water boils at 100°C. The curves of Fig. 17.5 show how the boiling points of four familiar liquids vary with pressure. (For water the boiling point is tabulated as a function of pressure in Appendix Table B.3.)

The influence of pressure on the boiling point can be shown by filling a flask half full of water and boiling it vigorously for some time to remove the air from above the water. Remove the flask from the flame and immediately insert a stopper, rendering the flask airtight. Invert the flask, and pour cold water on the bottom. This cold water causes some of the vapor in the flask to condense, and the pressure on the hot water in the flask is reduced sufficiently to allow the water to begin boiling again.

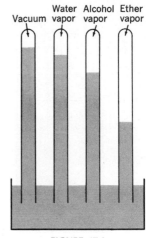

FIGURE 17.4
When a small volume of liquid is introduced into the Torricelli vacuum above a mercury column, the mercury level is depressed below the vacuum level by the saturated vapor pressure of the liquid.

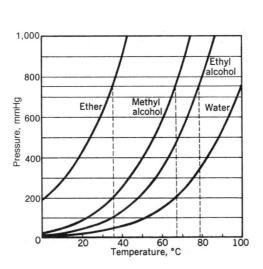

FIGURE 17.5
The saturated vapor pressure of four liquids as a function of the temperature. The dashed lines indicate the boiling points at 1 atm (760 mmHg) pressure.

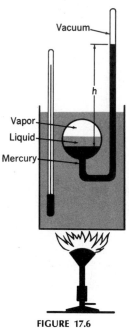

FIGURE 17.6
Apparatus for measuring the saturated vapor pressure of a liquid as a function of temperature. The saturated vapor pressure in millimeters of mercury is the height *h* in millimeters.

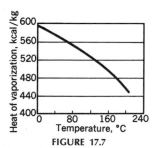

FIGURE 17.7
The heat of vaporization of water decreases as the temperature is raised.

The effect of pressure on the boiling point of water is well illustrated in the action of geysers. If it is 30 m down from the surface of a narrow column of water to a cavity surrounded by hot rocks, the pressure at these rocks is about 4 atm, three of the four being due to the column of water. At this pressure water boils between 140 and 150°C. Once water in the cavity begins to boil, bubbles rush up the column carrying quantities of water with them, thus reducing the pressure at the cavity and encouraging more rapid boiling until the temperature of the water in the cavity is lowered to the point at which boiling ceases.

In a pressure cooker the increased pressure results in a higher boiling point for water and a higher cooking temperature. On the other hand, sirups or milk to be evaporated may be placed in "vacuum pans" so that boiling can be carried out at a reduced temperature, thereby eliminating undesirable changes which would occur at the normal boiling temperature.

When a ship is driven by a rapidly rotating propeller, the pressure on one side of the propeller decreases as the angular velocity is increased. When the pressure falls below its saturated vapor pressure, the water flashes into vapor, producing cavities in the liquid. This phenomenon is known as *cavitation*.

17.6 HEAT OF VAPORIZATION

Just as a certain amount of heat is required to convert a unit mass of ice into water without changing its temperature, so a certain amount of heat is required to change a unit mass of water into steam without changing its temperature. When the steam condenses, this heat is liberated.

The heat of vaporization of a substance is the heat necessary to change a unit mass of the liquid to the vapor state without any change in the temperature and pressure. For water at 100°C and 1 atm pressure, it is 540 kcal/kg (970 Btu/lb). It takes more than 5 times as much heat to change 1 g of water at 100°C into steam at 100°C as it takes to heat that same gram of water from 0 to 100°C. The heats of vaporization of several other liquids are listed in Appendix Table B.2.

The heat of vaporization depends on the temperature (and pressure) at which the change takes place. The higher the temperature, the easier it is to evaporate the liquid; i.e., the higher the temperature, the lower the heat of vaporization (Fig. 17.7).

The following experiment may be used to determine the heat of vaporization of water. Fill a calorimeter with a known mass of water. Pass dry steam from a boiler into the calorimeter. Measure the temperature of the water in the calorimeter at the beginning and after steam has been condensed in it. The mass of condensed steam is found by weighing the calorimeter after the steam has been condensing long enough to produce the desired rise in temperature. The latent heat of vaporization of water can now be calculated as illustrated below.

□ **Example** Dry steam at 100°C is condensed in a large calorimeter of water equivalent 0.25 kg, which contains 2.8 kg of water at 5°C. The final temperature of 27°C is reached after 110 g of steam has been condensed. Find the latent heat of vaporization *V* as given by these data.

Heat gained = heat lost

$$2.8(1)(27 - 5) + 0.25(27 - 5) = 0.11V + 0.11(1)(100 - 27)$$
$$61.6 + 5.5 = 0.11V + 8.03$$
$$V = 537 \text{ kcal/kg} \qquad \text{(about 0.5\% low)} \quad \square$$

When a liquid is evaporating, heat is taken from the remaining liquid and the surrounding bodies. The withdrawal of this heat may cause their temperature to decrease. For this reason the evaporation of water sprinkled on the sidewalk cools the air. Evaporation of perspiration from the skin lowers the temperature of the body.

17.7 BOILING POINT OF A SOLUTION

When water has some substance such as sugar dissolved in it, the temperature at which it boils is raised. The amount the boiling point is raised is proportional to the amount of the substance dissolved in the water. In making candies or sirups, the temperature is used as a means of determining the concentration of the solution. By observing the temperature at which the solution is boiling, it is possible to determine when the candy or sirup has boiled long enough. When vegetables, fruits, meats, etc., are boiled in water, some of their contents dissolve in the water. This raises the boiling point slightly, so that the water boils above 100°C. The effect of a dissolved substance on the vapor pressure and the boiling point is shown in Fig. 17.8. The solid curve gives the relation between the temperature and the vapor pressure of the pure solvent, and the dashed curve gives the same relation for the solution.

When a solid is dissolved in a liquid, the vapor that rises above the liquid does not contain the dissolved substance. To obtain pure water from water containing solids in solution, it is only necessary to evaporate the water and condense the vapor. The solids are left behind.

17.8 SUBLIMATION

If a substance such as solid camphor or solid iodine is left for some time in a closed vessel, vapor given off by the solid condenses on the sides of the vessel to form small crystals. The escape of molecules directly from a

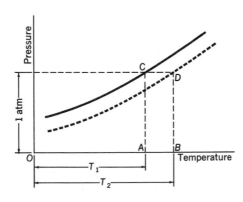

FIGURE 17.8
The saturated vapor pressure of a solution (dashed line) is lower than that of the solvent (solid line) at any temperature.

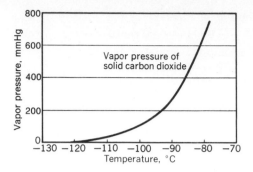

solid without passing through the liquid phase is called *sublimation*. In like manner, vapor may go directly to the solid state. For most solids, the vapor pressure is very small, but for some our sense of smell tells us that vapor is being given off. The vapor pressure of ice at 0°C is 4.6 mmHg; as the temperature decreases, the vapor pressure decreases.

The heat of sublimation is the heat necessary to change a unit mass of a substance from the solid to the vapor state without change of temperature. A vapor-pressure curve (Fig. 17.9) showing the vapor pressure when a solid is in equilibrium with its vapor can be plotted for a solid just as for a liquid; it is called a *sublimation curve*.

17.9 TRIPLE POINT

Under certain conditions of temperature and pressure a liquid may be in equilibrium with its vapor. On a temperature-pressure diagram these conditions are represented by the vapor-tension curve *OC* (Fig. 17.10). Similarly, a solid may be in equilibrium with its vapor along the sublimation curve *OA*. In like manner the melting curve *OB* can be drawn showing the relation between the temperature and pressure when the solid is in equilibrium with its liquid. When all these curves are drawn for a substance, they intersect at *O*, which is called the *triple point*. The triple point indicates the one temperature and pressure at which the solid, vapor, and liquid can exist together without any of them gaining in mass at the expense of the others. At this temperature and pressure, the solid, liquid, and vapor are in equilibrium.

Some substances, such as water, have several solid phases which would be represented by domains to the left of *AOB*. Water is unusual also in the sense that the melting curve *OB* leans slightly to the left of vertical, indicating that an increase in pressure lowers the melting point. The triple point for water can be reached by the following experiment. Some water is placed on a watch glass under a bell jar. Evacuation of the bell jar produces boiling, the heat for which is removed from the water, which is cooled. Eventually the temperature falls to the point where water is simultaneously boiling and freezing. With care one can reach the triple point for water, which is at a temperature of 0.0100°C and a pressure of 4.6 mmHg. If evacuation is continued and the pressure is reduced below 4.6 mmHg, only the solid and vapor phases can exist in equilibrium. A wafer of ice remains on the watch glass long after all the liquid has disappeared.

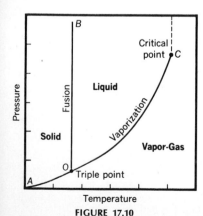

FIGURE 17.10
Equilibrium between the solid, liquid, and vapor phases of a material occurs at the *triple point O*.

17.10 THE CRITICAL POINT

If an *unsaturated* vapor is enclosed in a cylinder *HFEG* (lower part of Fig. 17.11) and piston *M* is moved in while the *temperature of the vapor remains constant*, the pressure at first increases as the volume is decreased, as represented by the curve *AB*. Near *A*, Boyle's law is almost followed, but deviations increase as *B* is approached. At *B* the vapor is *saturated*, and liquefaction begins. Along *BC* the volume decreases while the pressure remains constant (Boyle's law does not apply to saturated vapors). Meanwhile, the quantity of vapor decreases, and that of liquid increases. At *C* liquefaction is complete. A further application of pressure to the piston causes a small decrease in the volume of the liquid, and the curve *CD* representing the relation between the pressure and the volume of the liquid becomes very steep. The curve *ABCD* shows the relation between the volume and the pressure when the *temperature is kept constant;* it is called an *isothermal* curve.

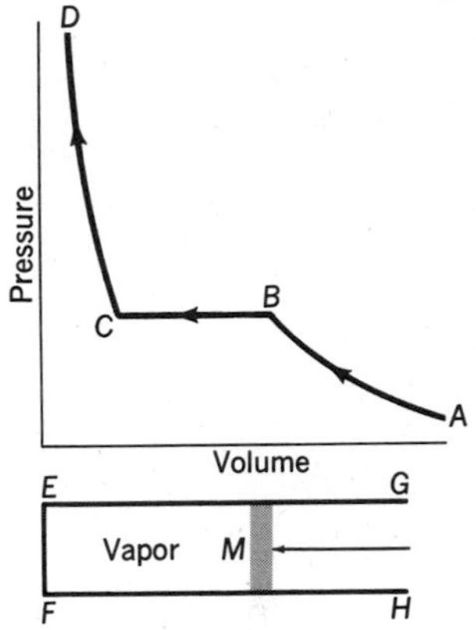

If this process is carried out at higher and higher temperatures, the horizontal part of the curve becomes shorter (Fig. 17.12) until it finally disappears and is replaced by a slight bend in the curve. The point *P* at which the horizontal part of the isothermal first disappears is known as the *critical point*, and the temperature corresponding to the isotherm through *P* is called the *critical temperature*. The pressure and specific volume (volume per unit mass) of the vapor at the point *P* are known as the *critical pressure* and *critical volume*, respectively. At temperatures below the critical temperature, it is possible to liquefy the vapor by the application of pressure *alone;* at temperatures above, it is impossible to liquefy the vapor, however great the pressure. Thus the critical temperature is the highest temperature at which a gas can be liquefied. On Fig. 17.10 the critical temperature and pressure are indicated by point *C*.

The isothermals of Fig. 17.12 refer to carbon dioxide. Its critical temperature is 31.1°C, and its critical pressure 77 atm. If a heavy-walled glass tube closed at both ends is partly filled with liquid carbon dioxide, the remainder of the tube being filled with the saturated vapor of carbon dioxide, at room temperature there is equilibrium between the liquid and its vapor. As the temperature of the tube is increased, the density of the liquid decreases while the density of the vapor increases. As a result, the meniscus marking the boundary between liquid and vapor becomes less

FIGURE 17.11
When the vapor in the cylinder (below) is compressed at constant temperature, the vapor begins to liquefy at *B* and the pressure remains constant at the saturated vapor pressure as liquefaction proceeds from *B* to *C*.

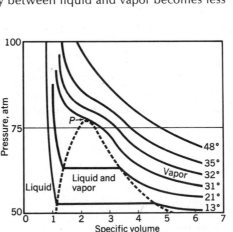

FIGURE 17.12
Isothermal pressure-volume curves for carbon dioxide at six temperatures. Below the critical temperature of 31.1°C the vapor can be liquefied by application of pressure alone.

TABLE 17.1
Critical Temperatures, Boiling Points, and Freezing Points of Common Gases

Gas	Freezing point, °C	Boiling point, °C	Critical temperature, °C	Critical pressure, atm
Helium	−272	−268.9	−267.9	2.26
Hydrogen	−259	−252.8	−239.9	12.8
Argon	−189	−186.0	−122.0	41.3
Nitrogen	−210	−195.8	−147.1	33.5
Oxygen	−219	−183.0	−118.8	49.7
Water (steam)	0	100.0	374	218

distinct. At the critical temperature the densities of liquid and vapor become equal and the meniscus disappears.

To liquefy a gas, it is necessary to cool it below its critical temperature and to apply sufficient pressure to produce liquefaction (Table 17.1). Most of the so-called "permanent" gases have low critical temperatures and require the application of large pressures. The boiling and freezing points quoted are for a pressure of one standard atmosphere.

QUESTIONS

1. Can heat be added to a substance without changing its temperature? Under what circumstances?
2. Why does 100°C steam produce a much worse burn than 100°C water?
3. Why do crystalline solids have sharp melting points while amorphous solids usually soften and melt over a broad temperature range?
4. Many foods are cooked in boiling water; why do they cook faster in a pressure cooker? Would increasing the rate at which heat is supplied speed up the cooking in an open pan?
5. How can pails of water in a small storage room help prevent potted plants or fruit from freezing during a cold night?
6. How is regelation involved in packing a snowball? Why is very cold snow powdery?
7. If 100 cm³ of saturated pure ether vapor at 25°C exerts a pressure of 530 mmHg, what pressure is exerted if the volume is reduced to 50 cm³ without changing the temperature? Explain.
8. What is the distinction between a vapor and a gas? How does the pressure of a gas vary as a function of volume when the temperature is kept constant? Of a saturated vapor? Of an unsaturated vapor?
9. How does the pressure of air above a water surface affect the saturated vapor pressure of the liquid and the rate of evaporation?
10. What role does surface tension play in the violent boiling, called *bumping*, which often occurs when water is heated to boiling in a smooth-walled container? Why is bumping reduced by placing rough, inert crystals in the container?
11. Why does the heat of vaporization increase as the temperature is lowered?
12. What determines the time between eruptions of a geyser? Explain what conditions must be satisfied for an eruption to occur.

PROBLEMS

1. A 150-g piece of iron is heated in an oil bath and then placed on a block of ice, causing 25 g of ice to be melted. What was the temperature of the iron? *Ans.* 121°C
2. A specimen of copper is heated to 150°C and placed on a block of ice. If 36 g of ice is melted, what is the mass of the copper specimen?
3. If the specific heat of snow is 0.5 kcal/(kg)(C°), how much heat energy is required to turn 4 kg of snow at −20°C into steam at 100°C? *Ans.* 2920 kcal
4. In an experiment designed to measure the heat of fusion of ice, 25 g of ice at −10°C is dropped into 200 g of water at 30°C contained in a copper calorimeter of mass 100 g. The final temperature is 17.5°C. What value of the heat of fusion do these data yield? Take the specific heat of ice to be 0.5 kcal/kg.

5. How many kilocalories are required to heat 300 g of mercury from 20 to 357°C and to vaporize it at that temperature? *Ans.* 23.7 kcal

6. A steel sphere with a mass of 90.23 g is placed in an atmosphere of steam at 100°C; after it has reached equilibrium, it is weighed with the water condensed on it. The observed mass is 91.17 g. What was the original temperature of the sphere?

7. Four ice cubes at 0°C, each of 35 g mass, are dropped in a 50-g glass containing 390 g of water at 20°C. If losses are negligible, how many grams of ice are left when the water and glass have cooled to 0°C? *Ans.* 40 g

8. How much heat is required to change 10 lb of ice at 12°F to steam at 320°F? The specific heats of both ice and steam are approximately 0.5 Btu/(lb)(F°).

9. What power is required to melt 3 kg of ice at 0°C and to raise the temperature of the water to 100°C in 450 s? *Ans.* 5.02 kW

10. Ice has a specific gravity of 0.917. How much heat is required to melt 0.2 m³ of ice at 0°C?

11. Three ice cubes, each of mass 30 g and at 0°C, are placed in a 50-g glass which contains 240 g of water. If the glass and the water are initially at 24°C, find the final temperature of the system and the mass of ice remaining, if any. *Ans.* 0°C; 15 g

12. If 70 g of ice at −20°C is placed in a calorimeter of 25 g water equivalent which initially contains 275 g of water at 40°C, and if the specific heat of ice is 0.5 kcal/(kg)(C°), find the heat required to warm the ice to 0°C, the heat required to melt the ice at 0°C without changing its temperature, and the final temperature of the calorimeter and its contents.

13. The condenser at an electric generating plant receives steam at 100°C and changes it to water at 60°C. The cooling is done by water, which enters at 10°C and leaves at 40°C. How much cooling water is required for each kilogram of steam condensed? *Ans.* 19.3 kg

14. When 55 g of ice at 0°C is dropped into 190.7 g of water at 30°C contained in a 100-g copper calorimeter, find the final temperature.

15. Steam at a temperature of 100°C and with a mass of 20 g is passed into an aluminum calorimeter containing 367 g of water. The initial temperature of the water is 6°C, and the mass of the calorimeter is 300 g. Find the final temperature in the calorimeter. *Ans.* 34°C

16. Steam at 100°C enters a radiator, and water at 70°C leaves. (*a*) How many grams of steam are required per hour if the radiator delivers 3 × 10³ kcal/h? (*b*) What fraction of the heat is associated with the cooling of the water?

17. When 20 g of steam at 100°C is condensed in a 200-g calorimeter containing 360 g of water initially at 10°C, the final temperature is 40°C. Find the heat lost by the steam, the heat gained by the calorimeter, and the specific heat of the calorimeter. *Ans.* 12 kcal; 1.2 kcal; 0.2 kcal/(kg)(C°)

18. A calorimeter contains 350 g of water and 60 g of ice at 0°C. The mass of the calorimeter is 150 g, and its specific heat is 0.12 kcal/(kg)(C°). How many grams of steam at 100°C must be introduced into the water to make the final temperature 40°C?

19. A calorimeter of 30 g water equivalent contains 270 g of water at 10°C. Into this calorimeter 200 g of ice at −20°C is dropped; then 15 g of steam at 120°C is introduced. The specific heats of steam and ice are, respectively, 0.48 and 0.50 kcal/(kg)(C°). Find the final temperature of the system. What remains in the calorimeter? *Ans.* 0°C; 419.3 g water and 65.7 g of ice

20. How many pounds of coal must be burned to produce 600 lb of steam at 300°F if one starts with water at 50°F and if 80 percent of the heat of combustion of the coal goes into the water? Take the heat of combustion of coal to be 11,000 Btu/lb and the specific heat of steam to be 0.48 Btu/(lb)(F°).

21. If 500 g of ice at 0°C and 60 g of steam at 100°C are mixed with 580 g of water at 30°C in a calorimeter with a water equivalent of 20 g, find the final temperature. *Ans.* 14.1°C

22. When 50 g of steam at 100°C was passed into a 400-g copper calorimeter, the calorimeter contained 600 g of water and a certain amount of ice, all at 0°C. The final temperature was 15°C. How much ice was in the calorimeter?

23. A 200-g copper calorimeter contains 500 g of water at 10°C. How many grams of steam at 100°C must be added to raise the temperature to 65°C? *Ans.* 49.6 g

24. How much steam at 120°C must enter a calorimeter to turn 200 g of ice at −10°C to water at 25°C? The specific heats of ice and steam are both approximately 0.5 kcal/(kg)(C°). Assume that negligible heat is needed to raise the temperature of the calorimeter.

25. It takes 540 kcal, or 2.26 × 10⁶ J, to change 1 kg of water at 100°C to steam at the same temperature at one standard atmosphere pressure. At 100°C, 1 kg of water occupies 0.001 m³, and 1 kg of steam 1.671 m³. Find the external work done in

evaporating 1 kg of water at 100°C and 1 atm pressure. Find the change in the internal energy of the system during the evaporation.

Ans. 1.69×10^5 J; 2.09×10^6 J

26. If the specific heat of water vapor at constant pressure is 0.48 kcal/(kg)(C°) and the heat of vaporization at 100°C is 540 kcal/kg, calculate the approximate heat of vaporization of water at 90°C. (*Hint:* By conservation of energy, it should take the same amount of energy to change 1 g of water at 90°C to 1 g of water vapor at 100°C whether the evaporation is done first or last.)

Compare your value with that given in Appendix Table B.4. Why is your value only approximate?

27. One possible way to store solar energy to heat a house is to liquefy Glauber's salt (sodium sulfate decahydrate, $Na_2SO_4 \cdot 10HOH$) when the sun is shining and to use the heat released as the fluid solidifies. Glauber's salt has heat capacities of 0.46 and 0.68 kcal/(kg)(C°) in the solid and liquid phases, respectively, a heat of fusion of 58 kcal/kg, a density of 2,680 kg/m³, and a melting point of 32.4°C. If a house requires 2.5×10^5 kcal/day on a cold day, find the volume of Glauber's salt required to store a 2-day requirement when the salt is heated to 50°C and is allowed to cool to 25°C. What volume of water operating between 50 and 25°C would be required to store the same 2-day energy requirement? *Ans.* 2.54 m³; 20 m³

In Chap. 6 we learned that energy is the ability or capacity to do work. A body at a high temperature has internal energy. How can part of this internal energy be converted to useful work? What fraction of the energy is available? The answers to the first of these questions—and there are several—tell us how heat engines operate. The maximum possible efficiency of any particular heat engine is limited by the answer to the second question. In this chapter we consider the operation and efficiencies of various types of heat engines.

18

THERMODYNAMICS

18.1 HEAT AND WORK

The heat engines that play such a major role in modern life depend on the transformation of heat energy into work. The hot gases of a jet engine push forward on a moving airplane and thereby do work. An automobile engine drives an automobile only when it is supplied constantly with heat from burning gasoline in the cylinders.

Practically all heat engines depend upon the work done by expanding gases to achieve mechanical work output. The simplest situation to discuss, although a difficult one to use for a heat engine, is that in which a gas expands at *constant pressure* (Sec. 15.8). In this case the work done by the gas is the product of the pressure p and the change in volume $V_2 - V_1$, where V_1 and V_2 represent the volume occupied by the gas before and after the expansion, respectively. If the pressure varies during the expansion, the process can be described by a pV diagram such as Fig. 18.1. The work W done during a small volume increment ΔV is given by Eq. (15.5):

$$W = p \, \Delta V \qquad (18.1)$$

where p is the average pressure appropriate for the small change ΔV. This work is represented by the hatched area in the figure. Clearly, any volume change $V_2 - V_1$ can be regarded as the sum of many small ΔV increments, and the total work done by the gas in producing the volume change $V_2 - V_1$ is represented by the area $ABba$.

The work performed by the expanding gas comes from one or both of two sources: heat supplied to the gas and the internal energy of the gas. In the *thermodynamics* of heat engines we are interested in energy transformations involving the heat supplied to the "working substance" of the engine (usually a gas), the work done by this substance, and changes in its internal energy.

18.2 FIRST LAW OF THERMODYNAMICS

When the principle of conservation of energy is used to include heat energy specifically, we may state it in the following form: *When heat energy is added to (or removed from) a system, an equal amount of some other form of energy appears (or disappears).* In the thermodynamics of heat engines we are concerned primarily with transformations between heat, internal energy, and work done by the system. Thus, the *first law of thermodynamics is the law of conservation of energy* with heat energy specifically included.

In most heat engines the working substance is a gas. In this case the first law of thermodynamics requires that

Heat added to gas = increase in internal energy of gas

+ mechanical work done by gas

It is understood that any of the quantities of this relation may be negative. Thus, if heat energy is removed from the gas, the heat "added" is negative. By Eq. (15.2) the increase in internal energy of the gas is the product of the mass m, the specific heat at constant volume c_V, and the change in temperature ΔT of the gas; ΔT is positive for an increase in the internal energy and negative for a decrease. By Eq. (18.1) the work done *by* the gas is $p\,\Delta V$ when ΔV is positive; when ΔV is negative, $|p\,\Delta V|$ gives the work done *on* the gas.

If we let ΔQ represent a small amout of heat *added* to the gas ($-\Delta Q$ represents heat removed), we can write the first law of thermodynamics for any process

$$\Delta Q = mc_V\,\Delta T + p\,\Delta V \tag{18.2}$$

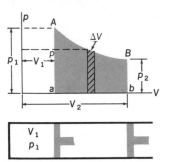

FIGURE 18.1

The external work done by a gas in expanding a small amount ΔV is $p\,\Delta V$, where p is the pressure; when the gas in the cylinder below expands at variable pressure, the work done by the gas is the shaded area of the pV diagram.

18.3 HEAT ENGINES

Before we continue with theoretical considerations, it will be instructive to discuss how some familiar heat engines use heat energy to perform mechanical work by repeating a series of processes over and over again. Each complete series of processes which restores the working substance to its initial condition is called a *cycle*.

Gasoline Engines Perhaps the most common heat engine is the four-stroke-cycle gasoline engine found in automobiles all over the world and in many other uses. It typically has four, six, or eight cylinders, in each of which a piston moves up and down (or back and forth), describing roughly simple harmonic motion. Gasoline or some other fuel is burned in the cylinder, and the resulting hot gases do work on the piston. The four strokes associated with each cylinder are illustrated in Fig. 18.2. The *intake stroke* begins when the piston, at its greatest penetration in the cylinder, is drawn back by the rotating crankshaft at the right. The intake valve A is opened, and a mixture of gasoline vapor and air is drawn into the cylinder. Near the end of this stroke valve A is closed, and the volume of the mixture is reduced to perhaps one-eighth to one-tenth of its maximum value. When this *compression stroke* is near its end, a spark ignites the mixture. The oxidation of the fuel releases heat energy, resulting in a great increase in the temperature and pressure of the gas in the chamber. As a consequence, there is a large force driving the piston outward during the *power stroke*. When the piston reaches the end of its range, valve B is open and the *exhaust* stroke begins. The combustion products are forced out, and at the end of the exhaust stroke valve B is closed. The system has now been restored to the condition which existed when the intake stroke began. One cycle has been completed, and the engine is ready to repeat.

For each cycle only one of the four strokes is a working stroke, during which heat is transformed into work. The working strokes of other cylinders or a flywheel provide the energy to carry the piston through its other three strokes. On the compression stroke of a gasoline engine, the temperature of the air and gasoline vapor increases rapidly because external work is done in compressing the gas. If the gas is compressed so

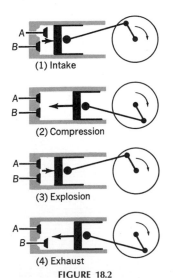

FIGURE 18.2

The four strokes in the operation of a typical gasoline engine.

that its final volume is less than about one-tenth its original volume, the gasoline may ignite spontaneously because of the high temperature. When such ignition occurs before the cylinder is near the top of its stroke, *knocking* results, with serious loss of efficiency. The ratio of the initial volume of gas to the final volume in the cylinder is called the *compression ratio*. The greater the compression ratio the greater the efficiency of the engine, but the greater the difficulty in bringing the explosive mixture safely to the point at which the spark is desired. Modern developments in high-octane fuel have permitted substantial increases in the compression ratios of automobiles, but a serious practical limitation remains.

Diesel Engines The cycle for the diesel engine closely resembles that of the gasoline engine in that both have four strokes per cycle and both are *internal-combustion engines*, i.e., the fuel is burned in the same cylinder in which the external work is performed. In the diesel engine only air is introduced during the intake stroke. There is no preignition problem when air without fuel vapor is compressed, so that compression ratios of 16 to 1 or more can be obtained; the temperature of the air rises to 500 or 600°C during the compression when work is done on the gas with no heat added or removed [see Eq. (18.2)]. When the desired compression ratio is reached, fuel is injected into the cylinder, where it ignites spontaneously because of the high temperature. Thus, no spark is required. The diesel engine can use ordinary fuel oil or even powdered coal. Because the diesel engine is more efficient than an ordinary gasoline engine, and because it can burn lower-grade fuel, large internal-combustion engines are usually diesels.

Steam Engines In a steam engine (Fig. 18.3) a closely fitting piston moves in a cylinder that is connected to the steam chest by two pipes, A and B. These pipes, which are provided with valves, serve alternately as inlet and exhaust for the steam. Steam entering through A produces a high pressure at the piston, which moves to the right, forcing the previously used steam out through B; this steam is exhausted to the atmosphere. When the piston has moved through about one-fourth of its stroke, the slide valve closes off A. The steam continues to expand and pushes the cylinder to the right during the remainder of the stroke. During this expansion the steam is doing work on the piston; meantime its pressure is decreasing and its temperature falling. Part of the internal energy of the steam is thus converted into useful work.

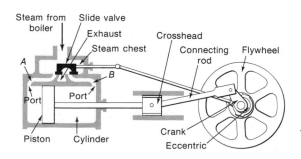

FIGURE 18.3
The steam engine was one of the first heat engines to be developed.

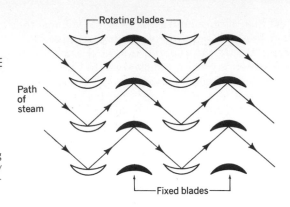

FIGURE 18.4
In the steam turbine steam recoiling from the moving blades is redirected by fixed blades against the next set of rotating blades.

When the piston has reached the end of its stroke, the slide valve opens *A* to the exhaust and connects *B* to the steam chest, and the return stroke begins. Thus there are two essentially identical strokes to each cycle.

Steam Turbines Modern electric generators are often driven by steam turbines, in which steam impinges upon a set of rotating blades (Fig. 18.4). The steam is deflected from the rotating blades to a set of fixed blades, where its direction is reversed once more so that it impinges upon a second set of rotating blades fastened to the same shaft as the first set. In a modern steam turbine a number of sets of rotating and fixed blades may be used (Fig. 18.5). As the steam transfers energy to the blades, its temperature drops, the pressure falls, and the volume of the steam increases. To allow for this, the wheels upon which the blades are assembled increase in diameter from the high-pressure end to the exhaust end. The steam may be superheated to above 560°C and supplied at a pressure of 240 atm. Under such circumstances the efficiency may be about 40 percent.

Jet Engines If a firecracker is set off in a tube closed at one end, the tube is driven forward by the increased pressure on the closed end. At the open end the hot gases escape. The momentum transferred to the tube is equal to the momentum of the escaping gases in the opposite direction. In jet propulsion it is the reaction from the escaping gases themselves,

FIGURE 18.5
A modern steam turbine. (*General Electric Company.*)

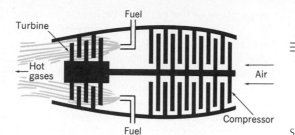

Fuel
Turbine
Hot gases
Air
Compressor
Fuel

FIGURE 18.6
Schematic diagram of a turbojet engine.

and not the interaction of the escaping gases with the air, which provides the thrust. This is why jet motors are sometimes called *reaction motors*. For many years jet motors have been used in fireworks for rockets, pinwheels, and other displays. Only in recent years have they been adapted for propelling aircraft and large missiles.

The principles underlying rocket propulsion have been discussed in Sec. 7.9. Rocket engines can operate in a vacuum, since they carry with them all the materials needed for combustion. Other jet engines require oxygen from the air for burning fuel. In the *turbojet engine* (Fig. 18.6) fuel is oxidized in a combustion chamber. The resulting hot gases escape through a turbine which drives a compressor to provide the air necessary to burn fuel. Many modern aircraft use turbojet engines. They are particularly suited to aircraft flying at high speeds and high altitudes, where air resistance is low. It is precisely in this region that propellers are not able to exert substantial thrust against the rare air. At low speeds and low altitudes, propeller aircraft are more efficient and have a number of superior features. A combination of the desirable (and undesirable) features of the propeller and jet engine can be achieved by operating a propeller from the shaft of the turbojet engine. Such a combination is called a *turboprop engine*.

18.4 THERMAL EFFICIENCY OF A HEAT ENGINE

In practical heat engines, which achieve mechanical work output through the expansion of hot gases, the variables pressure, volume, and temperature change simultaneously and in a complex manner. To gain quantitative understanding of the transformation of heat into mechanical energy, it is simplest to imagine an engine that uses an ideal gas as working substance and holds one of these variables constant during each portion of its heat cycle.

As a particular example of such a cycle, we consider first an idealized heat engine in which four steps are involved in each cycle (Fig. 18.7). Here we have a mass m of gas at pressure p_1, volume V_1, and temperature T_1 in a cylinder which is closed by a movable piston; the initial state is represented by point A of the figure. With the *pressure kept constant*, we add heat Q_1 sufficient to make the volume increase from V_1 to V_2. In this *isobaric* (constant-pressure) process the first law of thermodynamics [Eq. (18.2)] requires that the heat added be equal to the sum of the work performed by the gas and the increase in the internal energy of the gas. The work done by the gas is $p_1(V_2 - V_1)$ by Eq. (15.5); it is represented by the area $ABba$ in the figure. By Eq. (14.9), $p_1V_1/T_1 = p_1V_2/T_2$; therefore the

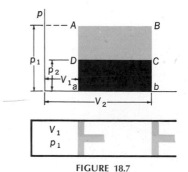

FIGURE 18.7
The work done on the piston *by the gas* expanding from A to B is taken as positive and is proportional to the area $ABba$; the work done *on the gas* in compressing it at pressure p_2 is negative and proportional to the area $DCba$.

temperature of the gas must be increased as we carry out the expansion. The increase in the internal energy of the gas is $mc_V \Delta T$, where in this case $\Delta T = T_2 - T_1 = (V_2 - V_1)T_1/V_1$.

Next, holding the volume of the gas constant, we remove some heat Q_2 from the gas, thereby lowering the temperature to T_3. A process in which the volume is held constant is called *isochoric* or *isovolumic*; since ΔV is zero, no work is done and the heat removed comes entirely from the internal energy of the gas. This process is represented by BC in Fig. 18.7.

The third process in our cycle, indicated by CD in the figure, is a compression of the gas at constant pressure p_2. During this isobaric step more heat Q_3 must be removed from the gas to reduce the temperature to T_4 as the volume decreases and to make up for the work done *on* the gas $p_2(V_2 - V_1)$, which is represented by the area $DCab$. Finally, keeping the volume constant at V_1, we increase the pressure from p_2 to p_1 by adding heat Q_4 and thus increasing the temperature of the gas from T_4 back to T_1.

We have now completed one cycle, and the internal energy of the gas is exactly what it was at the beginning; so also are the pressure, volume, and temperature. During the cycle we have supplied heat $Q_1 + Q_4$ and removed the smaller quantity of heat $Q_2 + Q_3$. During this same cycle the gas has done work proportional to $ABba$ and has had work $DCab$ done upon it. In each cycle the *net work output ABCD* is the difference between the *work done on the piston by the gas* and the *work done on the gas by the piston*. In this particular cycle the net work output of our heat engine is less than the total heat energy supplied because some heat energy is withdrawn (or rejected). This is true for all heat engines.

In general, a heat engine is a thermodynamic system which (1) absorbs heat energy from some higher temperature source, (2) converts some of this energy into mechanical work, and (3) rejects the remainder of the energy to some lower-temperature sink. For example, in a gasoline engine the high-temperature source is produced by burning fuel in the cylinder, mechanical work is done against the piston, and the rejected heat energy is carried off by the exhaust gases. In any cyclic heat engine the system returns to its original thermodynamic state at the end of each cycle.

For purposes of discussion it is convenient to think of a cyclic engine operating between two heat reservoirs: a source at a higher temperature T_h and a sink at the lower temperature T_c. In each cycle heat Q_h is absorbed from the source, mechanical work W is performed, and heat Q_c is rejected to the sink (Fig. 18.8). By the law of conservation of energy, the heat absorbed Q_h is equal to the sum of the mechanical work W and the heat rejected Q_c. While some heat engines convert a larger fraction of the energy supplied Q_h to work than others, we can never make a heat engine which converts all the heat energy supplied into useful work.

An important characteristic of any heat engine is its efficiency: *The thermal efficiency of a heat engine is the ratio of the net external work done by the engine to the total heat energy supplied.* In any heat engine only a fraction of the heat developed by the combustion of fuel is converted into work. In a steam engine much of the heat passes out with the exhaust steam. Gasoline engines are more efficient, but even so, provision must be made for removing the heat not converted into mechanical work.

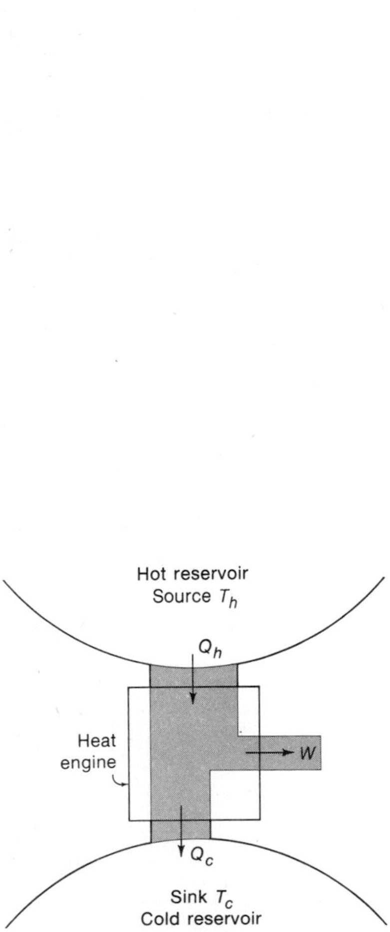

FIGURE 18.8
In an ideal heat engine the heat energy supplied by the hot source at temperature T_h is converted partly into mechanical work, the remainder being delivered to the colder sink at temperature T_c.

□ **Example** If 0.3 kg of coal, which has a heat of combustion of 7600 kcal/kg, is burned in an engine which raises 2,000 kg of water 40 m, what fraction of the heat from the coal has been converted into useful work?

$$\text{Heat supplied} = Q_h = (0.3 \text{ kg})(7600 \text{ kcal/kg})(4{,}186 \text{ J/kcal})$$
$$= 9.55 \times 10^6 \text{ J}$$

$$\text{Work done} = W = \text{weight} \times \text{height} = mgh$$
$$= (2{,}000 \text{ kg})(9.8 \text{ N/kg})(40 \text{ m}) = 7.84 \times 10^5 \text{ J}$$

$$\text{Efficiency} = \frac{\text{mechanical work performed}}{\text{heat supplied}} = \frac{W}{Q_h}$$
$$= \frac{7.85 \times 10^5}{9.55 \times 10^6} = 0.082 = 8.2\% \qquad \qquad □$$

18.5 ISOTHERMAL AND ADIABATIC PROCESSES

Consider a gas in a cylinder with walls which are good conductors of heat and with a movable piston at one end. Let the gas expand slowly, maintaining a constant temperature. To do this heat must be supplied; otherwise the temperature falls as the gas expands. By the first law of thermodynamics, $\Delta Q = mc_V \Delta T + p \Delta V$. Here we are keeping $\Delta T = 0$, so we must supply heat ΔQ which is equal to the work $p \Delta V$ done by the gas.

A process which is carried out in such a way that the temperature of the working substance remains constant is called an isothermal process. When a gas is compressed isothermally, work is done on the gas, and heat must be removed during the process. The relation between the pressure p and volume V of a gas during isothermal expansion or compression is shown in Fig. 18.9 by the line ST. The equation of this curve is given by Boyle's law: $pV = $ constant.

If a gas is compressed in such a way that no heat is allowed to enter it or to escape it, the temperature increases. On the other hand, the temperature falls during expansion. An expansion or compression during which no heat enters or escapes may be realized experimentally by enclosing the gas in a cylinder surrounded by nonconducting materials. *A process in which there is no exchange of heat between the substance and its surrounding is called an adiabatic process.*

In terms of Eq. (18.2), an adiabatic process is one for which $\Delta Q = 0$; no heat is added or removed. In this case the external work done by the gas is equal to the decrease in its internal energy: $p \Delta V = -mc_V \Delta T$. The relation between the volume and the pressure of a gas during an adiabatic expansion is shown by the curve LM in Fig. 18.9. The curve for the adiabatic process is steeper than the corresponding curve for the isothermal process. Consider two identical gas samples, one of which undergoes an adiabatic expansion and the other an isothermal expansion. If the pressure, temperature, and volume of the two samples are the same at the beginning, and if they expand to the same final volume, the temperature and pressure of the sample expanded adiabatically are less than the temperature and pressure of the gas expanded isothermally. *When a gas expands adiabatically, the product pV^γ is constant*, where γ is the ratio of the specific heat of the gas at constant pressure to the specific heat at constant volume (Sec. 15.8).

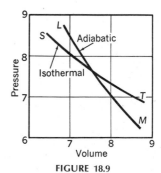

FIGURE 18.9
The slope of the adiabatic curve LM through any point on the pV diagram is negative and greater in magnitude than the slope of the isothermal curve ST through the same point.

18.6 THE CARNOT CYCLE

To analyze the fundamental principle involved in heat engines Carnot in 1824 discussed an idealized heat engine in which some *working substance* undergoes a sequence of four processes during which the substance takes heat from a hot *source*, performs some external work, and delivers some heat to some lower-temperature *sink*. At the end of the cycle the working substance and all parts of the engine are in the same state as they were in at the beginning. A great advantage of a Carnot cycle is that two of the four processes are isothermal and two are adiabatic. During the adiabatic processes ΔQ in the first law of thermodynamics [Eq. (18.2)] is zero, so the only energy transformations involved are between internal energy and work. Similarly, during the isothermal processes the internal energy of the working substance remains constant, so $\Delta Q = p\,\Delta V$.

We shall describe a Carnot cycle using an ideal gas as working substance. Consider a cylinder filled with gas and closed by a movable piston. Suppose that initially the gas in the cylinder has a volume and pressure represented by point A in Fig. 18.10 and that it is at the temperature T_h of an energy source. We now carry out the sequence of four processes which constitute a Carnot cycle and bring the gas back to point A so it is ready for another cycle.

1. The gas expands *isothermally* until point B is reached. To keep the temperature at T_h the gas must be in good thermal contact with the source. As the volume increases, external work is done by the gas; this work is represented by the area under the curve AB. Since $\Delta T = 0$ for this process, Eq. (18.2) requires that the heat Q_h supplied by the source be equal to the external work performed.

2. The cylinder is then insulated so that no heat is added or removed as the gas expands *adiabatically* until point C is reached. During this expansion external work represented by the area under BC is done by the gas; the energy is supplied by a reduction in the internal energy of the gas, so its temperature falls from T_h to the lower value T_c.

3. Excellent thermal contact is now made with a heat sink, which has temperature T_c, and the gas is compressed *isothermally* at T_c until it reaches point D. During this process work (represented by the area

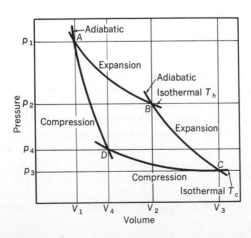

FIGURE 18.10
In a Carnot cycle there is an isothermal expansion at T_h of the working gas followed by an adiabatic expansion in which the temperature falls to T_c. The gas is then compressed isothermally at T_c and finally compressed adiabatically to complete the cycle.

under *CD*) is done *on* the gas, and heat Q_c must flow from the gas to the sink to keep the temperature constant. The point *D* was so chosen that the cycle can be closed by an adiabatic compression along the curve *DA*.

4. The gas is insulated once more and is then compressed adiabatically until point *A* is reached. During this adiabatic process work (represented by the area under *DA*) is done *on* the gas, so that its internal energy is increased. When point *A* is reached, the temperature is again T_h. The cycle is now complete.

The net work output of the cycle is represented by the area enclosed in *ABCD*. It is the difference between the work done *by the gas* during the isothermal expansion *AB* and the adiabatic expansion *BC* and the work done *on the gas* during the isothermal compression *CD* and the adiabatic compression *DA*.

During the isothermal expansion along *AB* at temperature T_h, Q_h units of heat flow into the gas, while during the isothermal compression along *CD*, Q_c units of heat are rejected by the gas at temperature T_c. During the adiabatic compression along *DA* and the adiabatic expansion along *BC*, no heat flows out of the gas or into it. In the full cycle $Q_h - Q_c$ units of heat are transformed into mechanical energy. If *W* denotes the work done by the engine during this cycle,

$$W = Q_h - Q_c \tag{18.3}$$

An important characteristic of an ideal Carnot engine is that it is *reversible;* i.e., we could drive it backward so that it removes energy from some reservoir at a lower temperature T_c and rejects energy to a second reservoir at a higher temperature T_h. To do this work *W* must be done *on* the engine (Fig. 18.11), which is now acting as a refrigerator. In this case the energy the engine receives each cycle $Q_c + W$ is equal to the energy delivered Q_h. Note that we are discussing ideal engines here in which there are *no losses* of any kind due to dissipative forces such as friction.

18.7 THE EFFICIENCY OF A CARNOT ENGINE

The efficiency of a Carnot engine depends only on the temperatures between which it works. By definition, for a Carnot engine,

$$\text{Eff} = \frac{W}{Q_h} = \frac{Q_h - Q_c}{Q_h} = 1 - \frac{Q_c}{Q_h} \tag{18.4}$$

where Q_h and Q_c are the heats transferred in a Carnot cycle at absolute temperatures T_h and T_c, respectively.

In establishing his absolute thermodynamic temperature scale, Lord Kelvin showed that

$$\frac{Q_h}{T_h} = \frac{Q_c}{T_c} \tag{18.5}$$

If we replace Q_c/Q_h in Eq. (18.4) by T_c/T_h from Eq. (18.5), the efficiency of a Carnot engine becomes

$$\text{Eff} = \frac{T_h - T_c}{T_h} \tag{18.6}$$

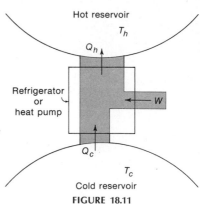

FIGURE 18.11
In a heat pump or a refrigerator external work *W* is used to remove heat Q_c from the cooler reservoir at temperature T_c and to deliver heat Q_h to the warmer reservoir at temperature T_h.

where T_h is the temperature of the hot source providing heat Q_h and T_c the temperature of the cold sink to which the heat Q_c is delivered.

No real engine can have an efficiency greater than that of a Carnot engine working between the same two temperatures. If it could, we could use it to drive a Carnot engine in reverse and transfer net energy from the cold to the hot body with no other change. This would violate the second law of thermodynamics (Sec. 18.8).

Similar reasoning shows us that *the efficiency of any ideal reversible heat engine operating between the same two temperatures T_h and T_c is the same as that of a Carnot engine.* To prove the truth of this theorem of Carnot's, suppose that some other reversible heat engine had a higher efficiency than the Carnot engine. Then we could use it to drive a Carnot engine backward and again transfer net energy from the cold reservoir to the hot one with no other change. By using a Carnot engine to drive the other one backward we can show that the other engine cannot have a lower efficiency; thus *any two ideal reversible engines operating between the same temperatures must have the same efficiency.*

☐ **Example**   Find the efficiency of an ideal engine working between temperatures of 127 and 77°C.

$$\text{Eff} = \frac{T_h - T_c}{T_h} = \frac{400 - 350}{400} = 12.5\% \qquad \square$$

To obtain high efficiencies, we want heat engines to have T_h as high and T_c as low as is practical. Modern electric generating plants using steam turbines often employ lakes or rivers to condense the steam, thereby obtaining the lowest practical T_c. When a very large amount of energy Q_c must be rejected at temperature T_c, its delivery to a lake or river may result in *thermal pollution*, an ecologically unacceptable increase in the temperature of the water. In such cases, some other type of sink, such as a cooling tower, is a possible alternative.

18.8   SECOND LAW OF THERMODYNAMICS

In the Carnot engine, one of the most efficient conceivable, a quantity of heat Q_h is taken from the hot reservoir at a temperature T_h. Some of this heat is rejected to a cold reservoir at temperature T_c, and the rest of the energy is converted into work. Only if we have a heat source which is at a temperature T_h greater than that of its surroundings can we get useful work from thermal energy. This idea was formalized by Kelvin in a statement of what he called the second law of thermodynamics: *There is no natural process the only result of which is to cool a heat reservoir and do external work.* This law follows directly from the fact that *heat by itself flows only from bodies at higher temperature to bodies at lower temperature.* Indeed, we may state the second law in the form: *Heat can be made to go from a body at lower temperature to one at higher temperature only if external work is done.* In refrigerators we compel heat to go from a colder to a warmer body, but work must be done to produce this unnatural heat flow in accord with the second law of thermodynamics.

Although there is a vast store of thermal energy in the ocean, we cannot simply take this thermal energy and convert it into work to operate a ship. Only if we have some body at a lower temperature than the ocean to which we can transfer part of the energy can we make an engine operate from the ocean's heat energy. Even then we can convert into work only the *difference* between the heat energy provided at the source and the heat energy rejected to the colder reservoir.

On the other hand, we can always convert mechanical energy into heat energy completely. Thus, the second law of thermodynamics indicates a certain *irreversibility* of natural processes. For example, we can always convert the kinetic energy of an automobile into heat in the brake drums and tires, but we cannot reconvert all this heat energy to kinetic energy. In Sec. 15.2 we discussed the fact that when a fast-moving bullet is brought to a stop in a piece of wood, the organized mechanical kinetic energy $\frac{1}{2}mv^2$ is converted into *random* mechanical internal energy. Here again, in a natural process net energy is transformed from the organized, available form into the random internal form; only by performing external work could we restore the system to its original state.

If we have 50 g of water at 0°C and 50 g at 100°C, we can always mix them to obtain 100 g at 50°C. But, if we have 100 g of water at 50°C, we cannot pour half of it into one vessel and have it at 0°C and the other half into a second vessel and have it at 100°C. This would be in violation of the second law of thermodynamics, which tells us that many natural processes occur only in one direction. The world is filled with irreversible processes; people age, eggs rot, and zinc dissolves in acid.

If the second law of thermodynamics is applicable throughout the universe, and if heat always flows from bodies at higher temperatures to those at lower temperatures, eventually the entire universe will come to the same temperature. Then, although the total energy in the universe remains the same, none of the energy will be available for doing mechanical work. Thus, the second law of thermodynamics tells us that although the total energy is constant, the available energy becomes always less. This implication of the second law of thermodynamics has aroused wide philosophical interest and leads to what philosophers of the nineteenth century called the *Wärmetod* or *heat death*.

18.9 THE REFRIGERATOR AND THE HEAT PUMP

The process of reducing the temperature of a body below that of its surroundings is essentially the reverse of the process employed in a heat engine. One may imagine a Carnot cycle operated in reverse so that the gas is expanded isothermally at a lower temperature T_c and then compressed isothermally at a higher temperature T_h.

In a typical electric refrigerator (Fig. 18.12) liquid Freon is pumped into the cooling coils, where the pressure is reduced. As a consequence, the liquid vaporizes and the gas expands. Both processes remove heat from the surroundings, which are cooled. The gas is pumped out of the cooling chamber, and in an external set of coils it is compressed and liquefied. In both of these processes large amounts of heat are given up. This heat is removed by water cooling in large systems or by air cooling in small systems. The liquid is then ready to repeat the cycle. The work required to transfer heat from lower to higher temperature is supplied by the "heat pump." This pump is usually operated by an electric motor, although a gas flame can also be used.

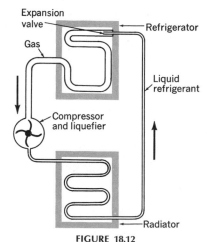

FIGURE 18.12
Schematic diagram of a refrigerator with refrigerant circulated and compressed by an electrically operated pump.

Some modern homes are heated by means of *heat pumps* which remove heat from groundwater or some other heat reservoir and transfer it into the house by processes similar to those in the refrigerator. Let us consider a system using water in a well. The heat pump takes heat from the water, which may be at a lower temperature than the house, and transfers or "pumps" it into the house. Just as in the refrigerator, the cycle may involve the vaporization and expansion of a liquid in coils in the water. In the house a compression and liquefaction of the vapor is accomplished with the release of heat there. One of the major advantages of the heat pump is that in the summer it can be used to reduce the temperature inside the house by transferring heat to the water in the well. In this case the vaporization and expansion occur in the house, thereby removing heat, and the compression and liquefaction occur in the well, where the heat is transferred to the water.

QUESTIONS

1. Can a kitchen be cooled indefinitely by leaving the door of an electric refrigerator open? Explain.
2. An electric refrigerator transfers heat from the cold cooling coils to the warm surroundings. Does this violate the second law of thermodynamics? Why not?
3. What is an adiabatic process? An isothermal process? An isobaric process?
4. Two identical samples of nitrogen are compressed until the volume is one-half the original volume, one isothermally and the other adiabatically. Which sample has the higher final pressure? Why is the pressure higher?
5. Why are the efficiencies of heat engines so low?
6. What are some examples of common irreversible processes in nature?
7. The internal-energy content of the ocean waters is enormous. Is there some way a ship could use this energy for propulsion?
8. What is gained by using a high compression ratio in an internal-combustion engine? What are the disadvantages?
9. What are the advantages of jet engines? Why weren't they used more in the past?

PROBLEMS

1. An ideal Carnot engine takes 800 kcal from a heat source at 227°C, does some external work, and delivers the balance of the energy to a heat sink at 127°C. Find the efficiency of the engine, how much work is done, and how much heat is delivered to the sink.   *Ans.* 20%; 670 kJ; 640 kcal
2. Steam is injected into a turbine at 327°C and exhausted at 77°C. If the turbine is 70 percent as efficient as an ideal heat engine, find its overall efficiency.
3. Compute the efficiency of an ideal heat engine operating between 207 and 87°C. If this engine takes 50 kJ from the source, how much work does it perform?   *Ans.* 25%; 12.5 kJ
4. A steam engine develops 2,000 hp. How many Btu of heat must be supplied per minute if the overall efficiency is 12 percent? If the engine operates between 240 and 140°F, what fraction of the efficiency of an ideal Carnot engine does this engine have?
5. An ideal Carnot engine has a source temperature of 277°C and sink temperature 127°C. Find the efficiency of this engine and find the heat received from the source and the heat released to the sink when 50 kJ of external work is done.   *Ans.* 0.273; 183 and 133 kJ
6. A Carnot refrigerator takes 40 kcal from a freezing chamber at −23°C and rejects heat at 27°C. How much work must be done? How many calories are rejected?
7. If it were possible to make an ideal refrigerator utilizing a Carnot cycle, how many joules of mechanical energy would be required to remove 10 kJ from the cold compartment at −13°C and deliver it to the outside air at 37°C? How much work would be required to remove this same 10 kJ from a cold compartment at −63°C and reject it at 37°C?   *Ans.* 1.92 kJ; 4.76 kJ
8. A heat pump takes heat from a water reservoir at 7°C and delivers it to a series of pipes in a house at 27°C. Assuming that the energy needed to operate the pump is twice that of an ideal Carnot pump, how much mechanical energy is needed to supply the house with 5×10^4 kcal?
9. In a heat pump or a refrigerator work is done to extract heat from a cold source and deliver it to a

hot sink. The ratio of the heat extracted to the work done is called the *coefficient of performance.* (*a*) Show that for an ideal Carnot engine the coefficient of performance is given by $T_c/(T_h - T_c)$. (*b*) Find the coefficient of performance for an engine operating between -13 and $37°C$. *Ans.* 5.2

10. The refrigeration capacity of air conditioners is typically rated in *tons*, where a capacity of 1 ton implies a rate of extraction required to freeze 1 ton of water at $32°F$ per day. Find the rate of heat extraction for a 1-ton system in terms of Btu's per hour, kilocalories per hour, and joules per second.

11. A four-cylinder gasoline engine makes $1,200 \, r/min$. Each piston has an area of $60 \, cm^2$ and a length of stroke of 140 mm. If the average gauge pressure during a power stroke is $6 \times 10^5 \, N/m^2$, find the power developed in watts and in horsepower. *Ans.* 20.2 kW; 27 hp

12. Show that the power output of a reciprocating engine (such as a steam, gasoline, or diesel engine) is given in watts by the relation

$$\text{Power} = pLAn$$

where p is the average pressure in pascals on the piston during the power stroke, L is the length of the piston stroke in meters, A is the area of the piston in square meters, and n is the number of power strokes per second.

13. A gasoline engine makes $1,200 \, r/min$. It has six cylinders and develops 60 hp. The cylinder is 4 in. in diameter, and the length of the stroke is 5.5 in. Find the average net pressure that is developed during each working stroke. *Ans.* 95.5 lb/in.^{2}

14. A brake is applied to the driving shaft of an engine that develops 4 kW. The brake and the shaft are immersed in a calorimeter so that all the work done against friction is transformed into heat and goes to increase the temperature of the water in the calorimeter. If the mass of the water in the calorimeter is 50 kg, how much will its temperature rise per minute?

15. At the beginning of the compression stroke of an automobile engine, the gas occupies a volume of 0.6 L at atmospheric pressure. At the end of the compression the pressure is 12.9 atm and the volume 0.1 L. What is the compression ratio?

What is the temperature according to the general gas law if the original temperature was $27°C$? *Ans.* 6; 372°C

16. Find how much fuel oil with a heat of combustion of 45 MJ/kg an electric generating plant consumes in 1 h if its power output is 5 MW and the overall efficiency of the plant is 38 percent.

17. Air is admitted to the cylinder of a diesel engine at atmospheric pressure and $27°C$. If the gas is compressed adiabatically to one-fifteenth its original volume, find the pressure and temperature after the compression. *Ans.* 44.3 atm; 613°C

18. How much heat must be added to 28 g of nitrogen gas in a cylinder originally at $0°C$ to double the volume if the pressure is kept at 1 atm? How much external work is done? What happens to the difference between the energy supplied and the work done?

19. If 8 L of helium ($\gamma = \frac{5}{3}$) at $0°C$ and $760 \, mmHg$ pressure is compressed adiabatically to a volume of 2 L, find the new pressure and temperature. *Ans.* 7,660 mmHg; 608 K

20. A gas subject to a constant pressure of $1.5 \times 10^5 \, N/m^2$ is supplied with 30 kcal by an electric heater. If it expands from 0.35 to 0.55 m^3, what is the increase in the internal energy of the gas? What is γ for this gas?

21. A quantity of oxygen ($\gamma = 1.4$) is compressed to occupy a volume of 3 L at 10 atm pressure and $27°C$. If it is allowed to expand adiabatically to 1 atm pressure, find the new volume and temperature. *Ans.* 15.5 L; 155 K

22. Given that pV^γ is constant for the adiabatic expansion of a given mass of ideal gas from a state characterized by pressure p_1, volume V_1, and temperature T_1 to a state characterized by p_2, V_2, and T_2, show that

$$T_1 V_1^{\gamma-1} = T_2 V_2^{\gamma-1}$$

and
$$\frac{T_1^{\gamma/(\gamma-1)}}{p_1} = \frac{T_2^{\gamma/(\gamma-1)}}{p_2}$$

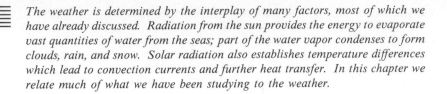

19

ATMOSPHERIC PHYSICS

The weather is determined by the interplay of many factors, most of which we have already discussed. Radiation from the sun provides the energy to evaporate vast quantities of water from the seas; part of the water vapor condenses to form clouds, rain, and snow. Solar radiation also establishes temperature differences which lead to convection currents and further heat transfer. In this chapter we relate much of what we have been studying to the weather.

19.1 WEATHER AND THE ATMOSPHERE

Nowhere in our daily lives do we meet heat phenomena on a more grandiose scale than in our contacts with the weather. Any understanding of the weather and its patterns has its foundations in the principles of mechanics and thermodynamics. In the solar radiation we have our prime source of energy, part of which establishes the giant convection currents we call *winds* and another part of which goes into evaporating water from the surfaces of lakes and oceans. The water vapor, transported by horizontal and vertical winds to distant places where temperature and other conditions are very different, may condense to form clouds, rain, or snow. The interplay of many energy transformations, including phase transitions, leads to that assemblage of atmospheric conditions which we call the *weather*. The weather at any one point is determined by a myriad of conditions, some general and readily understood, others local and of a very special nature. In the sections which follow we shall try to outline some of the principal considerations, but before we can discuss weather patterns in any detail, we must learn something about the atmosphere of the earth, how the sun's radiation produces daily and seasonal variations in temperature, and how the amount of water vapor in the air influences the situation. The science of the atmosphere is *meteorology*, a major area of physical science that is devoted to the study of atmospheric phenomena.

It is only in the lower reaches of the earth's atmosphere, called the *troposphere*, that clouds and storms occur. The troposphere extends to a height of about 11 km in middle latitudes, 18 km in equatorial regions, and 8 km above the poles. The greater height at the equator is due to the higher average temperature there and to the rotation of the earth. Above the troposphere is the *stratosphere*, which extends to a height of approximately 80 km. The stratosphere is particularly well suited for long-distance flying, since the meteorological conditions are more constant there. The stratosphere gets its name from the idea that the region is stratified in layers with roughly constant properties. Above the stratosphere is the *ionosphere*, a region where many molecules are ionized and radio waves are strongly reflected.

19.2 THE SUN'S RADIATION

On a clear day a surface placed normal to the sun's rays receives approximately 1,400 W/m² from the sun. If the rays from the sun strike the surface of the earth at an angle other than 90°, the energy in a given bundle of rays is spread out over a larger area (Fig. 19.1). Specifically, the amount of power received is proportional to the intensity of the beam and to the cosine of the angle between the rays and the perpendicular to the surface. Figure 19.2 shows the importance of the angle at which the

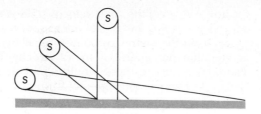

FIGURE 19.1
The position of the sun radically affects the energy received each second by any particular area of the earth's surface.

sun's rays reach the earth; much more energy reaches a unit area in a day in summer than in winter. This is the primary factor which establishes our seasons. In addition to the change in the area over which a given bundle of rays is spread, another factor which helps to diminish the amount of heat from the sun's rays as the sun approaches the horizon is the greater distance the rays must travel in the earth's atmosphere. Consequently, there is somewhat greater absorption.

19.3 GENERAL CIRCULATION OF THE AIR; THE CORIOLIS ACCELERATION

The inequalities of solar radiation received in different regions of the earth's surface are in part responsible for the general circulation of the atmosphere. If the earth were not in rotation, one might expect a relatively simple circulation pattern in which air in equatorial regions would be heated and rise. The denser air at greater latitudes would then push underneath, and we would have a pattern with air at the surface of the earth flowing toward the equator. The warmed air would rise and flow at high altitudes toward the polar regions, where it would cool and descend. This oversimplified picture is complicated by the rotation of the earth, the effects of which we approach by discussing a far simpler situation.

Consider a large merry-go-round, indicated schematically in Fig. 19.3a. When it is at rest, a boy at P_1 can throw a ball to a girl at P_2 by giving it a velocity v along the line P_1P_2 which establishes the x axis for a coordinate system on the merry-go-round. However, when the merry-go-round is rotating about an axis through O with a constant angular velocity ω, the problem of throwing a ball from P_1 to P_2 becomes more complex. Now if

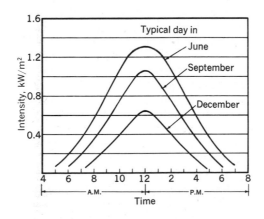

FIGURE 19.2
How the solar radiation received each second on a horizontal surface varies with time of day at a latitude of approximately 40°N.

(a)

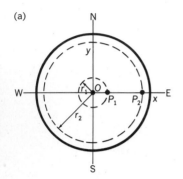

(b)

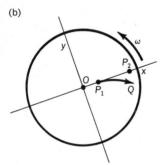

(c)

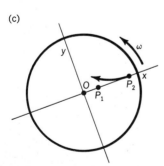

FIGURE 19.3
The horizontal surface of a large merry-go-round. (a) When it is at rest, a boy at P_1 can throw a ball to a girl at P_2 by aiming along the x axis. (b) When the merry-go-round is rotating, a ball thrown along the x axis from P_1 follows the path P_1Q, while (c) one thrown toward the origin from P_2 also curves toward the right as viewed by the thrower.

the boy at P_1 gives the ball a velocity v along the line P_1P_2, the ball traces a path such as P_1Q and passes the girl at P_2, as shown in Fig. 19.3b. To an observer on the merry-go-round the ball *appears* to have undergone an acceleration perpendicular to its velocity, although an observer in a stationary reference frame would report that the ball had traveled in a straight line and would correctly attribute the failure of the ball to reach P_2 to the rotation of the merry-go-round. *The apparent acceleration seen by an observer on the rotating merry-go-round is called the Coriolis acceleration* after G. G. de Coriolis (1792–1843), who worked out the equations of motion in the rotating system.

In a qualitative way it is not hard to see why the path of the ball on the merry-go-round is curved. When the merry-go-round is rotating and in the position shown in Fig. 19.3a, the point P_2 is moving north relative to the fixed frame *WSEN* with a speed ωr_2, while P_1 is moving north with a smaller speed ωr_1. Since there is no net force acting northward on the ball, there is no acceleration in that direction and when the ball gets a distance r_2 from the axis, it is moving north more slowly than the point on the merry-go-round directly below the ball. Relative to the merry-go-round the ball has experienced an apparent acceleration to its right.

Similarly, a ball thrown from P_2 toward P_1 undergoes a deflection toward its right when the merry-go-round is rotating. It is easy to understand this deflection (Fig. 19.3c) when we realize that when the ball is released at P_2, its true velocity perpendicular to the line P_1P_2 is greater than the velocity of P_1 perpendicular to P_1P_2.

In general, in a rotating coordinate system there is an apparent Coriolis acceleration a_c given by

$$a_c = 2\omega v \sin \phi \tag{19.1}$$

where ω is the angular velocity of the rotating axes relative to a fixed set of axes, v is the velocity of the body in question measured in the rotating system, and ϕ is the angle between the vectors $\boldsymbol{\omega}$ and $\mathbf{v}$. The acceleration is at right angles to both $\boldsymbol{\omega}$ and $\mathbf{v}$; it is in the direction in which a right-handed screw would move if it were rotated from $\mathbf{v}$ to $\boldsymbol{\omega}$.

The earth upon which we live and relative to which we observe the motions of bodies forms a rotating coordinate system with an angular velocity $\omega = 2\pi/86{,}164 = 7.3 \times 10^{-5}$ rad/s. The Coriolis acceleration in the Northern Hemisphere is to the right for an observer facing in the direction of the velocity and in the Southern Hemisphere to the left. It is important to take this apparent acceleration into account in the accurate calculation of the impact point of a long-range ballistic missile.

The Coriolis acceleration plays a vital role in our understanding of the circulation of the atmosphere. The earth revolves about its axis once each day, carrying a point at the equator eastward at a speed of 470 m/s and a point at latitude 40° at 360 m/s. At the poles there is no eastward velocity. If a mass of air in the Northern Hemisphere starts moving toward the south, its eastward velocity component is less than that of points on the earth's surface farther south. As it approaches these points, it appears to come from the north *and east*. Similarly, a mass moving northward has a greater eastward velocity component than points on the earth's surface farther north. When this air reaches these northern points, it is traveling eastward faster than the ground; therefore, the wind comes from the south *and west*. The Coriolis acceleration affects not only the motion of air masses but also the circulation of water. It is one of the factors involved in explaining the Gulf Stream, the Japan Current, and other persistent flows in the ocean system.

The combination of unequal heating in equatorial and polar regions and the rotation of the earth leads to the general atmospheric circulation indicated in Fig. 19.4. At any given point on the earth's surface there may be other factors, such as mountains or seas, which profoundly influence the wind distribution.

19.4 HUMIDITY

The next important factor governing the weather is the humidity, which is a measure of the amount of water vapor in the air. *The absolute humidity is the mass of water vapor per unit volume in the atmosphere.*

From the point of view of weather the actual mass of water vapor per unit volume is considerably less important than how nearly saturated the air is. At a high temperature the air can hold vastly larger quantities of water vapor per unit volume than at a low temperature. We can expect water to condense out of the atmosphere only when the air is saturated or supersaturated in some region. The ratio of the actual mass of water vapor per unit volume to the mass per unit volume which the air could hold at the same temperature if saturated is the relative humidity:

$$\text{Relative humidity} = \frac{\text{mass of water vapor per unit volume}}{\begin{array}{c}\text{mass of water vapor per unit volume if air}\\ \text{were saturated at same temperature}\end{array}}$$

Since the pressure exerted by the water vapor is proportional to the mass per unit volume, the relative humidity is also given as the ratio of the vapor pressure of the water to the saturated vapor pressure at the same temperature (Appendix Table B.4).

One method of determining the absolute humidity is the chemical hygrometer. In this instrument a measured volume of air is passed through a series of U-tubes filled with a substance that absorbs water vapor. These tubes are weighed before and after passing the known volume of air through them. The absolute humidity is obtained by dividing the mass of water absorbed in the tubes by the volume of air passed through.

The relative humidity of the air can be measured with a number of different types of instruments. Some of these depend on the hygroscopic properties of hair or of a thin strip of an appropriate material. When the moisture content of hair changes, its length also changes. Instruments based on this property are convenient but ordinarily not very accurate. The wet- and dry-bulb hygrometer consists of two thermometers, one of which reads the true temperature of the air. The other thermometer is covered with a wet cloth and reads a temperature that depends on the rate of evaporation of the water into the air. If the air is saturated, the rate of evaporation is no greater than the rate of condensation, and the wet bulb has the same temperature as the dry one. The drier the air, the greater the evaporation from the wet bulb, and the lower the temperature indicated by that thermometer. Charts showing the relative humidity in terms of the readings of the two thermometers have been prepared and may be found in many handbooks.

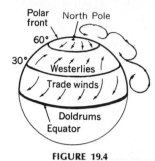

FIGURE 19.4
The general atmospheric circulation in the Northern Hemisphere, where winds deflect to the right; in the Southern Hemisphere the Coriolis effect leads to a leftward deflection.

The moisture of the air is important to our health and comfort. When the relative humidity is high in hot weather, perspiration evaporates slowly from the skin, and we have "sticky" weather. If the relative humidity is low at the same high temperature, water evaporates rapidly from the skin and the body is cooled because it supplies the heat of vaporization. Extremes of heat and cold are felt less when the humidity is relatively low. The relative humidity has a marked effect on the physical condition and behavior of many materials, such as wood and wool. Lack of moisture may cause furniture to shrink and come unglued. An excess of moisture results in swelling. In many forms of manufacturing it is necessary to control the humidity. One of the important functions of any good air-conditioning system is to maintain a desirable humidity.

19.5 CLOUDS, FOG, AND DEW

On a humid summer day water condenses on the outside of a glass filled with ice water. The temperature of the outside glass surface is low, and the air in close proximity is cooled. As the temperature of air falls, the amount of water vapor it can hold is reduced. Eventually air in contact with the glass becomes saturated. If the temperature is lowered still more, water vapor condenses on the cold surface. *The temperature at which the air is saturated is called the dew point.*

Water condenses on any surface which has a temperature lower than the dew point. One way of determining the dew point is to lower the temperature of a polished metal container by evaporating ether from it. When the dew point is reached, a thin film of water is formed on the polished surface.

When the sun goes down and the ground begins to cool, some objects lose heat more rapidly than others. When warm, moist air comes in contact with these objects, some of the moisture is deposited in the form of *dew*. If the temperature is below the freezing point, water vapor is deposited as *frost*. On a typical clear night exposed objects cool faster than the air. As soon as their temperatures fall below the dew point, moisture condenses on them.

If the air cools until it reaches the dew point, moisture condenses out on minute particles in the air. Such a collection of water particles is called a *fog* if the particles are sufficiently small (less than about 30 μm in diameter) so that they do not fall to the earth as rain. Fog often forms in low areas during the night. In the morning, the water particles reevaporate as the air warms. If fog is formed in a region above us, we speak of it as a *cloud*. Whether we use the term fog or cloud to describe the formation depends on whether we are in it or observing it from some distance. Not all clouds, of course, are formed of water droplets. The high fleecy-white cirrus clouds are composed of tiny ice crystals; they exist where the temperature is below freezing.

A cloud is often found above a tropical island. During the day the sunlight produces greater warming of the island than of the surrounding ocean. As the warm air above the island rises, its temperature falls because the pressure becomes less and the gas expands adiabatically. In a standard atmosphere this expansion results in a drop of temperature of 1 C° for every 110-m increase in altitude. Eventually the air above the island is cooled to the point at which it becomes saturated; here a cloud begins to form. A cloud of this type is characterized by a flat bottom, which suggests that the temperature at that particular height is uniform over a wide area.

19.6 LOCAL TEMPERATURE DIFFERENCES

Along shores and coastal areas movements of air, known as land and sea breezes, are associated with diurnal variations in the temperature. As the sun shines on land adjacent to the sea, the land becomes warmer than the water. Convection currents are set up in the air over the land. As the air rises, the pressure over the land is decreased and there is a breeze from water to land (Fig. 19.5a). At night the land cools more rapidly than the water. Convection currents are again established, but now they are from shore to water (Fig. 19.5b).

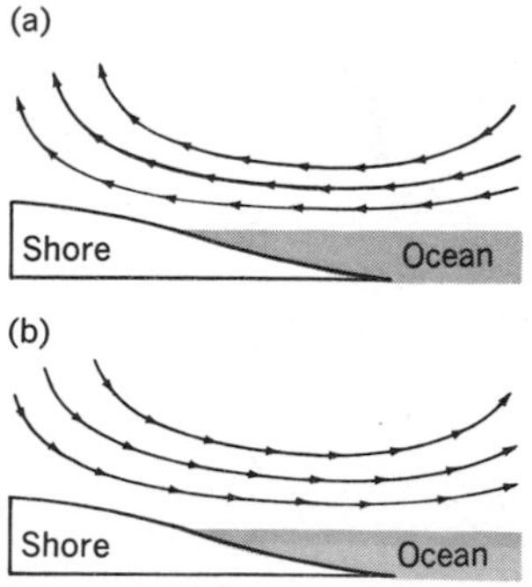

19.7 THE WEATHER MAP

As a consequence of unequal heating of various parts of the earth's surface and of the rotation of the earth, huge convection currents, which we call winds, are established. In regions where excessive heating has occurred there are updrafts and reduced pressure, while in cold regions one expects increased density and pressure. Areas with above-average pressure are known as *highs*, and regions of subaverage pressure as *lows*. These regions are shown on typical weather maps by drawing lines known as *isobars*. An isobaric line connects all neighboring places at which the pressure is the same. Such lines are shown in Fig. 19.6 with the pressure given in millibars (1 mbar = 100 Pa = 100 N/m²). The characteristic pattern of highs and lows in the United States moves from west to east with a speed of roughly 800 km/day.

FIGURE 19.5
Wind directions depend in part on relative temperatures. (*a*) Sea breezes blow from ocean to shore when temperatures are higher on land, and (*b*) land breezes blow from shore to ocean at night, when temperatures are lower on land.

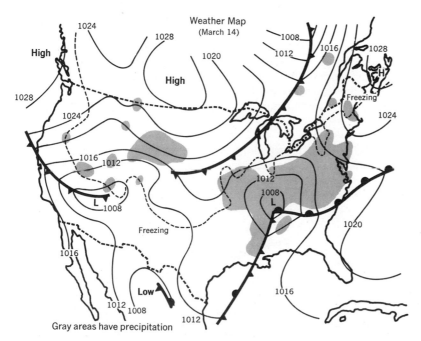

Gray areas have precipitation

FIGURE 19.6
Weather map showing a well-defined low over Tennessee with associated warm front extending toward the east, cold front to the southwest, and an extensive rain area. The numbers give the pressure in millibars.

Barometers measure changes in atmospheric pressure, which may vary from hour to hour and from day to day. These variations are important for weather prediction. When the air contains an unusually large fraction of water vapor, the pressure is ordinarily below normal, since water vapor is lighter than air. Thus a falling barometer is often associated with a storm, while a high atmospheric pressure is characteristic of dry air.

Over fairly extended regions we have large bodies of air which are nearly homogeneous in the horizontal plane with respect to both temperature and pressure. Such a body is called an *air mass*. It is, of course, never homogeneous in the vertical plane; even in the horizontal plane there are gradual variations. An air mass may cover thousands of square miles and extend to heights of many kilometers. It acquires its characteristics by moving for some time over an area where conditions are reasonably uniform. One which has developed over northern Canada is called a *polar continental* or *polar Canadian* air mass, while one which has originated in the northern regions of the Pacific Ocean is called a *polar Pacific* or *polar maritime* air mass. One which has originated over a tropical land mass is described as *tropical continental*, and one over tropical water as *tropical maritime*.

As an air mass moves, it is modified by heating or cooling, by the addition or removal of moisture, and by mixing with other air. Weather conditions are determined to a great extent by the characteristics of air masses and by the direction of their movements.

Where air masses at different temperatures and pressures meet, the surface of separation is called a *front*. Across a front there is a sharp transition in weather conditions, which may include temperature, pressure, humidity, and wind velocity. Several different types of fronts are recognized (Fig. 19.7).

A *warm front* is a surface on which the direction of motion of the air is such that warm air replaces cold air. On the weather map, warm fronts are represented by black semicircles pointing in the direction toward which the warm air is moving. Figure 19.8 shows a warm front advancing. Because the warm air is less dense, it rides over the colder air, and in the early stages of the warm front high clouds are formed by the condensation of moisture in this invading mass of warmer air. As the warm front moves in, lower clouds are formed, and if the warm front involves air of high relative humidity, rain is produced. It ordinarily takes about a day from the first signs of approaching warm air before the warm front passes on the earth's surface.

A *cold front* is a surface along which cold air is replacing warm. It is represented by small black wedges along the front, pointing in the direction in which the cold air is moving. When cold air moves in under warmer and humid air, upward currents of warm air may develop into

FIGURE 19.7
Symbols for fronts.

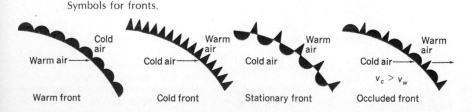

Warm air⟶   Cold air   Warm air

Warm front   Cold front   Stationary front   Occluded front

$v_c > v_w$

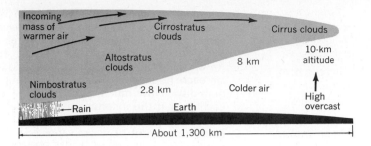

FIGURE 19.8
An advancing warm front.

violent thunderheads. Ordinarily, a cold front moves more rapidly than a warm front, and there is little warning before it arrives. As the cold front passes, there is often a hard shower, which removes much water vapor from the air and leaves crisp bright weather in its wake. Cold fronts move about 40 km/h and are steeper and faster than warm fronts. When a cold front overtakes a warm front, the result is an *occluded front*, represented by a combination of cold-front and warm-front symbols pointing in the direction of motion of the front.

A *stationary front* is a surface separating cold and warm air when the surface is not moving.

19.8 CYCLONES AND ANTICYCLONES

An atmospheric condition characterized by a low-pressure region surrounded by closed isobars is called a *cyclone*. As air from the south moves northward (Fig. 19.9), it is deflected to the east in the Northern Hemisphere, while colder air from the north is deflected toward the west. Thus the circulation about a cyclone is counterclockwise in the Northern Hemisphere. Often warmer air moving from the south forms a warm front, and colder air from the north a cold front, as indicated in the figure.

A high-pressure region on the weather map is known as an *anticyclone*. About this high-pressure region winds circulate clockwise in the Northern Hemisphere. A high-pressure area is ordinarily characterized by bright, clear weather. The pattern of cyclones and anticyclones which migrates across the country is largely responsible for exchange of heat between high and low altitudes.

19.9 TORNADOES

Tornadoes are gigantic whirling funnels of air. The motion of the air is characterized by a cyclonic upward spiral that causes rapid expansion, cooling, and condensation. This condensation forms the dark cloud of the tornado funnel. The width of the storm may vary from 30 m to 2 km, and the forward velocity may vary from 30 to 60 km/h. The tapering end of the funnel may skip over one area and descend on another. It often has a whirling and serpentine appearance. The whirling velocity of the air causes a decrease in pressure at the center of the funnel. The wind velocity is excessive, both in horizontal and vertical directions. Vertical wind speeds may be as much as 300 km/h, and horizontal speeds may exceed this figure.

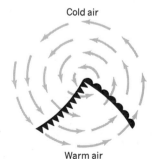

FIGURE 19.9
Cyclonic circulation about a low in the Northern Hemisphere, showing typical warm and cold fronts.

QUESTIONS

1. Why does the barometer fall as moist air moves into a region?
2. Why do the trade winds typically blow from east to west?
3. Why are the prevailing winds westerlies in the continental United States?
4. Why should the temperature fall less than the ordinary 1°C for every 110-m increase in altitude if the air is unusually moist?
5. Do raindrops cool down or warm up (*a*) as they fall and (*b*) as they land?
6. Why do ground fogs appear less often on cloudy overcast nights than on clearer nights?
7. How can an air-conditioning system reduce the relative humidity in a building by forcing the air through a spray of water?
8. Why is cyclonic circulation counterclockwise in the Northern Hemisphere and clockwise in the Southern Hemisphere?
9. Why are satellites put into roughly equatorial orbits by firing them eastward rather than westward?
10. Why are the bases of neighboring cumulus clouds often at about the same level? How are such clouds formed?

PROBLEMS

See Appendix Table B.4 for the properties of saturated water vapor.

1. A room has a temperature of 30°C. When a surface in the room is cooled to 10°C, moisture just begins to condense on it. Find the dew point, the approximate absolute humidity, and the relative humidity. *Ans.* 10°C; 8.76 g/m³; 29%
2. A chemical hygrometer gains 0.85 g in mass when 70 L of air at 23°C is passed through it. Find the absolute humidity of the air and the relative humidity if saturated air at 23°C contains 20.6 g/m³.
3. How great is the vapor pressure of water in a room at 20°C if the humidity is 45 percent? What mass of water vapor is present in each cubic meter of air? *Ans.* 1.04 kPa; 7.74 g/m³
4. Find the relative humidity and the absolute humidity on a day when the temperature is 30°C and the dew point is 20°C.

5. A room is 10 m long, 5 m wide, and 3 m high. The temperature is 20°C. How much water must be evaporated in the room to raise the relative humidity from 15 to 50 percent? *Ans.* 903 g
6. An air-conditioning system delivers 5,000 m³ of air each hour at 20°C and 40 percent relative humidity. How much water must be removed each hour if the incoming air has a relative humidity of 75 percent at 30°C?
7. Find the amount of water vapor contained in a room which is saturated at 10°C if the room has a volume of 4,000 m³. If the air is heated to 20°C, no water vapor is added, and the pressure remains the same, find the relative humidity. *Ans.* 37.5 kg; 52.7%
8. An air-conditioning system delivers 2,900 m³ of air each hour at 17°C and 60 percent relative humidity. To do so, it takes in 3,000 m³ of air at 27°C and 80 percent relative humidity. Find the absolute humidity of the air taken in and the mass of water vapor removed per hour. For saturated air:

Temp., °C	11	13	15	17	19	21	23	25	27
Absolute humidity, g/m³	10	11	12.8	14.5	16.3	18.4	20.6	23.0	25.8

9. Air that is saturated at 30°C rises vertically until its volume is doubled. At the same time, its temperature decreases until it is 10°C. Find the number of grams of water that will condense out of each cubic meter of the air, measured at the original temperature and pressure. *Ans.* 11.4 g
10. Assume that the density of the atmosphere is 1.29 kg/m³ regardless of the height (of course, the density of the atmosphere does depend on what is above it). If a barometer reads 760 mmHg at sea level, what would the height of this imaginary constant-density atmosphere be? Convert your answer into miles.
11. The relative humidity of an auditorium that has a volume of 6,000 m³ is 30 percent at the beginning of a concert and 75 percent at the end of the concert. If the temperature of the room is assumed to remain at 20°C throughout the concert, how many kilograms of water have been added to the room? *Ans.* 46.4 kg
12. The solar constant is 1,400 W/m². How much energy falls each minute on 1 m² of a lake at 39.5° north latitude on a clear day at high noon in June, when the sun is 23.5° north of the equatorial plane? In December when the sun is 23.5° south of the equatorial plane?

13. One mole of dry air at 27°C and 1 atm pressure expands adiabatically as it rises to 5,400-m altitude, where the pressure is $\frac{1}{2}$ atm. Calculate the temperature at 5,400 m, the decrease in the internal energy of the gas, and the work done by the gas. On the average, how many meters did the air have to rise to be cooled 1 C°? (Compare your result with the value of 1 C° for each 110 m quoted in Sec. 19.5 for a standard atmosphere.)
   Ans. 246 K; 1,120 J; 1,120 J; 100 m

14. A body dropped from a height h at the equator will strike the ground a distance $(8 h^3/g)^{1/2}\omega/3$ from the spot directly below the point of release if air effects are negligible. In what direction is the impact point displaced? If a steel ball is dropped from a captive balloon 300 m high, what is the distance between the impact point and the spot directly below the release point? Find the Coriolis acceleration just before the ball strikes the ground.

15. An iceberg with a mass of 10^7 kg is moving eastward 15 km/day at latitude 75°N. Find the Coriolis acceleration. What resultant force would be required to produce this acceleration in an inertial frame of reference?
   Ans. 2.5×10^{-5} m/s² south; 250 N

THREE

WAVE MOTION AND SOUND

20

WAVE MOTION

We have studied several types of motion which can be performed by "particles" of various kinds. Associated with each particle there is an instantaneous position in space as well as mass, kinetic energy, and momentum. A gas molecule or a baseball can be idealized as a point mass in many situations although each occupies a volume in space. A fundamental property of classical particles is that no two particles can occupy the same point in space at the same time. Now we turn to a study of classical waves, which have properties very different from those of classical particles.

20.1 WAVES

The transmission of energy and information from one point to another by wave motion is of the utmost importance to us. Our sense of sight is activated by the electromagnetic waves we call *light* falling on our eyes; our sense of hearing results from sound waves reaching our ears. Radio, television, air-traffic control, radar systems, and x-ray examinations are a few of the vital uses of electromagnetic radiation which make use of other electromagnetic waves. We communicate with each other through sound waves, and we use acoustical waves in other frequency regions to measure the depth of the ocean or detect a submerged submarine. As we saw in Chap. 16, electromagnetic radiation from the sun is our principal source of energy; coal, oil, and natural gas are all energy sources derived from the capture and utilization of solar radiation in the past.

There are many kinds of wave motion; in each kind a disturbance of some sort is propagated. In sound waves the disturbance consists of small fluctuations in pressure; in electromagnetic waves there are rapidly oscillating electric and magnetic fields; in water waves there are changes in the position of the water surface; in stretched strings the disturbance arises from displacements of particles perpendicular to the line of the undistorted string. When mechanical waves pass through the medium, an individual particle oscillates up and down or back and forth but does not progress with the wave. Energy is transferred through the medium although the particles themselves are not transmitted. In a typical wave motion a vibrating center produces motion in particles of the medium immediately in contact with it; these particles in turn impart motion to their neighbors. In many cases, but by no means always, the particles describe simple harmonic motion about their equilibrium positions, with phases varying along the wave.

Of the many kinds of waves, one is the familiar water wave. When a stone is dropped into a pool of still water, the surface is covered with circular wavelets, which widen out from the point where the stone entered the water. The water itself does not move outward from this central point; instead it rises and falls. That this is the case can be seen by observing a floating cork. It moves up and down and back and forth in a roughly elliptical path. The water on which the cork rests goes through this same kind of motion.

20.2 TRANSVERSE AND LONGITUDINAL WAVES

It is convenient to classify waves in terms of how the motion of the individual particles of the medium is related to the movement of the wave itself. In waves produced in a stretched rope (Fig. 20.1) the individ-

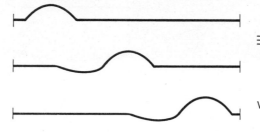

FIGURE 20.1
Wave pulse moving along a stretched string.

ual particles move up and down at right angles to the direction in which the wave propagates. A wave of this kind is known as a *transverse wave* because the particles in the medium move perpendicular (*trans* means "across") to the direction of the wave motion. In the simplest form of transverse wave each particle vibrates in simple harmonic motion with its displacement at right angles to the propagation direction. Light and other forms of electromagnetic radiation show behavior characteristic of transverse waves in many experiments. Waves in stretched strings are an example of a transverse wave motion. *A transverse wave is one in which the vibrations take place at right angles to the direction in which the wave travels.*

If a series of equal masses are joined together by springs (Fig. 20.2) and particle *A* is moved toward particle *B*, a wave is propagated down the system. The motion of *A* toward *B* compresses the spring between them. *B* is accelerated toward *C*, which in turn moves toward *D*. A compressional wave is produced. Similarly, when mass *A* is pulled away from *B*, the spring between them is stretched and *B* is accelerated toward *A*. If *A* is moved back and forth with simple harmonic motion, a wave motion is propagated along the system with each particle describing simple harmonic motion along the axis of the system. The various masses do not all vibrate in phase. *A wave motion in which the individual particles vibrate back and forth along the direction in which the wave travels is called a compressional or a longitudinal wave.* Sound exhibits the characteristic properties of longitudinal waves.

In some wave motions individual particles do not move back and forth along a single line. For example, in water waves, water molecules move forward and back as well as up and down. When a particle is on a crest *A* (Fig. 20.3), it moves in the direction in which the wave is moving. In a trough *B* a particle is moving in the opposite direction. The paths are ellipses or circles. We have in this case a mixed wave, neither longitudinal nor transverse. We can, of course, resolve the motion of the particle into components perpendicular to and parallel to the propagation direction and treat the mixed wave as a combination of a longitudinal and a transverse wave.

There are many other kinds of waves besides those in material media in which the disturbances arise out of the displacement of particles. If the temperature of one end of a metal rod is first raised gradually, then lowered, raised again, etc., a succession of changes in the temperature of

FIGURE 20.2
A series of masses joined by springs transmits a longitudinal wave when the particles move along the line connecting the masses.

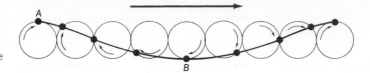

FIGURE 20.3
The particles in a water wave describe circles or ellipses.

the rod is set up. These changes travel forward in the rod as a wave of temperature. The daily heating and cooling of the surface of the earth, as it turns toward and away from the sun, produce waves of temperature that go down into the earth a short distance.

When one end of an ocean cable or telephone wire is suddenly joined to a battery, the change in potential produced in the wire is rapidly transmitted along the conductor. Thus, electric waves may travel through the conductor.

20.3 WAVELENGTH, FREQUENCY, AND VELOCITY

When waves spread out from the center of a disturbance, a surface marking the points which the disturbance has reached is called a *wavefront* (Fig. 20.4). When a drop of water falls on the surface of a pond, the wavefront is a circle that expands continuously. When a small balloon bursts in the air, the wavefront of the sound produced is a sphere with the balloon as the center. The waves of light from a distant star have spherical wavefronts of such large radius that small portions of them may be considered plane.

In a wave the distance from crest to crest (or trough to trough) is called the *wavelength* (Fig. 20.5a). In general, *the wavelength is the distance in the propagation direction between two successive points which are in the same phase of vibration.* Let ν be the frequency of the wave, or the number of complete vibrations made by each particle in unit time. If each wave is of length λ, the total distance traversed by the disturbance in 1 s is just the number of vibrations per second multiplied by the wavelength, or

$$V = \nu\lambda \tag{20.1}$$

where V is the velocity of the wave motion. This relation, applicable to all types of wave motion, is one of the most widely used relationships in physics.

□ **Example**   If the velocity of a disturbance in a steel rod is 5,000 m/s and the frequency is 2,500 Hz (or 2,500 vibrations per second), find the wavelength.

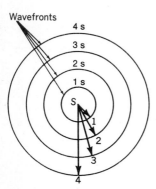

FIGURE 20.4
Wavefronts from a point source S generated at intervals of 1 s.

(a)

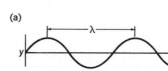

(b)

FIGURE 20.5
Displacement of particles at a given instant as a function of the distance from the source for (*a*) a plane wave for which the amplitude is constant and (*b*) a spherical wave in which amplitude decreases with distance.

$$\lambda = \frac{V}{\nu} = \frac{5,000 \text{ m/s}}{2,500 \text{ vib/s}} = 2.000 \text{ m} \qquad \square$$

Frequency and wavelength are among the important attributes of a wave; other properties are the velocity of propagation, the amplitude of the disturbance, and the energy and momentum transferred. For the frequency of an ideal wave to be measured with great precision the wave must have great extension in space. A classically ideal wave has well-defined frequency and amplitude but no specific position; by contrast, a classical particle has no frequency, but its position can be well defined.

20.4 REPRESENTATION OF WAVES

When each particle of a medium through which a wave is traveling describes simple harmonic motion, a common way of representing the wave graphically is to choose two axes (Fig. 20.5) and to plot on the vertical axis the displacement of a particle at a given instant and on the horizontal axis the distance of that particle from the source of the disturbance. Such a curve is an instantaneous "snapshot" of the wave, showing the displacement for all particles of the medium along the direction of propagation. In Fig. 20.5a the amplitude is constant, and the intensity of the wave remains the same as we go away from the source. In Fig. 20.5b the amplitude and intensity decrease with distance.

For a complex wave with a number of frequencies involved, it is possible to consider the motion of any particle to be the resultant of a number of simple harmonic motions. Thus, a very complex motion may be treated as the sum of a number of simple ones. Although many waves have a far more complicated picture than that of Fig. 20.5b, Fourier showed that all waves can be represented in terms of sine and cosine waves similar to that of Fig. 20.5a. A Fourier analysis of a complex wave into simple harmonic waves is beyond the scope of this book, but the fact that such an analysis can be made means that the arguments and developments which we base on simple sinusoidal curves can be extended to complex waveforms.

When a wave passes through a medium, the individual particles describe simple harmonic motion, for which the displacement y of a particle can be represented by

$$y = R \sin 2\pi\nu t \qquad (9.6)$$

where R is the amplitude, ν the frequency, and t the time, with its zero chosen as the instant at which the particle passes through the equilibrium position in the positive direction. The displacement of the particle as a function of time is shown at the right of Fig. 20.6a. Making use of the idea of the circle of reference (Sec. 9.3), we can think of this displacement as the projection of the uniform circular motion of a particle P on a vertical diameter. At $t = 0$, P is at the point represented by the heavy dot on the horizontal diameter. Imagine P moving with the constant angular velocity of ω rad/s. At the time t_1, P has an angular displacement ωt_1, and its projection is indicated by P_1. As P moves around the circle of reference, its projection describes one full cycle. The period T of the harmonic motion is $2\pi/\omega$, and the frequency $\nu = \omega/2\pi$.

Figure 20.6a showed the displacement of a particular particle as a function of time. Now consider how the displacement at a particular instant varies from one particle to another in the direction of propagation; this corresponds to a snapshot of the wave. Figure 20.6b shows the displacement as a function of x at a particular instant when the particle described by Figure 20.6a has zero displacement and is moving upward; this corresponds to the situation denoted by the two heavy dots on the axis. Further we locate our cartesian axes so that this particle has $x = 0$. The displacement y_1 associated with point P_1 will reach $x = 0$ at time t_1 after our snapshot; at $t = 0$ its position is at $-x_1$, a distance Vt_1 to the left of the origin, where V is the speed of the wave. Similarly, the trough now at point c, one-fourth wavelength beyond the origin, passed the origin one-fourth period before our snapshot. This trough will pass point d half a period after it passed the origin. Thus, as we move to the right in Fig. 20.6b, each successive particle is a little later in phase than the preceding one.

At point e, a distance x_2 from the origin, each crest and trough arrives a time x_2/V after it passed the origin. Let the displacement of the particle at $x = 0$ be given by $y = R \sin 2\pi v t$ [Eq. (9.6)]. Then the displacement at e is the same as the displacement at the origin a time x_2/V earlier. In general, for a wave moving in the $+x$ direction, the displacement of a point at distance x from the origin is given by

$$y = R \sin 2\pi v \left(t - \frac{x}{V} \right) \tag{20.2}$$

Observe that Fig. 20.6a is a plot of y as a function of t when $x = 0$, while Fig. 20.6b shows y as a function of x when $t = 0$.

Although in our plot of displacement as a function of position we obtain a sine curve which suggests a transverse wave motion, this same curve represents a longitudinal wave equally well since the displacement

FIGURE 20.6

(a) The instantaneous displacement y of a particle at a fixed x as a function of time t is given by the vertical projection of a point describing uniform circular motion on the circle of reference at the left. At time t_1 the displacement is y_1 and the angular displacement on the circle of reference is $\omega t_1 = 2\pi v t_1$. (b) The instantaneous displacement of particles along a wave as a function of x. If the wave is moving to the right with a speed v, the displacement at $x = x_2$ is the same as the displacement at $x = 0$ a time x_2/V earlier.

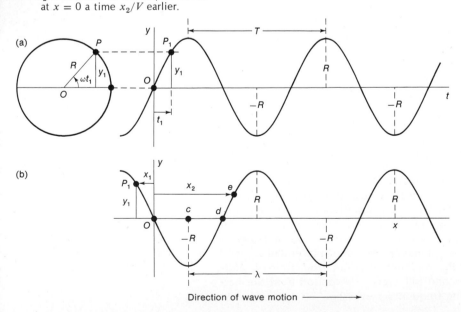

Direction of wave motion ⟶

y obeys exactly the same equations. The only difference is that the displacement for longitudinal wave motion is in the direction of propagation while for a transverse wave motion it is perpendicular.

20.5 THE FLOW OF ENERGY IN A MEDIUM

The energy transmitted in unit time through a unit area perpendicular to the direction of propagation of a wave is called the *intensity* of the wave motion. *The intensity is proportional to the square of the frequency and to the square of the amplitude of the vibrations of the oscillating particles,* as shown below.

For a particle describing simple harmonic motion, energy is transformed from kinetic energy into potential and back into kinetic. The energy of the particle is $\frac{1}{2}mv^2_{\max} = 2\pi^2\nu^2R^2m$ by Eq. (9.11). If we consider a wave propagated in the *x* direction (Fig. 20.7), the wavefront sets in motion every particle it passes. The intensity of the wave is then given by the total energy supplied to all particles included in a rectangle having a base of unit area and a length equal to the distance traversed by the wave in unit time. The number of particles included in this tube is equal to *nV*, where *n* is the number of particles per unit volume and *V* is the velocity of the wave. Each particle receives energy $2\pi^2\nu^2R^2m$, and the total power transmitted per unit area is the intensity, given by

$$I = 2\pi^2\nu^2R^2mnV \qquad (20.3)$$

If a disturbance produces waves which are sent out equally in all directions, the wavefronts are spheres with the source as center (Fig. 20.4). If we assume that there is no transformation of the wave energy, the total energy per unit time which passes through any one sphere surrounding the source is exactly equal to the energy per unit time which passes through any other sphere. The energy per unit time passing through a sphere is just the product of the area of the sphere and the intensity *I*. Therefore, $4\pi R_1{}^2 \times I_1 = 4\pi R_2{}^2 \times I_2$, where I_1 and I_2 are the intensities at the spheres of radii R_1 and R_2, respectively, and

$$R_1{}^2I_1 = R_2{}^2I_2 \qquad (20.4)$$

For a spherical wave the intensity varies inversely as the square of the distance from the source, providing there is no energy loss.

When the energy cannot spread out freely in all directions, the intensity does not vary inversely as the square of the distance. A cheerleader's megaphone is designed to reflect sound waves toward the cheering section. In this case the intensities at equal distances from the source depend strongly on direction. If there is absorption between the source and a receiver, the intensity falls off more rapidly with distance than the inverse-square law suggests, even though the energy may be radiated equally in all directions.

20.6 HUYGENS' PRINCIPLE

Wavefronts from a point source are spherical if wave speed is the same in all directions. If we place a series of obstacles in the medium through which the wave is traveling, these obstacles distort and change the wavefronts. How can we predict where the new wavefront will be a time

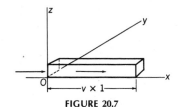

FIGURE 20.7
The energy transmitted per unit area in 1 s across the plane *zOy* occupies a volume of unit cross-sectional area and of length numerically equal to the speed of the wave motion.

Δt after the wavefront has struck an obstacle? The answer was found by Huygens, who observed that in a typical wave motion each particle is set into vibration by a neighboring particle. This led him in 1678 to postulate that *every point on a wavefront acts as a new source sending out secondary wavelets.*

Consider a spherical wavefront (Fig. 20.8*a*). Let us, following Huygens, assume that each point on this wavefront is a source of secondary spherical wavelets and ask where the wavefront corresponding to *AB* will be a time Δt later. About each point 1, 2, 3, 4, . . . on the initial wavefront we draw little spheres of radius $V \Delta t$, where V is the velocity of the waves. To find the position of the new wavefront, we find the surface which is tangent to all the secondary wavelets. This surface, *CD* in Fig. 20.8*a*, is called the *envelope* of the wavelets. When a plane wave passes through a medium, we can find the new wavefront by applying Huygens' method. The new wavefront is again plane, as shown by Fig. 20.8*b*.

Huygens' principle can be summarized as follows: *Every point on a wavefront due to any wave motion may be regarded as a secondary source of Huygens wavelets, which spread out with the velocity of the primary wave. To find the wavefront at any time Δt later, we find the forward surface which is tangent to all these secondary wavefronts. This surface gives the new position of the primary wavefront.*

At a large distance from a point source waves are almost plane; there are also other ways of producing plane waves. In water, for example, plane waves can be generated by a straightedge which oscillates up and down through a water surface. Such a plane-wave generator was used to produce the waves arriving from above in each part of Fig. 20.9. In the top picture plane waves are incident on a partition with a broad opening through which they pass. They continue as plane waves, except that there is some bending into the shadow region, as predicted by Huygens'

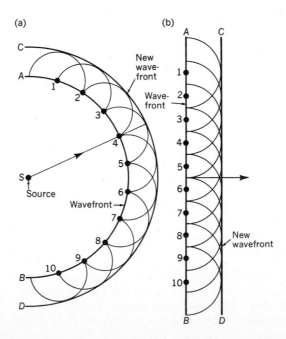

FIGURE 20.8
Huygens' principle applied to (*a*) a spherical wavefront and (*b*) a plane wavefront.

principle. In the other pictures the width of the opening is successively reduced. The transmitted beam of plane waves becomes narrower and the spread of waves into the shadow becomes more apparent. In the lowest picture only a narrow passage remains, and the wave pattern closely resembles that from a point source; there is little to suggest that the curved wavefronts arose from plane waves.

The bending of waves into the shadow region of obstacles and related phenomena are called *diffraction*. All types of waves exhibit diffraction, but it is most prominent when the wavelength is large compared with the size of the opening or of the obstacle. Diffraction is less apparent when λ is very small relative to obstacles or openings. Sound waves, with λ of the order of a meter, diffract readily around typical obstacles, while light, with λ a few hundred nanometers, does not. It is because sound diffracts prominently while light does not that we can talk with a person through an open doorway even though we can not see her.

For points on a wavefront which are far from an edge, the Huygens wavelets traveling to the side overlap in such a way as to cancel each other. However, for points near the edge of the wavefront there are insufficient neighbors to produce such destructive interference on one side. As a result, the Huygens wavelets from these points combine to form a wave spreading into the shadow region.

Early opponents of Huygens' principle raised the question: Why isn't there a backward wave if every particle on the wavefront sends out wavelets with the same amplitude in all directions? This question is based on an incorrect assumption; the wavelets do not have the same amplitude in all directions. About 1815 Fresnel showed that the amplitude of the Huygens wavelet varies according to the obliquity factor $(1 + \cos\theta)$, being greatest in the forward ($\theta = 0$) direction and diminishing to zero in the backward ($\theta = \pi$) direction. At right angles to the propagation direction the amplitude falls to one-half that in the forward direction and the intensity to one-fourth.

20.7 THE REFLECTION OF WAVES

An important application of Huygens' principle is to the reflection of waves by an obstacle. When a spherical wave strikes a plane surface, it is reflected and the curvature of the wavefront is reversed (Fig. 20.10).

Consider a plane wave (Fig. 20.11) falling upon a plane reflecting surface. Let A_1B_1, A_2B_2, and A_3B_3 be consecutive positions of the plane wavefront. Let us apply Huygens' construction to the wavefront A_3B_3 to find the position of the new wavefront. We choose the instant at which the end B_3 of the wavefront reaches the plane surface as the time at which we wish to locate the new wavefront. Let Δt be the interval required for the wave to move from B_3 to the surface. We observe from the figure that the new wavefront is given by RS. Since $V\Delta t$ is perpendicular to the wavefront, and since the distance A_3S is the same for the triangles A_3SR and A_3B_3S, the two triangles are similar and equal. Therefore, the angle of reflection r is equal to the angle of incidence i.

In dealing with the reflection of waves, it is often more convenient to follow the path of a ray which moves in the direction of propagation of the wave than to observe the wavefront. For isotropic media the ray is perpendicular to the wavefront. If N represents a normal to the surface

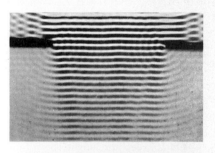

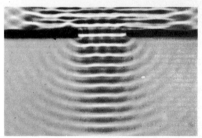

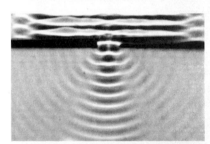

FIGURE 20.9
Plane water waves from a straight-wave generator fall on four openings of different widths. The bending (diffraction) into the shadow region, as predicted by Huygens' principle, is clearly evident in all cases but is most prominent from the narrowest (bottom) slit. (*From "Project Physics," Holt, Rinehart and Winston, New York, 1975; by permission.*)

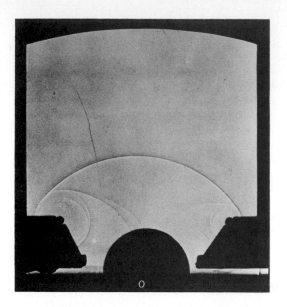

FIGURE 20.10
Reflection and diffraction of sound waves can be studied by photographing the wavefronts created by an electric spark. Here the spark originated at point O on the stage of Royce Hall at the University of California at Los Angeles. The picture is taken looking down at the stage, only a small fraction of which shows; the dark semicircle is a shield to keep the bright spark from fogging the sensitive film. Note the complex wavefronts arising from the proscenium at the left of the stage. (*Courtesy of Vern O. Knudson and L. P. Delasso.*)

(Fig. 20.11), the angle of incidence is the angle between the incident ray and N, while the angle of reflection is the angle between the reflected ray and N. The reflected ray is in the plane determined by the incident ray and the normal. This leads us to the two laws of reflection:

1. *The angle of incidence is equal to the angle of reflection.*

2. *The incident ray, the normal, and the reflected ray lie in the same plane.*

20.8 REFRACTION OF WAVES

When waves pass from one medium in which they have a speed V_1 to a second in which their speed is V_2, the directions of the wavefronts and rays change at the boundary between the two media. This phenomenon is known as *refraction*. Figure 20.12 shows the refraction of waves as they pass from one region to another in which their speed is decreased.

Refraction at the interface between two media is readily explained by use of Huygens' principle. In Fig. 20.12 a plane wavefront ABC is just reaching the interface between two media. Let V_1 be the speed of the waves in the medium in which the waves are incident and V_2 their speed in the second medium. While the wavefront goes from BC to EF in the first medium, the wavelets from A travel only as far as D in medium 2. When the wavelets from F reach J, the wavefront is entirely in the second medium, as is represented by GHJ. If Δt is the time required for the wavefront to travel from A to G in medium 2, we have $AG = V_2 \Delta t$ and $CJ = V_1 \Delta t$. Now $\sin \theta_1 = CJ/AJ$, and $\sin \theta_2 = AG/AJ$, where θ_1 is the angle of incidence† and θ_2 the angle of refraction. Therefore $(\sin \theta_1)/(\sin \theta_2) = (V_1 \Delta t)/(V_2 \Delta t) = V_1/V_2$. The ratio $(\sin \theta_1)/(\sin \theta_2)$ is

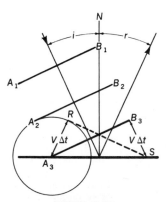

FIGURE 20.11
Huygens' construction for a plane wave reflected from a plane surface.

† Note that here we are designating the angle of incidence by θ_1 rather than by i. This has advantages because the path of rays are reversible. For a ray going from G to A in medium 2, θ_2 would be the angle of incidence and θ_1 the angle of refraction.

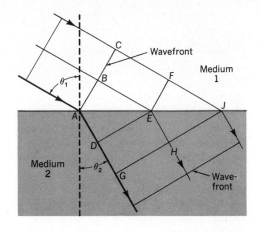

FIGURE 20.12
Refraction at a plane surface.

called the *index of refraction of medium 2 relative to medium 1;* we represent it by *n*. Thus

$$n = \frac{\sin \theta_1}{\sin \theta_2} = \frac{V_1}{V_2} \qquad (20.5)$$

For refraction:

1. *The ratio of the sine of the angle of incidence to the sine of the angle of refraction is a constant n, known as the index of refraction (of medium 2 relative to medium 1).*

2. *The incident ray, the normal, and the refracted ray lie in the same plane.*

Water waves become slower as the water becomes shallower. In this case there is no sharp bending, but the wavefronts bend continuously as the speed changes. On a sloping beach the waves bend so that the wavefronts are almost parallel to the shore regardless of the direction they came from. In general, a ray bends toward the normal to an interface as a wave goes from a region of higher speed to one of lower speed.

20.9 SUPERPOSITION AND INTERFERENCE

It is common to have sound waves from different sources moving through a room at the same time. *When two or more wave motions pass a given point at the same instant, the displacement is the resultant of the displacements which would be produced by each of the waves acting separately.* This is a statement of a very important law of nature, sometimes called the *principle of superposition.* It is valid not only for sound but for all other kinds of waves, provided the amplitudes are not too great.

We shall at the moment apply the principle of superposition to the particular case in which two waves are traveling in the same direction at the same time with the same amplitude and the same frequency. If the waves arrive in such a way that crests meet crests and troughs meet troughs (Fig. 20.13), the displacements due to the two waves add and the resultant is a wave of double the original amplitude. These two waves show *constructive interference.* On the other hand, if the two waves arrive at the same point in such a way that crest meets trough and trough meets crest, they cancel each other and we have *destructive interference.*

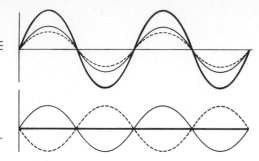

FIGURE 20.13
Constructive and destructive interference.

Two waves of unequal amplitude may still interfere; the resulting amplitude is the resultant of the individual amplitudes. Interference is a fundamental property of wave motions. Indeed, if it is uncertain whether some phenomenon has wave properties, the test which resolves the question in favor of the wave position is one which produces interference between two "rays."

20.10 STANDING WAVES

A particular type of interference of great importance in connection with musical instruments is interference between two waves traveling in opposite directions. Consider a long elastic cord which is fixed at one end (Fig. 20.14) while the other end is held in the hand. If the cord is stretched tight and the free end is moved up and down, disturbances are set up in the cord which travel to the fixed end, where they suffer reflection and travel back to the hand. In Fig. 20.14 a trough *A* is moving to the right. When it reaches the fixed end, it is reflected as a crest *B*. In general, there is a 180° phase shift from reflection at a fixed end; a crest is reflected as a trough and a trough as a crest.

In many physical situations, including stringed instruments and organ pipes, we typically have two trains of waves moving in opposite directions. When two trains of sinusoidal waves with the same amplitude and frequency are traveling in opposite directions along a cord, in certain situations a stationary pattern may be formed so that to the eye the waves appear to be standing still rather than progressing (Fig. 20.15). The cord ceases to have the appearance of being traversed by trains of waves, instead vibrating in one or more segments. At the ends of the segments there are points called *nodes* where there is no motion of the cord. At intermediate points each particle performs simple harmonic motion transversely, with the amplitude of vibration increasing as we go away from a node. The amplitude reaches a maximum halfway between the nodes at a point called an *antinode* or *loop*. When the frequency is high, the eye sees a characteristic blur between the stationary nodes. This appearance has given rise to the names *standing waves* or *stationary waves*.

FIGURE 20.14
When a wave in a stretched string is reflected at a wall, a trough is reflected as a crest and a crest is reflected as a trough; there is a 180° phase shift at this reflection.

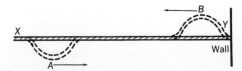

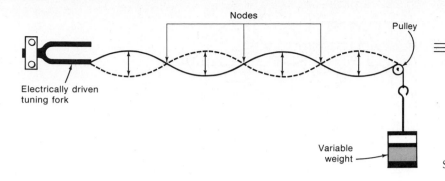

Nodes

Electrically driven
tuning fork

Pulley

Variable
weight

FIGURE 20.15
Standing wave in a stretched string.

In order to set up stationary waves in the cord of Fig. 20.15 certain conditions must be met. Let us assume that the electrically driven fork vibrates with a fixed frequency and that initially the tension in the cord, determined by the hanging weight, is less than that pictured. Then small-amplitude waves will be seen running back and forth along the cord, but there is no well-defined pattern of vibration. If the tension is now gradually increased by adding small weights, at some critical tension a sharp standing-wave pattern will evolve, perhaps with six segments. If we continue to increase the tension by small increments, the stationary wave disappears and then at some well-defined tension a five-segment standing wave develops. Higher critical tensions can be found to produce successively the four-segment pattern shown in the figure, and then standing waves with three, two and one segments. In Chap. 21 we shall learn how to calculate the critical tensions required to put the cord into *sympathetic* (or *free*) vibrations in resonance with the driving fork.

To understand the formation of stationary waves better consider Fig. 20.16a; at L the crests of two component waves are approaching each other. When the two crests coincide, the resulting displacement is maximum. One-quarter of a period later the crest of one wave meets the trough of the other. At this instant the cord is straight. As the waves travel farther in opposite directions, the portion of the string $N'LN$ is depressed below the horizontal, and after another quarter of a period it has its maximum displacement in the negative direction. At the points N and N' there is never any displacement. At these nodes the two waves traveling in opposite directions always interfere destructively. All the particles of the cord between two adjacent nodes are moving in the same direction at any given instant, but two adjacent segments of the cord are

FIGURE 20.16
(a) Standing waves result when two waves of the same amplitude and frequency travel in opposite directions, interfering constructively at some points and destructively at others. (b) Successive positions of incident and reflected waves producing standing waves.

(b)

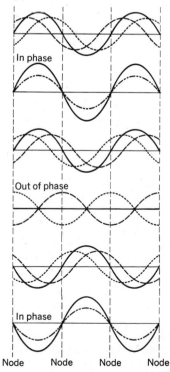

In phase

Out of phase

In phase

Node   Node   Node   Node

(a)

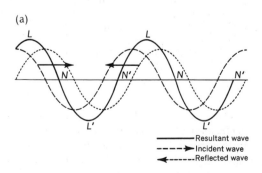

——— Resultant wave
----▶ Incident wave
◀----- Reflected wave

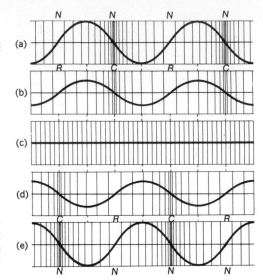

FIGURE 20.17
Successive stages of standing sound waves during one-half cycle, in the course of which nodes (*N*) representing rarefactions (*R*) and compressions (*C*) in (*a*) and (*b*) reverse to become compressions and rarefactions in (*d*) and (*e*).

always displaced in opposite directions. The length of the segment between two consecutive nodes is one-half wavelength.

Standing waves may result from any kind of wave motion. The basic essential for their production is that two waves of the same frequency and amplitude travel in opposite directions in the medium. Figure 20.17 represents standing waves in a column of air. The short vertical lines represent layers of air displaced as shown. At each place marked *N* there is a node in the displacement; here the air is alternately compressed and rarified. Consequently, these regions are called *compressions* and *rarefactions*, as we shall see in Sec. 21.2. The pressure at a compression is slightly greater than the ambient pressure, while it is slightly less at a rarefaction. Thus the antinodes in the pressure coincide with the nodes in the displacement and vice versa. In Fig. 20.17 the heavy sinusoidal line and the nodes are for the displacement; the corresponding quantities for pressure are 90° (or one-quarter wavelength) out of phase with those for displacement.

QUESTIONS

1. Energy can be transported by particles and by waves. How can one distinguish experimentally between these means of energy transport? What characteristics differ between waves and particles?
2. What general properties must a medium have to be able to transmit longitudinal waves? Transverse waves? Explain why liquids and gases do not transmit transverse mechanical vibrations.
3. If the intensity of any kind of wave falls off as $1/r^2$ [Eq. (20.4)], how does the amplitude vary as a function of *r*?

4. As water waves spread out over the surface of a lake, how does the intensity vary with distance from the source if essentially no energy is dissipated? How does the amplitude vary? Why isn't the inverse-square law followed?
5. Two sources produce waves of identical frequency, but one produces waves having 3 times the amplitude of the others. What is the ratio of the two intensities? If the frequency of a wave motion is doubled while the amplitude is kept constant, how does the intensity of the new wave compare with that of the old?
6. When two waves interfere, is there a change in intensity? Is there a change in energy? Explain.

7. What sort of experiments could you perform to prove that energy is associated with a wave motion?
8. What happens to a wave when it comes to an interface between two media?
9. What conditions must be satisfied to establish a standing-wave pattern?
10. What points remain stationary in a standing wave? Why don't they move?

PROBLEMS

Unless otherwise stated, take the speed of sound in air to be 340 m/s or 1,100 ft/s and the speed in water to be 1,450 m/s.

1. It is 12 m from crest to crest in a system of water waves. If 20 waves pass a given point each minute, find the speed of the waves.   *Ans.* 4 m/s
2. Water waves are observed passing a certain point at a speed of 7 m/s with 31 m between crests. What is the frequency of the waves?
3. A tuning fork with a frequency of 700 Hz sends out waves which travel 350 m/s. What is the wavelength of the sound waves? How many vibrations does the fork make in the time required for the sound to travel 250 m?
Ans. 0.5 m; 500
4. Waves travel at 250 m/s through a medium. If 40 waves pass a given point each second, find the wavelength and the period.
5. When a sound wave is transferred from one medium to another, its frequency does not change. If the wavelength of a disturbance is 0.68 m in air, find its wavelength in water, steel, and brass if the speed of sound in steel is 5,000 m/s and in brass 3,500 m/s.   *Ans.* 2.9 m; 10 m; 7 m
6. A sounding source with a frequency of 800 Hz sends out waves that travel from air into water. Find the wavelength in each medium.
7. A set of circular ripples is produced on the surface of a ripple tank by an oscillating point source. At a certain instant, the first crest is 0.40 m from the point where the needle point hits the water, and the sixth crest is 0.15 m from the same point. What is the wavelength of the disturbance? If the speed of the waves is 0.3 m/s, what is the frequency of the source?
Ans. 50 mm; 6 Hz
8. A violin string emits a sound with a frequency of 445 Hz. What is the wavelength of the disturbance that passes through the air? What is the period of the vibration?
9. The speed of electromagnetic waves through vacuum or air is approximately 3×10^8 m/s. Find the frequency associated with green light with a

wavelength of 540 nm and the wavelength of radio waves from a station broadcasting at 900 kHz.   *Ans.* 5.56×10^{14} Hz; 333 m
10. A phonograph record rotates $33\frac{1}{3}$ r/min. If a sound of 1,200 Hz is produced when the needle is in a groove 86 mm from the axis, find the distance between the wiggles in the groove at that point.
11. A swimmer is under water when the starter of a race fires his gun. If the sound ray to the swimmer's ear approaches the water surface at an angle of incidence of 10°, what is the angle of refraction in water?   *Ans.* 47.8°
12. A sound wave from an underwater explosion approaches the surface at an angle of incidence of 70°. What is the angle of refraction in air? The angle of reflection in water?
13. Standing waves are produced in a stretched rope. If the distance between successive nodes is 0.8 m, what is the wavelength? If the waves travel with a speed of 12 m/s what is the frequency?
Ans. 1.6 m; 7.5 Hz
14. What is the speed of waves in a stretched cord if a frequency of 220 Hz produces standing waves with 365 mm between adjacent nodes?
15. A sound source producing compressional waves at 1 kHz sends waves down a tube closed at the far end by a movable piston. Standing waves are produced when the piston is moved 90 mm into the tube. Then the piston is moved farther into the tube, and standing waves are next produced when the piston is 265 mm into the tube. Find the speed of the sound waves in the tube.
Ans. 350 m/s
16. Compressional waves with a frequency of 1,200 Hz are propagated through the air in a tube, the far end of which is closed by a movable piston. Standing waves are produced. The reflected wave reinforces the source at two successive positions of the piston differing by 144 mm. What is the speed of the waves in the tube?
17. A wave is represented by $y = 0.5 \cos 6\pi (t - 0.0035x)$, where y and x are in millimeters and t is in seconds. For this wave find the amplitude, period, frequency, wavelength, and speed.
Ans. 0.5 mm; 0.333 s; 3 Hz; 95.2 mm; 285.8 mm/s
18. If a certain particle has a displacement in millimeters given by $5 \sin 2\pi \nu t$, where $\nu = \frac{1}{10}$ Hz, find the displacement at times 1, 2, 3, . . . , 9, 10 s.

Plot the particle's displacement as a function of time.

19. The displacement of a particle in micrometers is given by $y = 8 \sin 2\pi\nu t$. If $\nu = 10$ Hz, find the displacement and velocity of the particle at times t of 0.02 and 0.06 s.

   Ans. 7.61 μm, 155 μm/s; -4.70 μm, -407 μm/s

20. If the displacement along a wave is given by the relation $y = 9 \sin 2(t - 0.006x)$, where x and y are in millimeters and t is in seconds, what are the amplitude, frequency, wavelength, and speed of the wave motion?

21. A transverse wave in a string is represented by $y = 0.02 \sin 4\pi(2x - 15t)$, where y and x are in meters and t is in seconds. For this wave find the amplitude, the frequency, the period, the wave-length, the wave speed, and the maximum speed of a particle of the string.

   Ans. 0.02 m; 30 Hz; 0.0333 s; 0.25 m; 7.5 m/s; 3.77 m/s

22. If the displacement along a wave is given by $y = 0.02 \sin (40\,\pi t - 0.2\,\pi x)$, where distances are in meters and times in seconds, what are the amplitude, frequency, wavelength, and speed of the wave motion?

23. A transverse wave in a stretched wire is described by the equation $y = 0.04 \sin (100\pi t - 0.4\pi x)$, where x and y are in meters and t is in seconds. What are the frequency, amplitude, period, wavelength, and velocity of the wave?

   Ans. 50 Hz; 0.04 m; 0.02 s; 5 m; 250 m/s

24. A wave $y_1 = R \sin (kx - \omega t)$ traveling to the right and a wave of the same amplitude and frequency $y_2 = R \sin (kx + \omega t)$ traveling to the left are superimposed to give a resultant $y = y_1 + y_2$. Show that the resultant is a standing wave $y = 2R \sin kx \cos \omega t$. [*Hint:* Use the formula for $\sin (\theta \pm \phi)$ given in Sec. A.1 of Appendix A.] Show that the wavelength of the resultant wave is given by $2\pi/k$.

In Chap. 20 we discussed many of the general characteristics of wave motions and introduced the vocabulary of waves with words such as wavelength, frequency, amplitude, refraction, reflection, and interference. In this chapter and the next we treat sound waves, their sources, and their properties. First we are concerned with the speed of acoustic waves and how they originate. We discuss how various kinds of sound sources produce the periodic changes in pressure we hear as sound. For some of the simpler sources we can predict both the fundamental frequency and the prominent overtones from the dimensions of the source.

SOUND SOURCES

21.1 THE NATURE OF SOUND

Sound has its source in vibrating bodies. Consider the tuning fork of Fig. 21.1. As the tine *A* swings toward the right, it pushes air molecules and produces a region in which the molecules are crowded together. Such a region is called a *compression* or *condensation*. When the tine swings to the left, a region of reduced pressure called a *rarefaction* is produced. The next swing to the right produces another compression, and so forth. Thus, the sound wave consists of a series of alternate compressions and rarefactions. The molecules move back and forth along the line of propagation, which establishes the *longitudinal* nature of the waves. When the compressions and rarefactions reach the eardrum of a listener, they produce small inward and outward motions of the eardrum, and this starts the physiological processes of hearing.

Even in solids and liquids a sound wave involves variations in pressure traveling through the medium. Associated with the pressure changes are small longitudinal displacements of the molecules. These displacements and the associated changes in density are very much smaller than they are in gases.

21.2 THE VELOCITY OF SOUND

The velocity of sound in any medium depends upon its density and its elastic properties. In a wire or rod of a solid material the velocity of sound waves is given by

$$V = \sqrt{\frac{Y}{d}} \tag{21.1}$$

where Y is Young's modulus and d the density. In fluids the velocity is given by

$$V = \sqrt{\frac{B}{d}} \tag{21.2}$$

where B is the appropriate bulk modulus. For a gas B is the *adiabatic bulk modulus* given by γp, where p is the pressure and γ is the ratio of two specific heats for the gas (Sec. 15.8). It is 1.67 for monatomic gases, 1.40 for diatomic gases, and 1.28 for many triatomic gases. For any gas

$$V = \sqrt{\frac{\gamma p}{d}} \quad \left(\text{for air, } V = \sqrt{\frac{1.4p}{d}} \right) \tag{21.3}$$

The speed of sound in air at 0°C is 331.45 m/s (1,087 ft/s). In hydrogen,

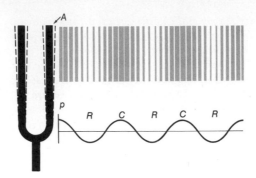

which has a low molecular weight and therefore low density, the speed of sound at 0°C is about 1,270 m/s.

According to Boyle's law, the ratio of the pressure to the density of a gas is constant if the temperature of the gas remains unchanged. Since the velocity depends on the *ratio* of pressure to density, it is constant for a given gas so long as the temperature is constant. However, the ratio p/d does depend on the temperature. For 1 kmol of gas of molecular weight M, $d = M/V$ while, by the general gas law (Sec. 14.7), $pV = RT$. Therefore, $d = pM/RT$ and

$$V = \sqrt{\frac{\gamma p}{d}} = \sqrt{\frac{\gamma RT}{M}} \tag{21.3a}$$

Since the density at constant pressure is inversely proportional to the absolute temperature, the speed of sound in a gas increases with temperature according to the equation

$$V_t = V_0\sqrt{\frac{T}{273}} = V_0\sqrt{1 + \frac{t}{273}} \tag{21.4}$$

where t is the Celsius temperature and V_0 is the speed at 0°C.

□ **Example**   Find the velocity of sound in dry air at 100°C.

$$V_{100} = V_0\sqrt{1 + \frac{t}{273}} = 331\sqrt{\frac{373}{273}}$$

$$= 386 \text{ m/s} \qquad\qquad □$$

The reason the speed of sound in a gas increases with temperature is that the velocity of the molecules increases. In a sound wave the energy is transferred from one molecule to the next; in a gas the speed of the transfer cannot exceed the speed of the molecules. The speed of sound in a diatomic gas is about 74 percent of the average speed of the molecules.

21.3   FREQUENCIES AND WAVELENGTHS OF AUDIBLE SOUNDS

A typical young human ear can hear frequencies from 20 and 20,000 Hz. There is substantial difference in range between individuals. Since the velocity of sound waves is given by the product of the frequency and the wavelength, and since sounds of all different frequencies have the same velocity in air, the range of wavelengths we hear runs from about 17 m to 17 mm. That all frequencies travel with essentially the same speed is clearly shown by the fact that the sounds from all instruments of an orchestra reach a listener at the same time. If velocity depended on frequency, the music would sound quite different to a person far from the orchestra and one close by.

Sound waves of higher frequency than the human ear can hear, known as *ultrasonic* sounds, are readily produced and are emitted by many animals. Ultrasonic frequencies can be heard by dogs, birds, and many other species; indeed, the bat locates obstacles and finds its way about by emitting ultrasonic frequencies, which it detects as they are reflected from various obstacles.

21.4 SOUNDING BODIES

Ordinary sound waves are set up by vibrating bodies. A typical source in vibration sends a series of alternate compressions and rarefactions through the air. Most sound sources do not produce a single frequency but a combination of many frequencies. The lowest prominent frequency is the *fundamental;* the higher frequencies are called *overtones.* In vibrating strings and in organ pipes the overtones are usually harmonics of the fundamental. *A harmonic is an overtone with a frequency that is an integral multiple of the fundamental frequency.* The frequency of the third harmonic is 3 times that of the fundamental, and so forth.

Most of the overtones of bells, drums, and vibrating plates are not harmonically related to the fundamental. A few sound sources are especially designed to produce as nearly as possible a single frequency, or a *pure tone.* A tuning fork is an example of such a source. It is essentially a rod bent in the form of a U with the central region reinforced to suppress overtones; even so, careful observation often reveals the presence of undesired frequencies.

In most musical sounds, the same frequency is emitted for an appreciable time, during which the amplitude does not decrease rapidly. In a noise, on the other hand, we usually have a sudden burst of a wide range of frequencies with no regularity and a marked drop in intensity during the emission of a single vibration. Of course, the distinction between noises and musical sounds is not sharp and definite. What is music for one person may be noise to another.

21.5 WAVES IN WIRES AND STRINGS

Many musical instruments make use of stretched strings as sound sources. These strings may be set into vibration by striking, plucking, bowing, or strumming. When the string is displaced at some point and then released, the disturbance is passed from one element of the string to the next and transverse waves proceeding in both directions in the string are produced. These waves are reflected at the fixed ends of the string, return in the opposite direction, and go on to the other ends, where they are again reflected. Of the many frequencies into which the initial disturbance can be analyzed, only those which are suitable for establishing standing waves in the string are maintained for any length of time.

The velocity V of a transverse wave along a flexible stretched string

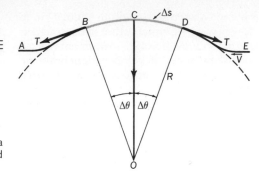

FIGURE 21.2

Grossly enlarged representation of a transverse-wave pulse on a stretched cord.

depends on the tension T and on the mass per unit length μ (Greek letter mu) of the string. It is shown below that

$$V = \sqrt{\frac{T}{\mu}} \tag{21.5}$$

Consider a wave traveling toward the right in the cord AE of Fig. 21.2. Assume that while the pulse, or wave, is moving toward the right with a velocity V, the cord is made to move toward the left with an equal velocity. As a result of these superposed velocities, the wave pulse remains at rest. For simplicity, assume that the small arc $BD = \Delta s$ is circular in form. The component of the tension T along the radius CO is $T \sin \Delta\theta$, so the net force on Δs toward O is $F = 2T \sin \Delta\theta = 2T \Delta\theta$ (approximately). Since $2 \Delta\theta = \Delta s/R$, $F = T \Delta s/R$. This force provides the centripetal acceleration required if the mass $\mu \Delta s$ is to traverse the circular path. By Eq. (8.2),

$$\mu \Delta s \frac{V^2}{R} = F = T \frac{\Delta s}{R}$$

from which $V = \sqrt{T/\mu}$.

Of the standing waves established in a stretched string, the one with the longest wavelength is that for which the string vibrates as a single segment (Fig. 21.3). The longest wavelength corresponds to the lowest frequency and therefore to the *fundamental* of the string. The standing wave of the next longest wavelength is that for which the string vibrates in two segments. In this case the length of the string is equal to the wavelength. The third longest standing wave which can exist in the

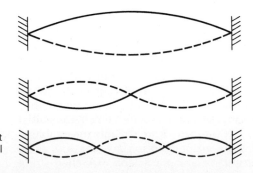

FIGURE 21.3

Vibrations in a stretched string fixed at both ends, showing the fundamental mode and the first two overtones.

stretched string occurs when the string vibrates in three segments, the next in four segments, etc.

The allowed modes of vibration for a stretched string are easy to determine if one remembers that both ends of the string are fastened. These ends cannot move and must therefore act as nodes for any possible standing wave. We are dealing with a situation in which we impose a condition on the ends of the strings, namely, that they remain at rest. Such conditions, known as *boundary conditions*, play a very important role in determining the types of vibrations allowed in various kinds of systems. Once we have established the boundary conditions that there be nodes at the ends of the stretched strings, we can readily find the fundamental by finding the longest wave we can produce with nodes at both ends. The first overtone corresponds to the next longest possibility, and so on.

If a string is plucked, it vibrates in such a way that both the fundamental and overtones are present. The overtones are all harmonics, and the note from a string consists of the fundamental together with several harmonics.

21.6 FUNDAMENTAL AND OVERTONES OF A STRING

Let L be the length of the string. Then, for the fundamental, the wavelength is $2L$ (Fig. 21.3). Since for any wave motion $V = \nu\lambda$, for the fundamental frequency ν_1 we have

$$\nu_1 = \frac{1}{2L}\sqrt{\frac{T}{\mu}} \qquad (21.6)$$

The frequency of the fundamental depends on length, tension, and mass per unit length. A musical instrument such as the piano affords an illustration of how these factors apply. The strings for the bass notes are ordinarily heavy, thus having large μ, and they are also long. Strings for the high notes are short, light, and tightly stretched. In tuning a piano, the tension is varied. In instruments like the violin, with only four strings, the strings are of the same length, but they have different masses per unit length and different tensions. To play different frequencies on the same string, violinists vary the length of the string by placing a finger at the point where they wish to create a node.

The first overtone for the vibrating string occurs when $\lambda = L$ (Fig. 21.3), and therefore the frequency of the first overtone is given by

$$\nu_2 = \frac{1}{L}\sqrt{\frac{T}{\mu}} \qquad (21.7)$$

The first overtone has a frequency twice that of the fundamental and is therefore the second harmonic. For the second overtone $\lambda = 2L/3$, and the frequency is

$$\nu_3 = \frac{3}{2L}\sqrt{\frac{T}{\mu}} \qquad \text{(second overtone = third harmonic)} \qquad (21.7a)$$

In general, *the allowed overtones of stretched strings consist of all harmonics.*

Most harmonics blend well with the fundamental to produce pleasing tones. The seventh harmonic is an exception. In order to suppress the seventh harmonic, piano strings are struck at a point where the seventh harmonic would have a node. This assures that there cannot be a perfect node at this point. Violins are often bowed in the vicinity of a node for the seventh harmonic in order to suppress this relatively unpleasant overtone. The superior tonal quality of an expert violinist is associated with the ability to produce not only the desired fundamental but also a combination of overtones which are pleasing. The relative intensities of overtones can be varied over substantial ranges by an expert.

21.7 CLOSED ORGAN PIPES

In an organ pipe the vibrating body is a column of air. At one end of a pipe, the column of air is set into vibration by sending a narrow jet of air toward a thin edge or lip (Fig. 21.4). This end of the pipe is always open so that sound waves can be transmitted from the pipe into the surrounding air. In a closed pipe the other end is blocked off. Thus, a closed organ pipe is closed at one end and open at the other. When air first strikes the lip, a compression starts down the pipe, is reflected at the closed end, and returns to the lip as a compression. When it reaches the lip, it pushes the airstream outside the lip, and this starts a rarefaction down the tube. The rarefaction is reflected at the closed end, returns to the lip as a rarefaction, and draws the stream into the pipe again, thereby starting a new compression. Thus the airstream is made to move back and forth across the lip with its period determined by the time required for a compression to travel up and back and a rarefaction to travel up and back. The period is 4 times the time required for sound to traverse the length of the tube.

An alternative approach to determining the fundamental frequency involves the application of boundary conditions. At the closed end of the pipe an air particle is unable to move forward when a compression comes down the tube. We have a *displacement node at the closed end of an organ pipe*. On the other hand, a particle at the open end is not restrained, and the amplitude of the vibration of a particle is a maximum there. We find a *displacement antinode at the open end of a pipe;* the displacement there reaches its largest value. However, as we saw in Sec. 20.10, the particle displacement and the pressure variations are 90° (one-fourth wavelength) out of phase. In an organ pipe the pressure has an antinode at a closed end and a node at an open end. In our treatment we shall focus attention on the boundary conditions for the displacement.

The possible standing waves which can be set up in a closed organ pipe are those for which we have a node in the displacement at the closed end and an antinode at the open end. The longest wavelength which can fit into the organ pipe is indicated in Fig. 21.4. In this case the wavelength is 4 times the length L of the pipe. The frequency ν_1 of the fundamental is given by the ratio of the velocity V of the wave motion, which in this case is the velocity of sound in the gas in the organ pipe, to the wavelength λ.

FIGURE 21.4
Displacement pattern (dashed line) for the fundamental in a closed organ pipe.

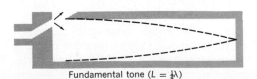

Fundamental tone ($L = \frac{1}{4}\lambda$)

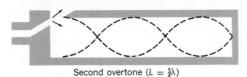

First overtone ($L = \frac{3}{4}\lambda$)

Second overtone ($L = \frac{5}{4}\lambda$)

Therefore,

$$\nu_1 = \frac{V}{4L} \tag{21.8}$$

There are other possible modes of vibration for this column of air which have a node at the closed end and an antinode at the open end. The next lowest frequency which can be emitted involves the vibration in which there is one additional node in the pipe (Fig. 21.5). For this case the wavelength is $4L/3$; the *first overtone is the third harmonic*, which has a frequency

$$\nu_3 = \frac{3V}{4L} \tag{21.9}$$

The second overtone is also shown in Fig. 21.5. Its frequency, $\nu_5 = 5V/4L$, is that of the fifth harmonic. The closed pipe can emit any frequency which is an odd integer times the fundamental frequency. Harmonics which have frequencies 2, 4, 6, . . . times the fundamental frequency cannot be sustained in a closed organ pipe. *Closed organ pipes emit only odd harmonics.*

The fundamental depends on the velocity of sound in the gas in the tube. If hydrogen is placed in the tube instead of air, the fundamental frequency is almost quadrupled. For any gas the fundamental depends upon the temperature, since the speed of sound is a function of temperature. As the temperature is raised, the frequencies emitted go up.

21.8 OPEN PIPES

The boundary condition for an open end requires that the oscillation amplitude be maximum. Since both ends of an open pipe are open, there is an antinode at each end. The wavelength of the longest possible standing wave is twice the length of the pipe, as shown in Fig. 21.6. Since $V = \nu\lambda$, the frequency ν_1 of the fundamental is

$$\nu_1 = \frac{V}{2L} \tag{21.10}$$

The fundamental of an open pipe has twice the frequency of the fundamental for a closed pipe of the same length. The next possible mode of vibration (Fig. 21.6) is that in which the pipe has a length equal to one wavelength. This gives rise to the first overtone (second harmonic), with a frequency

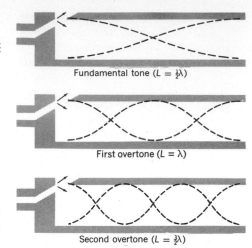

Fundamental tone ($L = \frac{1}{2}\lambda$)

First overtone ($L = \lambda$)

FIGURE 21.6
Displacement patterns for fundamental and first and second overtones of an open organ pipe.

Second overtone ($L = \frac{3}{2}\lambda$)

$$\nu_2 = \frac{V}{L} \tag{21.11}$$

For the second overtone, $\nu_3 = 3V/2L$, which is the third harmonic. The frequencies of the harmonics of the open pipe are in the ratio of the integers 1, 2, 3, 4, ... For an open pipe it is possible to have all harmonics; only the odd harmonics are emitted by a closed pipe.

Throughout this discussion we have assumed that an antinode occurs exactly at the open end of the pipe. Actually the antinode lies outside this open end by an amount which depends on the diameter of the pipe if round and on its shape and size if it is not round. The length must be corrected slightly for exact calculation of the frequencies.

21.9 RESONANCE

If two identical tuning forks are placed a few feet apart and one of them is set into vibration, the second fork also begins to vibrate (Fig. 21.7). This can be shown by stopping the vibrations of the first fork by grasping the tines. Energy has been transferred through the air from the first fork to the second. Because the second fork has the same frequency as the first, the compressions set out by the first fork arrive at just the right time to

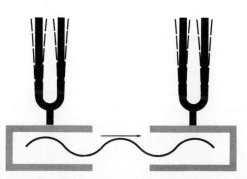

FIGURE 21.7
Resonance occurs when two oscillators with the same frequency interact so that the vibrations in one induce sympathetic vibrations in the other.

build up the amplitude of vibration in the second fork. The two forks are said to be in *resonance*. When two bodies are in resonance, a substantial amount of energy can be transferred to one of the bodies through a series of very small, but perfectly timed, impulses from the other. The amplitude is built up just as the amplitude of a child's swing can be built up by small, well-timed pushes.

When marchers cross a bridge, they are often commanded to break step because if a bridge structure happened to be resonant to the frequency of the steps, large and perhaps destructive vibrations could be set up. In automobiles very annoying noises are sometimes produced at certain speeds when some loose object happens to be resonant to a small impulse which is received with the proper timing. The designers of aircraft must be careful to avoid flutters and oscillations which may build up to intolerable levels by resonance.

On the other hand, resonances are sometimes very desirable. If we want a tuning fork to produce a louder sound, we can hold it over a tube which we gradually fill with water until the air column is resonant to the frequency emitted by the tuning fork. This occurs when the air column, closed at one end by the water, satisfies Eq. (21.8).

21.10 THE HUMAN VOICE

When we speak, the lungs, by their bellows action, force a stream of air between the vocal cords, two tightly stretched membranes in the throat, thereby setting them into vibration. The frequency of vibration can be changed by varying the tension in the vocal cords. The vibrations start a train of sound waves through the vocal passages. The tongue, the lips, and the cavities of the chest, nose, and throat interact with this wave-train. To prove that the quality of the sound is affected by resonant columns in the head, one need only hold one's nose closed while speaking.

The differentiation of speech sounds is nearly all accomplished by the mouth and by positioning the lips and tongue (Fig. 21.8). The *voiced sounds* include all the vowel and consonant sounds except *p, t, ch, k, f, s, th* (thin), and *sh*. The vocal cords do not enter into the production of these speech sounds, which arise from vibrations set up in the mouth itself.

Vocal sounds are transmitted through the air by exceedingly complex pressure waves. The amplitudes and frequencies of the various components present in speech sounds vary from one voice to another, but average speech includes frequencies from about 60 to 6,000 Hz. Most of the speech energy is carried by the vowel sounds. The power output of the normal human voice is only about 10 μW.

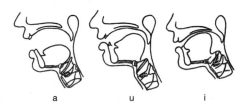

a   u   i

FIGURE 21.8
The position of vocal organs in uttering the vowels *a*, *i*, and *u*.

QUESTIONS

1. What evidence is available from everyday life to support the assertion that the speed of audible sounds is independent of frequency?
2. Upon what properties of a transmitting medium does the speed of sound depend? Explain qualitatively why these properties affect the speed.
3. How are stringed instruments tuned? Explain.
4. Why are the bass strings of a piano wrapped with a close helical winding of other wire?
5. When a liquid is being poured into a bottle, why does the pitch of the sound increase as the liquid level rises?
6. If the wave in an organ pipe is a sound wave, how does the sound get from within the organ pipe to your ear?
7. A violin and a pipe organ are tuned together; what happens to the frequencies if the temperature of the room rises from 20 to 27°C?
8. What sound sources have overtones which are not harmonics?
9. What is required to observe resonance between two mechanical systems?
10. Why do marchers break step when crossing a light bridge?
11. If a person takes a deep breath of helium and then speaks, how will the characteristics of his voice be altered? What would be the effect of inhaling carbon dioxide?
12. How are different frequencies produced in a bugle, which has an open air column of constant length?
13. How can the speed of sound in air be independent of the pressure when Eq. (21.3) tells us the speed is given by $\sqrt{1.4\,p/d}$?
14. How can the position of a gun be determined by using three listening posts and measuring the time of arrival at each post with accuracy?
15. Why is the speed of sound in a gas a little less than the average speed of the molecules?

PROBLEMS

Take the speed of sound in air to be 340 m/s or 1,100 ft/s.

1. What is the speed of sound in neon at 0°C and 1 atm pressure if 20.2 g of neon under standard conditions occupies a volume of 22.4 L? For neon $\gamma = 1.67$. *Ans.* 433 m/s

2. Find the speed of sound in hydrogen ($\gamma = 1.4$) at 17°C if the molecular weight of H_2 is 2.
3. Sound travels in fresh water at the rate of 1,450 m/s. What is the bulk modulus of elasticity of water. *Ans.* 2.1×10^9 N/m²
4. At what temperature would the speed of sound in air be 350 m/s if it is 331.5 m/s at 0°C?
5. What is the speed of sound waves in a brass rod for which Young's modulus of elasticity is 9.8×10^{10} N/m² and the density is 8,600 kg/m³? *Ans.* 3.38 km/s
6. What is the speed of sound waves in a steel rod for which Young's modulus of elasticity is 24×10^{10} N/m² and the density is 7,800 kg/m³?
7. A ship sends signals to a neighboring ship. The sound waves travel by two paths, one in air, and the other in seawater. The signals are heard on the neighboring ship at intervals 7 s apart. How far is it from one ship to the other if the speed of sound in seawater is 1,500 m/s? *Ans.* 3,080 m
8. An air-driven turbine breaks up the stream of air into 50 pulses per revolution. What frequency of sound will be heard when the turbine is rotating at the rate of 4,800 r/min? What is the wavelength of the sound waves generated?
9. (*a*) A timer sets his watch by the report of a gun 100 m away. What is the error due to the time required for the sound to travel from the gun to his ear? (*b*) A track worker strikes the steel rail of a railroad track with a hammer. The sound thus produced reaches an observer through the rail and through the surrounding air. The difference in time is 5 s. If the speed of sound in steel is 5,000 m/s, how far is the observer away from the worker? *Ans.* (*a*) 0.29 s; (*b*) 1,824 m
10. In pounding a spike a worker strikes a steel railroad track. A person 1.2 km along the track hears two reports from the pounding, one traveling through air and one through the rail. What time interval separates the reports if the speed of sound in steel is 5,000 m/s?
11. What is the velocity of a transverse wave in a string 3 m long with a mass of 2.5 g and subject to a tension of 4,800 N? What is the fundamental frequency of this string? *Ans.* 2,400 m/s; 400 Hz
12. A steel wire has a mass per unit length of 2.5 g/m and is 1.2 m long. It is stretched by a force of 784 N. Find the speed of transverse waves in this wire. What is the frequency of its fundamental vibration?
13. A string with a mass per unit length of 0.002 kg/m is stretched by the application of 320 N. What length of string will be required to produce a fundamental frequency of 600 Hz? What tension is required to give this string a fundamental frequency of 800 Hz? *Ans.* 0.333 m; 569 N
14. A stretched wire 2.4 m long has a fundamental

frequency of 300 Hz. What is the velocity of the wave in the wire? What is the wavelength of the fundamental vibration? What is the frequency of the first overtone? What is the frequency of the third overtone? What is the frequency of the third harmonic?

15. A steel piano wire is 0.6 m long and has a fundamental frequency of 440 Hz. If the mass per unit length is 2 g/m, find the tension in the wire. What is the speed of transverse waves in the wire? If the amplitude at the center of the wire is 0.5 mm, find the maximum acceleration of the midpoint. *Ans.* 558 N; 528 m/s; 3,820 m/s^2

16. A stretched string made of steel is vibrating at its fundamental frequency of 2,000 Hz. What is the fundamental frequency of a second string made from the same steel but with a diameter twice that of the original and a length half that of the original if it is stretched by twice the force of the original?

17. A copper wire is stretched between two points 0.5 m apart. It has a fundamental frequency of 400 Hz for transverse vibrations. If the wire has an area of 0.2 mm^2, find (*a*) the stretching stress in the wire, (*b*) the elongation of the wire due to this stress, and (*c*) the frequency of the first overtone. *Ans.* (*a*) 1.43 GN/m^2; (*b*) 5.96 mm; (*c*) 800 Hz

18. Water is poured into a long glass tube closed at one end. What length of air column must be left above the water level if the column is to resonate at a frequency of 256 Hz?

19. A glass tube open at both ends is so placed that one end is under water. The tube is adjusted until there is resonance when a sounding tuning fork is held above the open end. If the tuning fork has a frequency of 440 Hz, what is the shortest length of the tube for resonance? *Ans.* 0.193 m

20. A sounding tuning fork is held over a vertical glass tube into which water is poured slowly. The remainder of the tube is filled with air. What is the frequency of the tuning fork when the shortest column of air for resonance is 170 mm?

21. A closed organ pipe is 0.25 m long. Find the frequency of the fundamental and the frequency of the first two overtones. *Ans.* 340 Hz; 1,020 and 1,700 Hz

22. Calculate the lengths in meters of open and closed pipes for a fundamental frequency of 600 Hz.

23. An open organ pipe is 425 mm long. What is the wavelength of its fundamental vibration? What is the frequency of its fundamental vibration? What is the frequency of the first overtone? What is the frequency of the third overtone? What is the frequency of the third harmonic? *Ans.* 850 mm; 400 Hz; 800 Hz; 1,600 Hz; 1,200 Hz

24. Repeat Prob. 19 for a closed organ pipe 425 mm long.

25. The whistle of a steamer is in the form of a closed pipe 0.85 m long. Calculate the frequencies of the fundamental and the first three overtones. *Ans.* 100 Hz; 300 Hz; 500 Hz; 700 Hz

26. An open organ pipe is 1.65 ft long. Find the frequency of the fundamental and the wavelength of the first overtone. If this pipe is now closed at one end, find the frequency of the fundamental and the wavelength of the first overtone.

27. (*a*) Find the frequencies of the fundamental and first two overtones of an open pipe 340 mm long. (*b*) Repeat for a closed pipe. *Ans.* (*a*) 500 Hz; 1,000 Hz; 1,500 Hz; (*b*) 250 Hz; 750 Hz; 1,250 Hz

28. Show that the ratio of the speed of sound in a gas to the effective (rms) speed of the molecules is $\sqrt{\gamma/3}$.

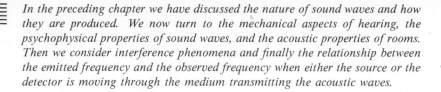

22

SOUND WAVES
AND HEARING

In the preceding chapter we have discussed the nature of sound waves and how they are produced. We now turn to the mechanical aspects of hearing, the psychophysical properties of sound waves, and the acoustic properties of rooms. Then we consider interference phenomena and finally the relationship between the emitted frequency and the observed frequency when either the source or the detector is moving through the medium transmitting the acoustic waves.

22.1 PITCH, LOUDNESS, AND QUALITY

The word *sound* is used with two related but separate and distinct meanings. Sometimes we use the word to mean the *sensation due to the stimulation of the auditory nerve centers* and at other times to mean the *longitudinal waves transmitted through elastic media*. To the psychologist "sound" is usually used in connection with hearing, whereas much of the time in physics "sound" is a type of wave motion, even though it may not be heard. For example, if we oscillate a wire at a frequency of 30,000 Hz, human ears cannot hear the waves produced. In the sense that longitudinal waves exist in the air there is a sound, but from the point of view of hearing there is none.

To a listener sound is ordinarily characterized by the psychophysical properties *pitch*, *loudness*, and *quality*. These subjective characteristics are, of course, related to physical properties of the waves, namely, the frequencies present, their amplitudes, and their phase relations.

In general, the pitch of a sound depends upon the frequency of the *fundamental*, or the lowest frequency present in the sound wave. As the fundamental frequency of a sound source is increased, the pitch is raised. A simple experiment showing a direct relation between pitch and frequency can be performed with the aid of a siren which consists of a disk with a number of holes uniformly spaced on concentric circles. If a jet of air is directed against the holes while the disk is in rotation, a puff goes through each hole as it passes the jet. When the number of puffs per second is increased by speeding up the disk, the pitch becomes higher. Although pitch is determined primarily by the frequency, it depends to some extent on loudness and quality as well.

Loudness describes the strength of the auditory sensation produced by a sound. In addition to depending on the intensity and the frequency composition of the sound, it also depends on the auditory acuity and experience of the individual listener. For a given frequency the loudness of a sound is closely related to the intensity; it increases roughly as the logarithm of the intensity. For a given intensity the loudness depends rather sensitively on the frequency.

Sounds which have the same loudness and pitch may have very different *tone qualities*. For example, if a piano and a trumpet play the same note at the same loudness, it is easy to distinguish between them. The reason is that neither sound source produces a single frequency; each sends out a group of frequencies. As we have seen, the lowest frequency is called the *fundamental*; it is this frequency which primarily determines the pitch. In addition to the fundamental there are present higher frequencies, called *overtones*. When two notes differ in quality, or *timbre*, the frequencies, phases, and relative intensities of the various overtones are typically different; further the temporal evolution of the tone is important in determining its quality. Figure 22.1 shows how the charac-

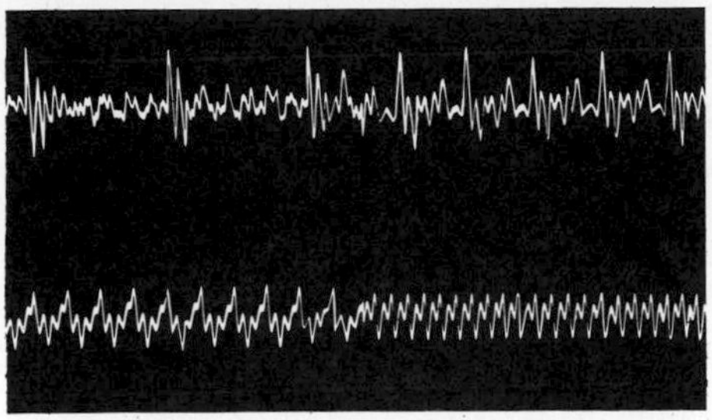

FIGURE 22.1
Sound waves emitted by a trumpet
blown in different ways.

teristics of the sound emitted by a trumpet change according to the way
the musician blows it.

22.2 THE EAR

The process of hearing begins when sound waves impinge upon an ear.
The human ear (Fig. 22.2) can be divided into three distinct parts: the
external ear, the middle ear, and the inner ear. The external ear collects
the sound waves and directs them along the auditory canal to the *eardrum*, which is caused to vibrate. These vibrations are carried across the
middle ear by three small bones, called the *hammer*, the *anvil*, and the
stirrup because of their shapes. Vibrations of the eardrum are passed
from hammer to anvil to stirrup, which in turn is attached to the "oval
window" separating the middle ear from the inner ear. The principal
parts of the inner ear are the *cochlea* and three *semicircular canals*. The
cochlea, shaped somewhat like the shell of a snail, is filled with fluid,
which is set into vibration by the movement of the oval window. Inside
the cochlea are about 2.5 turns of the basilar membrane, to which are
connected the roughly 30,000 nerve endings which initiate signals to the
brain. Figure 22.3 indicates the location of the nerve endings which show
maximum response to a pure tone of 700 Hz and moderate loudness. The
semicircular canals are not part of the hearing mechanism but are vital in
keeping one's balance.

The human ear is a remarkably sensitive detector of sound waves. At

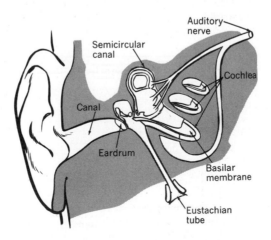

FIGURE 22.2
Anatomical structure of the human ear.

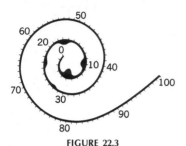

FIGURE 22.3
Parts of the cochlea where maximum response occurs for a pure tone of frequency 700 Hz at moderate loudness.

3,000 Hz the faintest sound the ear can hear has pressure variations of about 20 μN/m², which corresponds to a displacement amplitude of less than 10 pm, or about one-tenth the diameter of a molecule! Yet it can also hear sounds with pressure variations a million times this great.

The loudness of a sound, as judged by the ear, is approximately proportional to the logarithm of the intensity. This fact has led physicists to measure the intensity levels of sounds on an arbitrary logarithmic scale. The intensity level n of a sound wave *in decibels* (dB) is defined by the equation

$$n = 10 \log \frac{I}{I_0} \qquad (22.1)$$

where I_0 is a reference level, taken as 1 pW/m², which is roughly that of the weakest sound which can be heard. For a sound which has an intensity level $n = 60$ dB, the logarithm of I/I_0 is 6, so $I = 10^6 I_0$. In other words a 60-dB sound level corresponds to an intensity 1 million times that of the reference level. For ordinary speech, n ranges from 30 to 70 dB, while for loud music it may reach 100.

The lowest sound level which can produce an audible sound depends on the frequency. The range of intensity levels and the frequencies which can be heard by a typical ear are shown in Fig. 22.4. A sound more intense than that corresponding to the upper curve produces pain rather than hearing.

22.3 MUSICAL SCALES

Some combinations of sounds are pleasing to the ear; others are not. What is felicitious depends in no small measure on the training of the listener. As long ago as 530 B.C. Pythagoras experimented with stretched strings and found that simultaneous notes from strings were harmonious when the ratios of the frequencies were the ratios of two small integers. When one frequency is double another, we say they differ by one *octave;* this ratio of 2 to 1 is a pleasing one. On the other hand, ratios such as 100 to 81 are ordinarily displeasing, at least in part because of the existence of unpleasant beat notes. Modern occidental music is written with scales which consist of notes with frequencies related by fairly simple ratios. We call the ratio of the frequencies of two sounds the *musical interval* of the two notes. The diatonic C major scale is an example of a typical scale

FIGURE 22.4
Chart of intensity levels, showing the threshold of audibility and the threshold of feeling. Only 1 percent of a large number of people tested had the auditory acuity indicated by the lower solid line, while 50 percent could hear intensities represented by the dashed line.

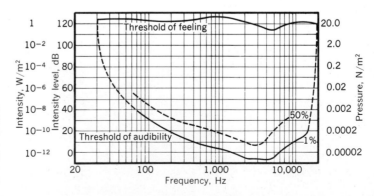

TABLE 22.1
Diatonic C Major Scale, Frequency in Hertz (Based on A = 440 Hz)

Major scale note	C	D	E	F	G	A	B	C'
Name	Do	Re	Mi	Fa	Sol	La	Ti	Do
Frequency	264	297	330	352	396	440	495	528
Interval relative to C	1	$\frac{9}{8}$	$\frac{5}{4}$	$\frac{4}{3}$	$\frac{3}{2}$	$\frac{5}{3}$	$\frac{15}{8}$	2
Musical interval		$\frac{9}{8}$	$\frac{10}{9}$	$\frac{16}{15}$	$\frac{9}{8}$	$\frac{10}{9}$	$\frac{9}{8}$	$\frac{16}{15}$
Musician's interval		Major tone	Minor tone	Semi-tone	Major tone	Minor tone	Major tone	Semi-tone

which is rather commonly used. The relationships between the notes in this scale are shown in Table 22.1.

A piano can be tuned to play a diatonic scale in any one key, but then the frequencies are not satisfactory for other keys. In order for a piano to be able to play in a number of different keys, it is customary to make certain compromises in which the piano is not tuned perfectly for any key. The compromise usually adopted is known as the *equal-tempered scale*. In one octave of this scale there are 13 notes and 12 intervals; there is a constant ratio between the frequencies of adjacent notes. This ratio is the twelfth root of 2, or 1.05946. The standard frequency of this scale is A = 440 Hz, so that this note agrees with that on the diatonic C major scale in Table 22.1. No other note in the octave has exactly the same frequency on the two scales.

The physics of music and musical instruments has many fascinating sides, delightfully presented in A. H. Benade, "Horns, Strings, and Harmony" (Anchor Books, Doubleday & Company, Inc., Garden City, N.Y., 1960).

22.4 THE REFRACTION OF SOUND

When sound waves pass from one medium to another, there is usually a change in the velocity of the sound. If the incident wavefronts meet the surface of separation obliquely, i.e., the angle of incidence is not zero, the direction of propagation changes as the waves enter the second medium, in accord with the law of refraction (Sec. 20.8).

Since the speed of sound in warm air is greater than it is in cold air, the direction of propagation of the sound changes continuously as it passes from air at one temperature to air at a different temperature. If the air is at rest and the temperature and density are uniform, a wavefront from a point source on the surface of the earth is spherical and the sound travels in straight lines. When the air at the ground is warmer than it is at higher altitudes, the speed of the sound is greater at the surface of the earth and a wavefront is no longer spherical. Since the direction of propagation is perpendicular to the wavefront, the sound is deflected upward (Fig. 22.5) and it cannot be heard for as great distances as it could without this distortion. When air at the ground is colder than at higher altitudes, the sound travels more slowly near the ground. The sound is deflected downward, and the distance at which the sound can be heard is increased. This sometimes happens over a lake at the end of a hot day.

When the wind is blowing, the speed of the sound with respect to the

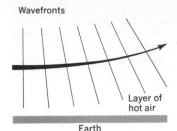
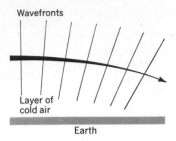

FIGURE 22.5
Wavefronts are deflected upward when air at the ground is warmer than air higher up and downward when air at the ground is cooler.

earth is decreased in the direction from which the wind comes and increased in the direction toward which the wind is blowing. Usually near the earth, the higher the altitude, the greater the velocity of the wind and the greater the change in the speed of the sound with respect to the earth. This unequal change in speed causes a distortion in the wavefront. On the windward side of the source the speed is greater at the ground than at points above the ground (Fig. 22.6). This inequality causes the wavefront near the ground to be inclined to the vertical and the line of propagation to be directed upward from the earth. On the side of the source toward which the wind is blowing, the speed near the ground is less than at higher altitudes. In this case, the direction of propagation is bent toward the ground, making it possible for the sound to be heard at greater distances.

22.5   REFLECTION OF SOUND

If an observer stands some distance in front of a cliff and produces a sound, the sound is returned to him with its characteristics essentially unchanged. If the observer is 340 m from the cliff, it takes about 2 s for the sound to return to him. We call the reflected sound an *echo*. The roll of thunder is due to the reflection of the original sound by various bodies including clouds. These echoes reach the observer at different times and produce the rolling continuation of the sound.

When an orchestra is to play out of doors, a large reflecting shell is often provided behind it so that sound waves are reflected toward the audience, thus greatly increasing the loudness.

Sound waves are reflected when they reach the walls of a speaking tube, which thereby prevents them from spreading out. Consequently, the intensity of the sound does not decrease appreciably as the wave advances, and the sound can be heard with only slightly diminished intensity at the other end of the tube many yards from the speaker. In an ear trumpet the waves entering the wide end are gradually diminished in area by reflections by the wall until, at the small end, most of the energy of the incident waves is concentrated over a small area. As a consequence, the intensity is greatly increased.

FIGURE 22.6
Wavefronts change shape and propagation direction in a wind.

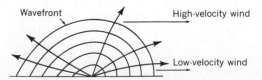

22.6 ARCHITECTURAL ACOUSTICS

When sound waves are produced in a room, they spread out until they strike the walls, ceiling, or floor. Here they are partly reflected, partly absorbed, and partly transmitted. A hard smooth wall reflects most of the sound; it transmits and absorbs little. On the other hand, a porous, soft material absorbs most of the sound energy and reflects little.

If a steady sound source is maintained for some time in a room, the sound level in the room builds up, as shown in Fig. 22.7. After a short time a state of equilibrium is reached in which the energy lost each second through absorption is equal to the sound power provided by the source. When the source is shut off, the intensity of the sound dies down, as indicated in the second half of the curve in Fig. 22.7.

The echoing and reechoing of sounds in a room because of repeated reflections is known as *reverberation. The reverberation time of a room is defined as the time required for the sound level to fall 60 dB after the sound source is shut off.* A drop of 60 dB means that the intensity falls to one-millionth of its original value. If the reverberation time of a room is too long, the sound waves bounce back and forth many times; a speaker's words become blurred because these reflected waves blend with the new sound waves. Such a blending effect is familiar to anyone who has sung in a hard-walled shower room with a long reverberation time. On the other hand, if the reverberation time is too short, sounds are thin and weak; we say the room is "dead." In general, a somewhat longer reverberation time is desirable for music than for speech. Reverberation times of about 1 s for a small room and 1.5 to 2 s for a larger one are desirable for ordinary uses. For making phonograph recordings, a very dead room with virtually no reverberation is used because the desired reverberation properties are supplied by the room in which the listener sits.

The reverberation time of a classroom or auditorium can be estimated by the use of a formula due to Sabine, a pioneer in the field of architectural acoustics. He found that the reverberation time t_R in seconds for a typical room was given by

$$t_R = 0.16 \frac{V}{A} \tag{22.2}$$

where V is the volume of the room in cubic meters and A is the total *absorbing power* of the room and its contents. The total absorbing power of the room is found by adding the contributions of all sound-absorbing surfaces, the contribution of each surface being the product of the

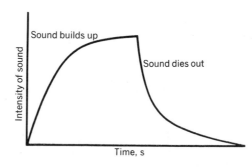

FIGURE 22.7
Rise and decay of sound in a room.

exposed area in square meters and the appropriate absorption coefficient. The absorption coefficient, in turn, is the fraction of the incident sound wave energy removed either by absorption or transmission. It is 1.00 for an open window because all the incident energy goes on through and is thus removed from the room. The absorption coefficients of a number of typical materials are listed in Table 22.2.

□ **Example**   A shower room has the dimensions 5 by 4 by 3 m. All the walls are of tile, and the door has the same absorption coefficient as tile. Find the approximate reverberation time if one man is showering and singing in this room.

$A = (20 + 20 + 12 + 12 + 15 + 15) \times 0.03 +$ the absorbing power of the man, which we shall take to be equivalent to 0.25 m² of open window (at least half the absorption of a person is associated with his clothing).

$$A = 2.82 + 0.25 = 3.07$$

$$t_R = 0.16 \frac{V}{A} = 0.16 \times \frac{60}{3.07} = 3.1 \text{ s} \qquad \square$$

□ **Example**   A classroom has a volume of 400 m³. There are exposed 600 m² of tile and 500 m² of wood in chairs, doors, and woodwork. There are 30 people in the room. Estimate the reverberation time.

$$A = (600 \times 0.03) + (500 \times 0.05) + (30 \times 0.5 \times 1)$$
$$= 18 + 25 + 15 = 58$$
$$t_R = 0.16 \times \tfrac{400}{58} = 1.1 \text{ s} \qquad \square$$

22.7   INTERFERENCE OF SOUND WAVES

When two trains of waves pass the same point, the displacement of a particle at this point is the resultant of the displacements which each of the two wavetrains would produce if it acted alone. If the two trains have the same wavelength and direction, the resultant amplitude depends on the amplitude of the two waves and on the phases of the two component disturbances. If the two arrive at the same phase, they reinforce each other, and we have constructive interference; if they arrive one-half wavelength out of phase, the resultant amplitude is the difference between the individual amplitudes. If these amplitudes are equal, the waves cancel each other and we have total destructive interference.

The interference of sound waves can be demonstrated with the apparatus of Fig. 22.8. Sound waves from a source O travel to point D by separate paths ACD and ABD. If the lengths of the two paths are equal,

TABLE 22.2
Sound-Absorption Coefficients

Open window†	1.00
Brick wall	0.02
Clay tile	0.03
Concrete	0.02
Glass	0.03
Hair felt	0.40
Heavy curtains	0.50
Perforated acoustic ceiling	0.60
Plaster	0.03
Wood	0.05

† Absorbing power of typical person is 0.5 m² of open window.

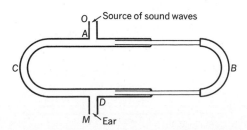

FIGURE 22.8
Apparatus for showing interference of two sound waves which have traveled paths of different lengths.

the waves arrive at the ear in phase and we have constructive interference. If the path *ABD* is increased in length, the waves travel different distances and the resultant amplitude is less than when the interference was constructive. When path *ABD* is one-half wavelength longer than path *ACD*, the waves are one-half wavelength out of phase and destructive interference results. As the path *ABD* is increased still more, the amplitude of the resultant disturbance grows until we have constructive interference once more when path *ABD* is one wavelength greater than *ACD*. In general, there is destructive interference when path *ABD* exceeds *ACD* by an odd number of half wavelengths and constructive interference when the paths differ by an integral number of wavelengths.

22.8 BEATS

When two steady sound sources of the same frequency are activated together, the intensity of the sound at a given listening point remains constant. If the frequencies are slightly different, however, the intensity of the sound fluctuates. There are bursts of louder sound with comparative silence between them. Each burst occurs when the disturbances from the two sources reinforce one another, while the periods of relative silence occur when the two waves interfere destructively. The fluctuations in intensity when two sound sources of slightly different frequency interfere are called *beats*.

The origin of beats is as follows. Suppose that at a certain instant compressions from both sound sources arrive simultaneously (*A* of Fig. 22.9). The amplitude of the resultant disturbance is large, and the intensity relatively high. A short time later the more rapidly vibrating source is one-half vibration ahead of the other, and a compression from one source arrives at the same time as a rarefaction from the other (*B* of Fig. 22.9). The two disturbances interfere destructively, and a minimum intensity results. A little later the more rapidly vibrating source has picked up a full vibration, and once again an intensity maximum is observed.

For two sounds differing in frequency by 1 Hz there is reinforcement once each second, and we observe one beat per second. As the frequency difference is increased, a higher beat rate is produced. In general, the number of beats observed per second is equal to the difference between the frequencies of the two sound sources:

$$\text{Number of beats per second} = N = \nu_1 - \nu_2 \qquad (22.3)$$

where ν_1 and ν_2 are the frequencies of the two sources.

Beats are used in tuning string instruments. As two strings are brought more and more nearly into unison, the number of beats per second

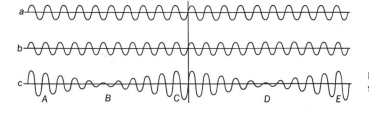

FIGURE 22.9
Beats arise from the superposition of sound waves differing slightly in frequency.

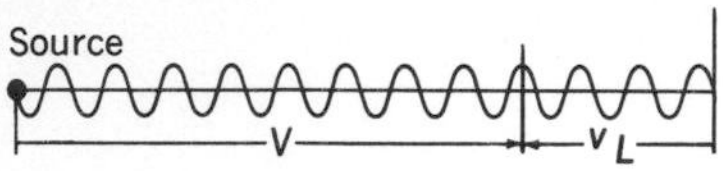

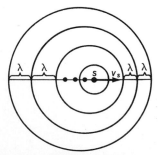

FIGURE 22.10

A listener moving with speed v_L toward a sound source passes in 1 s all the waves included in the distance numerically equal to $V + v_L$, where V is the speed of the sound.

becomes less; when no beats are observed, the strings have the same frequency.

22.9 THE DOPPLER EFFECT

As an automobile traveling at high speed and sounding its horn passes, the pitch heard by a pedestrian drops sharply. When a source of sound is moving toward an observer or an observer toward a sound source, the pitch heard is higher than the normal pitch. When the source moves away from the observer or the observer from the source, the pitch is lowered. The fact that the frequency received by an observer is different than the frequency emitted by the source whenever either is moving relative to the other is known as the *doppler effect*, named in honor of the Austrian physicist Christian Doppler, who proposed the theory in 1842. As Doppler correctly foresaw, light waves also exhibit the doppler effect although the mathematical relationships for light are not quite the same as the ones he derived for sound waves (see Sec. 29.4).

Consider first the case in which the observer approaches a sound source at rest relative to surrounding air. Then the observer passes more waves each second than he would if he were at rest (Fig. 22.10). Let V represent the speed of sound, and v_L the velocity of the listener toward the sound source. In 1 s the source sends out ν waves. In this same time the observer passes all the waves included in a distance $V + v_L$. The frequency ν_L heard by the listener is the number of waves he passes each second; thus $\nu_L = (V + v_L)/\lambda$, where λ is the wavelength. Since $\nu = V/\lambda$,

$$\frac{\nu_L}{V + v_L} = \frac{\nu}{V} \tag{22.4}$$

If the observer is moving away from the source, v_L is negative and the observer frequency ν_L is lower than ν.

If the source is moving (relative to the surrounding air) toward a stationary listener (Fig. 22.11), the wavelength of the sound waves in the air is different in different directions. Let ν be the frequency of the waves passing a point in the air and ν_S the frequency emitted by the source. In 1 s the sound waves moving to the right go a distance $V - v_S$ relative to the source, which has emitted ν_S waves in this second. Therefore the wavelength in air is $(V - v_S)/\nu_S = V/\nu$, and

$$\frac{\nu_S}{V - v_S} = \frac{\nu}{V} = \frac{\nu_L}{V} \tag{22.5}$$

since the listener hears the frequency of the waves passing in the air. When the source is moving away from the listener, v_S is negative and the pitch heard is reduced. If both source and observer are moving relative to the air, we write

$$\lambda_{\text{air}} = \frac{V}{\nu} = \frac{V - v_S}{\nu_S} = \frac{V + v_L}{\nu_L} \tag{22.6}$$

or

$$\frac{\nu_L}{V + v_L} = \frac{\nu_S}{V - v_S} \tag{22.6a}$$

Note that both v_S and v_L are measured *relative to the air; both are positive when the source is moving toward the observer and the observer toward the source.*

FIGURE 22.11

When a sound source is moving relative to the air, the wavelength is smaller in the direction of motion and larger in the opposite direction.

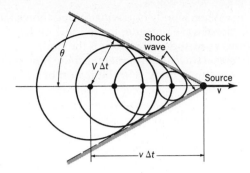

FIGURE 22.12
Wavefronts produced by a source moving faster than sound.

22.10 SUPERSONIC VELOCITIES AND SHOCK WAVES

When a body travels with a speed greater than that of sound, it is said to be *supersonic*. Figure 22.12 shows a source moving faster than the speed of sound. We observe that the object is beyond the spherical sound waves which it sent out a short time before. However, there exists a surface tangent to all these sound waves which, by Huygens' principle, gives us the position of a compressional wave. This wave accompanying a body traveling at a speed in excess of the speed of sound is called a *shock wave*. Figure 22.13 shows a shock wave accompanying a bullet traveling at supersonic speed. Whenever a body travels through a medium at a speed greater than that with which the resulting disturbance is propagated, a similar wave is observed. A common example is the bow wave from a speedboat.

The angle between the direction of motion of the source and the shock wave permits us to compute the velocity of the source, provided the velocity of sound in the medium is known. From Fig. 22.12 we see that $\sin \theta = V/v$, where v is the velocity of the source and V the speed of the shock wave or *shock front*.

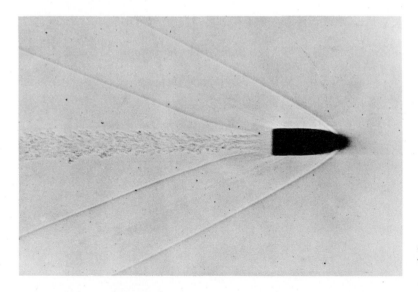

FIGURE 22.13
Compressional shock wave traveling with a bullet. (*Olin Mathieson Chemical Corporation.*)

When a body is moving through the air at subsonic speeds, a compressional wave precedes it and some of the air particles are moved out of the way. At supersonic speeds the body is traveling faster than the compressional wave. Consequently, there is no preparation in the medium for the oncoming body, and the region of sudden compression which we have called the *shock wave* is produced. A great deal of energy may be associated with this shock wave—energy which comes from the object passing through the air. An aircraft traveling at supersonic speed builds up a substantial shock wave. Where the shock wave reaches the earth, it may break windows and cause other damage.

Because of the buildup of a shock wave as an aircraft reaches the speed of sound, the drag on the aircraft increases markedly in this region. In discussing the drag under these conditions, it is customary to compare the speed of the aircraft with the local speed of sound. The latter depends, of course, on the temperature and is substantially lower at high altitudes, where the temperature is low. The ratio of the velocity of the body to the local velocity of sound is called the *Mach number*. Mach 5 means 5 times the velocity of sound. The speeds of missiles and high-velocity aircraft are often quoted in Mach numbers.

If a missile or aircraft is passing through the air, the drag can be written (see Sec.13.9)

$$F_{\text{drag}} = C_D S(\tfrac{1}{2}dv^2) \tag{22.7}$$

where S is the cross-sectional area normal to the airstream, d the density of the air, v the velocity of the object, and C_D an empirical factor called the *drag coefficient*. A typical plot of drag coefficient as a function of Mach number for a missile is shown in Fig. 22.14. The drag coefficient depends on the velocity of the moving body and on its size, shape, and smoothness. As one approaches Mach 1, the drag increases rapidly and much faster than v^2. This region of rapid increase in drag coefficient is called the *transonic region* because it occurs where the speed passes from below to above sonic velocity. The drag coefficient decreases above Mach 1.5, which indicates that once again the drag force is increasing less rapidly than v^2.

When an atomic bomb is exploded in the air, a tremendous shock wave is set up. The shock front represents a moving wall of highly compressed air which may blow down buildings and cause other damage. It has been reported that at a distance of 500 m from the detonation point of a nominal atomic bomb similar to that used at Nagasaki, the pressure of the

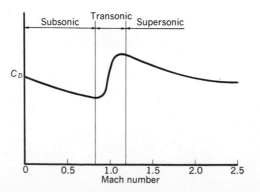

FIGURE 22.14
The drag coefficient of a particular missile as a function of the missile speed in Mach numbers.

shock wave is more than 4 times atmospheric and the speed of the shock wave is about 650 m/s. The tremendous compressional shock wave is followed by a rarefaction or *suction phase*, which lasts much longer than the pressure phase.

QUESTIONS

1. Why is thunder often so prolonged even though the single lightning flash which produced it is almost instantaneous?
2. Why doesn't the inverse-square law apply exactly to the decrease in sound intensity with distance? Is it possible for the intensity to decrease less rapidly than the inverse-square law predicts? Under what circumstances?
3. Why is it often true that sounds can be heard for particularly great distances at night or after a rain?
4. Draperies and furniture often improve the acoustics of a room. Why? On what factors do the acoustics of the room depend?
5. How may beats be used in tuning a musical instrument such as a piano or a harp? How is such an instrument tuned?
6. Beats are sometimes called *difference tones*. Why is that term used?
7. If two sounds are an octave apart, how are their frequencies related?
8. What are the advantages and disadvantages of the equal-tempered scale?
9. What is meant by the statement that the sound level in a room is 40 dB? 80 dB?
10. Sound waves with intensities greater than 1 W/m² often cause damage to human ears. To how many decibels does this intensity correspond?
11. If two tuning forks have the same frequency, how could you arrange to produce a fluctuating beat frequency from them?
12. How many times more intense is a 60-dB sound than a 40-db sound?
13. Does wind velocity play any role in the doppler effect? Explain quantitatively.
14. If an observer or a source is moving perpendicular to the line joining them, will there be a doppler shift in the frequency of acoustic waves? Explain.
15. Can an airplane designed to fly at supersonic speed get its thrust from a propeller? Explain.

PROBLEMS

Take the speed of sound in air to be 340 m/s or 1,100 ft/s. The logarithm of 2 on base 10 is 0.30103,

so doubling the intensity of a sound corresponds to increasing its level by 3 dB to a good approximation.

1. Find the intensity in watts per square meter corresponding to sound levels of 30, 43, 66, and 80 dB relative to $I_0 = 10^{-12}$ W/m² as the zero reference level.
Ans. 10^{-9} W/m²; 2×10^{-8} W/m²; 4×10^{-6} W/m²; 10^{-4} W/m²
2. The threshold of pain involves an intensity level of roughly 120 dB and higher intensities damage ears, so earmuffs are desirable. Find the intensity in watts per square meter 30 m from a jet engine, where the sound level is 133 dB.
3. If a loudspeaker is regarded as a point source radiating equally in all directions, what acoustic power must it develop to produce a sound level of 70 dB at a distance of 100 m? *Ans.* 1.26 W
4. What is the level (relative to the reference level of 10^{-12} W/m² as 0 dB) of a sound of intensity 10^{-4} W/m²? 2×10^{-2} W/m²?
5. Acoustic treatment of certain surfaces in a foundry reduced the sound level by 23 dB. How many times greater was the intensity before the treatment than after? *Ans.* 200
6. If the intensity due to five violins is 5 times that due to a single violin, how many decibels is the sound level raised by having five violins playing rather than a single one? (The logarithm of 5 to the base 10 is 0.699.)
7. A major diatonic scale is based on F = 360 Hz. Find the frequencies of the other seven notes. Show that for this scale the intervals D/C = $\frac{10}{9}$ and E/D = $\frac{9}{8}$. How do these intervals compare with those for the C major scale of Table 22.1? In view of the fact that the intervals for the key of C and for the key of F are different, how does one make a piano which can play reasonably well in both keys?
Ans. 405, 450, 480, 540, 600, 675, and 720 Hz
8. Find the notes of a major diatonic scale based on C = 256 Hz.
9. The vertical walls of a canyon are 540 m apart. A girl in the canyon fires a gun and hears the echo from the farther wall 2 s after the echo from the nearer wall. How far is she from the nearer wall?
Ans. 100 m
10. A lecture room has a volume of 15,000 m³. If it has 1,100 m² of acoustic ceiling, 2,400 m² of plaster, 1,000 m² of concrete, and 4,000 m² of wood,

find the reverberation time when the room is empty and when it holds 400 people.

11. An auditorium is essentially a rectangle 50 by 30 by 12 m. It has a perforated acoustic ceiling, concrete floor, and plaster walls. It contains 800 seats, each of which is equivalent to 1 m² of wood. Find the approximate reverberation time of the auditorium empty and full.

Ans. 2.8 s; 2.0 s

12. Both loudspeakers of a stereo hi-fi system are sending out, in phase, a pure frequency of 800 Hz. A listener originally equidistant from the speakers moves to one side until the note fades to a minimum loudness. If he is then 2.98 m from the closer speaker, how far is he from the farther one?

13. An oscillator emits a sound with a wavelength of 0.17 m which is divided into two parts in the apparatus of Fig. 22.8. The two parts travel the same distances, so that constructive interference is experienced. In order to eliminate the sound, one path is increased. What is the shortest distance this path can be lengthened to satisfy this condition? If it is lengthened more, the intensity increases and then decreases. What total change in path is necessary to obtain the second case of destructive interference? What total change in path is necessary to obtain the third case of destructive interference?

Ans. 85 mm; 255 mm; 425 mm

14. A sound wave travels through two branches of a tube which are 1.10 and 1.35 m long. List the three lowest frequencies which would suffer destructive interference if the waves were recombined after traveling through the branches.

15. Two speakers of a hi-fi set are 3 m apart. When they are playing a pure single frequency, there is a minimum intensity 4 m immediately in front of each speaker. As one moves along the line 4 m from the line of the speakers, the intensity reaches a single maximum between these two minima. Calculate the frequency. *Ans.* 170 Hz

16. Two strings *A* and *B* originally produced the same frequency. The tension of *B* was released slightly, thereby reducing the frequency, and the strings then produced seven beats per second when sounded together. If the frequency of *A* is 440 Hz, what is the frequency of *B*?

17. Two open organ pipes of lengths 340 and 345 mm are sounded simultaneously. How many beats per second are produced? *Ans.* 7.2

18. Two closed organ pipes 0.5 and 0.51 m long are sounded together and produce 3.5 beats per second. What is the velocity of sound in the air with which the organ pipes are filled?

19. Two automobile horns emit the same note frequency 250 Hz. If one of these horns is on a car approaching an observer at 108 km/h (30 m/s) and the other is on a car moving away from the observer at 108 km/h, calculate the frequency heard by the observer in each case.

Ans. 274 Hz; 230 Hz

20. A passenger on a train running at 34 m/s listens to a siren which has a frequency of 440 Hz. What is the apparent frequency of the sound when the train is approaching the siren? When it is going away from the siren?

21. A sounding object emits a sound of frequency 300 Hz. An observer heard it at a frequency of 280 Hz. (*a*) If the source is moving, what is its speed? (*b*) If the observer is moving, what is his speed? *Ans.* (*a*) 24.3 m/s; (*b*) 22.7 m/s

22. Find the frequency heard by a listener at rest when a sound source emitting a frequency of 1,000 Hz goes away at a speed of 25 m/s.

23. At what speed in meters per second must a source of sound approach an observer in order for the pitch of each note to be raised by a half tone, i.e., to $\frac{16}{15}$ of the original frequency?

Ans. 21.3 m/s

24. What is the apparent frequency heard by an observer at rest toward whom a police car is approaching at 30 m/s with its siren emitting a frequency of 240 Hz?

25. A passenger standing on the rear platform of a train notes that a warning bell at a grade crossing rings with an apparent frequency of 546 Hz as the train approaches and 474 Hz as the train moves away. Find the speed of the train and the actual frequency of the bell. *Ans.* 24 m/s; 510 Hz

26. A car moves away from a wall toward an observer with a speed of 30 ft/s. Find the beat frequency which the observer hears between a 180-Hz horn on the car and its echo from the wall.

27. A train approaching a tunnel in a vertical cliff perpendicular to the tracks sounds its horn, which has a frequency of 250 Hz. If the speed of the train is 25 m/s, what is the frequency reflected by the cliff? What frequency does the engineer hear reflected? *Ans.* 269.8 Hz; 289.7 Hz

28. A siren emitting a frequency of 1,000 Hz moves away from you toward a cliff at a speed of 20 m/s. Find the frequency of the sound you hear coming directly from the siren and the frequency you hear reflected from the cliff.

29. Find the angle between the shock wave and the path of a bullet which has a speed of 425 m/s.

Ans. 53.1°

30. Estimate the angle θ in Fig. 22.13 and calculate the approximate speed of the bullet.

FOUR

LIGHT

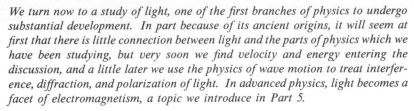

23
PROPAGATION AND REFLECTION OF LIGHT

We turn now to a study of light, one of the first branches of physics to undergo substantial development. In part because of its ancient origins, it will seem at first that there is little connection between light and the parts of physics which we have been studying, but very soon we find velocity and energy entering the discussion, and a little later we use the physics of wave motion to treat interference, diffraction, and polarization of light. In advanced physics, light becomes a facet of electromagnetism, a topic we introduce in Part 5.

One of our principal contacts with the world around us is through light. Not only are we personally dependent on light to convey visual information, but most of what we know about the stars and the solar system is derived from light waves impinging on our eyes and on optical instuments. Similarly, our concept of the structure of atoms comes largely from observing the radiations they emit. We have many instruments which use light to operate; important among them are the microscope, the telescope, and many kinds of spectrometers.

23.1 THE NATURE OF LIGHT

Much of the early knowledge of light dealt with its general behavior when it strikes materials: how it is reflected and how it is bent when it goes from one medium to another. The ultimate nature of light is not important in studying these phenomena, and the great science which we know as *geometrical optics* was well established before an understanding of the nature of light evolved.

Most objects emit no visible light; they are visible by light which is reflected from them. The sun is our chief source of light and heat, but many other bodies are *luminous* and serve as light sources. Any body heated to a sufficiently high temperature becomes self-luminous through *incandescence*. A gas through which an electric discharge is passed also emits light, the color depending on the gas involved. Entire panels of certain solid materials can be electrically excited to produce light through *electroluminescence*. Other solids emit light when bombarded by electrons or ultraviolet radiation; the television screen is an example of such a *fluorescent* source. A few chemical reactions result in the emission of visible light, even though the chemicals themselves are at a low temperature; the firefly is a *chemiluminescent* source.

During the eighteenth century supporters of Huygens argued that light was a wave motion, while followers of Newton held that light was a stream of particles. Early in the nineteenth century decisive experiments (Chaps. 27 and 28) showed that light exhibits the wave properties of interference, diffraction, and polarization, and for a hundred years it seemed that light was a classical wave phenomenon. Then evidence from blackbody radiation, the photoelectric effect, and x-rays established that light also has properties characteristic of classical particles. Now we hold that light is neither a pure wave phenomenon nor a pure particle phenomenon but electromagnetic radiation with both wave and particle properties. We also know now that the fundamental constituents of matter (such as electrons), once assumed to be classical particles, similarly display such wave characteristics as interference and diffraction.

The wave properties of visible light are associated with transverse electromagnetic disturbances having wavelengths between roughly 380 and 780 nanometers (1 nm = 10^{-9} m; the nanometer was formerly called the millimicron, abbreviated mμ). In some experiments the wave characteristics of the light dominate; in others the particle properties do. This *wave-particle duality* was a source of great confusion until the development of modern quantum mechanics.

In addition to visible light, there are radiations characterized by longer and shorter wavelengths than the eye can see but with physical properties identical with those of visible light. Sometimes the term *light* is used to denote only radiations which we can see, while at other times we use *light* in a broader sense to mean electromagnetic radiation, regardless of visibility. From the psychophysical point of view, "light" is radiant energy capable of producing a standard response in the human eye. In the broader physical sense, "light" is visible light plus invisible radiations having the same fundamental properties.

The relation between frequency ν, velocity c, and wavelength λ is the same for light as for sound:

$$c = \nu\lambda$$

Yellow light has been found to have a wavelength equal to about 590 nm. Taking the speed of light to be 3×10^8 m/s, the frequency of yellow light is

$$\nu = \frac{c}{\lambda} = \frac{3 \times 10^8 \text{ m/s}}{590 \times 10^{-9} \text{ m}} = 5.08 \times 10^{14} \text{ Hz}$$

The wavelength of light is usually expressed either in nanometers or in angstroms (1 Å = 10^{-8} cm = 10^{-10} m).

23.2 THE VELOCITY OF LIGHT

Light travels through empty space and through air with a speed of approximately 3×10^8 m/s (186,000 mi/s). In 1 s light travels a distance more than 7 times around the earth at the equator. It takes light a little over 8 min to reach us from the sun and 4 years from the next nearest star. Several methods have been devised for determining the speed of light.

The first evidence that the speed of light is finite came from the work of the Danish astronomer Römer about 1675. He studied the periods of Jupiter's moons, four of which are visible in a small telescope. The innermost of these moons revolves around Jupiter in about 42.5 h. Römer found that as the earth moved from *B* to *C* (Fig. 23.1) the measured periods of this moon were all somewhat longer than average; the periods were shorter than average while the earth moved from *G* to *H*. He concluded correctly that the longer periods were due to the fact that while the moon revolved around Jupiter, the earth was moving farther away. The shorter periods occurred when the earth moved closer to Jupiter during the revolutions of the moon. From his measurements Römer estimated that 22 min is required for light to travel the diameter *AD* of the earth's orbit. Subsequent measurements show that this time is 1,000 s.

About this same time the French astronomers Richer and Cassini measured the diameter *AD* of the earth's orbit. In 1678 Huygens divided the Richer-Cassini value of *AD* by Römer's transit time to obtain the first value of the speed of light.

One of the most famous determinations of the velocity of light was that of Michelson. Figure 23.2 is a simplified diagrammatic representation of

Jupiter and moon

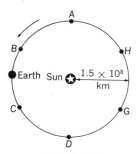

FIGURE 23.1
By studying the eclipses of Jupiter's moons Römer was able to estimate the time required for light to travel the diameter of the earth's orbit.

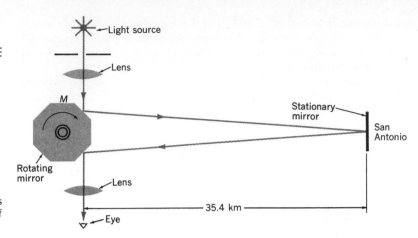

FIGURE 23.2
Simplified representation of Michelson's apparatus for measuring the speed of light.

his experiment. Light from an intense source is focused by a lens and reflected from an octagonal mirror which, for the moment, is at rest. This mirror is adjusted so that the beam of light travels to a stationary mirror and is reflected back to the octagonal mirror and finally to the eye of an observer. In Michelson's determination a stationary mirror was placed on Mount San Antonio and the octagonal mirror on Mount Wilson, in California. The distance between mirrors was carefully surveyed and found to be 35.4 km. Then the octagonal mirror was put into rapid rotation. Each time one of the eight sides passed the position indicated in the figure, a flash of light was sent to the stationary mirror on Mount San Antonio and returned. If, when it returned, the mirror was in some different position, the light was reflected in some direction other than that of the eye. However, when the speed of rotation of the mirror was made great enough for the mirror to rotate through exactly $\frac{1}{8}$ revolution (r) while the light made the trip to Mount San Antonio and returned, the light was reflected to the eye and observed. When this condition was satisfied, the time in which the mirror made $\frac{1}{8}$ revolution was exactly equal to the time for the light to travel 70.8 km.

□ **Example** In Michelson's experiment the light path from a rotating mirror to a fixed mirror (Fig. 23.2) is 35.4 km. If an octagonal mirror is rotated at the rate of 31,800 r/min, the light flashes are observed for the first time. Find the speed of light.

$$\text{Time for 1 revolution} = \frac{1}{31,800} \text{ min} = \frac{60}{31,800} \text{ s}$$

$$\text{Time for } \tfrac{1}{8} \text{ revolution} = \frac{60}{31,800 \times 8} \text{ s} = 2.36 \times 10^{-4} \text{ s}$$

In this time light travels 70.8 km, so that

$$c = \frac{70.8 \text{ km}}{2.36 \times 10^{-4} \text{ s}} = 3 \times 10^8 \text{ m/s} \ (186,000 \text{ mi/s}) \qquad \square$$

Many other methods of determining the speed of light have been devised, several of which are capable of far higher precision than the

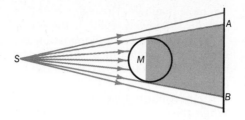

FIGURE 23.3
An opaque body casts a sharp shadow when illuminated by a point source. (There are weak diffraction effects, discussed in Chap. 27.)

Michelson measurement. The results of many determinations by many different physicists using a wide variety of methods lead us to believe that the speed of light in free space is 2.997925×10^8 m/s.

23.3 RECTILINEAR PROPAGATION OF LIGHT

Under ordinary circumstances light travels in straight lines, not bending appreciably around objects. Indeed, we depend on light to tell us when a meterstick is straight. The rectilinear (straight-line) propagation of light is shown by the fact that a small source of light casts sharp shadows. A point source S (Fig. 23.3) illuminates all points on a screen above A and below B. No light arrives at the screen between A and B because it is stopped by the opaque body M. The fact that the boundary of the shadow is sharp shows that light from S does not bend appreciably into the shadow of the opaque body.

When the source of light is not small, the boundary of the shadow is no longer sharp (Fig. 23.4a). Points on the screen above A and below D are fully illuminated by S. The screen between B and C receives almost no light. The screen between A and B and between C and D is partially illuminated and in a less dense shadow than the part between B and C. This outer shadow gradually shades off from complete shadow at B to complete illumination at A, giving the blurred appearance that character-

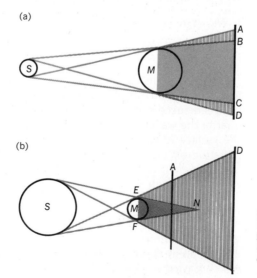

FIGURE 23.4
(a) The shadow of an opaque body illuminated by a source of significant angular dimension consists of an umbra and a penumbra. (b) Shadow cast by an object smaller than the source of the illumination.

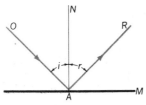

FIGURE 23.5

When a light ray is reflected from a plane mirror, the angle of incidence *i* is equal to the angle of reflection *r*.

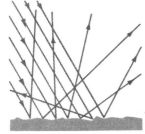

FIGURE 23.6

Irregular, or diffuse, reflection.

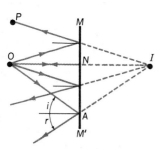

FIGURE 23.7

The image *I* of a point source *O* formed by reflection from a plane mirror is as far behind the mirror as the object *O* is in front.

izes most shadows. The region of total shadow is called the *umbra* and the partially shadowed region is the *penumbra*.

When the source of illumination is larger than the object casting the shadow (Fig. 23.4*b*), the only region of complete shadow is the cone *ENF* having the object *M* as a base. If the screen is placed at *A*, there is a small umbra surrounded by a large penumbra; if at *D*, the umbra is absent.

The best illustration of shadows on a large scale is found in eclipses. When the moon is interposed between the sun and the earth and its shadow falls on the earth, the sun is wholly or partly obscured by the moon and there is a total or partial eclipse of the sun. If the earth is sufficiently near the moon and passes through the complete shadow represented by the cone *FNE* (Fig. 23.4*b*), the eclipse is *total*. If, however, the earth is farther from the moon and passes through only the partial shadow, the eclipse is said to be *annular*. Sometimes the moon passes through the shadow of the earth; there is then an eclipse of the moon.

The pinhole camera illustrates the rectilinear propagation of light. If the pinhole is circular, the light from each point on the object strikes the photographic plate in the form of a small circle (or ellipse). For any other shape of hole each object point illuminates a small area shaped similar to the hole. If the pinhole is large, the areas overlap badly and the image is blurred. The smaller the hole, the sharper the image (unless the hole is so small that diffraction becomes important) but the longer it takes to expose the picture. Note that the image is *inverted*.

23.4   LAWS OF REFLECTION

When a beam of light traveling in a homogeneous medium comes to a second medium, part of the light is reflected. At a silvered surface the fraction of the light reflected is almost 100 percent, while at the surface of clear glass it is only a few percent. In Fig. 23.5, *OA* is a ray incident on a plane mirror and *AR* a reflected ray. The angle between *OA* and *AN*, the perpendicular or normal to the surface, is called the *angle of incidence;* the angle *NAR* between normal and reflected ray is the *angle of reflection*. The laws governing reflection (Sec. 20.7) are:

1. *The angle of incidence is equal to the angle of reflection.*

2. *The incident ray, the reflected ray, and the normal to the surface lie in the same plane.*

When light falls on a rough, opaque surface (Fig. 23.6), the incident light is scattered in all directions. If the surface is so smooth that the distances between successive elevations on the surface are less than about one-quarter the wavelength of the light, there is little random scattering, and the surface is said to be *polished*. Thus, a surface may be polished for radiation of long wavelength but not polished for light of short wavelength. A polished reflector or *mirror* is said to exhibit *regular* (or specular) *reflection*. Rough surfaces produce *diffuse reflection*.

23.5   THE PLANE MIRROR

When a luminous object is placed in front of a plane mirror *MM'* (Fig. 23.7), a point *O* on the object sends light in all directions. Several rays which leave this point and strike the mirror are shown in the figure. To an observer in front of the mirror *all* the reflected rays appear to come from the point *I* behind the mirror. The observer therefore sees a bright spot which appears to be behind the mirror and which we call the *image* of the object. For every point *O* on the luminous source there is a corresponding point *I* on the image. The image behind the mirror is a *virtual image*, because the light rays do not actually come from *I*. However, every ray of light which leaves a point on the object and is reflected from the mirror *appears to come* from the corresponding point on the image.

The ray *ON* (Fig. 23.7) falls normally on the mirror and is reflected directly back on itself; from the figure triangles *ONA* and *INA* are similar and equal. Therefore, the image point is as far behind the mirror as the object point is in front.

Reflection may also be considered in terms of wavefronts and Huygens' principle (Secs. 20.6 and 20.7). Such a treatment leads to the same conclusions so far as the location of the image of an object point is concerned. In terms of this description, spherical waves diverge from each point on the object, have their curvatures reversed (Fig. 20.10) at the reflecting surface, and thereby appear to have originated at the corresponding image point.

Figure 23.8 shows an object in front of a mirror. It is evident from the figure that the image is as far behind the mirror as the object is in front of the mirror and that image and object have the same size. However, *the image is reversed with respect to the reflecting plane*. This is shown more clearly in Fig. 23.9, in which the image *O'X'Y'Z'* of a set of axes *OXYZ* is indicated. Note that the axis *O'X'* points in a direction opposite to that of the axis *OX*. Such an image is *perverted*. Object and image are related in the same way that a right hand is related to a left. If a person salutes with his right hand before a plane mirror, his image appears to salute with the left hand.

23.6 THE CONCAVE SPHERICAL MIRROR

A concave spherical mirror is part of a spherical shell with its inner surface polished. The center *C* of the sphere from which the mirror was taken is the *center of curvature* of the mirror; the radius of the sphere is called the *radius of curvature* of the mirror. The middle point *V* of the mirror *M* (Fig. 23.10) is known as the *vertex*, and the straight line *CV* through the vertex of the mirror and its center of curvature *C* is the *principal axis*.

Consider a ray of light *AB* approaching the mirror parallel to the principal axis. At the mirror the ray is reflected with angle of reflection equal to angle of incidence. Since *AB* is parallel to the axis, angles *i* and ϕ

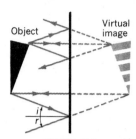

FIGURE 23.8
The image of an extended object formed by a plane mirror is the same size as the object and an equal distance from the reflecting surface.

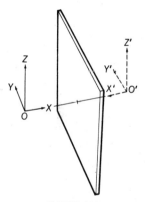

FIGURE 23.9
The image formed by a plane mirror is reversed with respect to the reflecting plane.

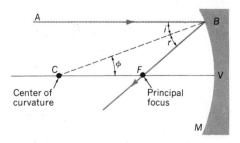

FIGURE 23.10
An incident ray parallel to the principal axis of a concave mirror is reflected so that it crosses the axis at the principal focus.

are equal. Triangle *FBC* is isosceles; therefore *BF* is equal to *FC*. If *B* is not too far from *V*, *BF* and *VF* are nearly equal. Hence, *VF* is nearly equal to *FC*; therefore, *F* is halfway between the mirror and the center of curvature *C*. This point *F* is known as the *principal focus* of the mirror. *The distance from the vertex V to the principal focus F is the focal length f of the mirror.* The focal length of a mirror is one-half the radius of curvature *R*:

$$f = \frac{R}{2} \tag{23.1}$$

Any ray of light which comes to the mirror parallel to the principal axis is reflected so that it passes through the principal focus F, provided the angle ϕ in Fig. 23.10 is small. If a ray of light passes along the line *FB*, it is reflected back in the direction *BA*. It is true in general that if a ray takes a certain path through an optical system in one direction, a ray sent backward along the path on which the original ray leaves traverses the same path and comes out along the line on which the original ray entered.

If an object *OP* is placed before a concave mirror (Fig. 23.11), we can locate an image of point *O* by drawing a suitable ray diagram. The ray *OA* parallel to the principal axis is reflected back through *F*, the principal focus. The ray *OF* strikes the mirror at point *B* and is reflected back parallel to the principal axis *CV*. A ray from *O* through *C* comes to the spherical surface along the radius and is reflected directly back on itself. All three reflected rays pass through a common point *I*. Indeed, if we draw any number of rays which leave point *O* and are reflected by the mirror, we find that *every ray which leaves point O and strikes the mirror passes through point I.* The point *I* is the image of point *O*. In this case it is a *real image*, since the rays of light actually cross at this point. If we put a piece of white paper at *I*, we see a real image of *O*. A similar construction can be made for other points on the object; for each object point we can thus locate the corresponding image point. In this case the image is *inverted*.

If an object were placed at *IQ*, the image would be formed at *OP*. The points *P* and *Q* are called *conjugate points;* an object at one has an image at the other.

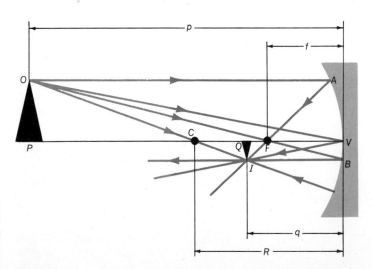

FIGURE 23.11
A concave mirror forms a real, inverted image of an object outside the principal focus. The ray reflected at the vertex is shown in addition to the three standard rays.

Let p, q, and f denote, respectively, the distances of object, image, and principal focus from the mirror. These three distances are related by a simple formula which can be derived with the aid of Fig. 23.11 as follows. The triangles OPV and IQV are similar; hence

$$\frac{OP}{IQ} = \frac{PV}{QV} = \frac{p}{q} \qquad (23.2)$$

Also, triangles OPF and BVF are essentially similar, so

$$\frac{OP}{BV} = \frac{PF}{FV} = \frac{p-f}{f} = \frac{p}{f} - 1 \qquad (23.2a)$$

Since $BV = IQ$, the left-hand members of Eqs. (23.2) and (23.2a) are equal and

$$\frac{p}{q} = \frac{p}{f} - 1$$

If we divide through by p and rearrange, we obtain

$$\frac{1}{p} + \frac{1}{q} = \frac{1}{f} \qquad (23.3)$$

□ **Example** An object is situated at a distance of 80 cm from a concave mirror of radius of curvature 60 cm. Find the position of the image.

$$p = 80 \text{ cm} \qquad f = \frac{60 \text{ cm}}{2} = 30 \text{ cm}$$

$$\frac{1}{p} + \frac{1}{q} = \frac{1}{f}$$

$$\frac{1}{80} + \frac{1}{q} = \frac{1}{30}$$

$$q = \frac{2{,}400}{50} = 48 \text{ cm} \qquad □$$

When an object is at a great distance, $q = f$, and the image is at the principal focus. As the object is moved toward the mirror, the image moves away. When the object distance is equal to the radius of curvature, the image distance is also the radius of curvature; object and image are identical in size. As the object is moved still closer to the mirror, the image moves away, until it reaches infinity when the object is at the principal focus. From Fig. 23.11 we observe that since the triangles OPV and IQV are similar,

$$\frac{\text{Height of image}}{\text{Height of object}} = \frac{\text{image distance}}{\text{object distance}} = \frac{q}{p} \qquad (23.4)$$

The ratio of the height of the image to the height of the object is known as the (lateral or transverse) *linear magnification*.

□ **Example** Find the size of the image of a body 25 mm high when it is placed 500 mm in front of a concave mirror with a focal length of 200 mm.

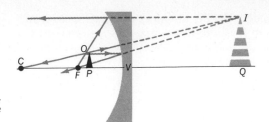

FIGURE 23.12
A concave mirror forms a virtual, erect,
magnified image of an object inside the
principal focus.

$$p = 500 \text{ mm} \qquad f = 200 \text{ mm}$$

$$\frac{1}{p} + \frac{1}{q} = \frac{1}{f}$$

$$\frac{1}{q} = \frac{1}{200} - \frac{1}{500}$$

$$q = 333 \text{ mm}$$

$$\frac{\text{Height of image}}{\text{Height of object}} = \frac{333}{500} = 0.667$$

$$\text{Height of image} = 0.667 \times 25 = 16.7 \text{ mm} \qquad \square$$

If an object lies inside the principal focus of a concave mirror, a ray diagram (Fig. 23.12) indicates that the image lies behind the mirror. It is *upright, virtual,* and *enlarged.* When we solve for q for the case in which p is less than f, we find that it is *negative.*

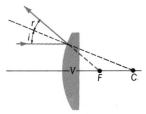

FIGURE 23.13
A convex mirror has a virtual principal focus.

23.7 THE CONVEX MIRROR

In a convex mirror (Fig. 23.13) a ray of light parallel to the principal axis is reflected away from the principal axis. However, if we extend the reflected ray backward, the line crosses the principal axis at point F, the principal focus for this convex mirror. The principal focus is *virtual.* Rays parallel to the principal axis are reflected away from the axis in such a way that they appear to come from the principal focus. The mirror is a *diverging* mirror, because rays initially parallel diverge after reflection.

If we place an object in front of a convex mirror (Fig. 23.14), the rays of light which leave a point O on the object and are reflected at the mirror never actually cross, but if we extend these rays behind the mirror, the extensions cross at point I. In a convex mirror a real object forms an *erect, virtual* image, *reduced* in size.

Equation (23.3) is applicable to the convex mirror, provided that we

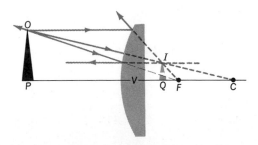

FIGURE 23.14
A convex mirror forms a virtual, erect,
reduced image of the object at the left.

assign a negative sign to the focal length. Further, for a virtual image formed behind the mirror, the image distance q is negative.

☐ **Example** A bright spot situated 600 mm in front of a convex mirror forms a virtual image 200 mm behind the mirror. Find the focal length of the mirror.

$$\frac{1}{p} + \frac{1}{q} = \frac{1}{f}$$

$$p = 600 \text{ mm} \qquad q = -200 \text{ mm}$$

$$\frac{1}{600} - \frac{1}{200} = \frac{1}{f}$$

$$f = -300 \text{ mm} \qquad\qquad ☐$$

23.8 THE STANDARD RAYS AND THE SIGN CONVENTION

The same formula can be used for both concave and convex mirrors if we are careful to use the proper signs for various quantities. We establish conventions for choosing these signs by considering a line parallel to the principal axis connecting a point on the object with the mirror.

1. If a ray of light along this line goes from the object to the mirror, the object distance p is positive.

2. The focal length f is positive if this ray parallel to the principal axis strikes the mirror and is reflected so that it passes through the principal focus. If instead of passing through the principal focus after reflection it diverges from the principal axis, the focal length is negative.

3. If, after reflection, this ray proceeds toward the image, the image distance q is positive. If, after reflection, the ray goes away from the image, the image distance is negative.

It is of great help in solving optics problems to draw scaled ray diagrams. In such a diagram, the intersection of two rays which leave a point on the object and are reflected from the mirror is sufficient to establish the corresponding point on the image. It is desirable to use three rays and thus guard against errors. We shall place emphasis on three rays, which we call the *standard rays* and which can be drawn with a ruler. These three standard rays and their paths are as follows:

1. The ray which goes from a point on the object along the line parallel to the principal axis leaves the mirror along a line which passes through the principal focus.

2. A ray which leaves a point on the object along a line connecting that point with the principal focus is reflected along a line parallel to the principal axis.

3. A ray from a point on the object through the center of curvature of the mirror is reflected directly back on itself.

23.9 SPHERICAL ABERRATION AND PARABOLIC MIRRORS

If the width of a mirror (Fig. 23.15) is comparable to its radius of curvature, parallel rays after reflection do not all meet at the single point F, the

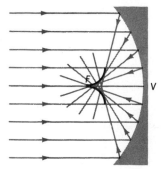

FIGURE 23.15
Spherical aberration occurs because rays parallel to the principal axis of a spherical mirror of large aperture are reflected so that the outer rays do not cross at the principal focus but over a small region called the *caustic surface.*

principal focus. Rays that are reflected from a limited region of the mirror in the neighborhood of its vertex V are brought to a focus at F, but the rays that strike the mirror at points far from the vertex cross the axis at points nearer to the mirror than F. The effect of this *spherical aberration* is to destroy the sharpness of the image formed by the mirror.

If the section of the mirror is a parabola rather than a circle, incident rays parallel to the principal axis are all brought to focus at a single point F, regardless of the size of the mirror. Similarly, all reflected rays which originated at a point source at F leave the parabola parallel to the principal axis. For this reason parabolic mirros are often used for searchlights. If a light source is placed inside the focus of a parabolic mirror, the rays of light diverge after reflection. If it is outside the focus, the rays converge after reflection. In automobile headlights one filament is located near the principal focus to give a *bright*, high beam, and a second filament is placed above and inside the principal focus to give a broad, low beam.

QUESTIONS

1. What is the shortest plane mirror in which a man 1.8 m tall can see himself completely?
2. Light waves travel through vacuum, but sound waves do not. Why don't sound waves propagate through an evacuated region?
3. If someone asked you why a mirror reverses right and left but not up and down, what would your answer be?
4. Is it possible to photograph a virtual image? Under what circumstances? If you wish to take a photograph of yourself in a plane mirror and you stand 2 m from the mirror, for what distance should the camera be set? Why?
5. In Michelson's method of measuring the speed of light, what was the advantage of using eight- and twelve-sided mirrors rather than four- or six-sided ones? What was the disadvantage?
6. When a street light is viewed by reflection in a river, the surface of which is rippled, the light appears to be elongated. Why?
7. A trick mirror makes a person's head look large and hips small. How is this mirror made?
8. A burning candle stands between two plane mirrors which touch along a vertical line. How many images of the candle does an observer see if the angle between the mirrors is (a) 90°, (b) 60°, or (c) 30°?
9. An object stands before a spherical mirror which may be either concave or convex. Under what conditions will the image be (a) erect, (b) larger than the object, and (c) real?
10. How could you locate the center of curvature of a concave mirror with only a sharpened pencil and your eye?
11. If a ground-glass surface is wetted with water, it becomes more transparent. Why?

PROBLEMS

Draw a suitable ray diagram for each mirror problem.

1. A standard radio broadcasting station has an assigned frequency between 535 and 1,605 kHz. The vhf (*very-high-frequency*) television stations (channels 2 to 13) have frequencies between 54 and 216 MHz, while the uhf (*u = ultra*) stations (channels 14 to 83) have frequencies between 470 and 890 MHz. What is the wavelength corresponding to each of the frequencies mentioned?
   Ans. 561, 187, 5.56, 1.39, 0.638, and 0.337 m
2. Find the frequencies associated with electromagnetic waves of wavelength (a) 550 nm (visible light), (b) 55 pm (x-rays), (c) 55 μm (infrared), and (d) 5.5 km (radio).
3. An astronomical unit of distance is the light-year, which is the distance light travels in a year. Compute this distance in kilometers. The nearest star is at a distance of 4 light-years. To how many kilometers does this correspond?
   Ans. 9.47×10^{12} km; 3.79×10^{13} km
4. If the separation of the mirrors in a Michelson-method determination of the speed of light is 30 km and the rotating mirror has six sides, how rapidly must the mirror be turning when light is first reflected to the eye?
5. What minimum speed of rotation is necessary for an eight-sided mirror used in measuring the speed of light with Michelson's arrangement (Fig. 23.2) if the distance from the fixed mirror to the rotating one is 35.4 km? *Ans.* 529.7 r/s
6. Fizeau devised a method of measuring the speed of light in which a beam of light passed between two teeth of a toothed wheel, was reflected back by a mirror, and arrived back at the wheel just in time to pass through the next slot between the

teeth. Fizeau's wheel had 720 teeth, and the mirror was 8.6 km from the toothed wheel. He observed no light returning through the wheel when it was rotating at 12.6 r/s and maximum brightness of returning light at 25.2 r/s. Find Fizeau's value for the speed of light.

7. A pinhole camera with film 75 mm high is to be used to take a picture of a tree 10 m high. The film is 150 mm from the pinhole. How far should the camera be from the tree to include the full height of the tree? *Ans.* 20 m

8. A plane mirror lies face up, making an angle of 15° with the horizontal. A ray of light shines down vertically on the mirror. What is the angle of incidence? What will the angle between the reflected ray and the horizontal be?

9. An object 12 mm high is placed 0.5 m from a concave mirror with a radius of curvature of 0.2 m. Find the focal length and the location, height, and orientation of the image.
Ans. 0.1 m; 125 mm; 3 mm, inverted

10. As the position of an object reflected in a spherical-shell mirror of 0.25 m focal length is varied, the position of the image varies. Plot the image distance as a function of the object distance, letting the latter change from $-\infty$ to $+\infty$. Where is the image real? Where virtual?

11. An object 24 mm high is 0.72 m from a concave mirror with a radius of curvature of 0.16 m Locate the image. Is it real or virtual? Is it erect or inverted? What is its size?
Ans. 90 mm; real, inverted; 3 mm

12. An object 18 mm high is located 60 cm in front of a concave mirror which has a focal length of 20 cm. Find the position, size, and character of the image.

13. An object 8 mm high is located 125 mm in front of a concave mirror which has a focal length of 200 mm. Find the position, size, and character of the image.
Ans. -333 mm; 21.3 mm; erect and virtual

14. What magnification will be obtained by using a concave mirror with a focal length 18 in. if the mirror is held 12 in. from the face?

15. A man is shaving with his chin 0.4 m from a concave magnifying mirror. If the linear magnification is 2.5, find the radius of curvature of the mirror. *Ans.* 1.33 m

16. Where must an illuminated object be placed with reference to a concave mirror with a radius of curvature of 2 m in order to have its image focused on a screen 4 m from the mirror?

17. A doctor looks through a small hole at the vertex of a concave mirror to examine a sore throat. If the radius of curvature of the mirror is 462 mm and the light source is 1 m from the mirror, how

far from the throat should the mirror be if the light source is to be imaged on the inflamed area? *Ans.* 0.300 m

18. To give a magnified image of a cavity a dentist holds a small mirror with a focal length of 12 mm a distance of 9 mm from a tooth. What is the linear magnification obtained?

19. A magnifying mirror has a radius of curvature of 0.9 m. How far must a face be from a mirror if the image is to be erect, virtual, and with linear dimensions 3 times those of the face?
Ans. 0.300 m

20. An incandescent lamp is located 2.5 m from a wall. It is desired to throw on the wall an image magnified 3 diameters, using a concave mirror. What must the radius of curvature of the mirror be? Where must it be placed?

21. An object is placed in front of a concave mirror having a radius of curvature of 0.3 m. If you want to produce first a real image and then a virtual image 3 times as high as the object, find the object distance required in each case.
Ans. 0.2 m; 0.1 m

22. A fortune-teller uses a polished sphere of 8 in. radius. If her eye is 10 in. from the sphere, where is the image of the eye?

23. An object 28 mm high is 0.48 m from a convex mirror with a radius of curvature of 0.32 m. Locate the image. Is it real or virtual? Is it erect or inverted? What is its size?
Ans. -0.12 m, virtual, erect, 7 mm

24. An object is placed 3 ft from a convex mirror which has a focal length of 1.2 ft. Find the image distance and the ratio of the height of the image to the height of the object.

25. A man's eye is 175 mm from the center of a spherical reflecting Christmas tree ornament which is 100 mm in diameter. Find the image position and the linear magnification.
Ans. -20.8 mm; 0.167

26. An object is 375 mm from a concave mirror of 250 mm focal length. Find the image distance. If the object is moved 5 mm farther from the mirror, how far does the image move?

27. About 125 B.C. Hipparchus determined that the distance from the earth to the moon is approximately 30 times the diameter of the earth. His method, proposed by Aristarchus more than a century earlier, involved measuring the angle

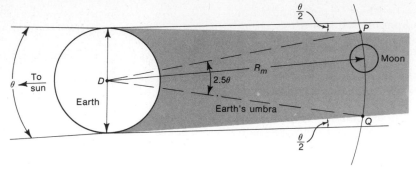

PROBLEM 27

subtended at the earth by the sun's diameter and the length of time the moon spent in the earth's umbra during a lunar eclipse (see accompanying figure). From this time and the period of the moon he found that the angle subtended by the earth's umbra (arc PQ) at the moon's orbit is 2.5 times the angle θ subtended by the sun (and the moon) at the earth. Given that $\theta = 0.53°$ $= 0.0093$ rad, find the ratio of the radius of the moon's orbit R_m to the diameter D of the earth as given by these data.

Ans. 30.7 (Actually the moon's orbit is not exactly circular, and 29.8 is a better modern value.)

28. Show that a ray of light reflected from a plane mirror rotates through an angle 2θ when the mirror is rotated through an angle θ about an axis perpendicular to both the incident ray and the normal to the surface.

29. A short match of length L lies along the axis of a concave spherical mirror of focal length f with its most distant point a distance p from the vertex. Calculate the length L' of the image and find the *longitudinal magnification* L'/L. Show that if $L \ll p - f$, the longitudinal magnification is equal to the square of the lateral linear magnification [Eq. (23.4)].

30. Show that the path of the reflected ray from O of Fig. 23.7 which passes through point P after reflection is the shortest possible path which reaches the mirror. (This is an example illustrating *Fermat's principle of least time*, which states that the actual path of a light ray between two points is such that it takes less time for the light to traverse this path than it would to traverse any other path which varies slightly from the actual path.)

In Chap. 23 we discussed what happens when light is reflected at a surface, and we learned how images are formed by spherical reflectors. Next we examine the bending of light rays as they pass through an interface which separates one medium from another. It is this phenomenon of refraction *which we must understand if we are to appreciate how lenses form images and how optical instruments such as microscopes and cameras operate.*

24

REFRACTION AND DISPERSION

24.1 REFRACTION

When light passes obliquely from one medium to another, there is a change in the direction of propagation. *The bending of light rays as they pass from one medium to another is called refraction.* As a consequence of refraction, a ruler dipped obliquely into water appears to bend sharply at the surface.

Figure 24.1 shows a ray OA incident obliquely on a plane water surface RS. Upon reaching the surface, the ray undergoes a sharp change in direction, proceeding along the line AI. The angle OAN between the incident ray and the normal NN' to the surface is the angle of incidence θ_1, while the angle IAN' is the angle of refraction θ_2. (Note that here we are using θ_1 for the angle of incidence rather than i.) When a ray of light is refracted at a surface, two laws are obeyed:

1. *The ratio of the sine of the angle of incidence to the sine of the angle of refraction is a constant independent of the angle of incidence.*

2. *The incident ray, the refracted ray, and the normal to the surface lie in the same plane.*

The first law of refraction is called *Snell's law* in honor of its discoverer.

The ratio $(\sin \theta_1)/(\sin \theta_2)$ depends on the media on the two sides of the interface at which refraction occurs. In order to compare the refraction associated with various media, we refer refraction measurements to a common medium—a vacuum.

The index of refraction of a medium is the ratio of the sine of the angle of incidence measured in vacuum to the sine of the angle of refraction measured in the medium.

When we say the index of refraction of glass is 1.50, we are automatically implying that the light is incident from a vacuum and falls upon the glass. Table 24.1 lists the indices of refraction of a number of common transparent materials. The index of refraction of air differs little from unity; in most situations it makes no significant difference whether light is incident from vacuum or from air.

The correct explanation for refraction was given by Huygens (Sec. 20.8). Whenever a wave motion passes from one medium in which its speed is V_1 to another in which its speed is V_2, the wave pattern changes its direction in such a way that $V_1/V_2 = (\sin \theta_1)/(\sin \theta_2)$, where θ_1 is the angle of incidence and θ_2 the angle of refraction.

In general, for any medium the index of refraction n is given by

$$n = \frac{\sin \theta_1 \text{ (in vacuum)}}{\sin \theta_2 \text{ (in medium)}} = \frac{c}{V} \qquad (24.1)$$

where c is the speed of light in vacuum and V the speed in the material.

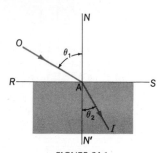

FIGURE 24.1
Light ray refracted at the interface *RS*
between air and water.

TABLE 24.1
Indices of Refraction
(For light of $\lambda = 589$ nm;
sodium D radiation)

Substance	n
Air	1.00029
Carbon dioxide	1.00045
Canada balsam	1.53
Ethyl alcohol	1.36
Carbon disulfide	1.63
Diamond	2.42
Glycerin	1.47
Water	1.333
Crown and plate glass	1.52
Flint glass	1.63
Heavy flint glass	1.66
Quartz	1.54

24.2 RELATIVE INDICES OF REFRACTION

In defining the index of refraction of a material we chose to have light incident from a vacuum. Often an interface separates two transparent media of substantial density, such as glass and water. The laws of refraction are valid in such cases; the ratio $(\sin \theta_1)/(\sin \theta_2)$ is called the *index of refraction of the second medium relative to the first.* In general, the ratio of the sine of the angle of incidence to the sine of the angle of refraction is equal to the ratio of the speed of light in the first medium V_1 to the speed of light in the second medium V_2:

Index of refraction of medium 2 relative to medium 1

$$= n_{2,1} = \frac{\sin \theta_1}{\sin \theta_2} = \frac{V_1}{V_2} \quad (24.2)$$

When light goes from one medium to another in which its speed is smaller, it is bent toward the normal; when it goes from one medium to another in which its speed is greater, it is bent away from the normal.

If light moves from medium *A* into medium *B*, it is convenient to write the law of refraction in the form

$$n_A \sin \theta_A = n_B \sin \theta_B \quad (24.3)$$

where n_A and n_B are the indices of refraction and θ_A and θ_B are the angles between the normal and the rays in medium *A* and medium *B*, respectively. This relationship applies whether the ray is going from *A* to *B* or from *B* to *A*.

If the light travels from *A* to *B*, the index of refraction of *B* relative to *A* is

$$n_{B \text{ relative to } A} = \frac{\sin \theta_A}{\sin \theta_B} = \frac{V_A}{V_B} = \frac{n_B}{n_A} \quad (24.3a)$$

On the other hand, if the light goes from *B* to *A*, the index of refraction of *A* relative to *B* is

$$n_{A \text{ relative to } B} = \frac{\sin \theta_B}{\sin \theta_A} = \frac{V_B}{V_A} = \frac{n_A}{n_B} \quad (24.3b)$$

24.3 REFRACTION THROUGH SLABS WITH PARALLEL FACES

If a ray of light (Fig. 24.2) falls on a plate of glass with parallel faces, it is bent toward the normal as it enters the glass. As it leaves the glass, it is bent away from the normal. Since the faces of the plate are parallel to

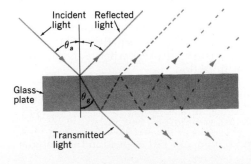

FIGURE 24.2
Refraction and reflection at parallel surfaces.

each other, the normals to the first and second faces are parallel to each other, and therefore the angle θ_g between the normal and the ray in the glass is the same at both surfaces. Consequently, the direction of the ray emerging from the glass is the same as its direction on entering; however, the ray is displaced laterally.

When light goes from one transparent medium to another with different optical properties, usually there is a reflected beam as well as a refracted one. Figure 24.2 shows schematically that the angle of reflection r is equal to the angle of incidence θ_a. There is reflection at the lower surface of the plate as well as at the upper. Light which undergoes several reflections is said to be *multiply reflected*.

When a ray of light from air traverses several parallel layers of transparent substances (Fig. 24.3), it is refracted at each surface. If it returns to air, its direction is unchanged but it is laterally displaced from its original path.

24.4 TOTAL REFLECTION

When a ray of light passes from an optically denser medium such as water to a rarer medium such as air, it is bent away from the normal so that the angle of refraction is greater than the angle of incidence (Fig. 24.4a). If the angle of incidence is made larger, the angle of refraction increases until it reaches 90°. In this case the refracted ray travels along the surface of separation between the two media. *The angle of incidence for which the angle of refraction is 90° is called the critical angle,* which we shall designate by θ_C.

□ **Example**   The index of refraction of water is 1.33. What is the critical angle for light going from water to air?

Application of Eq. (24.3) yields

$$n_{\text{water}} \sin \theta_C = n_{\text{air}} \sin 90°$$
$$1.33 \sin \theta_C = 1.00 \times 1.00$$
$$\sin \theta_C = 0.75$$
$$\theta_C = 49° \qquad \square$$

When the angle of incidence exceeds the critical angle, light no longer enters the rarer medium and all the incident light is reflected. This type of reflection, known as *total reflection*, takes place at a surface separating an optically rarer from a denser medium when the light comes from the denser to the rarer medium and the angle of incidence exceeds the critical angle. The prism ABC (Fig. 24.5) produces total reflection of the ray OL, since the angle of incidence of 45° exceeds the critical angle for glass.

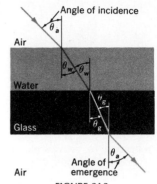

FIGURE 24.3
Path of a light ray through parallel layers of different materials.

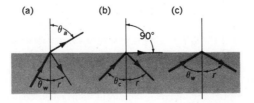

FIGURE 24.4
When light goes from water to air, (a) the refracted ray bends away from the normal, (b) at the critical angle of incidence the refracted ray is parallel to the surface, (c) the light is totally reflected when the angle of incidence exceeds the critical angle.

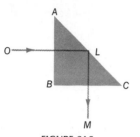

FIGURE 24.5
Total reflection by an isosceles right-angle prism.

It would be erroneous to believe that when the critical angle of incidence is reached, there is a sudden transition from all light refracted to all light reflected. At any angle of incidence there is always some reflected light.† At normal incidence the fraction of the incident intensity reflected is given by $[(n_2 - n_1)/(n_2 + n_1)]^2$, where n_1 and n_2 are the indices of refraction of the first and second media. As the angle of incidence is increased, a greater fraction of the incident light is reflected and a smaller fraction refracted. It is easy to observe with a pane of glass that as the angle of incidence increases, the fraction of the light reflected becomes larger.

There are many applications of total reflection. Light can be "piped" down a bent tube of glass or clear plastic. As the light travels along the tube, it strikes the surface at angles of incidence greater than the critical angle and is totally reflected. Thus the light is transmitted down the tube until it reaches the end, where the angle of incidence is less than the critical angle, and the light emerges. Such "light pipes" are used in surgical operations to illuminate places where it is difficult to operate a lamp because of the heat. Light can also be piped in streams of water; colored fountains depend on total reflection to keep the light in the water stream until it finally strikes at an angle of incidence smaller than the critical angle and escapes.

24.5 APPARENT THICKNESS Of A TRANSPARENT BODY

One consequence of refraction is that a body immersed in a transparent medium appears nearer to the surface than it actually is. Let OA and OB represent rays from an object under water (Fig. 24.6); both are bent away from the normal when they reach the surface. An eye at E sees light which appears to come from the image at I; an eye at D would "see" the object at J.

The index of refraction of the water is

$$n = \frac{\sin EAN}{\sin AOV} = \frac{\sin AIV}{\sin AOV} = \frac{AV/IA}{AV/OA} = \frac{OA}{IA}$$

If the eye receives only rays very near the normal to the surface, we can write OV for OA, and IV for IA. Then

$$n = \frac{OV}{IV} = \frac{\text{actual depth}}{\text{apparent depth}} = \left| \frac{p}{q} \right| \qquad (24.4)$$

The greater the inclination at which the object is viewed, the less its apparent depth.

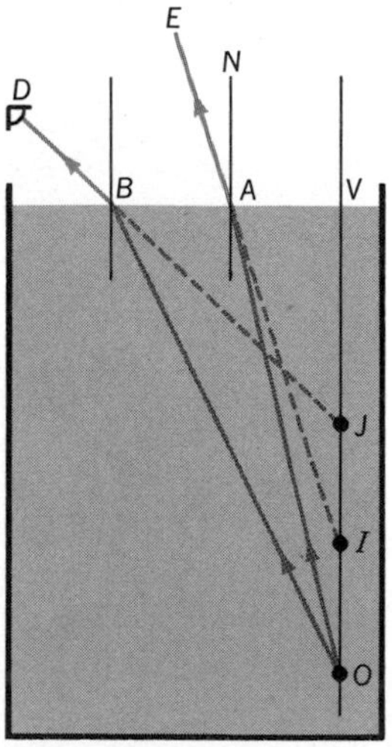

24.6 ATMOSPHERIC REFRACTION AND THE MIRAGE

Although the index of refraction of air under standard conditions deviates from unity by less than 0.03 percent, there are situations in which atmospheric refraction is far from negligible. One of the most interesting occurs in the *mirage*. On hot sunny days there may be a layer of very hot air in contact with the ground. This air has a lower density and a slightly smaller index of refraction than the air above it. Light from a distant mountain or

FIGURE 24.6
The apparent depth of an object O in water is less than the actual depth by an amount which depends on the angle of incidence.

---

† This remark applies to ordinary light; the reflection of polarized light is discussed in Sec. 28.7.

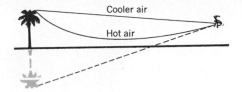

FIGURE 24.7
A mirage may be observed when the air
near the ground is substantially hotter
than the air higher up.

treetop approaches this layer of hot air at a large angle of incidence and is
totally reflected (Fig. 24.7). The light appears to have come from the image
just as though it had been reflected from the surface of a lake.

When the air next to the ground is substantially colder than higher air,
rays of light are deviated downward (Fig. 24.8). An image of a ship may
appear above the ship itself. This phenomenon is called *looming*. It is
rather common when an observer is looking over a snowfield or over a
body of water which is substantially colder than the air several feet above
the surface. Cases have been reported of a lighthouse being seen at a
distance of 60 km when the curvature of the earth would have completely
cut it off had there been no looming.

When one looks at an object over the hot burner of a stove or over a
hot pavement, one may observe a wavy, shimmering effect. This arises
from the bending of the light as it passes from colder to warmer to colder
air. The twinkling of stars is in part due to similar phenomena in that the
light travels through unstable layers in the atmosphere where the index of
refraction is changing significantly.

A ray of light from the sun (or a star) entering the earth's atmosphere
obliquely is refracted continuously as it enters denser air. For this reason
the sun appears to be at a higher altitude angle than it actually is. When
the sun is at the horizon, the correction is about $\frac{1}{2}°$, roughly the angle
subtended at the earth by the sun's diameter. As a consequence, the sun
is actually below the horizon when we watch it rise or set. When the sun
(or moon) is near the horizon, the rays from the lower edge are bent more
than those from the upper edge. This produces a shortening of the
vertical diameter, so that the sun appears elliptical.

24.7 REFRACTION AT A SPHERICAL SURFACE

Consider a ray OP in medium A which is incident upon a spherical surface
of medium B (Fig. 24.9) with center of curvature at C. This ray is refracted
at the surface; by Eq. (24.3),

$$n_A \sin \theta_A = n_B \sin \theta_B$$

By definition,

$$\sin \phi = \frac{z}{R} \qquad \tan \alpha = \frac{z}{p + x} \qquad \tan \beta = \frac{z}{q - x}$$

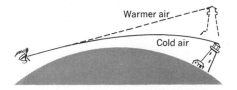

FIGURE 24.8
Looming occurs when the air near the
surface is substantially colder than the
air above.

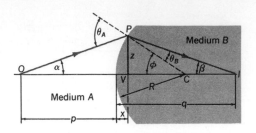

FIGURE 24.9
Refraction at a spherical surface.

Since $\theta_A + \angle OPC = 180° = \angle OPC + \alpha + \phi$, we have $\theta_A = \alpha + \phi$. Similarly, $\phi = \theta_B + \beta$.

We now confine our derivation to the case in which θ_A, θ_B, α, and ϕ are all small angles. If an angle is small, its sine, its tangent, and the angle itself (in radians) are essentially equal. For this case x is negligible compared with p and q, and we write

$$\sin\theta_A = \theta_A = \alpha + \phi = \frac{z}{p} + \frac{z}{R}$$

and

$$\sin\theta_B = \theta_B = \phi - \beta = \frac{z}{R} - \frac{z}{q}$$

Then, by Eq. (24.3),

$$\frac{n_A z}{p} + \frac{n_A z}{R} = \frac{n_B z}{R} - \frac{n_B z}{q}$$

or

$$\frac{n_A}{p} + \frac{n_B}{q} = \frac{n_B - n_A}{R} \qquad (24.5)$$

The line OC is the principal axis for this optical system. Equation (24.5) is valid only for paraxial rays, i.e., rays which make small angles with the axis. The radius R is positive when it is measured from the surface to the center of curvature in the direction in which the light is traveling. Thus R is positive for a surface convex to the incident light and negative for a surface concave to the incident light. Similarly, q is positive if the image is on the side of the surface toward which the light is propagating and negative if on the side from which the light approaches.

24.8 REFRACTION BY A PRISM

A wedge-shaped portion of a refracting medium bounded by two plane surfaces is called a *prism*. If the prism is optically denser than the surrounding medium, a ray of light incident on one of the faces is bent toward the normal to that face on entering the prism. On emerging from the opposite face, the ray is bent away from the normal to that face. The angle D (Fig. 24.10) through which the ray is deflected in passing through the prism is called the *angle of deviation*. When the angle at which the ray enters one face is equal to the angle at which it leaves the opposite face, the angle of deviation has its least value and is known as the *angle of minimum deviation*.

By measuring the angle of minimum deviation and the angle between the faces of the prism, it is possible to find the index of refraction of the material of which the prism is made. Let A be the angle of the prism, D the

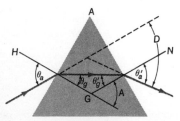

FIGURE 24.10
Refraction by a prism.

angle of minimum deviation, and n the index of refraction. Since the angle between the faces of the prism is the supplement of the angle between the normals HG and NG (Fig. 24.10),

$$A = \theta_g + \theta'_g$$

The deviation at the first face is $\theta_a - \theta_g$ and at the second face $\theta'_a - \theta'_g$. Hence, the total deviation is

$$D = \theta_a - \theta_g + \theta'_a - \theta'_g = \theta_a + \theta'_a - A$$

The angle of deviation is minimal when the angle of incidence is equal to the angle of emergence. In this case, $\theta_a = \theta'_a$, $\theta_g = \theta'_g$, $A = 2\theta_g$, $D = 2\theta_a - A$, and

$$\theta_a = \frac{A + D}{2}$$

From the law of refraction,

$$\text{Index of refraction} = n = \frac{\sin \theta_a}{\sin \theta_g}$$

or

$$n = \frac{\sin [(A + D)/2]}{\sin (A/2)} \qquad (24.6)$$

□ **Example** For a 60° glass prism, the angle of minimum deviation is found to be 48°. What is the index of refraction of the prism?

$$\text{Index of refraction} = \frac{\sin [(A + D)/2]}{\sin (A/2)}$$

$$n = \frac{\sin 54°}{\sin 30°} = \frac{0.809}{0.500} = 1.62 \qquad \square$$

24.9 THE DISPERSION OF LIGHT BY A PRISM

When a very narrow beam of white light passes through a prism (Fig. 24.11), it is spread out into a band of colors known as a *spectrum*. Violet light, which has the shortest wavelength of the visible, is bent most, while red, with the longest wavelength, is bent least. *The separation of white light into its component colors is called dispersion.*

Since different colors are deviated different amounts by the prism, it is clear that the index of refraction and the speed of light in the prism must be different for the various colors. In ordinary crown glass, the speed of red light is approximately 1 percent greater than the speed of violet light.

The question may well be raised of why we did not notice dispersion in connection with experiments performed earlier with the deviation of light

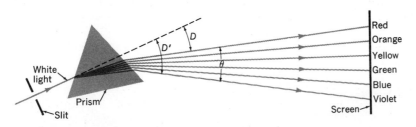

FIGURE 24.11
A prism refracts longer wavelengths less than shorter ones, resulting in the dispersion of white light into a spectrum.

by a prism. In these experiments a much wider beam of light was used, and the level of illumination in the room was great. Because of the width of the beam, the blue from one part of the beam fell on the red from another, on the orange from still another, and so forth. This led to a white core, but at the edges of the deviated beam there were doubtless evidences of dispersion. However, the colors were so dim compared with the bright central image that they probably were not noticed.

24.10 ACHROMATIC PRISM

When light of more than one color falls on a prism, the emerging beam is not only deviated but spread out into a spectrum. If a second prism is placed just beyond, with its vertex at the base of the first, the second prism tends to gather the different colors together again and combine them into white light. The net dispersion produced by the two prisms is the difference between the dispersions produced by the individual prisms. The deviation produced by the second prism is opposite that produced by the first prism, and the net deviation is the difference between the two. If the two prisms were identical, the net effect would simply be that of passing the light through a parallel-faced plate—no dispersion and no deviation; but if the prisms are made of different kinds of glass, they may have the same net dispersion but quite different deviations. A flint-glass prism which produces the same dispersion as a crown-glass prism of larger central angle gives a smaller deviation. By placing together crown- and flint-glass prisms which have the same dispersion, it is possible to construct a prism that deviates the light without spreading it into a spectrum (Fig. 24.12). Such a prism is called an *achromatic* prism.

If the angle of the flint-glass prism is increased until the average deviation it produces is equal to that produced by the crown-glass prism, the dispersion of the flint-glass prism is substantially greater. When two such prisms are put together, the combination produces *dispersion without net deviation*. When white light passes through such a combination of prisms, called a *direct-vision spectroscope*, it is dispersed into a spectrum, but its general direction is unchanged.

24.11 THE RAINBOW

No doubt the first evidence of dispersion seen by man was the rainbow. The formation of the rainbow was explained by Descartes about 1637. Let us consider a situation in which the sun is in the west and it is raining in the east. Rays of sunlight enter water droplets (Fig. 24.13), where they are refracted and dispersed. If a ray enters a drop at the proper place, it will

FIGURE 24.12
Achromatic prism combination producing deviation without dispersion. The red from the incident ray shown recombines with the yellow, green, blue, and violet components of neighboring rays to give white light once more.

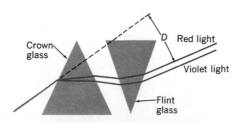

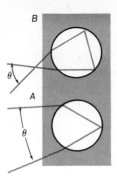

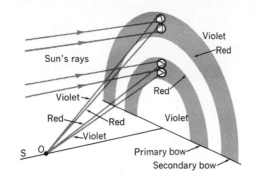

FIGURE 24.13
Formation of a rainbow.

be partly reflected at the internal surface of the drop and leave the drop as shown in insert *A* of the figure. Descartes showed that when the angle θ is 42°, red light is strongly reflected while violet light appears prominently at 40°. The other colors of the spectrum are intensely reflected at angles between 40 and 42°. Each color is strongly reflected at that angle for which it can pass through the droplet with minimum deviation. If an observer stands at *O*, and if the line *SO* represents the direction of the incident sunlight, all those raindrops which lie on the surface of a cone of angle 40° about *SO* send an excess of violet light to the observer, while those which lie on a conical surface of angle 42° about *SO* return red light.

In addition to the primary bow which we have just discussed, there is sometimes a secondary bow in which the colors are reversed, with violet on the outside and red on the inside. This secondary bow is produced by refraction, dispersion, and two internal reflections inside the raindrop, as shown in insert *B*. For the secondary bow, the angle between the incident and emergent rays is about 51° for red and 54° for violet.

When exceedingly tiny drops are involved, as may be the case in a garden spray, diffraction effects may become large and the structure of the bow much more complicated.

QUESTIONS

1. When a light beam goes from one medium to another, which of the following quantities always remain the same: the wavelength, the frequency, the direction of propagation, the speed? Which sometimes remain the same and sometimes change? Explain.
2. A clear cylindrical glass filled with water rests on a table. Explain why you cannot see the table through the water surface and the side of the glass but can see your fingers when they are in contact with the glass.
3. What are the advantages of using a totally reflecting prism (Fig. 24.5) rather than a silvered mirror in an optical instrument such as binoculars or a periscope? What are the disadvantages?
4. Why does the air shimmer over a hot stove? Why do stars twinkle?

5. Very little light is transmitted through chipped ice in a clear glass. When water is added, the system becomes much more transparent even though no significant melting has occurred. Explain why.
6. A swimming pool of clear water is flat where the depth is 1.2 m. If a person stands at the center of this region, why does it appear to her that she is standing at the deepest point?
7. Why are optical systems using totally reflecting prisms usually designed so the light enters and leaves the surfaces along a normal?
8. If you were standing waist deep in clear water and wished to spear a fish swimming nearby, would you aim above or below the apparent position of the fish? Why?
9. How and why does the earth's atmosphere alter the apparent shape of the sun and moon when they are near the horizon?

10. Why are colors observed in the light from a cut diamond?
11. A coin lies at the bottom of an empty cylindrical silver cup. If you move your head, you can find a position in which the coin disappears behind the wall of the cup. Now if water is poured into the cup, the coin comes into view even though you do not move your head. Explain, using a ray diagram.
12. A fish lies at the bottom of a clear pool. Describe and explain what it sees as it looks in various directions.
13. Why is it that when light goes from one medium to another, the wavelength changes but the frequency does not?
14. What properties should a material have if it is to be particularly useful as a light pipe?

PROBLEMS

1. Light of a certain color has 1,800 waves to the millimeter in air. What is its frequency? Find the number of waves per millimeter when the light is traveling in water and in plate glass. What is the wavelength in each medium?
   Ans. 5.4×10^{14} Hz; 2,399 mm^{-1}; 2,736 mm^{-1}; 556 nm; 417 nm; 365 nm
2. The index of refraction of *n*-propyl alcohol is 1.39. Find the speed of light in that medium and the angle of refraction if light comes from air with an angle of incidence of 55°.
3. The index of refraction of a flint glass is 1.64. Find the speed of light in the glass, the angle of refraction in glass if light is incident from water at an angle of 44°, and the critical angle of incidence for a glass-water interface.
   Ans. 1.83×10^{8} m/s; 34.4°; 54.4°
4. Find the sine of the critical angle of incidence for an air-water interface and the angle of refraction in water if light is incident from air at an angle of 38°.
5. If the speed of light in ice is 2.3×10^{8} m/s, find its index of refraction. What is the critical angle of incidence for light going from ice to air?
   Ans. 1.304; 50.1°
6. What is the speed of light in water? Find the angle of refraction of light incident on a water surface at an angle of 48°.
7. The bottom of a glass vessel is a thick plane plate

that has an index of refraction of 1.50. The vessel is filled with water. What is the largest angle of incidence in the water for which a ray can pass through the glass bottom and into the air below? How does this compare with the critical angle of incidence for a water-air interface which is 48.6°? What is the angle of refraction in the glass?
   Ans. 48.6°; same; 41.8°
8. Light is passing from air into a liquid and is deviated 19° when the angle of incidence is 52°. Under what conditions will total reflection occur at this interface?
9. A glass dish with a plane parallel bottom and refractive index 1.51 is half filled with water. Then carbon disulfide is poured on top of the water. Finally, a flat cover of the same type of glass is placed on top of the dish. A beam of light making an angle of 50° with the vertical is incident on the horizontal cover. Find the angles which the beam makes with the vertical as it passes through glass, carbon disulfide, water, glass, and air. *Ans.* 30.5°; 28.0°; 35.1°; 30.5°; 50°
10. A ray of light passes from crown glass to water. What is the critical angle of incidence? If the angle of incidence in the glass is 55°, what is the angle of refraction?
11. A coin lies at the bottom of a swimming pool at a depth of 2 m. Find its apparent depth when it is seen from directly above. *Ans.* 1.5 m
12. A fish is 1.2 m below the surface of a pool. What is its apparent depth when viewed from above?
13. A beam of white light is incident on a block of flint glass at an angle of 55°. What is the angular separation in the glass of two rays of light, one of wavelength 486 nm and the other of wavelength 656 nm, if their respective indices of refraction are 1.670 and 1.650? *Ans.* 0.392°
14. A ray of light is passed through a prism having a refracting angle of 50°. Rotation of the prism causes the ray to be deviated various amounts, the least of which is 30°. Determine the index of refraction of the glass of which the prism is made.
15. What is the index of refraction of a 60° glass prism which produces an angle of minimum deviation of 38°? *Ans.* 1.51
16. A ray of light makes an angle of incidence of 40° with a glass prism of index of refraction 1.52 and refracting angle $A = 56°$. Through what angle is the ray deviated by the prism? Does this represent minimum deviation?
17. Given a 60° prism of flint glass, find the angle of minimum deviation. At what angle of incidence must the light strike the prism? Find the deviation if the angle of incidence is 5° less.
   Ans. 49.2°; 54.6°; 49.8°
18. A hollow prism is made of plane parallel plates of

glass. The angle between the faces is 45°. What is the angle of minimum deviation when the prism is filled with carbon disulfide and sodium D radiation with a wavelength of 589 nm is passed through it?

19. Experiment shows that the index of refraction of heavy flint glass is 1.717 for D light (yellow) and 1.742 for F light (blue). Find the angle of dispersion of these two colors produced by a 60° prism if the light strikes the prism with an angle of incidence of 50°.   *Ans.* 5.0°

20. Show that Eq. (24.4) follows directly from Eq. (24.5) if the refracting surface is plane, i.e., has a radius of curvature of infinity.

21. If light is incident on face *AB* of the prism of Fig. 24.5 at an angle of 45° (midway between *AB* and *OL*), what must be the index of refraction of the prism if the refracted ray is incident on *BC* at the critical angle of incidence? What happens to the refracted ray at *BC* if the index of refraction is greater than this value?

                    Ans. 1.225; totally reflected

22. Show that when a light ray passes through a parallel plate of thickness t (see Fig. 24.2), it is displaced sideways by an amount $d = [t \sin (\theta_1 - \theta_2)]/(\cos \theta_2)$.

23. A clear plastic ball with a diameter of 80 mm has a flaw which is actually 28 mm under the surface. The index of refraction of the plastic is 1.40. Find the position of the image of the flaw when it is viewed along a diameter passing through the flaw from (*a*) the near side and (*b*) the far side.
                    Ans. (*a*) −25 mm; (*b*) −59.1 mm

24. Show that if a ray of light is incident on a plane parallel glass plate of thickness d at a small angle of incidence θ, it emerges from the plate parallel to its original direction but displaced a distance $\theta d(n - 1)/n$, where n is the index of refraction of the glass. (Recall that for small angles $\sin \theta = \theta$ in radians.)

25. A fish is swimming in a spherical bowl 444 mm in diameter. If the fish is actually 125 mm from the point on the bowl closest to the observer, what is the apparent distance to the fish from that point? (Assume the glass of the bowl to be of uniform and negligible thickness.)   *Ans.* 109 mm

26. Show that the deviation produced by a thin wedge of transparent material of angle A is given approximately by $D = (n - 1)A$ if the angle of incidence is small. *Hint:* For small angles, $\sin \theta = \theta$ in radians.

27. A glass ashtray has an upper surface which is approximately spherical and concave with a radius of curvature of 180 mm. If the glass has bubbles in it and an index of refraction of 1.60 find the apparent depth of a bubble 17.6 mm below the surface.   *Ans.* 10.6 mm

25

LENSES

Our knowlege of how light rays are bent when they pass from one material to another can now be applied to lenses, which are the basic elements of most optical instruments. When light rays pass through a lens, they undergo refraction at both surfaces, and, except in very special cases, they emerge traveling in a direction different from that in which they were incident. In this chapter we see how a lens deviates rays to form an image of an object. Much of what we learned about image formation by spherical mirrors is applicable to lenses as well.

25.1 SIMPLE LENSES

Lenses are bodies of transparent material shaped to converge or diverge a beam of light. Simple lenses are bounded by faces which are small sections of spheres.

Consider a beam of parallel rays falling on the prism in Fig. 25.1*a*. The rays are deviated by the prism but remain parallel to each other. If we wish to bring these rays together at a point, we must deviate the uppermost ray more than the lowest one. This can be accomplished by grinding the surfaces so that they have the cross section indicated in Fig. 25.1*b*. If the surfaces of a glass blank are small sections of spheres, they will deviate the upper ray more than any of the others; the higher the ray, the greater the deviation. Of course, this does not guarantee that all incident parallel rays will pass through the same point; but if we do not use too large a section of the spherical surfaces, the emerging rays all pass close to the same point.

If a lens brings a bundle of parallel rays together at a point focus, the lens is said to be *converging*. If it separates such a bundle of parallel rays, the lens is *diverging*. When glass or plastic lenses are used in air, they are converging if they are thicker in the middle than at the edges, diverging if they are thinner at the middle. Figure 25.2 shows six possible types of spherical lenses with their names.

Consider the lens of Fig. 25.3. The left surface is a section of a spherical surface having point C_1 as its center, while the right surface is a portion of a sphere having point C_2 as its center. *The principal axis of this lens is the line connecting the centers of curvature of the two lens surfaces.* If one of the surfaces is plane, the principal axis is the line through the one center of curvature which is perpendicular to the plane surface.

Rays which approach a converging lens parallel to the principal axis are deviated so that they pass (Fig. 25.4*a*) through a common point, the *principal focus*, on the principal axis. A lens has *two principal foci*, one for light incident from the left, which we denote by *F*, and the second *F'* for light incident from the right. For thin lenses the principal foci are equidistant from the optical center of the lens. By a thin lens we mean one whose thickness is negligible compared with the distance to the principal foci and to the objects and images concerned. In this text we confine our attention to thin lenses.

When the rays incident on a lens are parallel, the incident wavefront is a plane perpendicular to the rays. The part of the wavefront which goes through the edges emerges ahead of the central part because it has less distance to go in glass, where the speed is reduced. By Huygens' principle it can be shown that the emerging wavefront is spherical and converges on the principal focus.

In the case of a diverging lens, rays parallel to the principal axis are bent as shown in Fig. 25.4b. These rays never intersect, but they all appear to come from the virtual principal focus F.

25.2 STANDARD RAYS FOR LENSES

In Fig. 25.5 an object OP is a distance p from a converging lens. Every ray of light which leaves point O on the object and goes through the lens passes through point I. In the figure only three rays are shown, but any other ray from point O which traverses the lens proceeds through point I. Therefore, at point I we have an image of point O. For the entire object OP we find an image IQ, which in this case is real and inverted.

Whenever we are concerned with locating the image I of an object point O by ray tracing, we shall use three standard rays. These rays, analogous to those discussed for mirrors, propagate as follows:

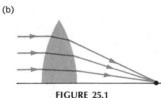

1. A ray from O parallel to the principal axis is deviated so that it passes through the principal focus of a converging lens, or diverges as though it came from the principal focus of a diverging lens.

FIGURE 25.1
(a) Parallel rays deviated by a prism remain parallel. (b) A lens with spherical surfaces brings parallel rays to a focus.

2. The ray from O which approaches the lens along the line through the principal focus F' for parallel rays coming from the opposite side of the lens is deviated so that it leaves the lens parallel to the principal axis.

3. A ray through the optical center of the lens passes through the lens undeviated. For a thin lens we may take the optical center to be the center point of the lens.

In Fig. 25.5 these three standard rays are, respectively, $OAFI$, $OF'I$, and OCI.

25.3 SINGLE LENSES

Let p be the distance from the object to the lens and q that from the image to the lens. The distance from the lens to either principal focus is the *focal length f.* It is easy to show that the equation

$$\frac{1}{p} + \frac{1}{q} = \frac{1}{f} \tag{25.1}$$

applies as well to lenses as to mirrors.

Equation (25.1) can be derived as follows. The triangles OPC and IQC (Fig. 25.5) are similar, and hence $OP/IQ = CP/CQ = p/q$. Also, triangles ACF and IQF are similar, so that $AC/IQ = FC/FQ = f/(q-f)$. Since $AC = OP$, we have $AC/IQ = OP/IQ = p/q = f/(q-f)$. Therefore,

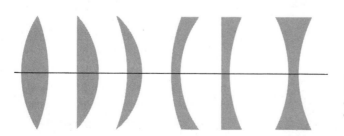

FIGURE 25.2
Lens shapes (from left to right) are double-convex, plano-convex, concavo-convex (converging meniscus), convexo-concave (diverging meniscus), plano-concave, and double-concave.

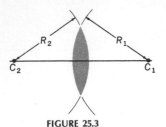

FIGURE 25.3
The principal axis of a spherical lens is the line connecting the centers of curvature of the two lens surfaces.

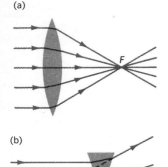

FIGURE 25.4
(a) A converging lens deviates rays parallel to the principal axis so that they pass through the principal focus. (b) A diverging lens deviates rays incident parallel to the principal axis so that they appear to come from a virtual focus.

$pq - pf = qf$. If we divide by pqf and rearrange, we obtain $1/p + 1/q = 1/f$.

We note further that $IQ/OP = q/p$. Since IQ is the height of the image and OP the height of the object, we have

$$\text{Linear magnification} = \frac{\text{height of image}}{\text{height of object}} = \frac{q}{p} \qquad (25.2)$$

For the lens of Fig. 25.5, an object at IQ has its image at OP. Points P and Q are *conjugate points* of the lens—an object at either point has an image at the other.

When an object is inside the principal focus of a converging lens, the image formed is *erect, virtual,* and *enlarged* (Fig. 25.6). In this case, the image distance is negative.

A real object placed before a diverging lens has an erect, virtual, and reduced image (Fig. 25.7). In applying the general lens formula to the diverging lens, the focal length is taken as *negative*. In Fig. 25.7 the image distance q is also negative.

In calculations involving the lens relation [Eq. (25.1)], one must be careful that the proper signs are taken for object distance, image distance, and focal length. The sign convention used in this text for both lenses and mirrors is as follows:

1. If the direction of an incident ray is from an object point O toward the lens (or mirror), the object distance is positive; if its direction is from the lens toward the object, p is negative.

2. The focal length is positive for a converging lens (or mirror) and negative for a diverging lens (or mirror).

3. The image distance is positive if a ray from O goes from the lens (or mirror) toward the image. It is negative if the ray goes away from the image when it leaves the lens (or mirror).

25.4 COMBINATION OF LENSES

When light passes first through one lens and then through a second, the position of the final image can be calculated by repeated use of the general lens equation. *The object for the second lens is the image formed by the first.* If the second lens is interposed before the rays actually cross (Fig. 25.8), its object is still taken to be the image which the first lens would have formed had the second not been present. In this case the object distance for the second lens is negative, since the light is deviated

FIGURE 25.5
Formation of a real image by a converging lens.

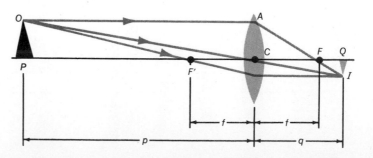

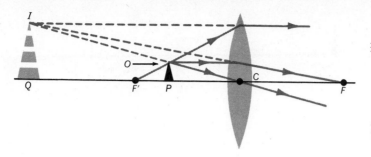

FIGURE 25.6
A converging lens forms an erect, virtual, enlarged image of an object inside the principal focus.

by this lens before its "object" is formed. If there is a third lens, the image formed by the second lens acts as its object. The final image formed by a very complicated optical system, such as that of a submarine periscope, can be located by successive applications of the relations governing simple lenses and mirrors. In every case the object for any single component is the image formed by the preceding one. Great care must be taken to assign the proper sign to each distance.

☐ **Example** An object 50 mm high is 300 mm from a lens of 100 mm focal length. A diverging lens of focal length −80 mm is placed 90 mm beyond the converging lens (Fig. 25.8). Find the position and height of the final image.

For the first lens,

$$\frac{1}{p} + \frac{1}{q} = \frac{1}{f}$$

$$\frac{1}{300} + \frac{1}{q} = \frac{1}{100} \qquad \frac{\text{Height of image}}{\text{Height of object}} = \frac{150}{300}$$

$$q = 150 \text{ mm} \qquad \text{Height of image} = 25 \text{ mm}$$

For the second lens $p = -60$ mm and

$$\frac{1}{-60} + \frac{1}{q} = \frac{1}{-80}$$

$$q = 240 \text{ mm from second lens}$$

$$\frac{\text{Height of image}}{\text{Height of object}} = \frac{240}{60}$$

Height of final image = 100 mm ☐

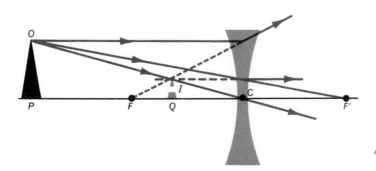

FIGURE 25.7
An erect, virtual, reduced image of a real object is formed by a diverging lens.

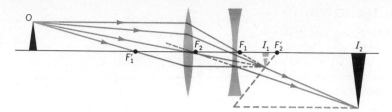

FIGURE 25.8
Image formed by a pair of separated
lenses.

□ **Example** If the diverging lens of the preceding example had been only
40 mm from the first lens, where would the final image have been located?

The image which would be formed by the first lens is still 150 mm from
the first lens. The object distance for the second lens is now −110 mm.

$$\frac{1}{p} + \frac{1}{q} = \frac{1}{f}$$

$$\frac{1}{-110} + \frac{1}{q} = -\frac{1}{80}$$

$$q = -293 \text{ mm} \qquad \qquad □$$

□ **Example** An object is 250 mm from a diverging lens of 150 mm focal
length. A converging lens of focal length 100 mm is 50 mm beyond the
diverging lens (Fig. 25.9). Find the position of the final image.

For the first lens,

$$\frac{1}{p} + \frac{1}{q} = \frac{1}{f}$$

$$\frac{1}{250} + \frac{1}{q} = \frac{1}{-150}$$

$$q = \frac{-750}{8} \text{ mm}$$

For the second lens,

$$p = \frac{750}{8} + 50 = \frac{1,150}{8} \text{ mm}$$

$$\frac{1}{p} + \frac{1}{q} = \frac{1}{f}$$

$$\frac{8}{1,150} + \frac{1}{q} = \frac{1}{100}$$

$$q = 330 \text{ mm} \qquad \qquad □$$

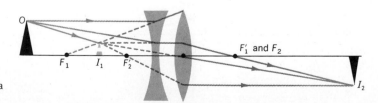

FIGURE 25.9
Image formation by a combination of a
diverging lens and a converging lens.

25.5 THE TELEPHOTO LENS

An interesting application of two lenses in series is found in the tele-photo lens (Fig. 25.10). In order to produce a large image of a distant object with a single lens, the focal length must be large. This requires a camera of inconvenient length, since the length of the camera must be somewhat greater than the focal length of the lens. This difficulty can be avoided by a telephoto lens, which consists of a combination of a converging lens A and a diverging lens B placed at a distance d from each other. The converging lens A would form an image of a distant object just outside its focus F_1, but the lens B causes the image to be formed at F^*. The image formed at F^* is larger than the one that would have been formed at F_1. The magnification of the lens system is increased without increasing the length of the camera too much.

□ **Example** The telephoto lens of Fig. 25.10 is used for taking an action shot at a football game. Lens A has a focal length of 120 mm, and lens B a focal length of -60 mm; they are separated by a distance of 80 mm. If the action is 100 m away, find the position of the final image and the magnification of the system. What focal length would be necessary in a single lens to produce an image of the same size? How long would the camera have to be?

For lens A,

$$\frac{1}{100} + \frac{1}{q} = \frac{1}{0.120}$$

$$q = 0.1202 \text{ m} = 120.2 \text{ mm}$$

and the size of the image is 0.120/100 that of the object.
For lens B, $p = -40.2$ mm and $f = -60$ mm, and

$$\frac{1}{-40.2} + \frac{1}{q} = \frac{1}{-60}$$

$$q = 122 \text{ mm}$$

$$\text{Height of second image} = \frac{122}{40.2} \text{ height of first image}$$

$$= \frac{122}{40.2}\frac{0.120}{100} \text{ height of original object}$$

$$\text{Magnification} = 0.00364$$

The film is $80 + 122 = 202$ mm from lens A.
For a single-lens camera to produce the same image size, $p = 100$ m and $q = 0.00364p = 0.364$ m.

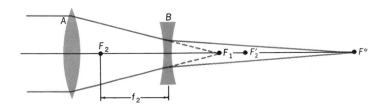

FIGURE 25.10
A telephoto-lens arrangement.

$$\frac{1}{p} + \frac{1}{q} = \frac{1}{f}$$

$$\frac{1}{100} + \frac{1}{0.364} = \frac{1}{f}$$

$$f = 0.362 \text{ m}$$

The minimum length of camera is the distance from lens to film, which is 364 mm; therefore this camera is almost twice as long as the one using the telephoto lens. ☐

A telephoto lens has a fixed effective focal length. For motion pictures and television it is particularly desirable to have a lens system of variable focal length so that the camera can be focused on a scene and then "zoom in" on an item of interest. Such a system, known as a *zoom lens*, requires many lenses as components. The variable focal length is achieved by the axial movement of several lenses simultaneously. They must be moved in such a way that the image is kept in focus and retains the desired brightness. This requires a complicated set of motions of some lenses within the zoom-lens system. A relatively simple zoom lens may have five achromatic doublets (Sec. 25.7) with two of the inner doublets moving together as the effective focal length is varied. A more complex zoom lens may have 18 or more simple lenses, several of which move internally. Modern computers have played an important role in the development of zoom lenses.

25.6 THE LENS-MAKER'S EQUATION

The focal length of a lens depends on the material of which the lens is made and on the radii of curvature of the two surfaces. It is shown below that the focal length is given by

$$\frac{1}{f} = (n - 1)\left(\frac{1}{R_1} - \frac{1}{R_2}\right) \tag{25.3}$$

where n is the index of refraction of the lens material relative to its surroundings, R_1 is the radius of curvature of the surface upon which the light is incident, and R_2 is the radius of curvature of the surface through which the light leaves the lens. R_1 *is positive when the surface is convex* and *negative when the surface is concave to the incident light.* The same rule applies to R_2. By this convention the radius of curvature of a surface is positive when measured from the surface to the center of curvature in the direction the light is leaving the surface. Thus, for a double-convex lens, the first surface has a positive radius of curvature and the second a negative one.

Rays from an object point O (Fig. 25.11) are refracted at the first surface of the lens and would form an image at I_1 if everything to the right of the first surface were medium B. By Eq. (24.5),

$$\frac{n_A}{p_1} + \frac{n_B}{q_1} = \frac{n_B - n_A}{R_1} \tag{A}$$

I_1 serves as object for the second surface; since the light goes from the second surface to I_1, the object distance is negative. Thus $p_2 =$

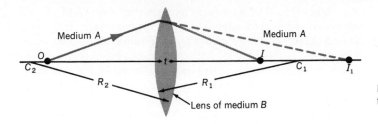

FIGURE 25.11
Light rays from object point O are refracted at both surfaces of the lens to produce an image at I.

$-(q_1 - t)$, where t is the lens thickness. For a thin lens t is negligible compared to q_1, so $p_2 = -q_1$, and for the second surface

$$\frac{n_B}{-q_1} + \frac{n_A}{q_2} = \frac{n_A - n_B}{R_2} \qquad \text{(B)}$$

Adding Eqs. (A) and (B) yields

$$\frac{n_A}{p_1} + \frac{n_A}{q_2} = (n_B - n_A)\left(\frac{1}{R_1} - \frac{1}{R_2}\right)$$

or

$$\frac{1}{p} + \frac{1}{q} = \left(\frac{n_B}{n_A} - 1\right)\left(\frac{1}{R_1} - \frac{1}{R_2}\right) \qquad (25.3a)$$

which is equivalent to Eq. (25.3).

□ **Example** A double-convex lens is made of glass having an index of refraction of 1.5. The front surface has a radius of curvature of 100 mm, and the back surface a radius of curvature of -200 mm (Fig. 25.11). Find its focal length.

$$\frac{1}{f} = (n - 1)\left(\frac{1}{R_1} - \frac{1}{R_2}\right)$$
$$= (1.5 - 1)[\tfrac{1}{100} - (-\tfrac{1}{200})] = \tfrac{1}{2}(\tfrac{3}{200}) = \tfrac{3}{400}$$
$$f = \tfrac{400}{3} = 133 \text{ mm} \qquad \square$$

The reciprocal of the focal length of a lens in meters is the dioptric power of the lens in diopters. Thus, a lens of focal length 500 mm has a power $P = 1/0.50 = 2$ diopters. The powers of lenses used in typical eyeglasses range in magnitude from about 0.2 to 10 diopters.

One advantage of using the powers of lenses is associated with the fact that when two thin lenses are in contact the power of the combination is equal to the sum of the individual powers. If f represents the focal length of the combination of two thin lenses in contact,

$$\frac{1}{f} = \frac{1}{f_1} + \frac{1}{f_2} \qquad (25.4)$$

where f_1 and f_2 are the focal lengths of the two lenses.

25.7 DEFECTS OF LENSES

If we had an ideal lens, every ray of light from any given point on the object would cross at exactly the same point on the image. Further, the image would be similar to the object in every respect. Any circle on

Object

Pincushion

Barrel

FIGURE 25.12
The image of an object may suffer from pincushion or barrel distortion.

the object would be a perfect circle on the image, and any square on the object would lead to a perfect square on the image. If the object were all in one plane, the image would be all in one plane. Actually no such ideal lens exists. For a single lens, the rays from a point on the object ordinarily fall over a small region of the image. If most of the rays from a point on the object cross at the desired image point but some of them cross off on one side, a defect called *coma* occurs. With some lenses the image of a rectangular object (Fig. 25.12) is distorted so that it is *barrel-shaped;* in other cases it may look like a *pincushion.* The detailed explanation of these defects is beyond the scope of this book, but there are three types of image defect which can be readily explained.

Spherical Aberration   When rays of light parallel to the principal axis of a spherical lens pass through zones near its edges, they cross the axis nearer the lens than those rays which pass through nearer to the center. Instead of converging to a sharp focal point, they cross the axis at slightly different points and form a blurred image. This defect, known as *spherical aberration,* can be minimized by using only a small portion of the lens; for a lens of small aperture and long focal length, spherical aberration becomes small. Spherical aberration can be reduced by proper choice of the curvatures of the lens surfaces. When the bending is roughly the same at each lens surface, deviation is minimum, and so is spherical aberration.

Chromatic Aberration   When rays of white light parallel to the principal axis pass through an ordinary lens, they are refracted in such a way that different colors are brought to focus at different distances from the lens (Fig. 25.13). Since violet rays are bent more than the red, the focal length for violet light is smaller. A single lens cannot bring all the rays of white light from a point on the object to a single point on the image. As a result the image is not sharp and is likely to have colored rings or markings. This image defect is called *chromatic aberration.* It can be corrected to a large extent by combining a converging lens of crown glass with a

FIGURE 25.13
Chromatic aberration (top and middle) is corrected by combining two lenses with compensating dispersion but differing deviations to form an achromatic doublet (bottom).

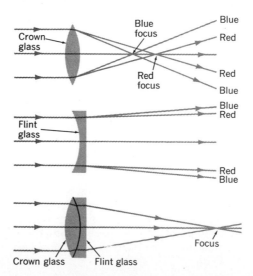

diverging lens of flint glass. These lenses are chosen so that the dispersion of one is equal and opposite to that of the other. In this way no major separation of light into colors occurs. On the other hand, the bending or refractive power of one lens exceeds that of the other; thus there is a resultant bending of the rays in spite of the fact that the lenses deviate the rays in opposite directions. Lenses which are free from dispersion are called *achromatic lenses.* Two lenses combined to form an *achromatic doublet* can bring only two colors to exactly the same focus; the other colors are not perfectly corrected. For better correction of chromatic aberration, more lenses can be used.

Astigmatism   Rays of light from an object point far from the principal axis pass through the lens obliquely and do not converge to a common image point. Figure 25.14 shows side and top views of several rays from the object point O. Rays in the vertical plane converge on I_1, while those in the horizontal plane converge on I_2. Thus, the rays from this single point at O focus in two lines I_1 and I_2, called *focal lines.* At a point midway between I_1 and I_2, a roughly circular patch of light corresponds to the closest available approximation to a point image. In quality optical systems astigmatism is corrected by the use of two or more lenses with the proper separation between them.

In a high-quality camera or microscope a combination of several lenses is utilized to perform the function of a single *ideal* lens. The different kinds of glass used in these lenses, their radii of curvature, and their separations are carefully chosen by the optical designer to minimize image defects.

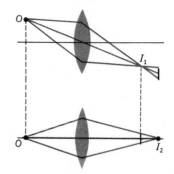

FIGURE 25.14
Astigmatism and astigmatic focal lines. The upper figure shows rays in a vertical plane, and the lower figure shows rays in a horizontal plane.

QUESTIONS

1. What happens to the image formed by a lens if the upper half of the lens is covered? The lower half?
2. Under what conditions is a converging spherical lens thicker at the edges than at the center?
3. If the object for a diverging lens is real, can the image formed by the lens ever be inverted? What arrangement could you set up so that a diverging lens produces a real, inverted image?
4. What happens to the focal length of a lens when it is immersed in water? Does it matter whether the lens is converging or diverging?
5. Why does a lens have two focal points and a mirror only one?
6. Can an empty almost spherical fishbowl focus a beam of light? A full one? A full water glass?
7. If a cylindrical lens is used to form an image of a point source, what will the image look like? Is it really an image in the sense of the discussion in Sec. 25.2?
8. As a self-luminous object moves closer to a lens, does its image get brighter or dimmer? Why? Explain with equations.
9. A mercury thermometer is often made in such a way that the mercury column is viewed through a cylindrical lens. Is the lens converging or diverging? How does it function?
10. If a plano-convex lens is used to form an image of a distant object, should the plane or the convex surface be on the side of the object? Why?

PROBLEMS

Draw a ray diagram whenever appropriate, showing the paths of the three standard rays.

1. A converging lens has a focal length of 18 cm. An object 12 mm high is placed 45 cm in front of the lens. Find the position and height of the image.
   Ans. 30 cm; 8 mm
2. An object 16 mm tall is placed 40 cm from a lens of 60-cm focal length. Find the image distance and the height of the image.
3. A converging lens has a focal length 150 mm. Find the image distance for an object at each of the following distances from the lens: 50, 100, 150, 200, 250, and 500 mm.
   Ans. −75; −300, ∞, 600, 375, 214 mm
4. An object 240 mm from a lens results in a real

image 180 mm from the lens. Find the focal length of the lens. If the image is 12 mm high, how high is the object?

5. An object is placed 125 mm in front of a lens of 150 mm focal length. Find the image position. Is the image real or virtual? Erect or inverted? What is the linear magnification?

      Ans. −750 mm; virtual; erect; 6

6. The image formed by a converging lens is erect, virtual, and 3 times as high as the object. If the focal length of the lens is 144 mm, find the object and image distances.

7. A candle is placed at a distance of 0.6 m from a diverging lens with a focal length of −0.3 m. Where is the image located? How high is it relative to the object?   *Ans.* −0.2 m; one-third

8. An object 24 mm high stands 405 mm from a diverging lens of −135 mm focal length. Find the position and height of the image.

9. An illuminated arrow 15 mm high is placed 585 mm from a diverging lens of focal length −195 mm. Find the position and the height of the image.   *Ans.* −146 mm; 3.75 mm

10. An object 9 mm high is 0.09 m from a diverging lens with a focal length of 0.45 m. Calculate the image distance and the height of the image.

11. A lens made of glass with index of refraction 1.5 has surfaces with radii of curvature of 200 and 600 mm. If this is a double-convex lens, what is the focal length? If double-concave, what is the focal length? What is the power in each case?
Ans. 300 mm; −300 mm; 3.33 and −3.33 diopters

12. A double-convex lens is made of glass (index of refraction 1.5) and has radii of curvature of 225 and 300 mm. Find the focal length of this lens in air and in water.

13. A plano-convex lens is made of glass with index of refraction 1.5. The curved surface has a radius of 375 mm. An object is placed 2.25 m in front of the lens. Where is the image? If the object and lens were both immersed in clear water, where would the image be formed?
      Ans. 1.125 m; −9 m

14. An object 16 mm high is located 20 cm in front of a plano-convex lens of 30 cm focal length. The lens is made of glass with index of refraction 1.55. Find the image distance, the height of the image, and the magnitude of the radius of curvature of the convex side of the lens.

15. A plano-convex lens is made of flint glass, which has an index of refraction of 1.69 for violet light and 1.64 for red light. If the radius of curvature of the curved surface of the lens is 205 mm, what is the difference between the focal lengths of the lens for these two colors?   *Ans.* 23.2 mm

16. A glass lens having an index of refraction of 1.52 and radii of curvature of 375 and −250 mm is cemented to another glass lens having an index of refraction of 1.66 and radii of curvature of −250 and −750 mm. Find the focal length of the combination and its power.

17. A plano-convex lens has a radius of curvature of 125 mm and an index of refraction of 1.60. A plano-concave lens has the same radius of curvature and an index of refraction of 1.50. What is the focal length of the combination if the two lenses are placed in contact to form a plane slab? What is the power?   *Ans.* 1.25 m; 0.80 diopter

18. An object 16 mm high stands 0.75 m in front of a converging lens of 125 mm focal length, and 6 cm beyond the first lens there is a diverging lens with focal length of −150 mm. Find the image distance from the second lens and the height of the final image.

19. An object 35 mm high is 1.5 m from a converging lens of 0.5 m focal length. A diverging lens with a power of −1 diopter is placed 0.35 m beyond the converging lens. Find the position and height of the final image.   *Ans.* 0.667 m; 29.2 mm

20. An object is 12 cm from a lens of 20 cm focal length, and 15 cm beyond the first lens is a second lens with a focal length of 30 cm. Where is the final image relative to the second lens? What is the magnification of the system?

21. A beam of sunlight falls on a diverging lens of −165 mm focal length; 235 mm beyond this is placed a converging lens of 240 mm focal length. Where should a screen be placed to receive the final image of the sun?

     Ans. 0.6 m beyond second lens

22. Two converging lenses, each of 375 mm focal length, are 250 mm apart. An object 12 mm high is placed 0.5 m from one of the lenses. Where is the final image and what is its height?

23. A converging lens with a focal length of 25 cm is placed 0.5 m in front of a diverging lens having a focal length of −40 cm. An object is placed 30 cm in front of the converging lens. Find the position and lateral magnification of the image produced by the two lenses.

     Ans. −0.667 m; 3.33

24. A converging lens and a diverging lens are placed 145 mm from each other. Where will this combination of lenses produce an image of a distant object if the converging lens has a focal length of 361 mm and the diverging lens a focal length of −72 mm?

25. Two double-convex lenses have focal lengths of 0.4 m. They are separated by 0.3 m. An object 45 mm high is placed 1.2 m in front of the first lens. Find the position of the image formed by the system of two lenses. How high is the image?
Ans. 171 mm; 12.8 mm

26. A converging lens with a focal length of 355 mm touches a diverging lens with a focal length of −435 mm. What are the focal length and power of the combination?

27. An object is 0.9 m from a converging lens of 0.3 m focal length. A diverging lens of −0.6 m focal length is placed 0.35 m beyond the converging lens. Find the position of the final image and the ratio of its height to that of the object.
Ans. 0.12 m; 0.6

28. Prove that when two thin lenses are in contact, the power of the combination is the sum of the powers of the two lenses. *Hint:* Assume a distant object and use the fact that the image formed by the first lens acts as object for the second.

29. A telephoto-lens system (Fig. 25.11) consists of a converging lens of 75 mm focal length and a diverging lens of −40 mm focal length separated by a distance of 50 mm. If the system is used to photograph an object 50 m away, how far must the film be from the converging lens? How far would the film be from the lens in a camera using

a single lens if the final image were the same size in both cases?
Ans. 117.5 mm; 203 mm

30. An object and a screen are a distance L apart on an optical bench. A converging lens of focal length f can be at either of two positions to give a sharp image of the object on the screen. Show that the distance of the lens from the object is given by

$$ p = \frac{L}{2} \left(1 \pm \sqrt{1 - \frac{4f}{L}} \right) $$

What is the largest value of f for which a sharp image is formed?

31. Show that if $s = p - f$ (Fig. 25.6) is the distance of an object from the principal focus for parallel rays coming from the opposite side of a lens and $s' = q - f$ is the distance of the image from the other principal focus, then $ss' = f^2$. (This is known as the *newtonian form* of the lens equation.)

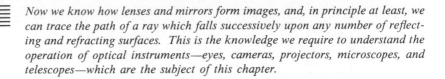

26

OPTICAL INSTRUMENTS

Now we know how lenses and mirrors form images, and, in principle at least, we can trace the path of a ray which falls successively upon any number of reflecting and refracting surfaces. This is the knowledge we require to understand the operation of optical instruments—eyes, cameras, projectors, microscopes, and telescopes—which are the subject of this chapter.

26.1 THE PHOTOGRAPHIC CAMERA

One major application of lenses is in the photographic camera (Fig. 26.1). In a quality camera a combination of lenses acts as a single ideal lens and produces a real image of an external object on a film. The distance between lens and film can be altered to focus the image. In some cameras this adjustment is made by a bellows, and in some by sliding one tube inside another.

The image of a distant object is formed in the focal plane, which contains the principal focus. Therefore, if two cameras of different focal lengths are set up to take the same picture, the lengths of the images are directly proportional to the focal lengths. If both cameras are using the same kind of film, the total amount of light which must reach unit area on each film for proper exposure is the same. The luminous flux entering the camera is directly proportional to the area of the lens opening, which in turn is proportional to the square of the diameter of the circular opening. Consequently, the exposure times of the two cameras are the same if the ratios of focal length to diameter of the opening are the same. *The ratio of focal length to diameter of the opening is called the f number.* When a camera is set at $f/3$, the diameter of the lens opening is one-third the focal length. If the opening is reduced to $f/6$, for the same exposure a time 4 times as long is required. When a picture must be taken in a very short time, a lens of small f number must be used, and this in turn means a large opening. To produce a lens with a large aperture for which lens defects have been satisfactorily corrected is an expensive process. For this reason a camera with an $f/1.5$ lens may cost several times as much as one with an $f/3$ lens.

The smaller the opening through which light is admitted, the greater the depth of focus for a camera. A smaller cone of rays is used to form the image of a point when the opening is small; therefore the circle on which they fall is smaller when a point is not quite in focus. On the other hand, a smaller opening calls for a longer exposure. If one is taking pictures of fast action such as that in a football game, a compromise between short exposure time and great depth of focus must be made.

If the regular lens of a camera is replaced by a wide-angle lens, i.e., one of shorter focal length, a larger area of the object field is focused on the emulsion because the image size of any given object is smaller, as we saw above. On the other hand, a telephoto lens with its longer effective focal length produces a larger image of an object at the cost of a reduced object field.

26.2 THE PROJECTION LANTERN

In the projection lantern the fundamental optical parts are essentially those of a camera. The lantern produces a large image at some distance of a small object close by. All the light which forms the image must come

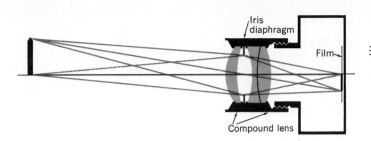

FIGURE 26.1
A photographic camera.

from the object. Since the image has a much greater area than the object, the object must be strongly illuminated if a bright image is desired. Basically, the projection lantern (Fig. 26.2) consists of a powerful source of light, large condensing lenses, and a projecting lens. The condensing lenses collect light from the source and send it through the slide, so that this slide is brilliantly illuminated. The objective lens then projects a real image of the slide on the screen. Since the slide is just outside the principal focus of the projecting lens, the image on the screen is enlarged, real, and inverted.

☐ **Example** Find the focal length of the lens which must be used in a lantern to produce the image of a slide 8 cm square upon a screen 3 m square at a distance of 10 m from the lantern. Assume that the image covers the entire screen.

$$\text{Linear magnification} = \frac{q}{p} = \frac{300}{8}$$

Since $q = 1,000$ cm,

$$\frac{q}{p} = \frac{1,000}{p} = \frac{300}{8} \qquad \text{and} \qquad p = 26.7 \text{ cm}$$

$$\frac{1}{p} + \frac{1}{q} = \frac{1}{f}$$

$$\frac{1}{26.7} + \frac{1}{1,000} = \frac{1}{f}$$

$$f = 25.9 \text{ cm} \qquad\qquad ☐$$

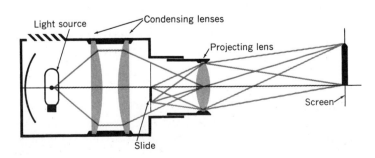

FIGURE 26.2
A slide projector.

Aqueous
humor

Vitreous humor

Ciliary
muscle

Retina

Lens

Fovea

Iris

Cornea

Optic
nerve

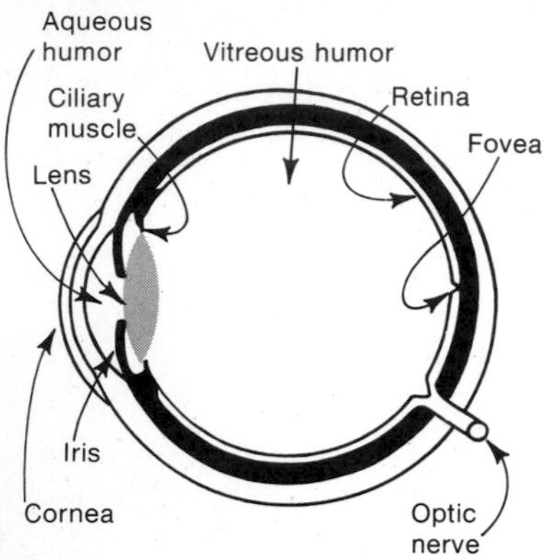

FIGURE 26.3
Right human eye, viewed from above.
The fovea is the area of most acute
vision; it is about 0.75 mm in radius and
subtends an angle of about 5° at the
lens. There is a blind spot where the
optic nerve joins the retina.

26.3 THE EYE

Like the camera, the eye can be looked upon as a lighttight enclosure
having a lens at one end and a sensitive film of nerve fibers at the other
end. The eyeball is roughly a hollow sphere of dense fibrous tissue (Fig.
26.3) about 15 mm in diameter. Light enters through a transparent tissue
called the *cornea* into a clear fluid, the *aqueous humor*. It then passes
through the *lens*, a flexible structure with index of refraction somewhat
greater than that of the aqueous humor and of the jellylike *vitreous humor*
on the other side. The light is brought to focus on the *retina*, which plays
the role of a sensitive screen. The amount of light which enters the eye is
regulated by a circular diaphragm, the *iris*, which is responsible for the
color of the eye. It is provided with muscles which contract or dilate the
pupil, the opening through which the light enters the lens.

The principal bending of light occurs at the cornea, which has a radius
of curvature much less than the remainder of the eyeball. The cornea
provides roughly two-thirds of the bending required to focus the image
of a distant object on the retina, while the lens provides a power of about
20 diopters. When a normal eye is adjusted for very distant objects, it is
relaxed, and the lens has its thinnest shape with surfaces of maximum
radii of curvature. To bring the image of a near object in focus on the
retina, the muscles of the ciliary body which encircles the lens contract,
thereby making the radii of curvature smaller, particularly that of the
front surface. Thus the focal length of the lens is reduced, and the image
focused on the retina.

The remarkable ability of the eye to adjust its focal length is called
accommodation. The greatest and shortest range at which a given eye can
see distinctly are called the *far point* and *near point*, respectively. For a
normal eye the far point is at infinity; the near point may be 60 mm for a
child and usually increases with age.

26.4 DEFECTS OF VISION

If the image formed by a distant object falls in front of the retina, an eye
is *nearsighted* or *myopic* (Fig. 26.4a). This may occur because the lens has
too short a focal length, because the eyeball is too long, or for some other
reason. Such an eye may see a near object distinctly, since the image

(a)

FIGURE 26.4
How eyeglasses correct vision. (*a*) In
nearsightedness the image of a distant
object falls in front of the retina; a diverging lens enables the eye to form a
sharp image on the retina. (*b*) In farsightedness the image of a nearby object falls behind the retina; a converging
lens provides correction.

(b)

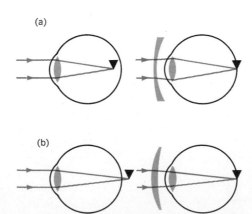

moves back as the object approaches. To correct for *myopia*, a diverging lens is placed in front of the eye. The image formed by this lens serves as the object for the eye.

☐ **Example** A nearsighted eye cannot see objects clearly when they are more than 3 m distant. Find the power of the weakest correcting lens which will just allow this eye to form a sharp image of the moon on the retina.

The image formed by the lens serves as object for the eye. The lens desired must form an image of a distant object 3 m from the eye. Therefore, $p = \infty$ and $q = -3$ m.

$$\frac{1}{p} + \frac{1}{q} = \frac{1}{f}$$

$$\frac{1}{\infty} + \frac{1}{-3} = \frac{1}{f}$$

$$f = -3 \text{ m}$$

$$\text{Power} = \frac{-1}{3 \text{ m}} = -0.33 \text{ diopter} \qquad \qquad ☐$$

In the *farsighted* or *hyperopic* eye the image of a nearby object is formed behind the retina (Fig. 26.4*b*). This defect may arise because the lens is too flat or the eyeball too short. As the object is moved farther away, the image moves nearer to the retina, so this eye may see distant objects clearly. To correct for farsightedness, a converging lens is placed in front of the eye. This makes the equivalent focal length of the eye shorter.

☐ **Example** The near point of an eye is 900 mm. What is the lowest power lens which will permit this eye to see print clearly at a distance of 250 mm?

The image formed by the lens acts as object for the eye. The lens must produce an image of the print at 250 mm at a distance of 900 mm in front of the eye.

$$p = 250 \text{ mm} \qquad q = -900 \text{ mm}$$

$$\frac{1}{p} + \frac{1}{q} = \frac{1}{f}$$

$$\frac{1}{250} + \frac{1}{-900} = \frac{1}{f}$$

$$f = 346 \text{ mm}$$

$$\text{Power} = \frac{1}{0.346} = 2.89 \text{ diopters} \qquad \qquad ☐$$

Another common defect of the eye is *astigmatism*, which occurs when the lens or the cornea does not have a truly spherical surface. The curvature is not the same in different planes containing the axis of the eye. Then the focal length is not the same in different planes. When light from an object is received by the normal eye, there is an attempt by the eye to adjust so that light from all parts of the object forms sharp images. With an astigmatic eye this is not possible, because the optical

(a)   (b)

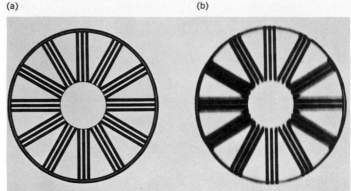

FIGURE 26.5
To the normal eye the spokes of the wheel appear equally sharp (*a*), but to the astigmatic eye some of the spokes are blurred (*b*). (Image *b* was formed with a cylindrical lens in the system.) (*Courtesy of Glenn A. Fry, Ohio State University.*)

system has different focal lengths for light in different planes. Consequently, the image formed is indistinct (Fig. 26.5). This defect can be overcome with lenses that have cylindrical surfaces.

26.5 LUMINOUS FLUX AND THE EYE

The eye has two mechanisms by which it adjusts for changes in the level of illumination. The first is varying of the diameter of the *pupil* through the motion of the iris. In bright light the pupil closes down; in dim light it opens. However, this adjustment covers only a relatively minor range, and the total amount of light entering the eye can be changed by the iris by only a factor of about 16. The eye, however, can see in brilliant sunlight and in dim moonlight, a range of illumination which covers a factor of millions. The principal mechanism which the eye uses to adapt for changes in illumination is that of changing the sensitivity of the retinal surface. To understand this, we must consider the retina in more detail.

Just under the surface of the retina is a mosaic of photosensitive cells, of which there are two kinds, named *rods* and *cones* after their shapes. Both rods and cones have higher indices of refraction than the surrounding material. Light is transmitted down their length by total internal reflection. There are no rods or cones where the optic nerve leaves the eye. The image formed on this region, called the *blind spot*, is not seen. The cones are responsible for color vision; the rods give only achromatic perceptions of white, gray, and black. The rods are grouped in bunches, with each bunch having a connection to the brain, while each cone usually has a private line to the brain. Cones are largely concentrated in the central region of the retina, the *fovea*. As we go away from this central region, the number of cones per unit area decreases. The number of rods per unit area increases to a maximum at 20° from the optic axis and then decreases slowly.

The cones of the foveal area give us acute and color vision under conditions of high to adequate illumination. The solid line of Fig. 26.6 shows how cones respond to equal intensities of various wavelengths. For equal amounts of light energy a yellow-green light of $\lambda \approx 560$ nm

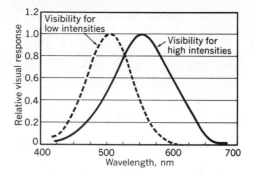

FIGURE 26.6
Relative sensitivity of the eye to equal intensities of monochromatic radiation. The solid curve is for cones, the dashed curve for rods.

appears brightest to the normal light-adapted eye. At this wavelength 1 W of radiant energy corresponds to 680 lumens of luminous flux (Sec. 28.1). Under illuminances lower than about 0.5 lm/m² the cones do not respond well. The rods are sensitive to very small amounts of radiant energy. The dashed curve of Fig. 26.6 shows that the rods have maximum sensitivity at $\lambda \approx 510$ nm; they are relatively more sensitive to blues and less sensitive to reds than the cones. This is one reason for using blue lights for aisles of theaters and darkened railroad trains.

On the rods the eye builds a photosensitive pigment called *rhodopsin* or *visual purple*. When light strikes a rod, rhodopsin is bleached, and a signal goes to the brain. In bright daylight little rhodopsin is present. In darkness the rhodopsin is built up in the rods, and the eye becomes *dark-adapted*. It takes about 30 min to build up a maximum concentration. When this has been accomplished, the eye is very sensitive to light. This accounts for the blinding effect of bright headlights from an approaching car at night; these same lights would produce no problem at all in daytime.

When the eyes are focused on a distant object, the visual axes are parallel. For viewing a near object there is a movement of the eyeballs which causes the visual axes to converge. Such a movement is necessary for the image in both eyes to fall on the proper area of the retina. Because of the distance between them, the eyes view an object from slightly different angles. Each retina receives a slightly different picture; this helps us judge distance, solidity, and depth.

26.6 SIMPLE MICROSCOPE

For most distinct vision, an object must be about 250 mm from a "normal" eye. If the object is placed at a greater distance, the image on the retina is smaller and its details are not seen so distinctly. When the object is placed too near, the image on the retina is blurred unless an additional converging lens is used to aid the eye. A lens used in this way constitutes a *simple microscope* or *magnifying glass*. When the object *OP* (Fig. 26.7) is inside the principal focus, the lens forms an erect, virtual, magnified image *IQ* which in turn serves as object for the eye, placed just to the right of the lens. To obtain the greatest advantage, the eye should be as near as possible to the lens. In this way, the field of view is made as large as possible.

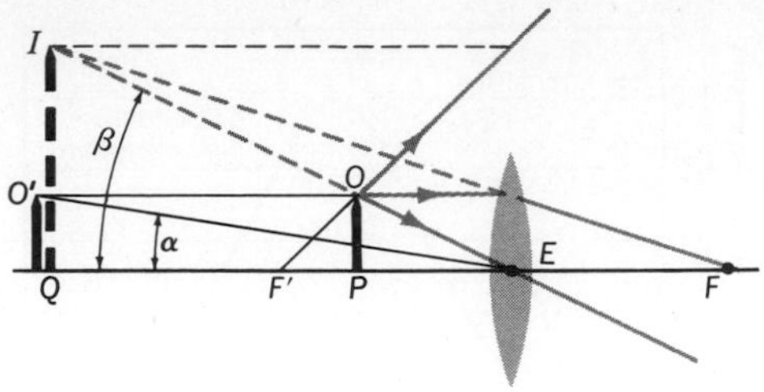

FIGURE 26.7
A simple microscope produces a virtual, erect, magnified image.

When an object is examined with the aid of a simple microscope, the object is brought nearer to the eye than would be possible for distinct vision without the magnifying glass. In Fig. 26.7 the angles subtended at the center of the lens by the object OP and the image IQ are the same. If QE is the distance of most distinct vision, the eye could not see OP distinctly without the lens until it was removed to $O'Q$. Hence, by using the lens, the usable visual angle subtended by OP has been increased from α to β. The *angular magnification M* of a microscope is the ratio of the angle subtended at the eye by the image formed by the microscope to the angle subtended by the object when it is at the point of most distinct vision. For the simple microscope of Fig. 26.7 the angular magnification, also known as the *magnifying power*, is given by

$$M = \frac{\beta}{\alpha} = \frac{IQ/QE}{O'Q/QE} = \frac{IQ}{OP} = \frac{QE}{PE}$$

If we take $QE = 250$ mm, the distance of most distinct vision for a normal eye, and observe that PE is the object distance p, we have, by Eq. (25.1), $1/p + 1/(-250 \text{ mm}) = 1/f$, and

$$M = \frac{250 \text{ mm}}{p} = \frac{250 \text{ mm}}{f} + 1$$

for an eye placed as close as possible to the lens.

It is usually less tiring to use a simple microscope if the final image is at infinity, since then the eye muscles are relaxed. In this case

$$\beta = \frac{OP}{f} \qquad \text{and} \qquad \alpha = \frac{OP}{250}$$

so that
$$M = \frac{250 \text{ mm}}{f}$$

With a single converging lens used at the eye, angular magnifications up to about 4 are possible before aberrations become a serious problem. By using eyepieces composed of two lenses it is possible to correct for aberrations and increase the angular magnification to 15 or 20.

26.7 COMPOUND MICROSCOPE

To obtain great magnification, a compound microscope is used. It utilizes a group of lenses acting as an ideal single converging lens L of short focal length (Fig. 26.8). This lens, called the *objective*, produces a real, magnified image IQ of a small object placed just outside its focus. The image IQ is viewed through the eyepiece L', which acts as a simple microscope. This *ocular* produces an enlarged, virtual image $I'Q'$ of the real image IQ.

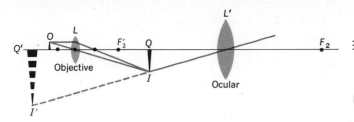

FIGURE 26.8
Lens scheme for a compound microscope (not to scale).

The magnification of a compound microscope depends on the focal lengths of objective and eyepiece. The objective might have a focal length of 5 mm and form a real image 200 mm from the objective. By Eq. (25.1), $1/p + \frac{1}{200} = \frac{1}{5}$, or $p = \frac{200}{39}$ mm. The linear magnification produced by the lens is then

$$\frac{q}{p} = \frac{200}{\frac{200}{39}} = 39$$

If now the eyepiece has a focal length of 25 mm and is adjusted for a final image at infinity, the angular magnification is $250/f = 10$. The overall angular magnification is $39 \times 10 = 390$.

26.8   ASTRONOMICAL TELESCOPE

The principle of the astronomical telescope (Fig. 26.9) is the same as that of the compound microscope. The objective lens is modified so that the instrument can be used to view distant objects. The objective, a large converging lens of long focal length, forms a real, inverted image IQ of a distant object OP. This real image is formed in the focal plane of the ocular, by which an enlarged virtual image $I'Q'$ of the real image IQ is produced at infinity.

Because of the great distance of the object from the telescope, the rays from any point are essentially parallel when they reach the objective. Hence, the image formed lies essentially in the focal plane of the objective. The angular magnification of the telescope is determined by the angles that the object subtends and appears to subtend at the eye (Fig. 26.9). The angle α that the object OP subtends at the eye when no lenses are present is $OEP = IEQ$. The angle β that the image $I'Q'$ subtends at the eye is $I'GQ' = IGQ$. For a distant object, with the ocular adjusted to place $I'Q'$ at infinity, F_1, Q, and F'_2 coincide. Then the length of the telescope is $f_0 + f_e$, and

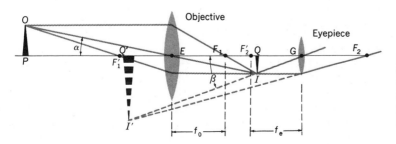

FIGURE 26.9
Lens arrangement of an astronomical telescope (not to scale). Actually IQ is formed at the principal focus of the objective lens, and the final image $I'Q'$ is placed at infinity by adjusting the eyepiece so that IQ is at its principal focus F'_2.

$$\text{Angular magnification} = \frac{\beta}{\alpha} = \frac{QI/f_e}{QI/f_0}$$

$$= \frac{f_0}{f_e} = \frac{\text{focal length of objective}}{\text{focal length of eyepiece}}$$

The image formed by an astronomical telescope is inverted. For terrestrial uses of the telescope it is desirable that the image be erect. This condition is realized by inserting a third convex lens between the objective and the eyepiece in such a way that the image formed by the objective is again inverted before it is viewed by the eyepiece.

☐ **Example**   The focal length of the objective of a telescope is 1,500 mm, and the focal length of the eyepiece is 20 mm. Find the angular magnification of the telescope for distant objects.

$$\text{Angular magnification} = \frac{\text{focal length of objective}}{\text{focal length of eyepiece}}$$

$$= \frac{1,500}{20} = 75 \qquad\qquad ☐$$

In very large telescopes the objective lens is replaced by a concave mirror. The great mirror in use at the Mount Palomar Observatory has a diameter of 200 in. (5.08 m).

26.9   PRISM BINOCULARS

A terrestrial telescope of reasonable field and magnification becomes inconveniently long. The prism binocular has largely replaced the old "spy glasses." Each side of the binocular achieves a long optical path by making the light traverse almost three times the length of the binoculars (Fig. 26.10). The beam of light OB from the objective is reflected internally at B and C by a right-angled prism. In this way its direction is reversed, and it travels back to a second right-angled prism which is placed at right angles to the first. Here it is again reflected internally at D and E and then passes through the eyepiece. The image, after the reflec-

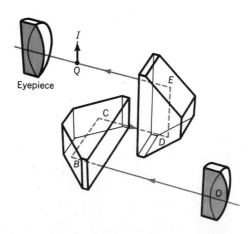

FIGURE 26.10
Optical scheme of one telescope of prism binoculars.

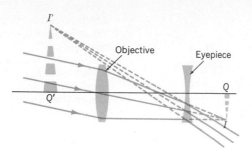

FIGURE 26.11
Optical scheme of an opera glass. The final image $I'Q'$ is usually placed at infinity, in which case Q coincides with one of the principal foci of the eyepiece.

tions by the two right-angled prisms, is restored completely to the upright position. The eyepiece, serving as a simple microscope, magnifies this image.

26.10 OPERA GLASS OR GALILEAN TELESCOPE

The opera glass uses an objective which converges the rays from a distant object toward an image IQ (Fig. 26.11). Before they reach this image they pass through a diverging lens which deviates rays that were converging to make them diverge. To an eye on the right-hand side of the eyepiece, the rays appear to have come from the virtual, erect image $I'Q'$.

To focus the opera glass, the eyepiece is adjusted so the rays emerging from it are parallel and the final image is at infinity. Then,

$$\text{Angular magnification} = \frac{\text{focal length of objective}}{\text{focal length of eyepiece}} = \frac{f_o}{f_e}$$

The length of the opera glass is $f_o - |f_e|$; thus the glasses need not be as long as a telescope of comparable magnifying power. A major disadvantage of the opera glass is its small field of view.

QUESTIONS

1. Can you photograph a virtual image? Could you do it by placing a film at the position of the image? Explain.
2. How is the camera similar to the eye? How different?
3. Why is the depth of focus of a camera at $f/11$ greater than the depth of focus of the same camera at $f/3.5$?
4. Does a wide-angle lens for a camera have a longer or a shorter focal length than the standard lens for the camera? Why? How does the focal length of a telephoto lens compare with the focal length of the standard lens? What are the advantages and disadvantages of the wide-angle and telephoto lenses?
5. If the screen is moved closer to a slide projector, which way must the lens be moved to bring the image back into sharp focus?
6. How does the eye accommodate to see objects which are at different distances?
7. When swimmers are under clear water, why can't they see nearby objects clearly? How do goggles help swimmers see better under water?
8. There is a blind spot in each eye where the optic nerve leaves the retina. Can you devise a simple experiment to locate the blind spot approximately? Why isn't a person more conscious of the blind spots and bothered by them?
9. How does the eye adjust to permit it to function under conditions of radically different illumination?
10. Why is a simple microscope useful although its image gets farther away at the same rate the linear dimensions of the image increase?
11. Why do inexpensive microscopes for children form images with colored edges? How is this avoided in more expensive microscopes?

12. Why does the magnification of a telescope increase with the focal length of the objective while the magnification of a compound microscope decreases with increasing focal length?

PROBLEMS

1. A camera has a lens of 50 mm focal length. When it is set at $f/2$, what is the diameter of the lens opening? If the setting is changed to $f/9$, how much more exposure time is required? If the camera is originally set to photograph distant objects, how far must the lens be moved to focus on a flower 0.6 m away?
                    Ans. 25 mm; 20.3 times; 4.55 mm
2. The lens of an aerial camera has a focal length of 0.5 m. What will be the dimensions on the film plate of a square of the earth's surface 3 km on a side, photographed from a height of 8 km?
3. A camera lens is set at $f/3$ for a picture which includes a distant point source of light. If the film is actually 0.5 mm too far from the lens for the point source to be in focus, what is the diameter of the circle of light on the film? If the lens had been at $f/11$, what would the diameter have been?   *Ans.* 167 μm; 45 μm
4. The lens of a camera has a focal length of 125 mm. It is used to photograph an object that is 3 m from it and then an object that is 30 m from it. How much must the lens of the camera be moved?
5. A camera has a regular lens of 50 mm focal length and a telephoto lens of 150 mm focal length. If each is used to take a picture of a statue 3 m high from a distance of 20 m, find the height of the image on the film in each case.
                    Ans. 7.5 mm; 22.7 mm
6. An enlarging camera has a lens with a focal length of 125 mm. It is used to enlarge a negative of dimensions 28 by 35 mm. How far from the lens must the negative be placed for the enlargement to be 120 by 150 mm?
7. A projection lantern is desired which will throw an image 1.5 m wide of a slide 35 mm wide on a screen 6 m from the lens. What must the focal length of the projecting lens be?   *Ans.* 137 mm
8. The lens of a projection lantern is to be 20 m from a screen on which an image with a height of 2 m is desired. The height of the slide is 6 cm. What should the focal length of the projecting lens be?
9. A farsighted eye cannot form a sharp image of an object which is closer than 375 mm. Find the focal length and the power of a lens which will permit the eye to form a sharp image of an object 200 mm distant but not of an object closer than this.   *Ans.* 429 mm; 2.33 diopters
10. A farsighted eye has a near point of 426 mm. Find the focal length and the power of the lens which will bring the near point to 142 mm. Where would the near point for this eye be with a lens having a power of 4 diopters?
11. A nearsighted person cannot see objects clearly when they are more than 1.25 m from his eyes. Find the power of the weakest lens which will permit him to see distant objects distinctly. If he puts on a pair of −0.6-diopter glasses, where is his new far point?   *Ans.* −0.8 diopter; 5 m
12. A myopic (nearsighted) eye has a far point of 667 mm. What is the power of the spectacle lens required for this eye to see distant objects clearly? What would be the far point for this eye with a lens of −1.2 diopters?
13. The near point of an eye is 0.5 m. Find the smallest distance at which a slide rule can be held and clearly read if a spectacle lens of +4 diopters power is worn.   *Ans.* 167 mm
14. A double-convex lens of 62.5 mm focal length is used as a simple microscope. If it is held close to the eye to form an image 25 cm from the eye, where should the object be located? What is the angular magnification? What would the angular magnification be if the final image were at infinity?
15. A farsighted eye has a far point of infinity and a near point of 750 mm. A plano-convex lens made of glass with an index of refraction of 1.5 is used to enable the eye to see an object 250 mm away. What is the largest radius the curved surface may have? Where would the near point and the far point be if this eye were using a 5-diopter lens?
                    Ans. 188 mm; 158 mm; 200 mm
16. A thin converging lens of 10 cm focal length serves as a simple microscope. By drawing the three standard rays find the image of a real object 75 mm from the lens graphically. Check your result by finding the position of the image analytically.
17. A metric scale is placed 250 mm from the eyes and observed with one eye unaided. The other eye observes a closer similar scale through a converging lens placed close to the eye. A magnified millimeter division appears to be the same size as 4 mm divisions seen with the naked eye

when the image formed by the lens is also 250 mm from the eyes. Find the focal length of the lens. *Ans.* 83.3 mm

18. A converging lens used as a reading glass has a focal length of 55.6 mm. What is the angular magnification if the lens is used to produce an image 250 mm from the eye?

19. A source emits monochromatic light of 560 nm wavelength. If the radiant flux (energy per unit time) through an aperture is 0.3 W, what is the luminous flux through this aperture? If the source had emitted light of 600 nm wavelength, what would the flux have been (Fig. 26.6)?
Ans. 204 lm; ~120 lm

20. A compound microscope has an objective lens of focal length 5 mm which forms an image 150 mm from the lens. The eyepiece produces a magnification of 8. What is the total magnification?

21. A crude microscope consists of an objective of 15 mm focal length and an eyepiece of 50 mm focal length. The lenses are placed 170 mm apart. A person uses the microscope to form an image at infinity. Where must the object be placed? What is the linear magnification of the objective? The angular magnification of the eyepiece? The overall magnification?
Ans. 17.1 mm; 7; 5; 35

22. The objective of a compound microscope has a focal length of 5 mm, and its eyepiece has a focal length of 30 mm. If the distance between the objective and the eyepiece is 230 mm, what is the linear magnification of the objective when the image is at infinity? Where must the object be placed? What is the overall angular magnification?

23. An 8-mm microscope objective forms an image at a distance of 168 mm. The eyepiece has a focal length of 25 mm. Find the lateral magnification of the objective and the angular magnification of the microscope, assuming the final image is at infinity. If an eye can see two dots as separate

when they are 100 μm apart, how close together could two dots be and still be resolved by this microscope? *Ans.* 20; 200; 0.5 μm

24. The world's largest refracting telescope at the Yerkes Observatory has an objective lens approximately 1 m in diameter with a focal length of 18.9 m. If the image formed by this objective is viewed through an ocular (eyepiece) of 75 mm focal length, find the angular magnification for viewing at great distances.

25. A telescope which consists of an objective that has a focal length of 0.8 m and an eyepiece with a focal length of 25 mm is used to view an object that is 16 m from the objective. What must the distance between the eyepiece and the objective be for a final image at infinity? What is the overall angular magnification? *Ans.* 0.867 m; ~34

26. What is the diameter of the image of the moon formed at the prime focus by the 200-in. (5.08-m) reflector on Mount Palomar if the mirror has a prime focal length of 16.8 m and the moon subtends an angle of 0.009 rad at the earth?

27. The objectives of a pair of binoculars have apertures of 35 mm diameter and focal lengths of 240 mm. The oculars (eyepieces) have focal length of 34 mm. Find the angular magnification when the final image is placed at infinity. Why are these binoculars designated as 7 × 35?
Ans. 7.06; ang. mag. × objective aperture

28. The eyepiece of a telescope has a focal length of 44 mm. When the final image is viewed at infinity, the objective and the eyepiece are 1.264 m apart. Find the angular magnification of the telescope and the focal length of the objective.

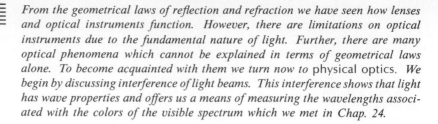

From the geometrical laws of reflection and refraction we have seen how lenses and optical instruments function. However, there are limitations on optical instruments due to the fundamental nature of light. Further, there are many optical phenomena which cannot be explained in terms of geometrical laws alone. To become acquainted with them we turn now to physical optics. *We begin by discussing interference of light beams. This interference shows that light has wave properties and offers us a means of measuring the wavelengths associated with the colors of the visible spectrum which we met in Chap. 24.*

27 INTERFERENCE AND DIFFRACTION

27.1 DOUBLE-SLIT INTERFERENCE

In 1801 Thomas Young performed a celebrated experiment which led to the general acceptance of the wave theory of light. Huygens and others had advanced the wave theory much earlier, but their arguments were inconclusive until Young showed that light beams produce interference phenomena.

In order to understand Young's experiment, consider a source of light placed behind a screen containing a narrow slit S (Fig. 27.1). A second screen having two narrow slits A and B is placed in front of the first screen in such a way that the openings are equally illuminated by the light from S. If the illumination on a third screen DF is examined, it is found that a series of light and dark bands (Fig. 27.2b) result. At point D, where we might expect the center of a shadow, there is a bright line. This, of course, means that light from the source S arrives at the screen at a point which it could not reach if light travels in perfectly straight lines. Thus, any explanation requires acceptance of the fact that light bends, to some extent at least, around obstacles. *The bending of light when it passes an obstacle is called diffraction.*

To explain the pattern of Fig. 27.2b, we accept the tentative hypothesis that light is a wave motion and that Huygens' principle is applicable. Then we can understand the origin of the bright and dark lines by the following reasoning. A wavefront from S reaches slits A and B. Each point on this wavefront acts as a new source of Huygens' wavelets. Since A and B are equidistant from S, the Huygens' sources at A and B both send out crests at the same time, and half a cycle later both send out troughs. The waves from A and B arrive at the screen DF in such a way that at some points crests meet crests (Fig. 27.2a) and troughs meet troughs, thereby giving constructive interference and a bright line. At other points the troughs meet crests and crests meet troughs, resulting in destructive interference and darkness. The bright and dark bands on the screen are called *interference fringes.*

The experiment we have just described with slits differs from that of Thomas Young in that he used pinholes in the screens rather than slits. In his experiment, interference fringes were also formed, but they were not straight lines.

Let us now see how we can determine the wavelength of light from the interference pattern observed. Note that a bright fringe is produced at D (Fig. 27.1), which is in the heart of the shadow of the slit. This bright line arises from the constructive interference of the light coming through slits A and B. The waves from these two slits arrive in phase, since the light

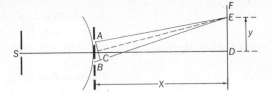

FIGURE 27.1
When light from a source S passes through two parallel slits A and B, an interference pattern is formed on the screen FD.

travels exactly the same distance in both cases. For the first bright fringe above D, the light must travel one wavelength farther from slit B than from slit A. For the nth bright fringe above D, the light from B goes n wavelengths farther than that from A. Suppose for the moment that the nth bright fringe is formed at point E. For this particular case the distance BC is n wavelengths if $CE = AE$. On the other hand, for a dark fringe at E the path difference BC must be some odd number of half wavelengths so that the waves interfere destructively.

We can calculate the path difference BC as follows. Let d represent the distance between slit centers, X represent the distance from the double slit to the screen DF, and y represent the distance ED. Then

$$(AE)^2 = X^2 + \left(y - \frac{d}{2}\right)^2 \quad \text{and} \quad (BE)^2 = X^2 + \left(y + \frac{d}{2}\right)^2$$

$$(BE)^2 - (AE)^2 = X^2 + y^2 + yd + \frac{d^2}{4} - X^2 - y^2 + yd - \frac{d^2}{4} = 2yd$$

$$(BE - AE)(BE + AE) = 2yd$$

(a)

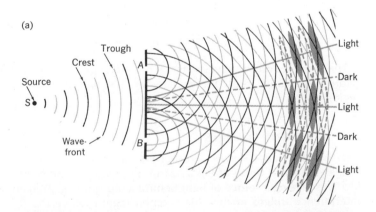

(b)

FIGURE 27.2
Interference of light from the parallel slits A and B (a) gives rise to interference fringes (b).

Then
$$BE - AE = \frac{2yd}{BE + AE} \approx \frac{2yd}{2X}$$

since BE and AE are almost exactly equal to X. Therefore, we have

$$\text{Path difference} = BC \approx \frac{yd}{X} \tag{27.1}$$

The central bright fringe at D occurs for zero path difference. If we count this fringe as zero, the nth bright fringe in either direction comes when the path difference BC is $n\lambda$, which we know from Eq. (27.1) to be yd/X. For the dark fringes, on the other hand, the difference in distance traversed from the two slits must be some odd integral number of half wavelengths. The first dark fringe occurs for $yd/X = \lambda/2$, the second for $yd/X = 3\lambda/2$, and so forth.

We have required that our slit S be illuminated with monochromatic light. If we use white light, the red wavelengths interfere constructively at different places than the blue. As a result, we observe white light at the central point D; then there is a spectrum on each side of D in which the path difference is one wavelength, then another spectrum in which the path difference is two wavelengths, etc. However, in a short distance the spectra overlap seriously, and the illumination appears uniform.

□ **Example** The distance between slits A and B in Young's interference experiment is 2.5 mm. The distance X from the slits to the screen on which the fringes are formed is 1,000 mm, and the distance from the central bright fringe to the third dark one is 0.59 mm. Find the wavelength of the incident light.

For the third dark fringe the path difference BC must be 2.5λ.

$$2.5\lambda = \frac{yd}{X} = \frac{(0.59 \text{ mm})(2.5 \text{ mm})}{1,000 \text{ mm}}$$

$$\lambda = 0.00059 \text{ mm} = 590 \text{ nm} = 5900 \text{ Å} \qquad \square$$

The eye responds to waves ranging in wavelength roughly from 0.38 μm (violet) to 0.78 μm (red). Other detectors respond to other wavelength regions.

27.2 COHERENT SOURCES

Shortly after Young's work, Fresnel was able to produce interference fringes similar to those of Fig. 27.2b in other ways. One of his methods involves placing a source of light behind a biprism (Fig. 27.3) and finding interference fringes in the crosshatched region where the light which went through the upper prism and the light through the lower prism overlap. Still another method devised by Fresnel involved the use of two mirrors set so that the angle between their surfaces is very nearly zero. The Fresnel double-mirror arrangement is shown in Fig. 27.4.

In Young's experiment and in the two interference experiments of Fresnel, light must come from a single source. No interference would be observed in Young's experiment, for example, if one light source were placed behind slit A and another behind slit B. Each source would send out a very large number of waves with completely random phases. At one

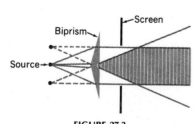

FIGURE 27.3
Fresnel biprism experiment; light from source S passing through the biprism appears to come from the virtual sources shown, and interference fringes are formed in the gray region.

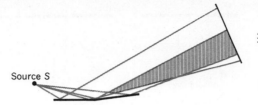

Source S

FIGURE 27.4
Fresnel double-mirror method for producing interference.

instant the crest of one wave might meet the trough of another, but an instant later crest might be meeting crest. The light reaching any area would be the sum of the contributions of each of the sources. When two sources send out waves in such a way that there is no regular phase relationship between these waves, the sources are *incoherent*. If there is a regular phase relationship between the waves, the sources are *coherent*.

In the interference experiments which we have discussed, light comes from a single source. A single wavefront is separated into two parts which travel different paths. When the parts are recombined, interference results. In order to produce interference, there must be two "apparent" sources initiating light waves with a well-defined phase relationship. This cannot be achieved with independent light sources. For electromagnetic waves in general it can be done in one of two ways: (1) we can start with a single source, send portions of the wavefronts by different paths, and recombine them in some region or (2) we can drive a source in such a way that it radiates coherently either with a master source or with another source controlled by the same master. In the *laser* excited atoms in a source are led to emit coherent radiation by stimulating emission from the atoms (Sec. 47.9).

27.3 LLOYD'S MIRROR AND PHASE CHANGE AT REFLECTION

Lloyd was able to obtain interference fringes with a light source and a single mirror in the manner indicated in Fig. 27.5. Light coming directly from the source and light reflected from the plane mirror overlap in the crosshatched region. If a screen is placed at the end of the mirror, one might expect a bright fringe at point P where the screen and mirror come in contact. However, a dark fringe is observed at this point. This fact is explained by a change in phase of one-half wavelength (or 180°) which occurs when light is reflected at the interface between two media if the light approaches the interface through the medium in which it has the higher speed. *When light in a medium is reflected at the surface of a second medium which has a greater index of refraction, there is a half-wavelength phase shift; there is no phase shift when the second medium has a lower index of refraction.*†

It may be helpful to consider an analog to this phase shift at reflection. If a small mass m is set into vibration as a pendulum bob, the x component of the displacement as a function of time is given by the sine wave of Fig. 27.6a. If as m swings through the equilibrium position it collides with a larger mass M, the laws of conservation of momentum and energy lead

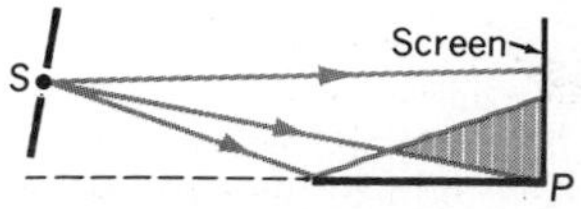

†This is an oversimplification which leads to correct results in any situation discussed in this text.

S

Screen

P

FIGURE 27.5
Lloyd's mirror experiment.

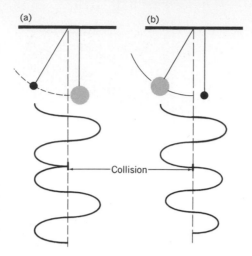

FIGURE 27.6
Mechanical analog of the 180° phase shift in reflection. The smaller mass reverses direction in collision with a larger mass (*a*), while the latter continues on after collision with the smaller mass (*b*).

to a recoil of *m*. From the figure we can see that the small mass has missed one-half of an oscillation; one-half wavelength is missing. On the other hand, if a large mass *M* serves as the bob of the pendulum and it has a collision with a smaller mass *m* as it swings through the equilibrium position (Fig. 27.6*b*), the larger mass continues in its original direction with somewhat reduced amplitude. There is no phase shift when the larger mass collides with the smaller one, but a half-wavelength change in phase when the smaller mass collides with the larger.

27.4 INTERFERENCE IN THIN FILMS

Consider the thin air film between two optically plane pieces of glass when the plane surfaces touch at one edge and are slightly separated at the other end (Fig. 27.7). When monochromatic light falls normally from above on this air wedge, part of the light is reflected at the upper surface of the wedge and part of the light passes to the bottom, where some of the light is reflected, this time with a change in phase of one-half wavelength. If the air wedge is viewed from above, a series of light and dark bands (Fig. 27.8) is observed.

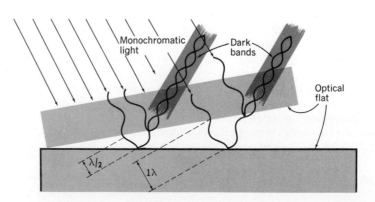

FIGURE 27.7
When a thin air wedge is illuminated from above, interference fringes are produced through the superposition of light reflected at the two surfaces of the wedge.

FIGURE 27.8
Interference by air films between an optical flat and gauge blocks. The parallel, equispaced fringes show that the block at the left is plane, while the curved fringes show that the block on the right is worn. (*Do-All Company.*)

In the region in which the two plates touch there is a dark band because although the two reflected waves go the same total distance, there is a change in phase for the reflection in which the light goes from air to glass. The first light fringe occurs when the air wedge is one-fourth wavelength thick. The wave reflected from the lower plate goes one-half wavelength farther than the wave reflected at the upper surface, and there is a half-wavelength phase shift at the lower reflection. Therefore, the waves reflected at the two sides of the air wedge meet in phase, and a light band results. When the thickness of the air wedge is one-half wavelength, the wave reflected at the lower plate goes one wavelength farther, but the half-wavelength change in phase on reflection again puts the two waves out of phase by one-half wavelength, and a dark band results. When the air wedge is three-quarters of a wavelength thick, constructive interference occurs once more, and so forth. There is a bright band whenever the thickness of the air wedge is an odd number of quarter wavelengths and a dark band whenever the thickness of the air wedge is an even number of quarter wavelengths. If n represents the number of a bright fringe, counting the first bright fringe after the point of intersection of the planes as 1, the thickness of the wedge at the nth bright fringe is given by $(2n - 1)\lambda/4$.

Vernier and micrometer calipers in precision machine shops can be checked by making certain that they read the width of a working gauge block correctly. To make sure that this block has not been worn down, it is compared with a master block by using the interference of light waves. The two blocks are placed next to each other on a flat surface, and an optically flat glass plate is placed on top of them (Fig. 27.9). If the two are exactly the same height and the optical flat is in contact with both blocks across their entire upper surfaces, no interference fringes are observed. If the working block is shorter than the master block, parallel fringes are observed, and the amount by which the working block is low can be calculated.

□ **Example** When the optical flat of Fig. 27.9 is viewed from above with helium yellow light of $\lambda = 587.6$ nm, there is a dark fringe where the flat

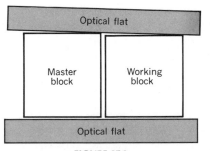

FIGURE 27.9
Use of interference to test gauge blocks.

touches the working block, a second dark fringe one-third the way across the block, a third one two-thirds the way across, and a fourth one just before the master block is reached. How much shorter is the working block than the master?

At the line of contact the separation is a negligible fraction of a wavelength. At the second dark fringe the air wedge is $\lambda/2$ thick, at the third λ, and at the last $3\lambda/2$. Therefore, the working block is $3\lambda/2$ shorter than the master block.

$$\frac{3\lambda}{2} = \frac{3}{2} \times 587.6 = 881 \text{ nm} \qquad \square$$

When a plano-convex lens is placed on an optical flat and illuminated with monochromatic light from above, a series of light and dark rings is observed. This pattern is known as *Newton's rings*. These light and dark rings are analogous to the light and dark straight fringes observed with plane surfaces. In the region where the lens touches the optical flat, there is destructive interference. Where the thickness of the air film is an odd number of quarter wavelengths, a bright circular fringe is seen, while there are dark fringes where the thickness is an integer times $\lambda/2$.

Consider a very thin film of transparent material with surfaces parallel to each other (Fig. 27.10). If one of these surfaces is illuminated by a beam of light of a single wavelength, light is reflected from the upper and lower surfaces. An incident ray such as AB is partly reflected from the upper surface, but most of the light enters the film, and a small part of it is reflected at C. In a similar manner, part of the ray DE is reflected at E, and the remainder of the light enters the film. Some of it is reflected at F, etc. If rays parallel to AB and DE illuminate the upper surface XY of the film, parallel rays EK, GH, etc., are reflected from the upper surface. There are also rays which are parallel to EK, GH, etc., which are reflected from the lower surface MN of the film. The rays reflected from the lower surface MN are superposed on those reflected from the upper surface XY. These two sets of reflected beams have nearly the same brightness but differ in phase, because those which are reflected from the lower surface of the film have traveled a distance significantly greater than those reflected at the upper surface while the latter have undergone a phase shift of one-half wavelength. The two sets of beams may interfere constructively or destructively, depending on the angle of incidence, the wavelength of

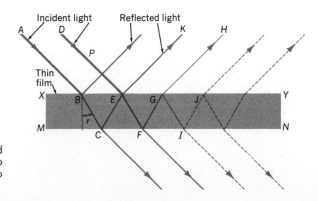

FIGURE 27.10
Interference fringes may be produced when light is reflected from the two surfaces of a thin film and the two beams recombine.

the light, and the thickness and refractive index of the film. The condition for destructive interference does not occur at the same place for different wavelengths. Hence, when the film is illuminated by white light, certain wavelengths reinforce each other where light of other wavelengths destroy each other. The result is a series of colored fringes. The colors of thin films of oil illustrate this type of interference.

27.5 THE MICHELSON INTERFEROMETER

An interesting example of a device in which part of a wavefront is sent along one path and part along a different one is the Michelson interferometer (Fig. 27.11). Light emerging from a source S falls on a glass plate A, one surface of which is thinly silvered so that it reflects half the light and transmits the remainder. The reflected part goes to mirror C, by which it is reflected, and returns along its original path. The part which went through plate A passes through a second glass plate B, is reflected at mirror D, and returns to mirror A. There it is reflected in such a way that its

(a)

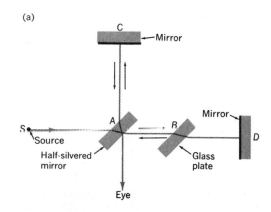

(b)

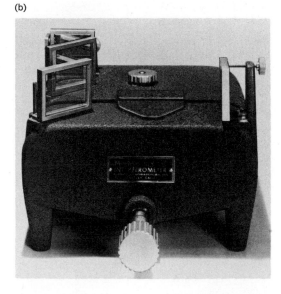

FIGURE 27.11
(a) Optical paths in a Michelson interferometer and (b) a commercial interferometer. (*Atomic Laboratories, Inc.*)

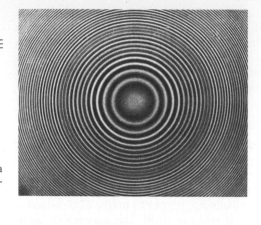

FIGURE 27.12
Interference fringes produced with a Michelson interferometer. (*Atomic Laboratories, Inc.*)

FIGURE 27.13
When light from a distant monochromatic light source passes through a keyhole aperture in an opaque screen and the shadow is observed some distance beyond the screen, the diffraction pattern below is found. Above is the shadow predicted on the assumption of rectilinear propagation. (*Courtesy of John M. Stone.*)

direction coincides with the direction of the ray which was reflected at mirror C and which subsequently passed through the glass plate A. The wavefront from source S has been split into two wavefronts. Both are received by the eye, one of them after reflection at C, and the other after reflection at D. The plane parallel glass plate B is introduced to compensate for the extra thickness of glass through which the ray reflected at C has passed in reaching the eye.

If the distance from the half-silvered surface at A to mirror C is the same as the distance from A to mirror D, the two rays of light travel the same distance, and they are therefore in phase. Under these conditions, the central rays reinforce each other. If, however, the distance AD is greater than the distance AC by one-quarter of a wavelength of the light, the central rays are out of phase and destroy each other. Rays which make small angles with the central ray travel slightly different distances and interfere alternately constructively and destructively (Fig. 27.12). If mirror D is moved along the line AD, the light and dark circles interchange position for each quarter wavelength the mirror is moved.

27.6 DIFFRACTION

When light passes an obstacle, it does not proceed in exactly straight lines but spreads out somewhat into the geometrical shadow. Ordinarily this effect is small for light waves. However, light does bend slightly around corners in much the same way that water waves bend around obstacles. The effect is prominent when the wavelength is large compared with the size of the obstacle; it is small when the wavelength is short compared with the size of the obstacle. Thus, radio waves and sound waves, which have relatively long wavelengths, bend readily around objects. The wavelengths of visible light are small compared with the sizes of ordinary obstacles; therefore the bending is not ordinarily conspicuous for light waves.

If light always traveled in perfectly straight lines and obeyed only the laws of geometrical optics, a picture like that of the upper half of Fig. 27.13 would be formed by light from a distant monochromatic source passing through a small keyhole aperture some distance from the film. The lower half of Fig. 27.13 shows an actual picture. Not only has there been some spreading of light into the region of the geometrical shadow, but interference between wavelets from various portions of the wavefront passing the aperture has led to a series of light and dark fringes. *The spreading of a wave motion into the geometrical shadow of an object and the other deviations from the predictions of geometrical propagation are called diffraction.* A definition due to Sommerfeld states that diffraction is "any deviation of light rays from rectilinear paths which cannot be interpreted as reflection or refraction."

To account for the fact that light bends around obstacles, we need only invoke Huygens' principle, which tells us that for any kind of wave motion we may regard each point on a wavefront as a new source. Whenever we are dealing with diffraction phenomena, we are treating a case in which various parts of the same wavefront act as coherent sources and the secondary wavelets from these coherent sources interfere. For interference in a Young's double-slit experiment we take two limited sections of a wavefront, send them by different paths, and combine them to get constructive and destructive interference. In diffraction much larger sections of the wavefront may be involved. The general theory calls for finding the net interference effect at a given point for a large number of waves with a continuous variation in phase. The mathematical treatment of such diffraction phenomena is more difficult than that for two-slit interference. For this reason we discuss diffraction in a qualitative way, except for the case of the diffraction grating.

27.7 DIFFRACTION GRATING

If a series of very fine equidistant parallel slits is ruled on a plate of glass with a fine diamond point, we have a *diffraction grating* Where the diamond point has made a furrow on the glass, the light cannot pass regularly. Between the furrows, where the surface of the glass is undisturbed, the glass is transparent. The plate of glass is then somewhat like a picket fence. In effect, there are strips through which the light can pass separated by strips through which it cannot pass.

Let AB (Fig. 27.14) represent such a grating on which parallel light is

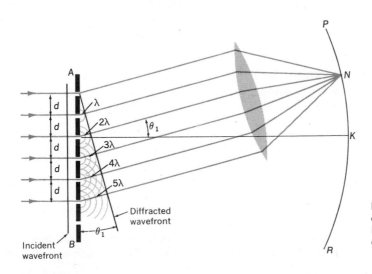

FIGURE 27.14
Diffraction grafting. Reinforcement occurs when $n\lambda = d \sin \theta_n$, where n is an integer and d is the grating space (or the distance between the centers of adjacent slits).

falling so that the direction of the rays is perpendicular to the plane of the grating. Figure 27.14 shows the Huygens wavelets that start out from the slits. These wavelets destroy or reinforce each other, according to whether they are in phase or out of phase at a given point.

Through the slits in the grating come beams of parallel light; these beams come from light falling on a narrow slit S in the focal plane of a converging lens. The rays emerging from the lens are parallel to the principal axis. If these rays, incident in the direction perpendicular to the plane of the grating, are brought to a focus on the screen PR by the lens, a bright line is formed at K. This line is the image of the slit S, and it is identical to the image which would be formed in the absence of the grating except that it is much less bright. This image, formed by straight-through rays, is called the *central maximum;* it is also known as the *zero-order maximum* because the optical-path length is the same for all rays.

If the light is viewed in a direction making an angle θ with the normal to the grating, parallel rays of light emerging from the slits in this direction travel unequal distances to reach the screen and when they are brought to a focus by the lens they may either reinforce or destroy each other. If the angle θ is made such that a ray of light from one slit is one-half wavelength behind the corresponding ray from the adjacent slit, the rays from one slit are out of phase with the rays from the adjacent slit and they destroy each other. Hence, there will be darkness. If the angle between the normal to the grating and the ray is increased, the rays from the lower slits will be farther and farther behind the rays from the upper slits. When the difference in path between corresponding rays from adjacent slits amounts to one wavelength, wavelets from all the slits are again in phase and interfere constructively. If these rays are focused by the lens, a bright image of the slit is formed on the screen at N. Whenever

$$\sin \theta = \frac{n\lambda}{d} \qquad n \text{ an integer}$$

the wavelets from different slits reinforce each other and the screen is illuminated.

The angle θ_1 at which reinforcement is first obtained after leaving the central image K can easily be measured on the divided circle of a spectrometer. For this first reinforcement each ray travels one wavelength farther than from the adjacent slit below, so that

$$\lambda = d \sin \theta_1 \tag{27.2}$$

This image is called the *first-order maximum.* An identical image is formed at angle θ_1 below the axis.

Reinforcement also occurs at other angles θ_n which satisfy the equation

$$n\lambda = d \sin \theta_n \tag{27.3}$$

The integer n gives the *order* of the spectrum. When $n = 2$, the waves from each slit go two wavelengths farther than those from the adjacent slit. In the third order $n = 3$, and the path difference is 3λ. The number of orders which can be seen is limited by the fact that $\sin \theta_n$ cannot exceed 1. Therefore the maximum number of orders which can be observed is given by the largest integer in the ratio d/λ.

When white light is incident on a grating, it is spread out in a spectrum with the first-order blue diffracted through a smaller angle than first-order red. There may be other spectra corresponding to second and third orders, but they begin to overlap. Note that the third-order maximum for $\lambda = 400$ nm (blue-violet) coincides with the second-order maximum for $\lambda = 600$ nm (yellow-orange).

Since the grating space d and the angle θ_n in Eq. (27.3) can be measured for a spectral line, the wavelength λ of the light can be calculated. Measuring wavelengths with a diffraction grating can be a very accurate method since both d and $\sin \theta_n$ can be determined with high precision.

□ **Example** For yellow light from a spectral source the angular separation between the central image and the first-order spectrum produced by a plane grating is 17°. The grating has 500 rules to the millimeter. What is the wavelength of the light? How many orders can be observed for this wavelength with this grating?

$$\text{Wavelength} = d \sin \theta_1$$

$$\lambda = \frac{1 \text{ mm}}{500} \times \sin 17°$$

$$= \frac{1 \text{ mm}}{500} \times 0.292$$

$$= 0.000584 \text{ mm} = 5840 \text{ Å} = 584 \text{ nm}$$

$$\frac{d}{\lambda} = \frac{2 \times 10^{-6} \text{ m}}{584 \times 10^{-9} \text{ m}} = 3.4 \qquad n_{\text{max}} = 3 \qquad \square$$

27.8 DIFFRACTION BY A STRAIGHT EDGE

Suppose that light is diverging from a narrow slit L (Fig. 27.15a) and that it passes by the straight edge of an opaque screen. If the light were propagated exactly in straight lines, there would be uniform illumination on the screen above the line LO and complete darkness below it. However, the illumination does not become zero immediately below O but fades away continuously. There is almost complete darkness a small distance below O. Immediately above O the illumination is not uniform but shows a series of bright and dark bands parallel to the edge. The appearance of the fringes thus produced is seen in Fig. 27.15b.

Consider now the case of a fine wire AB (Fig. 27.16a) placed in front of a narrow slit L. The shadow of this wire on the screen MN is found to be

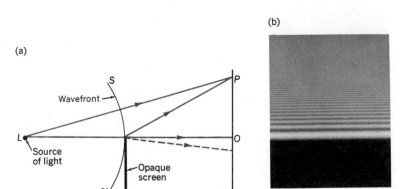

(a)

(b)

FIGURE 27.15
Light passing a straight edge (a) produces a diffraction pattern (b) with monotonically decreasing intensity below O and a series of fringes in the region OP. (*Courtesy of John M. Stone.*)

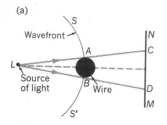

(a)

(b)

FIGURE 27.16
Light passing a fine wire is diffracted (*a*), producing the diffraction pattern (*b*). (*Courtesy of John M. Stone.*)

bounded on each side by a system of parallel fringes (Fig. 27.16*b*). Each edge of the wire behaves like a straight edge. Further at the very center of the shadow there is constructive interference of waves diffracted around each side of the wire, so that the center is bright relative to the main part of the shadow, though still dark compared with regions *CN* and *DM*.

27.9 DIFFRACTION BY A CIRCULAR APERTURE

One of the most important and striking illustrations of diffraction occurs when light passes through a circular aperture such as the pupil of a human eye, the iris diaphragm of a camera, a pinhole, or the lens of a telescope. If plane waves from a distant point source pass through the circular aperture and are then brought to focus by a lens, the image formed by even a perfectly aberration-free lens is never a point. Instead, the image consists of a central bright circular spot, called the *Airy disk*, surrounded by a series of dark and light rings (Fig. 27.17), which rapidly become dimmer as we go outward from the center. The radius of the central Airy disk decreases as the diameter of the aperture increases.

When an image is formed by an optical instrument, even the most perfect optical parts cannot produce a point image of a point source; at best the image appears as a diffraction pattern consisting of the bright Airy disk surrounded by much fainter rings. Ultimately these diffraction patterns limit the angular separation of two sources which can be distinguished as separate. Lord Rayleigh proposed a simple and convenient criterion for determining when two point sources are resolved. According to the Rayleigh rule, two equal point sources are just resolved when the center of the Airy disk of the diffraction pattern of one source falls on the first dark circle of the diffraction pattern of the second. Figure 27.18 shows overlapping images for two sources which come close to satisfying the Rayleigh criterion. Clearly, if the sources were much closer together, one would not be able to tell that there are two sources. (However, a

FIGURE 27.17
When parallel rays of light are incident on a small circular aperture, the diffraction pattern consists of a central Airy disk surrounded by light and dark rings. (*From J. W. Goodman, "Introduction to Fourier Optics," copyright 1968 by McGraw-Hill, Inc. Used with permission of McGraw-Hill Book Company.*)

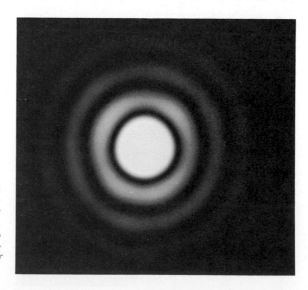

skilled observer can recognize two sources as separate when they are a little closer than the Rayleigh criterion allows.) A good optical system with a larger aperture would produce smaller Airy disks and could resolve sources closer together. The angular separation θ of two sources which can be resolved by a lens or mirror of diameter a is given by the relation $\theta_{min} = \sin\theta_{min} = 1.22\lambda/a$, where λ is the wavelength of the radiation involved. There are double stars in the sky which cannot be resolved by the unaided human eye, but which are resolved by a small telescope with a 100-mm aperture.

When one drives at night and can see a distant car approaching, the images of the two headlights on the retina overlap so much that one sees only a single image; we say the two headlight images are not resolved. As the car draws nearer, the Airy disks of the images overlap less and less. Once the Rayleigh condition is met, one can recognize that there are two headlights and, as the cars approach, the images on the retina separate from each other. For the human eye the angle between two point objects which can just be resolved is about 3.4×10^{-4} rad, or 1 minute of arc. Thus, headlamps 2 m apart can be resolved at a distance of about 6,000 m. A typical person 3 m from a television set cannot resolve two point sources separated by less than 1 mm and thus does not recognize that the picture is formed by a huge number of tiny discrete sources emitting different colors. At a distance of 30 cm we cannot resolve point sources closer than 0.1 mm, and thus we ordinarily do not perceive that a typical newspaper photograph actually consists of black dots on the whiter paper although a magnifying glass to give higher resolution makes it clear.

FIGURE 27.18
Overlapping images of two neighboring point sources. If the sources were closer or if a smaller aperture were used, the images would overlap more and it might not be possible to recognize that there are two sources. (*Courtesy of E. Hecht, from "Optics" by E. Hecht and A. Zajac, Addison-Wesley Publishing Company, Reading, Mass., 1974; by permission.*)

QUESTIONS

1. What happens to the distance between the interference fringes from two slits as the slits are moved closer together?
2. In the interference pattern of Fig. 27.2 light from two slits overlaps in a region to produce light and dark bands. What happens to the energy of two beams of light which interfere destructively to give the dark lines?
3. If the beams from two flashlights overlap on a screen, is an interference pattern produced? What conditions must be satisfied to produce an interference pattern?
4. Why is it easy to observe interference effects with thin films but difficult with thick ones?
5. How do the colors observed in Newton's rings, oil films, and soap bubbles arise? Explain.
6. Newton's rings can be observed by transmitted light as well as by reflected light. Would the point of contact be light or dark by transmitted light? Show that the transmitted rings are dark where the reflected ones are bright and vice versa.

7. An observer sees a green interference fringe at a given point in an oil film. Would other observers looking at the same point necessarily see green there? Explain.
8. Why is the diffraction of sound waves far more frequently observed in daily experience than the diffraction of visible light?
9. How do the spectra of sunlight formed by diffraction gratings differ from those formed by prisms? Which color is deflected most by prisms? By gratings?
10. Would a diffraction grating with 500 lines/mm be useful for studying infrared spectra with wavelengths around 3 μm? Explain.
11. Is the angular separation between a blue line and a red one greater in second order than in first? Why?
12. A tie clasp has a plane surface on which a large number of equally spaced parallel grooves have been ruled, so this surface constitutes a reflection-type diffraction grating. How would its appearance differ from that of one with no grooves?
13. Two spectral lines of different colors are observed by means of a grating, and it is seen that

the third-order image of one line coincides with the fourth-order image of a second line. What is the ratio of the wavelengths of the two colors?

14. Why do coated lenses look purple by reflected light?

PROBLEMS

1. The two parallel slits used for a Young's interference experiment are 0.3 mm apart. The screen on which the fringes are projected is 1.2 m from the slits. What is the distance between the fringes for monochromatic green light of wavelength 550 nm (5500 Å)? How far is the third dark fringe from the central bright one?
    Ans. 2.2 mm; 5.5 mm

2. Yellow light of wavelength 589 nm (5890 Å) from a single source is passed through two narrow slits separated by 0.25 mm. On a screen 1.1 m away an interference pattern is formed. Calculate the separation of intensity maxima in this pattern.

3. Monochromatic light from a narrow slit illuminates two parallel slits 0.22 mm apart. On a screen 1 m away interference bands are observed 2.6 mm apart. Find the wavelength of the light.
    Ans. 572 nm (5720 Å)

4. Light is incident normally on a grating which has 450 lines/mm. Find the wavelength of a spectral line for which the deviation in first order is 15°. Find the angle of deviation for the second order.

5. If 420-nm (4200-Å) radiation falls normally on a grating ruled with 500 lines/mm, how many orders can be observed on each side of the direct beam? Determine the angles at which the orders are observed. *Ans.* 4; 12.1°; 24.8°; 39.1°; 57.1°

6. A glass grating is ruled with 520 lines/mm. Light striking the grating normally forms a second-order image diffracted at an angle of 39° with the normal. What is the wavelength of the light?

7. The fourth-order spectrum contains a certain color diffracted at an angle of 55°. If the grating is ruled with 450 lines/mm, what is the wavelength of the light? *Ans.* 455 nm (4551 Å)

8. Compute the longest wavelength which can be observed in third order by a transmission diffraction grating with 580 lines/mm.

9. A grating is ruled with 600 lines/mm. Calculate the angles of diffraction for first-order red and blue light, the wavelengths being, respectively, 710 nm (7100 Å) and 380 nm (3800 Å). Determine the angular separation in radians. A lens of 525 mm focal length is placed in the path of light just beyond the grating. Determine the linear separation of the red and blue lines. (This is essentially the basis for a grating spectrometer.)
    Ans. 25.2°; 13.2°; 0.21 rad; 110 mm

10. The grating in a grating spectroscope is ruled with 460 lines/mm. What is the angle of diffraction of the second-order spectrum of light of wavelength 520 nm?

11. Two flat glass plates which are almost parallel produce interference fringes by successive reflections from the surfaces of normally incident light. If the light has a wavelength of 480 nm, what difference in thickness would be indicated in passing from one bright fringe to the next? If the bright fringes are 1.2 mm apart along the plates, what is the angular separation between the surfaces? *Ans.* 240 nm; 2×10^{-4} rad

12. Two glass optical flats are placed together with a piece of paper between the two at one edge. When the air film is illuminated and viewed from above with sodium yellow light ($\lambda = 589$ nm or 5890 Å), there is a dark fringe at the line of contact and a series of light and dark fringes. Find the thickness of the air wedge at the third bright fringe counted from the dark line of contact. Explain your solution in terms of path differences and phase shifts.

13. A thin air wedge is formed between two optical flats by slipping a piece of paper under one edge of the upper flat. Forty bright fringes are observed across the upper flat, which is 60 mm long. What is the thickness of the paper if the light is monochromatic with a wavelength of 620 nm? What is the angular separation of the surfaces? *Ans.* 12.2 μm; 2.04×10^{-4} rad

14. Two pieces of plane glass are placed together with a piece of paper between them at one edge. When the air film is illuminated from above with cadmium blue light of $\lambda = 480$ nm (4800 Å) and is viewed from above, there is a dark fringe along the line of contact and a series of light and dark fringes. Find the thickness of the air film at the fourth bright fringe from the line of contact. If the dark fringes are 1.8 mm apart, find the angle between the glass surfaces.

15. Two plane pieces of optical glass are pressed together at one edge and separated by a fine wire at the other edge. The distance between the wire and the edges of glass that are in contact is 125 mm. When light of wavelength 500 nm (5000 Å) is incident normally on the surface of one of the pieces of glass, interference fringes are observed. If the light fringes are 2 mm apart, what is the diameter of the wire? What is the

thickness of the air wedge at the second bright fringe from the line of contact of the plates?

Ans. 15.6 μm; 375 nm

16. Newton's rings are observed by reflected light of wavelength 520 nm. The central area is dark and is surrounded by light and dark circles. Find the thickness of the air film at the first, third, and fifth bright circles.

17. Newton's rings are observed by reflected light of wavelength 480 nm. The central area is dark and is surrounded by light and dark circles. Find the thickness of the air film at the second and fourth dark circles. Explain your method.

Ans. 480 nm; 960 nm

18. Newton's rings are formed in reflected light by a plano-convex lens balanced upon an optical flat. If the light incident from above has a wavelength of 600 nm, find the thickness of the air film at the second bright ring. In terms of wavelengths how much farther did light reflected from the flat travel than the light reflected from the spherical surface?

19. The angular separation between two point sources which can just be resolved by an ideal lens with aperture diameter d is $\theta = 1.22\lambda/d$, where λ is the wavelength of the light involved. If the human eye has an angular resolution of 3.4×10^{-4} rad (≈ 1 minute of arc) at a wavelength of 550 nm, find the minimum diameter of the pupil of the eye.

Ans. 2 mm

20. When the movable mirror of a Michelson interferometer is moved 25 μm, how many fringes pass the reference mark if light of wavelength 5890 Å is used?

21. One arm of a Michelson interferometer utilizing light with $\lambda = 500$ nm contains a cell 95 mm long between plane parallel windows. How many fringe shifts would be observed if all the air were evacuated from this cell? The index of refraction of the air was initially 1.00029.

Ans. 110.2

22. When a Michelson interferometer was used to measure the wavelength of monochromatic light, it was found that 280 fringes passed the observing microscope when the movable mirror was displaced 65 μm. What was the wavelength of the light?

23. Light of wavelength 560 nm is incident normally on a wedge-shaped soap film which has index of refraction of $\frac{4}{3}$. By reflected light the top of the film is dark, and below come a series of light and dark lines. What is the thickness of the film at

the uppermost light band? Explain how your answer was obtained, mentioning path differences, phase shifts, etc.

Ans. 105 nm

24. A lens having an index of refraction of 1.60 has a nonreflecting coating with an index of refraction of 1.30. Find the minimum possible thickness of this coating to give destructive interference in the reflected beam for light of wavelength 550 nm.

25. What is the thinnest film of oil (index of refraction 1.40) floating on water in which green light ($\lambda = 540$ nm in air) is essentially eliminated by destructive interference from a beam of white light incident normally on this film? If the oil film were on glass (index of refraction = 1.50), what thickness would be required?

Ans. 193 nm; 96.4 nm

26. Oil with index of refraction 1.40 rests on glass (index of refraction 1.50). Yellow light of $\lambda = 590$ nm is incident normally from above (angle of incidence = 0°). (a) Compute the wavelength of the light in the oil and in glass, and (b) calculate the thickness of the thinnest layer of oil from which this yellow light would be essentially eliminated from the reflected beam by destructive interference.

27. A film of oil with an (unrealistically low) index of refraction 1.20 rests on a puddle of water. When green light of $\lambda = 520$ nm is incident normally on the film and the reflected light is viewed from above the film, destructive interference is observed. What are the three smallest thicknesses the oil film could have? What are the three smallest thicknesses the film could have for constructive interference?

Ans. 108, 325, and 542 nm; 0, 217, and 433 nm

28. Show that if AB and DE of Fig. 27.10 are two rays associated with a single wavefront, then destructive interference occurs for reflected light when $2nt \cos r$ is an integer number of wavelengths. Here t is the thickness of the film, n is its index of refraction, and r is the angle of refraction in the film. Show that constructive interference takes place where $2nt \cos r$ is an odd number of half wavelengths.

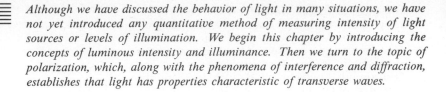

28

ILLUMINATION AND POLARIZATION

Although we have discussed the behavior of light in many situations, we have not yet introduced any quantitative method of measuring intensity of light sources or levels of illumination. We begin this chapter by introducing the concepts of luminous intensity and illuminance. Then we turn to the topic of polarization, which, along with the phenomena of interference and diffraction, establishes that light has properties characteristic of transverse waves.

28.1 STANDARD SOURCES AND LUMINOUS FLUX

To compare light sources and illumination, we make measurements in terms of arbitrary standards. In the early days the standard source was the *British standard candle*, made of spermaceti and burned at the rate of 120 grains per hour. At that time the *lumen* (lm), which is the unit of *light flux*, was the light emerging through an opening made by cutting an area of one square foot out of a spherical surface of radius one foot when a standard candle emitted light at the center of the sphere. These standards are not of current scientific value, but modern standards have been chosen in such a way that we still measure light flux in *lumens* and the luminous intensity of a light source in *candelas* (cd), a unit which evolved from the old candle and is sometimes called the new standard candle.

The *candela* is based on the following arbitrary but highly reproducible operation involving a standard source developed by the National Bureau of Standards and adopted by the International Commission of Weights and Measures in 1948. A schematic diagram of the standard source, basically a glowing blackbody cavity at the temperature of melting platinum, is shown in Fig. 28.1. A cylindrical tube of thorium oxide, surrounded by pure platinum at its melting point and with powdered thorium oxide in the bottom, acts as an ideal blackbody radiator. The tube is closed at the top except for a small hole, the area of which is $\frac{1}{60}$ cm². When this source, operating at the freezing temperature of platinum, is observed from above, its luminous intensity of one candela is essentially that of the old standard candle. *The candela is the luminous intensity in the perpendicular direction of a surface of $\frac{1}{60}$ cm² of a blackbody at the temperature of freezing platinum (1773.5° C) under a pressure of 101,325 N/m².* Luminous intensities of light sources are measured in candelas.

Imagine an ideal point source S of radiation at the center of a sphere (Fig. 28.2) emitting light equally in all directions and having a luminous intensity of I cd. Actually such an isotropic source is not something we can make, but we can discuss its properties. The total luminous flux F emitted is arbitrarily chosen to be $4\pi I$ lm. Thus an isotropic 1-cd source emits 4π lm of flux, and *one lumen is the luminous flux which falls upon one square meter of the area of a sphere of one meter radius when an isotropic source of one candela is at the center of the sphere.*

The solid angle subtended by 1 m² of the surface of a sphere of 1 m radius is called a *steradian*. The solid angle subtended by an area A of the surface of a sphere of radius R (Fig. 28.2) is given by $\omega = A/R^2$. Thus the entire sphere subtends an angle of 4π steradians. A one-candela source emits one lumen per steradian.

Typical light sources, such as fluorescent and incandescent lamps, emit different luminous fluxes in different directions. Consequently, the apparent source intensity depends on the direction from which it is viewed, just as with the standard source. In Sec. 28.3 we describe a means of measuring the intensity of a practical source.

28.2 ILLUMINANCE

When luminous flux strikes a surface, the surface is illuminated. We define the *illuminance* (or *illumination*) E of the surface as the *luminous flux incident per unit area*. In the metric system we measure the illuminance in lumens per square meter, or *luxes*; in the British system in lumens per square foot, sometimes called *foot-candles*. Consider light proceeding from a point source S of I cd. As the light advances, it is distributed over a larger and larger area. If the medium through which the light travels does not absorb any energy, the total light energy passing any spherical surface concentric with S is constant. The light falling on a unit area of a sphere of radius R is given by the ratio of the luminous flux F emitted to the area $4\pi R^2$ of the sphere:

$$E = \frac{F}{A} = \frac{4\pi I}{4\pi R^2} = \frac{I}{R^2} \qquad (28.1)$$

For concentric spheres (Fig. 28.3), the luminous flux F is constant; thus

$$E_1 = \frac{F}{4\pi R_1^2} \quad \text{and} \quad E_2 = \frac{F}{4\pi R_2^2}$$

The amount of light per unit area arriving from a point source varies inversely as the square of the distance from the source.

When the light rays are not incident along a normal to the surface but make an angle i with the normal (Fig. 28.4), the illuminance is given by

$$E = \frac{I \cos i}{R^2} \qquad (28.2)$$

If the eye were equally sensitive to radiant energy of all wavelengths, we could measure the illuminance E of a surface in terms of the amount of energy falling on a unit area of the surface in unit time. However, the eye is more sensitive to some wavelengths than to others (Sec. 26.5). Therefore, if we are interested in light from the point of view of vision, we must compare illuminances on surfaces by use of the eye or an instrument deliberately constructed to respond like the eye. One such instrument is similar to an exposure meter used by photographers. The output of a photoelectric cell is adjusted so that it reads the illuminance directly. The instrument differs from an exposure meter in that suitable filters over the cell give it a response which is proportional to the luminous flux. If the eye and the camera film had identical responses to all wavelengths, the same instrument could be used for both purposes.

Table 28.1 lists a few typical illuminances. The figures are approximate; one reason is that different surfaces reflect vastly different fractions of the incident light. A gray surface receiving an illuminance of 50 lm/m² does not appear as bright as a white one under the same illuminance, since more light is reflected back from the white surface. Hence, one may find a lower illuminance acceptable for reading a book printed on good white paper than for reading a newspaper.

The *luminous efficiency* of a light source is the ratio of light flux emitted by the source to the power supplied to the source. Table 28.2 lists some

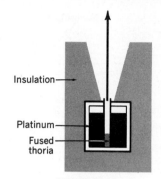

FIGURE 28.1
International standard light source.

FIGURE 28.2
Ideal point source S at the center of a sphere of radius R. The solid angle subtended by an area A of the spherical surface is A/R^2 steradians.

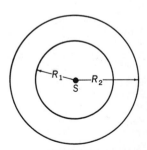

FIGURE 28.3
The total light flux from point source S which passes through each spherical shell is the same though it falls on different areas. The flux per unit area is inversely proportional to the square of the distance from the source.

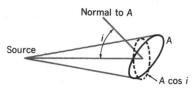

FIGURE 28.4
The illuminance falling on a surface at an angle of incidence i is proportional to $\cos i$.

TABLE 28.1
Typical Illuminances

Situation	Illuminance, lm/m²
Brilliant day, sun overhead	100,000
Overcast day	8,000
Minimum recommended for fine work	500
Night football or baseball	400
Office, classrooms, reading room	300
Street lights	5
Full moonlight	0.2
Starlight	0.0003

TABLE 28.2
Typical Luminous Efficiencies

Source, W	Luminous flux, lm	Efficiency, lm/W
Tungsten lamp:		
10	80	8
40	470	12
100	1,600	16
1,000	22,000	22
Fluorescent lamp:		
6	240	40
10	450	45
30	1,800	60
40	3,000	75

typical luminous efficiencies of modern light sources. For an incandescent lamp, the higher the temperature at which the source operates, the greater the luminous efficiency.

28.3 THE PHOTOMETER

A photometer is a device for comparing the intensities of two sources. To make this comparison, the distances of the sources from a screen are adjusted until they produce the same illuminance. Under these circumstances, if I_1 is the intensity of one source, I_2 that of the other source, and R_1 and R_2 their respective distances from the screen when they produce equal illuminance,

$$\frac{I_1}{R_1{}^2} = \frac{I_2}{R_2{}^2} \tag{28.3}$$

□ **Example**   A standard 32-cd lamp at a distance of 0.6 m from a screen gives the same illumination as a lamp of unknown intensity at a distance of 1.2 m from the screen. Find the intensity of the unknown lamp.

$$\frac{I_1}{R_1{}^2} = \frac{I_2}{R_2{}^2}$$

$$\frac{32 \text{ cd}}{(0.6 \text{ m})^2} = \frac{I_2}{(1.2 \text{ m})^2}$$

$$I_2 = 128 \text{ cd} \qquad \square$$

If two sources have different colors, it is impossible to obtain proper balance with a simple photometer. However, if an arrangement is made whereby a screen is illuminated first by one source and then by a second, it is possible to find a frequency of alternation such that the color difference disappears. Then the observer can adjust the source distances until the sources produce equal illuminances. A photometer constructed for this process is called a *flicker photometer*.

28.4 POLARIZATION

When a beam of ordinary light passes through a Polaroid sheet, the unaided eye recognizes only that the intensity has been reduced. If the Polaroid is rotated about an axis normal to its surface, the transmitted beam appears to differ in no way as the rotation proceeds. However, if this transmitted radiation falls on a second Polaroid, a dramatic change in the intensity beyond the second Polaroid occurs as either Polaroid is rotated through 360° (Fig. 28.5). At some angle, and again 180° later, no light is observed beyond the second Polaroid, but 90° from those positions essentially all the light which goes through the first Polaroid passes also through the second. At intermediate angles the intensity is between these extremes. It is evident that light, in passing through the first Polaroid, acquires properties that ordinary light does not possess: the light is *plane-polarized*.

To understand this experiment, consider a stretched string (Fig. 28.6) in which the particles are vibrating perpendicular to the length of the string. If a block of wood with a slot in it is placed over the string, the

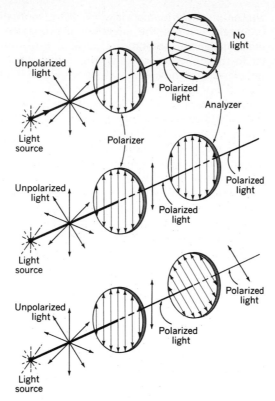

FIGURE 28.5
Light is plane-polarized by passing it through one Polaroid. All, part, or none of this light passes through a second Polaroid, depending on its orientation to the first. The arrows on the Polaroids indicate the directions of the components of the electric vectors which are allowed to pass.

vibrations are not affected when the slot is parallel to the direction of vibration but when the slot is at right angles to this direction, the vibrations do not pass. If the slot makes various angles with the direction of vibration of the string, the component of the vibratory motion parallel to the slot passes through. If a second slot is placed over the string, the vibrations that pass the first slot also pass the second when the two slots are parallel. When the slots are perpendicular to each other, the vibrations that pass the first slot do not pass the second.

Visible light and other electromagnetic radiation have many of the properties of classical transverse waves. The wave model of light involves

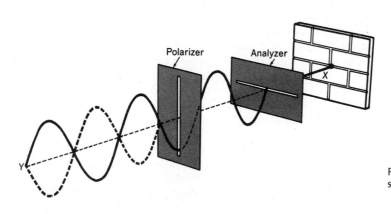

FIGURE 28.6
Polarization of waves in a stretched string is analogous to the polarization of light.

alternating electric and magnetic fields mutually perpendicular both to each other and to the direction of propagation. In ordinary light the electromagnetic vibrations occur in all directions in a plane perpendicular to the direction in which the light is traveling (Fig. 28.5); in plane-polarized light the electric vibrations are all in one direction perpendicular to the propagation direction, while the magnetic oscillations are in the third direction which is perpendicular both to the electric ones and to the direction of propagation. In Fig. 28.5 the lines on the polarizer and analyzer indicate the direction of the electric field which is allowed to pass through. (There are no lines or slits in polarized filters!) In discussing polarized light, it is ordinarily desirable to focus attention on the *electric vibrations,* since most common optical phenomena are due to the interaction of the electric vector with the charged particles in matter. The *plane of vibration* is defined as the plane determined by the electric vibrations and the direction of propagation.

When ordinary light passes through a Polaroid sheet, one component of the vibrations is absorbed and the other transmitted. Consequently the emerging beam differs from ordinary light in that all the electric vibrations are in one direction and the beam is *plane-polarized.* If it falls on a second Polaroid, that Polaroid transmits only those vibrations which are parallel to its transmission direction. When ordinary light falls on a polarizing agent and plane-polarized light is produced, we call the agent a *polarizer.* To determine whether a beam is polarized or not, we pass the light through a second polarizing agent. When the latter is rotated, there is no change in transmitted intensity if the light is unpolarized, while the intensity goes from maximum to zero if the beam is plane-polarized. A polarizing agent used in this way is called an *analyzer.* If the incident light is partially plane-polarized, the intensity is maximum for one orientation of the analyzer and minimum for another but there is no zero. When two polarizing agents are arranged with their axes perpendicular so that they cut out all light, they are said to be *crossed.*

28.5 PLANE-POLARIZED LIGHT

When a polarizing agent of large aperture is desired, Polaroid sheets are ordinarily used (Fig. 28.7). Several kinds of Polaroid sheet have been developed by E. H. Land. The first type (J-sheet) is a plastic sheet containing ultramicroscopic polarizing crystals of iodosulfate of quinine (also known as *herapathite* after Herapath, who studied the polarizing properties in 1852). The tiny needlelike crystals are aligned with their axes parallel by subjecting them to a strong electric field as the plastic in which they are embedded solidifies. Actually a pair of crossed Polaroids is not perfectly effective for eliminating the far-red and far-violet radiations, so that one can see a little deep purple through crossed Polaroids.

Other materials are also known to produce plane-polarized light; an example of primarily historical interest is a crystal of the mineral *tourmaline.* Tourmaline crystals are generally somewhat colored and are seldom used in modern studies of polarization phenomena. Nicol prisms (Sec. 28.8) are excellent polarizers and analyzers; their disadvantages are high cost and limited aperture.

For discussing polarization we look upon a beam of ordinary light as consisting of a large number of waves, each with its own plane of vibra-

FIGURE 28.7
Polaroid sheet crossed on itself to show
how transmitted light is absorbed in
varying degrees. (*Polaroid Corporation.*)

tion, with every direction of vibration perpendicular to the rays having
equal probability. We represent such a beam, looking head on, by the
end on view of Fig. 28.8a. When we pass such a beam through a Polaroid,
it comes out plane-polarized. Only an infinitesimal part of an unpolarized light beam contains vibrations in any single direction. However, if
we have a polarizer set to transmit only vertical vibrations, approximately
half the incident intensity is transmitted. Vibrations which are not exactly vertical also contribute to this beam, since a polarizer transmits all
components which are parallel to its select direction. We may look on a
beam of unpolarized light as being composed of two equal beams
plane-polarized at right angles to each other. Any vibration can be
resolved into two components, one in each of these directions. Hence,
we sometimes represent unpolarized light by showing only two of the

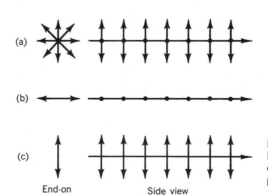

End-on   Side view

FIGURE 28.8
Pictorial representation of (*a*) ordinary
light, (*b*) plane-polarized light with
electric vibrations horizontal, and (*c*)
plane-polarized light with the electric
vector vertical.

many directions of vibration. We use a dot to represent vibration perpendicular to the plane of the paper and arrows to indicate vibrations in this plane (Fig. 28.8).

When a beam of plane-polarized light falls on an analyzer set to transmit a plane of vibrations making an angle ϕ with that of the incident beam, a portion of the incident beam is transmitted. To calculate the fraction transmitted, we proceed as follows. We resolve (Fig. 28.9) the vibration amplitude A of the incident beam into two components, one parallel to the preferred plane of the analyzer and the other perpendicular. The polarizer passes the parallel component and rejects the perpendicular one. The amplitude transmitted A_T is given by

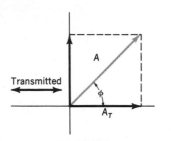

FIGURE 28.9
Resolution of the electric vector into rectangular components.

$$A_T = A \cos \phi \tag{28.4}$$

Since the intensity is proportional to the square of the amplitude, the fraction of the incident intensity I transmitted is given by

$$\frac{I_T}{I} = \frac{A_T{}^2}{A^2} = \cos^2 \phi \tag{28.5}$$

This relation is called the *law of Malus*, after the French engineer who discovered it in 1809.

□ **Example**   An analyzer is set to transmit vibrations at 37° to those of an incident plane-polarized beam. Find what fraction of the incident intensity is transmitted.

$$A_T = A \cos \phi = A \cos 37° = 0.8A$$
$$\frac{I_T}{I} = \frac{A_T{}^2}{A^2} = 0.64 \qquad\qquad □$$

28.6   POLARIZATION BY REFLECTION

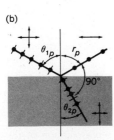

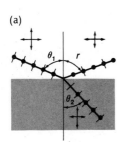

FIGURE 28.10
Partial polarization by reflection and refraction occurs at most angles (*a*), but at the polarizing angle the reflected beam is completely plane-polarized (*b*).

When a beam of unpolarized light is reflected at the surface of glass or water, the reflected beam is unpolarized for normal incidence and for grazing incidence. For any other angle of incidence, the reflected ray contains more vibrations parallel to the reflecting surface, while the transmitted beam contains more vibrations in the plane of incidence (Fig. 28.10*a*). If the angle of incidence is about 45°, an analyzer reveals that some reflected light is transmitted for every orientation of the analyzer, but the intensity is maximum when the analyzer passes electric vibrations parallel to the reflecting surface.

When the angle of incidence is such that the angle between the reflected and refracted rays is 90° (Fig. 28.10*b*), the reflected beam is completely plane-polarized. We call the angle of incidence the *polarizing angle* θ_{1p}. With light incident at the polarizing angle, it is clear from Fig. 28.10*b* that $r_p + 90° + \phi_{2p} = 180°$, where r_p is the angle of reflection. Since $r_p = \phi_{1p}$, by the law of reflection $\phi_{1p} + \phi_{2p} = 90°$; consequently the angle of refraction ϕ_{2p} is the complement of ϕ_{1p}, so sin $\phi_{2p} = \cos \phi_{1p}$. If n is the index of refraction of the reflecting material relative to the first medium, we have

$$n = \frac{\sin \phi_{1p}}{\sin \phi_{2p}} = \frac{\sin \phi_{1p}}{\cos \phi_{1p}} = \tan \phi_{1p} \tag{28.6}$$

The relation $\tan \phi_{1p} = n$ is known as *Brewster's law*.

Although the reflected beam is completely plane-polarized for incidence at the polarizing angle, the refracted beam is only partially plane-polarized. Most of the intensity is associated with the transmitted beam, which includes all vibration components in the plane of incidence and part of those perpendicular to this plane.

Light reflected from metallic and other conducting surfaces is not polarized. A well-silvered mirror may reflect 98 percent of the incident light; both types of polarization are reflected. In order to have plane polarization by reflection, we must have a dielectric medium, which is not a good conductor of electricity.

Sunlight reflected from the surface of a lake or from an asphalt pavement is partially polarized. Glasses equipped with polarized lenses oriented so they transmit only vertical vibrations cut out most of the reflected light and only half the unpolarized light. It is for this reason that Polaroid glasses reduce glare which is due primarily to reflected light.

28.7 DOUBLE REFRACTION

When a crystal of calcite (often called *Iceland spar*) is placed over a page of print, two images are seen in the calcite. If these images are examined through an analyzer, it is found that each of the images is produced by plane-polarized light, with the planes of vibration at 90° to one another. Thus, ordinary light entering the crystal is split into two plane-polarized beams which travel through the crystal in different directions. This phenomenon, known as *double refraction,* is shown by many transparent crystals.

Suppose that a flash of light is created at point *P* inside some medium and that we look at the wavefront a time Δt later. In glass or air the wavefront is a sphere with radius $V \Delta t$, where V is the speed of light in the medium. In calcite there are two wavefronts (Fig. 28.11)! One, due to the *ordinary* wave, is spherical; the other, due to the *extraordinary* wave, is an ellipsoid of revolution. There is one direction, known as the *optic axis,* in which the two waves travel at the same speed; in other directions they have different speeds. At any point both waves are plane-polarized, the ordinary with electric vibrations perpendicular to the plane determined by the point and the optic axis through *P*, and the extraordinary with the electric vector in that plane.

For a crystal to show double refraction it must be *anisotropic;* i.e., it must have different properties in different directions. Calcite and quartz are examples of anisotropic crystals. Water and glass are *isotropic,* i.e., have the same properties in all directions. However, a piece of glass can be made anisotropic if it is subjected to stress.

For calcite the optic axis of a typical cleavage crystal is oriented as shown in Fig. 28.11. If we use Huygens' principle to trace light through a calcite crystal, we find that the Huygens wavelets for the extraordinary wave are ellipsoidal rather than spherical (Fig. 28.12). To find the new wavefront for the extraordinary wave, we find the surface which is tangent to these wavelets. This is a plane in the crystal parallel to the ordinary wavefront, but the light which enters the crystal at point *A* in the figure moves along the line *AX* which represents the path of the *extraordinary ray.* The construction for the ordinary ray is just as it would be for an isotropic medium. Thus, an ordinary ray goes straight through the crystal, while the extraordinary ray moves off to one side. If we place a

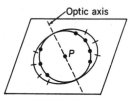

FIGURE 28.11
In calcite a flash of light at *P* produces two wave surfaces, a spherical one, with electric vibrations perpendicular to the plane containing the optic axis through *P* and the point of observation, and an ellipsoidal one, with electric vibrations in this plane.

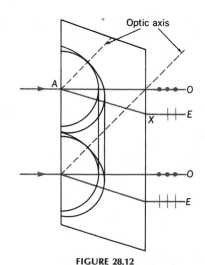

FIGURE 28.12
Double refraction in calcite showing the Huygens wavelets and the paths of the ordinary and extraordinary rays.

calcite crystal over a dot on a piece of paper and rotate the crystal, the image formed by the ordinary ray stands still, while the image due to the extraordinary ray rotates about it.

28.8 NICOL PRISM

One method of separating the ordinary from the extraordinary ray is by use of a *Nicol prism* (Fig. 28.13). A rhomb of calcite *AMBN* is cut into two parts by a plane *MN*, which makes an angle of about 22° with *MB*. When the cut surfaces have been polished, they are cemented together with Canada balsam, the index of refraction of which is less than that of calcite for the ordinary ray and greater than that of calcite for the extraordinary ray. The refractive indices are Canada balsam 1.55, ordinary ray in calcite 1.658, and extraordinary ray 1.468.

The ray of light *LC* entering the face of the rhomb at *C* is broken into the ordinary and extraordinary rays, which are polarized at right angles to each other as indicated by the dots and crosslines in the figure. When the ordinary ray reaches the surface of separation of the Iceland spar and the Canada balsam at *D* at an angle greater than the critical angle, it is totally reflected and emerges from the rhomb in the direction *OO'*. This surface of the rhomb is ordinarily painted black, and the ray is absorbed in this black coating. The extraordinary ray *CEE'* emerges from the rhomb at the face *BN*. In this way, a beam of plane-polarized light is obtained.

28.9 POLARIZATION BY SELECTIVE ABSORPTION

We have introduced polarization phenomena with the aid of tourmaline crystals and Polaroid sheets. These materials produce plane-polarized light by the phenomenon of *selective absorption*, which is exhibited by a number of minerals and organic compounds. Materials which yield polarized light by selective absorption are not only *anisotropic* and double-refracting, but in addition they strongly absorb one of the polarized beams while transmitting the other.

28.10 POLARIZATION BY SCATTERING

When light rays pass through the air, they interact with air molecules and dust particles. Part of the energy is scattered. Particles with dimensions much smaller than the wavelength of the radiation scatter short wavelengths more strongly than long ones. Indeed, the scattering is proportional to the fourth power of the frequency. Since the violet part of the light from the sun has a frequency approximately double that of red

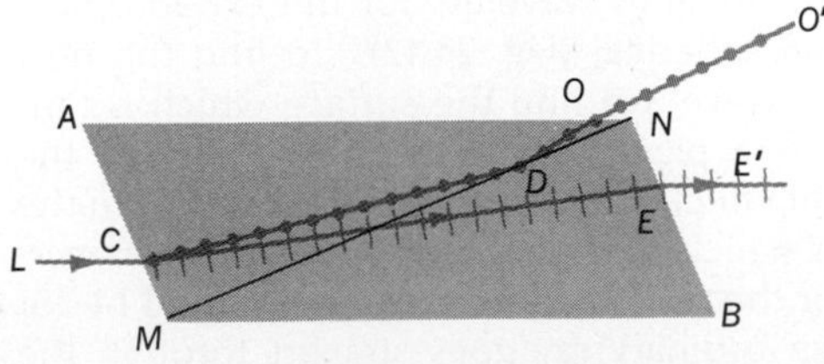

FIGURE 28.13
Total reflection eliminates the ordinary ray in a Nicol prism, leaving only the plane-polarized extraordinary ray in the emerging beam.

waves, the violet rays are scattered about $2^4 = 16$ times as strongly as the far red.

Light from the sun is scattered in passing through the atmosphere, and some of this scattered light is rescattered to our eyes. Since this scattered light is much richer in blues and violets than in the longer wavelengths, the sky is blue. By the same token the sun, even at midday, looks far more orange than it would if it were viewed from a satellite above the earth's atmosphere. Much of the blue and violet has been scattered out before the light reaches the earth's surface. At sunrise and sunset the sun appears redder because its rays go through far greater thicknesses of atmosphere.

If the light of the sky is examined through a Polaroid, it is found that this light is partially plane-polarized if the sky is viewed from almost any angle. Light scattered at an angle of 90° with the incident rays is plane-polarized.

28.11 INTERFERENCE WITH POLARIZED LIGHT

If plane-polarized light is incident on a thin sheet of some double-refracting material, the beam is spread into ordinary and extraordinary rays, which travel with different speeds. When they emerge from the double-refracting material, the vibrations are likely to be out of phase. If the light is now passed through an analyzer which transmits only horizontal vibrations, this analyzer selects only the horizontal components of the vibrations from both ordinary and extraordinary rays. These horizontal components may interfere with one another, either constructively or destructively. In particular, if highly convergent light is passed through the double-refracting crystal, the paths of various rays in the crystal differ in length and there is then constructive interference for a given wavelength for some directions and destructive interference for others. Consequently, interference patterns are formed (Fig. 28.14). If the original light contains all the wavelengths in the visible spectrum, the interference pattern is vividly colored. If either the polarizer or analyzer is rotated through 90°, the colors change to the complementary colors, because those for which the interference was initially constructive then have destructive interference, and vice versa.

When a sheet of cellophane is placed under stress, it becomes double-refracting. Interference colors can be observed when such a sheet is placed between two Polaroids. It is possible to determine where the stresses are and how great they are by careful analysis of such an interference pattern. One of the most useful methods of studying the stresses in various structural shapes involves making a plastic model, subjecting it to external stresses, and studying the interference patterns produced (Fig. 28.15). Such studies permit a fairly complete analysis of the stress distributions. Once one knows what stress distribution is most likely to cause difficulty, one can change the shape to produce a better distribution.

(a)

(b)

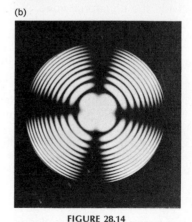

FIGURE 28.14
Interference patterns from crystals of (a) mica and (b) quartz placed in converging monochromatic polarized light.

28.12 ROTATION OF THE PLANE OF POLARIZATION

When a beam of monochromatic plane-polarized light passes through certain substances, the plane of polarization is rotated. Thus, if a plate of

FIGURE 28.15
Stress distribution in an open-end wrench made of Photo-Stress plastic, photographed by polarized light. (*Instrument Division, Budd Company; Zandman Method.*)

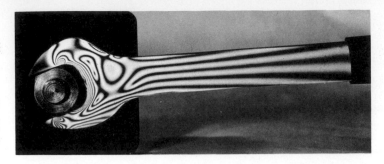

quartz cut so that the faces are perpendicular to the axis of the crystal is placed between crossed Nicol prisms, the light is no longer extinguished by the second Nicol prism. If, however, the second Nicol prism is rotated, a new position can be found at which the light is again extinguished. Rotation of the plane of polarization is called *optical activity*.

The angle through which the plane of polarization is rotated depends on the kind of substance interposed between the Nicol prisms, the thickness of the substance, the wavelength of the light, and the temperature. The rotation may be either clockwise or counterclockwise. In this respect there are two kinds of quartz. One kind rotates the plane of polarization clockwise; the other rotates it counterclockwise. Some liquids and gases also cause a rotation of the plane of polarization. Molecules in which a carbon atom is attached to four different atoms or groups of atoms often show optical activity. Optical activity is a matter of importance in organic chemistry and is fundamental to stereochemistry.

QUESTIONS

1. The sun is closer to the earth in December than it is in June. Nevertheless, the illuminance on a horizontal surface is greater on a clear summer day than on a clear winter day everywhere outside the equatorial zone. Why?
2. How, in principle, would you calculate the illuminance at a point illuminated by a large, spread-out source such as a fluorescent lamp?
3. Is the source intensity of an incandescent lamp the same in all directions? What do we mean by source intensity in this case?
4. Do a photometer and a photographic light meter read the same physical quantity? If not, how do they differ?
5. Why is the sky blue? Is the light from the blue sky polarized or partially polarized? Explain.
6. Why is the sun red at sunset? Is this red light polarized or partially polarized? Explain.
7. How should the plane of vibration be oriented in Polaroid sunglasses if they are to be effective in cutting the glare of reflected light from a lake?
8. Proposals have been made to use Polaroid for reducing the dangers of night driving by eliminating the glare of blinding headlights. Explain how this could be done. (Remember that the driver must still see reflected light from his own headlights.) What are the disadvantages of such a system?
9. How could you make an estimate of the index of refraction of polished black marble by use of a piece of Polaroid and a reflected beam?
10. Why do only anisotropic media exhibit double refraction?

PROBLEMS

1. An arc lamp 15 m from a screen produces the same illumination as a 200-cd lamp 2.5 m from the screen. What is the luminous intensity of the arc lamp?   *Ans.* 7,200 cd
2. A lamp of unknown luminous intensity is placed 3 m from a 60-cd standard. The proper placement of a photometer screen placed between the sources is found by experiment to be 1.4 m from the standard lamp. What is the luminous intensity of the other lamp?

3. A 50-cd lamp is placed 0.6 m in front of a screen. How far from the screen must a 180-cd lamp be placed for the illuminance on the screen to be 3 times as great as it was with the first lamp?

Ans. 0.657 m

4. Two lamps of 80 and 20 cd are placed 2 m apart. At what position between them will the illumination on both sides of an interposed screen be equal?

5. It is desired to have an illuminance of 250 lm/m² at the working area of a desk. If the light comes from a source 0.9 m away and the angle of incidence is 40°, find the luminous intensity of a source which produces the desired illuminance.

Ans. 264 cd

6. If a 100-cd incandescent lamp sends out light uniformly in all directions, how many lumens per square meter are received on a screen 2 m from the lamp when the angle of incidence is 30°?

7. A screen is placed 1.4 m from a lamp. When a sheet of smoked glass is placed between the screen and the lamp, the lamp must be moved so that it is 0.85 m from the screen in order to produce the same illuminance on the screen. Calculate the percentage of light transmitted by the sheet of glass.

Ans. 36.9 percent

8. A small screen is 4.5 m from a 90-cd source. The normal to its surface makes an angle of 15° with a line drawn from the source. What is the illuminance?

9. A football field is illuminated at night from six towers, each with twenty 2-kW lamps with luminous efficiencies of 25 lm/W. If one-fourth of the luminous flux reaches the used area, which is 120 by 75 m, find the average illuminance at the field.

Ans. 167 lm/m²

10. A photocell actuates a relay which controls the opening and closing of the door of a supermarket. At least 0.06 lm must reach the photocell through an opening with an area of 5 cm² if the relay is to operate. What source intensity would be required if a point source is to be used 1.2 m from the relay? (In an actual installation one would use a lamp of much lower candlepower. How would the installation differ from the setup of the problem?)

11. A lens of 45 mm diameter is 0.6 m from a lamp which is essentially a point source with a luminous intensity of 160 cd. Find the illuminance at the lens. How many lumens are incident on the lens if the lamp lies on the axis of the lens?

Ans. 444 lm/m²; 0.707 lm

12. The luminous efficiency of a 200-W light bulb is 20 lm/W. If such a bulb is used as a street light, what is the illumination on a horizontal street (*a*) 6 m directly below the lamp and (*b*) 8 m from the point directly below the lamp? Assume that there is no reflector and that light is emitted equally in all directions.

13. Light of a certain wavelength passes through a Nicol prism and is thus polarized. It then passes through a second Nicol prism whose plane of vibration makes an angle of 55° with the plane of vibration of the first prism. What percentage of the light incident on the second Nicol prism is transmitted?

Ans. 32.9 percent

14. Two Polaroids are crossed. If the analyzer is now rotated through an angle of 40°, what percentage of the plane-polarized light from the polarizer is transmitted, assuming the Polaroid is a perfect transmitter of one polarization and a perfect absorber of the other? What percentage of the ordinary light incident on the polarizer is transmitted through the analyzer?

15. Ordinary light of intensity I_0 falls on a Polaroid P_1 with its characteristic polarizing direction vertical. An intensity I_1 passes through and falls on a second Polaroid P_2, the polarizing direction of which makes an angle of 53.1° with the vertical. An intensity I_2 passes through P_2. If the Polaroids pass all components parallel to the polarizing direction and none perpendicular, find the ratios I_1/I_0, I_2/I_1, and I_2/I_0.

Ans. 50, 36, and 18 percent

16. Ordinary light of intensity I_0 falls on a Polaroid P_1 with its characteristic polarizing direction vertical. An intensity I_1 passes through and falls on a second Polaroid P_2, the polarizing direction of which makes an angle of 36.9° with the vertical. An intensity I_2 passes through P_2 and strikes Polaroid P_3, with polarizing direction horizontal. If the Polaroids pass all components parallel to the polarizing direction and none perpendicular, find the ratios I_1/I_0, I_2/I_1, I_3/I_2, and I_3/I_0. How can any light get through P_3 when it is crossed relative to P_1?

17. Two Polaroids are crossed so no light passes the analyzer. If a third Polaroid is placed between these two with its axis making an angle of 40° with that of the polarizer, find the fraction of the plane-polarized light incident on the middle Polaroid (*a*) which is transmitted and (*b*) which passes through the analyzer.

Ans. (*a*) 0.587; (*b*) 0.242

18. Two Nicol prisms have their planes of vibration parallel. One of the prisms is then turned so that its plane of vibration makes an angle of 25° with

that of the other. What fraction of the amplitude incident on the second Nicol prism is transmitted? What percentage of the light incident on the second Nicol prism is transmitted?

19. Find the angle of incidence for which light reflected from water of refractive index 1.333 is plane-polarized.   *Ans.* 53.1°

20. Determine the angle of incidence (polarizing angle) for which light reflected from flint glass of index of refraction 1.63 is plane-polarized.

21. Light reflected from the surface of heavy flint glass is plane-polarized when the angle of incidence is 59°. What is the index of refraction of the glass?   *Ans.* 1.66

22. Determine the critical angle of incidence for the ordinary ray passing from calcite to Canada balsam in a Nicol prism.

23. A calcite crystal is 15 μm thick and cut so that its faces are parallel to the optic axis. If the indices of refraction of calcite for $\lambda = 550$ nm are those given in Sec. 28.8, find the number of ordinary O waves and the number of extraordinary E waves in the crystal for light incident normally. What is the approximate difference in phase for the O and E waves leaving the crystal?

   Ans. 45.22 and 40.04; 10.36π or 0.36π rad

24. The index of refraction of a plate of glass is 1.52. (*a*) Find the polarizing angle of this glass. (*b*) Find the critical angle of incidence for light going from the glass to air.

OPTICAL SPECTRA

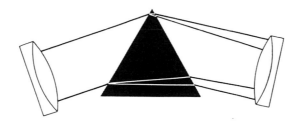

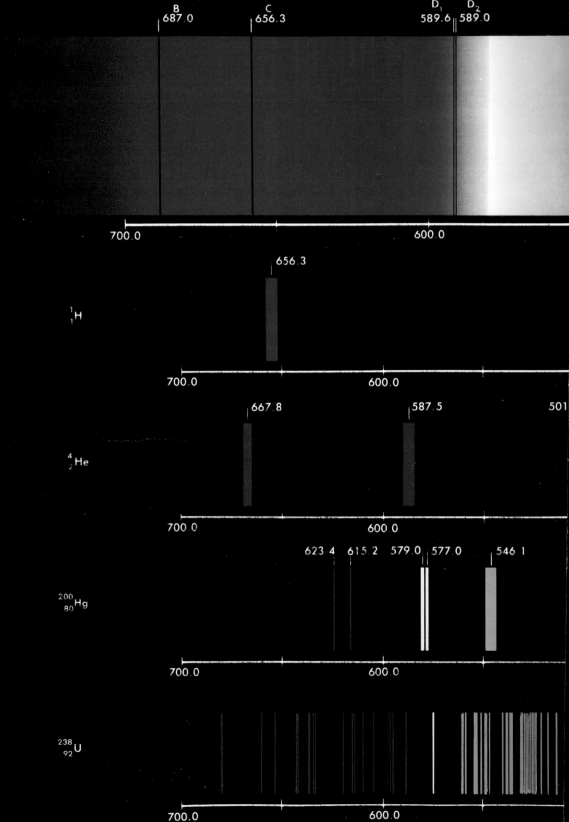

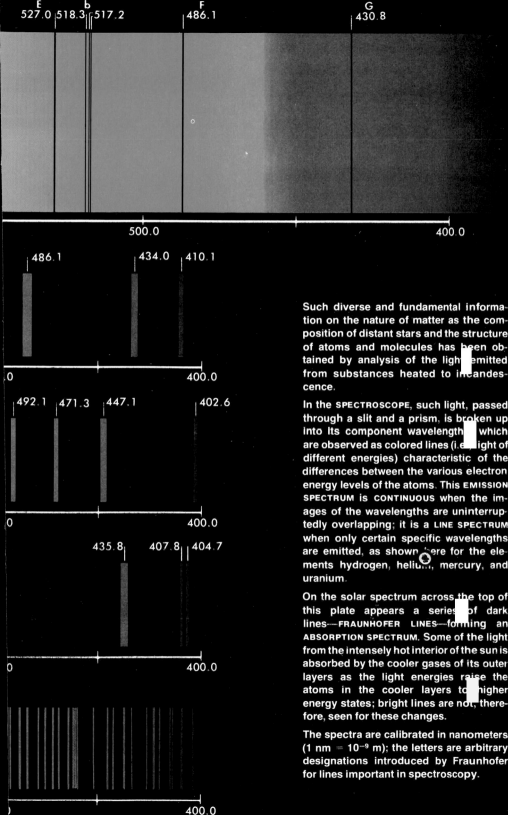

E 527.0   b 518.3 517.2   F 486.1   G 430.8

500.0   400.0

486.1   434.0   410.1

0   400.0

492.1   471.3   447.1   402.6

0   400.0

435.8   407.8   404.7

0   400.0

0   400.0

Such diverse and fundamental information on the nature of matter as the composition of distant stars and the structure of atoms and molecules has been obtained by analysis of the light emitted from substances heated to incandescence.

In the SPECTROSCOPE, such light, passed through a slit and a prism, is broken up into its component wavelengths which are observed as colored lines (i.e. light of different energies) characteristic of the differences between the various electron energy levels of the atoms. This EMISSION SPECTRUM is CONTINUOUS when the images of the wavelengths are uninterruptedly overlapping; it is a LINE SPECTRUM when only certain specific wavelengths are emitted, as shown here for the elements hydrogen, helium, mercury, and uranium.

On the solar spectrum across the top of this plate appears a series of dark lines—FRAUNHOFER LINES—forming an ABSORPTION SPECTRUM. Some of the light from the intensely hot interior of the sun is absorbed by the cooler gases of its outer layers as the light energies raise the atoms in the cooler layers to higher energy states; bright lines are not, therefore, seen for these changes.

The spectra are calibrated in nanometers (1 nm = 10^{-9} m); the letters are arbitrary designations introduced by Fraunhofer for lines important in spectroscopy.

The radiation from any light source can be spread out in a spectrum by using a prism or a diffraction grating. The study of spectra has been rewarding in many ways, and various branches of spectroscopy continue to employ many physicists in research, in process controls, and in the identification of elements and molecules. When white light is spread into a spectrum, the eye finds color everywhere. Color vision still has its mysteries, but many interesting facts about color are well established.

29
SPECTRA AND COLOR

29.1 EMISSION SPECTRA

In 1666 Newton darkened a room and made a small hole in his "window-shuts" through which a beam of sunlight could enter. He placed a prism at the hole and produced a spectrum on the far wall of the room. This discovery and the subsequent investigation of the colors produced represent one more of Newton's great contributions to physics.

When a beam of white light defined by a narrow slit is passed through a prism (Fig. 29.1), the light is dispersed into a spectrum containing all the colors of the rainbow, with red light deviated the least and violet the most. In 1800 Sir William Herschel held a thermometer in each of the colors obtained by passing sunlight through a prism and found that the temperature rose as he moved toward the red end of the spectrum. The temperature increased still further when the thermometer was slightly outside the visible region. The invisible radiation responsible for this temperature rise was described as *infrared*. One year after Herschel's discovery Ritter showed that there is also radiation just beyond the violet end of the visible spectrum; it is designated as *ultraviolet*.

If we put salt (NaCl) in the flame of a bunsen burner, the light emitted is a yellow characteristic of sodium atoms. A lithium salt in the flame produces a bright red. A systematic examination of the spectra emitted by various elements was initiated by Kirchhoff about 1859. The spectra of elements gave us one of the most important clues in the development of modern atomic theory. There are several ways in which emission spectra may be classified. We shall group them as follows.

Continuous Spectra   When the spectrum from an incandescent solid or liquid is examined, it shows no regions of darkness (Fig. 29.1). Such spectra, characteristic of an incandescent lamp and a carbon arc, are known as *continuous spectra*, since they contain light of every wavelength over a broad region.

Bright-Line Spectra   When atoms in the gaseous state are excited to emit radiation, the spectrum consists of a number of *narrow bright lines* with wavelengths characteristic of the element that emits them. Optical spectra of hydrogen, helium, mercury, and uranium are shown in the preceding color insert. Such spectra can be excited by passing an electric discharge through the gas or by raising the temperature of the atoms sufficiently. Each element has its own characteristic *bright-line* spectrum.

Band Spectra   When radiation is emitted by excited molecules, the spectrum often appears in the form of bands of light with regions of darkness between. Such molecular spectra are often referred to as *band spectra*. However, when band spectra are studied in spectrometers of

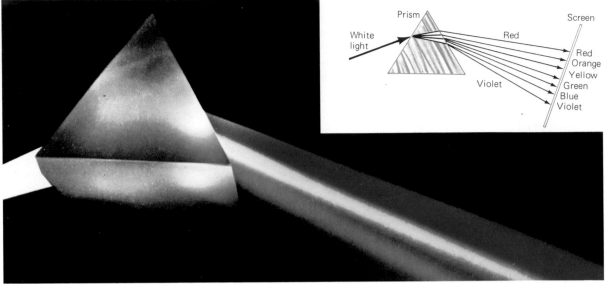

FIGURE 29.1
A visible spectrum is formed when white light from a narrow slit is dispersed into its component colors by the prism. (*From Konrad B. Krauskopf and Arthur Beiser, "The Physical Universe," 3d ed., McGraw-Hill, New York, 1973.*)

great resolving power, it is found that a band results from a large group of lines very close together. Overlapping of the lines gives rise to the band structure.

29.2 SPECTROSCOPY

In spectral measurements we are concerned directly with the wavelengths of the radiation, which range from many kilometers to far less than a billion billionth of a meter. The observation and interpretation of spectra underlies the science of *spectroscopy*, which deals not only with visible light but also with electromagnetic waves of longer and shorter wavelengths including radio, radar, infrared, ultraviolet, x-rays, and gamma rays. The wavelength regions of these radiations are suggested in Fig. 29.2. Actually there is substantial overlapping of regions; the boundaries are not sharp. Except for their inability to excite vision, radiations with wavelengths a little longer than 780 nm (near infrared) or a little shorter than 380 nm (near ultraviolet) have properties very close to those of visible radiation. Infrared and ultraviolet rays are reflected and refracted like visible rays and can expose suitably prepared photographic film. For wavelengths far from those of the optical region, some phenomena occur which are not apparent with visible light.

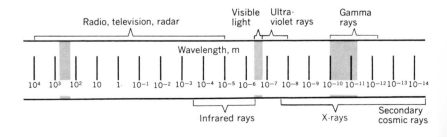

FIGURE 29.2
The electromagnetic spectrum.

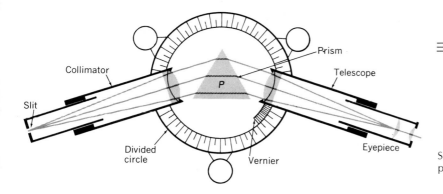

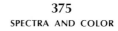

FIGURE 29.3
Spectrometer showing how rays made parallel by a collimator are brought to a focus by the telescope.

As an example of how potent a tool spectroscopy is, consider the element *helium;* its spectral lines were observed in light from the sun some 25 years before helium was discovered on the earth. Thanks to study of the spectra of the stars, we know a great deal about their chemical composition, even though no sample has ever been placed in the hands of a chemist. To this day spectroscopy is one of the most active branches of physics and chemistry; both research and industrial applications are legion.

Since each element has a unique spectrum, an examination of the light emitted by a vaporized substance gives direct evidence of its composition. This method is very useful in detecting small quantities of many elements. It is a rapid and sensitive method of analysis. The spectrum of an element can be excited in a variety of ways, among which are passing an electric discharge through a gas containing the element, heating the element in a furnace, and introducing it into an electric arc. The origin of spectral lines is discussed further in Chap. 45.

For the study of optical spectra, a spectroscope or a spectrometer is used. One part of a typical spectrometer (Fig. 29.3) is the collimator, which consists of a tube with a slit at one end and a converging lens at the other. The slit is located at the principal focus of the lens so that the light rays which emerge from the lens are parallel. These rays pass through a grating or a prism which disperses the light to a telescope mounted so that it can rotate about the vertical axis of the instrument. The angle through which the telescope is rotated is read on a divided circle.

29.3 DARK-LINE SPECTRA; FRAUNHOFER LINES

The spectrum of the sun is crossed by a number of fine dark lines known as *Fraunhofer lines* (see color insert following page 372). These lines are produced by absorption in the atmospheres of the sun and the earth. The core of the sun emits white light, which then passes through an atmosphere of less hot vapors and gases. Certain wavelengths corresponding to the spectral lines emitted by various atoms are absorbed and reradiated in random directions in these less hot layers. For example, excited sodium atoms emit strongly two yellow lines, named the D lines by Fraunhofer. Sodium vapor in the solar atmosphere absorbs these wavelengths, which are almost absent when the solar spectrum is examined on

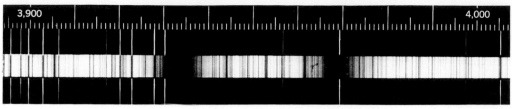

FIGURE 29.4
The correspondence of many Fraunhofer lines with the absorption lines of iron vapor is apparent in these emission spectra between 3900 and 4000 Å for iron and the sun (center). (*Courtesy of Mt. Wilson and Mt. Palomar Observatories.*)

the earth. From the Fraunhofer lines it is possible to identify several elements which are relatively abundant in the sun's outer layers. Figure 29.4 compares a portion of the emission spectrum of iron with the corresponding part of the solar spectrum, showing that iron vapor is present in the solar atmosphere.

A dark-line spectrum can be produced in the laboratory by passing an intense beam of continuous radiation through the vapor of some element which absorbs from the continuous spectrum certain wavelengths found in the bright-line spectrum emitted by this element. For example, when light from a carbon arc is sent through sodium vapor, there is strong absorption of the wavelengths of the yellow sodium D lines.

29.4   THE DOPPLER EFFECT

When a sound source is moving toward or away from an observer, the frequency observed is different from the source frequency (Sec. 22.9). A similar effect is observed in light. The doppler effect provides a means of determining the motions of distant stars. If a star is moving away from the earth, the frequency of a spectral line observed at the earth is lower than the emitted frequency, so that the observed wavelength is greater than normal; if the star is approaching the earth, the wavelengths are shifted toward smaller values. By measuring the displacements of the lines, the speed of the moving star can be determined.

The doppler effect in light differs in several important respects from that in sound waves. In sound waves it is the velocity of the source (or of the observer) *relative to the air* which is of importance. In the case of light waves there is no comparable medium. When source and observer are approaching each other along the line connecting them with relative speed v,

$$\frac{\nu_{\text{observed}}}{\nu_{\text{source}}} = \frac{\sqrt{c + v}}{\sqrt{c - v}} \tag{29.1}$$

where c is the speed of light.

The doppler effect is used for measuring speeds of space vehicles and of automobiles. For example, a radar transmitter on a police car sends out very-high-frequency waves, which are reflected by oncoming automobiles. The reflected waves have a different frequency and are "beat" (Sec. 22.8) against the initial frequency, yielding a *beat frequency* equal to the difference between the two. In this case the source is the image of the transmitter reflected from the vehicle, which acts as a plane mirror. Therefore, the speed of the source is twice the speed v of the oncoming vehicle. Let the frequency emitted by the radar be ν, and that received be ν_r. Then Eq. (29.1) yields

$$\frac{\nu_r}{\nu} = \frac{\sqrt{c + 2v}}{\sqrt{c - 2v}} \approx 1 + \frac{2v}{c}$$

since $2v$ is very small compared to c. The beat frequency F is given by

$$F = \nu_r - \nu \approx \frac{2v\nu}{c}$$

□ **Example** A radar transmitter sends out a signal at 100 MHz. An automobile is approaching, and the beat frequency observed is 19 beats/s. Find the speed of the approaching car.

$$F = \frac{2v\nu}{c}$$

$F = 19 \text{ beats/s} \qquad \nu = 10^8 \text{ Hz} \qquad c = 3 \times 10^8 \text{ m/s}$

$$19 = \frac{2v \times 10^8}{3 \times 10^8}$$

$v = 28.5 \text{ m/s} = 103 \text{ km/h} \ (64 \text{ mi/h})$ □

29.5 COLOR CLASSIFICATION

In studying the spectrum of an element, we might well use a diffraction grating to disperse the radiation into spectral lines and then determine the wavelengths of these lines and their relative intensities. From this information it would almost surely be possible to identify the element. However, an experimental spectroscopist might well be able to identify the element just by looking at the light from the source. Relatively little experience is necessary before a typical student can identify the light from an electric discharge through sodium vapor or mercury or hydrogen with reasonable confidence. Each has a characteristic color which is readily recognized. In this case, as for sound waves, there are psychophysical properties of the radiation which are related to, but not identical to, the physical properties. There are three psychophysical characteristics in terms of which it is convenient to describe the color of an object, *hue*, *saturation*, and *lightness* (or, for a self-luminous object such as a light bulb, the *brightness*). The *hue* is specified in terms of the color in a hue circle (Fig. 29.5) which the object most closely resembles. In the system of color notation developed by Munsell, the complete hue circle consists of 100 hues.

The term *saturation* has to do with the extent to which the hue is influenced by the addition of other colors. If blue is the only color present, the saturation approaches unity. On the other hand, if other colors are present but there is enough blue for the object to have a blue hue, the saturation is low (Fig. 29.6). In mixing paints, for example, a paint of lower saturation can always be made by mixing white with the color. Thus, it is easy to produce a wide range of variations in saturation for the same hue. As we go from 0 to 100 percent in saturation, we go from whites and grays to pure colors. Whites and grays are *achromatic* and correspond to zero saturation.

Lightness is connected with the relative amount of light which reaches the eye from the object. If the object is illuminated by white light and reflects all the incident light, it has maximum lightness and appears

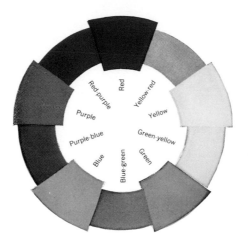

FIGURE 29.5
Hue circle.

white; if it reflects none of the light, it is black. It is possible to arrange achromatic grays in a scale of brightness and to match against these grays samples of a given hue which reflect the same fraction of the incident light (Fig. 29.6).

One way of representing colors is in a three-dimensional color ellipsoid or color solid (Fig. 29.7) in which the lightness (or brightness) is plotted, going from black to grays to brilliant white along an axis. Around this axis can be drawn concentric circles, with the various hues given by the angles, and the saturations by the distance from the central white-black axis.

Ordinary "white" light from the sun or a lamp consists of a mixture of all the visible wavelengths. When it is dispersed by a grating or prism, a continuous spectrum is formed. However, it is not necessary that all visible wavelengths be present to excite the sensation of white light in an observer. Indeed, only two properly chosen wavelengths need be pres-

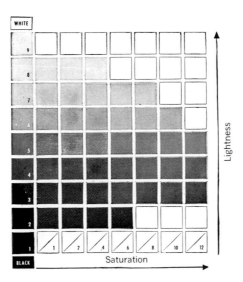

FIGURE 29.6
Variations in blue hue from a Munsell color atlas.

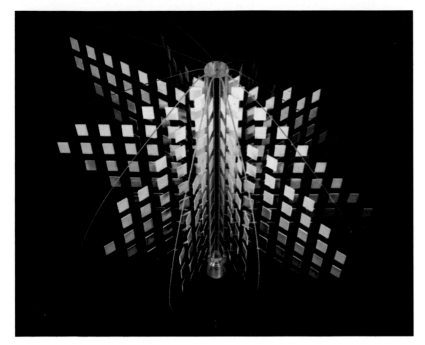

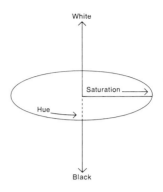

FIGURE 29.7
A color solid formed by arranging colors in three dimensions vividly shows the relationship between lightness, hue, and saturation. Along the vertical axis lightness increases from the least intense at the bottom to the most intense at the top. The various hues are arranged around the circle with saturation increasing with distance from the axis.
(*National Bureau of Standards.*)

ent, and a wide variety of pairs is possible. Two wavelengths (or two colors) which when mixed together give white light are said to be *complementary*. The physical, chemical, and psychological processes associated with vision are highly complex, and what an observer "sees" depends not only on what wavelengths are present but also on many other factors. The seeing process can make distinctions of amazing sublety and may display a color for some object in the visual scene even though no wavelengths ordinarily associated with that color are present. Colors seen in complex images depend not only on the wavelengths incident on the eye but also on intricate feedback mechanisms in the visual chain.

29.6 THE MEASUREMENT OF COLOR

Physicists interested in quantitative measurements of colors have found that it is possible to make reproducible measurements by matching colors projected on two neighboring areas of a reflecting screen. Suppose a projector (Fig. 29.8) throws on area S_1 an arbitrary color X. Let us illuminate the adjacent area S_2 by means of three lanterns, each of which sends out a light of a single narrow wavelength region, A at 425 nm (blue), B at 551 nm (green), and C at 650 nm (red). For almost any color we can find one and only one combination of relative intensities of these three lights which exactly matches the color X. We can then describe color X by three specific numbers representing the amounts of light from A, B, and C.

Although we can match most colors in this way, there are some which cannot be matched by adding everything on S_2. These colors are said to be outside the *gamut* of these *primaries*. However, if we move the proper

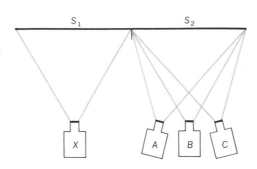

FIGURE 29.8
Color of almost any hue or saturation thrown on screen S_1 can be matched on screen S_2 by blending (adding) three monochromatic primaries in specific amounts.

one of the lamps to the other side and add its light to the unknown color, it is possible to achieve a match. For example, if we try to match a pure green of $\lambda = 500$ nm with the lamps, we find that there is too much red, but if we move lamp C to the other side, we find that we can match $\lambda = 500$ nm plus some red from C with light from lamps A and B. One way of saying this is to say we can match $\lambda = 500$ nm with light from A and B and a *negative* amount of light from C. This, however, introduces the complication of negative numbers for specifying some colors.

We have chosen for our illuminants blue, green, and red, because with them we can match the widest gamut of colors. In this sense red, green, and blue are *primary* colors. However, if we use any three different wavelengths, we can still match any color if we allow the use of negative contributions and assume that the sources A, B, and C can be varied in intensity over an infinite range.

To avoid the difficulties associated with negative numbers, and for other reasons, the International Commission on Illumination agreed in 1931 on a standard *operational procedure* for determining the color of a surface. (This method can be shown to be equivalent to the type of experiment in which we match the color with three primary sources, except that the sources chosen do not exist in the realm of *real colors.*) The surface is placed in a spectrophotometer. This instrument contains an optical system in which the light from a standard lamp is dispersed into a spectrum by a prism. One narrow range of wavelengths at a time is allowed to fall upon the surface under test, and the amount of light reflected is compared with the amount of the same wavelength range reflected from a standard white surface. A plot is made of the relative reflectance of the surface under test as a function of wavelength. Several such *spectral-reflectance* curves are shown in Fig. 29.9. From the curve it is

FIGURE 29.9
Spectral-reflectance curves for various objects.

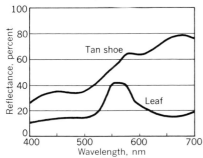

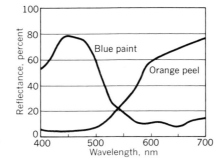

possible to calculate the coordinates x and y for the color of the surface on the ICI *chromaticity diagram* (Figs. 29.10 and 29.11).

The horseshoe-shaped boundary (Fig. 29.10) is called the *spectrum locus*, since here we find the pure colors of the continuous spectrum. Point B on the diagram represents a white surface illuminated by sunlight at noon, while point C represents average daylight from an overcast sky. The other points on the inner curve represent the colors of blackbodies at the temperatures indicated on the diagram. Point C is ordinarily chosen as the illumination standard.

Any color above the dotted lines in Fig. 29.12 can be regarded as a mixture of illuminant C and a single spectral frequency. Thus the point N in this figure could be reached by mixing daylight and sodium D radiation. The line CN extended to the spectrum locus gives us 589.3 nm, which is called the *dominant wavelength* for the color specified by N. Since the yellow of point N lies between C and the spectrum color, it is not *saturated yellow*, as the spectrum color is. The distance from C to N divided by the distance from C to the spectrum locus gives the *purity* of the color. For point N the purity is 40 percent.

Any two colors on opposite sides of a line through C on the chromaticity diagram are complementary. The point P below the dotted line is a purple and not a spectral color. If we extend the line CP, it reaches the spectral curve at 500 nm. Point P is said to have the coordinate $500c$, where the c (= complementary) is added to indicate that the point P is on the opposite side of C from the spectrum locus. We can assign a purity to the color represented by P if we use a straight line connecting the endpoints of the visible region as the analog of the spectrum locus.

By use of the ICI chromaticity diagram it is possible to specify measurable physical quantities, dominant wavelength and purity, which correspond to the psychological *hue* and *saturation*. From the spectral-reflectance curve for a surface it is possible to compute the *luminous reflectance*, where the word *luminous* means that the value takes into account the visual response of a standard observer and the color characteristics of the light source. The luminous reflectance is the physical analog of *lightness*.

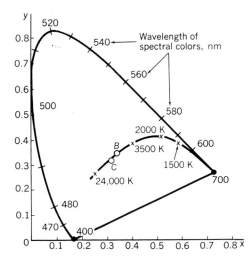

FIGURE 29.10
Chromaticity diagram shows the spectrum locus and the colors of blackbodies at the temperatures indicated. Point C represents average daylight from an overcast sky and is chosen as the illumination standard.

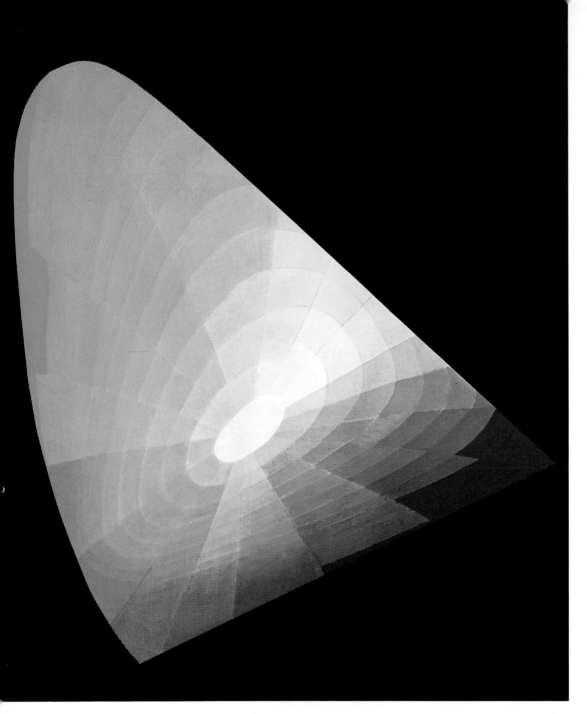

FIGURE 29.11 Chromaticity diagram has radial boundaries between segments to separate the various hues, with oval boundaries representing lines of constant saturation. The horseshoe-shaped outer boundary corresponds to the positions of pure spectrum colors. (*Courtesy of L. M. Condax and the Eastman Kodak Company.*)

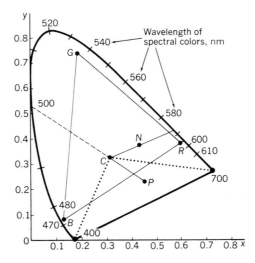

FIGURE 29.12
Chromaticity diagram showing method of specifying a color in terms of dominant wavelength and purity. The phosphors of color television screens glow in the primaries *R*, *B*, and *G*, permitting mixtures to reproduce any color in the triangle *RBG*, a close approximation to the full spectrum of visible color.

For color television, phosphors are available which give the primaries *R*, *G*, and *B* of Fig. 29.12 when they are bombarded by an electron beam. The screen is made up of these phosphors. By appropriate variations of intensities, all the colors in the triangle *RBG* can be obtained. Although this triangle does not include all visible colors, it covers a sufficiently wide range for rich color communication to be possible.

29.7   COLOR VISION

Thus far we have discussed the color of a surface by matching it with a neighboring surface from which adjustable amounts of three pure standard wavelengths are reflected. In this way it is possible to develop an *operational* definition of the color of a surface which permits us to specify the location of this color on a chromaticity diagram. A reasonable explanation for the ability to match any color with three spots of light is built on the hypothesis, proposed by Young in 1802 and developed by Maxwell and Helmholtz, that there are three types of color receptors in the eye, each of which is excited to some extent by almost the whole visual spectrum.

Modern research has established that there are indeed three color-sensitive pigments segregated in three different kinds of receptor cells in the cones of the retina. One of these, primarily responsible for sensing the blue-violet portion of the spectrum, has a peak sensitivity at 440 nm (Fig. 29.13); a second, detecting green, is most sensitive at 535 nm; and the third has maximum sensitivity at 565 nm (yellow) but still responds well to the red region. When all three types of cells are aroused about equally, the sensation of white light is produced. It is not necessary that all wavelengths be present to get the sensation of white light. As we have seen, properly chosen amounts of radiation of only two wavelengths can excite a sensation of perfect white light, since such a pair can excite all three responses roughly equally. Purples arise from excitation of the blue and red sensations without any substantial amount of green. Purple is not a spectral color but a mixture of spectral colors. Yellow is a true spectral color; nevertheless, the simultaneous excitation of the red and green sensations in roughly equal amounts gives the eye the sensation of yellow, even though none of the "yellow wavelengths" are present.

The three-color theory is highly successful in dealing with spots of light and is the foundation upon which both color television and color photography have been based. However, matching spots of light is a far cry from seeing colors in complete images under normal conditions. Then the colors of images arise from the interplay of longer and shorter wavelengths over the total visual field. For example E. H. Land has shown that if a photographic transparency of a colored scene taken with red light is projected on a screen with yellow light of $\lambda = 599$ nm, the resulting image is yellow with various areas lighter and darker than others. Now if the image of a second transparency of the same scene taken with green light and projected with yellow light of $\lambda = 579$ nm is superposed, the resulting image includes reds, greens, blues—a broad range of colors, paler and less saturated than the original scene, but still a reasonably faithful reproduction. Thus it appears that color is not determined uniquely by the wavelengths reaching the eye; instead the various wavelengths bear information by which the eye assigns colors to various objects. Apparently, to see colors the eye needs information concerning long and short wavelengths in a scene, but this information can be brought to the eye by as few as two narrow bands of wavelengths. Further, even when the source of illumination shifts from broad daylight to the relatively red-rich tungsten-lamp filament, the eye does not experience any substantial change in the color of such familiar objects as apples, bananas, oranges, lettuce, and blue pottery.

To explain the relative constancy of colors in shifting and uneven

FIGURE 29.13
Spectral-sensitivity curves (above) for the three cone pigments involved in color vision, with the colors associated with the various wavelengths (below).

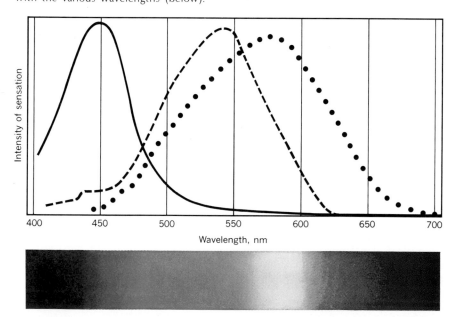

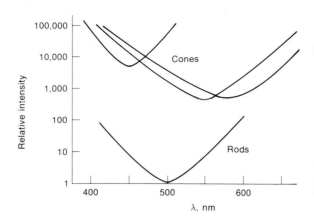

FIGURE 29.14
Threshold response curves for rods and cones in the retina. It requires 1,000 times as great an intensity to activate the cones as is needed for the rods.

illumination, Land has proposed the *retinex theory* of color vision† based on a number of significant experiments which he and his associates have performed. They have established that the eye has a remarkable ability to determine lightness values independent of the radiant flux received. As we saw in Sec. 26.5, the retina has both rods and cones as receptors; the cones have a threshold almost 1,000 times higher than the rods (Fig. 29.14). At very low intensity levels only the rods are excited; they have no color response, giving only a range of lightnesses from white to black. If one observes an assortment of light and dark papers at these low intensity levels and places an additional light close enough to a dark paper for it actually to send more light to the eye than a white paper, the dark paper will continue dark and the white one will continue to be lighter. Thus it is not the total flux received which determines the lightness but the fraction of the incident radiation returned.

If a dark-adapted eye is looking at a series of colored papers illuminated with 550-nm light so weak that only the rods are excited, lightnesses are observed but no colors. However, if a weak beam of 656-nm radiation is added and its intensity raised until the long-wavelength receptor system is activated, a remarkable range of colors can be recognized. This experiment shows that the lightness information collected at two wavelengths by separate receptor systems is not averaged area by area but is somehow compared. When one now goes to full illumination, all three color responses (plus the rod response) are activated. To determine the color of an object the reflectances for each of the sensor systems are compared in some way not yet understood. Thus the colors of illuminated objects may be thought of as a code for the determination of the reflectances of objects in three different but overlapping wavelength regions. Of course, for luminous objects such as a television screen or a colored light we are not concerned with reflectances but with the flux reaching the eye from the luminous sources.

† Color vision is a fascinating topic. Reports of definitive studies in this area may be found in four articles in *Scientific American*: E. H. Land, The Retinex Theory of Color Vision, December 1977, p. 108; W. A. H. Rushton, Visual Pigments and Color Blindness, March 1975, p. 64; E. F. MacNichol, Jr., Three-Pigment Color Vision, December 1964, p. 68; and E. H. Land, Experiments in Color Vision, May 1959, p. 84.

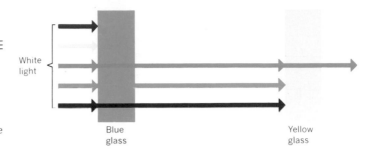

FIGURE 29.15
Transmission and absorption of white
light by yellow and blue glass.

29.8 MORE ON COLORS

A piece of glass which appears blue may actually reflect and transmit other colors such as the neighboring green and violet (Fig. 29.15). Similarly, a yellow glass plate may transmit red and green. If both glasses are placed in a beam of white light, the only color which can pass through both is green. This is an example of color by the *subtractive* process, in which we start with white light and remove various wavelength regions by absorption. The mixing of pigments is another example of a subtractive process. Mixing suitable yellow and blue pigments may give rise to a green paint just as green light passes through the combined filters of Fig. 29.15.

Although the eye holds the color of objects remarkably stable under varying illuminations, it does depend on the light received from the object and to a lesser extent on light from neighboring areas to judge color. Consequently, the apparent hue of an object is affected by its surroundings (Fig. 29.16) and, indeed, by the immediate past exposure of the retina. When any area of the retina is subjected to continued stimulation by intense red light and then subjected to white light, the area "sees" blue-green. In general, a retinal area strongly stimulated by any color responds to subsequent white light with a *fatigue image* dominated by the color complementary to the original stimulus.

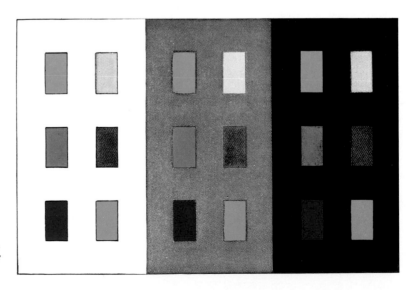

FIGURE 29.16
Background color affects identical hues.
(*From F. A. Osborne, "Physics of the Home."*)

If a white surface is illuminated with only pure red light, it appears red because it reflects only red light to the eye. However, if any other wavelengths are present, the eye can readily distinguish the white surface from a red one returning the same red flux. An object that appears red in white light is also red in red light, but it is black in any light which it does not reflect or transmit. Two bodies which have the same color in lamplight may not have the same color in daylight, although they usually do. Daylight is far richer in blues and violets than light from incandescent lamps. Further sunlight contains enough ultraviolet radiation to produce fluorescence (Sec. 29.9) in some dyes.

29.9 LUMINESCENCE

When a substance absorbs energy in some form and then emits visible (or near visible) radiation, the phenomenon is called *luminescence*. It involves a minimum of two processes: (1) the excitation of the electronic system of the substance and (2) the subsequent emissions of radiation. Crystalline solids used as luminescent sources are referred to as *phosphors*. When the emission of radiation takes place immediately (within 10 ns or less of excitation), the process is referred to as *fluorescence;* if the emission takes place more than 10 ns after the excitation, the process is called *phosphorescence*, or afterglow; for some phosphorescent materials the delay time may be minutes or even hours.

The screen of a television tube or of a cathode-ray oscilloscope is coated with phosphors which emit visible radiation after excitation by a beam of high-energy electrons. For color TV each element of the picture has three different neighboring phosphors, each of which emits in a different wavelength region (see Fig. 29.12). In fluorescent light sources ultraviolet radiation from mercury vapor excited by an electric discharge is absorbed in phosphors which coat the inside of the tube. By a suitable choice of phosphors a wide range of colors of fluorescent light is available. When the excitation of luminescence is by electromagnetic radiation (photoluminescence), the *wavelength of the emitted radiation is longer than the wavelength of the exciting radiation.*

QUESTIONS

1. What is meant when we say two colors are complementary?
2. The most common type of color blindness is one in which reds and greens appear the same. In terms of the theory of color vision discussed in Sec. 29.7, how might this difficulty arise?
3. A "daylight" incandescent bulb has a blue glass envelope. Why?
4. Green color can be seen in (*a*) a green leaf, (*b*) a mercury arc, (*c*) a shirt, and (*d*) a stained-glass window. What is the origin of the green in each case?
5. Pieces of cloth appear red, yellow, green, and blue-violet in daylight. How is their appearance modified when seen under candlelight? Under an incandescent light bulb?
6. Why does each of the following appear red? (*a*) A piece of iron heated to 1300 K; (*b*) a sunset observed through an atmosphere with a high concentration of very small particles; (*c*) a piece of white paper seen through a piece of red glass; (*d*) a ripe tomato; (*e*) a red shoe.
7. Why does a piece of black marble look black and a piece of white paper look white even when the marble is so intensely illuminated that the luminance of the marble exceeds that of the paper?
8. Why does a blue suit appear almost black by candlelight?
9. How can only three colors (and black) be used to

print color pictures? Can any color be reproduced in this way?

10. How does the doppler effect enable us to determine the angular velocity of the sun? To determine the component along the line of sight of the velocity of a distant star?

PROBLEMS

1. Find the wavelength of monochromatic light complementary to each of the following: (a) blue light with $\lambda = 470$ nm; (b) sodium yellow light with $\lambda = 589$ nm; and (c) red light with $\lambda = 610$ nm. In each case assume the standard illuminant C as white light and use Fig. 29.12.

Ans. (a) 571 nm; (b) 486 nm; (c) 492 nm

2. A stationary police radar unit operating at a frequency of 2 GHz observes a beat frequency of 356 as an automobile speeds away from the unit. Find the speed of the automobile.

3. If a police radar operates at a frequency of 2.5 GHz, find the beat frequency observed from a car approaching at a speed of 30 m/s.

Ans. 500 beats/s

4. Show that if $v \ll c$, the Doppler shift in wavelength is given approximately by $\Delta\lambda = -(v/c)\lambda$, where v is the velocity of approach of source and observer. *Hint:* If x is very small, $(1 + x)^n = 1 + nx + \cdots$.

5. For the galaxy Cygnus A the observed doppler shift toward longer wavelengths (lower frequencies) is such that $\Delta v/v = -0.06$. Estimate the speed at which Cygnus A is receding relative to the earth.

Ans. 1.8×10^7 m/s

FIVE

ELECTRICITY AND MAGNETISM

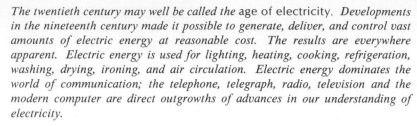

30
ELECTRIC CHARGES AT REST

The twentieth century may well be called the age of electricity. *Developments in the nineteenth century made it possible to generate, deliver, and control vast amounts of electric energy at reasonable cost. The results are everywhere apparent. Electric energy is used for lighting, heating, cooking, refrigeration, washing, drying, ironing, and air circulation. Electric energy dominates the world of communication; the telephone, telegraph, radio, television and the modern computer are direct outgrowths of advances in our understanding of electricity.*

Less obvious but no less important is the role of electricity in our economic development. The automobile and aluminum industries, to mention only two, could not have evolved in their present form until man learned how to generate and control electric energy. So important is electricity in our everyday life that our standard of living would be tremendously lowered without it. Electrical servants have replaced human servants in the average home. In less than a century the science of electricity developed from a curiosity to a dominant position in our technology.

Although most of the common uses of electrical phenomena involve electric charges in motion, the fundamental concepts were first developed in electrostatics, *the science of stationary electric charges.*

30.1 CHARGES BY CONTACT AND SEPARATION

As early as 600 B.C. it was known that amber rubbed with fur has the interesting property of attracting light pieces of straw or paper. A rubber comb run through the hair may exhibit this same property. Such bodies are said to be *electrified*, a term derived from the Greek word for *amber*. Experiments reveal that it is possible to electrify any kind of material by rubbing it with a suitable second material and then separating the two. An electrified body is said to bear an *electric charge*. When a charged body and an uncharged one are brought together, a portion of the charge is likely to be transferred to the uncharged body.

If a small pith ball is charged by contact with an electrified glass rod, the rod and ball repel each other. Similarly, if a pith ball is charged by contact with an electrified rubber rod, they repel each other. However, a ball charged by contact with the electrified *glass* rod is attracted by the electrified *rubber* rod, while the charged *glass* rod attracts a pith ball charged by contact with the electrified *rubber* rod. This simple experiment suggests (see Fig. 30.1) that there are *at least* two different kinds of electric charge. Exhaustive experiments with many substances have led to the conclusion that there are *only* two kinds of charges. The charge on a glass rod rubbed with silk is called *positive*, while that on a rubber rod rubbed with cat's fur is *negative*. From a number of experiments it is possible to establish the qualitative laws of electrostatic reaction: *Like charges repel each other; unlike charges attract each other.*

30.2 THE ELECTRICAL STRUCTURE OF MATTER AND THE FERMI LEVEL

The idea that all matter is composed of tiny particles called *atoms* is well known. Although the word *atom* means "indivisible," atoms have a complex structure. Practically all the mass of an atom is concentrated in a tiny core called the *nucleus*. The nucleus bears a *positive* charge, the magnitude of which depends on the atom in question. All nuclei of a given element bear the same charge. The nucleus with the smallest charge, *1 atomic unit* (u), is that of the hydrogen atom; the nucleus of ordinary hydrogen has a special name—*proton*. Every other type of atom has a nucleus bearing some integral number of atomic units. All oxygen nuclei have 8 atomic charges, all copper nuclei 29 charges. Indeed, the atomic number of the atom is a measure of how many atomic units of charge its nucleus bears, i.e., of how many protons the nucleus contains.

Outside the nucleus of a neutral atom are *electrons*, the number of electrons being equal to the number of positive charges of the nucleus. Each electron bears one fundamental atomic unit of negative charge, exactly equal in magnitude to the positive charge of the proton. It is sometimes convenient to think of an atom as a submicroscopic solar system, with the nucleus as the sun and the electrons revolving about it much as the planets revolve around the sun (Fig. 30.2). The concept of sharply defined orbits must be abandoned in a correct wave-mechanical treatment, but nevertheless the orbit model is a widely used reasoning aid which is capable of yielding plausible values for the average distance of an electron from its nucleus and of the allowed energies of the electrons.

In all atoms more massive than those of helium some of the electrons are closer to the nucleus than others and much more tightly bound. It is the outermost electrons which interact most strongly with neighboring atoms and participate in chemical bonds. Just as water runs downhill, so electrons move to lower energy states. When neutral atoms approach each other, a stable combination may be formed if the energy of the system is reduced by a sharing of outer electrons or of the transfer of an electron from one atom to another.

At 0 K electrons occupy the lowest possible energy states, and even at room temperature all but a minuscule fraction of the electrons in a substance are in the lowest possible energy states. For every material there is a *Fermi energy level* E_F such that the probability that an allowed state of energy E is occupied is given by†

$$f = \frac{1}{e^{(E-E_F)/kT} + 1} \tag{30.1}$$

where e (= 2.71828) is the base of the natural logarithms, T is the absolute temperature, and k is the Boltzmann constant (Sec. 14.7). At 0 K, f is 1 for E less than E_F and 0 for E greater than E_F; thus all levels below E_F are filled at 0 K and all levels above E_F are empty.

Consider water filling a swimming pool. The floor and walls play the role of the block of material, while the water is analogous to the electrons and the surface plays the role of Fermi level. When nothing disturbs the surface, all available energy states below this level are filled with water and all higher states are empty. Raising the temperature of a material corresponds to setting up small waves on the water surface. Then at troughs water is missing from levels below the Fermi level and at crests we find water above the Fermi level. When the absolute tem-

† Use of this formula involves mathematics beyond the scope of this course. As a consequence there are no problems using it, and you are not expected to remember it.

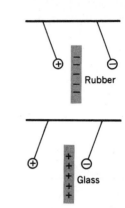

FIGURE 30.1
Like charges always repel, and unlike always attract.

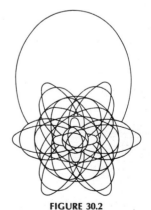

FIGURE 30.2
Electron orbits of a copper atom. The nucleus and two innermost electrons are not shown at the center of the atom.

perature is nonzero, f is exactly $\frac{1}{2}$ when $E = E_F$. *The Fermi energy E_F is the energy for which the probability of occupancy of an allowed energy state at E_F is $\frac{1}{2}$.*

30.3 EXCHANGE AND CONSERVATION OF CHARGE

When two electrically uncharged bodies of different composition are brought into intimate contact, it is almost certain that the Fermi level of one will be initially higher than that of the other. Then electrons in the first can achieve a lower energy by transferring to the second. Just as water runs down hill, so electrons by themselves move to states of lower energy. In order to move an electron from a lower state to a higher one, energy must be supplied in one of its many possible forms, e.g., thermal energy, electromagnetic radiation, or acoustic energy. When a substance with a higher Fermi energy is brought into contact with a second material with a lower Fermi energy, electrons move from the first to the second. For example, in Sec. 30.1, we observed that when a rubber rod is rubbed with cat's fur, the rod becomes negatively charged and the fur positively charged. In this case the Fermi energy of the rubber is lower than that of the cat's fur, so that electrons move from the fur to the rubber. Similarly, the Fermi level in silk is lower than the Fermi level in glass, so that when glass is rubbed with silk, the glass gives up electrons, thereby becoming positive, while the silk gains electrons and becomes negative.

When electrons move from a substance with higher Fermi level to a substance with lower Fermi level, the transfer lowers the Fermi level in the first and raises it in the second. When equilibrium is achieved, *the Fermi levels coincide* and there is no further net exchange of charge. It is not difficult to understand why this occurs. As the substance with lower Fermi level accumulates electrons, its excess negative charge repels other electrons; meanwhile the substance with the initially higher Fermi level has excess positive charge and attracts electrons.

Fermi levels in charge exchange between two substances are directly analogous to water levels in the exchange of water between two interconnected beakers and to temperature levels in the exchange of heat between two interacting bodies. In all three cases there is a transfer of something from the reservoir with higher level to the one with a lower level. Further, when full equilibrium is established, the levels are the same in both reservoirs. Since electrification by contact and separation always involves a transfer of charges from one body to another, the amount of positive charge which appears on one is equal to the negative charge which appears on the other. Whenever a certain positive charge appears on one body, an *equal* negative charge must appear on some other body or bodies.

Charge is one of the fundamental properties of matter. Whenever one body exchanges charge with another, the charge gained by one is exactly equal to the charge lost by the other. Consequently, *the total net charge of any isolated system never changes.* This statement expresses the *law of conservation of charge,* one of the basic laws of nature. This law and the laws of conservation of energy, of momentum, and of angular momentum are of great importance in the interactions of atoms and in the reactions of atomic nuclei.

30.4 CONDUCTORS AND INSULATORS

In some materials, notably the metals, a small fraction of the electrons are not bound to any one nucleus but are free to wander among the atoms. Materials in which charges are free to move about are called *conductors*. In other materials each electron is held by one or two atoms. A material in which charges are not free to move about is called a *nonconductor* or *insulator*. As might be expected, the line between conductors and insulators is not a sharp one. For example, a piece of damp wood is neither a good conductor of electricity nor a good insulator.

In metallic conductors the transfer of charge is by movement of electrons, but there are many situations in which charges are not conducted by electrons. A solution of sodium chloride in water is a good conductor of electricity because the molecules break into two parts, a sodium atom with an electron missing and a chlorine atom with one extra electron. These charged particles are called *ions*. The conduction takes place by the movement of both positively charged sodium ions and negatively charged chlorine ions. In liquids and gases, the transfer of charge by ions, both positive and negative, is common.

The early observations of electric charges were made on materials such as amber and glass, which are nonconductors. A piece of glass rubbed with silk obtains a positive charge which does not escape readily because the glass is a good insulator. If a copper rod is rubbed with silk, it also becomes charged, but unless it is carefully insulated the charge is rapidly conducted away. If one holds the copper rod in one's hand, the charges have no difficulty in traveling along the copper conductor and escaping through the body, since the human body is a fairly good conductor, especially when the skin is moist.

It is possible to charge one end of a hard rubber rod positively and the other end negatively by rubbing one end with a material of great electron affinity and the other end with a material of low electron affinity. Because the rubber is a good insulator, the charges remain on the ends. However, if one tries to do this to a metal rod, the charges do not stay on the ends; instead there is an immediate flow of electrons. Similarly, if a wire of conducting material is connected between a positively charged metal sphere and a negatively charged sphere, there is a flow of charges. Such a flow is called an *electric current*. In sending electric energy from one point to another, it is standard practice to produce a flow of charges; since this requires some sort of conducting medium, wires of good conductors, usually copper or aluminum, connect homes with power-generating stations.

30.5 COULOMB'S LAW

In 1785 the French physicist Coulomb used a torsion balance (Fig. 30.3) to measure the force between two small charged spheres as a function of the distance between them. He found that *the force between two charges Q_1 and Q_2 is directly proportional to the product of the charges and inversely proportional to the square of the distance r between the charges:*

$$F = k\frac{Q_1 Q_2}{r^2} \tag{30.2}$$

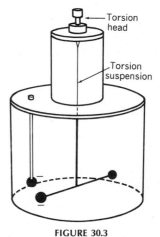

Torsion head

Torsion suspension

FIGURE 30.3
Coulomb measured the force exerted by the charge of a fixed sphere on the charge of one of the movable spheres of the torsion balance.

where the proportionality constant k depends on the units in which the variables of Eq. (30.2) are measured. Coulomb's law is directly applicable to charged bodies only in situations in which the distance r between charges is large compared with the dimensions of the charged bodies. When two charged objects are very small compared with the distance between them, we may consider the charges as concentrated at points and we refer to them as *point charges.*

Charge may be measured in a number of units, four of which have been widely used. We shall work with only the first two.

1. The *atomic unit of charge* (u) is the charge on a single proton; the charge on the electron has the same magnitude but is negative in sign. The atomic unit of charge is exceedingly minute, far too small to be convenient for use in most practical problems in electricity. It is, however, a truly basic unit in that the nucleus of every atom bears an integral number of positive atomic units and every electron exactly one negative atomic unit. Further, every charged particle† discovered through 1977 carries an integral number of atomic charges.

2. The *coulomb* (C) is the practical unit of charge. We shall take it as a new fundamental unit which, with the meter, kilogram, and second, will permit us to formulate the laws of electricity and magnetism in a self-consistent manner. The formal definition of the coulomb in terms of the operations required for a precise measurement of charge involves the use of concepts which are developed in subsequent chapters (Sec. 36.8). We can, however, compare the coulomb with the atomic unit of charge:

$$1 \text{ C} = 6.242 \times 10^{18} \text{ atomic units}$$

or $$1 \text{ atomic unit} = 1.60219 \times 10^{-19} \text{ C}$$

In electrostatics we seldom meet charges as large as 1 C. Typically we shall be dealing with charges of the order of a microcoulomb (1 μC = 10^{-6} C) or even a picocoulomb (1 pC = 10^{-12} C).

3. The *electrostatic unit of charge,* sometimes called the *statcoulomb,* is especially convenient for dealing with electrostatic problems. It is defined in terms of Coulomb's law as follows: The electrostatic unit of charge is that charge which repels an identical charge one centimeter away in vacuum with a force of one dyne (10^{-5} N).

$$1 \text{ C} = 3 \times 10^9 \text{ electrostatic units of charge}$$

4. The *electromagnetic unit of charge,* sometimes called the *abcoulomb,* is defined in terms of the magnetic field produced at the center of a circular coil by a current.

$$1 \text{ C} = 0.1 \text{ electromagnetic unit of charge}$$

Complete systems of electrical units have been built around both the electrostatic and the electromagnetic units of charge. These systems are particularly convenient for handling certain special problems, but since it is difficult to learn three distinct systems of electrical units, we shall use only the practical system in the problems of this text.

† In 1962 Gell-Mann advanced the hypothesis that there are particles named *quarks* which have charges of $\frac{1}{3}$ and $\frac{2}{3}$ u. Experiments have found evidence of quarks.

When the charges Q_1 and Q_2 are in coulombs, the force in newtons, and the distance r in meters, the constant k of Eq. (30.2) for a vacuum has the value 8.9878×10^9 N-m^2/C^2, which in turn is $10^{-7}c^2$, where c is the speed of light in free space in meters per second. It is customary to write Coulomb's law in the form

$$F = \frac{1}{4\pi\epsilon_0}\frac{Q_1Q_2}{r^2} \qquad (30.2a)$$

where $\epsilon_0 = 8.8543 \times 10^{-12}$ C^2/N-m^2 is called the *permittivity of free space*. From the point of view of Coulomb's law, replacing k with $1/4\pi\epsilon_0$ appears to offer no advantage; however, we shall see in Chap. 33 that Eq. (30.2a) leads to simpler expressions for capacitances; the advantages become still greater in advanced electromagnetic theory.

As we have seen, in vacuum $1/4\pi\epsilon_0 = k$ is almost 9×10^9 N-m^2/C^2. In air the constant in Coulomb's law is less than in vacuum by the factor $1/1.0006$. For our problems we shall use

$$F = (9 \times 10^9 \text{ N-m}^2/\text{C}^2)\frac{Q_1Q_2}{r^2} \qquad (30.2b)$$

for charges in either air or vacuum.

When several charges are present in one region, the force on any one of the charges is equal to the vector sum of the forces which each of the other charges would exert on the first one if it acted independently.

□ **Example** Charges A, B, and C of $+25$, $+20$, and $-8\,\mu$C, respectively, are arranged as shown in Fig. 30.4. Find the magnitude of the force on charge A.

$$F_{AB} = 9 \times 10^9\frac{(20 \times 10^{-6})(25 \times 10^{-6})}{25} = 0.180 \text{ N}$$

$$F_{AC} = -9 \times 10^9\frac{(25 \times 10^{-6})(8 \times 10^{-6})}{9} = -0.200 \text{ N}$$

The vertical component of F_{AB} is $\frac{3}{5} \times 0.180 = 0.108$ N; the horizontal component of F_{AB} is $\frac{4}{5} \times 0.180 = 0.144$ N.

The vertical component of the resultant force on A is $-0.200 + 0.108 = -0.092$ N. The horizontal component of the resultant is 0.144 N. The magnitude of the resultant is then

$$\sqrt{(0.144)^2 + (0.092)^2} = 0.17 \text{ N} \qquad \square$$

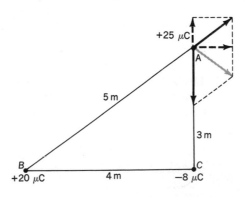

FIGURE 30.4
Charges of $+25$, $+20$, and $-8\,\mu$C at the corners of a 3-4-5 right triangle.

30.6 ELECTRIC FIELD INTENSITY

Any region in which electric forces can be detected is called an electric field. The intensity (or strength) E of an electric field at any point is defined as the ratio of the force F acting on a small test charge q at that point to the charge q:

$$E = \frac{F}{q} \tag{30.3}$$

A very small test charge should be chosen, since a large test charge would change the charge distribution which creates the field and thus distort the very thing it is being used to measure. Clearly, the electric intensity is a vector quantity. Its direction is the direction of the force on a *positive* charge.

The intensity of the electric field at a distance r from an isolated point charge Q is obtained by performing an imaginary experiment in which a small test charge q is placed at the point where the intensity is desired. By Coulomb's law the force on this test charge is $F = Qq/4\pi\epsilon_0 r^2$. Since $E = F/q$, we conclude that the electric intensity at a distance r from a point charge Q is given by

$$E = \frac{Q}{4\pi\epsilon_0 r^2} \tag{30.3a}$$

If an electric field is due to two or more charges, the electric intensity E at any point is given by the resultant of the electric intensities due to each charge taken individually.

☐ **Example** Find the magnitude of the electric intensity at A of Fig. 30.4 due to the charges at B and C.

We have found the magnitude of the force on a charge of 25 μC at this point to be 0.17 N. By Eq. (30.3),

$$E = \frac{F}{q} = \frac{0.17\ N}{25\ \mu C} = 6{,}800\ N/C \qquad\qquad ☐$$

☐ **Example** Find the electric field strength at the point P of Fig. 30.5, which is 10 cm from a charge of $+6\ \mu$C and 40 cm from a charge of $-8\ \mu$C.

The electric field due to the $+6$-μC charge has a magnitude $(9 \times 10^9 \times 6 \times 10^{-6})/(0.1)^2 = 54 \times 10^5$ N/C. The field due to the -8-μC charge has a magnitude $(9 \times 10^9 \times 8.0 \times 10^{-6})/(0.4)^2 = 4.5 \times 10^5$ N/C. The fields due to both charges are in the same direction—to the right. Therefore, the resultant has a magnitude of 58.5×10^5 N/C and is directed toward the right. ☐

30.7 LINES OF FORCE

The electric field in the vicinity of one or more charged bodies is frequently represented by drawing *lines of force* to help us visualize the

FIGURE 30.5
Electric field strength **E** at point P is the resultant of the fields due to the $+6$- and -8-μC charges.

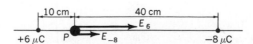

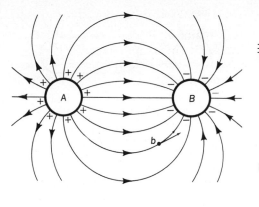

FIGURE 30.6
Lines of force for a pair of equal and
opposite charges.

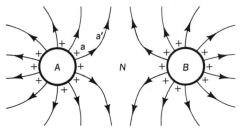

FIGURE 30.7
Lines of force for two identical positive
charges; at the neutral point N the field
intensity is zero.

field. *An electric line of force is a line drawn so its tangent at every point has the direction of the electric intensity at that point.* Some of the lines of force associated with a pair of equal and opposite charges are shown in Fig. 30.6, while some due to two identical positive charges are indicated in Fig. 30.7. In electrostatics lines of force always begin on positive charges and terminate on negative charges. The lines of Fig. 30.7 terminate on negative charges which might, for example, be on the walls of the room.

We may draw as many lines of force as we wish in picturing an electric field. A reasonable number of lines can give us a visualization of the direction of the force on a small positive test charge any place in the field. The line of force through a point tells us the direction of the electric intensity at that point but nothing of its magnitude. However, a qualitative estimate of the magnitude of the field can be obtained from a reasonably detailed plot which shows the lines of force approaching each other as the field becomes stronger. (Why does this happen?) Indeed, in an ideal three-dimensional plot, the number of lines of force per unit area perpendicular to the lines would be directly *proportional* to the electric intensity.

30.8 ELECTROSTATIC INDUCTION

If an uncharged conductor B (Fig. 30.8) is brought into the electric field of a positively charged conductor A, the attractive forces due to the excess positive charge on A cause the electrons in B to be pulled toward end C, leaving the farther end D with a deficit of electrons and therefore charged positively. Since B was originally neutral, i.e., contained as much positive as negative electricity, the positive charge on end D is just equal to the negative charge on end C. However, by Coulomb's law, the

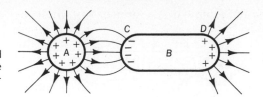

FIGURE 30.8
Equal and opposite charges are induced at ends C and D when a positive charge A is brought near the uncharged conductor B.

attractive force on the charges at C is greater than the repulsive force on those at D because the former charges are closer to A. Although conductor B bears no net charge, it is attracted by conductor A.

In the example of the paragraph above, body B was an uncharged conductor. What would happen if we replaced it by an insulator of similar shape? In this case there would be no free electrons to move to end A, but now for each atom body A would exert an attractive force on all atomic electrons and a repulsive force on all the positive nuclei. As a result the electrons would undergo slight displacements toward A and the nuclei slight displacements in the opposite direction. Again, by Coulomb's law, there is an attractive force between A and the body B. Thus in general, a charged body exerts an attractive force on *any* neighboring uncharged body whether the latter is a conductor or an insulator. The attractive forces between a charged comb and uncharged bits of paper arise from the induction process described above. The bits of paper are *polarized* by the charge on the comb; polarization of nonconductors is discussed further in Sec. 32.2.

Let us now return to the situation shown in Fig. 30.8, where body B is a good conductor. If B is connected to the earth by a wire, enough electrons come from the earth to B to neutralize the positive charge at D. Meanwhile the electrons on end C are held fast by the attractive force due to the positive charge on A. If the connection to the earth is broken and B removed from the presence of A, B will have an excess of electrons; it has been negatively charged by electrostatic induction.

A charged sphere in the neighborhood of the surface of the earth (Fig. 30.9) induces a charge on the surface of the earth below it. A charged antenna in the neighborhood of the surface of the earth has a similar effect.

Another illustration involving electrostatic induction is seen in Fig. 30.10. If a metal sphere charged positively is introduced into an insulated uncharged hollow sphere, some of the electrons of the hollow sphere are drawn to its inner surface, leaving the outer surface charged positively. If

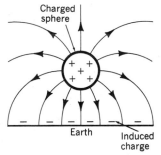

FIGURE 30.9
A positively charged object induces a negative charge on the surface of the earth below it.

FIGURE 30.10
Faraday's ice-pail experiment shows that the charge induced on the inside wall of the cavity is equal and opposite to the inducing charge on sphere A, since the external field is not affected by touching A to the inside of the pail and thereby discharging A.

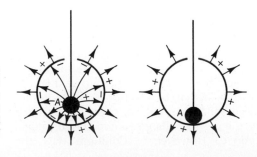

the metal sphere A is placed in contact with the inner surface of the hollow sphere, the electrons from the inner surface of the hollow sphere go over to the sphere A and just compensate the deficit of electrons. This leaves both sphere A and the inner surface of the hollow sphere without a charge, while the outer surface of the hollow sphere is charged positively. If the outer surface is now connected to the earth, it gains enough electrons to compensate for its deficit and it is left uncharged. By experiments of this type Faraday was able to show that the charge induced on the outside of an almost closed hollow conductor is equal to the inducing charge on the inside. As his hollow conductor Faraday used a small ice pail, and the operations described in this paragraph are often referred to as the Faraday ice-pail experiment.

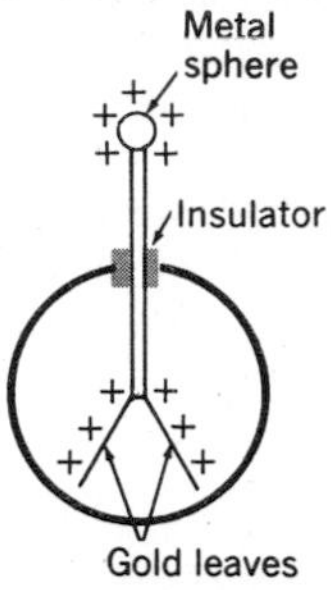

30.9 THE ELECTROSCOPE

A useful instrument for studying electrostatic phenomena is the electroscope, one form of which consists of two thin leaves of gold foil attached to one end of a metal rod which is terminated at the other end by a metal sphere (Fig. 30.11). When the metal sphere is charged, part of the charge goes to the gold foils, which repel each other and therefore diverge. The greater the charge on the leaves, the greater the divergence.

An electroscope can be charged positively or negatively by touching the knob with either a positive glass rod or a negative rubber rod, but there is a substantial danger of transferring so much charge that one gold leaf will be torn off. A safer and easier way to charge an electroscope is by induction. To charge an electroscope positively by induction, a negatively charged rod is brought near the knob, as shown in Fig. 30.12a. Electrons in the knob are repelled by the rod and descend to the leaves. The knob is then grounded by touching it (Fig. 30.12b), and the excess electrons on the leaves run off. The connection to ground is then removed, leaving a net positive charge on the electroscope knob. When the negative rod is removed, these charges rearrange themselves so that both the knob and leaves are positively charged (Fig. 30.12c).

When a negatively charged rod is brought near a positively charged electroscope, some of the electrons on the metal sphere are repelled and move to the leaves, where they reduce the net positive charge and hence the divergence of the leaves (Fig. 30.13). As the negative rod is brought

Metal sphere

Insulator

Gold leaves

FIGURE 30.11
A gold-leaf electroscope.

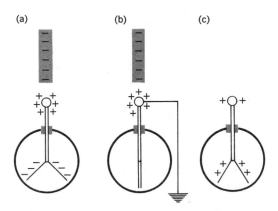

(a)   (b)   (c)

FIGURE 30.12
Charging an electroscope by induction.

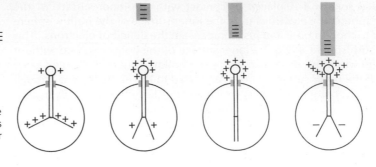

FIGURE 30.13
Charge distribution on an electroscope changes but the total charge remains constant as a charged rod comes nearer and nearer.

still nearer, more electrons go to the leaves, which eventually lose all their charge and converge. If the negative rod is brought still closer, even more electrons go to the leaves, causing them to diverge once more. In this case, although the total charge on the electroscope is positive, the leaves bear a negative charge.

QUESTIONS

1. In what respects are electric and gravitational forces similar? In what respects are they different?

2. How can one prove that the force between two charges is proportional to the product of the charges? If someone refused to believe this, what experiment could you devise to convince him?

3. The nuclei of carbon, tin, and uranium bear 6, 50, and 92 fundamental positive units of charge respectively; each unit corresponds to 1.6×10^{-19} C. How many electrons does a neutral atom of each type have?

4. When a charged rubber rod is brought near small pieces of cork, some bits of cork may at first cling tightly to the rod, but soon some bits may jump off the rod. What has happened? Why don't all the bits of cork fly off?

5. A small charged pith ball is brought near to an uncharged metal sphere, and an attractive force is evident. Why? If the sphere is grounded, the attraction is still greater. Why?

6. What does it mean to say that charge is always conserved? After all, you can have a system of several uncharged bodies and make one in which every body is charged.

7. Why do electrostatic experiments often work poorly in hot, humid rooms?

8. Does it make any difference which end of conductor B in Fig. 30.8 is grounded by the wire? Justify your answer with an explanation.

9. How could you charge an electroscope positively with a negatively charged rod? Explain what happens during each step of the process.

10. In one form of Cottrell precipitator for capturing smoke particles, a highly negative wire is suspended vertically in a rising column of polluted air. The wire produces a strong inhomogeneous electric field. Why does such an arrangement remove smoke and soot?

11. Can two lines of force ever cross? Explain your answer carefully.

12. Three point charges are equally spaced along a line. Q_2 and Q_3 are equal in magnitude but opposite in sign. What is the magnitude of Q_1 if the net force on Q_3 is zero? Is the sign of Q_1 the same as that of Q_2 or of Q_3?

PROBLEMS

1. Find the force on a charge of 5 nC which is 50 mm from a charge of 20 nC.   *Ans.* 360 μN

2. The force on a small test charge is 5 μN when the charge is placed in an electric field of intensity 1.5 MN/C. Find the magnitude of the charge. How many electrons would be required to neutralize this charge?

3. How many electrons must be removed from a small pith ball to give it a charge of 3 pC? What is the electric field strength 25 mm from the pith ball?   *Ans.* 1.87×10^7; 43.2 N/C

4. In a Bohr model of the hydrogen atom, an electron revolves in a circular orbit of radius approximately 5.29×10^{-11} m about a nucleus which

bears a positive charge equal in magnitude to the electronic charge. Find the force on the electron which provides the centripetal acceleration for the uniform circular motion. How large is the centripetal acceleration?

5. Three charges of $+5$, -3, and $+6\,\mu C$ are placed in the same straight line 15 mm apart. What force acts on each charge because of the other two?
   Ans. 300 N; 120 N; 420 N

6. Two identical metal spheres have a mass of 1 kg each. How many electrons must be transferred from one sphere to the other in order to make the electrical attraction between the spheres equal to the gravitational attraction when the spheres are 1 m apart?

7. Charges of -6 and $+4\,\mu C$ are 3 m apart. Find the force they exert on each other and the electric intensity midway between them.
   Ans. 0.024 N; 40 kN/C toward $-6\,\mu C$

8. A point charge of 25 nC is placed 0.3 m from a point charge of 10 nC. What force is exerted on each charge? Find the electric field strength at the point midway between the charges.

9. A point charge of 8 nC is located 0.6 m from a point charge of 12 nC. Find the force exerted on each charge by the other and the electric field strength at a point 0.2 m from the larger charge on the line connecting the two charges.
   Ans. 2.4 μN; 2,250 N/C toward smaller charge

10. Two point charges repel each other with a force of 6×10^{-4} N when they are 0.3 m apart. Find the force if the distance between them is reduced to 0.15 m. If one of the charges is 500 pC, what is the other?

11. The electric intensity in the region between two deflecting plates of a cathode-ray oscilloscope tube is 25 kN/C. Find the force on an electron passing between these plates. What acceleration does the electron experience if it has a mass of 9.11×10^{-31} kg?
   Ans. 4×10^{-15} N; 4.39×10^{15} m/s^2

12. Calculate the ratio of the coulomb electric force to the newtonian gravitational force between two electrons if the mass of an electron is 9.11×10^{-31} kg. What electric intensity is required to balance the gravitational force on an electron?

13. Two small pith balls, each weighing 0.002 N, are hung from a common point by nylon threads 125 mm long. When the pith balls are given equal negative charges, they repel each other and stand 150 mm apart, so that each of the supporting threads makes an angle of 36.9° with the vertical. Find the charge on each pith ball.
   Ans. -6.12×10^{-8} C

14. Two small equally charged spheres, each with a

mass of 0.24 g, are suspended from the same point by silk fibers 0.8 m long. The repulsion between them keeps them 128 mm apart. What is the charge on each sphere?

15. An unknown charge and a charge of 60 nC are 2 m apart. The electric field strength is zero at a point 0.8 m from the unknown charge on the line connecting the two charges. Find the unknown charge.
   Ans. 26.7 nC

16. A small test charge of 5 pC experiences a force of 2.5 μN when it is 1.5 m from a point charge of unknown magnitude. Find the electric intensity at the test charge due to the unknown charge and the magnitude of the unknown charge.

17. A charge of 4 pC is placed in a downward-directed electric field of intensity 30 kN/C. Find the work required to move the charge (a) 0.5 m to the east, (b) 0.5 m upward, and (c) 2 m upward at an angle of 36.9° with the vertical.
   Ans. (a) none; (b) 60 nJ; (c) 192 nJ

18. Find the force on a charge of 6 pC if it is placed in a uniform field of intensity 45 kN/C. How much work is done *by the field* if the charge moves 0.2 m in the direction of the field? Perpendicular to the field?

19. A charge of 24 nC is placed at the origin of a cartesian coordinate system, and a charge of -50 nC is placed at $x = 2$ m, $y = 0$. Find the electric intensity at the point (a) $x = 4$ m, $y = 0$ and (b) $x = 2$ m, $y = 2$ m.
   Ans. (a) 99 N/C toward origin; (b) 95.3 N/C at $-78.5°$ with x axis

20. Charges of 60 nC are placed at opposite ends of the hypotenuse of a right triangle with sides 3, 4, and 5 m. Find the magnitude of the electric intensity at the other vertex.

21. Charges of -40 and $+70$ nC are placed at two of the vertices of an equilateral triangle with sides 0.2 m in length. Find the magnitude of the electric intensity at the third vertex and of the force which would act on a charge of 2 nC at that vertex.
   Ans. 13.7 kN/C; 27.4 μN

22. Three charges A, B, and C are located on the same straight line. The distance from A to B is 30 cm, and that from B to C is 60 cm. The charge at A is 50 nC, that at B is 80 nC, and that at C is 60 nC. What force is exerted on A by the charges at B and C if all the charges are positive? What is the electric intensity midway between A and B?

23. Charges of $+20$, -40, and $+30$ nC are placed on the corners of a square with sides 2 m long. Find the magnitude of the electric field strength at the fourth corner, which is diagonally opposite the -40 nC charge. *Ans.* 38 N/C

24. An electric dipole consists of a charge $+Q$ sepa-rated by a distance d from a charge $-Q$. Find the torque on this dipole when it is placed in a uniform field of intensity E (a) with its axis per-pendicular to E and (b) with its axis making an angle θ with E.

25. Identical 50-nC charges are placed at opposite ends of a diagonal of a rectangle 0.30 by 0.40 m. Find the values of the charges Q_1 and Q_2 which must be placed at the other two corners to make the resultant force on one of the 50-nC charges zero. Will the force on the other 50-nC charge then also be zero? If not, what will its magnitude be? *Ans.* -25.6 nC; -10.8 nC; no; 85 μN

From mechanics we know that the best approach to many problems is through the concept of energy and its conservation. By using the energy concept we avoid the need of knowing and working with force, which may be changing rapidly in space or in time. In electricity practical problems are conveniently solved by considering changes in the energy of a charge as it moves from place to place rather than focusing our attention on the force experienced by the charge at each point in its path. In this chapter we use Coulomb's law of force between electric charges to find the potential energy of a small test charge in the field of a point charge. This leads us to the concept of potential, *one of the key ideas of electricity.*

31

POTENTIAL

31.1 POTENTIAL ENERGY IN AN ELECTRIC FIELD

When a small test charge $+q$ is moved about in the field of a fixed charge $+Q$ (Fig. 31.1), work must be done to carry the test charge closer to Q. This work goes into increasing the potential energy of the test charge. On the other hand, as q moves away from Q, the electric field does work and the potential energy of q decreases. In discussing the movement of electric charges, the concept of electric potential energy plays an important role, since many problems in electricity involve the law of conservation of energy and electric potential energy is one of the forms which must be taken into consideration.

In expressing the electric potential energy of a charge quantitatively, it is necessary to specify some point at which the potential energy is zero, just as it is necessary to specify a zero configuration in dealing with any other form of potential energy. For example, the potential energy of a given mass relative to the earth (Chap. 8) is proportional to the elevation h. In this case it is necessary to decide from what level h is measured. The potential energy of a 1-kg mass 2 m above a table depends on whether one wishes to use the tabletop or some other level to establish the zero for potential energy. In dealing with the particular situation illustrated in Fig. 31.1, it is common to consider the potential energy of q to be zero when it is infinitely far from Q. This choice is arbitrary but conventional.

Having chosen the configuration for zero potential energy, we define the potential energy of a charge $+q$ at point A as the work necessary to move the charge from the point at which its potential energy is zero to the point A. If the potential energy of a charge q at a given point is P joules, the potential energy of a charge $3q$ at this same point is $3P$ joules, and, in general, for charge nq the potential energy at this point is nP joules. Since the potential energy of a charge at a given point is directly proportional to the charge itself, it is convenient to introduce the concept of *potential energy per unit charge*. The ratio of the potential energy of a charge at a given point to the charge is called the *potential at the point*. In electrostatics the zero of potential is usually chosen to be at infinity. For this choice *the potential at a point A is the ratio W/q, where W is the work required to move a small test charge $+q$ from infinity to the point A.*

The practical unit of potential, the *joule per coulomb*, is called the *volt* (V), in honor of Alessandro Volta, whose pioneering work with batteries was a major contribution to the infant science of electricity in 1800. *The potential at a point is one volt when it requires one joule of work to move a positive charge of one coulomb from a point of zero potential to the point in question.*

FIGURE 31.1

Work must be done to move a small test charge $+q$ from point A to point B against the electric intensity of charge $+Q$.

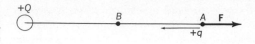

31.2 THE POTENTIAL DUE TO A POINT CHARGE

The potential at a distance r from a point charge of Q coulombs is the *work per unit charge* required to bring a positive test charge from an infinite distance to the point in question. If q is the test charge being transferred, the repelling force exerted on it by Q is $Qq/4\pi\epsilon_0 r^2$. To move q against this force we must apply a force of this same magnitude toward Q. The element of external work ΔW required to move q a distance Δr is given by $\Delta W = (-Qq/4\pi\epsilon_0 r^2)\,\Delta r$ (Fig. 31.2). If Δr is positive (r increasing), this work is negative: Q provides the required force, and no positive external work need be performed. However, when Δr is negative (r decreasing), external work must be done against the electric field of Q and ΔW is a positive quantity.

Since the repelling force on q varies with r in the same way as the gravitational attractive force on a small mass in the earth's gravitational field (Sec. 8.10), we may use the reasoning of that section to conclude that the work† required to move q from infinity to a distance r from charge Q is

$$W = \frac{Qq}{4\pi\epsilon_0}\left(\frac{1}{r} - \frac{1}{\infty}\right) = \frac{1}{4\pi\epsilon_0}\frac{Qq}{r} \tag{31.1}$$

This is the potential energy of the test charge relative to zero at $r = \infty$. The potential at distance r from Q is then

$$V = \frac{W}{q} = \frac{1}{4\pi\epsilon_0}\frac{Q}{r} = (9 \times 10^9 \text{ V-m/C})\frac{Q}{r} \tag{31.1a}$$

The potential 1 m from a charge of 1 C in vacuum is 9×10^9 V.

□ **Example**   Find the potential 2 m from a charge of $+6\,\mu$C.

$$V = \frac{9 \times 10^9 Q}{r} = \frac{(9 \times 10^9)(6 \times 10^{-6})}{2} = 27{,}000 \text{ V} \qquad \square$$

When there are a number of charges in a region, the potential at any point is the sum of the potentials due to each charge acting alone. Thus if there are three charges Q_1, Q_2, and Q_3 in a region, the potential at a point which is at distance r_1 from Q_1, r_2 from Q_2, and r_3 from Q_3 is simply given by

$$V = \frac{1}{4\pi\epsilon_0}\left(\frac{Q_1}{r_1} + \frac{Q_2}{r_2} + \frac{Q_3}{r_3}\right) \tag{31.2}$$

†It is assumed that all operations in this chapter are carried out in air or *in vacuo*.

FIGURE 31.2

The external work ΔW required to move a small test charge $+q$ a distance Δr in the field of $+Q$ is given by $\Delta W = -(Qq/4\pi\epsilon_0 r^2)\,\Delta r$. When q is moved from A to B, Δr is negative and ΔW is a positive quantity.

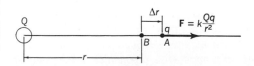

The contribution of a negative charge to the potential is negative; the electric field does work in bringing a positive test charge toward a negative charge. A great advantage of working with potential in preference to electric intensities arises from the fact that *potential is a scalar* while the *electric intensity is a vector.*

31.3 THE ISOLATED SPHERE

Consider a metal sphere of radius R, far away from all other bodies and bearing a charge of Q coulombs. The electric field associated with this charged sphere is everywhere radial and, for all points outside the sphere, is indistinguishable from the electric field which would be associated with a point charge Q located at the center of the sphere. Thus, a uniformly charged sphere behaves, *so far as all external points are concerned,* exactly as though all its charge were concentrated at the center. The work necessary to bring a unit charge from infinity to the surface of the sphere is the same as the work which would be required to bring the same unit charge from infinity to the distance R from a point charge Q. The potential of the sphere is therefore given by

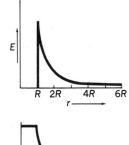

$$V = \frac{1}{4\pi\epsilon_0}\frac{Q}{R} = (9 \times 10^9 \text{ V-m/C})\frac{Q}{R} \tag{31.3}$$

Within the sphere the electric field is zero (see Sec. 31.7). Therefore, no work is required to carry a small test charge from any point on the surface of the sphere to any point within the sphere. *The potential at all points inside the sphere is the same as the potential at the surface.* The electric field and the potential are plotted as a function of r in Fig. 31.3.

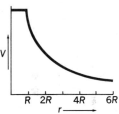

FIGURE 31.3
The electric intensity E and potential V at a distance r from the center of a uniformly charged conducting sphere of radius R.

31.4 SURFACE CHARGE DENSITY

When a sphere of radius R bears a charge of Q C, the charge per unit area is given by

$$\sigma = \frac{Q}{4\pi R^2} \tag{31.4}$$

where σ is known as the *surface charge density.*

Consider a large sphere and a small one, both charged to the same potential V and sufficiently far apart for the electric field of either to be negligible at the other. If the charges on the two spheres are then Q and q and the radii of the two spheres are R and r, respectively, V is given by

$$V = \frac{Q}{4\pi\epsilon_0 R} = \frac{q}{4\pi\epsilon_0 r}$$

Let σ_R and σ_r be the charge densities on the respective spheres. Then

$$V = \frac{4\pi R^2 \sigma_R}{4\pi\epsilon_0 R} = \frac{4\pi r^2 \sigma_r}{4\pi\epsilon_0 r}$$

and

$$R\sigma_R = r\sigma_r \tag{31.5}$$

For isolated spheres at the same potential, the charge densities are inversely proportional to the radii.

The surface of any conductor may be regarded as composed of a large number of spherical segments (Fig. 31.4) of different curvatures. While

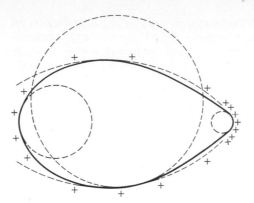

FIGURE 31.4
As a first approximation, the surface of an arbitrary curved conductor is assumed to be made up of sections of spherical surfaces; charge density becomes greater as the radius of curvature becomes smaller.

we may not regard these segments as portions of isolated spheres in order to apply Eq. (31.4) to calculate relative charge densities, it remains qualitatively true for convex surfaces such as those of Fig. 31.4 that where the surface has greatest curvature, i.e., smallest radius, the surface charge density is greatest. When a conductor terminates in a sharp point, the surface density at the point is so great that the molecules of air in the neighborhood of the point may become charged with electricity. Since like charges repel, the charged molecules are repelled from the point and the body to which the point is attached is discharged. The ionized molecules may emit light which is readily visible in a darkened room. Such a discharge is called a *corona discharge*. If the field is sufficiently great, flashover occurs (Fig. 31.5).

The fact that charged points allow electricity to escape is used in electrostatic machines, where rows of metallic points conduct electricity from moving to fixed parts of the machine. This fact is also important in the design of lightning rods. The pointed conductors on the rod bring about a silent and gradual discharge from the rod to the clouds. The

FIGURE 31.5
Flashover on a sealed bushing. (*General Electric Company.*)

escape of electricity from these points prevents the accumulation of enough electricity on the building on which the lightning rod is mounted to result in a dangerous disruptive discharge.

31.5 POTENTIAL REFERRED TO THE EARTH

In dealing with potential, it is sometimes convenient to assign zero potential to some conducting body, just as it is convenient to refer the height of a building to the ground level, while the height of a mountain is ordinarily referred to mean sea level. In many situations the potential of the earth is taken as zero. When one speaks of a 110-V lighting circuit, one ordinarily implies that the earth is taken as zero potential. In a radio or television receiver the chassis is usually regarded as being at zero potential; often it is actually tied to the ground by a direct conductor. In dealing with the electric circuits of an automobile, it is convenient to regard the frame of the car as being at zero potential, although it is ordinarily insulated from the earth by the tires.

In general, one may assign the potential zero to any convenient point. Then the potential at any other point is the ratio W/q, where W is the *work* required to move a small test charge q from the point at zero potential to the point in question.

31.6 POTENTIAL DIFFERENCE

An important concept in electrical theory is that of the *potential difference* between two points A and B, which is simply the potential at A minus the potential at B, just as the name *potential difference* implies. Thus, $V_{AB} = V_A - V_B$. Alternatively, *the potential difference between points A and B is the external work per unit charge done against the electrostatic field in moving a small positive test charge from B to A.* The potential difference is independent of the zero of potential. For example, the potential difference between the positive and negative terminals of an ordinary dry cell is 1.5 J/C, regardless of whether the earth or some other point is taken as the zero potential reference. The potential of the positive terminal relative to the negative terminal is $+1.5$ V, while the potential of the negative terminal relative to the positive is -1.5 V.

Let W be the external work done against an electric field in moving a small positive charge q from point B to point A. The potential difference V_{AB} between these two points is, by the definition above, the ratio of W to q.

$$V = \frac{W}{q} \qquad (31.6)$$

Because of its great importance, we rephrase our definition: *The potential difference between two points is the work per unit positive charge required to move a small test charge from one point to the other.*

Potential difference, as well as potential, is commonly measured in volts (joules per coulomb). *The potential difference between two points is one volt if it requires one joule of external work to move one coulomb of charge from one point to the other.* The potential difference across the terminals of an ordinary automobile battery is 12 V; this means that 12 J of external work is required to transfer 1 C of positive charge from the negative terminal to the positive terminal. When the battery is being discharged, this energy is supplied by the chemical reaction within the cells, so that chemical energy is converted into electric energy in the battery.

For any electric field the direction of the field is from points at higher potential toward points at lower potential. External work must be supplied to move a positive charge from a point at a lower potential to a point at higher potential. Conversely, external work must be supplied to move a negative charge from a point at higher potential to one at lower potential. Suppose a small positive test charge is moved from B to A against an electric field and then returned to B. External work is required to transfer the charge from B to A; when the charge returns to B, work is done on the charge by the electric field. Since the force on a negative charge in an electric field is opposite that on a positive charge, the field moves negative charges from points of lower potential toward points of higher potential.

31.7 EQUIPOTENTIAL SURFACES

It is often convenient to represent the potential distribution in an electric field by means of equipotential surfaces. An equipotential surface is defined as a surface whose points are all at the same potential. Some equipotentials for a spherical charge distribution are shown in Fig. 31.6. In this particular case the equipotential surfaces are spheres, while the lines of force are radial. The lines of force are always perpendicular at every point to the equipotential surfaces. This follows immediately from the fact that it takes no work to move a small test charge from one point to another on the same equipotential surface. The electric field has no component in the direction of the surface and hence must be perpendicular to it.

In Fig. 31.6, A and B lie on the same equipotential surface, while C lies on another equipotential surface. No work is done in moving a small test charge fom A to B along the equipotential surface, since the displacement is at all points perpendicular to the force on the charge. However, if a charge is moved from C to A, work must be done; regardless of what path is chosen, the force due to the field must have a component in the direction of this displacement for at least some part of the movement. The work required to transfer a charge from one point to another is independent of the path. The work required to move a small test charge from C to A is the same whether one goes first along the line of force to the equipotential surface AB and then to A, or from C to A along a straight line, or to B and then over to A.

Lines of force and equipotential surfaces are shown in Fig. 31.7 for the region around two equal point charges of opposite sign.

In a conductor there is a flow of charge whenever there is a potential difference between any two points. This flow stops as soon as all points in the conductor are at the same potential. In electrostatics, charges are at rest, and therefore there is no potential difference between points in the conductor. The entire conductor is an equipotential surface. Since there can be no electric field within the conductor, there is no net charge within the conductor. *The entire net charge of the conductor resides on the surface.* Any excess or deficiency of electrons is at the surface. Since the surface of a conductor is an equipotential surface, and since lines of force

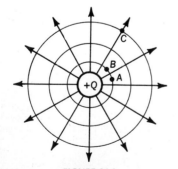

FIGURE 31.6
Lines of force (radial) and equipotential surfaces (spheres) for a spherical charge distribution.

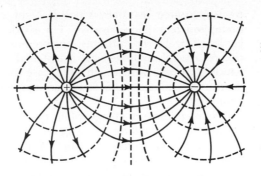

FIGURE 31.7
Lines of force (solid lines) and equipotential surfaces (dashed lines) near two equal but opposite charges.

are always perpendicular to equipotential surfaces, *lines of force always leave perpendicular to the surface of a conductor* when charges are at rest.

31.8 THE VAN DE GRAAFF ELECTROSTATIC GENERATOR

An important application of the fact that charges go to the outside of a conductor is the Van de Graaff generator (Fig. 31.8), a major tool of nuclear physics. A belt made of an insulating material carries charges at high speed from a charging source into a rounded metal shell (Fig. 31.9), where they are picked off; the charges then go to the outside of the shell. The potential of the high-voltage electrode builds up until charges are lost from the shell and gained at exactly the same rate. Often the entire system is placed in a large tank so that the pressure can be raised to several atmospheres to reduce the loss of charge by sparks and corona.

If the Van de Graaff generator is to be useful for physics experiments, one must utilize the high potential to accelerate some kind of charged particle, or *ion*. A Van de Graaff electrostatic accelerator consists of a Van de Graaff generator, an ion source, and an evacuated tube down which the ions are accelerated. When high-energy electrons are desired, the

FIGURE 31.8
The upper part of a 4-MeV Van de Graaff generator with the high-voltage shell raised for access to the terminal assembly, which includes both a deuterium source and an electron gun. This unique accelerator for radiobiological research accelerates positive ions down one tube or electrons down another. It was built in 1977 for the Gray Laboratory at Mount Vernon Hospital, Northwood, England. (*High Voltage Engineering Corporation.*)

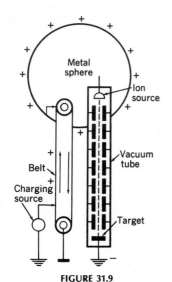

FIGURE 31.9
Schematic diagram of a Van de Graaff electrostatic accelerator.

electron source is usually a hot filament inside the high-voltage electrode, which is made negative. For exciting nuclear reactions by bombarding targets with protons, an ion source involving an electric discharge through hydrogen gas is installed in the high-voltage electrode, which is charged positive. In either case charged particles are accelerated down the evacuated tube to the grounded end, where they are focused on a target.

When a particle of charge q falls through a potential difference V in a vacuum, it gains a kinetic energy Vq. This energy is in joules if V is in volts and q is in coulombs. However, if we express q in atomic units of charge, the energy is given in electronvolts (eV). *An electronvolt is the kinetic energy gained by a particle bearing one atomic unit of charge in falling through a potential difference of one volt.* The negative charge on the electron and the positive charge on the proton both have the magnitude of 1 atomic unit of charge. Since the atomic unit of charge corresponds to 1.602×10^{-19} C,

$$1 \text{ eV} = (1.602 \times 10^{-19} \text{ C} \times 1.000 \text{ V})$$
$$= 1.602 \times 10^{-19} \text{ J}$$

Electrostatic generators are highly valuable sources of particles with energies from many thousands of electronvolts (keV) to several million electronvolts (MeV).

31.9   UNIFORM ELECTRIC FIELD

Consider two large parallel plates, one bearing a charge $+Q$ and the other a charge $-Q$ (Fig. 31.10). The electric field between the two plates (reasonably far from an edge) is constant in both magnitude and direction. Such a field is said to be *uniform*. When a charge q is carried from the lower plate to the upper one, an upward force of magnitude qE is required and the work done against the field is Eqd, where d is the distance between plates. The potential difference V between the plates is $W/q = Ed$. Solving for E yields

$$E = \frac{V}{d} \tag{31.7}$$

Electric field strengths may be expressed in units of potential difference divided by distance as well as in terms of the equivalent force divided by charge; they are often quoted in *volts per meter*. (An electric intensity of 1 V/m is exactly the same as one of 1 N/C.) If the test charge q is moved from X to Y (Fig. 31.10) the work done is given by

$$W = -(q\mathbf{E}) \cdot \mathbf{s} = -qEs \cos \theta = qEs \cos \varphi$$

and the potential difference between Y and X is $V_{YX} = Es \cos \varphi$.

The ratio of the potential difference ΔV between two neighboring points along a line of force to the displacement Δs between the points is a vector called the *potential gradient*. The electric intensity $\mathbf{E}$ is the negative of the potential gradient. The component of the potential gradient in any direction is the negative of the component of the electric intensity in that direction. The negative sign arises from the fact that the potential energy of a positive charge decreases when it is displaced in the direction of the electric vector $\mathbf{E}$.

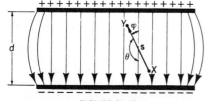

FIGURE 31.10
The electric field between two parallel charged plates is uniform except at the edges, where fringing occurs.

The uniform electric field between parallel charged plates was utilized by Millikan and his students to determine the charge on the electron. An extensive series of measurements performed between 1909 and 1917 produced evidence that every electron bears the same charge. Millikan's apparatus is shown schematically in Fig. 31.11. Two horizontal plates B and C are placed about 1 cm apart. An atomizer A shoots a fine spray of oil into the space above these plates. The drops of oil are so small that they do not settle for a long time. Eventually one or more of the drops finds its way through the opening O in the upper plate. A telescope is focused on this drop, and its movements are observed over an extended period. The rate at which the drop falls can be measured by means of a micrometer eyepiece in the telescope.

A beam of x-rays is next sent into the air between the horizontal plates. By means of these x-rays, electrons are detached from the atoms of air. One or more of these electrons may be captured by oil drops. When the upper plate B is charged positively and the lower plate C negatively, there is an upward force on a negatively charged oil drop. If the electric field is sufficiently strong, this upward force may equal or exceed the weight of the drop. Under such conditions the drop may be made to hang in the air for a substantial time. If the downward pull of gravity and the upward pull of electrostatic field leave the oil drop essentially at rest, we can write

$$w = Eq = \frac{Vq}{d}$$

where w is the weight of the drop, q the charge on the drop, E the electric field strength, V the potential difference between the plates, and d their separation. By determining w, V, and d it is possible to calculate the charge q. Frequently, a number of electrons may be attached to the drop, and, as a result, q is several times the charge on a single electron. Careful observations show that whatever charges are present, their magnitudes are always integral multiples of a single elementary charge. This charge e is the smallest known unit of electricity. It is the fundamental unit of electric charge. Every electron bears a charge $-e$, where

$$e = 1.602 \times 10^{-19} \text{ C}$$

The procedure just described is an oversimplification of the method used by Millikan, who actually measured the velocity of the oil drop as it

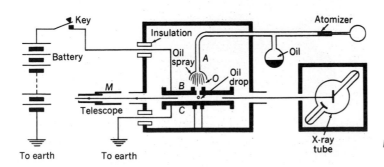

FIGURE 31.11
Millikan apparatus for measuring the charge on the electron.

moved upward under the influence of the electric field and the velocity of the same drop as it moved downward due to gravity with the electric field turned off. The calculations are straightforward but somewhat involved; they lead to the result quoted above.

31.11 ATMOSPHERIC ELECTRICITY AND LIGHTNING

On a clear day the surface of a level field freely exposed to the sky is negatively charged, so that there is an electric field downward. Between the earth and the ionosphere there is typically a potential difference of roughly 360 kV; in good weather, the electric field strength at the ground is about 100 V/m. This potential gradient changes continually. In addition to local variations, there are well-defined annual and diurnal variations, which differ at different parts of the surface of the earth. Above the earth's surface, the potential gradient diminishes, and at a height of about 10 km it approaches zero, a fact explained by the rapid increase in the conductivity of the atmosphere at higher altitudes.

The earth would require 500 kC of net negative charge to produce a potential gradient of 100 V/m over its entire surface. However, the gradient is not the same all over the earth, and it is actually reversed over substantial areas during lightning storms. The earth's negative charge is believed to come largely from lightning and point discharges under clouds. During thunderstorms the violent, ascending air currents in thunderheads (Fig. 31.12) break large raindrops into small ones. In this process the water droplets become positively charged and the air molecules negatively charged. Electrification occurs particularly strongly in the area of the thunderhead where the temperature is a few degrees below 0°C and tiny ice crystals abound. Water droplets are driven against ice crystals, and charging by contact and separation takes place, just as it does when a rubber comb is run through the hair. The ice crystals gain electrons and become strongly negative, while the water droplets are positively charged. Many of the smaller droplets are carried upward in the ascending air currents of the thunderhead to great elevations, where

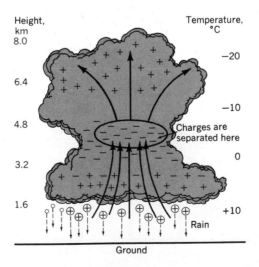

FIGURE 31.12
Clouds are electrified when ascending air currents in thunderheads produce charges by contact and separation.

they freeze. The result is that the top of the thunderhead has a large positive charge which eventually spreads out in the upper atmosphere. Many of the large droplets, also positively charged, fall to the earth as rain. Meantime the region near the 0°C isotherm is strongly negative. Most lightning flashes to earth are from this lower negative center to ground; these flashes help bring the earth its negative charge. Less frequently there may be flashes between the lower negative charge center and the upper positive region. Eventually the ice crystals in the negative charge center may melt and fall, bringing negatively charged raindrops to earth. When the thunderstorm is over, the earth is left with a negative charge and the upper atmosphere with a positive charge.

QUESTIONS

1. Why must any net charge on a perfect conductor reside at the surface?
2. Why must the electric intensity at the surface of an ideal conductor be perpendicular to the surface?
3. Electric lines of force never cross. Do equipotential surfaces ever cross? Explain.
4. Can charges be arranged so that both the electric intensity and the potential are zero at some point? Show an arrangement of charges which satisfies these conditions at some point.
5. It is sometimes stated that under static conditions all charges go to the surface of a conductor. What is implied? Do all the electrons and protons of the atoms go to the surface? How would you modify the initial statement to make it less misleading?
6. Is it possible for the potential at a point to be zero while there is a finite electric intensity at that point? Show a charge configuration which satisfies these conditions.
7. Is it possible for the electric intensity at a point to be zero while the potential at that same point is nonzero? Show a configuration of charge which satisfies these conditions at some point.
8. If a body bears a positive charge, is its potential necessarily greater than zero? Devise a situation in which a positively charged body has a negative potential relative to the earth.
9. Does an electron released at rest in an electric field move toward regions of higher or lower potential? Explain.
10. A metal cube bears a positive charge. What is the shape of the equipotential surfaces very close to the cube? Very far away from the cube? How does the charge distribute itself over the cube?
11. What is the theory behind the use of lightning rods? Explain how they may save a building from a stroke of lightning.

PROBLEMS

1. A conducting sphere of radius 25 mm bears a charge of 80 pC. Find the electric intensity and the potential at the surface of the sphere.
   Ans. 1,152 N/C; 28.8 V
2. Find the potential at a point 4 m from a charge of 60 nC. How much work is required to bring a 3-nC charge from infinity to this point?
3. Find the potential difference between two points, one 3 m and the other 1.5 m from a point charge of 4 nC. How much work is required to move a charge of 5 pC from the 3-m point to the other? *Ans.* 12 V; 60 pJ
4. The potential of an isolated conducting sphere of radius 75 mm is 3,000 V. Find the charge on the sphere and the potential at a distance of 1 m from the center of the sphere.
5. A water droplet with a radius of 400 μm bears a net charge of 1,000 electrons. Find its potential. If this droplet coalesces with an identical droplet, also bearing 1,000 excess electrons, find the potential of the new larger droplet.
   Ans. −3.6 mV; −5.71 mV
6. The potential difference between the terminals of an automobile battery is 12 V. How much work is done by the battery in transferring 1,500 C from one terminal to the other while starting the automobile?
7. Two points in the neighborhood of an isolated point charge differ in potential by 60 kV. How much work is required to carry 30 pC from the point of lower to the point of higher potential? If the points are 0.5 and 2 m from the original charge, find the magnitude of this charge.
   Ans. 1.8 μJ; 4.44 μC
8. How much work is required to move a charge of 20 nC from a point where the potential is −30 V to one where the potential is +150 V?
9. Suppose a solid copper sphere 10 mm in diameter had a positive charge arising from removing one

electron from each 10^{12} copper atoms. If the sphere contained 4.4×10^{22} atoms, what would be the charge on the sphere, the potential of the sphere, and the electric intensity at the surface?
Ans. 7.04 nC; 12.7 kV; 2.53 MN/C

10. How much work is required to move a charge of 4 nC from a point 2 m to a point 0.5 m from a point charge of 60 nC? What is the potential difference between these points?

11. A point charge of 80 pC is 0.2 m from a charge of −60 pC. Find the electric field intensity at the point P midway between the charges, the potential at this same point P, and the work required to bring a charge of 2 nC from infinity to point P.
Ans. 126 N/C; 1.8 V; 3.6 nJ

12. Air at atmospheric pressure breaks down and ionizes when the electric intensity is about 3 MV/m. Find the approximate maximum potential to which a smooth sphere of 125 mm radius can be charged in air. What charge is required for this potential?

13. Find the potential and electric intensity midway between point charges of +60 and −150 nC if they are separated by a distance of 3 m.
Ans. −540 V; 840 N/C

14. A charge of 160 nC is 0.6 m from a charge of 40 nC. Find the potential and the electric field strength at a point P midway between the two charges.

15. The plates of a cathode-ray tube are parallel and 5 mm apart. The potential difference between them is 12 kV. Find the force on an electron passing between the plates. What acceleration is produced if the electron has a mass of 9.11×10^{-31} kg?
Ans. 3.84×10^{-13} N; 4.2×10^{17} m/s^2

16. Find the kinetic energy and the speed of an electron which starts from rest and is accelerated through a potential difference of 12 V.

17. In a Millikan oil-drop experiment, the charge on the oil drop is that of five electrons. The oil drop is in the electric field arising from a difference of potential of 2,100 V between two plates that are 7.5 mm from each other. Find the force exerted on the oil drop by the electric field. Find the mass of the drop if it is almost in equilibrium in the field. *Ans.* 2.24×10^{-13} N; 2.29×10^{-14} kg

18. In a Millikan oil-drop experiment an oil drop of 1.31×10^{-11} g mass is almost exactly balanced by applying a potential difference of 1,200 V be-

tween plates which are 6 mm apart. How many elementary charges are on the oil drop?

19. Charges of −20, +30, −40, and +50 nC are placed at the corners of a square 0.3 m on a side. Find the potential at the center of the square.
Ans. 849 V

20. A rectangle is 4 by 3 m. If charges of +80 and −20 nC are placed at corners separated by 3 m, find the potentials of the other two corners and the potential difference between them.

21. Find the potential at the center of a uniformly charged thin ring of radius 30 mm which bears a charge of 60 pC. What is the potential on the axis of the ring 40 mm from the center? Find the electric intensity at this point.
Ans. 18 V; 10.8 V; 173 N/C along axis

22. Two large parallel plates are 25 mm apart. If the potential difference between them is 800 V, calculate the electric intensity between the plates and the force on a 30-nC charge placed anywhere between the plates.

23. A hollow copper sphere has a radius of 75 mm. If it bears a charge of 1.5 nC, find the potential and the electric intensity (*a*) inside the sphere, (*b*) at the surface of the sphere, and (*c*) at a point 2 m from the center.
Ans. (*a*) 180 V; zero; (*b*) 180 V; 2,400 N/C; (*c*) 6.75 V; 3.38 N/C

24. How much work is required to carry a charge of 50 pC from an infinite distance to a point midway between two identical 8-nC charges 0.2 m apart? How much work is required if one of the charges is −8 nC?

25. A 300-eV electron passes through a hole in a screen into a region where there is a uniform electric intensity of 900 N/C. If the velocity vector is in the direction of the field, how far does the electron move before it reverses its direction of motion? What was the initial speed of the electron? *Ans.* 0.333 m; 1.03×10^7 m/s

26. A proton with a charge of $+1.6 \times 10^{-19}$ C and a mass of 1.67×10^{-27} kg is accelerated through a potential difference of 3.29 MV in a Van de Graaff electrostatic accelerator. Find the final speed of the proton.

27. Electric charges of −12, +12, and +4 nC are placed at the vertices of an equilateral triangle with sides 1 m long. Sketch the configuration and find (*a*) the electric intensity and the potential at the center of the triangle and (*b*) the work required to assemble the configuration if the charges are brought from infinity.
Ans. (*a*) 571 N/C at 79° with line from 4 nC; 62.4 V; (*b*) −1.296 μJ

28. Charges of +60 nC are placed on diagonally opposite corners of a square with sides 0.5 m long, and a charge of −30 nC is placed at a third cor-

ner. (*a*) Find the potential energy of this configuration assuming it was assembled from charges initially at infinity. (*b*) Compute the potential at the fourth corner of the square, assuming $V = 0$ at infinity.

29. A charge of 6 nC is placed on a conducting sphere of radius 50 mm. This sphere is then connected temporarily by a copper wire to an uncharged conducting sphere of radius 25 mm. Find the charge on each sphere.

     Ans. 4 nC on larger sphere; 2 nC

30. A metal sphere having a radius of 125 mm carries a charge of 128 nC. It is temporarily connected by a conducting wire to a second uncharged metal sphere having a radius of 75 mm. Find the charge remaining on the smaller sphere.

31. A spherical conductor of 25 mm radius bears a charge of 20 nC. It is surrounded by a concentric sphere of 75 mm radius bearing a charge of -15 nC. Find (*a*) the potential difference between the two spheres and (*b*) the potential and electric intensity at $r = 50$ mm and $r = 125$ mm. *Ans.* (*a*) 4,800 V; (*b*) 1,800 V; 7.2 × 10⁴ N/C; 360 V; 2,880 N/C

32. A solid conducting sphere of 0.1 m radius which bears a charge of 12 nC is surrounded by a concentric sphere of 0.2 m radius bearing a charge of -8 nC. (*a*) Find the electric intensity E at points A, B, and C, which are respectively at distances of 0.05, 0.15, and 0.25 m from the center of the spheres. (*b*) Find the electric potential at these same three points, taking the potential as zero at an infinite distance from the spheres.

33. Find the electric intensity and the potential at the center of a cube with sides of length d (*a*) if there is a charge $+Q$ at each of the eight corners and (*b*) if one of the eight charges is changed to $-Q$. *Ans.* (*a*) zero, $4Q/\sqrt{3}\pi\epsilon_0 d$; (*b*) $2Q/3\pi\epsilon_0 d^2$ toward $-Q$, $\sqrt{3}Q/\pi\epsilon_0 d$

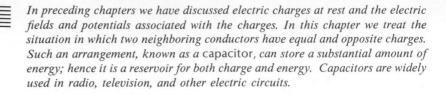

32

CAPACITANCE AND DIELECTRICS

In preceding chapters we have discussed electric charges at rest and the electric fields and potentials associated with the charges. In this chapter we treat the situation in which two neighboring conductors have equal and opposite charges. Such an arrangement, known as a capacitor, *can store a substantial amount of energy; hence it is a reservoir for both charge and energy. Capacitors are widely used in radio, television, and other electric circuits.*

32.1 CAPACITORS

A device on which electric charges can be stored is called a *capacitor*, formerly known as a *condenser*. Capacitors are important components in radio and television circuits, in the ignition systems of automobiles, and in other electrical equipment. The term *condenser* originated in the erroneous idea that electricity was a fluid which could be stored in a suitable container, such as the *Leyden jar* (a glass jar with a coating of tinfoil on the inside and another on the outside).

Capacitors have many forms. A typical capacitor consists of two conductors, one charged positively and the other negatively. Charging is accomplished by transferring the charge from one conductor to the other by means of a battery or other source of potential difference. The charge gained by one conductor is equal to that lost by the other. When we refer to the *charge on a capacitor*, we mean *the magnitude of the charge on either conductor*. Since the two conductors bear equal and opposite charges, the *net* charge is zero.

Let $+Q$ be the charge on the positive conductor, $-Q$ that on the negative conductor, and V the potential difference between the two conductors. *The capacitance C is defined as the ratio of the charge to the potential difference:*

$$C = \frac{Q}{V} \qquad (32.1)$$

From the definition it follows that capacitance can be measured in coulombs per volt. This unit has been named the *farad* (F) in honor of Michael Faraday. *One farad is that capacitance for which a charge of one coulomb will produce a potential difference of one volt.* The farad is a very large unit of capacitance; common capacitors seldom exceed a few microfarads (μF), and capacitances in the picofarad range ($1 \text{ pF} = 10^{-12} \text{ F}$) are common in radio and television circuits. Some low-voltage electrolytic capacitors (Sec. 32.7) have capacitances on the order of 0.1 F.

The capacitance of a pair of conductors is independent of the charge. When the charge is doubled, the potential difference between the conductors is also doubled. The capacitance depends on the size and shape of the conductors, on their relative positions, and on the character of the insulating material between them. Before we develop equations for the capacitances of various arrangements of conductors, it is desirable to consider how the insulating material affects the capacitance.

32.2 THE DIELECTRIC CONSTANT

When two parallel metal plates are insulated from each other and connected to an electroscope (Fig. 32.1), there is a potential difference between the plates if they are charged so that one bears $+Q$ and the

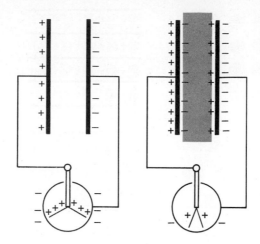

FIGURE 32.1
The potential difference between capacitor plates is reduced by inserting a
dielectric sheet because of the bound
surface charges induced on the dielectric.

other $-Q$ coulombs. The potential difference V can be measured in a variety of ways, e.g., by the divergence of the electroscope leaves. If a sheet of glass is inserted between the two plates, the potential difference between the two plates becomes smaller and the divergence of the electroscope leaves is reduced. Since no charge escaped from the plates while the potential was reduced, the capacitance of the system with the glass plate in place must be greater than the capacitance without the glass plate.

To understand this phenomenon, consider what happens to the glass plate. The glass is composed of atoms and ions, electrically neutral on the average. When it is placed in the electric field, electrons are attracted by the positively charged conductor and the nuclei by the negatively charged conductor. The electrons in the glass are not free, but they can undergo slight displacements toward the positive plate, while the nuclei undergo smaller displacements toward the negative plate. The net effect of these minute displacements throughout the glass is to produce a layer of positive charge adjacent to the negative plate and a layer of negative charge next to the positive plate. This negative charge layer reduces the potential of the plate (see Sec. 31.2). Similarly, the positive surface layer near the negative metal plates raises the potential of the latter. The potential difference between the metal plates is decreased. The glass plate increases the capacitance of the capacitor because it places a layer of negative charge close to the positive conductor and a layer of positive charge close to the negative conductor. In a very real sense it reduces the *effective* charge on the conductors, although the actual charge is not changed significantly.

A material is said to be *polarized* when the electric "center of charge" of the electrons and of the nuclei of a material do not coincide. This polarization may arise in two distinct ways. If the normal charge distribution of the molecules is initially symmetrical (Fig. 32.2a), the molecules are initially *nonpolar* and an applied electric field induces a polarization. If the molecules are already polar (Fig. 32.2b) but have random orientation, the external field produces at least a partial alignment of these polar molecules. In either case, the effect of the polarization is to leave the

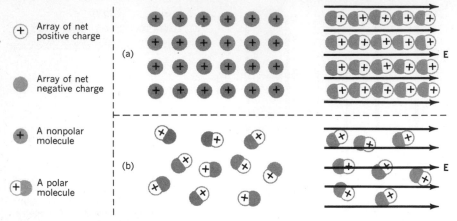

Array of net
positive charge

Array of net
negative charge

A nonpolar
molecule

A polar
molecule

(a)

(b)

E

E

FIGURE 32.2

Polarization effects in dielectrics. (*a*) Initially nonpolar molecules, shown here in a solid, have polarization induced by an applied field. Silicon, oxygen, and methane are nonpolar. (*b*) Polar molecules in a gas have a random orientation in a field-free region; application of an electric field introduces torques which favor alignment with the field. Thermal agitation prevents complete alignment. Water molecules and hydrogen chloride are polar.

interior of the dielectric uncharged but to produce a bound charge on surfaces normal to the field.

In the discussion above, we kept the charge on the two metal plates constant and found that insertion of a glass sheet resulted in a reduced potential difference between them. Next consider plates A and B (Fig. 32.3) connected to a battery which maintains a constant potential difference between them. If the space between the plates is filled by a sheet of glass, the induced surface charges on the glass lead to an increase of the charge $+Q$ on A and $-Q$ on B by a factor of perhaps 5. Since Q increases when the dielectric is introduced while V is held constant by the battery, the capacitance of the plates is increased by the insertion of the dielectric. *The ratio of the capacitance of a capacitor with a given material filling the space between conductors to the capacitance of the same capacitor when the space is evacuated is the dielectric constant† K of the material.* The dielectric constants of several materials are listed in Table 32.1.

32.3 PIEZOELECTRICITY

We have seen that when a slice of dielectric material is placed between two charged plates (or, more generally, in an electric field), there is a polarization of the medium. In some materials, such as quartz and rochelle salt, this displacement of electric charges is accompanied by small changes in the size and shape of the crystal slice, an effect called *electrostriction*. Electrostrictive distortions depend on the orientation of the crystal axes relative to the direction of the electric field.

In view of the fact that the shapes of some crystals change when the internal charges are displaced by an electric field, it is not surprising that changing the shape of a crystal may result in a redistribution of charges. When a thin slice of quartz is compressed (Fig. 32.4), one face becomes positive and the other negative. If the crystal is stretched instead of compressed, the charges on the faces are reversed. Compressing or elongating the crystal results in a potential difference between the faces;

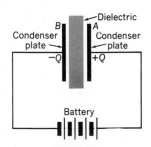

Dielectric

B A

Condenser
plate

Condenser
plate

$-Q$ $+Q$

Battery

FIGURE 32.3

Dielectric between the plates of a capacitor.

† Some authors prefer the name *specific inductive capacity*, and others use the name *relative permittivity*.

TABLE 32.1
Dielectric Constants and Dielectric Strengths

Material	K	Dielectric strength, MV/m
Vacuum	1	
Air (1 atm)	1.006	3
Ammonia (liquid)	22	
Ethyl alcohol (0°C)	28.4	
Transformer oil	2.1	5–15
Water (18°C)	81	
Amber	3	
Barium titanate (25°C)	1,200	
Glass	4.8–10	30
Mica	4.5–7.5	200
Paraffined paper	2	40
Polystyrene	2.6	20
Porcelain	6	15
Rubber (hard)	3	21

this potential difference may be hundreds or even thousands of volts. This phenomenon is known as *piezoelectricity* (*piezo* means "pressure").

If an alternating voltage is applied to a properly sliced quartz crystal, the crystal faces oscillate. By proper choice of the thickness of the slice, the mechanical oscillations can be made to have any desired frequency over a wide range. If a radio-frequency circuit has the same natural frequency as the mechanical oscillations of a quartz crystal, a sharp resonance can be obtained and the electrical oscillations can be accurately controlled by the mechanical frequency of the quartz crystal. Quartz crystals are often used to control the frequencies of radio and television transmitters.

32.4 THE CAPACITANCE OF AN ISOLATED SPHERE

Consider a single sphere of radius R in vacuum, removed sufficiently far from other bodies to permit their influence to be neglected. Let this sphere be charged with Q coulombs, presumably brought to the sphere from an infinite distance. (In this case the second conductor of the capacitor is a sphere of infinite radius, which now bears a charge $-Q$.) According to Eq. (31.3), the potential of the sphere is given by $V = Q/4\pi\epsilon_0 R$. Since $C = Q/V$, it follows immediately that

$$C = 4\pi\epsilon_0 R = \left(\frac{1}{9 \times 10^9} \text{ F/m}\right) R \qquad (32.2)$$

so that C is in farads when R is in meters.

If we imagine all space to be filled with a medium of dielectric constant K, we have

$$C = 4\pi\epsilon_0 KR = \left(\frac{1}{9 \times 10^9} \text{ F/m}\right) KR \qquad (32.3)$$

Obviously, the charge Q has not changed; therefore the increase in C

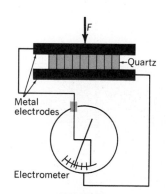

FIGURE 32.4
A piezoelectric cell in which pressure on the crystal faces produces a potential difference between them.

must arise from a decrease in V. The potential of the sphere under these conditions is given by

$$V = \frac{Q}{4\pi\epsilon_0 KR} = (9 \times 10^9 \text{ m-V/C})\frac{Q}{KR} \tag{32.4}$$

32.5 CAPACITANCE OF A SPHERICAL CAPACITOR

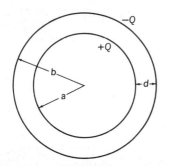

FIGURE 32.5
Capacitor formed by two concentric spheres.

Figure 32.5 shows a capacitor consisting of two concentric spheres. Let a be the radius of the inner sphere and b the radius of the outer sphere, which is connected to the earth. If the inner sphere bears a charge $+Q$, the outer one must have a charge $-Q$ if its potential is to be zero. (Remember the charge on the inner sphere raises the potential of the outer sphere, so a negative charge is required on the outer sphere to reduce its potential to zero.) If the region between the spheres is filled with a medium of dielectric constant K, the potential V_a of the inner sphere is the sum of the contributions of the charges $+Q$ on the inner sphere and $-Q$ on the outer sphere, so

$$V_a = \frac{Q}{4\pi\epsilon_0 Ka} - \frac{Q}{4\pi\epsilon_0 Kb} = \frac{Q}{4\pi\epsilon_0 K}\left(\frac{1}{a} - \frac{1}{b}\right)$$

Since the potential of the outer sphere is zero, the potential difference V between the spheres is just the potential of the inner sphere V_a and the capacitance of the two concentric spheres is

$$C = \frac{Q}{V} = \frac{4\pi\epsilon_0 Kab}{b - a} = \frac{Kab}{(9 \times 10^9)(b - a)} \tag{32.5}$$

32.6 CAPACITANCE OF TWO PARALLEL PLATES

If the radii of the two concentric spheres considered above are allowed to increase until they are very large while the difference $b - a$ remains constant, the surfaces of the spheres become approximately plane. The product ab becomes almost equal to a^2, since a and b are almost equal when both a and b are very large.

Let S be the area of the sphere of which a is the radius ($S = 4\pi a^2$), and let d ($= b - a$) be the distance between the spheres (kept constant) as a and b approach infinity. Then

$$C = \frac{4\pi\epsilon_0 Kab}{b - a} = \frac{4\pi\epsilon_0 Ka^2}{d} = \frac{\epsilon_0 KS}{d} = \frac{KS}{(9 \times 10^9)(4\pi d)}$$

If this relation is true for the entire spherical capacitor, it is true for a small portion of it, provided the area of that portion is used instead of the entire area of the sphere. If the spheres are very large, and if a small portion of area A is cut out of the spherical surfaces, the capacitor obtained in this artificial way consists of two parallel plates at a distance d apart, each plate with area A. The capacitance in farads of such a capacitor is

$$C = \frac{KA}{(9 \times 10^9)(4\pi d)} = \frac{\epsilon_0 KA}{d} \tag{32.6}$$

when A is in square meters and d is in meters.

□ **Example** Find the capacitance of a capacitor consisting of two parallel plates that are 5 mm apart. Each of the plates has an area of 100 cm². The space between the plates is filled with a medium whose dielectric constant is 3.0.

$$C = \frac{KA}{4\pi d(9 \times 10^9)}$$

$$= \frac{3.0 \times 0.01}{(0.005)(4\pi)(9 \times 10^9)}$$

$$= 54 \times 10^{-12} \text{ F} = 54 \text{ pF} \qquad \Box$$

32.7 PRACTICAL CAPACITORS

Most of the practical capacitors in everyday use represent some modification of the parallel-plate capacitor. An example is the familiar type used for tuning radio and television circuits (Fig. 32.6). The capacitance is varied by changing the effective area of the plates, which is that area close to a plate bearing the opposite charge. Alternate plates are connected together so that there are only two conductors; several plates are used to get a large area in a reasonable space. A further favorable factor in achieving a large area is that both sides of all but the outermost plates are used.

Another common form of capacitor consists of two thin foils of aluminum with a thin sheet of wax-impregnated paper between them. Since both the paper and the foil are flexible, the whole arrangement can be rolled up to form a small compact cylinder. From Eq. (32.6) it follows that the thinner the layer between the conducting plates, the greater the capacitance. However, there is a practical limitation on how thin the insulating separator can be, because if it is too thin, a spark may jump through it. The maximum potential difference which an insulating layer will stand can be computed from a knowledge of the thickness and of the *dielectric strength* of the material. *The dielectric strength is the potential*

FIGURE 32.6
Parallel-plate capacitor with variable capacitance.

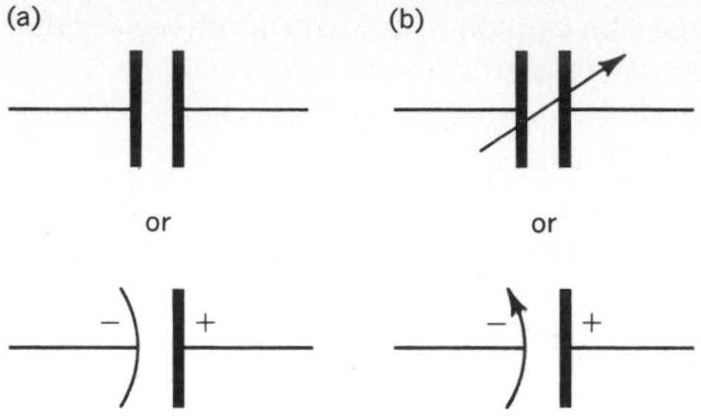

(a)   (b)

FIGURE 32.7
Symbols for (a) fixed and (b) variable capacitors.

gradient at which electrical breakdown occurs. For common insulators the dielectric strength (Table 32.1) is many millions of volts per meter, so a fraction of a millimeter is enough to support several thousand volts.

Among the materials which are used as insulators in capacitors are impregnated papers, mica, plastics, ceramic materials, glass, and oils. Air is the dielectric material in the variable capacitor of Fig. 32.6; because of the low dielectric constant, air capacitors have relatively small capacitance for a given area.

In the circuit diagrams of radios and other electric circuits, capacitors are commonly represented by the symbols of Fig. 32.7. An arrow drawn through the capacitor symbol indicates that it is variable; when no arrow is shown, it is implied that the capacitor has a fixed value.

The electrolytic capacitor is a common and inexpensive form. Such a capacitor has one plate of aluminum. The dielectric is a very thin coating of aluminum oxide on the surface of the plate, and the other conductor is an electrolyte. In such a capacitor d is exceedingly small, and therefore a relatively large capacitance can be provided in a rather small space. However, these capacitors break down at relatively low voltages; furthermore, it is important that the aluminum terminal be made positive, since the aluminum oxide layer conducts when the aluminum is negative.

32.8 ENERGY STORED IN A CAPACITOR

In charging a capacitor, it is necessary to do work to carry the electric charge from one conductor to the other. At the beginning, the two conductors of the capacitor are at the same potential. As charge is transferred from one plate to the other, the difference of potential between the two increases. Suppose that the final potential difference V between the terminals of the capacitor is attained after Q coulombs of electricity has been transferred from one conductor to the other. At the beginning of the charging, the potential difference is zero; at the end, the difference is V. The average potential difference during the charging is $V/2$. The work is equal to the product of the average difference of potential and the quantity of electricity transferred. The energy W stored in the capacitor is given by

$$W = \tfrac{1}{2}QV \tag{32.7}$$

This energy is released when the capacitor is discharged. If the capacitor is allowed to discharge through a wire, the energy is converted into heat in the wire.

□ **Example** A capacitor having a capacitance of $2\,\mu F$ is charged with 10^{-3} C of electricity. How much energy is stored in it?

$$\text{Energy} = \tfrac{1}{2}QV = \frac{1}{2}\frac{Q^2}{C}$$

$$= \frac{1}{2}\frac{10^{-3}\times10^{-3}}{2\times10^{-6}}$$

$$= 0.25\,\text{J} \qquad\qquad\square$$

32.9 CAPACITORS IN PARALLEL

When two or more capacitors are connected in such a way that all the positive conductors are at the same potential V^+ and all the negative conductors at potential V^-, the capacitors are said to be connected *in parallel*. When a number of capacitors (Fig. 32.8) are connected in parallel, the system has a capacitance C equal to the sum of the separate capacitances. This result can be proved as follows.

The capacitors are all charged to the same difference of potential. Let V denote this difference of potential, and let Q_1, Q_2, and Q_3 be the charges on capacitors C_1, C_2, and C_3, respectively. Let Q be the total charge on all the capacitors. Then

$$Q = Q_1 + Q_2 + Q_3$$
$$Q_1 = C_1V \qquad Q_2 = C_2V \qquad Q_3 = C_3V$$

and

$$Q = CV$$

Substituting yields

$$CV = C_1V + C_2V + C_3V$$

Dividing by V gives

$$C = C_1 + C_2 + C_3 \qquad\qquad (32.8)$$

Hence, the equivalent capacitance of a number of capacitors connected in parallel is the sum of the individual capacitances.

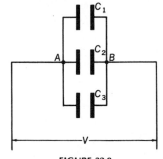

FIGURE 32.8
Capacitors connected in parallel.

32.10 CAPACITORS IN SERIES

Two or more capacitors are said to be *in series* when they are connected as shown in Fig. 32.9. When a potential difference V is applied between points A and B, a charge $+Q$ appears on the positive plate of C_1, and a corresponding charge $-Q$ on the negative plate of C_1. The electrons which produce the negative charge on C_1 must come from the plate of C_2 and leave it with a charge $+Q$. The negative plate of capacitor C_2 has a charge $-Q$ coming from the positive plate of capacitor C_3, which is left with a charge $+Q$. Finally, the negative plate of C_3 bears a charge $-Q$. Thus, *when capacitors are charged in series, the same charge is stored on each capacitor.* The potential difference across the combination of capacitors in series is equal to the sum of the potential differences of the individual capacitors.

Let Q be the charge on each capacitor, C the equivalent capacitance of the capacitors when joined in series, and V_1, V_2, and V_3 the differences of potential between the terminals of C_1, C_2, and C_3, respectively. Since the total difference of potential is equal to the sum of the separate differences of potential,

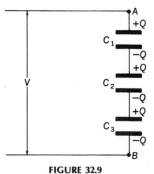

FIGURE 32.9
Capacitors connected in series.

$$V = V_1 + V_2 + V_3$$

When we divide both sides of this equation by Q and make use of the fact that $Q = Q_1 = Q_2 = Q_3$, we obtain

$$\frac{V}{Q} = \frac{V_1}{Q_1} + \frac{V_2}{Q_2} + \frac{V_3}{Q_3}$$

Since V/Q is the reciprocal of the capacitance,

$$\frac{1}{C} = \frac{1}{C_1} + \frac{1}{C_2} + \frac{1}{C_3} \tag{32.9}$$

When capacitors are connected in series, the reciprocal of the resultant capacitance is the sum of the reciprocals of the individual capacitances.

□ **Example** A capacitance of 4 μF is connected in series with one of 5 μF. What is the equivalent capacitance of the combination? If 100 V is the potential difference across the combination, find the potential difference across the 4-μF capacitor.

$$\frac{1}{C} = \frac{1}{C_1} + \frac{1}{C_2} = \frac{1}{4} + \frac{1}{5} = \frac{9}{20}$$

$$C = \tfrac{20}{9} = 2.22\ \mu F$$

$$Q = CV = 2.22 \times 100 = 222\ \mu C$$

$$V_4 = \frac{Q}{C} = \frac{222\ \mu C}{4\ \mu F} = 55.5\ V \qquad \square$$

32.11 GROUPS OF CAPACITORS

When several capacitors are connected in such a way that some are connected in parallel and others in series (Fig. 32.10), it is possible to replace any parallel group with the equivalent single capacitor of value given by Eq. (32.8). After this is done for the parallel group (or groups), one is left with a problem in series capacitors.

□ **Example** The four capacitors of Fig. 32.10a are 6, 8, 10, and 12 μF, respectively. They are charged to a potential difference V of 120 V. Find the charge and potential difference for each capacitor.

The capacitance of the three parallel capacitors is given by $C_{\parallel} = 6 + 8 + 10 = 24\ \mu F$. If we replace these capacitors with their equivalent, our circuit reduces to that of Fig. 32.10b. The capacitance of a 24- and a 12-μF capacitor in series is C_S, where

$$\frac{1}{C_S} = \frac{1}{24} + \frac{1}{12} = \frac{1}{8} \qquad \text{or} \qquad C_S = 8\ \mu F$$

The charge on each capacitor of Fig. 32.10b is therefore given by $Q = C_S V = 8 \times 120 = 960\ \mu C$. $V_{24} = \frac{960}{24} = 40\ V$, and $V_{12} = \frac{960}{12} = 80\ V$. Note that these potential differences add to 120 V. The 24-μF capacitor, which is equivalent to the 6-, 8-, and 10-μF capacitors in parallel, has a potential difference of 40 V, and so must the three individually. The charges are given as follows: $Q_6 = 6 \times 40 = 240\ \mu C$, $Q_8 = 8 \times 40 = 320\ \mu C$, and $Q_{10} = 10 \times 40 = 400\ \mu C$. Note that these charges add to 960 μC. $\qquad \square$

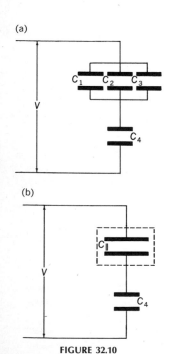

(a)

(b)

FIGURE 32.10
(a) Three capacitors in parallel are connected in series with a single capacitor. (b) Simple series circuit results when the three parallel capacitors in (a) are replaced with one equivalent capacitor $C_{\parallel}$.

1. Is the condition of a dielectric in a capacitor different when the capacitor is charged from when it is uncharged? In what way?
2. What determines the maximum potential difference which can safely be applied to a capacitor?
3. Why is mica often used as dielectric in capacitors when glass and porcelain are better insulators?
4. Why does a parallel-plate capacitor store more charge for a given potential difference when a dielectric (such as mica or glass) is inserted between the plates than when the space is evacuated?
5. The plates of a parallel-plate capacitor are moved together until their separation is half of its original value. What happens to the capacitance? Why?
6. Two parallel metal plates are separated by a 2-mm air gap. The plates are given equal but opposite charges. If a sheet of glass is placed between the plates, what happens to the potential difference between them? Why? What happens to the potential gradient? To the electric field intensity?
7. Suppose an insulated thin metal sheet is inserted between the plates of a parallel-plate capacitor. What happens to the capacitance if the metal sheet never touches either plate?
8. Why isn't distilled water used as a dielectric in capacitors? Its dielectric constant (relative permittivity) is 80, a very high value, and pure water is an excellent insulator.
9. If V is kept constant across the capacitors of Fig. 32.9 as a glass plate is inserted between the plates of C_1, what happens to the charge on each capacitor? To the total stored energy? Is work required to insert the plate? Why?
10. How is a capacitor analogous to a steel cylinder into which a gas such as nitrogen is pumped? What is analogous to V for a capacitor? To Q? To C?
11. Two identical capacitors are connected in parallel and charged to a potential difference of 100 V. If they are now removed from the circuit and the positive conductor of one capacitor is connected to the negative conductor of the second, what is the resulting potential difference? (Some voltage multipliers actually operate in this way.)
12. In how many *apparently* different units has the constant $4\pi\epsilon_0$ been expressed in this chapter? Show that they are all equivalent to $C^2/N\text{-}m^2$, in terms of which the constant was initially introduced with Coulomb's law.
13. Explain why the dielectric constant of the medium outside the two spheres makes no difference in the development of Eq. (32.5).
14. What is piezoelectricity? How can it be used?

PROBLEMS

1. What is the difference of potential between the terminals of a capacitor that has a capacitance of $5\,\mu F$ when the charge on the capacitor is 7.5×10^{-4} C? Find the energy stored in the capacitor.   *Ans.* 150 V; 56.3 mJ
2. A $4\text{-}\mu F$ capacitor is charged to a potential difference of 155 V. Find the charge on the capacitor plates and the energy stored.
3. The potential difference between two clouds is 8 MV. How much electric energy is dissipated if a lightning stroke of 25 C leaps from one cloud to the other? What is the capacitance of the two-cloud capacitor? Assume that the potential difference between the clouds decreases to zero and that the system behaves as though the clouds were the plates of a capacitor.
       Ans. 100 MJ; 3.13 μF
4. A charge of $360\,\mu C$ is stored in a capacitor at a potential of 900 V. What is the capacitance of the capacitor? What is the energy stored in the capacitor?
5. A parallel-plate capacitor has insulation which can withstand an applied potential of 12 kV. Its capacitance is $0.5\,\mu F$. What is the maximum energy the capacitor can store? The maximum charge? If the insulation is paraffined paper, what is its approximate thickness? (See Table 32.1.)   *Ans.* 36 J; 6 mC; 0.3 mm
6. What is the capacitance of a sphere of radius 0.3 m? If the dielectric strength for air is 3 MV/m, what is the largest charge which ideally could be placed on this sphere?
7. Find the capacitance of the earth if it is approximately a sphere of 6,400 km radius. If the electric intensity over the entire surface of the earth were 120 V/m directed downward, what charge would the earth have to have? What would the surface charge density be?
       Ans. 711 μF; -5.5×10^5 C; -1.06 nC/m²
8. A parallel-plate capacitor using mica as the dielectric is to have a capacitance of 40 pF and be able to withstand a potential difference of 5 kV. What is the minimum thickness of mica required? What is the minimum area the plates of the capacitor may have if the dielectric constant is 6?
9. A parallel-plate capacitor consists of two sheets of aluminum, each of area 0.15 m², separated by a thin layer of plastic insulation of dielectric con-

stant 3 and thickness of 0.1 mm. Find the capacitance and the charge stored when this capacitor is charged to a potential difference of 600 V.
Ans. 39.8 nF; 23.9 μC

10. A capacitor is made of two sheets of tinfoil in contact with a plate of glass of dielectric constant 6.5. If the area of each sheet of tinfoil is 0.0125 m² and the thickness of the glass is 1.6 mm, what is the capacitance of the capacitor? What charge is stored for a potential difference of 800 V?

11. A capacitor is made up of 50 sheets of tinfoil, each 800 by 55 mm. These sheets are separated by sheets of paraffined paper which are 0.12 mm thick and which have a dielectric constant of 2. What is the capacitance of the capacitor when alternate sheets of tinfoil are joined together?
Ans. 318 nF

12. A capacitor consists of two parallel plates which are separated by a sheet of mica 0.15 mm thick which has a dielectric constant of 5. The area of each plate is 0.035 m². The capacitor is charged to 400 V. Find the energy stored in the capacitor.

13. A parallel-plate capacitor has a capacitance of 4 pF with air between the plates and a capacitance of 26 pF with glass between the plates. (*a*) What is the dielectric constant of the glass in question? (*b*) If a charge of 500 pC is placed on the plates with glass between them, what is the energy stored? (*c*) If the glass is removed without affecting the charge, what energy is stored? (*d*) What is the source of the added energy?
Ans. (*a*) 6.5; (*b*) 4.81 nJ; (*c*) 31.3 nJ; (*d*) work done in pulling out glass

14. What is the joint capacitance of three capacitors of 4, 6, and 12 μF when they are connected in series? In parallel?

15. Capacitors of 30, 6, and 5 μF are arranged so that they can be connected in series or in parallel. What capacitance is obtained in each case?
Ans. 2.5 μF; 41 μF

16. Three capacitors, of capacitances 4, 9, and 11 μF, are connected in parallel across a 120-V potential difference. Find the total capacitance and the charge on the 9-μF capacitor.

17. A 12-μF capacitor is connected in parallel with an 8-μF capacitor across a 300-V potential difference. Find the charge on each capacitor and the total energy stored. *Ans.* 3.6 mC; 2.4 mC; 900 mJ

18. Two capacitors, of capacitance 6 and 3 μF, are connected in series across a 90-V battery. Find

the charge, the potential difference, and the energy stored for the 3-μF capacitor.

19. Two capacitors, of capacitances 24 and 6 μF, are connected in series across a potential difference of 150 V. Find the charge, potential difference, and energy stored for the 6-μF capacitor.
Ans. 720 μC; 120 V; 43.2 mJ

20. Three capacitors, of capacitances 20, 30, and 60 nF, are connected in series across a 120-V source. Find the charge and potential difference associated with each capacitor.

21. Three capacitors are connected as shown in the accompanying figure to a source of unknown potential difference *V* between *A* and *B*. If C_2 is a 10-μF capacitor charged to a potential difference of 60 V, find the potential differences and the charges for C_1 and C_3, which have capacitances of 30 and 15 μF, respectively.
Ans. $V_3 = 60$ V; $Q_3 = 900$ μC; $Q_1 = 1.5$ mC; $V_1 = 50$ V

22. Three capacitors are connected as shown in the accompanying figure. If the potential difference between *A* and *B* is 120 V, and if C_1, C_2, and C_3 are, respectively, 50, 7, and 18 nF, find the capacitance of the combination, the charge on C_3, and the potential difference across C_3.

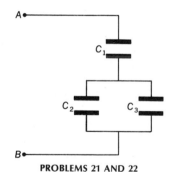

PROBLEMS 21 AND 22

23. A 3-μF capacitor is charged to a potential difference of 60 V. Find the charge. The capacitor is then disconnected from the source, but not discharged. Find the new potential difference if this capacitor is connected in parallel with an uncharged 15-μF capacitor. *Ans.* 180 μC; 10 V

24. A 5-μF capacitor is given a charge of 600 μC. (*a*) Find the potential difference between the conductors and the energy stored in the capacitor. (*b*) If an uncharged 7-μF capacitor is connected in parallel with the charged 5-μF capacitor, find the new potential difference.

25. Two identical air-dielectric parallel-plate capacitors with $C = 20$ pF are charged in parallel

to a potential difference of 240 V and then disconnected from the charging source. A glass
plate of dielectric constant 5 is inserted between
the plates of one of the capacitors, filling the
space completely. Find the new potential difference between the plates and the charge transferred from one capacitor to the other.

Ans. 80 V; 3.2 nC

26. When a capacitor that has a capacitance of 90 nF
and a charge of 18 μC is connected in parallel
with an uncharged capacitor, the resulting potential difference is 50 V. Find the capacitance of
the second capacitor.

27. The system shown in the accompanying figure is
charged by applying a potential difference V_{AB}
between points A and B. If the potential difference across C_1 is 20 V and if $C_1 = 5$ nF,
$C_2 = 20$ nF, $C_3 = 6$ nF, and $C_4 = 10$ nF, find V_{AB}.
What is the capacitance of the system?

Ans. 50 V; 5 nF

28. Find the capacitance of the system of the accompanying figure and the charge on capacitor C_2 if
C_1, C_2, C_3, and C_4 are, respectively, 60, 30, 20, and
10 nF and the potential difference between A and
B is 120 V.

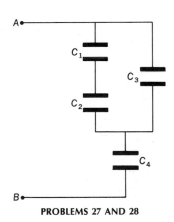

PROBLEMS 27 AND 28

29. Find the energy required to establish in a parallel-plate capacitor with transformer oil as dielectric a uniform electric field of 10^7 V/m over a
volume of 0.001 m³. *Ans.* 0.93 J

30. Show that the energy stored per unit volume in a
parallel-plate capacitor with a dielectric of con-
stant K is $\frac{1}{2}K\epsilon_0 E^2$. *Hint:* $E = V/d$, where d is the
thickness of the dielectric.

31. Capacitor $C_1 = 10$ μF is charged to a potential
difference of 45 V, capacitor $C_2 = 5$ μF to a potential difference of 60 V, and capacitor
$C_3 = 20$ μF to a potential difference of 50 V. They
are then connected as shown in the accompanying figure with the positive side of C_2 connected
to the positive side of C_1. Find the charge on C_3
after the key is closed. *Ans.* 900 μC

32. Capacitor $C_1 = 6$ μF is charged to a potential
difference of 100 V, capacitor $C_2 = 8$ μF is
charged to a potential difference of 80 V, and
capacitor $C_3 = 24$ μF is charged to a potential
difference of 30 V. They are then connected as
shown in the accompanying figure with the negative side of C_2 connected to the positive side of
C_1. Find the charge on C_1 after the key is closed.

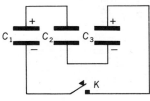

PROBLEMS 31 AND 32

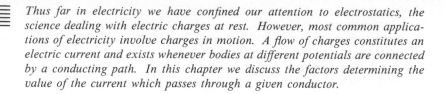

33

ELECTRIC CURRENT AND RESISTANCE

Thus far in electricity we have confined our attention to electrostatics, the science dealing with electric charges at rest. However, most common applications of electricity involve charges in motion. A flow of charges constitutes an electric current and exists whenever bodies at different potentials are connected by a conducting path. In this chapter we discuss the factors determining the value of the current which passes through a given conductor.

33.1 CURRENTS AND THEIR EFFECTS

When two metallic spheres, one charged positively and the other negatively, are connected by a copper wire, electrons flow until there is no longer a potential difference between the two spheres. Such a flow of charge is called an electric *current*. The magnitude of the current I is the charge per unit of time that passes any cross section of the wire.

$$I = \frac{Q}{t} \tag{33.1}$$

Current is commonly measured in coulombs per second, or *amperes*, named for the French physicist André Marie Ampère. *The ampere is the current when one coulomb per second passes any cross section of a conductor.*

Electric currents are of great practical importance because of the many ways in which we can use the three principal effects they produce: (1) heating, (2) chemical, and (3) magnetic effects.

Heating by electric currents is utilized in making the filaments of incandescent lamps luminous, in operating electric stoves, and in hundreds of other ways. There are cheaper ways to obtain thermal energy, but none is cleaner or easier to control and handle than electric energy.

Batteries and electroplating are applied electrochemical devices. Aluminum, high-purity copper and silver, and chlorine are among important industrial elements ordinarily obtained by electrolysis.

The magnetic effects of currents are used in giant electromagnets and in tiny electric relays in telephone circuits. The interaction between currents and magnetic fields is fundamental to electric motors and to reproducing the picture on a television receiver.

33.2 THE DIRECTION OF A CURRENT

The charges which are primarily responsible for the current in metallic conductors are negative electrons, but early in the nineteenth century there was no way to tell whether it was negative or positive charges (or both) which were in motion. In 1747 Benjamin Franklin discussed his experimental results in terms of the flow of an "electric fluid" which we now call *positive charge*. Relatively little work was done with electric currents in the eighteenth century, but in 1800 Volta developed the first successful battery, and in 1820 Oersted discovered that there is a magnetic field associated with an electric current. When news of Oersted's discovery was reported to the French Academy of Sciences, French physicists sprang into action; within 1 week Ampère was ready to discuss how the direction of the magnetic field was related to the direction of the current (Chap. 36), which he also took to be a flow of positive charge. The definitive work of Ampère and his French contemporaries brought into general use the convention which we follow in this book; according to this convention, *the direction of a current is the direction in which a positive charge would move under the influence of the electric field. By definition* then, the *conventional* current in a wire flows from a point at higher potential to a point at lower potential, as though the current represented a movement of *positive* charge. Actually, *in metallic conductors* the positive nuclei are not free to move, and the transfer of charge results from a flow of electrons in a direction opposite that of the conventional current. In liquid and gaseous conductors, both positive and negative ions are in motion. In some of the modern high-energy accelerators, such as Van de Graaff generators and cyclotrons, the current may be a movement of positive charges. Obviously no convention could be most convenient for handling every possible situation.

When a constant potential difference is maintained between two points in a conductor, a constant flow of charge results. The current is always in the same direction and is said to be a *direct current*. On the other hand, when the flow of charges is first in one direction and then in the opposite direction, the current is said to be *alternating*. The next five chapters deal primarily with direct currents.

33.3 OHM'S LAW FOR A RESISTOR

To produce a steady current through a conductor, such as the filament of a lamp, an electric field must be established in the conductor. Since the natural flow of charges is always such as to eliminate the electric field, a *steady* current can result *only* if some device such as a battery or generator maintains the field. In this agency, called a source of *electromotive force* (emf), some other form of energy is converted into the electric energy needed to keep the charges flowing. In batteries chemical energy is converted into electric energy, while in generators it is mechanical energy which is transformed to electric energy. These and other sources of emf are discussed in the chapters which follow.

Consider a conductor across which a constant potential difference is maintained by a battery, as indicated in Fig. 33.1. In this figure $\textcircled{A}$ represents an ammeter, an instrument for measuring electric current, while $\textcircled{V}$ represents a voltmeter, an instrument for measuring potential difference. The common forms of ammeters and voltmeters depend on the magnetic effects of electric currents (Chap. 36) for their operation. In use, the two terminals of the voltmeter are connected to the two points between which one wishes to measure the potential difference, while the ammeter is connected so that all the charges which pass through the device in which the current is to be measured also pass through the ammeter. (In Fig. 33.1 the ammeter passes not only the current through the wire but also the current through the voltmeter; the latter is assumed to be negligible in this case. It is not always negligible.) Let the current through the conductor be I_1 amperes (A) when the potential difference across it is V_1 volts. If the battery is replaced by a different one, the potential difference and current can again be read and found to be V_2 and I_2. When V_2 is twice V_1, I_2 is twice I_1. Indeed, so long as the temperature of the wire is constant, the ratio of the potential difference to the current is constant. The ratio

$$R = \frac{V}{I} = \frac{V_1}{I_1} = \frac{V_2}{I_2} \qquad (33.2)$$

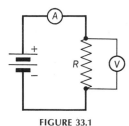

FIGURE 33.1
A current, read by ammeter A, is produced in resistor R (represented by the sawtooth symbol) by application of a potential difference, read by voltmeter V.

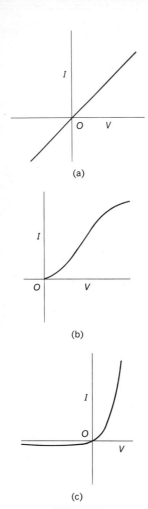

FIGURE 33.2
In an ohmic resistance (*a*) the current is directly proportional to the applied potential difference, but for many circuit elements, such as a vacuum-tube rectifier (*b*) or a solid-state junction rectifier (*c*), Ohm's law is not obeyed.

is called the *resistance* of the wire. The resistance is commonly expressed in *ohms*, where *one ohm is that resistance in which a potential difference of one volt produces a current of one ampere.* This unit is named in honor of Georg S. Ohm, who discovered (1825) that the current in a wire is proportional to the potential difference between the ends. Much of our knowledge about resistance is due to the pioneering work of Ohm, who made his own batteries, wires, and meters. The symbol Ω (Greek capital omega) is used as an abbreviation for ohms.

The relation

$$I = \frac{V}{R} \tag{33.2a}$$

is *Ohm's law for a single resistor.* This law is obeyed within wide limits for metallic and many other conductors. However, for many other nonmetallic conductors the current is *not* proportional to the potential difference. In Fig. 33.2 current is shown as a function of potential difference across (*a*) a metallic resistor, (*b*) a vacuum-tube rectifier, and (*c*) a solid-state rectifier. Ohm's law is applicable to the metallic resistor but not to the other two.

33.4 JOULE'S LAW OF HEATING

When a potential difference produces a current through a conductor of resistance R, electric energy is converted into thermal energy. The potential difference V represents the energy per unit charge converted from electric into thermal energy in the conductor. The charge Q transported through the conductor in time t is It, by Eq. (33.1). Hence the total energy W dissipated in the conductor in time t is

$$W = VQ = VIt \tag{33.3}$$

If Ohm's law is obeyed for this conductor, $V = IR$, and

$$W = I^2Rt \tag{33.3a}$$

The fact that the thermal energy produced by an electric current in a conductor is proportional to the square of the current, to the resistance, and to the time was reported by Joule, and Eq. (33.3a) is known as *Joule's law of heating.* This formula gives energy in joules when I is in amperes and R in ohms. (To obtain the energy in calories, use the relation 4.186 J = 1 cal.)

Power is defined as the ratio W/t; thus the power is equal to VI. The power is given in watts when the current is in amperes and the potential difference is in volts (volt-amperes = joules per coulomb × coulombs per second = joules per second = watts). With the aid of Ohm's law, we can express the power in the forms

$$P = VI = I^2R = \frac{V^2}{R} \tag{33.4}$$

It is desirable to have some sort of device to protect electric machines and appliances from excessive currents. One method of furnishing this protection is by fuses. A typical fuse consists essentially of a conductor that has a low melting point. When an excessive current passes through this fuse, the heat generated is sufficient to melt the conductor, thereby opening the circuit. Fuses are made to melt when the current becomes greater than some preselected amount; thus it is possible to buy a fuse appropriate to protect almost any circuit against a current too great for some component to carry. An alternative to a fuse is a *circuit breaker*, which opens a switch magnetically when the current is too great.

33.5 RESISTIVITY

In his studies of resistance Ohm made wires of various materials, lengths, and areas. He was able to show that for a wire of given material at constant temperature *the resistance is directly proportional to the length and inversely proportional to the cross-sectional area.* This fact is expressed by the equation

$$R = \rho\frac{l}{A} \tag{33.5}$$

where ρ is called the *resistivity* (or *specific resistance*) of the material. The resistivity depends on the material in question and on its temperature. In the metric system the resistivity of a material is numerically equal to the resistance of a piece of the material one meter in length and one square meter in cross-sectional area. The resistivities of several materials are recorded in Table 33.1.

In the British engineering system the *resistivity* of a material is numerically equal to the resistance of a piece of the material one foot long and one circular mil in area. A circular mil is defined as the area of a circle one one-thousandth of an inch in diameter. The units are usually written as *ohms per (circular) mil-foot*, which is dimensionally incorrect and misleading. The British engineering unit has an advantage of convenience for circular conductors in that the area of a wire d thousandths of an inch in diameter is d^2 circular mils.

33.6 TEMPERATURE COEFFICIENT OF RESISTANCE

The resistance of metallic conductors increases as the temperature is increased. Figure 33.3 shows how the resistivity of platinum varies with temperature. For many materials, such as carbon, the resistance de-

TABLE 33.1
Resistivities and Temperature Coefficients of Resistance
(Approximate values at 20°C)

Material	Resistivity,† Ω-m	Temperature coefficient per Celsius degree
Aluminum	2.6×10^{-8}	0.0040
Carbon	$3,500 \times 10^{-8}$	-0.0005
Constantan	49×10^{-8}	0.000002
Copper	1.7×10^{-8}	0.00393
Iron	9.7×10^{-8}	0.0058
Manganin	48×10^{-8}	0.0
Silver	1.6×10^{-8}	0.0038
Tungsten	5.5×10^{-8}	0.0047
Glass	$\approx 10^{13}$	
Quartz	$\approx 10^{17}$	

† To obtain ρ in ohms per mil-foot, multiply ρ in ohm-meters by 6×10^8.

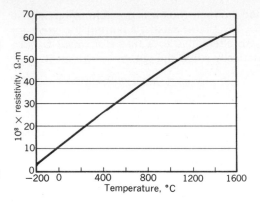

FIGURE 33.3
The resistivity (or specific resistance) of platinum rises with temperature.

creases as the temperature is increased. Very often the change in the resistance of a conductor is roughly proportional to the change in the temperature and to the original resistance R:

$$\Delta R = \alpha R \Delta t \tag{33.6}$$

where ΔR represents the change in resistance and Δt the change in temperature. The proportionality constant α is called the *temperature coefficient of resistance* and is defined as the change in resistance divided by the product of the original resistance and the change in temperature. Approximate values of α for several conductors are listed in Table 33.1.

If R_0 represents the resistance at $0°$ C and α_0 the temperature coefficient at $0°$ C, the resistance R_t at temperature t can be written in the form

$$R_t = R_0(1 + \alpha_0 t) \tag{33.7}$$

since in this case $\Delta R = R_t - R_0$, and $\Delta t = t$. The temperature coefficient of resistance is negative for materials which show a decrease in resistance as the temperature is raised.

For most pure metals the temperature coefficient of resistance is in the neighborhood of 0.004 per Celsius degree for small temperature changes near $0°$ C. Resistance is only approximately a linear function of temperature; if we define α as the slope of the curve showing resistivity as a function of temperature, pure metals behave much like platinum (Fig. 33.3), with α roughly proportional to the reciprocal of the absolute temperature. For most alloys α is much smaller than for metals. Indeed, for some alloys, e.g., constantan and Manganin, the temperature coefficients of resistance are very near zero. For this reason such alloys are often used for resistors when it is desirable that the resistance be independent of temperature.

When the resistance of a wire changes with temperature, it is possible to infer the change in temperature from measurements of resistance. A resistance thermometer using platinum wire can be used to determine temperatures over a wide range up to the melting point of platinum. When properly calibrated, such a thermometer gives high precision in the measurement of temperatures. Figure 33.4 shows the essential components of a resistance thermometer.

FIGURE 33.4
A resistance thermometer registers changes in resistance as the temperature varies.

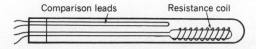

□ **Example** A resistance thermometer is made of platinum wire. Its resistance in a mixture of ice and water at 0°C was found to be 10 Ω; in a furnace of unknown temperature it was found to be 50 Ω. If the temperature coefficient of resistance of platinum has an average value of 0.0036 per Celsius degree over this range, what was the temperature of the furnace?

$$\text{Temperature of furnace} = t = \frac{R_t - R_0}{\alpha R_0}$$

$$= \frac{50 - 10}{0.0036 \times 10} = \frac{40}{0.036}$$

$$= 1100°C \qquad \qquad □$$

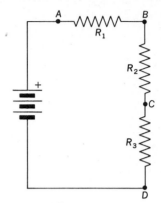

FIGURE 33.5
Three resistors in series.

33.7 RESISTORS IN SERIES

The battery and the resistors of Fig. 33.5 are connected *in series*. Two circuit elements are in series whenever all the charge passing through one of the elements also passes through the second. Every electron which passes through the battery of Fig. 33.5 passes through R_1, through R_2, and through R_3 and returns to the battery. The current through each resistor is the same. The current I is the same in all resistors connected in series.

The work necessary to move a coulomb from A to D is the work necessary to move it from A to B plus the work necessary to move it from B to C plus the work from C to D. Thus, the potential difference across the combination of resistors in series is the sum of the potential differences across the individual resistors:

$$V = V_1 + V_2 + V_3$$

The resistance of the combination is, by definition, the ratio of the potential difference V across the combination to the current I:

$$R = \frac{V}{I} = \frac{V_1 + V_2 + V_3}{I} = R_1 + R_2 + R_3 \qquad (33.8)$$

Thus, *the resistance of any combination of resistors connected in series is equal to the sum of the individual resistances.*

33.8 RESISTORS IN PARALLEL

The resistors of Fig. 33.6 are connected *in parallel*. When several conductors are connected between two points so that the current divides between them and then rejoins, they are said to be *in parallel*. An electron in going from B to A of the figure may pass through R_1 or R_2 or R_3. The potential difference across each resistor is the same, because the work required to move a charge from A to B is independent of the path chosen. If V is the potential difference between A and B, and V_1, V_2, and V_3 are the potential differences across R_1, R_2, and R_3, respectively, then $V = V_1 = V_2 = V_3$. The current I splits at point A, a part going through each of the parallel resistors. The total charge reaching B each second is equal to that leaving A each second. Therefore,

$$I = I_1 + I_2 + I_3$$

The resistance between A and B is, by definition, the ratio of V to I:

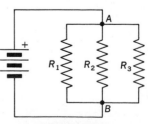

FIGURE 33.6
Three resistors in parallel.

$$R = \frac{V}{I} = \frac{V}{I_1 + I_2 + I_3}$$

or

$$\frac{1}{R} = \frac{I}{V} = \frac{I_1 + I_2 + I_3}{V} = \frac{I_1}{V_1} + \frac{I_2}{V_2} + \frac{I_3}{V_3}$$

and

$$\frac{1}{R} = \frac{1}{R_1} + \frac{1}{R_2} + \frac{1}{R_3} \tag{33.9}$$

When resistors are connected in parallel, the reciprocal of the total resistance is equal to the sum of the reciprocals of the individual resistances.

33.9 QUALITATIVE CONSIDERATIONS IN CONDUCTION

The high conductivity of metals was explained qualitatively by Drude in 1900, shortly after the discovery of the electron. He assumed that when a metal crystallizes in its lattice, one or more electrons are released from each atom. These electrons are free to gain energy from an applied electric field and are responsible for metallic conduction. In 1909 Lorentz extended Drude's theory, considering the electrons to behave like the molecules of a gas with an average kinetic energy of $\frac{3}{2}kT$ [Eq. (14.11)], but his theoretical results were only partially compatible with experimental data. In 1928 Sommerfeld applied wave mechanics and the Fermi distribution function [Eq. (30.1)] to the problem to develop a much better theory. In the Sommerfeld model even at 0 K the average kinetic energy of the "free" electrons is several electronvolts, corresponding to an average speed of the order of 1 Mm/s. When an electric field is applied, these "free" electrons are accelerated and a small drift velocity (≈ 1 mm/s) is superimposed on the high-speed random motions of the electrons. The current in the metal is thus a small electric "wind" superimposed on a very large random velocity distribution. As the temperature of the metal is increased, the average kinetic energy of the electrons is raised slightly but by a very small fraction of the electron kinetic energy at 0 K.

Ionized atoms of a pure metal are arranged in a crystalline lattice; because each atom has released one or more electrons, there is an abundant supply of nearly free electrons. When a current passes through the metal, energy is lost by electrons in collisions with the ion cores; this energy shows up as the Joule heat discussed in Sec. 33.4. As the temperature of the metal is increased, movements of the atoms become greater; the chances of an electron colliding with an atom increase, and so does the resistance. The resistance of an ideal crystal lattice would be zero if there were no motion of the atoms, and most of the resistance of pure metals at room temperature is associated with the movements of the ion cores, which are described in terms of *lattice vibrations*. In general, metals have low resistivities and high temperature coefficients of resistance.

Alloys typically have higher resistivities and lower temperature coefficients than pure metals. In alloys the crystal array is less favorable for the free movement of electrons; this leads to a higher resistivity, especially at very low temperatures. However, as the temperature increases, the thermal vibrations of the atoms in the alloy are almost as likely to move a given atom out of the way of a traveling electron as to move it into the way. Consequently, the resistance is considerably less temperature-sensitive than in the case of a pure metal.

Many nonmetallic solids are poor electric conductors; some are excellent insulators. At room temperature insulating materials have practically no free electrons or mobile ions available to carry current, but at sufficiently high temperature the internal energy becomes large enough to free electrons or conducting ions. For ordinary glass at elevated temperatures sodium ions, freed by virtue of the increased internal energy, play an important role in conduction, but for most insulators it is the internal liberation of electrons which produces conduction at high temperature.

Some materials which are excellent insulators near 0 K require so little energy to provide charge carriers thermally that at room temperature they are fair conductors. Such materials, known as *semiconductors*, typically have $\rho \approx 10^{-4}$ to 10^7 Ω-m, intermediate between metallic conductors ($\rho \approx 10^{-7}$ Ω-m) and good insulators ($\rho \approx 10^{16}$ Ω-m and above). The number of free electrons or other charge carriers increases rapidly with temperature for a typical semiconductor, with a resulting drop in resistance: copper oxide, for example, has a resistance at 70°C which is only one-tenth its resistance at 20°C. Since the 1950s there has been a dramatic expansion of the use of semiconductors in electronic circuits. Among the more common semiconductors are silicon, germanium, selenium, cuprous oxide, and lead sulfide. The physical characteristics of semiconductors, as well as those of metals and insulators, are discussed further in Secs. 47.4 to 47.7.

Carbon is widely used as an electric conductor. It has a negative temperature coefficient of resistance because the number of free electrons available for conduction increases with temperature. The coefficient of thermal conductivity also increases with temperature, and for the same reason; both the electric and thermal conductivities of the best room-temperature conductors are associated with free electrons. It is a well-known fact that good conductors of electricity are also good conductors of heat and that nonconductors of electricity are poor conductors of heat. Indeed, for all metals the *Wiedemann-Franz law* states that (except at low temperatures) the ratio of the thermal conductivity to the

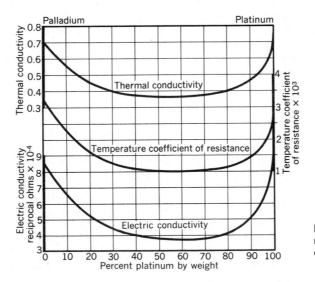

FIGURE 33.7
Relationship between electric and thermal conductivities and the temperature coefficient of resistance for palladium-platinum alloys.

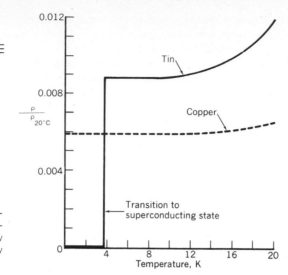

$\dfrac{\rho}{\rho_{20°C}}$

0.012

0.008

0.004

Tin

Copper

Transition to
superconducting state

0

4   8   12   16   20

Temperature, K

FIGURE 33.8
Resistivity-ratio curves for tin and copper, showing that tin becomes superconducting, losing all resistivity below 3.75 K. Ratios are of resistivity at low temperatures to those at 20°C.

electric conductivity is directly proportional to the Kelvin temperature, with the constant of proportionality independent of the particular metal.

For alloys of varying composition the thermal conductivity and the electric conductivity are closely related. Further, the temperature coefficient of resistance as a function of alloy composition varies in much the same manner as the two conductivities (Fig. 33.7).

33.10 SUPERCONDUCTIVITY

At very low temperatures the resistivities of all metals are very much smaller than at room temperature, but the resistivities of some materials drop suddenly to an immeasurably small value. The drop in resistance occurs over an exceedingly small temperature range, as is shown in Fig. 33.8. Metals in which the resistance vanishes at very low temperatures are called *superconductors;* when the resistance vanishes, they are said to be in the *superconducting* state. Not all metals become superconducting as the temperature approaches absolute zero. Metals such as gold, platinum, copper, sodium, and iron have resistivities which show no such abrupt changes. Among the elements which become superconductors are niobium, mercury, lead, tin, indium, and aluminum.

QUESTIONS

1. In what ways is the current in a conductor analogous to the flow of water through a pipe? At what points does the analogy fail? Do bends in a pipe affect the flow of water? Do bends in a wire affect the flow of current?

2. What do we mean when we say that two circuit elements are in series? What must be the same for both?

3. What do we mean when we say that several

circuit elements are in parallel? What is the same for all the elements? Why?

4. What are the advantages and disadvantages of strings of Christmas tree lights in series? In parallel?

5. What is the distinction between the *velocity* of an electron in a metal and the *drift velocity* of the electrons?

6. Why are good electric conductors also good conductors of heat?

7. The filament of a light bulb is a tungsten wire. If

the potential difference across the filament is measured as a function of the current, is Ohm's law obeyed? Why not? Under what conditions would Ohm's law be obeyed by a tungsten filament?

8. A 60-W light bulb and a 100-W bulb are designed to operate on 110 V. Which has the greater resistance? How do you reconcile your answer with the relation $P = I^2R$?

9. What is meant by a negative temperature coefficient of resistance? Explain why some materials have negative temperature coefficients.

10. Why is the variation of resistance with temperature so different for a pure metal and for a semiconductor?

PROBLEMS

1. A transmission line with a resistance of $2.5\,\Omega$ carries a current of 80 A. What is the potential difference between the ends of the line? How much power is dissipated in the line? What charge is transferred each minute?
   Ans. 200 V; 16 kW; 4,800 C

2. A 200-W incandescent lamp operates at 120 V. What current is drawn? What is the resistance of the lamp? How much charge passes through the filament each minute? How many electrons pass through each second?

3. Find the current drawn by, and the resistance of, a 660-W 110-V toaster. If electric energy costs 5 cents per kilowatt-hour, how much does it cost to keep the toaster in steady operation for 30 min? *Ans.* 6 A; 18.3 Ω; 1.65 cents

4. A potential difference of 24 V is applied across a resistance of $6\,\Omega$ for 4 min. What current is drawn? Calculate the charge which passes through the resistor. How many joules of energy does each coulomb lose in the resistor? What power is dissipated in the resistor?

5. A trouble lamp draws 2.5 A from a 12-V battery. What is the resistance of the lamp? The power it dissipates? The charge which passes through the lamp in 5 min? *Ans.* 4.8 Ω; 30 W; 750 C

6. An electric heater element is designed to dissipate 1,320 W when connected to a 110-V line. Find the current drawn, the resistance of the heater, the charge which passes through the heater in 20 s, and the energy transformed to heat in 20 s.

7. When the light switch of a car is turned on, two headlights, two taillights, and a dash light are connected in parallel with a 12-V battery of negligible internal resistance. If each headlight has a resistance of $3\,\Omega$, each taillight a resistance of

24 Ω, and the dash light a resistance of 48 Ω, find the current through the battery. *Ans.* 9.25 A

8. If three resistors have resistances of 50, 60, and 100 Ω, respectively, what is the resistance of the combination if they are connected (*a*) in series and (*b*) in parallel?

9. If in the accompanying figure $R_1 = 10\,\Omega$, $R_2 = 12\,\Omega$, and $R_3 = 24\,\Omega$, find the current in R_1 and the potential difference between A and B if the current in R_2 is 0.5 A. *Ans.* 0.75 A; 13.5 V

10. The resistors of the accompanying figure are a portion of a complete circuit, and $R_1 = 8\,\Omega$, $R_2 = 4\,\Omega$, and $R_3 = 12\,\Omega$. The current through R_3 is 0.5 A. Find the current in R_1 and the potential difference between A and B.

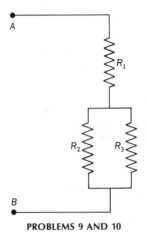

PROBLEMS 9 AND 10

11. Four 120-V 80-W resistors are connected in parallel. The group is then connected in series with a coil of 55 Ω resistance. How much power is dissipated in each resistor and in the coil of wire when the combination is connected across a 100-V battery? Assume that all the resistors retain the same resistance independent of the power dissipated. *Ans.* $P_R = 11.25$ W; $P_C = 55$ W

12. A copper bus bar is 15 mm thick, 35 mm wide, and 800 mm long. Find its resistance and the potential difference between its ends when it bears a current of 2 kA.

13. A ribbon of silver 75 mm long and 1.1 mm wide is to be made into a resistance of 0.1 Ω. How thick must it be? *Ans.* 10.9 μm

14. A silver wire of resistance of $0.7 \, \Omega$ is drawn through a die so that its length is tripled and its cross section is reduced to one-third of its previous value. Find the new resistance.

15. A piece of wire 10 m long and 0.5 mm in diameter has a resistance of $2 \, \Omega$. What length of wire of the same material 0.3 mm in diameter will have a resistance of $2.5 \, \Omega$? *Ans.* 4.5 m

16. If 1 g of copper and 1 g of aluminum are used to make uniform wires each 5 m long, find the resistances of both wires. What is the ratio of the resistance of the copper wire to that of the aluminum wire? What would be the ratio of the resistance of a copper wire to that of an aluminum wire of the same dimensions?

17. A platinum resistance thermometer is used to determine the temperature of an oven. If the average temperature coefficient of resistance of platinum is 0.0037 per Celsius degree and the resistance of the platinum coil is $80 \, \Omega$ at $0°C$, find the temperature when the coil has a resistance of $250 \, \Omega$. *Ans.* 574°C

18. A 15-Ω resistor has a temperature coefficient of 0.004 per Celsius degree. Initially the resistor is immersed in ice water to keep its temperature constant. (*a*) If a potential difference of 60 V is applied, find the current. (*b*) If the resistor is removed from the ice water and allowed to come to equilibrium with the 60 V still applied, the current falls to 3 A. Calculate the equilibrium temperature.

19. The electric resistance of a tantalum wire is $15 \, \Omega$ at $20°C$, and $20.4 \, \Omega$ at $120°C$. What is the temperature coefficient of resistance? What is its resistance at $160°C$ if the resistance of tantalum varies linearly with temperature? *Ans.* 0.0036 per C°; 22.6 Ω

20. The resistance of a tungsten lamp filament is $9 \, \Omega$ at $20°C$. If the filament operates at $2220°C$, find its resistance. How many times as great is the resistance at operating temperature as the resistance at room temperature?

21. The field coil of a motor draws a current of 1.65 A from a 110-V line when the motor is started and the coil is at $0°C$. What is the resistance of the coil at $0°C$? If the potential difference across the coil does not change, what current is drawn by the copper field coil at its normal operating temperature of $60°C$? *Ans.* 66.7 Ω; 1.34 A

22. The field coil of an electric motor has a resistance

of $49.2 \, \Omega$ at $40°C$ and $42.4 \, \Omega$ at $0°C$. What is its temperature coefficient of resistance?

23. If a lamp filament made of tungsten wire with a cross-sectional area of $6 \times 10^{-9} \, m^2$ is to have a resistance of $11 \, \Omega$ at $20°C$, how long must it be? Find the resistance of this filament at $2020°C$. *Ans.* 1.2 m; 114 Ω

24. Calculate the resistance between points A and B in the accompanying figure if $R_1 = 30 \, \Omega$, $R_2 = 20 \, \Omega$, $R_3 = 8 \, \Omega$, $R_4 = 60 \, \Omega$, $R_5 = 20 \, \Omega$, and $R_6 = 30 \, \Omega$.

25. Calculate the resistance between points A and B of the accompanying figure if $R_1 = 18 \, \Omega$, $R_2 = 9 \, \Omega$, $R_3 = 20 \, \Omega$, $R_4 = 24 \, \Omega$, $R_5 = 8 \, \Omega$, and $R_6 = 12 \, \Omega$. *Ans.* 30 Ω

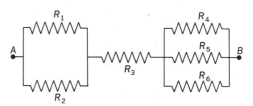

PROBLEMS 24 AND 25

26. An ammeter in series with a battery and a resistance R reads 5 A. When an additional resistance of $6 \, \Omega$ is inserted in series with the first, the reading of the ammeter is reduced to 3 A. Find the resistance R. The resistances of battery and ammeter are negligible.

27. The belt of a Van de Graaff generator is 0.75 m wide and moves 25 m/s. If half the charges sprayed on the belt reach the high-potential electrode, how many coulombs must be sprayed onto the belt each second to give a total current to the electrode of 2.5 mA? Find the charge sprayed on the belt per unit area. *Ans.* 5 mC/s; 267 $\mu C/m^2$

28. A 4-kW heating unit in a hot-water heater is on 2.5 h/day. If electric energy costs 5 cents per kilowatt-hour, how much does hot water cost each 30-day month? How many kilograms of water are heated from 10 to $80°C$ each day if all the heat dissipated in the heater is effective in warming the water?

29. An airplane deicer operates from a 24-V battery and is capable of melting 0.03 kg of ice per minute. Find the resistance and current when the deicer is in operation. *Ans.* 6.98 A; 3.44 Ω

30. An oven requires 9 A to heat it to the desired temperature when the applied voltage is 110 V. How much resistance must be inserted in series with the oven to keep it at the same temperature

if the potential difference is increased to 220 V? How much power is dissipated in this resistor?

31. An electric iron of 1.4 kg mass has an average specific heat of 0.10 kcal/(kg)(C°). The heating units takes 7 A from a 110-V line. If half the heat is lost by radiation, how long will it take to bring the iron to a temperature of 115°C when it is at 20°C originally? *Ans.* 144 s

32. Find the resistance between A and B of the ac-

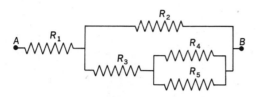

PROBLEMS 32 AND 33

companying figure if $R_1 = 9\,\Omega$, $R_2 = 24\,\Omega$, $R_3 = 8\,\Omega$, $R_4 = 6\,\Omega$, and $R_5 = 12\,\Omega$.

33. Find the resistance between A and B of the accompanying figure if $R_1 = 8\,\Omega$, $R_2 = 48\,\Omega$, $R_3 = 11\,\Omega$, $R_4 = 6\,\Omega$, and $R_5 = 30\,\Omega$. *Ans.* 20 Ω

34. If a 6-Ω resistor is placed along each of the 12 edges of a cube and connections are made at each corner, find the resistance between two diagonally opposite corners of the cube. *Hint:* By symmetry each of the remaining six corners has the same potential as two others.

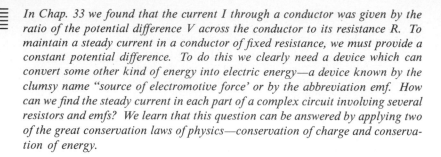

34

ELECTRIC
CIRCUITS

In Chap. 33 we found that the current I through a conductor was given by the ratio of the potential difference V across the conductor to its resistance R. To maintain a steady current in a conductor of fixed resistance, we must provide a constant potential difference. To do this we clearly need a device which can convert some other kind of energy into electric energy—a device known by the clumsy name "source of electromotive force" or by the abbreviation emf. How can we find the steady current in each part of a complex circuit involving several resistors and emfs? We learn that this question can be answered by applying two of the great conservation laws of physics—conservation of charge and conservation of energy.

34.1 ELECTROMOTIVE FORCE

To maintain a constant current in a conductor, it is necessary to maintain a steady potential difference across the conductor. This potential difference can be supplied only if some device converts some other form of energy into electric energy. Such a device is called a *source of electromotive force* (abbreviated emf).

There are many kinds of emf. In batteries chemical reactions occur, transforming chemical energy into electric energy. In the giant generators of our electric power plants mechanical energy is converted into electric energy. In a thermocouple it is heat energy, while in the photoelectric cell it is radiant energy which is transformed.

When a charge q receives an energy W in passing through a battery or some other source of electric energy, the emf $\mathcal{E}$ is given by

$$\mathcal{E} = \frac{W}{q} \tag{34.1}$$

When W is in joules and q in coulombs, the emf is in joules per coulomb, or volts.

When a charge of one coulomb receives one joule of energy upon passing through a source, the source is said to have an electromotive force of one volt. A 12-V battery delivers 12 J of energy to each coulomb which passes through it. Electromotive force and potential difference are measured in the same units. An emf is a particular kind of potential difference, namely, one which arises through the transformation of some other form of energy into electric energy. In contrast, the potential difference between two points A and B in the neighborhood of a charge Q (Fig. 31.1) is not an electromotive force and cannot be used to maintain a steady current in a conductor connecting these points.

In a source of emf not only may some other form of energy be transformed into electric energy, but the reverse process may also occur—electric energy can be converted into another form. For example, in charging a battery, charges are forced through the battery in a direction opposite that in which they go when the battery is discharging; these charges deliver electric energy to the battery, where it is converted into, and stored as, chemical energy.

34.2 THE CONSERVATION OF ENERGY IN A SIMPLE CIRCUIT

Consider the circuit of Fig. 34.1, in which a battery is connected in series with three resistors. A charge q which passes through the battery gains an amount of energy $\mathcal{E}q$ joules. In passing through the resistance R_1, this charge loses energy V_1q joules. In passing through resistors R_2 and R_3, this charge loses amounts of energy V_2q and V_3q, respectively. In going once around the complete circuit, the charge loses exactly the same amount of energy as it gains. Therefore,

$$\mathcal{E}q = V_1q + V_2q + V_3q \qquad \text{and} \qquad \mathcal{E} = V_1 + V_2 + V_3 \qquad (34.2)$$

or
$$\mathcal{E} = IR_1 + IR_2 + IR_3 \qquad (34.3)$$

From Eq. (34.3) we obtain

$$I = \frac{\mathcal{E}}{R_1 + R_2 + R_3} \qquad (34.4)$$

In a simple series circuit the current is equal to the ratio of the emf to the sum of the resistances in the circuit. This statement represents a simplified form of *Ohm's law for a complete circuit.*

34.3 THE RESISTANCES OF SOURCES OF ELECTROMOTIVE FORCE

Any source of emf, such as a battery or electric generator, has some internal resistance. As a consequence, a current through a battery produces some heating. When the battery is being discharged, the total energy given to a charge q is $\mathcal{E}q$, but a portion of this energy is converted into heat within the battery. If the internal resistance is r, the potential drop in the battery resistance is Ir. The potential difference between the terminals of the battery V_t is the net potential gain.

$$V_t = \mathcal{E} - Ir \qquad (34.5)$$

The terminal potential difference is $\mathcal{E}$ *only* when no current is being drawn.

The decrease in the terminal potential difference of a battery when the current drawn is changed is illustrated by the dimming of automobile headlights when the starter is activated. The starter draws a large current from the battery. Because of the increased Ir drop within the battery, the terminal potential is reduced, and the potential difference across the lamps of the car is lower. For an automobile battery of emf about 12 V, the internal resistance may be about 5 mΩ. The current drawn by the lights is approximately 6 A; thus, when current is being drawn only for the lights, the terminal potential difference of the battery is only a few hundredths of a volt less than the emf. If the current drawn from the battery is increased to 150 A by operating the starter motor, V_t becomes $12 - 0.75$ or 11.25 V. The brightness of the lamps is reduced by such a decrease in terminal potential difference.

For practical purposes we may assume that the real battery is made up of a pure emf and a series resistor, as suggested in Fig. 34.2. Ohm's law for a complete circuit is applicable, but the internal resistance of the battery must be included in the total resistance.

□ **Example** A 45-V battery has an internal resistance of 0.6 Ω. It is connected in a circuit as indicated in Fig. 34.3. Find the current in each resistor.

First, the series equivalent of the two parallel resistors must be found.

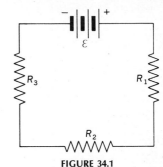

FIGURE 34.1
A coulomb gains an amount of energy in passing through the battery equal to the energy it loses in the resistors as it goes around the complete circuit.

FIGURE 34.2
A real battery may be treated as an ideal resistanceless battery and a series resistance r.

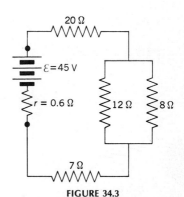

FIGURE 34.3
Circuit composed of a battery and four resistors, two of which are in parallel.

$$\frac{1}{R_{\parallel}} = \frac{1}{12} + \frac{1}{8} = \frac{5}{24}$$

$$R_{\parallel} = 4.8 \ \Omega$$

Now applying Ohm's law for the complete circuit yields

$$I = \frac{\mathscr{E}}{R} = \frac{45}{0.6 + 20 + 4.8 + 7} = \frac{45}{32.4} = 1.39 \ \text{A}$$

This is the current through the battery and the 20- and 7-Ω resistors. The potential difference across the parallel resistors is $1.39 \times 4.8 = 6.67$ V. Therefore, the current in the 8-Ω resistor is $6.67/8 = 0.83$ A, while that through the 12-Ω resistor is $6.67/12 = 0.56$ A. Note that the sum of these currents is 1.39 A.   □

34.4 CHARGING A BATTERY

The terminal potential difference of a battery as it is discharged is given by $\mathscr{E} - Ir$. While the battery is being charged, the terminal potential difference is greater than the emf by an amount Ir, as can be seen from consideration of the energy transformations. In charging the battery, electric energy is converted into chemical energy; the energy per unit charge is $\mathscr{E}$. The charging current produces heat in the battery, and the energy per unit charge required to produce this heating effect is Ir. Thus, the total electric energy which must be delivered to the battery per unit charge is

$$V_t = \mathscr{E} + Ir \tag{34.6}$$

Regardless of the direction of the current through a cell or other source of emf, Joule heat is always produced.

34.5 CELLS IN SERIES

When two or more sources of emf are connected in series, the net emf is the algebraic sum of the individual emfs. If two cells are connected in series in such a way that both would produce a current in the same direction, the emf is the sum of the two emfs; on the other hand, if the two are connected in series in such a way that they would send currents in opposite directions, the net emf is the difference between the two. In the first case the cells are said to be connected in *series aiding,* in the second case in *series opposing.* When a battery is to be charged, it must be connected in series opposing with some other source of emf which supplies electric energy to be transformed into chemical energy.

When several identical cells are connected in series, the total emf is equal to the number of cells multiplied by the emf of a single cell, while the resistance of the battery is equal to the resistance of an individual cell times the number of cells. The type of cell used in ordinary automobile batteries has an emf of approximately 2 V; in order to obtain an emf of 6 V, three cells are connected in series; for 12 V, six cells are required.

34.6 OHM'S LAW FOR A COMPLETE CIRCUIT

In Fig. 34.4 three batteries of negligible resistance are connected in series with three resistors. In this case $\mathscr{E}_2$, which is opposing $\mathscr{E}_1$ and $\mathscr{E}_3$, is

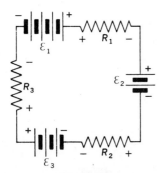

FIGURE 34.4
Series circuit containing three sources of emf.

smaller than the sum $\mathcal{E}_1 + \mathcal{E}_3$. Consequently, battery 2 is being charged by the other batteries.

For this circuit we again apply the fundamental principle that the total energy gained by a charge in going around a complete circuit is equal to the total energy lost. If a positive unit charge is carried clockwise around the circuit, the net change in potential energy for the complete circuit is zero, or

$$\mathcal{E}_1 - IR_1 - \mathcal{E}_2 - IR_2 + \mathcal{E}_3 - IR_3 = 0 \qquad (34.7)$$

It is often helpful to mark plus and minus signs at the appropriate ends of all resistors and batteries. The change in potential is positive when the positive test charge is moved from the minus terminal of a circuit element to the positive one and negative when the charge goes from plus to minus. It is not necessary that the charge be carried around the circuit in the direction in which the current flows. Indeed, if the direction of motion is reversed, all signs in Eq. (34.7) are changed but the sum is still zero. The current, which is the same in all parts of the circuit, can be found by solving Eq. (34.7) to obtain

$$I = \frac{\mathcal{E}_1 - \mathcal{E}_2 + \mathcal{E}_3}{R_1 + R_2 + R_3} = \frac{\Sigma \mathcal{E}}{\Sigma R} \qquad (34.8)$$

This equation represents Ohm's law for a complete series circuit, which can be stated as follows:

The current in any series circuit is given by the ratio of the algebraic sum of the emfs to the total series resistance of the circuit.

Ohm's law for a complete circuit is to be distinguished from Ohm's law for a single resistor. The law for a single resistor involves only the potential difference across the resistor, while the law for the complete circuit involves the algebraic sum of the emfs in the circuit and the total resistance of the series circuit. In a series circuit the algebraic sum of the emfs is equal to the sum of the IR drops in the resistors. This follows directly from the law of conservation of energy.

When one applies Ohm's law to a complete circuit in which there are parallel resistors, each group of parallel resistors is first replaced with the equivalent series resistor. Then Ohm's law for the circuit is applied directly.

☐ **Example**   A battery of emf 20 V and internal resistance 1 Ω is connected in series with a 5-Ω resistor, a second battery of emf 8 V and internal resistance 2 Ω which is in series opposing, and a group of three resistors of 12, 6, and 4 Ω resistance in parallel, as shown in Fig. 34.5. Find the current in each resistor and the terminal potential difference of each battery.

We first replace the three parallel resistors with the equivalent single resistor: from $1/R_{\parallel} = \frac{1}{12} + \frac{1}{6} + \frac{1}{4}$ we find that $R_{\parallel} = 2\,\Omega$. Next we apply Ohm's law for the circuit, which gives

$$I = \frac{20 - 8}{1 + 5 + 2 + 2} = \frac{12}{10} = 1.2 \text{ A}$$

for the current in each battery and in the 5-Ω resistor. The potential drop across $R_{\parallel}$ is given by $IR_{\parallel} = 1.2 \times 2 = 2.4$ V, which remains unchanged if we replace $R_{\parallel}$ with the original three resistors. The currents in

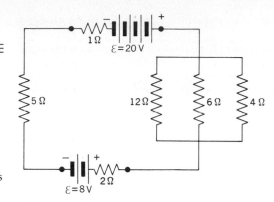

FIGURE 34.5
Circuit with two opposing batteries connected in series.

these resistors are given by V/R as follows: $I_{12} = 2.4/12 = 0.2$ A, $I_6 = 2.4/6 = 0.4$ A, and $I_4 = 2.4/4 = 0.6$ A. Note that the sum is 1.2 A, the current in the main circuit.

The terminal potential difference for the 20-V battery is $\mathcal{E} - Ir = 20 - 1.2 \times 1 = 18.8$ V, while the terminal potential difference for the 8-V battery which is being charged is $\mathcal{E} + Ir = 8 + (1.2 \times 2) = 10.4$ V. ☐

34.7 KIRCHHOFF'S LAWS

Circuits ranging from the simplest to very complex networks with many branches and many emfs can be handled by application of two fundamental principles known as *Kirchhoff's laws*. These laws, which apply once a steady state has been reached, have been used in previous discussions, although they have not been specifically named. Kirchhoff's two laws are:

1. *The sum of all currents arriving at any junction in a circuit is equal to the sum of the currents leaving that junction.*

2. *Around any closed loop, the sum of the potential rises is equal to the sum of the potential drops.*

The first law must apply if we are to avoid an accumulation of charge at any junction in the circuit or a continuing disappearance of charge there. If Kirchhoff's first law were not true and the sum of the currents reaching a junction exceeded the currents leaving it, the charge would build up at this point and the potential of the junction would change continuously.

Kirchhoff's second law is a special statement of the law of conservation of energy. If a charge gained more energy in going around a closed path than it lost, it would be able to gain more and more energy by repeated traversing of this path. This is obviously not permissible. Once the charge returns to its starting point, its potential energy must be exactly the same as when it started. When a man takes a hike in the mountains and eventually returns to his starting point, he has climbed up exactly as many meters as he has descended, since he ends at the same altitude at which he started. Kirchhoff's second law is the electrical analog of this mechanical illustration.

In applying Kirchhoff's laws to a problem, it is convenient to carry out the following steps in order:

1. *Assign a direction and a symbol to the current in each independent branch of the circuit.* It is not necessary to worry about which direction to assign the current in a given branch, since if an incorrect assignment is made, the current turns out to be negative.

2. *Place appropriate plus and minus signs at the terminals of every source of emf and every resistor in the circuit.* Remember that in a resistor the current is from the plus to the minus terminal; thus the choices of current directions in the first step determine the signs of the terminals of all resistors.

3. *Apply Kirchhoff's first law at enough junctions for each current to appear in an equation.* Be sure that each junction equation contains at least one current which has not appeared in earlier equations.

4. *Apply Kirchhoff's second law to closed loops until once again every current has been included in at least one equation.* When a positive charge goes through a circuit element from minus to plus, it gains potential energy and the change in potential is positive. When the positive charge goes from plus to minus, it loses potential energy and the change in potential is negative. Once again be sure that every new loop equation involves at least one current which has not appeared in a previous loop equation.

5. *Solve the equations for the desired unknowns.*

A familiarity with the use of Kirchhoff's laws can be obtained by studying one or more examples and then by practice on additional problems.

□ **Example** Find the current in all branches of the circuit of Fig. 34.6.

1. Let us designate the currents by I_1, I_2, and I_3 and assume them to be in the directions indicated by the arrows.

2. We place a plus on the higher-potential end of each circuit component and a minus on the lower-potential end. We treat each battery as a pure source of emf in series with a resistor equal to the internal resistance of the battery.

3. At point A, $I_1 + I_2 = I_3$, by Kirchhoff's first law.

4. If we start at point A and apply Kirchhoff's second law to the left loop going clockwise, we obtain

$$-15 + 2I_2 + 7I_2 - 3I_1 - I_1 + 40 = 0$$

By going clockwise from A around the right loop, we obtain

$$-8I_3 - 7I_2 - 2I_2 + 15 = 0$$

We now have the equations

$$
\begin{aligned}
I_1 + I_2 - I_3 &= 0 \\
4I_1 - 9I_2 &= 25 \\
9I_2 + 8I_3 &= 15
\end{aligned}
$$

The solutions are $I_1 = 4$ A, $I_2 = -1$ A, and $I_3 = 3$ A. The fact that I_2 is negative means that the current I_2 is in the direction opposite that assumed. The 15-V battery is being charged. □

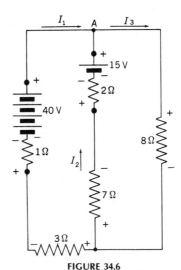

FIGURE 34.6
Circuit in which there are three branches, each with a different current.

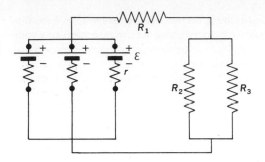

FIGURE 34.7
Circuit with three identical cells in parallel.

When several identical sources of emf are connected in parallel, the emf of the combination is the emf of a single source. The net resistance of this combination is equal to the resistance of a single cell divided by the number of cells, since, effectively, all the resistances are in parallel. Cells which are not identical are seldom connected in parallel. Although Kirchhoff's laws are valid in any situation in which currents are constant, it is easier to solve circuits such as that of Fig. 34.7 by the method illustrated below. Note that this method leads to correct results only when the cells are identical, having the same emf, the same internal resistance, and the same lead resistances connecting the sources to the external circuit.

□ **Example** Three identical dry cells with $\mathcal{E} = 1.5$ V and $r = 0.12\,\Omega$ are connected in parallel to the circuit of Fig. 34.7. Find the current in each cell and in each resistor if $R_1 = 5\,\Omega$, $R_2 = 30\,\Omega$, and $R_3 = 60\,\Omega$.

The emf of the three identical cells in parallel is 1.5 V, and the resistance of the combination is $0.04\,\Omega$ (three 0.12-Ω resistors in parallel). The resistance of the 30- and 60-Ω parallel combination is $20\,\Omega$. Application of Ohm's circuit law yields

$$I = \frac{1.5}{0.04 + 5 + 20} = \frac{1.5}{25.04} = 0.060 \text{ A}$$

The potential drop across the 30- and 60-Ω resistors is 20×0.060, or 1.2 V. Therefore, $I_{30} = 1.2/30 = 0.04$ A, and $I_{60} = 1.2/60 = 0.02$ A. The current through each of the three identical cells is one-third of 0.06 A, or 0.02 A.
□

34.8 THE WHEATSTONE BRIDGE

An accurate and simple method of measuring resistances employs the Wheatstone bridge. The circuit for this bridge is shown in Fig. 34.8. A and B are fixed resistors, the values of which are known. The resistance X whose value is to be determined is connected in the third arm of the bridge, while a known variable resistance R is connected in the fourth arm. The resistance R is varied until there is no current between c and d as indicated by the galvanometer G, an instrument for detecting small currents. When the galvanometer shows no current between c and d, the bridge is said to be *balanced*.

If there is no current in the galvanometer, the potential difference between points c and d must be zero. For this to be true, the potential drop across resistor A must be equal to the potential drop across X, since

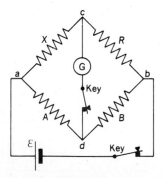

FIGURE 34.8
A Wheatstone bridge circuit.

one end of each of these resistors is at the potential of point a. If I_A is the current in A and I_X the current in X, the condition for no current in the galvanometer is that $V_{ac} = V_{ad}$ or $I_X X = I_A A$. Similarly, the fall of potential from c to b must be equal to the drop in potential from d to b, which gives $I_R R = I_B B$. The current in R and in X is the same when no current flows through the galvanometer. Similarly, the current in A is the same as that in B. Therefore $I_X X / I_R R = I_A A / I_B B$, or

$$\frac{X}{R} = \frac{A}{B} \qquad (34.9)$$

In the slide-wire Wheatstone bridge the resistors A and B are segments of a wire of uniform cross section and resistivity. The ratio A/B is determined by the position of the movable contact along this wire. In high-quality commercial Wheatstone bridges, the ratio of A to B can be set at one of several values and the final balance made by varying R.

34.9   THE POTENTIOMETER

The potentiometer occupies an important place in electrical measurements because it can be used to measure potential differences with great accuracy and *without drawing any current*. This feature is of great importance in working with sources of emf of high internal resistance and low current capabilities.

A schematic diagram of a potentiometer is shown in Fig. 34.9. The wire ab, which is of uniform cross section, carries a current maintained constant by the working battery B. There is a progressive drop in potential along the wire from a to b, directly proportional to the distance from a. To measure an emf whose value $\mathcal{E}_x$ is unknown, this emf is placed in series with a galvanometer and connected as shown in Fig. 34.9, in which point c represents a movable contact. It is important that the emf $\mathcal{E}_x$ oppose the current which the working battery B would produce if $\mathcal{E}_x$ were zero. If the potential difference between a and b is greater than $\mathcal{E}_x$, there is some point c at which the potential difference across ac is equal to $\mathcal{E}_x$. This point can be found by moving the sliding contact until the current through the galvanometer is zero. The potentiometer is then *balanced*.

In the usual application of the potentiometer a standard cell of known emf $\mathcal{E}_s$ is first used in place of $\mathcal{E}_x$, and the balance point is found at some point d. The unknown emf $\mathcal{E}_x$ is substituted for $\mathcal{E}_s$, and the balance point C is found. Then $\mathcal{E}_x$ can be calculated from the relation

$$\frac{\mathcal{E}_x}{\mathcal{E}_s} = \frac{ac}{ad}$$

where ac and ad represent the length of conductor from a to c and from a to d, respectively.

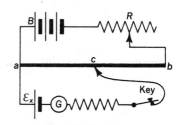

FIGURE 34.9
A simple potentiometer circuit.

QUESTIONS

1. What is the distinction between the terms *potential difference* and *emf*? Can a battery have an emf and still have no potential difference between its terminals? Under what circumstances?
2. Can the terminal potential difference of a battery exceed its emf? Can the terminal potential difference be negative in the sense that what is normally the positive terminal of the battery becomes the negative terminal? Explain.
3. What is the advantage of connecting identical cells in series?
4. What is the advantage of connecting identical cells in parallel?
5. Why are batteries which are not identical seldom connected in parallel?
6. Draw a Wheatstone bridge and derive the relationship between the resistances when the bridge is balanced.
7. Does interchanging the galvanometer and battery in a Wheatstone bridge affect the balance? Justify your answer with equations.
8. Draw a potentiometer circuit and explain how it operates. How could the potentiometer be used to calibrate an ammeter with a standard resistor? Draw a suitable circuit.
9. Why is it difficult to obtain a good measurement of resistance by using an ammeter and a voltmeter? Draw circuits to illustrate. Recall that ordinary ammeters and voltmeters have resistance too.
10. Why is a voltmeter rated at 20 kΩ/V usually more desirable and more expensive than one of the same range rated 5 kΩ/V?

PROBLEMS

1. A battery of emf 12 V and internal resistance of 0.2 Ω is connected in series with two resistors, one of 5.8 Ω and another of 18 Ω. Find the current, the potential difference across each resistor, and the terminal potential difference of the battery. *Ans.* 0.5 A; 2.9 V; 9 V; 11.9 V
2. A 15-Ω resistor bears a current of 0.6 A when it is connected in parallel with two other resistors of 18 and 25 Ω. Find the current in each of these two resistors.
3. A circuit has three parallel branches with resistances of 50, 40, and 70 Ω. When a current of 0.2 A

is flowing in the 50-Ω branch, how much current is flowing in each of the other branches?
Ans. 0.25 A; 0.143 A

4. A battery has an emf of 60 V and an internal resistance of 4 Ω. Find the power dissipated in a variable external series resistor set successively at the values 2, 3, 4, 5, and 6 Ω. Plot a rough curve of power dissipated in the load as a function of load resistance to confirm the fact that maximum power is supplied to the load when the load resistance is equal to the internal resistance of the source.
5. Four cells of a storage battery, each with a resistance of 0.01 Ω and an emf of 2.05 V, are connected in series with a coil having a resistance of 4.06 Ω. Find the current and the terminal potential difference of the battery. *Ans.* 2 A; 8.12 V
6. When an external resistance of 85 Ω is connected to the terminals of a battery, the current is found to be 0.5 A. When this resistance is increased to 145 Ω, the current drops to 0.3 A. Find the emf and internal resistance of the battery.
7. A battery has an emf of 3 V. Determine its internal resistance if a current of 20 A is drawn when an ammeter and leads with a total resistance of 0.01 Ω are connected across the terminals of the battery. *Ans.* 0.14 Ω
8. A battery having an internal resistance of 0.06 Ω and an emf of 6 V is used to send a current through an 8.7-Ω resistance connected in series with two resistances, one of 12 Ω and the other of 13 Ω, in parallel. Find the current in each of the resistances.
9. In the circuit of the accompanying figure, find the current in R_1 and the charge stored on the capacitor if $\mathcal{E} = 6$ V, $r = 0.5$ Ω, $R_1 = 14.5$ Ω, $R_2 = 25$ Ω, and $C = 2$ μF. *Ans.* 0.15 A; 7.5 μC
10. In the circuit of the accompanying figure, find the current in R_1 and the charge stored on the capacitor if $\mathcal{E} = 24$ V, $r = 1$ Ω, $R_1 = 19$ Ω, $R_2 = 52$ Ω, and $C = 8$ μF.

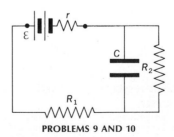

PROBLEMS 9 AND 10

11. A 72-V battery with an internal resistance of 1 Ω is to be charged at the rate of 8 A from a 120-V source. (*a*) What resistance must be connected in series with the battery? (*b*) What will be the

terminal potential difference of the battery during the charging process? (c) Assuming that the emf arises entirely from the conversion of chemical into electric energy, find the rate at which electric energy is converted into chemical energy. (d) At what rate is electric energy converted into heat in the battery? (e) In the series resistor?

Ans. (a) 5 Ω; (b) 80 V; (c) 576 W; (d) 64 W; (e) 320 W

12. A storage battery having an emf of 48 V and an internal resistance of 1.2 Ω is to be charged by connecting it to a 100-V generator. What resistance must be introduced in series with it in order for the charging current to be 5 A? What is the terminal potential difference of the battery?

13. In the accompanying figure, $\mathcal{E}_1 = 30$ V, $r_1 = 1$ Ω, $\mathcal{E}_2 = 10$ V, $r_2 = 2$ Ω, $R_1 = 6$ Ω, and $R_2 = 3$ Ω. Find the current, the terminal potential difference of each battery, and the potential difference between points A and B.

Ans. 1.67 A; 28.3 and 13.3 V; 18.3 V

14. In the accompanying figure, $\mathcal{E}_1 = 24$ V, $r_1 = 1$ Ω, $\mathcal{E}_2 = 6$ V, $r_2 = 2$ Ω, $R_1 = 5$ Ω, and $R_2 = 4$ Ω. Find the current, the terminal potential difference of each battery, and the potential difference between points A and B.

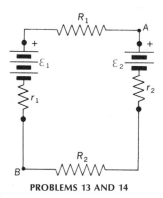

PROBLEMS 13 AND 14

15. Six identical cells, each with an internal resistance of 0.18 Ω, are connected in parallel to send a current through an external resistance of 2.97 Ω. How much current will be obtained in the external resistance if each cell has an emf of 1.5 V?

Ans. 0.5 A

16. Five storage batteries, each having six cells with an emf of 2 V and an internal resistance of 0.012 Ω per cell, are to be charged in series at the rate of 7 A from a 110-V line. How much resistance must be inserted in series with the batteries?

17. Find the current through the batteries and through R_3 of the accompanying figure if

$\mathcal{E}_1 = 36$ V, $r_1 = 1$ Ω, $\mathcal{E}_2 = 24$ V, $r_2 = 2$ Ω, $R_1 = 40$ Ω, $R_2 = 27$ Ω, $R_3 = 40$ Ω, and $R_4 = 120$ Ω. Find the terminal potential differences of the batteries and the potential difference between A and B. *Ans.* 0.6 and 0.45 A; 35.4 and 22.8 V; 4.8 V

18. Find the current through the batteries and through R_3 of the accompanying figure if $\mathcal{E}_1 = 50$ V, $r_1 = 1$ Ω, $\mathcal{E}_2 = 20$ V, $r_2 = 2$ Ω, $R_1 = 5$ Ω, $R_2 = 8$ Ω, $R_3 = 12$ Ω, and $R_4 = 36$ Ω. Find the terminal potential differences of the batteries and the potential difference between A and B.

19. In the accompanying figure R_4 is 8 Ω and bears a current of 1.5 A. If $R_3 = 4$ Ω, what current does R_3 carry? If $\mathcal{E}_1 = 24$ V, $r_1 = r_2 = 1$ Ω, $R_1 = 5$ Ω, and $R_2 = 3$ Ω, find $\mathcal{E}_2$. *Ans.* 3 A; 33 V

20. In the accompanying figure $\mathcal{E}_1 = 42$ V and $\mathcal{E}_2 = 6$ V, $r_1 = r_2 = 0.5$ Ω, $R_1 = 3$ Ω, and $R_4 = 8$ Ω. The current in R_3 is 2 A, and that in the main circuit is 3 A. Find R_3 and R_2.

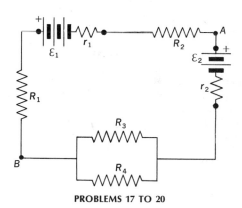

PROBLEMS 17 TO 20

21. Twelve identical cells, each with an internal resistance of 0.15 Ω, are connected so as to have three parallel groups of four cells in series. The combination sends a 1.25-A current through an external resistance of 7 Ω. What is the emf of each cell? *Ans.* 2.25 V

22. Twelve identical resistances of value R are connected to form the edges of a cube. If they are soldered together at the corners, find the effective resistance between opposite corners of one face of the cube.

23. In the accompanying figure, $I_2 = 2$ A, and

$I_3 = 6$ A. If $R_1 = 4\,\Omega$, $r_1 = 2\,\Omega$, $R_2 = 5\,\Omega$, $r_2 = 1\,\Omega$, $R_3 = 3\,\Omega$, and $R_4 = 5\,\Omega$, find $\mathcal{E}_1$ and $\mathcal{E}_2$.

Ans. 96 V; 36 V

24. In the accompanying figure, $I_1 = 4$ A and $I_2 = 1.5$ A. What is I_3? If $\mathcal{E}_1 = 50$ V, $r_1 = 1\,\Omega$, $\mathcal{E}_2 = 13$ V, $r_2 = 2\,\Omega$, $R_2 = 4\,\Omega$, and $R_4 = 5\,\Omega$, find R_1 and R_3.

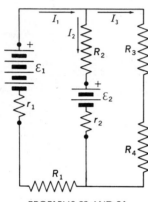

PROBLEMS 23 AND 24

25. In the accompanying figure $\mathcal{E}_1 = 64$ V, $r_1 = 1\,\Omega$, $R_1 = 5\,\Omega$, $\mathcal{E}_2 = 48$ V, $r_2 = 1.5\,\Omega$, $R_2 = 2.5\,\Omega$, and $R_3 = 6.667\,\Omega$. Find the current in each element in the circuit. *Ans.* 4 A; 2 A; 6 A

26. In the accompanying figure, $\mathcal{E}_1 = 64$ V, $r_1 = 0.5\,\Omega$, $R_1 = 20\,\Omega$, $\mathcal{E}_2 = 50$ V, $r_2 = 1\,\Omega$, $R_2 = 8\,\Omega$, and $R_3 = 4.6\,\Omega$. Find the current in each element in the circuit.

27. In the accompanying figure $R_3 = 20\,\Omega$ and $I_3 = 4$ A. If $\mathcal{E}_1 = 60$ V, $r_1 = 2\,\Omega$, $R_1 = 18\,\Omega$, $r_2 = 1\,\Omega$, and $R_2 = 3\,\Omega$, find I_1 and $\mathcal{E}_2$.

Ans. 1 A↓; 100 V

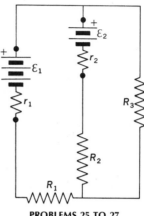

PROBLEMS 25 TO 27

28. The current in any element of a network (such as we have been considering) as the result of the simultaneous action of a number of emfs in the network is the sum of the currents which would exist in this element if each source of emf were considered separately, with all other sources of emf being replaced by their internal resistances. This is a statement of the *superposition theorem*. Find the currents in the circuit of Fig. 34.6 by applying this theorem.

In Chap. 34 we made use of the fact that batteries convert chemical energy into electric energy, but we made no effort to explain how this energy transformation was accomplished. We now treat some of the basic facts of electrochemistry, in terms of which the functioning of simple batteries can be understood. But batteries are not the only sources of emf in which we are interested. In this chapter we also discuss thermocouples, which convert thermal energy into electric energy.

35

CHEMICAL AND THERMAL ELECTROMOTIVE FORCES

35.1 ELECTROLYSIS

If two copper plates are inserted into a beaker filled with water and connected to the terminals of a battery (Fig. 35.1), an ammeter in the circuit shows no current; but if a little copper sulfate ($CuSO_4$) is poured into the beaker, there is a current through the solution. When the copper sulfate dissolves in water, many of the molecules split (or dissociate) into two charged particles called *ions*. One is a copper atom from which two electrons are missing, and the second is a sulfate (SO_4^{2-}) ion with two excess electrons. The net charge in atomic units carried by an ion is called the *valence* of the ion. Thus the valence of the Cu^{2+} ion is 2, while that of the sulfate ion is -2. In the solution the positively charged copper ions migrate to the negatively charged plate and the negatively charged sulfate ions to the positive plate.

After a current has passed between the two copper electrodes for some time, the *cathode* (the electrode by which the conventional current leaves) has gained weight and is bright. The *anode* (by which current enters) has lost weight. There has not only been a transfer of electricity through the solution, but copper has been carried from one plate to the other. The Cu^{2+} ions which reach the cathode obtain two electrons there and are deposited as neutral copper atoms, while each sulfate ion which goes to the anode gives it the two excess electrons and joins with an atom of copper. The copper sulfate so formed goes into solution, thus keeping the amount of $CuSO_4$ constant. The net effect is a gain of copper by the cathode and a loss of copper by the anode.

The addition of any acid, base, or salt to the pure water in the beaker makes it conducting, provided the solute dissociates into ions. When a current is passed between two platinum electrodes in a dilute solution of sulfuric acid in water (Fig. 35.2), hydrogen is released at the negative electrode and oxygen at the positive electrode. The water is decomposed into its constituents. The volume of hydrogen released is twice that of oxygen. A simplified explanation of the process by which the water is decomposed is as follows. A sulfuric acid molecule in solution splits into one H^+ ion and one HSO_4^- ion. The hydrogen ions are attracted to the cathode, where they receive electrons to form neutral hydrogen atoms. Two atoms promptly form a hydrogen molecule which rises to the top of the collecting tube. An HSO_4^- ion gives up an electron at the positive terminal and then unites with an atom of hydrogen to form sulfuric acid. These atoms of hydrogen are taken from the water, and oxygen is set free. Two atoms of oxygen unite to form a molecule of oxygen gas. The sulfuric acid formed at the anode goes into solution. The amount of sulfuric acid in the water does not change. The net result of passing current through the liquid is to decompose water into hydrogen and oxygen: $2H_2O \rightarrow 2H_2 + O_2$.

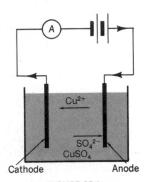

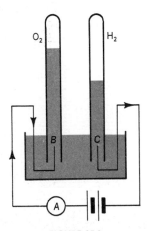

FIGURE 35.1
In the electrolysis of copper sulfate copper ions migrate to the cathode and sulfate ions to the anode.

FIGURE 35.2
The electrolysis of water produces hydrogen at the cathode and oxygen at the anode.

35.2 FARADAY'S LAWS OF ELECTROLYSIS

The quantitative laws of electrolysis may be introduced by considering a number of electrolytic cells connected in series (Fig. 35.3) with a battery so that the same current passes through each cell for the same time. Suppose cell *A* contains a solution of silver nitrate with silver electrodes, *B* a solution of sulfuric acid with platinum electrodes, *C* a solution of copper sulfate with copper electrodes, and *D* a solution of nickel chloride with nickel electrodes. Passage of a charge through the cells liberates, at the respective cathodes, silver, hydrogen, copper, and nickel. By determining the amount of substance liberated at each cathode as a function of the charge transported, it is possible to confirm two laws of electrolysis, which were discovered by Michael Faraday.

1. *The mass of any substance liberated is proportional to the charge which passes through the cell.* Hence, the mass liberated is proportional to the product of the current and the time.

2. *The masses of different elements liberated by a given charge are proportional to the ratios of atomic weight to valence.*

Suppose that the current exists until 1.008 g of hydrogen is liberated in cell *B*. Then, in cell *A*, 107.9 g of silver is deposited; in cell *C*, 31.77 g of copper; in cell *D*, 29.35 g of nickel. Whatever current is chosen, and whatever the length of time it exists, the masses deposited at these cathodes always bear the same ratio to each other and are proportional to the quotient obtained by dividing the atomic weight by the valence. The ratio of the atomic weight to the valence of an element is called the *chemical equivalent* or *combining weight.* When an element is monovalent, the chemical equivalent is equal to the atomic weight. If the element is divalent, the chemical equivalent is equal to one-half the atomic weight.

It requires 96,487 C to deposit one gram equivalent weight of any element, and 9.65×10^7 C to deposit one kilogram equivalent weight. A charge of 96,487 C is called one *faraday.* Faraday's two laws of electrolysis can be summarized by the relation

$$m = \frac{Q}{9.65 \times 10^7} \frac{A}{v} \tag{35.1}$$

where *m* is the mass of the element liberated in kilograms, *Q* is the total charge in coulombs, *A* is the atomic weight of the element, and *v* is its valence.

The mass deposited is directly proportional to the charge. *The electrochemical equivalent of any substance is defined as the ratio of the mass of the*

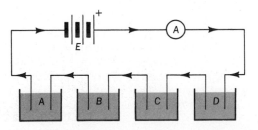

FIGURE 35.3
Four electrolytic cells connected in series.

substance deposited to the charge transferred. If Z represents the electrochemical equivalent, by Eq. (35.1)

$$Z = \frac{m}{Q} = \frac{1}{9.65 \times 10^7} \frac{A}{v} \tag{35.2}$$

One accurate method for measuring the electric charge which passes through a circuit is to insert in the circuit an electrolytic cell containing a solution of silver nitrate. The mass of silver deposited is a measure of the total quantity of charge passed. Indeed, for many years the ampere was defined as that current which would deposit 0.001118 g/s from a standard solution of silver nitrate. Since 1948 the ampere has been defined in terms of the interaction between two current-carrying conductors (Sec. 36.8).

35.3 APPLICATIONS OF ELECTROLYSIS

Electrolysis is of great commercial importance. The chromium plating of automobile parts and the silver plating of tableware are examples of the wide variety of commercial plating operations. Practically all our aluminum is produced by the electrolysis of aluminum oxide from a molten mixture. It takes 22 kWh of electric energy to produce a single kilogram of aluminum. Chlorine and many other commercially important elements are obtained by electrolytic processes.

By no means all electrolytic processes are desirable. Electrolysis is an important factor in limiting the life of underground pipes. When a steel pipe is laid near an electrified railroad, the current may find its way into the pipe instead of traveling from the generator to the motors directly through the track. At certain points the water pipe may be eaten away, like the anode in the electrolysis of copper sulfate. The electrochemical action is complex, but as part of the process iron is removed from the pipe in regions where it serves as anode. Electrolytic processes that lead to early disintegration must be carefully guarded against in many types of construction. Electrolytic action is a danger whenever different metals are electrically connected by a conducting solution. Vulnerable areas of some ships are protected by the use of "sacrificial" anodes; in one example an aluminum cable, charged positive relative to the hull, is dragged behind the ship to give electrolytic deposition on the ship in preference to the electrolytic removal of metal which would otherwise occur.

35.4 CHEMICAL ELECTROMOTIVE FORCES

Electric energy used to send electric current through a solution can produce chemical reactions such as the liberation of hydrogen and oxygen from water. The reverse processes, i.e., chemical reactions providing electric energy, also occur. This fact was discovered about 1800 by the Italian physicist Volta, who built the first batteries.

Most batteries are composed of several cells connected in series. In each cell a chemical reaction converts chemical energy into electric energy. One easily analyzed cell consists of a zinc plate and a lead plate immersed in a dilute solution of sulfuric acid (Fig. 35.4). Zinc ions leave the metal and go into solution. Each zinc ion bears two units of positive charge, having left two electrons with the metal. Thus, the zinc metal

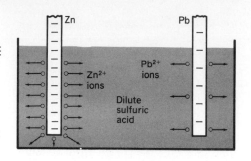

FIGURE 35.4
Diffusion of ions in a zinc–lead–sulfuric acid cell.

becomes negatively charged. As it becomes more negative, it attracts positive zinc ions in the solution. Eventually an equilibrium is reached in which the rate of loss of zinc ions is equal to the rate of return. When this occurs, the potential of the zinc is lower than the potential of the solution. Similarly, lead ions leave the lead plate and go into solution. They also bear two unit positive charges and leave the lead plate negative relative to the solution. However, equilibrium for the lead ions is established when the lead is less negative than the zinc. There is now a potential difference between the zinc plate and the lead plate, as shown in Fig. 35.5a. If a wire is connected between the zinc and lead plates, electrons flow through the wire from the zinc to the lead. When there is a current through the cell, there is a potential drop in the solution, owing to its resistance (Fig. 35.5b). As electrons leave the zinc plate, it becomes less negative; the equilibrium is disturbed, so that more zinc ions go into solution. At the same time the lead plate becomes more negative, and positive ions are attracted back.

Chemical cells can be made from a large number of different materials. The emf depends on the particular materials involved. Generally speaking, the chemically more active metal forms the negative terminal. Table 35.1 shows the electromotive series of metals. It indicates the potentials of various metals relative to a hydrogen electrode formed by bubbling hydrogen gas over a spongy platinum conductor. The values in the table are for standard ion concentrations.

An *ideal* cell utilizing any two of these elements as electrodes has an emf given by the difference between the voltages listed in Table 35.1 for the elements in question. For example, the potential of the copper-zinc cell can be predicted by observing that zinc, relative to hydrogen, has a potential of -0.76, while copper has a potential of $+0.34$. A copper-zinc cell has an emf of approximately 1.10 V.

The emf of a cell is determined primarily by the energy released in the

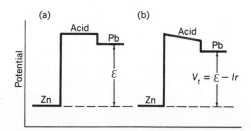

chemical reactions per coulomb of charge transferred through the cell. However, it would be erroneous to assume that the only energy transformations occurring in a cell are between chemical and electric energy. When some cells are discharged, part of the chemical energy is transformed into heat; when such cells are charged, this heat is retransformed into chemical energy. Other cells transform heat energy into electric energy during discharge and electric energy into heat energy when they are being charged.

Major research programs to produce new and improved batteries were initiated in the early 1970s in an effort to meet new demands. One objective is a battery which is satisfactory for powering electric road vehicles. A second major use envisioned for new batteries is to store electric energy generated by utilities during off-peak hours and to store energy obtained from solar sources. The fact that fossil fuels are becoming scarcer and more expensive has stimulated intense interest in large-capacity batteries. Meantime the demand for small, light-weight, long-life batteries has increased because they are used in hand-held calculators, semiautomatic cameras, hearing aids, and various other solid-state devices.

35.5 POLARIZATION OF CELLS

If a zinc strip and a copper strip are inserted into a juicy lemon, the juice serves as electrolyte and a potential difference can be observed between the zinc and the copper. Another simple cell can be made by immersing a zinc plate and a copper plate in a dilute solution of some acid such as H_2SO_4. Neither of these zinc-acid-copper cells performs very satisfactorily as a source of emf because, as current is drawn, hydrogen ions are deposited on the copper. This reduces the emf, since the positive electrode is now essentially hydrogen, rather than copper. Further, hydrogen bubbles on the surface of the plate form an insulating layer and greatly increase the resistance of the cell. This is an example of an effect called *polarization* of a cell. (This should not be confused with polarization of a dielectric or with polarization of light.) Polarization can occur in many ways in a cell, but its net effect is to reduce the observed emf to a value below that expected on the basis of the electrochemical series. This occurs because one or both of the terminals is coated with some less effective material.

35.6 PRIMARY CELLS

In most cases the chemical reaction in a cell cannot be reversed by changing the direction of the current. A cell in which the chemical reaction is irreversible is known as a *primary* cell, while one which can be charged and discharged repeatedly and reversibly is called a *secondary* cell. The simple zinc-acid-copper cell discussed in Sec. 35.4 is a primary cell. If one attempts to reverse the chemical process by making the zinc electrode positive, zinc ions do not plate out. Instead copper and hydrogen ions are deposited, and the zinc stays in solution. The reaction is not reversible.

Next we consider the chemical reactions which take place in several kinds of cells of practical or historical importance.

TABLE 35.1
The Electromotive (or Electrochemical) Series

Element	Potential difference,† V
Li	−2.96
Rb	−2.93
K	−2.92
Ca	−2.76
Na	−2.71
Mg	−2.40
Al	−1.70
Zn	−0.76
Fe	−0.44
Cd	−0.40
Ni	−0.23
Sn	−0.14
Pb	−0.12
H	0
Cu	0.34
Ag	0.80
Hg	0.80
Au	1.5

†All potential differences are referred to a standard hydrogen electrode taken as zero and are for a temperature of 25°C.

The Daniell Cell   One of the earliest practical sources of emf was the Daniell cell. It consists of a copper electrode immersed in copper sulfate and a zinc electrode in zinc sulfate. A porous partition separates the zinc sulfate from the copper sulfate so that charges can pass through but the chemicals do not mix readily. The zinc electrode becomes negative and the copper plate positive. When the plates of the Daniell cell are connected through a resistance, electrons flowing through the wire reduce the negative charge on the zinc terminal and permit more zinc to go into solution. The potential of the copper electrode is reduced, and copper ions are deposited from the solution. The net reaction converts zinc and copper sulfate into zinc sulfate plus copper. In this reaction chemical energy is transformed into electric energy.

In the early days of telegraphy the Daniell cell was the standard source of emf. Its output is about 1.1 V, but the terminal potential difference of a cell is somewhat less when current is drawn. The *volt* as the practical unit of potential difference originated with the use of Daniell cells in early telegraph circuits, when the "voltage" was just the number of Daniell cells connected in series.

Dry Cells   The Leclanché cell (Fig. 35.6) is a primary cell of interest because a modification, known as the *dry cell*, is widely used. A Leclanché dry cell consists of a zinc can containing a centered graphite (carbon) rod surrounded by a moist paste of manganese dioxide (MnO_2), zinc chloride ($ZnCl_2$), ammonium chloride (NH_4Cl), and carbon black. Manganese dioxide is a poor conductor, and the carbon black serves to reduce the cell resistance. A thin porous separator prevents direct electric contact between the zinc and the mixture of manganese dioxide and carbon black. The term "dry cell" originates from the fact that the electrolyte is absorbed in the porous materials. As the cell delivers current, zinc is oxidized at the negative terminal and Zn^{2+} ions go into solution. At the positive electrode MnO_2 is reduced through complex reactions which seem to vary depending on how much current is drawn from the cell. In the discharge of NH_4^+ ions at the inert carbon electrode, hydrogen is released in the reaction $2NH_4^+ + 2e^- \rightarrow 2NH_3 + H_2$, where e is the abbreviation for electron. The MnO_2 reacts with the hydrogen to produce water and perhaps Mn_2O_3. The NH_3 is immediately absorbed in the water. When too large a current is drawn, hydrogen accumulates rapidly and the cell becomes polarized with the terminal potential difference falling and the cell resistance rising. After a rest period the cell may regain its normal emf of about 1.5 V. Such a cell is well adapted to work in which it is used a short time and then allowed to stand.

The *magnesium dry cell* is constructed in a similar way except that the outer can is magnesium and the electrolyte includes magnesium bromide rather than ammonium chloride. The *mercury dry cell* has a zinc negative electrode, a positive electrode of solid mercuric oxide (HgO) mixed with carbon, and an electrolyte of potassium hydroxide (KOH) and zinc oxide. The overall reaction $Zn + HgO \rightarrow ZnO + Hg$ gives an emf of 1.35 V. This cell is widely used in cameras and hearing aids because it provides more energy and has lower internal resistance than a Leclanché cell of the same weight. A further advantage is that it will operate at lower temperature, so it is used in some missile and aviation applications.

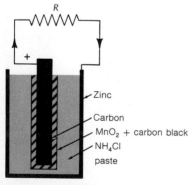

FIGURE 35.6
A Leclanché dry cell.

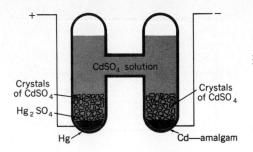

FIGURE 35.7
A Weston mercury-cadmium standard cell.

Weston Standard Cell Standard cells offer a means of obtaining definite, known, and constant potential differences. The most widely used of these standard cells is the cadmium, or Weston, cell (Fig. 35.7). Its emf changes very little with temperature, which is one of the reasons the Weston cell is considered the best standard available. At 20°C its emf is 1.0183 V, and the emf decreases about 40 μV for each Celsius degree of temperature increase.

35.7 SECONDARY CELLS

A few of the many possible chemical reactions which can be used for batteries are reversible. The lead storage cell and the Edison cell are the best known among those which can be discharged and recharged repeatedly.

Lead Storage Cell In the lead storage cell, both the positive and negative plates are made of heavy lead grids full of holes or grooves filled with the active material. The positive plates contain lead peroxide and the negative plates spongy lead. A cell is usually formed of a number of such plates, alternately negative and positive, covered with sulfuric acid. The negative plates are connected together and act as one terminal of the cell, and the positive plates are connected together to form the other terminal. It is customary to have one more negative plate than positive so that every positive plate lies between two negative ones. In this way, both sides of a positive plate are charged or discharged. During the process of recharging, there is a restoration of the peroxide accompanied by an increase in volume, causing a swelling of the plate. Since this swelling takes place equally on both sides of a positive plate, there is little tendency for the plates to warp. If the plates are close together and if a large number of plates are used so that the area is large, the cell has a small internal resistance. By increasing the number of plates and making the areas larger the current capacity of the cell is increased.

The overall chemical reaction taking place in the discharge of the lead storage cell is

$$Pb + PbO_2 + 2H_2SO_4 \rightarrow 2PbSO_4 + 2H_2O$$

When the cell is delivering energy, both the lead peroxide on the positive plates and the spongy lead of the negative electrode are converted into lead sulfate. As this occurs, the emf gradually decreases and the density of the electrolyte falls (Fig. 35.8). Consequently, it is possible to find the

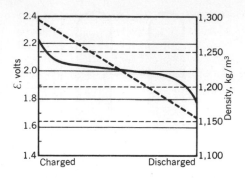

FIGURE 35.8
As a lead storage cell is discharged, the emf (solid line) and the density (dashed line) of the sulfuric acid both decrease.

state of charge of a battery by measuring the density of the sulfuric acid. When the battery is charged, the reactions are reversed.

Edison Storage Cell A storage battery composed of Edison cells is lighter, more rugged, and longer-lived than a lead storage battery. In the Edison cell the negative plate is a nickel-plated steel grid with a large number of pockets filled with powdered iron and iron oxide (FeO). The positive plate is a nickel-plated steel grid with perforated steel tubes filled with alternate layers of nickel oxides and flaked nickel. The electrolyte is a solution of potassium hydroxide. When the cell is discharging, the nickel oxides are reduced to lower oxides and the iron is oxidized. When the cell is being charged, the reaction is reversed. The electrolyte enters into intermediate reactions. Its effect in charging and discharging is to transfer oxygen from one plate of the cell to the other. The density changes only slightly during the reactions. The normal emf of an Edison cell is about 1.2 V, varying from 1.4 to 0.9 V as the cell discharges. This variation of emf and the relatively high cost are major disadvantages of Edison cells.

Other types of secondary cells have considerable use. The *nickel-cadmium-alkaline cell* gives a nominal potential difference of 1.34 V from the oxidation of cadmium and the reduction of nickel in $Ni(OH)_4$. The *zinc–silver oxide–alkaline cell* uses a potassium hydroxide electrolyte and has an emf of 1.86 V. In it zinc is oxidized while silver oxide is reduced to silver.

35.8 FUEL CELLS

In a battery chemical energy can be converted into electric energy until one or more of the active materials is used up, at which point the battery is ready for discard or, possibly, recharge. In a fuel cell active materials are supplied, and the reaction products removed, continuously; thus the fuel cell can operate as long as the required materials are provided.

In one type of fuel cell hydrogen gas and oxygen are supplied, and energy is released by the oxidation of hydrogen to form water; the process is the reverse of that which occurs in the electrolysis of water. In other kinds of fuel cells chemical energy from the oxidation of various hydrocarbons is converted into electric energy. A major advantage of the fuel cell is its high efficiency—vastly greater than that available when the fuel is burned in a heat engine and then converted to electric energy, chiefly because of the low Carnot efficiency (Sec. 18.6). A hydrogen-oxygen fuel cell at 25°C has a theoretical peak efficiency of 83 percent; other fuel cells have still higher theoretical efficiencies, and practical fuel cells can convert 60 to 70 percent of chemical energy to electric energy.

Although Grove produced the first successful fuel cell in 1839, it was not until the space program blossomed in the last half of this century that fuel cells became important sources of electric energy.

35.9 THERMOELECTRICITY: THE SEEBECK EFFECT

When two dissimilar materials are placed in intimate contact, electrons diffuse from the one with the higher Fermi level (Sec. 30.2) to the material with the lower Fermi level until the levels coincide. In electrostatics this phenomenon permits us to produce charges by contact and separation. In batteries the potential differences between metals and solutions give rise to the emfs. When two metals are placed in contact, the potential difference between them depends not only on the metals but also on the temperature of the junction.

If two wires of dissimilar metals, e.g., copper and iron, are joined together at the ends to make a closed circuit and one of the junctions thus formed is maintained at a temperature different from that of the other (Fig. 35.9), an electric current is established in the circuit. This effect was discovered in 1821 by Seebeck. The magnitude of the net emf producing the current depends on the two kinds of wire and on the temperatures of the two junctions. If the junctions are at the same temperature, the emfs established are equal and opposite, so that there is no net emf.

Consider first a thermocouple made of iron and copper (Fig. 35.9). If one of the junctions is kept at 0°C while the other is heated, a net thermoelectromotive force is produced. When the temperature of the hot junction is raised, the thermal emf first increases at a nearly uniform rate (Fig. 35.10). As the temperature is raised further, the rate of increase becomes less. When the temperature of the hot junction reaches 260°C, the emf reaches its maximum value and a further increase in temperature results in a decrease in the thermal emf. The temperature at which the emf reaches its maximum is the *neutral temperature,* so named because there is neither increase nor decrease of thermal emf with temperature. When the temperature of the hot junction is raised above the neutral temperature, the thermal emf decreases and finally becomes zero at approximately 520°C. If the temperature of the hot junction is still further increased, the current in the circuit reverses direction. The tem-

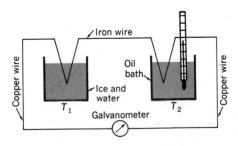

FIGURE 35.9
A thermal emf exists when the temperatures of the two copper-iron junctions are different.

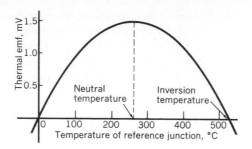

FIGURE 35.10
The thermal emf of a copper-iron thermocouple as a function of the temperature of the hot junction when the cold junction is at 0°C. The neutral temperature, the peak emf, and the inversion temperature depend to some extent on the purity of the metals.

perature at which the thermal emf passes through zero is called the *inversion temperature*. The curve obtained by plotting thermoelectromotive force as a function of the temperature of one junction (with the second junction kept at constant temperature) is parabolic.

For some materials, such as chromel and alumel, the emf increases continuously as the temperature of the hot junction is raised. For these materials the neutral temperature and the inversion temperature are reached by cooling the variable-temperature junction rather than by warming it.

Thermocouples are convenient for measuring temperatures, since the emf depends on the temperature difference between the junctions in a known way. Ordinarily, one junction is kept at a constant temperature, often in an ice bath at 0°C. The hot junction need not be anywhere near the cold junction. It can be installed at some inaccessible point, where it would be impossible to place and read a standard mercury thermometer. Thermocouples can be used for measuring temperatures roughly up to the melting points of the materials involved, which may be as high as 2900°F for some couples. Further, they can be made much smaller than glass or metal expansion thermometers, so that they can be used to measure the temperatures of small objects, such as insects or twigs of trees.

It is likely that one would want to make connection to a galvanometer by means of copper wires fastened to the ends of the thermocouple elements. Adding other materials in the circuit has *no effect so long as the junctions of the other metals are all at the same temperature.* Electromotive forces exist at the new junctions, but the algebraic sum of the emfs in the circuit remains the same. The lengths of the wires have no effect on the emf (although they do influence the resistance of the circuit).

The emfs involved in the Seebeck effect are ordinarily small, usually of the order of millivolts. To obtain higher emfs, we can connect a number of thermocouples in series, thus producing a *thermopile*. Thermopiles are often used for measuring radiant energy.

35.10 THE PELTIER AND THOMSON EFFECTS

In 1834 Peltier discovered that if a battery is connected in series with a thermocouple to produce a current in the circuit, one junction of the thermocouple is cooled while the other junction is heated. The cooling takes place in spite of the fact that some Joule heat is released all through the circuit. To explain the origin of this *Peltier effect* we recall that when two dissimilar metals are placed in contact, electrons diffuse from the metal with the higher Fermi level to the one with the lower Fermi level until the levels coincide (Sec. 30.3). The resulting potential difference, called the *Peltier emf*, depends on the metals in contact and upon the temperature of the junction. When the two junctions of a thermocouple are at the same temperature, the Peltier emfs are equal and opposite.

When a battery sends a current through a thermocouple, the Peltier emf at one junction aids the battery, while the Peltier emf at the second junction is in series opposing. At the junction where the Peltier emf is in opposition, the work done by the charges against the emf is transformed into internal energy and the temperature rises. At the junction where the Peltier emf is in series aiding, thermal energy is absorbed and converted into electric energy. As a consequence, the junction temperature falls even though a small amount of Joule heat is deposited there. Thus, if a current is sent through a thermocouple, heat is released at one junction and its temperature rises, while heat is absorbed at the other junction. The resultant cooling may be of the order of 80 C° for some metal-semiconductor junctions.

If the two ends of a wire of any material are at different temperatures, electrons are likely to diffuse more rapidly in one direction than in the other. This unequal diffusion produces a potential difference called the *Thomson emf*. The Seebeck effect, discussed in Sec. 35.9, arises as the result of Thomson emfs between opposite ends of wires of the same material which are at different temperatures and Peltier emfs at the junctions of the metals. Usually the Peltier emfs greatly exceed the Thomson emfs.

QUESTIONS

1. How does the mechanism of conduction in a sodium chloride solution differ from that in a copper wire?
2. Eight dry cells connected in series have an emf of 12 V. Could they be used to replace the 12-V lead-acid battery in an automobile? Explain.
3. What is the source of the electric energy supplied by a chemical cell?
4. Why should one never hold a lighted match near a storage battery?
5. A strip of copper and a strip of zinc are inserted in a lemon and connected by a wire. Why do electrons go to the zinc in the lemon while they go away from the zinc in the wire?
6. What factors determine the electromotive force of a battery?
7. What are the advantages of fuel cells over batteries? The disadvantages?
8. How does one get rid of the various metallic impurities in the electrolytic purification of copper?
9. A chromel-alumel thermocouple is connected to a galvanometer by copper wires. Under what condition does the presence of the copper wires not change the net emf in the circuit?
10. Explain why thermocouples and thermopiles are not more widely used as practical generators of electric energy.
11. Lithium-sulfur and sodium-sulfur batteries are being developed. What advantages are such batteries going to have over the conventional lead storage battery?

PROBLEMS

1. A copper plate of 107.392 g mass is placed in an electroplating bath. A steady current is sent through the bath for 30 min, and the mass of the plate is increased to 109.074 g. What total charge passed through the bath and what was the current in amperes, assuming copper is divalent?
   Ans. 5,108 C; 2.84 A
2. How long will it take for a current of 5 A to plate 1.2 g of silver on a knife?
3. An object that has a surface of 98 cm² is to be plated with silver. What will the average thickness of the silver be if a current of 0.5 A flows for 2 h? *Ans.* 39.1 μm
4. Find the electrochemical equivalent for lithium (monovalent), zinc (divalent), and indium (trivalent).

5. Find the electrochemical equivalent for sodium (monovalent), trivalent iron, and divalent cadmium.

   Ans. 0.238 mg/C; 0.193 mg/C; 0.582 mg/C

6. An electrolytic cell containing a solution of silver nitrate is connected in series with a cell containing copper sulfate. What mass of Ag^+ ions is deposited on the cathode of the first cell by a charge which deposits 2.5 g of Cu^{2+} ions on the cathode of the second cell? What charge has passed through the two cells?

7. A current of 2.25 A flows for 27 min through a series of cells containing nickel nitrate, copper sulfate, and silver nitrate. Find the masses of nickel (valence 2), copper (valence 2), and silver (valence 1) deposited. *Ans.* 1.11 g; 1.20 g; 4.07 g

8. A current of 1.6 A deposits 4.69 g of monovalent element in 24 min. Find the electrochemical equivalent for the element. What is the element?

9. Find the volumes of hydrogen and oxygen released under standard conditions (0°C; 1 atm pressure) by the electrolysis of acidulated water for a period of 500 s with a current of 2.4 A.

   Ans. 139 cm^3; 69.6 cm^3

10. An aluminum plant produces 5,000 kg/day by electrolysis in which trivalent Al ions are deposited. How many coulombs must pass through the electrolytic cells each day?

11. A battery storing at least 50 kWh of energy is needed to make certain electric-powered automobiles practical. Find the mass required for (*a*) a conventional lead-acid battery which stores 30 Wh/kg (the energy stored per unit mass is called the specific energy of the battery); (*b*) an improved lead-acid battery with a specific energy of 50 Wh/kg; (*c*) an improved nickel zinc with a specific energy of 70 Wh/kg; (*d*) a lithium-sulfur battery with a specific energy of 150 Wh/kg.

    Ans. (*a*) 1,670 kg; (*b*) 1,000 kg; (*c*) 714 kg; (*d*) 333 kg

Electric charges at rest exert forces on each other through coulomb interactions. Charges in motion also interact with each other through magnetic fields. Currents, of course, are charges in motion, and so magnetic interactions are of great importance whenever we deal with currents. Most of our meters, our electric motors, and the great generators of our electric companies depend on magnetic interactions for their operation. Before we can study the physics underlying these devices, we must gain some knowledge of magnetism and magnetic fields.

36

MAGNETIC FIELDS OF CURRENTS

36.1 MAGNETS

A piece of the mineral *magnetite* (Fe_3O_4), sometimes called a *lodestone*, has the ability to attract other pieces of the same mineral or small bits of iron. This magnetic iron ore is found in Magnesia in Asia Minor; the word "magnet" comes from the Greek for "stone from Magnesia." How early the remarkable properties of lodestones were discovered we do not know, but Thales of Miletus (624–546 B.C.) made the first known study of magnetism. Lodestones turn with the same side toward the north when they are suspended from cords or floated on corks in water. This property led to the invention of the mariner's compass before the twelfth century.

When a steel knitting needle is stroked from one end to the other with a piece of lodestone, the needle may acquire the property of attracting iron filings and of setting itself along a north-south line when suspended by a string. Such a needle is said to be *magnetized* and is commonly called a *magnet*. Most materials cannot be magnetized in this way; relatively few show attraction for a lodestone or a small magnet. Iron and some of its alloys are by far the best known of magnetic materials; cobalt, nickel, and a number of alloys also exhibit prominent magnetic properties.

One end of a magnetized needle suspended by a cord so that it is free to rotate in any direction normally points in a northerly direction. This end of the needle is commonly called the *north pole* (N pole), an abbreviation for the more fully descriptive *north-seeking pole*. The opposite end of the needle is called the *south pole* (S pole).

If this magnetized needle is dipped into soft-iron filings, the filings cling tenaciously to the ends; relatively few stick to the middle. It is often convenient to think of the magnetic properties of the needle as concentrated in the two ends, or poles, although this is an oversimplification.

When two magnetized needles are brought near each other, the two north poles repel each other, as do the two south poles. On the other hand, there is an attractive force between the north pole of one magnet and the south pole of the other. Such observations lead to the conclusion that *like poles repel, unlike poles attract.*

36.2 INTERACTIONS OF CURRENTS AND MAGNETS

In 1819 Oersted discovered that a magnet in the neighborhood of a current-bearing wire undergoes a deflection. A magnetized needle held above a straight wire carrying a current is deflected as shown in Fig. 36.1. If the magnet is held below the wire, the north pole is deflected in the opposite direction. The direction in which a north pole points can always be found by the application of a simple rule, illustrated in Fig. 36.2: *If the*

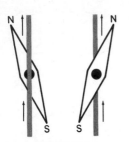

FIGURE 36.1
Magnetic field of a current deflecting a compass needle.

right thumb is pointed in the direction of the conventional current and the fingers are allowed to curl, the direction in which the fingers point is the direction in which the north pole of the needle is deflected by the current.

If current is passed through a cylindrical coil of wire (called a *solenoid*) produced by winding fine wire on a matchstick, the solenoid behaves like a magnet of the same shape. While current is passing through the solenoid, it will pick up small iron filings. It has a north and a south "pole," which exert forces on other poles. The fact that a current-bearing solenoid behaves like a magnet of the same dimensions led Ampère to suggest in 1820 that the forces between iron magnets arise from electric currents in the iron. This idea has been substantiated by later research and is discussed in Chap. 37.

Magnetic interactions occur between moving charges, whether they are currents in wires, a beam of electrons in a television tube, or the electrons in magnetic materials. The forces so produced are of tremendous practical importance. Electric motors depend on them, as do most of our meters for measuring electric currents and potential differences.

36.3 THE MAGNETIC FIELD

A magnetic field is any region in which forces can be observed to act on small magnets or on moving electric charges. *The direction of the magnetic field at any point is the direction of the force experienced by the north-seeking pole of a small test magnet when the test magnet is placed at the point.* A small current-bearing solenoid can be substituted for the test magnet.

One can readily explore the magnetic field associated with a large bar magnet by using a small compass needle to find the direction in which the north pole points at various places in the field. A plot of the magnetic field associated with a magnet is shown in Fig. 36.3*a*. In discussing magnetic fields, it is convenient to make use of magnetic lines of force, defined as lines whose tangents give the direction of the magnetic field at every point. It is customary to draw the lines close together where the field is strong and farther apart where the field becomes weaker. The lines of force for a bar magnet are similar to those for a suitably chosen current-bearing solenoid (Fig. 36.3*b*). For the magnetic field associated with a straight current-bearing conductor (Fig. 36.2) the lines of force are circles concentric with the wire.

The magnetic field at any point is characterized not only by a direction but also by a strength. The vector which describes the field at a point is called the *magnetic induction, magnetic flux density,* or *magnetic intensity.* It

FIGURE 36.2
When the outstretched right thumb is pointed in the direction of the conventional current, the fingers of the right hand curl in the direction of the magnetic field, which is the direction of the force on the north pole of a magnetic needle.

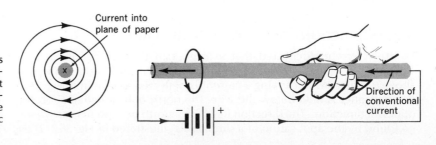

Current into plane of paper

Direction of conventional current

is represented by the symbol **B**. To measure **B**, we make use of the fact that there is a force on a current element in a magnetic field.

Consider first the special case of the uniform magnetic field between the pole pieces of the large magnet of Fig. 36.4. If a straight wire bearing a current *I* is placed perpendicular to the lines of magnetic force, and if *l* represents the length of the part of the wire in the magnetic field, there is a force **F** on the wire in the direction indicated. We define the magnetic intensity **B** with the equation

$$B = \frac{F}{Il} \tag{36.1}$$

In practical units **F** is measured in newtons, *I* in amperes, and *l* in meters. Then **B** has the dimensions *newtons per ampere-meter*, called *teslas* (T) in honor of Nikola Tesla.

In the preceding discussion we stated that the wire must be perpendicular to the direction of the magnetic field. This is very important; indeed, if the wire lies along the lines of force, there is no force at all. If the wire makes an angle θ with the field, the force is given by

$$F = IlB \sin \theta \tag{36.2}$$

Many magnetic fields are not uniform but vary from point to point. The field intensity **B** at any point is the limit of the ratio $\Delta F / I \Delta l$ as Δl is made smaller and smaller. Here ΔF represents the small force on a small element of length Δl. The result of using any finite length *l* of wire is to measure the average intensity over the region covered by *l*. In actual practice it is often difficult to determine the force on a small element of a current-bearing conductor; therefore other means are ordinarily used to measure **B**.

The force on a current-bearing wire is perpendicular both to the length of the wire and to the magnetic field. For the particular situation of Fig. 36.4 the direction of this force can be found as follows. The lines of flux associated with the current are concentric circles, as indicated. The magnetic field due to the current in the wire exerts a force on the north pole of the magnet in the direction opposite that indicated by **F**. Similarly, the force on the south pole due to the current is out of the paper (opposite to **F**). Since the current in the wire pushes outward on the magnet, the magnet exerts a force inward on the current-bearing wire in accordance with Newton's third law.

The origin of some magnetic fields, e.g., that of the earth, may not be obvious. In such a case one can find the direction of the force on the wire by observing that on one side of the wire the magnetic field due to the current (Fig. 36.5) reinforces the field already present while on the other side of the wire the two fields are opposite in direction. The force on the wire is directed from the stronger field toward the weaker one, or from the side on which the two fields reinforce to the side on which the two fields oppose.

36.4 MAGNETIC FORCE ON A MOVING CHARGE

The force exerted on a current-bearing conductor in a magnetic field is the resultant of the forces which act on individual moving charges in the conductor. Let us consider a conductor in which there are *n* charged

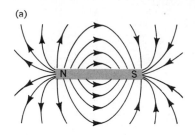

(a)

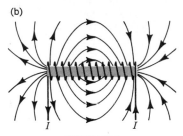

(b)

FIGURE 36.3
The magnetic fields are similar around (a) a bar magnet and (b) a current-bearing solenoid of similar shape.

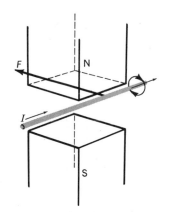

FIGURE 36.4
A current-bearing conductor lying perpendicular to a uniform magnetic field experiences a force which is perpendicular to both the magnetic field and its own length. (**B** is directed downward.)

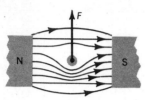

FIGURE 36.5
The force on a current-bearing conductor in a magnetic field is from the stronger toward the weaker resultant field.

particles per unit volume, each with charge $+q$ and velocity v perpendicular to a magnetic induction B (Fig. 36.6). The current through the plane PP' is the charge passing per second, which is

$$I = qnAv$$

where A is the cross-sectional area of the wire. The force on a length l of wire is out of the plane of the paper and is given by $IlB = qnAlvB$. Since the number of charges in the length l is nAl, the force on each particle is

$$F = qvB \qquad (36.3)$$

Equation (36.3) gives the force on a single charged particle moving with a velocity v perpendicular to a magnetic field of intensity B. But what is the force if the particle is not moving at right angles to **B**? In this case we resolve the velocity **v** into components $v \sin \theta$ perpendicular to **B** and $v \cos \theta$ parallel to **B**, where θ is the angle between **v** and **B**. There is no force exerted by a magnetic field on a charge moving parallel to the field; only the component of the velocity perpendicular to **B** contributes to the resulting force, which is given by

$$F = qvB \sin \theta \qquad \text{[in vector notation } \mathbf{F} = q(\mathbf{v} \times \mathbf{B})] \qquad (36.3a)$$

The force is perpendicular to the plane defined by **B** and **v**. Equation (36.3a) is one of the basic equations of charged-particle physics.

The deflection of charged particles by magnetic fields has many uses. The electron beam which sketches the pictures on a television tube is usually directed by magnetic forces. In cyclotrons, betatrons, and synchrotrons, accelerated charged particles are restrained to roughly circular paths by magnetic fields. Electron microscopes and mass spectrographs use magnetic forces to control beams of charged particles. Some of these important instruments are described in later chapters.

36.5 THE MAGNETIC MOMENT OF A COIL

Consider a rectangular loop of wire, bearing a current I, placed in a uniform magnetic field of intensity B with one side of the rectangle perpendicular to the field and another side lying along the field (Fig. 36.7). Let y represent the length of the side perpendicular to the magnetic field and x the length of the side parallel to the magnetic field. Under these circumstances there are no forces on the sides parallel to **B**, but the force on each of the wires perpendicular to the field is given by $F = BIy$. The lever arm about the axis OO' for one of the vertical sides is $x/2$, and the torque τ on this one side is given by $\tau = BIyx/2$. The net torque acting to rotate the loop is twice this, or $BIyx$. The product yx is

FIGURE 36.6
The force on a current-bearing conductor in a magnetic field results from the forces acting on the individual moving charges.

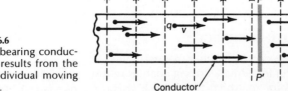

Conductor

equal to the area A of the rectangle. The torque τ on the loop is thus

$$\tau = BIA \qquad (36.4)$$

Current-bearing coils and magnetized materials in various shapes also experience torques in a magnetic field. The concept of magnetic moment, which we introduce now, is applicable to any object which is subject to a torque when it is placed in a uniform magnetic field. *The magnetic moment M of any object is the ratio of the torque experienced by the object when it is placed in a uniform magnetic field B in such a position that the torque takes on its maximum value to the magnetic induction B;* thus, $M = \tau_{max}/B$.

For the coil of Fig. 36.7 the torque, maximum when the plane of the coil is parallel to the magnetic field B, is BIA, and hence the magnetic moment is IA. By use of calculus it can be shown that Eq. (36.4) is applicable to a plane loop of any shape.

The magnetic moment M of a current-bearing loop is the product of the current I and the area A of the loop. The magnetic moment is a vector perpendicular to the plane of the coil in the direction in which a right-handed screw would advance if turned in the direction of the current I. It has the dimensions A-m². The magnetic moment associated with a current-bearing loop is shown in Fig. 36.8. The torque τ on a loop of magnetic moment M is given by BM when M is perpendicular to B. When M and B are parallel, the torque is zero. In general, if the angle between M and B is θ, the torque is given by

$$\tau = BM \sin \theta \qquad (36.5)$$

If there is no opposing torque, the loop (Fig. 36.7) rotates under the influence of the magnetic field until the magnetic moment is aligned in the direction of the field.

Thus far we have been discussing single loops. If a coil made by winding N turns in series bears a current I, each turn has a magnetic moment of magnitude IA and the resultant magnetic moment is N times that of a single loop. Thus, for a coil of N turns, $M = NIA$.

36.6 THE MOVING-COIL GALVANOMETER

If a small current is passed through a coil suspended between the poles of a permanent magnet (Fig. 36.9), the coil rotates until the restoring torque exerted by the suspension is equal to the torque due to the interaction of the current and the magnetic field. This latter torque is

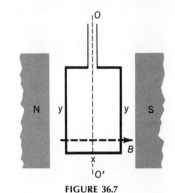

FIGURE 36.7
A rectangular loop of wire bearing a current is subject to a torque in a magnetic field unless its plane is perpendicular to the magnetic lines of force. The torque is maximum when the coil is in the position shown above.

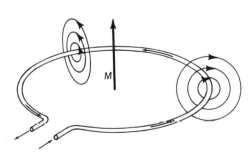

FIGURE 36.8
The magnetic field and the magnetic moment **M** associated with a current-bearing loop.

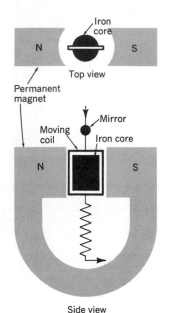

Side view
FIGURE 36.9
Schematic diagram of a moving-coil galvanometer.

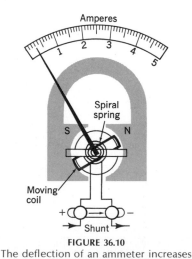

FIGURE 36.10
The deflection of an ammeter increases as the current through the coil rises.

proportional to the current. Therefore, the angular deflection of the coil is proportional to the current, provided the restoring torque is proportional to the angular displacement and the magnetic field is perpendicular to **M** for all displacements. This latter condition is reasonably well satisfied for angles up to 30° for the magnet of Fig. 36.9. Thus, a single loop of wire suspended by an elastic suspension in a magnetic field can be used to detect and measure currents. The sensitivity of the instrument can be increased by adding turns to the coil. Such a coil, suspended in a magnetic field, is called a *galvanometer* when it is used to detect and measure small currents. A light pointer or mirror attached to the coil can be used to measure the angular deflection.

36.7 AMMETERS AND VOLTMETERS

When a galvanometer and its scale are so adjusted that the scale readings indicate the current passing through some portion of a circuit in amperes, the meter is called an *ammeter*. A common form of ammeter is shown in Fig. 36.10. It consists of a coil of fine copper wire wound on a light frame which is mounted on jeweled bearings between the poles of a permanent magnet. When a current exists in the coil, it rotates between the poles of the magnet. Two spiral springs, one at the top and the other at the bottom, carry the current into and out of the coil and provide a restoring torque proportional to the angular displacement.

Since a small fraction of an ampere through the coil of an ordinary ammeter produces a full-scale deflection, it is necessary to use a low-resistance *shunt* to carry the remainder of the current. (A shunt is a resistance connected in parallel with a circuit element.) Because the current flowing in the movable coil is always a constant fraction of the full current entering the instrument, the scale can be calibrated so that the pointer indicates the entire current. Any galvanometer can be made to operate as an ammeter by use of a suitable shunt. By choice of some other shunt, the galvanometer can be made into an ammeter of different range.

A galvanometer can also be made into a *voltmeter* by the proper application of an additional resistance (Fig. 36.11). A voltmeter, as its name implies, is an instrument used to measure potential difference. By connecting a high resistance in series with the galvanometer coil, the current passing through the coil can be limited to a value which will not exceed full-scale deflection. Since for a given resistance the current is directly proportional to the potential difference across the instrument, the deflection of the pointer is proportional to the potential difference and the scale can be calibrated to read directly in volts. By suitable choice of the series resistance one can make a voltmeter which gives full-scale deflection for any desired potential difference across its terminals.

□ **Example**   A galvanometer requires 0.00015 A to produce a full-scale deflection. The coil has a resistance of 60 Ω. What shunt resistance is needed to convert this galvanometer into an ammeter reading 2 A full scale?

Of the 2 A which enter the ammeter for a full-scale deflection, 0.00015 A must pass through the galvanometer, and the remainder (2 − 0.00015 = 1.99985 A) through the shunt. Since the shunt and galvanometer are in parallel, their potential differences are the same.

$$V_G = V_S \quad \text{and} \quad I_G R_G = I_S R_S$$

$$0.00015 \times 60 = 1.99985 R_S$$

$$R_S = \frac{0.009}{1.99985} = 0.0045 \ \Omega \qquad \square$$

□ **Example** What series resistance R is needed to convert the galvanometer of the preceding example into a voltmeter reading 6 V full scale?

When 6 V is impressed across the voltmeter, a current of 0.00015 A must pass through the coil and the series resistance if the deflection is to be full scale. The total resistance of the voltmeter must be

$$R_V = \frac{6}{0.00015} = 40,000 \ \Omega$$

But R_V is equal to R plus the resistance of the galvanometer; therefore

$$R = 40,000 - 60 = 39,940 \ \Omega \qquad \square$$

36.8   THE MAGNETIC FIELD OF A LONG STRAIGHT WIRE

A long straight wire carrying a current has a magnetic field with the lines of force concentric circles about the wire (Sec. 36.2). When the wire lies in air (or some other nonmagnetic material), the magnetic intensity **B** at any point is proportional to the current I and inversely proportional to the distance r from the wire. It can be shown that $B = 2\kappa I/r$, where κ is a constant which has the value 10^{-7} N/A². The magnetic field strength B in teslas (N/A-m) at a distance r meters from a long straight wire carrying a current I amperes is given by

$$B = \frac{2I}{10^7 r} \tag{36.6}$$

When two long wires parallel to one another bear currents I_1 and I_2 in the same direction, there is an attractive force between them. If the wires are separated by a distance r, the magnetic field at the first wire due to the current in the second is given by $2I_2/10^7 r$. Therefore, the force on a length l of the first wire is given by

$$F = I_1 l B = \frac{2I_1 I_2 l}{10^7 r} \tag{36.7}$$

The force per unit length is $2I_1 I_2/10^7 r$. That the force is attractive can be seen by studying Fig. 36.12. When the directions of the currents in the two wires are opposite, the force between the wires is repulsive. In this

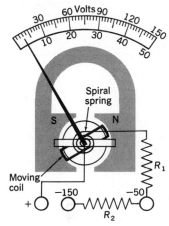

FIGURE 36.11
A two-scale voltmeter.

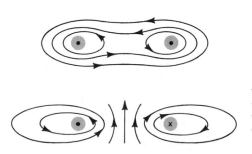

FIGURE 36.12
Two parallel wires attract each other through their magnetic interaction when they bear currents in the same direction and repel each other when the currents are in opposite directions.

figure we make use of the common practice of representing a current out of the paper with a dot (point of an arrow moving toward you) and a current into the paper with cross feathers (tail of an arrow moving away from you).

Although Eq. (36.7) has been derived for long straight wires, there is an attractive force between any two neighboring conductors which are carrying currents in the same direction. One of the most accurate methods of measuring current involves the determination, in an instrument known as the *current balance*, of the attractive force between two precisely made, parallel coils carrying the same current. In the national system of units, officially adopted by the United States (*National Bureau of Standards Circular* C459, 1947), the ampere is defined in terms of measurements made with a current balance. The coulomb is then derived from the ampere as the charge which is carried by a current of one ampere in one second. (For pedagogical reasons we have introduced the coulomb first and then defined the ampere; our units are consistent with those adopted by the National Bureau of Standards and with the SI units.)

When coils carry an alternating current, neighboring turns are attracted to one another by forces which change as the current increases and decreases. The turns may vibrate, moving together and then apart. This causes the hum associated with many coils carrying alternating currents.

36.9 THE MAGNETIC FIELDS OF OTHER CURRENT CONFIGURATIONS

The magnetic field due to any current configuration is the vector sum of the contributions due to the current in each minute element of length Δl. The contribution $\Delta \mathbf{B}$ to the magnetic field due to the current in a small element of length Δl (Fig. 36.13) has a direction given by Fig. 36.2 and a magnitude

$$\Delta B = \frac{\kappa I \, \Delta l \sin \theta}{r^2} \tag{36.8}$$

where θ is the angle between the element Δl and the vector $\mathbf{r}$ connecting Δl to the point at which $\Delta \mathbf{B}$ is to be measured and κ is 10^{-7} N/A². This equation is an expression of a fundamental rule known as *Ampère's law*. Equation (36.8) is applicable only if the wire is in a nonmagnetic medium. It should be emphasized that the contribution $\Delta \mathbf{B}$ from each element of length Δl is a vector which must be added vectorially to all the other $\Delta \mathbf{B}$'s if one wishes to find the resultant magnetic intensity.

The magnetic field at the center of a circular loop of wire (Fig. 36.8) of radius a is readily computed by use of Eq. (36.8). Since the contributions $\Delta \mathbf{B}$ of the tiny elements of length Δl are all in the same direction and are equal in size, addition of the components gives $B = Il/10^7 a^2$. The length of the wire l is $2\pi a$, so that this equation reduces to

$$B = \frac{2\pi I}{10^7 a}$$

If the circular loop is replaced by a circular coil having N turns, the contributions of these turns are all in the same direction. Accordingly, at the center of the circular coil the magnetic intensity $\mathbf{B}$ is given by

$$B = \frac{2\pi N I}{10^7 a} \tag{36.9}$$

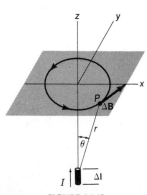

FIGURE 36.13
Ampère's law gives the contribution $\Delta \mathbf{B}$ to the magnetic field at point P due to the current-bearing element of length Δl.

One of the common current configurations in magnetic circuits is that of the solenoid, often used for windings on electromagnets. The solenoid provides a uniform field over its entire central section. If it is long compared with its diameter, the magnetic field is almost the same at all points inside the solenoid, except at distances less than one diameter from the ends. At the ends the field falls off. In the central section of the solenoid the magnetic field **B** is given by the relation

$$B = \frac{4\pi nI}{10^7} = \mu_0 nI \qquad (36.10)$$

where n is the number of turns per meter; the constant $4\pi/10^7$ J/A²-m is called the *permeability of free* space, designated by μ_0. It is important to note that n represents the *number of turns per unit length,* while N is used to represent the total number of turns; $n = N/l$, where l is the length of the solenoid. Equation (36.10) is applicable no matter what the cross-sectional shape of the solenoid may be. Solenoids with square or rectangular cross section are common.

The magnetic fields of a solenoid, a loop, and a single straight wire are shown in Figs. 36.3b, 36.8, and 36.2, respectively. In every case *magnetic lines of flux are continuous. They never begin or end;* instead they follow closed curves which encircle the current-bearing conductors. In this respect magnetic lines of force are different from electric lines of force, which begin and end on electric charges.

QUESTIONS

1. Can magnetic lines of force ever cross? Explain.
2. In what ways are magnetic fields analogous to electric fields? How are they different?
3. If there is no deflection of a beam of electrons as they move through a region in space, does this prove that there is no magnetic field in that region? Give two examples of situations in which there could be a field: one with no electric field present and the other with a transverse electric field.
4. How much work is done by a magnetic field on a stream of free electrons moving through it? Justify your answer. Prove that a constant magnetic field does no work on a charged particle passing through it.
5. A stream of electrons is moving toward the west. If the stream passes through a uniform magnetic field directed upward, in what direction is the electron beam deflected?
6. A current-bearing wire has no net static charge. Why can a magnetic field exert a force on the wire in spite of its uncharged condition?
7. In sensitive electronic circuits why are wires carrying equal and opposite currents often twisted together? How does this reduce their resultant magnetic field in neighboring regions?
8. How could you reverse the direction of motion of a beam of electrons without changing the speed of the electrons at any time?
9. In the equation $\mathbf{F} = q(\mathbf{v} \times \mathbf{B})$, which pairs of the three vectors are always mutually perpendicular? What pair may have any angle between them?
10. A long helical brass spring hangs vertically with its lower end just making contact with a pool of mercury. If a large current is passed through the mercury and the spring, what will occur? Why? Will the motion of the end of the spring be simple harmonic?
11. How can you convert a galvanometer into an ammeter of specified full-scale reading? Into a voltmeter?

PROBLEMS

1. Find the force on 85 mm of conductor bearing a current of 6 A if it lies perpendicular to a magnetic field of intensity 1.4 T (1 tesla = 1 N/A-m). What would the force be if the wire made an angle of 30° with the magnetic field?

Ans. 0.714 N; 0.357 N

2. A long wire bearing a current of 4 A lies perpendicular to a uniform magnetic field between the

poles of a large magnet. If 125 mm of the wire lies in the field, and if the force on this length is 0.15 N, find the magnetic intensity B.

3. An electron is moving perpendicular to a magnetic field with a speed of 4×10^6 m/s. If the intensity of the magnetic field is 0.12 T, what force acts on the electron? What is its acceleration? What is the radius of the circular path described by the electron in the field?
Ans. 7.68×10^{-14} N; 8.43×10^{16} m/s²; 190 μm

4. Find the force on and the acceleration of an electron moving 8×10^7 m/s at right angles to a magnetic field of intensity 0.9 mT in a television picture tube.

5. A rectangular galvanometer coil is 15 mm high and 12 mm wide and has 20 turns of wire. Find its magnetic moment when it bears a current of 80 μA. What torque is exerted on this coil in a uniform magnetic field of intensity 0.6 T when the plane of the coil (a) is parallel to the field, (b) is perpendicular to the field, and (c) makes an angle of 30° with the field?
Ans. 2.88×10^{-7} A-m²; (a) 1.73×10^{-7}, (b) 0, and (c) 1.50×10^{-7} N-m

6. A coil of 25 turns of wire has an area of 250 mm². It carries a current of 4 μA. Find the magnetic moment of the coil. If the magnetic moment is perpendicular to a magnetic field of intensity 0.8 T, what is the torque on the coil?

7. A coil of magnetic moment 2.4×10^{-8} A-m² experiences a torque of 1.44×10^{-8} N-m when it is placed in a magnetic field with the magnetic moment normal to the field. Find the magnetic intensity. If the coil has 15 turns and an area of 140 mm², what current does it bear?
Ans. 0.6 T; 11.4 μA

8. A galvanometer of 14 Ω resistance requires a current of 0.0015 A to produce full-scale deflection. What resistance is required to convert this galvanometer into an ammeter reading 2 A full scale? Into a voltmeter reading 30 V full scale?

9. A meter movement is a galvanometer with a resistance of 9 Ω, and it requires 200 μA for full-scale deflection. Find the resistance required to convert this galvanometer into (a) a voltmeter reading 5 V full scale and (b) an ammeter reading 0.6 A full scale. *Ans.* 24,991 Ω; 0.0018/0.5998 Ω

10. A two-scale voltmeter is wired as shown in Fig. 36.11, where the galvanometer has a resistance of 20 Ω and requires a current of 1.5 mA for full-scale deflection. Find R_1 and R_2 if the voltmeter is to read full scale for 40 and 150 V.

11. A galvanometer has a moving coil with a resistance of 6 Ω and a sensitivity of 1-mm deflection for 5×10^{-7} A. What shunt will be needed to produce a 40-mm deflection for 0.003 A in the main circuit? What series resistor is required to convert this galvanometer into a voltmeter reading 5 V for a 40-mm deflection?
Ans. 40.3 mΩ; 249,994 Ω

12. A 20,000 Ω/V voltmeter is connected in series with a 500 Ω/V voltmeter, both of which are calibrated to read 0 to 120 V. What will each read if the voltmeters in series are connected across a 100-V source?

13. The ohmmeter of the accompanying figure has a meter with resistance of 5 Ω, which deflects full scale for 1 mA, and a 1.5-V battery. If there is zero resistance between a and b, the meter reads full scale; when a and b are open, the meter reads zero. Find R_0. What resistance R between a and b would give a current of 0.75 mA (0.75 full-scale reading)? What current would $R = 250 \Omega$ give?
Ans. 1,495 Ω; 500 Ω; 0.857 mA

14. An ohmmeter circuit of a simple multimeter is shown in the accompanying figure. The meter G reads full scale for 0.4 mA and has a resistance of 7 Ω. The battery has an emf of 1.5 V. When there is zero resistance between a and b, the meter reads full scale; when a and b are open, the meter reads zero. What is the value of R_0? What deflection in terms of full scale would result if a 750-Ω resistor were connected between a and b?

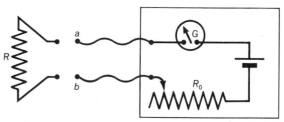

PROBLEMS 13 AND 14

15. A long straight wire bears a current of 5 A. At what distance from the wire is the magnetic field 3×10^{-5} T? *Ans.* 33.3 mm

16. Find the magnetic intensity a distance of 37.5 mm from a long straight wire bearing a current of 15 A.

17. Two long straight parallel wires are 75 mm apart. A current of 5 A passes through one wire and a current of 2 A through the other. If the two currents are in the same direction, what is the

magnetic intensity at a point midway between the wires? Find the attractive force per meter of length between the wires.

        Ans. 16 µN/A-m; 2.67×10^{-5} N/m

18. Find the force on the current loop of the accompanying figure if $I_1 = 10$ A; $I_2 = 7$ A, $a = 125$ mm, $b = 65$ mm, and $d = 35$ mm.

19. Find the force on the current loop of the accompanying figure if I_1 and I_2 are 5 and 9 A respectively and a, b, and d are 225, 55, and 20 mm, respectively.   *Ans.* 74.3 µN to right

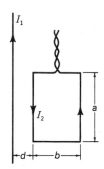

PROBLEMS 18 AND 19

20. A current in a circular loop of wire with a diameter of 65 mm produces a field strength of 40 µT at the center of the loop. What is the current? Find the magnetic moment of the loop.

21. Find the magnetic intensity at the center of a circular coil of 30 turns of 25 mm radius when the coil bears a current of 8 A. What is the magnetic moment of the coil?

        Ans. 6.03 mN/A-m; 0.471 A-m²

22. A current I of 8.8 A passes through a long straight wire which makes a semicircular bend of radius r about point C of the accompanying figure. If

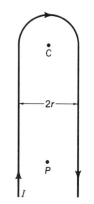

PROBLEMS 22 AND 23

$r = 44$ mm, find the magnetic intensity B at C and at point P far from the bend.

23. If the current in the wire of the accompanying figure is 10.5 A and the radius of the semicircular bend is 35 mm, find the magnetic intensity B at P, a point far from the bend, and at C, the center of curvature of the semicircle.

    Ans. 120 µN/A-m (or µT); 154 µT; both into the paper

24. A long horizontal wire is bent to include a circular loop of $r = 100$ mm (see accompanying figure). If a current of 8 A exists in this wire, what is the magnetic induction B at the center C of the loop? If the loop is twisted about the axis AA' so its plane is vertical, what is the magnitude of B at the center of the loop?

25. If the current in the wire of the accompanying figure is 21 A and the radius of the loop is 42 mm, find the magnetic induction B at the center of the loop. If the loop is now twisted about the axis AA' so its plane is vertical, what is the magnitude of B at the center of the loop?

    Ans. 214 µT; 330 µT

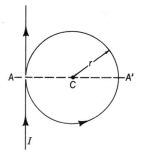

PROBLEMS 24 AND 25

26. A solenoid is 325 mm long. It is wound with 6,500 turns of copper wire. What current must be sent through the windings to produce a magnetic intensity of 4 mT?

27. A cylindrical solenoid of 2,000 turns of 125 mm diameter and 1.2 m long is placed with its axis parallel to the lines of the earth's magnetic field. If the latter has an intensity of 20 µT, what current must flow through the coil to make the magnetic field at its center zero?   *Ans.* 9.55 mA

28. How many turns of wire must there be in a sole-

noid 0.85 m long for a current of 5 A to produce a magnetic field of 0.006 T at its center?

29. In two concentric solenoids the currents flow in opposite directions. The inner one has 8 turns per millimeter and the outer one 3 turns per millimeter. What current in the outer coil will be necessary in order to have the field at the center zero when the inner coil is carrying 1.2 A? If no current passes through the outer coil, what is the magnetic intensity at the center?

Ans. 3.2 A; 12.1 mT

30. Find the magnitude and direction of the magnetic induction B at the center of curvature of the circular segment of the accompanying figure,

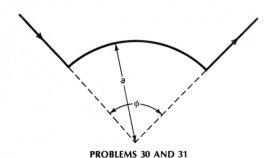

PROBLEMS 30 AND 31

which is bearing a current of 5.4 A. The radius a is 36 mm, and $\phi = 60°$.

31. Find the magnitude and direction of the magnetic induction B at the center of curvature of the circular segment of the accompanying figure if the segment is 68 mm long, the radius a is 62 mm, and the current is 10 A.

Ans. 17.7 μT into paper

32. Find the contribution to the magnetic induction at $a, b,$ and c due to the current I in a short length Δl of conductor (see accompanying figure) if I is 20 A and Δl is 5 mm.

33. Find the contribution to the magnetic induction at $a, b,$ and c due to the current I in a short length Δl of conductor (see accompanying figure) if I is 9 A and Δl is 2 mm. *Ans.* 80 mT; 27 mT; 0 T

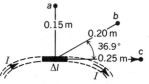

PROBLEMS 32 AND 33

34. The trajectory of a charged particle in a uniform magnetic field B_z in the z direction is given by $\mathbf{r}(t) = \mathbf{i}R \cos \omega t + \mathbf{j}R \sin \omega t + \mathbf{k}v \cos \theta\, t$, where $v \cos \theta$ is the z component of the velocity of the ion. This trajectory is a helix. Find R and ω in terms of B_z and $v \sin \theta$, the velocity perpendicular to the z direction.

We now know what a magnetic field is and how magnetic fields are associated with certain current distributions, but we have said little about the magnets of simple compasses and the familiar bar magnets children play with. What are the origin and nature of their magnetic fields? And why does a compass needle point roughly north? These are questions we discuss in this chapter.

37.1 THE MAGNETIC MOMENT OF A MAGNET

In Chap. 36 we saw that a coil placed in a magnetic field of strength **B** with its magnetic moment **M** perpendicular to the field experiences a torque equal to *BM*. Similarly, a current-bearing solenoid placed in a magnetic field with its axis perpendicular to the field experiences a torque equal to the resultant of the torques acting on each of the turns of the solenoid. These turns can be regarded as a series of individual coaxial coils. The resulting magnetic moment of the solenoid is *NIA*, where *N* is the total number of turns, *I* the current, and *A* the cross-sectional area of the solenoid.

If we place a small bar magnet (magnetic needle) with its axis perpendicular to a magnetic field (Fig. 37.1), it experiences a torque in a direction such as to align its axis with **B**. The magnetic moment **M** (Sec. 36.5) is the ratio of the torque τ to the field strength **B**. Since the *apparent* magnetic properties of the magnet are associated with the poles at the ends, it is often convenient to think of the torque as arising from the forces acting on the north and south poles, which are separated by a distance equal roughly to the length of the magnet *l*. Let us define the pole strength *p* of this magnetic needle as the ratio of the magnetic moment to the distance between the poles:

$$p = \frac{M}{l} \tag{37.1}$$

Since the dimensions of the magnetic moment are A-m², the dimensions of pole strength are ampere-meters. The force **F** on a pole *p* placed in a magnetic field of strength **B** has a magnitude

$$F = Bp \tag{37.2}$$

A pole has a strength of one ampere-meter if the force acting on it is one newton when the pole is placed in a magnetic field of intensity one tesla. The force on the north pole of the magnet is in the direction of **B**, while the force on the south pole is opposite to **B**.

If the axis of the magnet makes an angle θ with the magnetic field, the torque is less than if the magnet were perpendicular to the field. In general, the torque τ is given by

$$\tau = BM \sin \theta = Bpl \sin \theta \tag{37.3}$$

as can be seen from Fig. 37.2.

In many problems regarding magnets it is convenient to consider the forces and torques on the magnet as arising from the interaction between magnetic field and point poles, but it should be remembered that the pole picture is a simplification of the more general point of view that there are forces on current elements in magnetic fields. The torque on the magnet of Fig. 37.2 can be explained in terms of torques acting on elementary current loops. The magnetic moment **M** of the magnet is just

37

MAGNETS

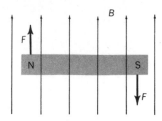

FIGURE 37.1
A bar magnet with its axis perpendicular to a uniform magnetic field experiences a torque which tends to align the magnet with the field.

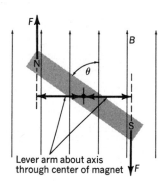

Lever arm about axis through center of magnet

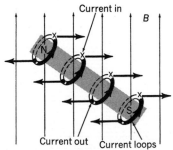

Current in

Current out   Current loops

FIGURE 37.2
The torque on a bar magnet whose axis makes an angle θ with a uniform magnetic field is the resultant of all the torques exerted on current loops of infinitesimally small dimensions.

that which would be produced by a current $I = M/A$ (where A is the cross-sectional area of the magnet) flowing as a current sheet around the magnet. The point of view of the current loops is rigorous and universally acceptable, but for many problems the force-on-pole concept is simpler and more convenient.

Magnets never† have single poles. The magnetic field arises from currents and, as we have seen in Chap. 36, magnetic lines of force are always continuous. The pole of a magnet is a region where a large fraction of lines of force enters or leaves the magnet. Since every line of force which enters a magnet also leaves, it is clear that no magnet can have a single pole.

37.2 COULOMB'S LAW

When the poles of bar magnets are brought close to each other, they interact. *Like poles repel each other, and unlike poles attract.* Coulomb studied the interaction between magnetic needles with torsion balances and found that the force between two poles varies as the product of the pole strengths and inversely as the square of the distance between the poles:

$$F = \kappa \frac{p_1 p_2}{r^2} \qquad (37.4)$$

When the force is in newtons, pole strength in ampere-meters, and r in meters, κ is 10^{-7} N/A², the same constant which appears in Ampère's law.

Coulomb's law is applicable whenever one can regard the magnetic properties of the needle as concentrated in point poles. This, in turn, requires that the distance between the two magnets be substantially greater than the thickness of either magnet, since otherwise a pole of a magnet cannot be correctly approximated by a point pole. If one wishes to calculate the force which one magnetic needle exerts on another, it is ordinarily necessary to calculate not only the forces between nearest poles but also the forces between the other poles. Each pole of one magnet exerts forces on both the north and south poles of the second magnet. Therefore, to calculate the total force exerted on one magnet by another, it is necessary to compute four forces and add them vectorially.

□ **Example** Two small identical magnets are 8 cm long and have pole strengths of 60 A-m. When they are restrained so they cannot rotate, it is possible to make one "float" 6 cm above the other (Fig. 37.3). Find the force which the lower magnet exerts on the upper one (which is the weight of the upper magnet, of course).

$$F_{NN} = \kappa \frac{p_1 p_2}{r^2} = 10^{-7} \frac{60 \times 60}{(0.06)^2} = 0.100 \text{ N}$$

$$F_{NS} = 10^{-7} \frac{60 \times 60}{(0.1)^2} = 0.036 \text{ N}$$

Similarly $F_{SS} = 0.100$ N, and $F_{SN} = 0.036$ N. The vertical component of

† The possibility that magnetic monopoles do exist has been treated theoretically, but extensive experimental searches have failed to find one. If a magnetic monopole is ever discovered, some of the statements above would obviously require modification.

F_{NS} is $0.036 \times 0.6 = 0.0216$ N, and the horizontal component is $0.036 \times 0.8 = 0.0288$ N. The horizontal component of F_{SN} exactly balances the horizontal component of F_{NS}. If we take the upward direction as positive, the vertical force on the upper magnet is given by

$$F = 0.100 + 0.100 - 0.0216 - 0.0216 = 0.1568 \text{ N} \qquad \square$$

Any pole has a magnetic field associated with it. To determine the field intensity **B** at any point, we may imagine that we bring a test pole p_t to the point in question and measure the force **F** on this pole. The intensity **B** is given by $\mathbf{F}/p_t$. To calculate the magnetic intensity due to any group of poles, we find the magnetic intensity which each pole would produce by itself and add all these fields together vectorially. The magnetic field due to one pole p can be calculated readily by Coulomb's law. The force on a test pole p_t at a distance r from the pole p is given by $F = pp_t/10^7r^2$, and

$$B = \frac{F}{p_t} = \frac{p}{10^7r^2} \qquad (37.5)$$

□ **Example** A magnet 15 cm long with poles of strength 250 A-m lies on a table. Find the magnitude of the magnetic intensity B at a point P 20 cm directly above the north pole of the magnet (Fig. 37.4).

The magnetic fields are given by

$$B_N = \frac{p_N}{10^7r^2} = \frac{250}{10^7(0.2)^2} = 6.25 \times 10^{-4} \text{ T}$$

$$B_S = \frac{p_S}{10^7r^2} = \frac{250}{10^7(0.25)^2} = 4.00 \times 10^{-4} \text{ T}$$

The vertical component of B_S is $B_S \cos \theta = 3.2 \times 10^{-4}$ T, and the horizontal component is $B_S \sin \theta = 2.4 \times 10^{-4}$ T. The vertical component of the resultant intensity is $(6.25 - 3.2) \times 10^{-4} = 3.05 \times 10^{-4}$, and the horizontal component is 2.4×10^{-4}. The resultant B is

$$\sqrt{(3.05)^2 + (2.4)^2} \times 10^{-4} = 3.9 \times 10^{-4} \text{ T} \qquad \square$$

37.3 THE MAGNETIC FIELD OF THE EARTH

The usefulness of the compass as a device for determining direction arises from the fact that the earth has a magnetic field which aligns the compass needle. At points high above the earth's surface this field is approximately that which would be produced by a tremendous bar magnet (Fig. 37.5) near the center of the earth with its axis making an angle of 11° with the earth's rotational axis, its south-seeking pole at 78.5° north latitude, 69° west longitude, and its north-seeking pole at 78.5° south latitude, 111° east longitude. At the surface of the earth this field is grossly distorted by magnetic materials of the earth's crust.

In view of the fact that the earth's magnetic axis does not coincide with its axis of rotation, a compass needle points to true north at relatively few regions on the earth's surface. The angle between a free horizontal compass needle and true north is called the *angle of declination* (or sometimes the *variation* or *deviation*). Figure 37.6 shows the declination at various places in the continental United States. A line on the earth's

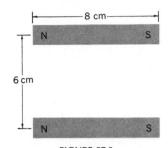

FIGURE 37.3
One bar magnet can be floated above another by magnetic repulsion.

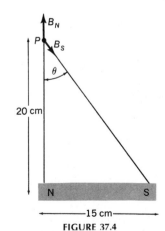

FIGURE 37.4
The magnetic intensity at any point P is the resultant of the fields of the two poles of the magnet.

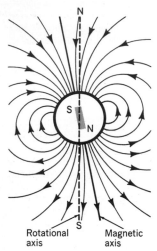

FIGURE 37.5
The earth's magnetic field.

surface along which a compass needle points to true north is called an *agonic line*. At points east of the agonic line in the United States, the declination is west; at points west of the agonic line, the declination is east. In New York City the declination is about 11.5° west, while in San Francisco it is roughly 18° east.

When a compass needle is free to orient itself along the lines of magnetic force, the needle points downward as well as to the north. The angle between the free magnetic needle and the horizontal plane is called the *inclination* or *angle of dip*. Figure 37.6 also shows lines of constant dip, which are called *isoclinic* lines. The inclination in various parts of the continental United States varies from about 50 to 80°.

The earth's magnetic field is a relatively weak one. The horizontal component varies approximately from 25 μT in the southern part of the United States to 15 μT in the northern part. The vertical component varies similarly from about 40 to 55 μT as one goes from south to north.

The magnetic isogonic and isoclinic lines, shown in Fig. 37.6 as smooth regular lines, actually have many local irregularities. One would expect the magnetic lines of the earth to be significantly altered by large deposits of iron ores and other magnetic materials, and indeed they are. One method of prospecting involves the careful measurement of the earth's magnetic field over large areas. The vertical component of the earth's field sometimes increases sharply above a body of magnetic ore and is often weaker than normal above an oil deposit. Geomagnetic surveys are an important tool for looking for materials hundreds of meters below the earth's surface. The largest known deposit of nickel was discovered by flying a sensitive magnetometer along a series of predetermined paths and then making a plot of magnetic intensity as a function of position.

The earth's magnetic field is not static but varies gradually. In 1576 a compass needle in London pointed 8° east of true north; by 1823 it had gradually shifted to point 24° west of north, and since then it has shifted back and now points a few degrees west of north. Data on the strength of the earth's field suggest that it has decreased about 6 percent in the past century. Even more surprising, studies of fossil magnetism show that

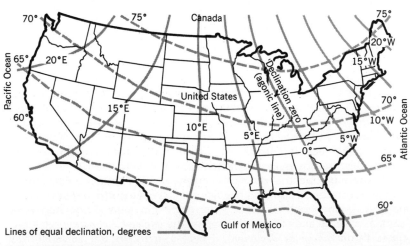

FIGURE 37.6
Geomagnetic declination (solid lines) and inclination (dashed lines) for the continental United States.

the direction of the earth's field has actually reversed itself several times. These reversals seem to be the result of the earth's field decreasing to zero over a period of thousands of years and then coming back with opposite polarity. It appears that the geomagnetic axis has always been fairly closely aligned to the rotational axis; the present angular separation of 11° is probably much larger than the average value over the past. The origin of these variations is explained in terms of a "dynamic theory" in which motions of the earth's fluid core are governed by a complex coupling of thermal, mechanical, electrical, and magnetic phenomena.

Roughly 99 percent of the earth's magnetic field is believed to be due to phenomena in the earth itself; the remainder has its origin in external currents in the ionosphere. In disturbed regions of the solar atmosphere, strong magnetic fields are accompanied by the emission of a tenuous stream of matter, chiefly protons and electrons. When the earth passes through such a stream, the moving charges interact with the earth's magnetic field and its ionosphere, producing a fluctuation in the field at the earth's surface. Magnetic storms of this kind can be partially correlated with sunspots and solar flares, as well as with northern lights (*aurora borealis*) and phenomena of a similar nature on the earth.

37.4 MAGNETIC MATERIALS

Classical studies of the magnetic properties of materials led to an assignment of each substance to one of three classes: *ferromagnetic, paramagnetic,* and *diamagnetic.* The assignment was made on the basis of a simple experiment in which a rod or needle of the substance was placed in a strong, nonhomogeneous magnetic field. A few materials interacted strongly with the field and were called *ferromagnetic* (*ferro* for "iron"). Of the common elements only iron, nickel, and cobalt ordinarily reveal prominent magnetic properties. Gadolinium and liquid oxygen are also strongly magnetic; so are a substantial number of compounds and alloys. All these are ferromagnetic in the classical nomenclature.

For the remaining vast majority of materials the magnetic effects were exceedingly small. However, in a strong, nonuniform field each material fell into one of two classes. Some became feebly magnetized in the direction of the field and aligned the axis of the needle with the field. Such substances were called *paramagnetic* (*para-* means "parallel"). Examples are palladium, manganese, and many metallic salts. Other materials became very feebly magnetized in the direction opposite the field. A needle of bismuth aligns itself at right angles to a strong nonuniform magnetic field. Such a material is called *diamagnetic* (*dia-* means "across"). Most elements and chemical compounds are slightly diamagnetic. Bismuth is the most diamagnetic element. For most practical purposes paramagnetic and diamagnetic materials can be regarded as magnetically inactive. They are not useful for magnets. Ferromagnetic materials are of very great practical importance and utility. They are vital for transformers, electric motors, electric generators, and electromagnets.

Twentieth-century research has led to a microscopic understanding of magnetic phenomena and further classification. We now know that some "atoms" have permanent magnetic moments arising from their structures, where "atom" implies the atom or ion as it exists in the material. For example, in magnetite (Fe_3O_4 or $FeO \cdot Fe_2O_3$) there are ferric (Fe^{3+}) ions

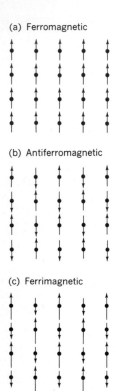

(a) Ferromagnetic

(b) Antiferromagnetic

(c) Ferrimagnetic

FIGURE 37.7
When the "atoms" of a material have magnetic moments large enough for the interaction with neighboring atoms to produce a repetitive orientation over a domain, a material is (a) *ferromagnetic* if all magnetic moments are aligned, (b) *antiferromagnetic* if the magnetic moments are arranged in two groups which contribute equal and opposite net magnetization, and (c) *ferrimagnetic* if they are arranged in two groups with opposite alignment but one group involves "atoms" with a greater magnetic moment than the other.

and ferrous (Fe^{2+}) ions; the latter contribute about eight-tenths the magnetic moment of the former. In other materials iron "atoms" may have electron configurations such that they contribute little or no net magnetic moment. While many compounds and alloys of iron are strongly magnetic, a number, including stainless steel, are not. On the other hand, the Heusler alloys (containing aluminum, copper, and manganese) are ferromagnetic, although none of the component elements are. Why some "atoms" are magnetic and others are not is a complex question, but a qualitative answer can be given in fairly simple terms. Magnetic effects are always associated with moving electric charges. As long ago as 1820, Ampère suggested that within magnetized iron there are circulating currents. It is obvious that these currents cannot be due to a flow of charges around the circumference of the iron core, since such currents would necessarily dissipate heat in the iron as a result of the resistance. Instead, the currents are due to the motions of the atomic electrons. The electrons associated with any given atom are in constant motion, and a moving charge produces a magnetic field.

An electron may establish a magnetic field in two ways: (1) we can picture the electron as traveling around the nucleus of the atom in an orbit crudely analogous to the orbit in which the earth revolves about the sun; (2) the electron may be thought of as rotating about its own axis, much as the earth rotates once each day about its axis. A single electron may establish a magnetic field by virtue of its orbital and its spin motions. In almost all atoms the *net* effect of the magnetic fields set up by various electrons is exceedingly small or zero because the magnetic field set up by one electron counterbalances the fields set up by other electrons. However, in iron, nickel, and cobalt the magnetic fields due to the spinning of the electrons do not cancel for certain of the electrons in an unfilled (M) shell. In these metals, individual "atoms" act as tiny magnets.

A magnetic field interacts with electrons in such a way as to induce a magnetic moment which opposes the applied field. This electronic diamagnetism occurs in all substances. If the "atomic" magnetic moments of the material are zero or very small, the substance is *diamagnetic*.

When "atoms" of a material have magnetic moments and the interaction between the moments of neighboring "atoms" is small, an applied field may align the moments and *paramagnetism* occurs. If the coupling between neighbors is large enough to order them in the absence of an applied magnetic field, there are three important classifications. When the "atomic" interactions align all the magnetic moments in a small domain, the substance is said to be *ferromagnetic* (Fig. 37.7a). However, the coupling may produce two (or possibly more) groups of moments, such that all neighbors of each group are aligned but the two groups are oriented *antiparallel*. If the "atomic" moments of the two groups are equal, the material is *antiferromagnetic* (Fig. 37.7b), while if the resultant magnetic moments of the groups is not zero, the substance is *ferrimagnetic* (Fig. 37.7c). Many ferrimagnets are poor electric conductors, a quality often desired for devices that operate at high frequencies.

It is unfortunate that the classical and current classifications use the same names but sometimes in different senses. For example, MnO is best described as *ferrimagnetic* in modern usage but is *paramagnetic* in the classical terminology. Similarly, the original magnetic material magnetite

(FeO·Fe$_2$O$_3$) is best described as *ferrimagnetic* in current language but is *ferromagnetic* classically.

37.5 MAGNETIZATION BY INDUCTION

In a ferromagnetic material, sometimes individual atoms are the elementary magnets, sometimes combinations or groups of atoms. Although the net magnetic field outside a piece of unmagnetized soft iron is zero, over small volumes of the order of 10^{-3} mm^3 most of the elementary magnets are aligned with their north poles in the same direction. There is no field outside the iron because these tiny *domains* are oriented at random. If this iron is placed in a magnetic field, it is magnetized *by induction*. This magnetization occurs through two processes: (1) Domains which are magnetized in the direction of the inducing field grow at the expense of neighboring domains which are magnetized in less favorable directions. (2) The direction of magnetization of an entire domain may be shifted by the simultaneous rotation of the elementary magnets which make up the domain.

Further evidence confirming that the nature of the magnetizing process is the alignment of elementary magnets are the following observations.

1. When a piece of soft iron lies in a weak field, its magnetization is increased by tapping. Tapping the iron gives the elementary magnets a better chance to align themselves with the magnetizing field. On the other hand, if a piece of magnetized material is not in a field, tapping usually disturbs the alignment of some of the domains and thereby reduces the strength of the magnet.

2. Increasing the temperature of the magnet tends to demagnetize it, because adding kinetic energy to the molecules by thermal agitation produces the same effects as tapping or dropping the magnet. Iron heated above 760°C loses its magnetic properties. Indeed, every magnetic material has a temperature above which it loses its magnetism. The temperature at which the magnetic properties of a given material disappear is called the *Curie point* of that material.

3. If a long permanent magnet is cut into two equal shorter pieces, the magnetized domains remain aligned. There are now two magnets, each half as long as the original. The pole strength of each new magnet is roughly equal to that of the original magnet. It is easy to understand why magnetic effects seem to be concentrated at the ends or poles when one remembers that throughout the body of the magnet the arrangement of elementary magnets leaves a north pole close to each south pole, so that their effects cancel one another. At one end of the magnet there is a concentration of elementary north poles, and at the other end of south poles. Consequently, the magnetic effects are strong at the ends.

When the north pole of a bar magnet is brought near an unmagnetized needle, the elementary magnets in the needle align themselves in the field. Thus, a south pole is produced near the north pole of the bar magnet. Since the attractive force between the south pole of the needle and the north pole of the bar magnet is greater than the repulsive force between the two north poles, there is a net attractive force. It should be

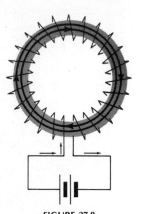

FIGURE 37.8

Magnetic lines of force exist within a toroid or ring solenoid; there is no external field.

noted that *this attractive force arises only in a nonuniform magnetic field.* In a uniform field the forces on opposite poles are equal and opposite.

While most common magnets have two poles, it is entirely possible for a magnet to have three or more. For example, if the middle of a piece of soft iron is brought up to the strong north pole of an electromagnet, it develops a north pole at each end and a stronger south pole at the middle. When steel gas or water pipes are explored with a small compass needle, several north and south poles are often found along a few feet of the pipe. Such poles are called *consequent* poles.

37.6 PERMEABILITY AND THE MAGNETIZING FIELD

If a long solenoid is bent in the form of a circle until the two ends touch (Fig. 37.8), it becomes a *ring solenoid* or *toroid.* Consider a toroid which has n turns per meter of length and bears a current I. Bending a long solenoid to form a toroid does not change the magnetic field inside; thus the magnetic field in a toroid is given by $B = 4\pi nI/10^7$, according to Eq. (36.10).

Suppose the entire core of the toroid is now filled with iron, and a current I is sent through the windings. How is the value of **B** inside the toroid affected by this iron? A measurement† of **B** shows that it is now several hundred times greater than it was before the iron was added.

The magnetic field in the iron is written

$$B = \frac{4\pi K_m nI}{10^7} \tag{37.6}$$

where K_m is called the *relative permeability.* For any material K_m is the ratio of magnetic intensity in the material to magnetic intensity in vacuum. If we fill the toroid with some other material, such as bismuth, silver, wood, or nickel, and measure the magnetic intensity for a given current in the windings, we can determine its relative permeability (Table 37.1). Materials with relative permeabilities less than unity are diamagnetic, those

† The measurement of **B** in the solid material can be made without disturbing the material in any way by wrapping a few turns of wire around the solenoid as indicated in Fig. 38.3 and observing the induced emf as the field is changed. The operation of such a measuring device, known as a fluxmeter, is described in Chap. 38.

TABLE 37.1
Relative Permeabilities of Various Substances (Vacuum = 1.000 by definition)

Material	Relative permeability	Material	Maximum relative permeability
Bismuth	0.99983	Cobalt	250
Silver	0.99998	Nickel	600
Copper	0.99999	Mild steel	2,000
Water	0.99999	Iron (0.2% impurity)	5,000
Air	1.0000004	Silicon iron†	7,000
Aluminum	1.00002	Permalloy	100,000
		Supermalloy	1,000,000

† Used in power transformers.

with permeabilities slightly greater than unity are paramagnetic, and those with permeabilities much larger than unity are ferromagnetic, according to the classical description.

The magnetic intensity in a toroid depends on two factors: (1) the properties of the medium in the toroid and (2) the influences which are inducing the magnetization of the medium. In this case the inducing field is provided by the current through the turns of the toroid. In other situations it may be provided by magnetic poles or other kinds of current distributions. It is called the *magnetizing field* and is represented by **H**. In the special case of the uniformly filled toroid we have continuous magnetization and no poles; the magnetizing force is the product of the current and the number of turns per unit length. Thus we have

$$H = nI \tag{37.7}$$

If nI is replaced by H in Eq. (37.6), we obtain

$$B = \frac{4\pi}{10^7} K_m H = \mu H \tag{37.8}$$

Here μ stands for $4\pi \times 10^{-7} K_m$ N/A^2 and is called the *absolute permeability* of the material. Since the relative permeability K_m of a vacuum is unity, the absolute permeability is $4\pi/10^7$ for a vacuum (and for air for practical purposes). It is represented by μ_0 and called the permeability of free space (Sec. 36.9).

B is different inside the toroid when it is filled with iron because of the magnetization of the iron which occurs when we apply the magnetizing field. We have seen (Sec. 37.1) that the magnetic moment M of a bar magnet of cross-sectional area A and length l is that which would exist if there were a current M/A flowing around the surface, i.e., a surface current M/lA per unit length. The ratio M/lA is the magnetic moment per unit volume, which we shall indicate with M_v. With magnetic material in the toroid the total magnetic intensity **B** comes from two contributions: μ_0**H** due to the conduction currents in the windings and μ_0**M**$_v$ due to the internal currents in the magnetic material.

$$\mathbf{B} = \mu_0(\mathbf{H} + \mathbf{M}_v) \tag{37.9}$$

In an *ideal* magnetic material **M**$_v$ is proportional to the magnetizing field **H**. From the equation $\mathbf{B} = \mu\mathbf{H} = \mu_0(\mathbf{H} + \mathbf{M}_v)$, we obtain the relation

$$\mu = \mu_0\left(1 + \frac{M_v}{H}\right) = \mu_0 K_m \tag{37.10}$$

□ **Example** A toroid of mean circumference 0.5 m has 500 turns, each bearing a current of 0.15 A. (*a*) Find H and B if the toroid has an air core. (*b*) Find B and M_v if the core is filled with iron of relative permeability 5,000. (*c*) Find the average magnetic moment per iron atom if the density of iron is 7,850 kg/m^3.

(*a*) By Eqs. (37.7) and (37.8),

$$H = \frac{500 \text{ turns}}{0.5 \text{ m}}(0.15 \text{ A}) = 150 \text{ A/m}$$

$$B = (4\pi \times 10^{-7} \text{ N/A}^2)(150 \text{ A/m})$$
$$= 1.88 \times 10^{-4} \text{ N/A-m}$$

(b) For iron with $K_m = 5,000$, by Eq. (37.8),

$$B = (4\pi \times 10^{-7})(5,000)(150 \text{ N/A-m})$$
$$= 0.94 \text{ N/A-m}$$

By Eq. (37.9),

$$0.94 \text{ N/A-m} = (4\pi \times 10^{-7} \text{ N/A}^2)(150 \text{ A/m} + M_v)$$
$$M_v = 7.5 \times 10^5 \text{ A/m (or A-m}^2/\text{m}^3)$$

(c) One kilogram atomic weight (55.85 kg) of iron has 6.02×10^{26} atoms. Therefore in 1 m³ there are $7,850 \times 6.02 \times 10^{26}/55.85$ atoms. Hence there are 8.48×10^{28} atoms in 1 m³ of iron, which has a magnetic moment M_v of 7.5×10^5 A-m²/m³. The average magnetic moment per iron atom is 8.9×10^{-24} A-m². □

For a material such as permalloy, which has a K_m of the order of 100,000, Eq. (37.10) shows that $\mathbf{M}_v$, the magnetic moment per unit volume due to the "internal currents," is about 99,999 times as great as the magnetizing field due to currents in the external windings.

Only when $\mathbf{M}_v$ is proportional to $\mathbf{H}$ is the permeability μ a constant. We shall see in Sec. 37.7 that for ferromagnetic materials μ is not constant over any large range of values of $\mathbf{H}$.

If we solve Eq. (37.9) for $\mathbf{H}$, we obtain

$$\mathbf{H} = \frac{\mathbf{B}}{\mu_0} - \mathbf{M}_v \qquad (37.11)$$

This is the defining equation† for $\mathbf{H}$ in the general case.

37.7 THE MAGNETIZATION CURVE

If a toroid is wound around an iron core and a fluxmeter provided to measure changes in magnetic intensity in the core, we can measure $\mathbf{B}$ as a function of $\mathbf{H}$ by starting with an unmagnetized core and no current in the windings and increasing the current in the windings step by step. If this is done, we obtain a *magnetization curve* in which the magnetic intensity B is plotted as a function of the magnetizing field $H = nI$. For a typical iron sample, a curve similar to that of Fig. 37.9a is obtained.

The permeability μ is the ratio of B to H, by definition. Figure 37.9b shows how μ varies with H for the particular magnetization curve of Fig. 37.9a. Observe that the permeability is not constant for a ferromagnetic substance, as it is for an ideal magnetic material. Instead it varies over a considerable range. It is not difficult to understand qualitatively why this is true. The iron core is composed of a large number of magnetic domains originally oriented at random. When a tiny current is passed through the windings, the magnetizing field H has a small effect on these domains compared with the influence of neighboring domains; therefore relatively small changes in magnetization occur. As the current is increased, the more favorably aligned domains grow at the expense of the others. The number of elementary magnetic dipoles oriented in the direction of the magnetizing field increases rapidly, as is shown by the steep portion of the magnetization curve. Once most of the elementary

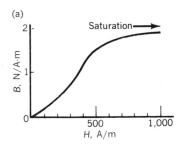

(a)

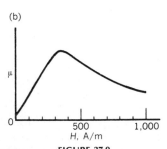

(b)

FIGURE 37.9

How (a) the magnetic intensity and (b) the permeability of a sample initially unmagnetized vary with the magnetizing field.

† $\mathbf{B}$, $\mathbf{H}$, and $\mathbf{M}_v$ are vectors and do not always have the same direction.

magnets are aligned, the region of "hard magnetization" begins. Here the magnetizations of the unaligned domains undergo rotations which bring them into more exact alignment with the applied field. When essentially all the elementary magnets are aligned, the iron is said to be *saturated.* From this point on, increases in H result in small increases in B. When the iron approaches saturation, the permeability (B/H) falls off rapidly as H is increased.

Observing a portion of the magnetization curve with instruments of great sensitivity will show that the curve is not perfectly smooth but is made up of a number of tiny jumps, called *Barkhausen steps;* they are caused by the fact that a large number of neighboring elementary magnets along a domain boundary align themselves with the applied field simultaneously, thereby producing sharp little jumps in the magnetization.

Although this picture of magnetism is simplified, it does present the qualitative features in reasonable perspective. During magnetization the sizes of domains change, and domains are reoriented. There may even be an observable change in the length of the magnet. Such a change in length with magnetization is known as *magnetostriction.* Magnetostriction oscillators are used to produce sound waves of very high frequency, particularly in liquid media.

37.8 HYSTERESIS

When an iron core is saturated and the magnetizing field is removed, the magnetization does not fall to zero because the magnetic domains have been aligned by the magnetization process and help to keep each other aligned. To demagnetize the specimen completely, a magnetizing field in the opposite direction must be applied. If a specimen is magnetized first in one direction and then in another, a plot of the relationship between B and H results in a *hysteresis curve.* Figure 37.10 shows hysteresis loops for three types of ferromagnetic material.

A material which makes a good permanent magnet has a very broad hysteresis loop, since this indicates a high retention of magnetism when the magnetizing field is removed. Such a material would not be good for an electromagnet, which should lose its magnetic properties when the magnetizing field is eliminated. In ac transformers, motors, etc., one ordinarily wants a magnetic material with a hysteresis loop of small area, because it takes energy to reverse the direction of the magnetization. The area of the hysteresis loop is a measure of the energy lost per cycle per unit volume of the material; this energy is converted into internal energy. A transformer made of steel with a broad hysteresis loop would be unsatisfactory because of the large amounts of energy which would be lost in magnetizing and demagnetizing it many times each second.

Soft iron demagnetizes rapidly when the magnetizing field is removed, so it is desirable for electromagnets and transformers. On the other hand, the alloy *alnico,* made of aluminum, nickel, and cobalt steel, is particularly high in retentivity and therefore makes a fine permanent magnet.

37.9 APPLICATIONS OF ELECTROMAGNETS

If a solenoid is wound around a soft-iron core, the iron becomes strongly magnetized when a current is passed through the solenoid and loses

(a)
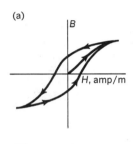

(b)
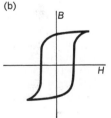

(c)
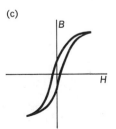

FIGURE 37.10
Hysteresis curves for (*a*) a piece of soft iron, (*b*) an alloy which can be made into a permanent magnet, and (*c*) an alloy suitable for use in a power transformer.

most of its magnetism when the current is stopped. There are many uses for such electromagnets. Large ones can pick up rails and operate switches on railroads, as well as lift and move scrap iron and other magnetic materials. Small electromagnets have been developed for extracting particles of iron from the eye and for opening and closing switches in telephone and automobile circuits.

A common use of electromagnetism is in the doorbell. Here a soft-iron vane is attracted by the electromagnet when the switch is closed. When the iron moves toward the poles of the electromagnet, the clapper strikes the bell and the electric connection is broken by the movement of the iron vane. The current ceases, and spring action moves the vane back toward its original position. Contact is reestablished, the electromagnet is activated once more, and another cycle begins. The same basic idea is also utilized in the electric horn, except that a diaphragm is set into vibration instead of a clapper. The loudspeakers of many radios, record players, and television receivers (Sec. 48.8) use electromagnets as fundamental components.

QUESTIONS

1. Why is an unmagnetized iron object (such as a nail) attracted by a permanent magnet?
2. Why can't one obtain an isolated north pole simply by cutting it off a bar magnet?
3. If a small permanent magnet is placed in a uniform magnetic field, is there any net force on the magnet? In a nonuniform field? Explain.
4. A steel post is driven into the ground in the United States. Is it likely to be magnetized? Why? At which end is the north pole likely to be? Why?
5. How can a magnetized watch or piece of iron be demagnetized?
6. One of two initially identical iron bars is magnetized and retains part of its magnetic moment. How can it be distinguished from the unmagnetized bar in a region in which there is no magnetic field due to the earth or other magnets?
7. What is meant by the "demagnetizing effect of poles"? Explain why a good permanent magnet usually has a "keeper" of soft iron placed between its poles when it is not in use.
8. Why do iron filings line up in a magnetic field?
9. What is the advantage of an iron core in an electromagnet?
10. Why is the size and shape of the hysteresis loop of great importance to the designer of electrical machinery?
11. Why are gyrocompasses and radiocompasses widely used in place of magnetic compasses in the navigation of aircraft and large ships?
12. A thin iron rod is deeply notched with a file and then magnetized. After the poles are marked, it is broken in halves. Why was it desirable to notch the rod before magnetizing it? Describe the poles in the two halves if the rod originally had a north pole at one end and a south pole at the other. How could you magnetize the rod so it had a south pole at each end? How would the halves be magnetized after you broke the rod this time?
13. What are the units of the product BH? Are they consistent with the statement about the area of the hysteresis loop in Sec. 37.8?
14. If a substantial quantity of magnetic ore lay beneath the earth's surface, how would the magnetic field of the earth be affected nearby? Consider both horizontal and vertical components.

PROBLEMS

1. A magnet has a pole strength of 50 A-m. It is placed at right angles to the earth's magnetic field of 20 μT and a torque of 60 μN-m acts on it. Find the magnetic moment of the magnet and the distance between the poles. What would the torque be if the magnet were rotated so its axis made an angle of 53.1° with the field?

    Ans. 3 A-m²; 60 mm; 48 μN-m

2. What torque is necessary to hold the axis of a magnet at an angle of 50° to the magnetic meridian where the horizontal component of the earth's magnetic field is 20 μT? The length of the magnet is 125 mm, and its pole strength is 40 A-m.
3. A bar magnet has poles of strength 12 A-m and a

magnetic moment of 0.96 A-m². This magnet is placed perpendicular to the earth's magnetic field, which has a horizontal component of 25 μT. (a) Find the torque acting to rotate the magnet about a vertical axis. (b) Find the magnitude of the resultant horizontal magnetic induction at point P on the south-north axis 0.2 m from the north pole of the magnet.

Ans. (a) 24 μN-m; (b) 29.0 μT

4. Two magnetic poles have strengths of 50 and 90 A-m, respectively. At what distance in air will the force of attraction between them be 1 mN?

5. Find the magnetic induction at a point 120 mm from the center of a magnet on the perpendicular bisector of the line joining the poles if the magnet is 100 mm long and its pole strength is 40 A-m. *Ans.* 182 μT parallel to axis

6. A long bar of cobalt steel with a mass of 40 g when placed horizontally over a similar bar, both being equally magnetized, remains suspended at a distance of 9 mm above it (Fig. 37.3). What is the pole strength at each end of each bar? *Hint:* Assume the bars are so long that only the nearest poles interact significantly.

7. Identical magnets 80 mm long with poles of 120 A-m strength lie 60 mm apart with axes parallel (Fig. 37.3). Find the repulsive force between the magnets if the south pole of one is nearest the south pole of the other. *Ans.* 0.627 N

8. The axes of two magnets are collinear. One has poles of strength 80 A-m separated by 125 mm, and the second has a magnetic moment of 12 A-m² with poles of strength 160 A-m. Find the attractive force between the magnets if the north pole of one is 45 mm from the south pole of the second.

9. A small test magnet has poles of 75 A-m strength separated by 25 mm. Find the force on each pole when this magnet is placed at the center of a circular loop with a radius of 375 mm in which a current of 12.5 A is flowing. (Assume the magnetic intensity is uniform over the volume occupied by the magnet.) What is the resultant force on the magnet? Find the torque on the magnet if its axis is perpendicular to the magnetic field.

Ans. 1.57 mN; zero; 39.3 μN-m

10. The magnetic declination in central Texas is 10° east. How far from a true-north course would a flier be after traveling 200 km following the compass without making a correction for declination? Would he be east or west of his course?

11. A bar magnet is 128 mm long, and each pole has a strength of 64 A-m. Find the magnitude of the intensity of the magnetic field at a point 96 mm from the south pole, measured at right angles to the axis of the magnet. *Ans.* 580 μT

12. A north pole of 60 A-m strength is placed 150 mm

from a south pole of 90 A-m strength. How far from the north pole, on a line drawn through the two poles, will the resultant field due to these poles be zero?

13. A bar magnet 75 mm long with a pole strength of 20 A-m is horizontal, at right angles to the earth's magnetic field, with its north pole pointing west. Find the intensity of the magnetic field in a horizontal plane at a point 225 mm west of the north pole. Take the horizontal component of the earth's magnetic field as 20 μT.

Ans. 26.4 μT 40.8°W of N

14. At a place where the horizontal component of the earth's magnetic field is 20 μT, a bar magnet 125 mm long with a pole strength of 4 A-m is horizontal, at right angles to the earth's field, with its north pole pointing toward the west. Find the direction and intensity of the field in a horizontal plane at a point 85 mm east of the south pole of the magnet.

15. The poles of a magnet are 95 mm apart, and each has a strength of 19 A-m. What is the magnitude of the magnetic intensity at a point 76 mm from one pole and 57 mm from the other?

Ans. 0.671 mT

16. A magnet with poles of 40 A-m strength separated by a distance of 25 mm is placed in the uniform field inside a solenoid with its axis at right angles to the lines of force. If the solenoid is 0.4 m long and has 600 turns bearing a current of 3.5 A, what torque does the field exert on the magnet?

17. Find the magnetizing field H and the magnetic flux density B at (a) a point 105 mm from a long straight wire bearing a current of 15 A and (b) the center of a 2,000-turn solenoid which is 0.24 m long and bears a current of 1.6 A.

Ans. (a) 22.7 A/m; 28.6 μT; (b) 13,333 A/m; 16.8 mT

18. An electromagnet has a solenoidal winding 225 mm long with a total of 900 turns. What is the magnetizing field H near the center of the winding and far from any poles if the current is 0.8 A? What is the magnetic induction B at this point if the iron has a relative permeability of 350?

19. An iron anchor ring is wound with 800 turns. If the current in this toroidal solenoid is 0.75 A and the relative permeability of the iron core is 400, what is the magnetic induction B, assuming that

the toroid has a mean circumference of 0.5 m? What is the magnetizing force H? Find the magnetic moment per unit volume of iron and the average contribution to this magnetization per iron atom.

Ans.   0.603 T;   1,200 A/m; 4.79×10^5 A/m; 5.65×10^{-24} A-m^2

20. A solenoid that is 0.6 m long is wound with 1,800 turns of copper wire. An iron rod having a relative permeability of 500 is placed along the axis of the solenoid. What is the magnetic intensity in the rod when a current of 0.9 A flows through the wire? What are the magnetizing field and the magnetic moment per unit volume of the iron?

Find the average contribution per iron atom to the magnetization.

21. A toroid with 1,500 turns is wound on an iron ring 360 mm^2 in cross-sectional area, of 0.75-m mean circumference and of 1,500 relative permeability. If the windings carry 0.24 A, find (*a*) the magnetizing field H, (*b*) the product of H and the length, called the *magnetomotive force*, (*c*) the magnetic induction B, (*d*) the product of B and the cross-sectional area of the ring, called the *flux*, and (*e*) the ratio of magnetomotive force to flux, called the *reluctance* of the circuit.

Ans.   (*a*) 480 A/m;   (*b*) 360 A;   (*c*) 0.905 T; (*d*) 3.26×10^{-4} N-m/A;   (*e*) 1.1×10^6 A^2/N-m

22. An iron ring has a cross section of 80 mm^2 and an average diameter of 0.25 m. It is wound with 900 turns of copper wire. If the iron in the core has a relative permeability of 250, what is the magnetic intensity in the iron when a current of 4 A exists in the windings?

In our discussion of electric circuits thus far, we have used batteries and thermocouples as sources of emf. Electric energy from either of these sources is relatively expensive. Now that we have learned some facts about magnetic phenomena, we are ready to develop the fundamental physics upon which the economic conversion of mechanical energy into electric energy depends. We recall that when an electric current is sent through a conductor, a magnetic field appears near the wire. If a current always produces a magnetic field, is there some way in which a magnetic field can produce a current? There is, and its discovery made feasible the broad use of electric energy in our daily lives.

38
INDUCED ELECTROMOTIVE FORCES

38.1 THE DISCOVERY OF INDUCED EMFS

The discovery by Oersted in 1819 that a current has an associated magnetic field led a number of physicists to search for some means by which a magnetic field might produce a current. The first observation of such a phenomenon was made in 1830 by Joseph Henry. He used a horseshoe-shaped electromagnet around which he wound a second coil, the terminals of which were connected to a galvanometer (Fig. 38.1). Henry found that when the current through the electromagnet is changed, either increased or decreased, there is a deflection of the galvanometer coil. When the current in the electromagnet is steady, there is no current through the galvanometer. When the current in the magnet is turned on, the deflection of the galvanometer is in one direction; when the current is stopped, the deflection is in the opposite direction.

Several months later Michael Faraday independently discovered the deflection of a galvanometer connected to one coil when the current in an adjacent coil was started or stopped. Faraday published his findings first and is therefore usually credited with the discovery.

Faraday made a thorough study of the phenomenon. He found that the galvanometer in series with the second coil deflected not only when the current in the first coil was started and stopped but also when the first coil carried a steady current and was moved nearer to or farther from the second. He brought a magnet near the second coil and then withdrew it. The galvanometer needle deflected during the motion of the magnet. When the north pole of a magnet was brought near the coil, the charges flowed in one direction. When this pole was withdrawn, the charges flowed in the opposite direction. If a south pole was brought toward the coil, the deflection was in the same direction as when the north pole was withdrawn. Whenever the magnetic lines of force linking the second coil were changed, there was an induced emf which produced a current through the galvanometer. This phenomenon is called *electromagnetic induction*. To develop quantitative relations between change in magnetic field and induced emf, we must become familiar with the meaning of the term *magnetic flux*.

38.2 MAGNETIC FLUX: THE WEBER

Consider an area A (Fig. 38.2a) with its plane perpendicular to a uniform magnetic field of intensity **B**. The magnetic flux Φ through this area is defined as the product of **B** and **A**:

$$\Phi = BA \qquad (38.1)$$

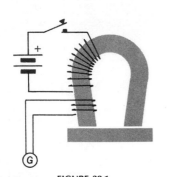

FIGURE 38.1
Apparatus with which Henry discovered electromagnetic induction.

The flux Φ is given in *webers* when B is in teslas (newtons per ampere-meter) and A is in square meters. The weber (Wb) is thus dimensionally equivalent to the newton-meter per ampere; it is named in honor of W. E. Weber (1804–1891).

In many situations it is easier to measure the flux through a coil than it is to measure **B** directly. If Φ and the area are known, the average value of **B** is given by Φ/A Wb/m². (The weber per square meter, the tesla, and the newton per ampere-meter are different names for the same unit.)

If **B** is not perpendicular to the plane of the area through which the flux is desired, $\Phi = BA \cos \theta$, where θ is the angle between **B** and the normal to the area (Fig. 38.2b). When **B** is not constant over the area, Eq. (38.1) is applicable if we take the *average* flux density for **B**.

An alternative point of view toward flux was developed by Faraday. We saw in Sec. 36.3 that a magnetic field can be represented by lines of force. If we draw many lines of force in regions where the field is strong and correspondingly fewer where the field is weak, we have a plot which shows not only the direction of the magnetic field at various points but also its intensity. If we agree to limit the number of lines of force so that we draw one line per square meter where **B** is 1 Wb/m², two lines per square meter where **B** is 2 Wb/m², etc., the number of lines through any large area is just equal to the magnetic flux through that area. Let us call lines drawn according to this convention *lines of flux* to distinguish them from lines of *force*, which give only the direction of the field. In terms of this physical picture, the *flux* through any area is equal to the number of *flux lines* which pass through the area.

38.3 FARADAY'S LAW OF ELECTROMAGNETIC INDUCTION

Faraday found experimentally that the magnitude of the induced emf in a single loop is directly proportional to the rate at which the flux linking the loop changes:

$$e = -\frac{\Delta \Phi}{\Delta t} \tag{38.2}$$

where $\Delta \Phi$ is the change in flux occurring in the time Δt. The emf e† is given in volts when $\Delta \Phi / \Delta t$ is in webers per second. An emf $\Delta \Phi / \Delta t$ is induced in each turn of a coil. If N turns are connected in series, the total emf in the coil is N times that induced in a single loop. The significance of the minus sign is discussed in Sec. 38.5.

38.4 THE FLUXMETER

In Sec. 37.6 it was assumed that we can measure **B** inside a toroid without cutting any holes in the material. We are now in a position to see how this can be done. Consider the iron-filled toroid shown in Fig. 38.3. Assume the iron is originally unmagnetized and the current in the winding is zero. If N turns of wire are wound around this toroid and connected to a suitable galvanometer, we can measure **B** as follows. We pass

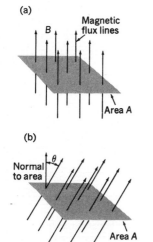

(a)

Magnetic flux lines

B

Area A

(b)

Normal to area

θ

Area A

FIGURE 38.2
The magnetic flux linking an area is the product of the area and the component of the magnetic intensity perpendicular to the area; $\Phi = AB \cos \theta$.

† Note that we have used e rather than $\mathcal{E}$ to indicate the induced emf in this case. In general, we use $\mathcal{E}$ and I to represent *constant* values and the lowercase letters e and i to indicate emfs and currents which *vary in time*.

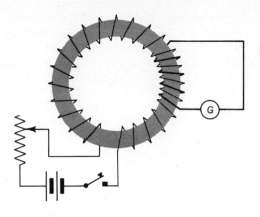

FIGURE 38.3
Iron-core toroid with a fluxmeter that
measures changes in magnetic flux.

a small current through the toroid windings. This changes the magnetic
intensity in the iron from zero to some value **B**. If A is the cross-sectional
area of the iron core, the flux in the iron goes from zero to BA in a time t.
According to Faraday's law of induction, there is an average induced emf
e in the fluxmeter windings given by

$$e = \frac{-N\,\Delta\Phi}{\Delta t} = \frac{-NAB}{t}$$

There is a current in the fluxmeter given by $i = e/R = NAB/Rt$. Now
$it = NAB/R$ is equal to the charge q passing through the fluxmeter. If we
use a suitable galvanometer, we obtain a deflection proportional to this
charge. If we know q, N, and A, we can calculate **B**.

38.5 LENZ'S LAW

The direction of the induced emf is readily predicted by application of a
rule due to Lenz. *The direction of an induced emf is always such that any
current it produces opposes, through its magnetic effects, the change inducing
the emf.*

To illustrate Lenz's law, consider a north pole that is being pushed (Fig.
38.4) toward a coil. The current induced in the coil is in such a direction

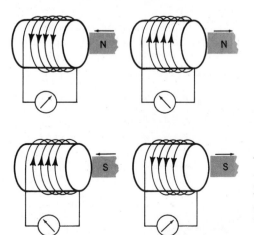

FIGURE 38.4
The current arising from an induced emf
always opposes, through its magnetic
effects, the change inducing the emf.
Whether the bar magnet is pushed into
the coil or pulled out of it, the magnetic
field of the induced current opposes the
motion of the magnet.

that its magnetic field opposes the motion of the magnet. In this case, it is directed toward the north pole of the magnet, which requires that the current be counterclockwise. When the north pole is withdrawn, the directions of induced emf and current reverse, since the induced magnetic effect opposes the act which creates it.

Actually, Lenz's law represents one of the many forms in which conservation of energy appears in physics. That this law follows from the conservation of energy can be seen by this reasoning. Changing a magnetic field induces an emf and a corresponding current if there is a conducting path. Thus, electric energy appears as the result of this act. It is the work done against the opposing magnetic field which is transformed into electric energy.

38.6 MOTIONAL ELECTROMOTIVE FORCE

If a wire of length **l** is moved with a velocity **v** perpendicular to a magnetic field of flux density **B** (Fig. 38.5), there is induced in this wire an emf

$$\mathcal{E} = vBl \tag{38.3}$$

The origin of this emf is intimately related to the force exerted on each individual charge moving through the magnetic field. This force, as we saw in Sec. 36.4, has magnitude qvB and gives rise to an electric field of strength $E = F/q = vB$. However, it is erroneous to think that the magnetic force does work on the charges to produce the emf. We recall that the magnetic force is perpendicular to the velocity and hence to the displacement of the charge. By Eq. (6.1), the work ΔW done by a force **F** in producing a displacement Δ**s** is $\Delta W = F \Delta s \cos \theta$; in this case $\theta = 90°$, and so no work is done by the magnetic force. This actual work on the charges producing the emf is done by the conductor in which the charges are confined, and the energy is supplied to the wire by work done in moving the current-bearing conductor through the perpendicular magnetic field. The potential difference across a wire of length l is given by the product of the field strength vB and the length l; $\mathcal{E} = vBl$. (In the event that **B**, **v**, and **l** are not mutually perpendicular, $\mathcal{E}$ is numerically equal to the volume of a parallelepiped with sides in the directions of **B**, **v**, and **l** and lengths proportional to the magnitudes of the respective quantities.)

In many electric generators the emf is induced by moving wires through a magnetic field at high speed. The direction of the induced emf is such that the force on the resulting current opposes the movement of

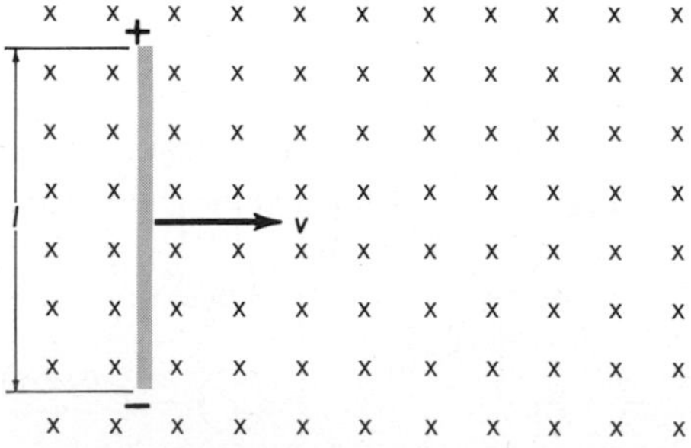

FIGURE 38.5
A motional emf is induced by moving a conductor across a magnetic field; in this case **B** is into the page, and the upper end of the wire is positive.

the wire across the field. From this it follows immediately that the magnetic field of the current produced by the induced emf strengthens the field on the side toward which the wire is moving and weakens it on the opposite side. Once we know the direction of the magnetic field associated with the induced current, we can let the fingers of the right hand curve along the lines of force; the right thumb then points in the direction of the induced emf (see Sec. 36.2). In the situation shown in Fig. 38.6, the induced emf produces a current into the paper when the wire is moved as indicated.

38.7 THE EMF IN A ROTATING LOOP

When a rectangular loop of wire rotates in a uniform magnetic field (Fig. 38.7), the conductors move through the field and a motional emf is induced. The conductors ab and cd cut across the flux lines, while conductors ad and bc do not. In conductor ab there is induced an instantaneous emf

$$e_1 = vBl \sin \theta$$

where θ is the angle between the velocity $\mathbf{v}$ of the wire and $\mathbf{B}$. The direction of the induced emf is from b toward a as ab moves downward. At the same time, in side cd there exists an emf of equal magnitude directed from d to c. The emfs in ab and cd are in series, and the net emf in the loop $badc$ is

$$e = 2vBl \sin \theta = \mathcal{E}_{max} \sin \theta \qquad (38.4)$$

where $\mathcal{E}_{max}$ is the maximum emf induced in the loop. For a flat coil consisting of n closely wound loops, both e and $\mathcal{E}_{max}$ are n times that of a single loop.

The instantaneous emf induced in the coil varies as θ changes. It is zero when the plane of the coil is vertical, since in this position the wires in the coil move parallel to the magnetic intensity $\mathbf{B}$ and $\sin \theta = 0$. When the plane of the coil is horizontal, the wires move perpendicular to the lines of flux, $\sin \theta = 1$ and the induced emf is maximum. The relation between the position of the coil and the induced emf is evident from Fig. 38.8. As the curve shows, the emf generated rises from zero to a maximum value at 90°, decreases to zero again at 180°, reverses its direction to reach its largest negative value at 270°, and returns to zero when the rotation is completed. An *alternating emf* is induced in the coil. If such an emf is applied to a circuit, the current alternates with the same frequency as the emf (Chap. 40).

38.8 MUTUAL INDUCTANCE

Consider the neighboring circuits of Fig. 38.9. When the key is pressed, the current in coil A rises and the associated magnetic lines of flux produce a change in the magnetic flux linking coil B. As a consequence, there is an induced emf and a resulting current in coil B. The current in B lasts only as long as the current in A is changing. In such a circuit, coil A is called the *primary* and coil B the *secondary*. When the current in the primary is increasing, the induced emf in the secondary produces a current which, by its magnetic effect, opposes the rise of current in the

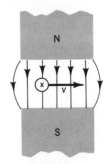

FIGURE 38.6
By moving the wire to the right, an emf is induced which results in a current into the page.

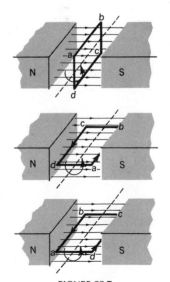

FIGURE 38.7
A varying emf is produced in a rectangular loop rotating in a uniform magnetic field with constant angular velocity.

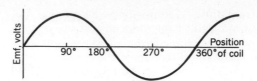

primary. Note that it is always the *change in current in the primary* which is opposed, not the current itself. Thus, when the current passes its maximum value and is being reduced, the magnetic effect due to the induced current in coil B is such as to keep the primary current at its previous value. The effect which results in an emf being produced in one circuit due to a changing current in another circuit is called *mutual induction*.

The emf e_2 induced in the secondary coil is directly proportional to the rate of change of current in the primary:

$$e_2 = M\frac{\Delta i_1}{\Delta t} \tag{38.5}$$

where the constant M is called the *coefficient of mutual inductance*. If the current in the secondary coil changes at the rate $\Delta i_2/\Delta t$, there is an emf in the primary given by

$$e_1 = M\frac{\Delta i_2}{\Delta t} \tag{38.5a}$$

The proportionality constant M is the same in both Eqs. (38.5) and (38.5a).

The coefficient of mutual inductance between two circuits is the ratio of the emf induced in the second circuit to the rate of change of current with time in the first circuit. When an emf of 1 V is induced in a secondary coil by a current change of 1 A/s in the primary, the coefficient of mutual inductance is said to be 1 *henry* (H).

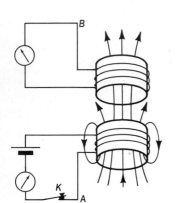

FIGURE 38.9
Mutual inductance between two circuits occurs when magnetic flux from circuit A links circuit B so that current changes in A induce an emf in B.

☐ **Example** The mutual inductance between two circuits is 0.4 H. Find the emf induced in the secondary at an instant when the current is changing at the rate of 90 A/s in the primary.

$$e_2 = M\frac{\Delta i_1}{\Delta t} = 0.4 \times 90 = 36 \text{ V} \qquad\qquad ☐$$

38.9 THE TRANSFORMER

If an iron core (Fig. 38.10) is wound with two separate coils, and if an alternating current is maintained in one of them, an alternating emf of the same frequency is induced in the other. Such a device is known as a *transformer*. The first coil is called the *primary* and the other the *secondary*. Power is transferred from the primary to the secondary by *mutual inductance*.

The change from one ac potential difference to another can be made very efficiently at low frequency by means of a transformer. For example, at 60 Hz the efficiency of a large transformer may be 99 percent, while smaller transformers may be 90 percent efficient. Some power is dissipated in hysteresis and eddy currents induced in the iron by the changing magnetic flux. Also, there is some Joule heating (I^2R) in the windings, but in a well-designed transformer the power delivered to the secondary

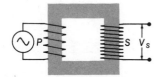

FIGURE 38.10
Transformer designed to step up the alternating potential difference across the primary.

is almost equal to the power supplied to the primary. If V_p and V_s are the primary and secondary potential differences† and I_p and I_s the corresponding currents,

$$V_p I_p \approx V_s I_s \tag{38.6}$$

Further, for this well-designed transformer the same magnetic flux links both the primary and the secondary windings. Consequently, the ratio of the emf induced in the secondary to that applied to the primary is equal to the ratio of the number of turns in the secondary N_s to the number in the primary N_p:

$$\frac{\mathcal{E}_s}{\mathcal{E}_p} = \frac{N_s}{N_p} \tag{38.7}$$

In an ideal transformer there are no losses, and the induced emfs in the primary and secondary are equal to the corresponding terminal potentials V_p and V_s. From $\mathcal{E}_p = V_p$, $\mathcal{E}_s = V_s$, and Eqs. (38.6) and (38.7), we can write

$$\frac{V_s}{V_p} = \frac{\mathcal{E}_s}{\mathcal{E}_p} = \frac{N_s}{N_p} = \frac{I_p}{I_s} \tag{38.8}$$

Where the emf is large, the current is small; where the emf is small, the current is large. By means of a transformer a small emf and a large current can be transformed into a large emf and a small current, or vice versa.

From Eq. (38.8) cross multiplication gives $V_s I_s = V_p I_p$. In words the power output of the secondary is equal to the power input of the primary in the ideal transformer. Since ordinary transformers have some energy losses to heat, the power input has to exceed the power output; however, transformers often have high efficiencies, as we saw above.

In small transformers, where the ratio of transformation is not large, economy of construction and efficiency of operation are obtained by using the same coil for both primary and secondary. Such a transformer is known as an *autotransformer*. The arrangement and connections of the coil are shown in Fig. 38.11. The entire coil AC is the primary of the transformer, and the part between B and C is the secondary. Equation (38.8) is applicable to autotransformers as well as to other types, provided always that heat losses are negligible.

At higher frequencies eddy-current and hysteresis losses in iron become prohibitive, so no iron is used. In such an air-core transformer often only a fraction of the magnetic flux from the primary links each turn of the secondary. Under these conditions the transformer equations above are no longer good approximations, and the power delivered to the secondary may be much less than the power supplied to the primary.

38.10 THE INDUCTION COIL

An illustration of mutual inductance is found in the induction coil, which is constructed as shown in Fig. 38.12. A primary coil made of a few turns of heavy copper wire is wound around an iron core. Insulated from this primary coil is the secondary coil, which is wound on the outside of the

---

† Here we may regard the values of the V's and I's to be either the maximum values attained during the cycle or the "effective" values as defined in Sec. 40.2. We must be consistent and use all maximum values or all "effective" values.

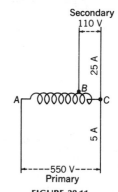

Secondary
110 V

25 A

B

A —〇〇〇〇〇〇〇〇〇— C

5 A

---- 550 V ----
Primary

FIGURE 38.11
Autotransformer designed to step down the potential difference from 550 to 110 V.

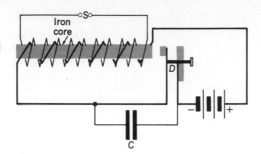

FIGURE 38.12
An induction coil.

primary. The secondary contains a large number of turns of fine, well-insulated wire. By making or breaking the current in the primary at D an emf is induced in the secondary. In order to make and break the current in the primary, an interrupter similar to that of the doorbell (Sec. 37.10) is connected in the circuit.

When the primary circuit is broken, the current is rapidly reduced to zero and an emf is induced in the secondary. This emf is large because the number of turns in the secondary is very large and the time in which the primary current is stopped is short. To get the greatest induced emf, the primary current must be stopped as quickly as possible. To effect this, a capacitor C is connected across the gap in the primary. It acts as a storage place into which the charge surges when the circuit is broken. Without this capacitor an arc would be established at the contacts when the circuit is opened. Such an arc pits the contacts and results in a slower stopping of primary current. There is also an emf induced in the secondary when the primary current is rising. This emf is much smaller, because the time required for the current to build up to its maximum value is long compared with the time required to stop the current. The primary current and the induced emf in the secondary are plotted as a function of time in Fig. 38.13.

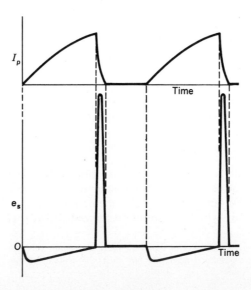

FIGURE 38.13
How the current I_p in the primary of an induction coil and the emf e_s induced in the secondary vary with time.

The spark coil of an automobile is an induction coil similar to the one just described, except that instead of a mechanical vibrator opening and closing the primary circuit a cam on the distributor shaft opens the "points" at the instant a high potential is needed to make a spark between the terminals of a spark plug. The secondary potential is many thousand volts, although only 12 V is available in the primary.

38.11 SELF-INDUCTANCE

When the current through a circuit like that in Fig. 38.14 is changing, the magnetic flux linking this circuit is also changing. So long as B is proportional to H, the flux Φ linking the circuit is directly proportional to the instantaneous current i; thus

$$\Phi = Li \qquad (38.9)$$

where L is a constant known as the *coefficient of self-inductance*. If, in a time Δt, the current changes by an amount Δi, the flux changes by $\Delta\Phi$ and there is, by Eq. (38.2), an induced emf

$$e = -\frac{\Delta\Phi}{\Delta t} = -L\frac{\Delta i}{\Delta t} \qquad (38.10)$$

The negative sign appears in Eq. (38.10) because, by Lenz's law, the direction of the induced emf is always such as to oppose the change in current.

Self-inductance is ordinarily measured in henrys. *The self-inductance of a circuit or component is one henry when there is induced an emf of one volt in that circuit or component at an instant when the current is changing at the rate of one ampere per second.*

The self-inductances of many dc circuits are negligibly small, but a coil with many turns, a large solenoid, or an electromagnet may have a large self-inductance. In ac circuits inductance is of great practical importance, since the current is constantly changing. The emf resulting from self-inductance always opposes the *change of current;* it operates to hold down the current when it is rising and to maintain the current when it is decreasing.

Consider the circuit of Fig. 38.15a. When the key K is closed, the current in the circuit begins to rise. If we apply Kirchhoff's second law to the circuit at any instant, we obtain

$$\mathcal{E} - Ri - L\frac{\Delta i}{\Delta t} = 0 \qquad (38.11)$$

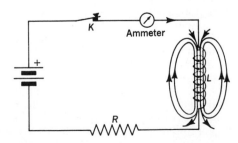

FIGURE 38.14
The flux linking the inductor L changes when the current varies, thereby inducing an emf which opposes the variation in current.

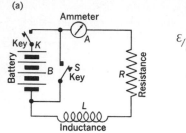

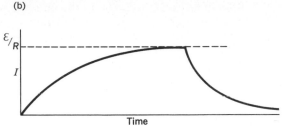

FIGURE 38.15

Inductor circuit (*a*) for studying the rise and decay in current (*b*).

At the instant the switch is closed, i is zero, and $\mathcal{E} = L\,\Delta i/\Delta t$. As time goes on, i increases and eventually the value $I = \mathcal{E}/R$ (Fig. 38.15*b*) is attained.

If the battery in Fig. 38.15 is suddenly shorted out of the circuit by closing the key S, the emf of self-induction keeps a current passing through the circuit until all the energy stored in the magnetic field of the inductor is dissipated. The current in the circuit is shown as a function of time after closing S in the right half of Fig. 38.15*b*.

When a current I passes through an inductor L, the energy W stored in the magnetic field is given by

$$W = \tfrac{1}{2}LI^2 \tag{38.12}$$

as can readily be shown by the application of integral calculus (see derivation in Appendix C.1). The energy calculated from Eq. (38.12) is expressed in joules when L is in henrys and I in amperes.

QUESTIONS

1. A current is induced in a coil by inserting a bar magnet. What is the source of the electric energy dissipated in the circuit?
2. How is Lenz's law related to the principle of conservation of energy?
3. How does one obtain the thousands of volts needed for the spark in the cylinder of an automobile when one starts with 12 V from a battery?
4. What happens to the work done against the back emf of an inductor in establishing a current?
5. A dc series circuit includes an inductor and a switch. If 10 A passes through the circuit, is there any limitation on how quickly the current can be reduced to zero by opening the switch? Why is there often a spark across the terminals of a switch in a dc circuit when the switch is opened?
6. When a switch in a series circuit consisting of a battery, a switch, and an inductor is opened, the induced emf is far larger than when the switch is closed. Why?
7. A strongly magnetized rod with its axis vertical is dropped down a long evacuated copper tube, which is vertical and just slightly bigger in diam-

eter than the rod. Why is it that even if it never touches the tube, the rod reaches a constant terminal velocity? What provides the upward force equal to the weight of the rod? What would happen if the rod were shot down the tube with an initial speed greater than the terminal speed reached after dropping?

8. What happens if a battery is connected across the primary of a transformer?
9. How can a wire-wound resistor be made so that it is essentially noninductive?
10. What factors determine the mutual inductance of two coils?
11. When a sensitive moving-coil galvanometer is to be shipped or carried about, it is customary to short-circuit the coil terminals. Why is this a good idea?
12. A closed conducting plane loop is moved perpendicular to a uniform magnetic field. Is there an induced emf when the plane of the loop is (*a*) parallel to the motion and (*b*) perpendicular to the motion? Discuss the proposal that the groundspeed of an airplane be measured by determining the emf induced in a horizontal loop fixed in the plane by the vertical component of the earth's magnetic field.

13. Why is there a capacitor connected in parallel with the breaker points in the ignition system of an automobile?

14. What kind of energy is stored in an inductor? Why don't we use a big inductor to store energy to start an automobile in preference to a storage battery?

PROBLEMS

1. A square loop of wire 75 mm on a side lies with its plane perpendicular to a uniform magnetic field of 0.8 T. (*a*) Find the magnetic flux through the loop. (*b*) If the coil is rotated through 90° in 0.015 s in such a way that there is no flux through the loop at the end, find the average emf induced during the rotation.
   Ans. (*a*) 0.0045 Wb; (*b*) 0.3 V

2. A small "search" coil with an area of 125 mm^2 has 50 turns of very fine wire. This coil is placed between the pole pieces of a small magnet and then suddenly jerked out. If the average induced emf is 0.07 V when the coil is pulled to a field-free region in 0.06 s, what is the magnetic intensity between the poles? What was the original flux through each turn?

3. A coil of 275 turns with an area of 0.024 m^2 is placed with its plane perpendicular to the earth's field and is rotated in 0.025 s through a quarter turn, so that its plane is parallel to the earth's field. What is the average emf induced if the earth's field has an intensity of 8 × 10^{-5} T? What was the original flux through each turn?
   Ans. 0.0211 V; 1.92 μWb

4. The secondary of an induction coil has 12,000 turns. If the flux linking the coil changes from 7.4 × 10^{-4} to 4 × 10^{-5} Wb in 1.8 × 10^{-4} s, how great is the induced emf?

5. The magnetic induction *B* in the core of a spark coil changes from 1.4 to 0.1 T in 0.21 ms. If the cross-sectional area of the core is 450 mm^2, find the original flux through the core. What is the average emf induced in the secondary coil if it has 6,000 turns? *Ans.* 630 μWb; 16.7 kV

6. A jet aircraft is flying due south at 300 m/s at a place where the vertical component of the earth's magnetic field is 8 × 10^{-5} T. Find the potential difference between wing tips if they are 25 m apart. Which tip has the higher potential?

7. An axle of a truck is 2.4 m long. If the car is moving due north at 30 m/s at a place where the vertical component of the earth's magnetic field is 9 × 10^{-5} T, find the potential difference between the two ends of the axle. Which end is positive? *Ans.* 6.48 mV; left

8. A horizontal wire 0.8 m long is falling at a speed of 5 m/s perpendicular to a uniform magnetic field of 1.1 T, which is directed from east to west. Calculate the magnitude of the induced emf. Is the north or south end of the wire positive?

9. An emf of 3.5 V is obtained by moving a wire 1.1 m long at a rate of 7 m/s perpendicular to a uniform magnetic field. What is the intensity of the field? *Ans.* 0.455 T

10. An electromagnet has a self-inductance of 8 H. How much energy is stored in the magnetic field when a current of 9 A exists in the coil? What average emf is induced if the current is reduced to zero in 0.15 s?

11. The current in a circuit changes from 24 A to zero in 0.003 s. If the average induced emf is 260 V, what is the coefficient of self-inductance of the circuit? How much energy was initially stored in the magnetic field of the inductor?
   Ans. 32.5 mH; 9.36 J

12. What back emf is induced in a coil of self-inductance 0.008 H when the current in the coil is changing at the rate of 110 A/s? What energy is stored in the inductor when the current is 6 A?

13. The coefficient of mutual inductance between two coils is 8 mH. What emf is induced in the second coil if the current is changing at the rate of 4 kA/s in the first coil? *Ans.* 32 V

14. Two circuits have a coefficient of mutual inductance of 16 mH. What average emf is introduced in the secondary by a change from 40 to 4 A in 6 ms in the primary?

15. When the current in the primary of a small transformer is changing at the rate of 600 A/s, the induced emf in the secondary is 8 V. What is the coefficient of mutual inductance? *Ans.* 13.3 mH

16. The secondary of an ideal transformer has 275 times as many turns as the primary. It is used in a 110-V circuit. What is the voltage across the secondary? If the current in the secondary is 50 mA, what is the primary current?

17. An ideal transformer has 550 turns on the primary and 30 turns on the secondary. What is the maximum output potential difference if the maximum input voltage is 3,300? If the transformer is assumed to have an efficiency of 100 percent, what maximum primary current is required if a maximum current of 11 A is drawn from the secondary? *Ans.* 180 V, 0.6 A

18. A constant potential difference of 60 V is suddenly applied to a coil which has a resistance of 30 Ω and a self-inductance of 8 mH. At what rate does the current begin to rise? What is the current at the instant the rate of change of current is 500 A/s? What is the final current?

19. A toroid with an iron core has an average circumference of 0.4 m and a cross-sectional area of 320 mm². When the current in the windings is 0.6 A, the magnetic induction B is 0.8 T. The relative permeability of the coil is 250. A four-turn fluxmeter connected to a galvanometer is used to determine B in the coil. (a) Calculate the flux through the solenoid. (b) Find the number of turns on the toroidal winding. (c) When the key is opened, the current in the winding and the flux drop to zero. Find the coefficient of mutual inductance between the fluxmeter and the toroidal winding.

   Ans. (a) 256 μWb; (b) 1,698 turns; (c) 1.71 mH

20. The circuit of Fig. 38.14 consists of a 50-V battery, a 20-Ω resistor, a 25-mH inductor, and a key. Find the rate at which current begins to rise when the key is closed, the current at the instant the rate of change of current is 400 A/s, and the final steady current.

21. The series circuit of Fig. 38.14 consists of a 30-V battery, a 6-Ω resistor, an 8-mH inductor, and a switch. Apply Kirchhoff's second law to the circuit to relate the current i and its rate of change $\Delta i/\Delta t$ to the emf, resistance, and inductance of the circuit at any time after the switch is closed. At what rate does the current begin to rise when the switch is closed? Find the back emf induced in the inductor when the current is 3 A and the energy stored in the inductor at that instant. What is the final current?

   Ans. 3,750 A/s; 12 V; 36 mJ; 5 A

22. A ballistic galvanometer with a resistance of 240 Ω gives a full-scale deflection for 5×10^{-4} C of electricity. A coil of 320 turns and 160 Ω resistance is to be constructed to study fields up to 1.4 T by observing deflections produced when the coil is suddenly removed from the field. What is the maximum area allowable for the coil if full-scale deflection is not to be exceeded?

23. A coil of 100 turns with a radius of 6 mm and a resistance of 40 Ω is placed between the poles of an electromagnet and suddenly removed. A charge of 32 μC is sent through a ballistic galvanometer connected to the coil. The resistance of the galvanometer is 160 Ω. What is the intensity of the magnetic field?   *Ans.* 0.566 T

24. Show that for a long narrow toroid of average circumference $2\pi r$, cross-sectional area A, number of turns N, and permeability of core $K_m\mu_0$, the coefficient of self-inductance is given by $L = N^2\mu_0 K_m A/2\pi r$ and the energy stored per unit volume in the magnetic field of the toroid is $\frac{1}{2}BH$.

25. A toroid of 0.5 m circumference and 480 mm² cross-sectional area has 2,500 turns bearing a current of 0.6 A. It is wound on an iron ring with relative permeability of 350. Find the magnetizing field H, the magnetic induction B, the flux Φ, the coefficient of self-inductance L, and the energy stored in the magnetic field.

   Ans. 3 kA/m; 1.32 T; 0.633 mWb; 2.64 H; 0.475 J

Mechanical energy can be converted into electric energy through the processes described in Chap. 38. In this chapter we are interested in practical arrangements for generating substantial emfs and transforming large amounts of mechanical energy into electric energy. In some cases the same apparatus can serve as a generator and can also be used to develop mechanical power from electric power. A device which performs this latter function is known as an electric motor.

GENERATORS AND MOTORS

39.1 INSTANTANEOUS ELECTROMOTIVE FORCE

When a single loop of wire is rotated at constant angular velocity in a uniform magnetic field (Fig. 39.1), the instantaneous emf induced in the loop varies sinusoidally in time (Sec. 38.7). One full cycle is completed each rotation. If ν represents the frequency (or number of cycles completed in unit time), the emf e at any instant is, by Eq. (38.4),

$$e = \mathcal{E}_{max} \sin \theta = \mathcal{E}_{max} \sin 2\pi\nu t \qquad (39.1)$$

Both the angle θ and the time t are measured from the instant the plane of the coil is perpendicular to the magnetic intensity B. In this equation and subsequent ones we use lowercase letters to represent the instantaneous values of quantities which vary in time and capital letters to indicate quantities which are constant.

39.2 COLLECTING RINGS

If the loop of Fig. 39.1 is opened at the axis and an outside circuit is connected to the two leads, the induced emf can be applied to the external circuit. To make continuous connection to the outside circuit, the ends of the wire forming the coil are fastened to rings (Fig. 39.2a) mounted on the axis of the rotating loop. Sliding connectors, called *brushes*, complete the circuit. The slip rings rotate with the loop, while the nonrotating brushes are pressed against the rings.

When an alternating emf is applied to a circuit which has resistance R only (no appreciable inductance or capacitance), the current i at any instant is simply the ratio of the applied potential difference to the resistance:

$$i = \frac{e}{R} = \frac{\mathcal{E}_{max} \sin 2\pi\nu t}{R} = I_{max} \sin 2\pi\nu t \qquad (39.2)$$

The current and the emf reach their maximum values at the same instant and pass through their zeros together. The current is an alternating one and is said to be *in phase* with the emf, since the phase (the argument of the sine function) is the same for both at any time.

39.3 THE COMMUTATOR

The emf induced in a loop rotating in a magnetic field is an alternating one. To obtain a current which is always in the same direction through an external circuit, the terminals of the rotating loop may be joined to a *commutator*, which is a ring divided into two segments, as shown in Fig. 39.2b. Against this divided ring press two brushes which are connected to

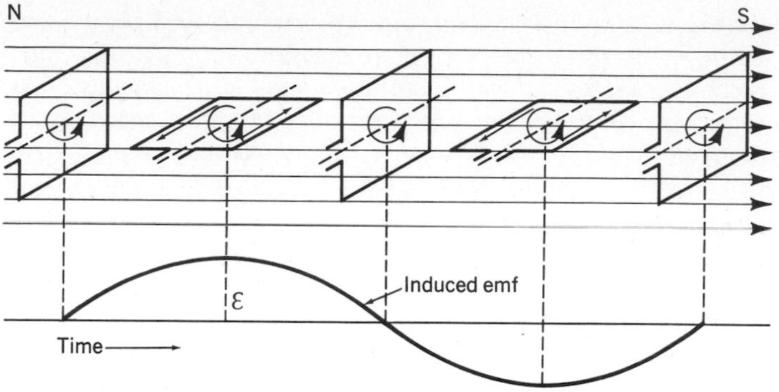

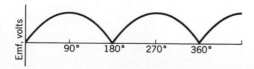

FIGURE 39.1
The motional emf induced in the rotating loop is greatest when the effective conductors are moving perpendicular to the magnetic intensity.

(a)

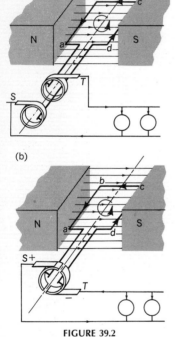

(b)

FIGURE 39.2
Connecting rings for generators. (*a*) Slip, or collection, rings provide continuous connection to an external ac circuit. (*b*) A split-ring commutator delivers unidirectional current to an external circuit.

the external circuit and so placed that they slip from one segment of the commutator to the other at the instant the emf of the revolving coil passes through zero.

When the loop of Fig. 39.2*b* is rotating counterclockwise, the wire *ab* which is moving downward has an emf in the direction of the arrow, while the current in the wire *cd* is in the opposite direction. Hence, current leaves the brush *S* and passes through the external circuit in the direction of the arrow from *S* to *T*. After the loop has made one-half revolution, the segment of the commutator which was in contact with brush *T* now makes contact with brush *S* and the other segment makes contact with brush *T*. The wire *ab* is now moving upward, and current in it is reversed. The wire *cd* is moving down, and current in it is also reversed. Because the segments of the commutator have reversed their positions, the current still leaves by brush *S* and passes through the external circuit in the direction in which it flowed originally. In the revolving coil the current alternates; in the external circuit the current and the potential difference are pulsating but *unidirectional* (Fig. 39.3).

39.4 PRACTICAL GENERATORS

In most practical generators several coils are rotated simultaneously in a magnetic field. These coils are connected in such a way as to give the desired emf and current.

Consider a dc generator with two coils whose planes are mutually perpendicular (Fig. 39.4). If the coils are connected in series through a suitable commutator arrangement, the fluctuation in the output emf is much smaller than the fluctuations in output of each coil separately, as shown in Fig. 39.5. By adding more coils, the resultant output can be made constant except for small fluctuations known as the *commutator ripple*.

The wires which rotate past the poles of a generator are ordinarily embedded in slots in an iron core. The rotating system involving the iron

FIGURE 39.3
The potential difference between the brushes of the split-ring commutator of Fig. 39.2*b* as a function of the angular position of the coil.

core and the wires is known as the *armature*. The purpose of the iron core is to increase the magnetic intensity **B** and thereby to increase the emf. There are emfs induced not only in the copper wires of the armature but in any material which moves in the magnetic field, including the iron of the armature. If the armature were a solid piece of iron, these induced emfs would lead to large currents circulating in the iron and much energy would be dissipated as heat by these current whirlpools, which are called *eddy currents*. To reduce the eddy currents, the iron core is constructed from thin sheets or *laminations* separated by thin insulating layers to make the resistance high.

Not only does the practical generator have several coils mounted on the same armature, but these coils often pass several magnetic poles during each revolution. Four- and six-pole generators are common; in some applications as many as twenty-four poles are passed in a single revolution. In a six-pole generator three cycles are completed in each revolution.

An emf may equally well be induced in a stationary coil by the varying of a magnetic field; one way of doing this is by rotating magnets.

39.5 EXCITATION OF THE FIELDS OF GENERATORS

Generators and motors are sometimes classified according to how their magnetic fields are produced. In one of the simplest generators the magnetic field is produced by a permanent magnet. An example of such a generator is the *magneto*, shown in Fig. 39.6.

Most large generators have fields which are produced by electromagnets. Such generators can be classified as self-excited or separately excited, depending on whether the generator produces its own magnetic field or uses some other source to excite the electromagnet. Self-excited generators may be series-, shunt-, or compound-wound. In the series-wound generator (Fig. 39.7*a*) the current to the external circuit goes through the field coil, which consists of a few turns of heavy wire. When the generator is not delivering current, the only magnetic field is that due to residual magnetism. The greater the current drawn by the external circuit, the greater the current in the field coils and therefore the greater the magnetic field in which the armature rotates. When the generator is operated at constant speed, the greater the current, the greater the emf generated. Series-wound generators are relatively uncommon because one does not ordinarily want an emf which rises rapidly as more current is drawn.

In the shunt-wound generator the field coils are in parallel with the external circuit (Fig. 39.7*b*). When such a generator is operated at constant speed, if the current in the external circuit is increased, the terminal potential difference V_t of the generator decreases for the following

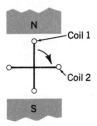

FIGURE 39.4
Generator with two coils with planes mutually perpendicular.

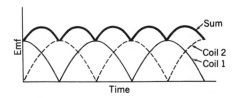

FIGURE 39.5
Connecting the commutated outputs of the two coils of Fig. 39.4 in series leads to a steadier output emf than either coil produces alone.

FIGURE 39.6
Distribution of the magnetic flux in a magneto.

reason. The terminal potential is the induced emf $\mathcal{E}$ minus Ir, where r is the resistance of the armature and I the armature current. As I increases, the terminal potential decreases, and so does the current in the field coils (given by V_t divided by the resistance of the field coils). This reduces the magnetic field and the induced emf. Therefore, a shunt-wound generator provides a potential which decreases as the current drawn is increased. Such a current-voltage response is not commonly desired.

When the terminal potential supplied by a generator is to be essentially independent of the current load, a compound-wound generator (Fig. 39.7c) is used. Such a generator has field coils which are composed of a series-wound part and a shunt-wound part. The characteristic of a series-wound field (increased emf with increasing current demand) and the characteristic of a shunt-wound generator (terminal potential decreasing with load current) can be combined to yield a generator with almost constant terminal potential independent of load current.

39.6 THE EFFICIENCY OF GENERATORS

The efficiency of a generator is the ratio of the electric energy output of the generator to the energy input, or, alternatively, the ratio of the electric power output to the power input. The power output of the generator is the product of the terminal potential and the current to the external circuit. The power input is greater, of course; it is equal to the power output plus the various losses, such as the I^2R heat loss in the armature and the field coils, hysteresis losses, eddy-current losses, and frictional losses.

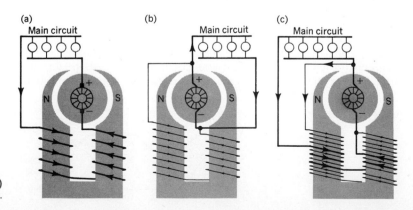

FIGURE 39.7
Generators may be (*a*) series-wound, (*b*) shunt-wound, or (*c*) compound-wound.

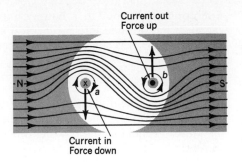

Current out
Force up

b

N→

×
a

S

Current in
Force down

FIGURE 39.8
The torque on a current-bearing coil arises from forces on the individual conductors.

39.7 ELECTRIC MOTOR

When a wire bearing an electric current I is placed in a magnetic field **B** so that the length **l** of the wire is perpendicular to the field, the wire experiences a force $F = IlB$ [from Eq. (36.2) with $\theta = 90°$]. This force is perpendicular to both the magnetic field and the length of the wire. On one side of the wire, the magnetic field due to the current in the wire adds to the field due to the external magnetic circuit, while on the opposite side the two fields oppose. The force on the wire is from the stronger toward the weaker field. If the current is reversed, the force is reversed.

Figure 39.8 shows the cross section of a single current-bearing loop of wire in a uniform magnetic field. Assume that the current enters at a and leaves at b. There results a force downward on wire a and upward on wire b. Thus, there is a torque in a direction such as to produce a counterclockwise rotation about the axis of the coil. In an electric motor there are ordinarily many conductors bearing currents perpendicular to a magnetic field. The resulting torque produces rotation of the armature.

There are many kinds of electric motors. That is, there are many ways to convert electric energy into mechanical energy. The underlying principle in most motors is that of a current-bearing conductor experiencing a force in a magnetic field.

39.8 BACK EMF IN A MOTOR

When the conductors in the armature of a motor rotate in a magnetic field, an emf is induced which opposes the current in the conductors. For this reason it is called a *back emf* or a *counter emf.*

If an incandescent lamp is connected in series with a small shunt-wound motor while the armature is held stationary, the lamp glows as if the motor were absent, since the armature resistance is small. When the armature is allowed to revolve, the lamp becomes dim because the emf generated in the armature opposes the impressed emf, thereby reducing the current in the circuit. In a constant magnetic field, the faster the armature revolves, the greater the back emf and the smaller the armature current.

For shunt-wound motors the potential difference across the armature is the potential V_t applied to its terminals. The terminal potential of the armature is equal to the back emf plus the potential drop across the armature resistance r_a due to the armature current I_a:

$$V_t = \mathcal{E}_b + I_a r_a \tag{39.3}$$

A motor is analogous to a battery under charge, except that electric energy is transformed into mechanical energy rather than chemical energy. If Eq. (39.3) is solved for the armature current, it yields

$$I_a = \frac{V_t - \mathcal{E}_b}{r_a} \tag{39.3a}$$

The electrical work done against the back emf appears as mechanical energy output of the motor. Indeed, the mechanical power output P of the motor is equal to the product of the back emf $\mathcal{E}_b$ and the armature current:

$$P = \mathcal{E}_b I_a \tag{39.4}$$

This follows directly from the fact than $\mathcal{E}_b$ is the electrical work per unit charge performed to produce motion of the conductors, while I_a is the charge per second passing through the conductors.

Since the armature current of a motor depends on the difference between the terminal potential and the back emf, the current is largest when the armature is at rest, for in this case the back emf is zero. In order to keep the current in a large motor from being excessive at the start, it is customary to introduce a series resistor that reduces the current flowing in the circuit. As the speed of the motor increases and the back emf becomes appreciable, this resistance is removed from the circuit. Such a resistance is called a *starting resistance* or a *starting box*.

39.9 SERIES-WOUND AND SHUNT-WOUND MOTORS

When large starting torques are required, motors having field coils which are in series with the armature are used. In such a series-wound motor the back emf is small when the motor is starting and the current drawn is large. Since this same current passes through the series field coils, the magnetic field is also large. The starting torque is proportional to IlB, the force on an individual conductor. As the motor gains speed, the back emf increases and both armature current and field strength decrease. Therefore, the torque is much smaller at high speeds. Series-wound motors are often used for rock crushers and other machines in which particularly large starting torques are desired.

In shunt-wound motors the field coils are in parallel with the armature coils. For a constant applied potential difference, the field strength is constant, and the torque is directly proportional to the armature current, whereas in the series-wound motor the torque is more nearly proportional to the square of the current. Most common small electric motors are shunt-wound.

The efficiency of a motor can be obtained by dividing the mechanical power output of the motor by the electric power input. The difference between the input and output power is dissipated in Joule heat, overcoming air friction, etc.

39.10 DYNAMOS

The fundamental components of many motors are identical with those of certain generators. The single term *dynamo* describes an electric device which can function as either motor or generator. When a dynamo is supplied with a potential difference and energy by an electric source and does mechanical work, it is a motor. When it is driven by a mechanical torque, it develops electric energy and acts as a generator. Not every motor can be operated as a generator, but many can.

There are many applications for a dynamo operating part-time as a motor and the rest of the time as a generator. An example is the dynamo used in diesel-electric locomotives. When the train is set in motion, diesel engines drive generators to supply current to the dynamos on the axles of the driving wheels. These dynamos exert large torques on the wheels and put the locomotive in motion. To slow the train down the dynamo is disconnected from its power supply, leaving the armature rotating in a magnetic field as long as the wheels turn. An emf is induced in the armature; if the terminals are connected through a resistor, a current flows. The dynamo is now operating as a generator, converting kinetic energy of the train into electric energy, and thus braking the train. In diesel-electric locomotives the electric energy is dissipated in heat, but in electric locomotives it may be fed back into the power lines.

39.11 WATT-HOUR METER

The Thomson form of recording watt-hour meter (Fig. 39.9) consists of a little shunt motor whose armature turns at a speed proportional to the power supplied to it. The armature is geared to dials which record the energy in kilowatt-hours. The stationary field coils L and M are connected in such a way that the field strength of the motor is proportional to the current to the load. The armature is connected across the line, as a voltmeter would be connected, so that the armature current is proportional to the potential difference. The torque which turns the motor is a constant times the product of the current in the armature and the magnetic field in which the armature turns. Hence, the torque which turns the armature is proportional to the product of the current supplied to the

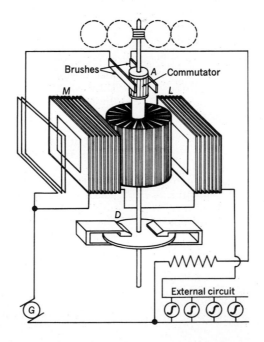

FIGURE 39.9
A watt-hour meter.

load and the potential at which it is supplied, i.e., to the power received.

To ensure that the motor does not run too fast and to enable it to stop as soon as the current ceases, an electromagnetic brake *D* is attached to it. This brake consists of an aluminum disk rotating between the poles of a permanent magnet. The eddy currents induced in the disk by its rotation between the poles of the magnet retard its motion and stop it as soon as the current ceases. This type of watt-hour meter can be used with either direct or alternating current.

QUESTIONS

1. Does an electric generator generate electricity? Explain exactly what it does generate.
2. How is electric energy used to do external work? Explain how an electric motor can accomplish this task.
3. Why is iron used in motors and generators? Why not replace the iron by a nonconducting plastic and thereby eliminate the eddy currents in the iron?
4. Does reversing the connections on a dc shunt-wound motor reverse the direction of rotation? Why? If it does not, how can the rotation direction be reversed?
5. Why does a dc generator ordinarily have many commutator segments rather than only two?
6. In electrically powered trains the same dynamo often serves as both motor and generator. Why do automobiles typically have both a generator and a starting motor rather than a single dynamo to perform both functions?
7. What is the response of a shunt-wound generator to an increase in load current? Of a series-wound generator? Of a compound-wound generator?
8. Why are series-wound motors used when very high starting torques are desired?
9. How could you make a generator which would have an emf constant in magnitude and direction?

PROBLEMS

1. Part of the windings of an electric motor is a segment of wire 165 mm long which lies perpendicular to the magnetic field of 1.1 T. Find the force on this segment when it bears a current of 4 A. What torque is produced if this wire is 45 mm from the axis of the motor?
   Ans. 0.726 N; 0.0327 N-m
2. In a motor a conductor 125 mm long carries a current of 3 A. If it lies perpendicular to a uniform magnetic field of 1.2 T, find the force on the conductor. What torque is produced if the conductor is 50 mm from the axis of the motor?

3. A rectangular loop of wire 8 by 15 cm bears a current of 6 A. The plane of the coil is parallel to a uniform magnetic field of 1.2 T, with its 15-cm length perpendicular to the lines of flux and the other side parallel to them. Find the force on each side of the loop. Using this force, find the torque on the loop. Find the magnetic moment of the loop and use its value to check the value of the torque.
   Ans. 1.08 N; 0; 0.0864 N-m; 0.072 A-m²
4. Find the torque on a rectangular loop of wire bearing a current of 9 A when its plane is parallel to the magnetic lines of a field with an intensity of 1.25 T. The dimensions of the loop are 0.25 m parallel to the flux and 12 cm at right angles to it.
5. The flux density between the poles of a certain motor is 1.1 T. A conductor carrying 8 A lies in this magnetic field so that it is perpendicular to the field. Find the force on it if 0.1 m of the conductor is in the field. If the wire moves 15 mm perpendicular to the field in 0.002 s, and if the field is uniform over this distance, find the work done on the wire. Show that the same work is predicted by finding the induced back emf and multiplying it by the charge transferred in the 0.002 s. *Ans.* 0.88 N; 0.0132 J; 0.825 V
6. A separately excited constant-speed generator that develops 118 V at no load furnishes only 112 V when a current of 50 A is drawn. What is the resistance of the armature?
7. The armature of a separately excited generator has a resistance of 0.16 Ω. When run at its rated speed, it yields 132 V on open circuit and 126 V on full load. What is the current at full load? How much power is delivered to the external circuit? What power is needed to drive the generator if its overall efficiency is 85 percent?
   Ans. 37.5 A; 4.73 kW; 5.56 kW or 7.45 hp
8. The brush potential of a separately excited generator when it is delivering 5 A is 125 V. When the generator delivers 15 A, the potential difference across the brushes falls to 122 V. What are the induced emf and the resistance of the armature?
9. The armature of a generator has a resistance of 0.2 Ω. When the current through the armature is 5 A, the terminal potential is 224 V. What will the

terminal potential be when the current is 40 A, assuming that the field strength and the speed remain unchanged? What is the emf induced?
Ans. 217 V; 225 V

10. The armature of a shunt-wound generator has a resistance of 0.12 Ω. The terminal potential difference of the generator is 118 V when the armature current is 60 A. Find the emf of the generator and the power dissipated in heat in the armature.

11. A shunt-wound generator delivers 48 A to an external load at a brush potential of 120 V. The field coils have a resistance of 60 Ω, and the armature has a resistance of 0.14 Ω. If the stray-power loss is 500 W, what is the efficiency of the generator? *Ans.* 84 percent

12. A shunt motor takes a total current of 9 A from 120-V mains. The resistance of the armature is 1.5 Ω, and that of the field coils is 240 Ω. Find the current in the field coils and in the armature, the back emf induced, and the mechanical power output of the motor.

13. A shunt-wound motor draws 2.9 A from a 120-V line. The field coils have a resistance of 300 Ω, and the armature resistance is 0.8 Ω. Find the armature current and the back emf. What is the mechanical power output of the motor? How much power is dissipated in I^2R heating in the motor? *Ans.* 2.5 A; 118 V; 295 W; 53 W

14. A motor running at full load on a 118-V line develops a back emf of 109 V and draws a current of 8 A through the armature. What is the mechanical power output of the motor, disregarding frictional losses? What is the armature resistance?

15. In the circuit of the accompanying figure a battery of 64 V emf and 1 Ω internal resistance is providing electric energy to operate a motor which has an armature resistance of 0.5 Ω. If $R_3 = 12\ \Omega$, $R_4 = 6\ \Omega$, $R_1 = 2.5\ \Omega$, and $R_2 = 3\ \Omega$, what is the back emf of the motor when the armature draws 2 A? What is the mechanical power output of the motor? *Ans.* 42 V; 84 W

16. In the circuit of the accompanying figure, $\mathcal{E}_a = 80$ V, $r_a = 1.5\ \Omega$, $R_2 = 3\ \Omega$, $\mathcal{E}_b = 32$ V, $r_b = 0.5\ \Omega$, $R_3 = 15\ \Omega$, $R_4 = 10\ \Omega$, and the current through the generator is 3 A. Find R_1 and the terminal potential differences of the generator and battery.

17. In the circuit of the accompanying figure, $\mathcal{E}_a = 100$ V, $r_a = 0.8\ \Omega$, $R_1 = 6\ \Omega$, $R_2 = 5\ \Omega$, $\mathcal{E}_b = 36$ V, $r_b = 0.2\ \Omega$, $R_3 = 20\ \Omega$, and $R_4 = 5\ \Omega$. Find the current in the main circuit and the terminal potential differences of the generator and battery. *Ans.* 4 A; 96.8 V; 36.8 V

18. In the circuit of the accompanying figure, $\mathcal{E}_a$ is the back emf of a motor. It has a value of 30 V. The armature resistance and the resistance of the

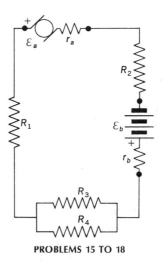

PROBLEMS 15 TO 18

battery are both 1 Ω, and all other resistances are 8 Ω. Find the emf of the battery if the motor draws 2 A. What are the terminal potential and the mechanical power output of the motor?

19. A shunt-wound motor with an armature resistance of 0.3 Ω operates on a 120-V circuit. (*a*) Find the starting resistance which must be temporarily connected in series with the armature to limit the starting current to 10 A. (*b*) Find the current drawn and the mechanical power output when the back emf generated is 118 V.
Ans. (*a*) 11.7 Ω; (*b*) 6.67 A; 787 W

20. A shunt-wound motor has an armature with a resistance of 0.2 Ω and operates from a 220-V line. What starting resistance is required if the armature current must be limited to 11 A? When the motor is operating normally with the starting resistor out, the armature draws 4 A. Find the back emf and the mechanical power output.

21. The armature of a motor connected to a 220-V line draws 25 A at the instant the connection is made, when the motor is at rest. There is a starting resistance of 8.5 Ω in the circuit. What is the resistance of the armature? What is the counter emf when the current drawn is 15 A (with the starting resistance shorted out)?
Ans. 0.3 Ω; 215.5 V

22. A series-wound motor has an armature with a resistance of 0.5 Ω and field coils with a resistance of 4.5 Ω. The motor draws 8 A at 120 V. Find the power supplied to the motor, the back emf induced in the armature, the power transformed into heat, and the mechanical power output.

40
ALTERNATING CURRENTS

When a coil is rotated in a uniform magnetic field, an alternating emf is generated. Because it is particularly easy to generate alternating emfs, and because transformers make it economically feasible to increase or decrease an alternating potential difference, most of the practical generation and distribution of electric power utilizes alternating currents. Many of the ideas we have treated for the simpler dc circuits can be carried over to ac circuits, but there are also many other factors which must be considered. In this chapter we discuss these factors.

40.1 THE ALTERNATING EMF

If the armature of a generator is rotated in a uniform horizontal magnetic field, the instantaneous emf induced in a single loop can be represented by a sine curve (Fig. 40.1) in which the angle through which the coil has turned from a vertical plane is plotted on the horizontal axis and the corresponding induced emf on the vertical axis (see Sec. 38.7). The instantaneous emf e is given by

$$e = \mathcal{E}_{max} \sin \theta = \mathcal{E}_{max} \sin 2\pi \nu t \tag{40.1}$$

where $\mathcal{E}_{max}$ is the maximum emf induced, θ is the angle through which the coil has turned from the position in which the conductors are moving parallel to the magnetic field, ν is the frequency, and t is the time.

It is often convenient to think of the sine wave of Fig. 40.1 as being generated by a vector of magnitude $\mathcal{E}_{max}$ rotating with frequency ν. The instantaneous emf is then represented by the vertical component of this vector, as shown in the figure.

The frequency ordinarily used in electric-power generation and distribution in the United States is 60 Hz. In Europe 50 Hz is a common frequency, while in aircraft and guided missiles 400-Hz generators are widely used. In telephones and in the audio amplifiers of radio and television sets the frequencies involved are the acoustical frequencies themselves. The frequencies used for radio transmission in the standard broadcast band range from 535 to 1,605 kHz; there are other bands at higher and lower frequencies. Television, radar, and microwave transmission is carried out at much higher frequencies.

40.2 EFFECTIVE VALUE OF ALTERNATING CURRENT

The power dissipated in heat in a resistor by a direct current is I^2R. For a direct current the maximum value, the average value, and the instantaneous value are all the same. For an alternating current, the *average* value of current over a whole cycle is zero; the current fluctuates between the maximum value in one direction and the same maximum value in the opposite direction. Nevertheless, an alternating current is effective in producing heat in a resistor. If the instantaneous value of the current is i, i^2 is always positive, even though i is negative during one-half of each cycle.

When an instantaneous potential difference $v = V_{max} \sin 2\pi \nu t$ is applied across a pure resistor R, the instantaneous i is $(V_{max} \sin 2\pi \nu t)/R = I_{max} \sin 2\pi \nu t$ (Sec. 39.2). The current is in phase with the applied potential difference, and Ohm's law is directly applicable to instantaneous current

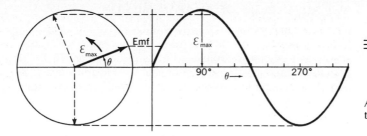

FIGURE 40.1
An alternating emf varies in time like
the vertical component of a vector rotating at the same frequency.

and maximum current, provided the appropriate potential difference is
used. Thus,

$$i = \frac{v}{R} \quad \text{and} \quad I_{max} = \frac{V_{max}}{R}$$

The power dissipated in a resistor R at any instant is vi, the product of
the instantaneous values of potential difference and current (Fig. 40.2).
Since $v = iR$, the instantaneous power is i^2R, and the average power
dissipated over each cycle is $\overline{i^2}R$, where $\overline{i^2}$ is the average value of the
square of the current. We call $I = \sqrt{\overline{i^2}}$ the *effective value* of the alternating current. *The effective value of an alternating current is the value of that
direct current which would produce heat at the same rate as the alternating
current in a given resistor.* An alternating current is said to have an effective value of 2 A when it generates heat in a 1-Ω resistor at a rate of 4 W,
the same rate at which a direct current of 2 A produces heat in a 1-Ω
resistor. Since the heating effect depends on the square of the current,
this is equivalent to saying that the average value of the square of the
alternating current is the same as the square of the corresponding direct
current. The effective value of an alternating current is found by taking
the average of the square of the instantaneous current and then extracting its square root. Sometimes the effective value of the alternating
current obtained in this way is called the *root-mean-square* (rms) current.
   The effective value of a sine-wave current is given† by

$$I_{eff} = \frac{I_{max}}{\sqrt{2}} = 0.707 I_{max} \tag{40.2}$$

where I_{max} is the maximum value of the current. Similarly, the effective
value of an alternating potential difference is 0.707 times the maximum;
i.e.,

$$V_{eff} = \frac{V_{max}}{\sqrt{2}} = 0.707 V_{max} \tag{40.2a}$$

Ordinary ac ammeters and voltmeters are calibrated to read effective
values. When we speak of a 110-V ac circuit or of an alternating current
of 5 A, we are specifying effective values; the maximum values are 156 V
and 7.07 A, respectively. Hereafter we use V and I without subscripts to
denote effective values.

---

† These relations and several others in this chapter can be obtained readily by use of calculus.
The derivations will be found in Appendix C.

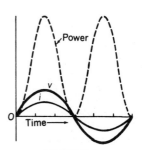

FIGURE 40.2
Instantaneous values of the potential
difference v, current i, and power dissipated in a resistor as functions of time
for an alternating potential difference.

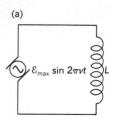

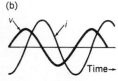

FIGURE 40.3

A circuit (*a*) containing a source of alternating emf and an inductor of negligible resistance has (*b*) current *i* that lags behind the applied potential difference by 90°.

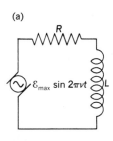

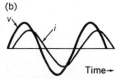

FIGURE 40.4

A series circuit (*a*) consisting of a source of alternating emf, resistance, and inductance has (*b*) current *i* that lags behind the applied potential difference *v* by an angle θ such that $\tan \theta = 2\pi v L/R$.

40.3 ADVANTAGES OF AC TRANSMISSION

One of the major reasons for the widespread use of alternating currents is that it is possible to change from one potential difference to another simply and efficiently by use of transformers (Sec. 38.9). This is highly desirable, since it is most economical to transmit electric power at potentials which are much too high to be used safely in a home or factory.

To illustrate why it is desirable to transmit power at high potentials, consider the following practical problem. A power company must deliver 1,100 kW to a distant city over a transmission line which has a resistance of 2 Ω. How much power is dissipated in the delivery line at 110,000 V and at 1,100 V? From $P = VI$ it is clear that the line must deliver 10 A at 110,000 V or 1,000 A at 1,100 V. The I^2R power loss in this line is 200 W at 110,000 V and 2,000,000 W at 1,100 V. In the latter case more power would be dissipated in the line than is delivered to the ultimate consumer!

If these arguments were the only ones involved, it would be desirable to transmit power at potentials of many million volts. However, as the potential increases, there are serious power losses due to corona, and the difficulties of insulation become severe. For these reasons high-voltage power transmission lines typically operate at potential differences of a few hundred thousand volts, although lines operating at about a million volts have been constructed.

40.4 CURRENT IN AN INDUCTIVE CIRCUIT

When an alternating potential of frequency v is applied to a large inductor with negligible resistance (Fig. 40.3a), the current is limited because as the current changes, a back emf is induced in the inductance. At every instant this emf opposes the *change in the current*. If we apply Kirchhoff's second law to this situation, we obtain for the instantaneous emf

$$e = \mathcal{E}_{max} \sin 2\pi v t = L\frac{\Delta i}{\Delta t}$$

We observe that $\Delta i/\Delta t$, the rate at which the current is changing with time, is greatest when e is maximum. When $\Delta i/\Delta t = 0$, e is also zero; this occurs when the current is either at its maximum value or at its minimum value (Fig. 40.3b). Consequently, when a potential difference is applied across a pure inductance, *the current is not in phase with the potential difference*. Instead it reaches its maximum value when the potential is passing through zero and is zero when the potential difference is at its maximum. The current lags 90° behind the potential difference. With calculus it can readily be shown that

$$i = -\frac{\mathcal{E}_{max}}{2\pi v L} \cos 2\pi v t = -I_{max} \cos 2\pi v t \qquad (40.3)$$

The potential-difference and current curves of Fig. 40.3b can be visualized as being generated by two mutually perpendicular vectors rotating together with the same frequency. The maximum current is $\mathcal{E}_{max}$ divided by the quantity $2\pi v L$, which is called the *inductive reactance* of the inductor and is indicated by X_L. The effective current is just $I = \mathcal{E}/2\pi v L = \mathcal{E}/X_L$. The dimensions of X_L are volts per ampere, or ohms.

When a circuit (Fig. 40.4a) contains both resistance and inductance, the current lags behind the potential by some angle less than 90° (Fig. 40.4b).

Kirchhoff's second law applied to the circuit yields

$$e = \varepsilon_{max} \sin 2\pi\nu t = Ri + L\frac{\Delta i}{\Delta t}$$

It is shown in Appendix C that in this case

$$I = \frac{\varepsilon}{\sqrt{R^2 + (2\pi\nu L)^2}} \qquad (40.4)$$

and that the current lags behind the potential by an angle θ such that $\tan\theta = 2\pi\nu L/R$. The quantity by which ε must be divided to get I is called the *impedance* of the circuit and is represented by Z. In this case $Z = \sqrt{R^2 + (2\pi\nu L)^2}$. It can be computed as the resultant of R and $2\pi\nu L$, where R and $2\pi\nu L$ are drawn as vectors at right angles to one another as shown in Fig. 40.5a. The angle θ by which the current lags behind the emf is the angle between Z and R.

An ac voltmeter connected across the resistor R would read IR. Connected across the inductor L, it would read IX_L. The arithmetical sum $IR + IX_L$ is not equal to the applied effective emf ε, since when the potential difference is maximal across the resistor, it is zero across the inductor. Instead we must add the potentials vectorially, with IX_L 90° ahead of IR as shown in Fig. 40.5b, to obtain ε. Obviously, the triangle in this figure is similar to the impedance triangle of Fig. 40.5a; each side is just I times the corresponding side in the impedance diagram.

When no current is being drawn in the secondary of an ordinary transformer, the current in the primary is limited by its inductive reactance. Since the primary of a large transformer ordinarily has a low resistance, the current is controlled almost entirely by the self-inductance when the secondary circuit is open and by the self-inductance and the mutual inductance when there is a current in the secondary.

40.5 THE CAPACITIVE CIRCUIT

When an alternating emf is applied across a capacitor, the current is zero when the applied emf is maximum because the capacitor is charged to the maximum potential difference. As the potential decreases, the capacitor begins to discharge. The current increases and reaches its maximum value when the emf passes through zero. Thus, the current and the emf are out of phase. In the case of a *pure capacitance* there is a 90° phase difference between the two, with the current leading the emf. The effective current I is given by

$$I = \frac{\varepsilon}{1/2\pi\nu C} = \frac{\varepsilon}{X_C} \qquad (40.5)$$

where C is the capacitance and the quantity $1/2\pi\nu C = X_C$ is called the *capacitive reactance*.

When a circuit (Fig. 40.6a) contains both capacitance and resistance, the current is given by

$$I = \frac{\varepsilon}{\sqrt{R^2 + (1/2\pi\nu C)^2}} \qquad (40.6)$$

In this case the current leads the emf (Fig. 40.6b) by the angle θ ($\tan\theta = X_C/R$). The effective potential difference across the resistance is

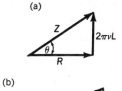

(a)

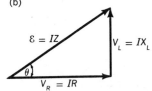

(b)

FIGURE 40.5
(a) Vector diagram for finding the angle by which the current lags behind the applied emf in an ac series circuit containing inductance and resistance. (b) The vector sum of the potential differences IR and IX_L is equal to the applied emf.

(a)

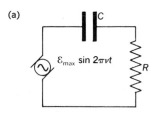

(b)

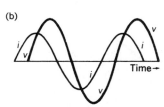

FIGURE 40.6
A series circuit (a) containing capacitance and resistance has (b) current which leads to applied emf by angle θ given by $\tan\theta = X_C/R$.

given by IR, while the effective potential difference across the capacitor is given by IX_C. The vector sum of IR and IX_C gives the applied emf $\mathcal{E}$. Note that instantaneous applied potential difference is the sum of the instantaneous potential differences across the two circuit elements but the effective potential difference applied is not equal to the sum of the effective potential differences because they are not in phase.

40.6 CIRCUIT CONTAINING INDUCTANCE, CAPACITANCE, AND RESISTANCE

In a series ac circuit containing inductance, capacitance, and resistance (Fig. 40.7a), the current is given by

$$I = \frac{\mathcal{E}}{\sqrt{R^2 + (2\pi\nu L - 1/2\pi\nu C)^2}} = \frac{\mathcal{E}}{Z} \tag{40.7}$$

Note that the impedance can be considered as the *geometric sum* of the resistance, the inductive reactance, and the capacitive reactance with due regard for the fact that the inductive reactance and the capacitive reactance shift the phase in opposite directions. The impedance can be computed by drawing a vector diagram like that in Fig. 40.7b. The angle by which the current lags behind the emf is θ. Clearly, $\tan\theta = (X_L - X_C)/R$, $\sin\theta = (X_L - X_C)/Z$, $\cos\theta = R/Z$. When X_C is larger than X_L, θ is negative and the current leads the applied potential difference.

(a)

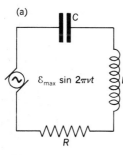

(b)

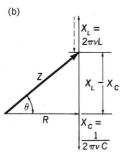

FIGURE 40.7
(a) Series circuit containing resistance, inductance, and capacitance and (b) its impedance, or phasor, diagram.

□ **Example** In the circuit of Fig. 40.7a, $\mathcal{E} = 24$ V, the total resistance $R = 10\ \Omega$, $L = 0.05$ H, and $C = 10^{-4}$ F. If the frequency is 60 Hz, find the impedance of the circuit and the current. What would an ac voltmeter connected across R read? Across L? Across C?

$$X_L = 2\pi\nu L = 6.28 \times 60 \times 0.05 = 18.9\ \Omega$$

$$X_C = \frac{1}{2\pi\nu C} = \frac{1}{377 \times 10^{-4}} = 26.5\ \Omega$$

$$Z = \sqrt{R^2 + (X_L - X_C)^2}$$
$$= \sqrt{(10)^2 + (18.9 - 26.5)^2}$$
$$= \sqrt{100 + 57.8} = 12.5\ \Omega$$

$$I = \frac{\mathcal{E}}{Z} = \frac{24\text{ V}}{12.5\ \Omega} = 1.9\text{ A}$$

$$V_R = IR = 19\text{ V}$$

$$V_L = IX_L = 1.9 \times 18.9 = 36\text{ V}$$

providing the resistance of the inductor is neglible. If it is not, the potential difference across the inductor is IZ_L, where $Z_L = \sqrt{R_L^2 + X_L^2}$.

$$V_C = IX_C = 1.9 \times 26.5 = 51\text{ V}$$

In this circuit the current leads the emf by an angle given by $\cos\theta = R/Z = 10/12.5 = 0.8$; $\theta = -36.9°$. (The minus sign means the current leads the emf.) □

40.7 RESONANCE IN SERIES CIRCUITS

Consider a series circuit with inductive reactance X_L, capacitive reactance X_C, and resistance R. When an emf of variable and increasing frequency is applied, the inductive reactance increases while the capacitive reactance decreases. At some frequency the inductive and capacitive reactances become equal, and the resultant reactance $X_L - X_C$ goes to zero. Then the circuit is said to be *resonant;* the frequency at which this occurs is called the *resonant frequency* ν_{res}. Clearly, when $X_L = X_C$, $2\pi\nu_{res}L = 1/2\pi\nu_{res}C$, or

$$\nu_{res} = \frac{1}{2\pi\sqrt{LC}} \tag{40.8}$$

At the resonant frequency the impedance Z of the circuit reduces to R and the current is in phase with the applied potential difference.

Physically, in a circuit at resonance, energy is stored during part of the cycle in the magnetic field of the inductor. A quarter cycle later this energy is stored in the electric field of the capacitor. In another quarter cycle it is once again stored in the magnetic field of the inductor. Thus energy is transferred back and forth between inductor and capacitor. The current is in phase with the applied emf because the effects of inductance and capacitance cancel each other. The only net energy supplied to the circuit is that dissipated as heat or radiated.

When an emf of constant amplitude but variable frequency is applied to the circuit under consideration, the current is greatest when the applied frequency is ν_{res}. If R is very small compared with X_L and X_C at the resonant frequency, the current at resonance may be many times greater than the current at slightly higher or slightly lower frequencies. At any frequency lower than ν_{res}, X_C exceeds X_L and the current leads the applied potential difference. Similarly, for a frequency greater than ν_{res}, X_L is larger than X_C and the current lags behind the applied emf.

40.8 POWER IN AC CIRCUITS

The power supplied at any instant in an ac circuit is obtained by multiplying the instantaneous current by the emf at the same instant:

$$p = ei \tag{40.9}$$

where p is the instantaneous power, i the instantaneous current, and e the instantaneous emf. The power varies from instant to instant, since both the current i and the applied emf e change with time.

When the current and the potential difference are in phase, the average power dissipated over one or more cycles is the product of the effective current I and the effective potential difference V (Sec. 40.2). The power is always positive (Fig. 40.2), since the current and the potential difference always have the same sign. When the current and potential difference are out of phase, the power is no longer given by IV. When an emf is applied across a pure capacitor, the power is negative as much as it is positive, and there is no *net* supply of energy by the generator; the generator supplies energy to charge the capacitor, and then the capacitor returns the energy. For a circuit in which the reactance is primarily

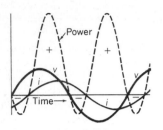

FIGURE 40.8
Instantaneous power supplied to a circuit in which applied potential difference v and current i are not in phase.

inductive, the power curve is similar to that in Fig. 40.8. In this case, the instantaneous power is negative during part of the cycle, which means that some of the energy stored in the inductance is being returned to the power source. In the general case, the power dissipated in a series ac circuit is given by the relation

$$P = \mathcal{E}I \cos \theta \tag{40.10}$$

where θ is the angle by which the current lags behind the applied emf. The vector diagram from which θ can be found is shown in Fig. 40.7b. Cosine θ is called the *power factor* of the circuit.

Power companies are eager to have a power factor as near unity as possible, because when the power factor is very different from unity, part of the energy supplied is returned each cycle. Since the power company has to provide the I^2R line losses which occur in delivering the power to the customer, this results in a disadvantageous situation. Most power companies charge higher rates for any consumer whose power factor is appreciably different from 1. By the suitable use of additional inductors or capacitors, it is possible to bring the power factors of most loads fairly close to unity.

40.9 OSCILLATORY DISCHARGE

In the three-branch circuit of Fig. 40.9a, there is a capacitor and an inductor. When key K_1 is closed, the capacitor is charged to a potential difference $\mathcal{E}$ and energy is stored in the capacitor. Let us now open K_1 and close K_2. The capacitor C begins to discharge, and a current is established through the inductor L. Energy is transferred from the electric field of the capacitor to the magnetic field of the inductor. When C is completely discharged, energy is stored in the inductor, which sustains the current until the capacitor has been given a charge opposite its original one. Now the capacitor discharges by a current in the opposite direction, thereby transferring energy to the inductor, which in turn uses this energy to recharge the capacitor to approximately its original condition.

If there were no resistance in this circuit, the capacitor would be charged to the potential difference with which it started. However, some of the energy is dissipated in heat in the inevitable resistance of the

(a)   (b)

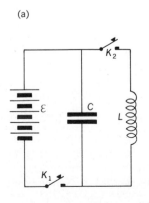

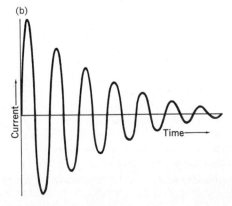

FIGURE 40.9
(a) Circuit for producing the oscillatory discharge of a capacitor through an inductor and (b) graph of damped oscillations.

circuit; thus the potential difference does not attain the value $\mathcal{E}$. The capacitor discharges again and again. The current rises and falls, as shown in Fig. 40.9b. After a number of oscillations, all the energy originally stored in the capacitor is dissipated.

The frequency of oscillation is the frequency to which the circuit is resonant. That is the frequency at which the inductive and capacitive reactances are equal, provided the resistance of the circuit is small compared with the reactances. By Eq. (40.8), the frequency of oscillation is given by

$$\nu = \frac{1}{2\pi\sqrt{LC}} \tag{40.11}$$

□ **Example** The oscillating circuit of a radio station has an inductance of 5 μH and a capacity of 0.01 μF. Neglecting the effect of resistance, find the natural (resonant) frequency and the wavelength of the electromagnetic waves.

$$\nu = \frac{1}{2\pi\sqrt{LC}} = \frac{1}{2\pi\sqrt{(5 \times 10^{-6})(1 \times 10^{-8})}}$$

$$= 712{,}000 \text{ Hz}$$

$$\text{Wavelength} = \frac{\text{velocity}}{\text{frequency}} = \frac{3 \times 10^8 \text{ m/s}}{712{,}000 \text{ Hz}} = 421 \text{ m} \qquad \square$$

40.10 ELECTROMAGNETIC WAVES

Classically, optics and electricity evolved as independent topics with no apparent interrelationships, but in 1864 James Clerk Maxwell published a great paper in which he developed what he called a "dynamical theory of the electromagnetic field." His work utilized the electrical researches of Faraday, introduced a missing concept in electric currents, and combined electricity and magnetism to bring forth a new, unified, and fruitful theory. An immediate consequence of Maxwell's differential equations was the prediction that energy could be transmitted by *electromagnetic waves* which should travel with a speed very close to the experimentally known speed of light. Faraday had suggested earlier that light might be an electromagnetic wave, and now Maxwell had a quantitative theory which supported this idea.

In 1887 Heinrich Hertz produced electromagnetic waves, as predicted by Maxwell's theory, with laboratory apparatus. He generated these waves with oscillating circuits, analogous to that of Fig. 40.9, in which spark gaps played the role of switches and an induction coil provided the source of emf. Using a detector which consisted of an incomplete loop of copper wire with a short air gap between the ends, Hertz showed that electromagnetic waves from his oscillator would produce a spark across the air gap of the detector as long as its dimensions were such that it was resonant at the frequency of the oscillator. Hertz's work established that electromagnetic disturbances have the wave properties predicted by the Maxwell theory; his experiments were in a sense the beginning of radio.

To understand the origin of the electromagnetic waves sent out by an oscillating current, we must be familiar with a conclusion of electromagnetic theory that an accelerated charge with available energy radiates an

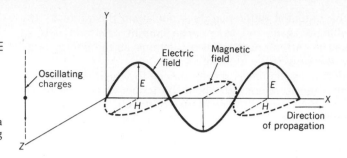

FIGURE 40.10
Electric and magnetic radiation fields a long distance from charges oscillating parallel to the *Y* axis.

electric field parallel to and proportional to the acceleration. Accompanying this electric field is a magnetic field *in phase and perpendicular to it*. Figure 40.10 shows schematically the radiation fields associated with a charge performing simple harmonic oscillations in the *y* direction. These radiation fields propagate away from the accelerated charge with the speed of light, decreasing in amplitude as $1/r$, where r is the distance from the radiating charge. (The radiation fields exist in addition to the electrostatic field of the charge and the magnetic field due to the velocity of the charge.) Thus the accelerated charge radiates electromagnetic energy, which is propagated at a speed of 3×10^8 m/s through free space. If the charge oscillates at a frequency in the neighborhood of 6×10^{14} Hz, the radiation is visible to the human eye as light.

QUESTIONS

1. Could one define the ampere of alternating current in terms of chemical effects? Why is the heating effect particularly appropriate for defining the ac ampere?
2. In Europe electric energy is supplied chiefly at approximately 220 V and at 50 Hz. What advantages and disadvantages does this have over the 60-Hz 110-V circuits common in the United States?
3. Why is a dc voltmeter unsuitable for measuring an alternating potential difference. How would such a voltmeter respond if the frequency across its terminals were slowly increased from 0.01 to 500 Hz?
4. A large inductor placed in series with an electric light bulb in an ac circuit results in a very dim lamp. A variable capacitor added in series can be adjusted to bring the lamp to almost normal brightness. How does this come about?
5. Why does the reactance of a capacitor decrease with frequency while that of an inductor increases? Offer physical reasons for this behavior.
6. What are the dimensions of RC? Verify by starting with the definition of resistance and capacitance and develop the dimensions. Show that $1/2\pi\nu C$ can be given in ohms. Is there any differ-

ence between these ohms and those in which resistance is measured? Explain.
7. What are the dimensions of LC? Of L/R. Of $2\pi\nu L$?
8. What happens to the resistance, the inductive reactance, and the capacitive reactance when the frequency applied to a series ac circuit is doubled?
9. What is the smallest value the power factor of an ac circuit can have? Under what conditions can this occur? What is the maximum value the power factor can have? Under what conditions does this value occur? Why does the power company prefer that its customers have high power factors?
10. Why do almost all transformers for 60-Hz alternating current use iron cores, while transformers for frequencies in the megahertz range do not?
11. Why are ac generators and transformers rated in kilovolt-amperes rather than kilowatts? What is the difference?
12. A good transformer automatically adjusts its primary current to take care of the load in the secondary circuit. How is this accomplished?
13. In what sense is Kirchhoff's second law obeyed for an ac series circuit such as that of Fig. 40.7? In what sense is it not obeyed?
14. An ac series circuit contains resistors, an induc-

tor, and a variable capacitor. If the frequency is constant, will increasing the capacitance increase or decrease the current? Explain carefully.

15. Radios are ordinarily tuned to various stations by changing the capacitance of an air capacitor. This is done by varying the effective area of the plates. If one wishes to tune to a station with a higher frequency, must the effective area be increased or decreased?

16. The 12 standard TV channels are located in the 54- to 72-, 76- to 78-, and 174- to 216-MHz bands, while the standard radio band is 535 to 1,605 kHz. With this in mind, give at least one reason why TV channels are switched by changing inductors while radio stations are switched by changing a single variable air capacitor.

PROBLEMS

Where appropriate, draw a phase diagram, roughly to scale, showing the impedances or voltages involved.

1. An alternating current passing through a resistance of 200 Ω produces heat at the rate of 50 W. What is the effective value of the current? Of the potential difference? What are the maximum values?   *Ans.* 0.5 A; 100 V; 0.707 A; 141 V

2. A pure 78-Ω resistor is connected to a 60-Hz power supply which has a 156-V amplitude. Find the maximum current, the rms (or effective) current, the average current, and the power dissipation.

3. An ac generating station generates 5 MW and transmits this power through a power line of 2 Ω total resistance. Find the power dissipated in the line if transmission is (*a*) at 500 kV and (*b*) at 5 kV.
          Ans. (*a*) 200 W; (*b*) 2 MW

4. An alternating emf has a frequency of 60 Hz and an effective value of 110 V. What is the maximum value of the emf? What is the instantaneous value of the emf at times $\frac{1}{720}$, $\frac{1}{360}$, $\frac{1}{240}$, $\frac{1}{180}$, and $\frac{1}{120}$ s after it passes the zero value?

5. An alternating emf of frequency 50 Hz has a peak value of 320 V. Find the instantaneous value of the emf at the instants 1, 2, 3, 4, 5, 8, and 10 ms after it passes through its zero value. What is the rms (or effective) value?
          Ans. 98.9, 188, 259, 304, 320, 188, 0 V; 226 V

6. Find the reactance of a 500-nF capacitor at 10 Hz, 10 kHz, 10 MHz, and 10 GHz. Note that a capacitor passes high frequencies readily and offers high reactance to low frequencies.

7. Find the reactance of a 0.15-H inductor and of a 3-μF capacitor at 60 Hz.   *Ans.* 56.5 Ω; 884 Ω

8. At what frequency would a 2-μF capacitor have a reactance of 80 Ω? What inductance would have this same reactance at this frequency?

9. A 2.5-μF capacitor and a 35-Ω resistor are connected in series with a 12-V 2,000-Hz emf. Find the current, the angle by which the current leads the emf, and the power dissipated.
          Ans. 0.254 A; 42.3°; 2.25 W

10. A current of 0.5 A is drawn from a 110-V 60-Hz line when a 100-Ω resistor is connected in series with a capacitor. Find the capacitance and the angle by which the current leads the potential difference.

11. A coil with a resistance of 15 Ω and an inductance of 10 mH is connected to a 28-V 400-Hz power source in an aircraft. Find the current through the coil, the angle by which the current lags behind the applied potential difference, and the power dissipated in the coil.
          Ans. 0.957 A; 59.2°; 13.7 W

12. An inductor has a resistance of 30 Ω and an inductive reactance of 60 Ω. It is connected in series with an emf of 120 V and a resistor of 50 Ω. Find the current, the potential difference across the inductor (do not forget its resistance), and the potential difference across the resistor.

13. The current in a solenoid is 2 A when it is connected to a 24-V battery and 1.2 A when it is connected to a 24-V 60-Hz power source. Find the resistance and the inductance of the coil.
          Ans. 12 Ω; 42.4 mH

14. A coil has an inductance of 127 mH and a resistance of 30 Ω. What are its reactance and its impedance at 50 Hz? By what angle does the current in this coil lag behind the potential difference? What resistance must be connected in series with this coil to give a total impedance of 100 Ω?

15. When a large coil is connected to a 50-Hz 220-V power source, a current of 0.4 A is drawn and the power dissipated in the coil is 32 W. Find the resistance of the coil, the power factor, and the inductance. If it is desired to increase the power factor to 1.00, what capacitance connected in series with the coil would achieve this objective?
          Ans. 200 Ω; 0.364; 1.63 H; 6.21 μF

16. An air-core coil bearing a 60-Hz current of 4 A dissipates heat at the rate of 320 W. If it has an inductive reactance of 50 Ω, find the inductance, the resistance, and the impedance of the coil and the applied potential difference.

17. A pure inductance of 0.3 H and a 60-Ω resistor are connected in series across a 220-V 50-Hz source.

Find the current and the angle by which it lags behind the source potential difference. What capacitance must be connected in series in this circuit to make it resonant at 50 Hz?

Ans. 1.97 A; 57.5°; 33.8 μF

18. Find the capacitive reactance of a 50-nF capacitor at 3 kHz. Compute the inductance required to produce series resonance with the capacitor at this frequency.

19. A television station operates at a frequency of 200 MHz. What inductance is needed with a capacitance of 1.5 pF to form a circuit resonant to this frequency? What is the wavelength of the radiation? *Ans.* 422 nH; 1.5 m

20. A radio station broadcasts on an assigned frequency of 630 kHz. Find the wavelength of the radio waves sent out. What inductance must be connected in series with a capacitance of 9 pF for resonance at this frequency?

21. Signals with a frequency of 1.2 MHz are sent out by a radio station. Find the wavelength. What capacitance is required to produce resonance at this frequency with an inductance of 25 μH?

Ans. 250 m; 704 pF

22. What capacitance is required with an inductance of 3 μH to form a resonant circuit for a wavelength of 125 mm. What is the frequency?

23. In a 110-V 60-Hz series circuit there is a pure capacitor with a reactance of 120 Ω, an inductor with 10 Ω resistance and inductive reactance of 40 Ω, and a 50-Ω resistor. Find (a) the impedance of the circuit, (b) the power factor, (c) the current, (d) the potential difference across the capacitor, (e) the potential difference across the inductor, and (f) the inductance L.

Ans. (a) 100 Ω; (b) 0.600; (c) 1.1 A; (d) 132 V; (e) 45.4 V; (f) 106 mH

24. In the ac circuit shown in Fig. 40.7, the effective emf is 130 V, the inductance is 0.31 H, the resistance is 50 Ω, and the capacitance is 1 μF. The frequency is 1,000/π Hz. (a) Find the inductive reactance, the capacitive reactance, and the impedance. (b) Find the current and the potential difference across each part of the system. (c) Find the angle by which the current lags the emf and the power factor. (d) Find the power dissipated in the circuit.

25. In the ac circuit shown in Fig. 40.7, the effective emf is 50 V, the inductance 0.19 H, the resistance 80 Ω, and the capacitance 4 μF. The frequency is 500/π Hz. (a) Find the inductive reactance, the capacitive reactance, and the impedance. (b) Find the current and the potential difference across each part of the system. (c) Find the angle by which the current leads the emf. (d) Find the power dissipated in the circuit.

Ans. (a) 190 Ω; 250 Ω; 100 Ω; (b) 0.5 A; $V_R = 40$ V; $V_L = 95$ V; $V_C = 125$ V; (c) 36.9°; (d) 20 W

SIX

MODERN PHYSICS

41

ELECTRONS AND PHOTONS

Before the end of the nineteenth century the electromagnetic nature of light was well established, and the atomic hypothesis of matter was strongly supported by chemical evidence. However, no satisfactory model of an atom and no real understanding of how atoms join to form molecules or solids were yet available. Then, beginning with the discovery of x-rays in 1895, major clues were rapidly uncovered, new ideas were advanced, and the era of modern physics *was born. Here "modern physics" implies the quantitative description and understanding of phenomena which could not be analyzed in terms of the "classical physics" we have been studying. In this chapter we discuss the discovery of the electron and the need for ascribing particle characteristics to electromagnetic radiation.*

41.1 THE EVOLUTION OF MODERN PHYSICS

Just as assembling a jigsaw puzzle usually proceeds by putting together a number of initially separate domains which are subsequently interconnected to give the final picture, so the development of physics has progressed, and continues to progress, by the simultaneous pursuit of apparently unrelated researches, the results of which often lead to a unifying denouement. For example, our present understanding of atomic structure arises from the synthesis of the outcomes of many experiments and calculations. Among the hundreds of important contributions which led to a useful model of the atom, we treat only a selected few which do not require the use of differential equations. In discussing these topics we shall find that experimental results in one area have often confirmed and extended ideas which arose in some very different realm. The sequence of presentation follows the chronological order of discovery in some places but departs from it when a more logical presentation seems available. Among the topics which are typically regarded as foundations of modern physics are three major experimental discoveries and three revolutionary theoretical developments which challenged the validity of some seemingly well-established ideas of classical physics. Of these six epic contributions five came in one eventful decade (late 1895 to late 1905). Each served to stimulate intense activity by scientists all over the world, and the result was a great step forward in our understanding of the physical universe. In the order of their appearance the three experimental milestones were:

1. The discovery of x-rays by Röntgen in 1895. Once this new type of radiation had been announced, hundreds of physicists began to work with x-rays. The origin, properties, and uses of x-rays were explored and x-rays became a valuable tool in areas as diverse as medicine, crystallography, and engineering.

2. The discovery of radioactivity in 1896 by Becquerel. Work with natural radioactive sources led Rutherford to propose the nuclear model of the atom in 1911. By the early 1930s accelerators were available for producing nuclear reactions, and nuclear physics had become one of the most active branches of physics; during the 1940s nuclear energy emerged as an important new source.

3. The discovery of the electron as one constituent of the atom. In 1896 Zeeman and Lorentz showed that certain features of the spectra of atoms (Sec. 46.7) indicate that the radiation comes from a negative charge, and they determined the approximate ratio of the charge e on

the electron to its mass *m*. In 1897 J. J. Thomson measured *e/m* by deflecting a beam of electrons by electric and magnetic fields (Sec. 41.5). About the same time Townsend determined the order of magnitude of the electronic charge *e*; by 1909 Millikan had developed a method of measuring *e* capable of good accuracy (Sec. 31.10).

The three milestones in the realm of theory were no less significant; they were:

1. The idea that electromagnetic radiation has *particle characteristics* as well as *wave properties*. In 1900 Planck proposed that harmonic oscillators radiate energy only in quanta of energy *hν* (Sec. 41.6). In 1905 Einstein extended Planck's idea to explain the photoelectric effect (Sec. 41.7). In 1913 Bohr combined the nuclear model, the electron, and the particle nature of radiation to provide a model of the hydrogen atom which correctly predicted the spectrum of atomic hydrogen (Secs. 45.2 and 45.3).

2. The theory of relativity proposed by Einstein in 1905 (Chap. 42), which revolutionized our thinking about space and time and which introduced the mass-energy relationship.

3. The idea that electrons and other classical "particles" have wave properties. This hypothesis was advanced by de Broglie in 1924, and the wave equation satisfied by the electrons was proposed by Schrödinger in 1926. Within a few years wave mechanics was used to solve many previously intractable problems and to give some understanding of the structures of atoms and of solids as well as an explanation of the periodic table of the elements.

In the short span of less than 40 years radically new ideas were established in physics, and several of the classical foundation stones had to be abandoned. For example, the law of conservation of mass and the idea that light was *purely* a wave phenomenon lost their fundamental positions and are now recognized to be useful approximations in many situations but not generally valid. Why should the third of a century starting in 1895 have been so fruitful in new discoveries and revolutionary ideas? Surely the presence of intellectual giants was a major factor, but so also was the fact that not until that time were many of the essential tools of scientific investigation available—reliable electrical equipment with high-voltage sources, good vacuum pumps, practical photographic emulsions, and so forth. The evolution of the heat engine, the Industrial Revolution, and the growth of higher education brought many minds to bear on the problems of physical science, and an explosive growth ensued.

Before we return to the great discoveries and theories in more detail, let us set the scene by describing some phenomena associated with the discharge of electricity through gases. It was research in this area which led to the discovery of x-rays and to the isolation of the electron.

41.2 GASEOUS IONIZATION

Molecules of air and other gases are normally electrically neutral. Un-ionized air is an excellent insulator, a fact demonstrated by the observation that the leaves of an electroscope can hold their charges for many hours.

However, an ionized gas is a good conductor. Indeed, if virtually every molecule of the gas has lost one or more electrons, the assemblage of positive ions and free electrons is an excellent conductor. A highly ionized gas is known as a *plasma*, sometimes called the fourth state of matter since it is not solid, liquid, or gas in the usual sense of these words. Actually the major portion of the matter in our universe is in the plasma phase, since this is the state of stars and most interstellar matter. On the earth, conditions of temperature and pressure are such that most matter is solid, liquid, or gaseous, but relatively little of the universe has anything similar to our terrestrial environment.

To ionize a gas molecule, one must provide the few electronvolts of energy needed to free an electron. This can be accomplished in a variety of ways, of which collision with an energetic electron or ion is perhaps the most familiar. This is the dominant process in fluorescent lamps, neon tubes, and similar devices in which positive ions are accelerated toward a negative electrode (cathode) and negative ions, chiefly electrons, toward a positive electrode (anode). If the ions have enough energy when they collide with neutral molecules, additional ion pairs are produced.

Charged particles emitted by radioactive nuclei produce ionization in any gas through which they pass. Many devices for detecting and measuring the radiation from fallout or other nuclear disintegrations are fundamentally electroscopes depending on gaseous ionization for their readings.

When gamma rays, x-rays, and ultraviolet radiation pass through gases, electrons are ejected from the atoms, thereby producing positive gas ions and free electrons. Gases drawn from a flame contain many ions. This ionization is greatly enhanced if the flame is fed with a salt such as sodium chloride, which readily breaks into ion pairs. Any gas can be ionized by raising the temperature sufficiently high. The temperature required for a specified fraction of ionized molecules depends on the gas involved. Even neutral molecules can ionize each other if they have enough kinetic energy.

Air at atmospheric pressure "breaks down" under electric fields of about 3,000 V/mm. At reduced pressures the breakdown occurs at lower field strengths. The corona glow from a lightning rod is an example of ionization by a strong electric field. Once such ionization begins, other ions are produced by collisions, and a conducting path between two charged objects is created.

41.3 DISCHARGE IN GASES AT LOW PRESSURES

When an electric discharge is maintained between two electrodes in a long glass tube filled with air (Fig. 41.1), some beautiful and interesting effects are observed as the air is pumped out of the tube. If the potential difference between the electrodes is not much greater than necessary to maintain the discharge, sparks jump between the electrodes at atmospheric pressure. As the pressure in the tube is lowered, a narrow pink streamer appears between the electrodes. As the pressure is reduced further, the streamer expands until it fills almost the entire volume of the tube. By the time the pressure is reduced to 1 mmHg, the entire gas glows.

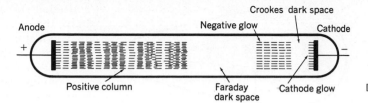

FIGURE 41.1
Discharge of electricity through a gas at
reduced pressure.

As the pressure is reduced still further, one reaches the stage of discharge indicated in Fig. 41.1. A velvety glow, known as the *cathode glow*,
covers the surface of the negative electrode. Outside this glow is a
region called the *Crookes dark space*. Beyond this is a luminous region
known as the *negative glow*, and then a second dark region called the
Faraday dark space. Then follows another luminous region, known as the
positive column, which reaches to the positive electrode. The positive
column is not perfectly continuous but may show alternate light and dark
layers called *striations* across the path of the discharge. If the distance
between the electrodes is increased, the appearance of the cathode
region is not much changed; the positive column increases in length to
fill the added volume. Practically all the potential difference across the
discharge tube occurs between the cathode and the negative edge of the
positive column.

The potential difference required to maintain the discharge in a tube
depends sensitively upon the pressure. When the air is at atmospheric
pressure, a high potential difference is required. As the pressure is reduced, the potential difference required to maintain the discharge decreases steadily (Fig. 41.2). At some relatively low pressure, a minimum in
the curve is reached; then, as the pressure is decreased still further, the
potential difference required to maintain the discharge increases sharply.
Practically all the ionization of the air in the discharge tube occurs as the
result of impacts. At high pressure the few residual ions (which may be
formed by cosmic rays or by radioactivity) are accelerated by the electric
field in the tube, but they travel short distances between collisions and
attain relatively low energies. In order to produce more ionization they
must have a certain minimum energy. If an ion is to attain this minimum
energy between collisions at high pressure, it is necessary that a high
electric field be maintained. As the pressure is reduced, the distance an
ion travels between collisions is increased and the ion has a greater
distance over which to acquire the energy needed to produce ionization
by collision. Eventually the pressure becomes so low that ions may go all
the way to the collecting electrode without colliding. Then a higher
potential difference is required to maintain the discharge; the ions may
eject electrons and other ions when they collide with the electrode.
These *secondary ions* aid in supporting the discharge.

41.4 CATHODE RAYS

As the pressure in a discharge tube is reduced beyond the stage pictured
in Fig. 41.1, the Faraday dark space lengthens and the positive column
shrinks. Eventually a greenish glow spreads over the entire tube. This is
due to fluorescence of the glass as it is bombarded with high-speed

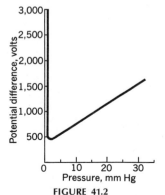

FIGURE 41.2
Potential difference required to maintain a discharge in a particular tube as a
function of the pressure.

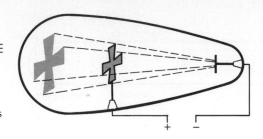

FIGURE 41.3
Sharp shadows show that cathode rays travel in straight lines.

electrons which, Crookes showed, have their origin at the cathode of the tube. These particles were named *cathode rays* before they were identified as electrons—indeed, before electrons were "discovered." Cathode rays have the following properties:

1. *The rays travel in straight lines in the absence of electric and magnetic fields.* This fact can be demonstrated by inserting some obstacle in the path of the rays and observing that the geometric shadow of the obstacle does not show fluorescence (Fig. 41.3).

2. *The rays are deflected by an electric field* (Fig. 41.4).

3. *The rays are deflected by a magnetic field.*

4. *The rays carry negative electric charge.* The direction of the deflection in both electric and magnetic fields supports this conclusion. Further, by applying a suitable magnetic field in the arrangement of Fig. 41.5, a cathode-ray beam can be deflected so that it enters the small cylindrical vessel at E, which is connected to a galvanometer G. The galvanometer shows that negative charge is collected in E when cathode rays enter the cylinder.

5. *The rays emerge normally from the surface of the cathode.* If the cathode has a concave surface, the cathode rays come to a focus near its geometric center. If the anode of the discharge tube is shielded by some obstruction or is off at one side of the tube, the cathode rays strike at points which one would expect to be hit if the rays left perpendicularly from the surface of the cathode.

6. *The rays penetrate small thicknesses of matter.* If a thin window of aluminum is inserted in the tube, the cathode rays pass through the aluminum and make themselves evident by forming luminous streamers in the air beyond the window.

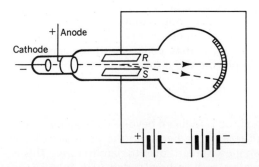

FIGURE 41.4
The deflection of cathode rays by an electric field reveals that they bear negative charge.

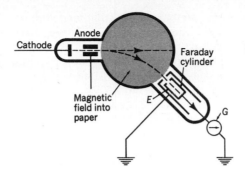

FIGURE 41.5
Cathode rays are deflected by a magnetic field in such a direction as to confirm that they are negatively charged.

41.5 THE RATIO OF ELECTRONIC CHARGE TO MASS

By using a tube similar to that in Fig. 41.4, J. J. Thomson was able to measure the velocity of cathode rays (electrons) in 1897. If the plate S of this cathode-ray tube is made positive, the beam is deflected downward. Now if a magnetic field uniform over the area between the plates R and S is applied out of the plane of the paper, it exerts an upward force on the moving electrons. It is possible to choose magnetic and electric fields in such a way that their effects exactly balance. The downward force due to the electric intensity E has a magnitude Ee, where $-e$ is the charge on an electron. The upward force due to the magnetic intensity B has magnitude evB if v is the speed of the electrons. When the beam is undeflected, $Ee = evB$, and

$$v = \frac{E}{B} \tag{41.1}$$

Thomson found that electrons which had been accelerated through a potential difference of 25 kV had a speed of about 8×10^7 m/s. From the deflection of the electrons by the electric field in the absence of a magnetic field and the speed of the electron, Thomson calculated e/m, the ratio of the charge of an electron to its mass, to be 1.7×10^{11} C/kg.

Thomson's pioneering experiments in deflecting beams of high-speed electrons by electric and magnetic fields laid the foundation for modern cathode-ray oscilloscopes and television picture tubes.

A more direct way to determine e/m is by measuring the radius of curvature of the path of an electron moving with known velocity perpendicular to a uniform magnetic intensity B. The centripetal force of magnitude mv^2/r required for the circular path is provided by the force evB exerted on the moving charge by the magnetic field. Therefore,

$$\frac{mv^2}{r} = evB \tag{41.2}$$

or

$$\frac{e}{m} = \frac{v}{Br} \tag{41.2a}$$

The result of many painstaking measurements is that

$$\frac{e}{m} = 1.759 \times 10^{11} \text{ C/kg}$$

When this value is combined with Millikan's much later determination of $e = 1.602 \times 10^{-19}$ C, we find that the mass m of the electron is 9.11×10^{-31} kg.

In Chap. 35 we found that it requires 9.649×10^7 C to deposit electrolytically 1 kg equivalent mass of any element. In the case of hydrogen this corresponds to 1.008 kg. If the cathode rays in a discharge tube bear the same magnitude of charge as a monovalent ion, we can calculate the ratio of the mass M of a hydrogen atom to that of the electron. For the hydrogen atom $e/M = 9.649 \times 10^7$ C/1.008 kg $= 9.575 \times 10^7$ C/kg, from which we have

$$\frac{\text{Mass of hydrogen atom}}{\text{Mass of electron}} = \frac{M}{m} = \frac{1.759 \times 10^{11}}{9.575 \times 10^7} = 1{,}837$$

41.6 PHOTOELECTRICITY AND THE PHOTON

In 1887, a decade before J. J. Thomson measured e/m for the electron, Heinrich Hertz was doing pioneer work in the study of radio waves. In the course of his experiments Hertz observed that when ultraviolet light fell on the electrodes of a spark gap, a spark jumped at a substantially lower potential difference than usual. A little later it was found that a charged sheet of zinc exposed to ultraviolet radiation loses its charge when negative but retains its charge when positive (Fig. 41.6). We now know that when ultraviolet light falls on a metallic surface, electrons are liberated from the metal. The energy required to eject an electron is provided by the incident electromagnetic radiation. When the metal is negatively charged, these *photoelectrons* are repelled. When the metal is positive, the photoelectrons are attracted back to the metal. No quantitative understanding of the *photoelectric effect* became available until 1905. It required the synthesis of the concept of the electron and the concept of a photon (or quantum), which we now develop.

Near the end of the nineteenth century Max Planck was endeavoring to develop an adequate theory for the wavelength distribution of blackbody radiation (Fig. 16.6). He discovered that he could account quantitatively for blackbody radiation curves if he assumed that energy is radiated in small packets, which he called *quanta* and for which G. N. Lewis later introduced the name *photons*. According to Planck's theory, the energy associated with each *quantum* or *photon* is given by

$$\text{Energy of photon} = W = h\nu \tag{41.3}$$

where ν is the frequency of the radiation and h is a universal constant equal to 6.626×10^{-34} J-s. In honor of its discoverer, h is called *Planck's constant*.

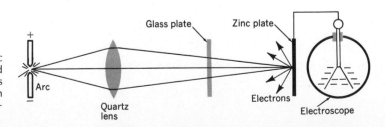

FIGURE 41.6
When ultraviolet light falls on a zinc plate, photoelectrons are ejected and the electroscope discharges. The glass plate absorbs the ultraviolet, and when it is inserted in the beam, the photoemission ceases.

It was in 1901 that Planck proposed his startling new idea, which now plays a vital role in our understanding of physics in atomic dimensions. Although Planck's theory gave excellent agreement with experimental measurements on blackbody radiation, many people regarded the agreement as fortuitous and considered the proposal that radiant energy comes in chunks or quanta as unacceptable. The objections were based on the misconception that light and other electromagnetic radiation were purely classical wave phenomena. In the early 1800s it had been established that light exhibits interference, diffraction, and polarization (Chaps. 27 and 28), which are characteristic wave properties. Virtually a century of experimentation had established that radiation does indeed have wave characteristics. In the 1860s Maxwell had developed the theory of electromagnetic fields and proposed that light consists of electromagnetic waves, a hypothesis which was strongly supported in 1887 by the experimental work of Hertz. Thus it is not surprising that many physicists regarded Planck's proposal that radiation had particle properties as erroneous and retrograde. According to Einstein's extension of Planck's idea, a quantum is emitted by one atom (or oscillator) and absorbed by another as though it were a particle with a well-defined position. On the other hand, a wave emitted from a source spreads out, and by the time it is a centimeter from the source, it is spread over a volume millions of times that of a single atom.

Now we believe that radiation is neither a classical wave phenomenon nor a classical particle phenomenon. It has wave characteristics under some conditions and particle properties under others. Classical waves and particles are idealizations which are extremely useful, but electromagnetic radiation is neither a pure wave motion nor exclusively particlelike. Instead it exhibits some aspects of each, as we shall see.

41.7 EINSTEIN'S PHOTOELECTRIC EQUATION

A theory like Planck's, which utilizes some new assumption to explain a phenomenon, is called an *ad hoc theory*. Such theories are usually viewed with skepticism until the radical assumption is found to have wider application. The first of much supporting evidence for the validity of Planck's hypothesis came in 1905, when Einstein used it to explain the photoelectric effect, a phenomenon which had been a puzzle to physicists since its discovery. By 1905 it had been established that electrons are emitted from all metals (and many other materials) when ultraviolet light is incident (but not visible light). On the other hand, there are several elements from which visible light ejects photoelectrons. To explain these facts, Einstein adopted the hypothesis that radiant energy comes in quanta of energy $h\nu$. In order to eject a photoelectron from a surface, a photon must have enough energy to free the electron from the atom or material with which it is associated. If it requires an energy W to remove an electron from the metal, and if the photon imparts to it an energy $h\nu$ which is greater than W, the difference in energy appears as kinetic energy of the ejected photoelectron. This explanation of the photoelectric effect in terms of photons was one of the early triumphs of the quantum idea. If we let W_{min} represent the smallest energy which can free an electron from the solid, the maximum kinetic energy of ejected electrons is given by Einstein's photoelectric relation:

$$K_{max} = h\nu - W_{min} \qquad (41.4)$$

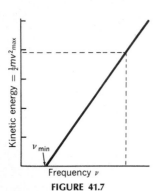

FIGURE 41.7

The maximum kinetic energy of ejected photoelectrons increases linearly with the frequency of the incident light.

The quantity W_{min} is called the *work function* of the material.

Many careful experiments have shown that the Einstein relationship is correct. If the frequency of the incident light is increased, the maximum kinetic energy of the ejected photoelectrons increases correspondingly (Fig. 41.7). Similarly, if the frequency of the incident photons is decreased, the energy of ejected photoelectrons decreases until the frequency $\nu_{min} = W_{min}/h$ is reached. For any lower frequency, no photoelectrons are emitted.

If the frequency of the incident light is kept constant while the intensity is changed by increasing the number of photons incident, the number of photoelectrons emitted increases. For a given frequency of radiation the number of photoelectrons ejected per second is directly proportional to the intensity of the incident radiation.

Among the metals from which photoelectrons can be ejected by visible light are barium, cesium, lithium, potassium, sodium, and rubidium. These are elements from which it is particularly easy to remove an electron in chemical reactions. Both the chemical and photoelectric properties of electropositive elements depend on the ease with which an electron can be removed.

□ **Example** Light of wavelength 555 nm is required to cause the emission of electrons from a potassium surface. What is the energy necessary to remove one of the least firmly bound electrons? Find the kinetic energy and velocity of such an electron if the potassium surface is bombarded with light of wavelength 400 nm.

$$\text{Energy of 555-nm photon} = h\nu = \frac{hc}{\lambda}$$

$$= (6.63 \times 10^{-34}\text{ J-s})\,\frac{3 \times 10^8\text{ m/s}}{555 \times 10^{-9}\text{ m}}$$

$$= 3.58 \times 10^{-19}\text{ J} = 2.23\text{ eV}$$

Since this is the lowest-frequency photon which causes photoemission, $W_{min} = 3.58 \times 10^{-19}$ J.

$$\text{Energy of 400-nm photon} = (6.63 \times 10^{-34})\,\frac{3 \times 10^8}{400 \times 10^{-9}}$$

$$= 4.97 \times 10^{-19}\text{ J} = 3.10\text{ eV}$$

$$K_{max} = h\nu - W_{min} = (4.97 - 3.58) \times 10^{-19}\text{ J}$$

$$= 1.39 \times 10^{-19}\text{ J} = 0.87\text{ eV}$$

$$\tfrac{1}{2}(9.11 \times 10^{-31})v_{max}^2 = 1.39 \times 10^{-19}$$

$$v_{max}^2 = 30.5 \times 10^{10}\text{ m}^2/\text{s}^2$$

$$v_{max} = 5.5 \times 10^5\text{ m/s} \qquad \square$$

41.8 PHOTOELECTRIC CELLS

Devices which utilize the photoelectric effect have many important applications. One is the exposure meter used by photographers. Here the rate of photoelectron ejection caused by the incident light is a measure of the illumination. Photocells are used to open store doors when people cut off a light beam, to sound burglar alarms, and to turn on street lights whenever the sun fails to provide adequate illumination. Photocells are used in physics, biology, and medicine for measuring the intensity of various kinds of electromagnetic radiation. In a television studio variations in light intensity on a photosensitive surface are used to convert the picture into electric signals in the image orthicon tube.

The sound track of motion-picture films permits varying amounts of light to pass through when the film is run between a light source and a photoelectric cell. The photoelectric cell converts the variations in light intensity into variations in electric current.

There are several different types of photoelectric cells. In some there is a light-sensitive surface which emits electrons and a second electrode, usually maintained at a fairly high positive potential, which collects the electrons. A circuit for such a photocell is shown in Fig. 41.8 with a galvanometer to measure the relative light intensity.

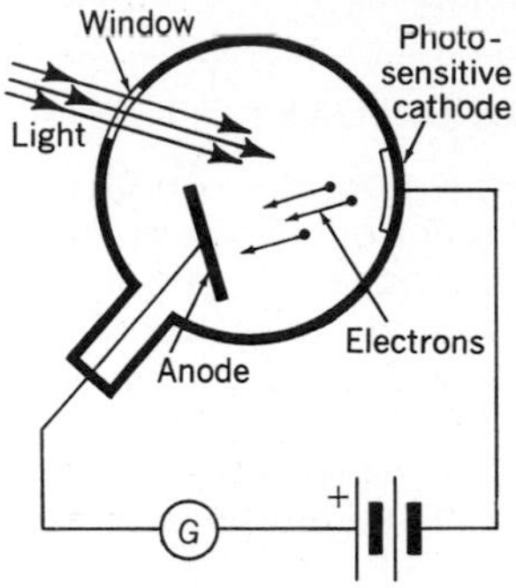

A second type is the *photovoltaic* or *barrier-layer photocell*, in which incident light produces an emf between two terminals. A typical photovoltaic cell (Fig. 41.9) contains a thin metal disk A on which there is a film of light-sensitive material B. In contact with B is a conducting ring or layer C. In one common type A is an iron plate and B is a thin film of selenium over which a thin gold coating is laid to serve as the conductor C. Another type of cell uses a copper plate as A, cuprous oxide as B, and a thin layer of silver as C. When the sensitive surface is illuminated, an emf is generated between the sensitive layer B and the metal disk A. Electrons are driven by the light from the layer B to the metal A across the thin insulating layer. This type of cell is used in exposure meters, since the light produces the emf required to deflect the galvanometer.

Solar cells, widely used to provide energy for satellites and space vehicles, are photovoltaic cells converting radiant energy into electric energy. A familiar form uses a very perfect silicon crystal which is treated with trace impurities to introduce extra positive charges on one side of a thin junction and extra negative charges on the other side (see Sec. 47.7). Radiant energy disturbs the electrical balance of the charges and thereby induces an emf which continues as long as light strikes the crystal. Such cells convert roughly 15 percent of the incident radiant energy into electric energy; efficiencies 2 or more times as great are theoretically possible.

A third type of photoelectric cell utilizes the photoconductivity effect. When photons are absorbed in a material, relatively few electrons leave the material. If a material which is normally an insulator or a poor conductor is irradiated, electrons originally bound to atoms are freed and become available for conduction through the material. The resistance of

FIGURE 41.8
A photoelectric cell.

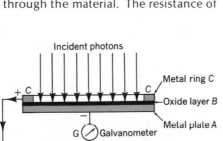

FIGURE 41.9
An electromotive force is produced when light is incident on a barrier-layer photoelectric cell.

a piece of germanium, for example, decreases greatly when it is subjected to high illumination. The conductivity of a selenium cell may be 25 or more times greater in bright light than in the dark. The conductivity of a thallium sulfide cell changes by an even larger factor.

QUESTIONS

1. How does the conduction of electricity in a gas differ from conduction in a copper wire? Discuss the similarities and differences.
2. What are the properties of cathode rays?
3. What is "recombination" in electric discharges? Why does it occur?
4. Why can visible light remove photoelectrons from sodium but not from zinc?
5. If visible light comes in photons, why don't we perceive the discontinuous structure in the light received by the eye?
6. Why does the existence of a cutoff frequency below which no photoelectrons are ejected support the photon theory rather than the wave theory of light?
7. What determines the energy of a photon? Can two photons have the same energy but different wavelengths in air?
8. When electromagnetic radiation of a given wavelength strikes a surface and ejects photoelectrons, these electrons may have a wide range of energies. Why don't they all have K_{max} of the Einstein equation?
9. Why are measurements of the work function W_{min} of a material by determining the energies of photoelectrons very sensitive to the preparation of the surface?
10. How can one reconcile the wave-theory explanation of interference and diffraction with the photon theory used by Einstein to explain the photoelectric effect?
11. Ordinary film and x-ray film can be developed in a room illuminated by a red safelight because such films are insensitive to red light. Why are they insensitive to red but not to blue light? Explain in terms of photons.
12. How can beams of electrons be focused? What types of electron lenses can you suggest?

PROBLEMS

1. An electron with an energy of 250 eV enters a uniform magnetic field of intensity 4×10^{-4} T which is perpendicular to the electron's velocity. Find the speed of the electron, the radius of the path of the electron, and the energy of the elec-

tron after it has traveled 25 mm in the magnetic field in a good vacuum.
   Ans. 9.38×10^6 m/s; 133 mm; still 250 eV
2. What speed must an electron have for its path to be a circle of radius 225 mm in a uniform magnetic field of 500 μT if the velocity is perpendicular to the magnetic field lines? What is the energy of the electron?
3. In determining the speed of electrons by Thomson's method it is found that a magnetic field of 50 mT is just adequate to compensate for an electric field of 375 kN/C. Find the speed.
   Ans. 7.5×10^6 m/s
4. A beam of cathode rays with energy 1,500 eV per electron carries a current of 24 μA. The beam strikes a thin metal foil. How many electrons strike the foil each second? If all the electrons are stopped in the foil, what power is dissipated in the foil?
5. A beam of 5-kV electrons in a television picture tube pass through a region 15 mm long where there is a uniform magnetic field perpendicular to the velocity of the electrons. What magnetic induction B is required to deflect the beam through an angle of 30°? *Ans.* 7.95 mT
6. An electron initially at rest is accelerated through a potential difference V. It then passes between two parallel plates, which are oppositely charged to produce a transverse electric intensity of magnitude E. If a magnetic field B is applied in the region between the plates and its intensity and direction are such that the electron is not deflected as it passes between the plates, show that for the electron $e/m = E^2/(2VB^2)$.
7. What is the current through a mercury discharge tube operated on a dc potential difference if each minute 2.5×10^{19} electrons and 7.5×10^{16} singly charged mercury ions pass a given cross section of the tube? What fraction of the current is carried by Hg^+ ions?
   Ans. 66.9 mA; 0.3 percent
8. A photon has an energy of 2.4 eV. Find its frequency and its wavelength in vacuum.
9. Determine the energy of a photon of sodium yellow radiation whose wavelength is 589 nm both in joules and in electronvolts. What is the frequency? *Ans.* 3.37×10^{-19} J; 2.11 eV; 509 THz
10. The human eye can detect as few as three photons of 550-nm light arriving together. How

much energy do these three photons collectively represent?

11. The eye can detect a flux of about 3 photons/mm²-min of green light ($\lambda = 500$ nm), while the ear can detect about 10^{-13} W/m² under optimum conditions. As a power detector, which is more sensitive and by what factor? *Ans.* The eye; 5

12. On a clear day, about 1,400 W of radiant energy from the sun strikes an area of 1 m² if that area is perpendicular to the direction of the sun. If this energy were all in the form of photons of green light of wavelength 5000 Å, how many photons would arrive per square millimeter in 1 s?

13. The work function W_{min} of mercury is 4.53 eV. Find the longest wavelength which can produce photoelectric emission from this material and the maximum kinetic energy of photoelectrons emitted when the material is irradiated with ultraviolet photons of 6 eV energy.
Ans. 274 nm; 1.47 eV

14. The work function W_{min} of uranium is 3.63 eV. Find the longest wavelength which can produce photoelectric emission from this element and the maximum kinetic energy of photoelectrons emitted when uranium is irradiated with photons of 5 eV energy.

15. It requires 7.00×10^{-19} J to remove one of the least tightly bound electrons from a chromium surface. Find the longest wavelength which will be effective in producing photoelectrons. What is the maximum energy of photoelectrons ejected from a chromium surface by ultraviolet radiation of $\lambda = 150$ nm? *Ans.* 284 nm; 3.90 eV

16. Electrons with a maximum energy of 3.51 eV are emitted from a cadmium surface radiated with light of $\lambda = 200$ nm. Find the energy of the incident photons and W_{min}, the smallest energy which can free an electron from the surface.

17. Find the maximum kinetic energy in electronvolts of the photoelectrons ejected from a sodium photoemitter for wavelengths of 200, 300, 400, and 500 nm if $W_{min} = 2.28$ eV. What is the longest wavelength which will eject photoelectrons?
Ans. 3.92; 1.85; 0.82; 0.2; 544 nm

42

RELATIVITY

By expressing the laws of electricity in terms of differential equations and introducing some new ideas, Maxwell showed that light is electromagnetic in origin and thus brought unity to optics and electricity. His theory eventually led to new questions about the propagation of light. It was found that while the classical physics we have been studying makes predictions in excellent agreement with observed results for macroscopic bodies and moderate speeds, it fails when one deals with objects of atomic dimensions and with speeds comparable with the speed of light. In 1905 Einstein advanced his celebrated special theory of relativity, linking space and time in a new way that resolved the dilemmas regarding the speed of light. The existence of this revolutionary theory has been widely publicized; in this chapter we develop several facets of Einstein's theory and its predictions.

42.1 ELECTROMAGNETIC WAVES AND THE ETHER

During the nineteenth century it was established that radio signals, visible light, and other electromagnetic disturbances have the properties of transverse waves, but the particle properties introduced in Chap. 41 were not yet known. Late in the nineteenth century physicists were speculating on the question: Through what medium are these waves propagated? Such a question seemed most reasonable; after all, acoustic waves are transmitted through air, transverse waves are transmitted by the strings of a violin, and water waves are propagated at a water-air interface. There was no obvious medium to which one could assign electromagnetic waves, since they were known to pass through a vacuum. However, physicists believed that some medium was necessary for energy to be transmitted through a distance, and this medium was given the name *luminiferous aether* (the modern spelling is *ether*). Does such an ether exist? If it exists, how can one detect it and determine its characteristics? These, too, were questions which haunted physicists during the latter part of the nineteenth century.

Early speculations on the ether revealed that it must have remarkable properties. First of all, to sustain waves with the tremendous speed of light, a material of high rigidity was required. At the same time this rigidity raised problems about how the earth could move through the ether with no apparent retardation; or if one assumed the ether to be transported along with the earth, there was the problem of how the ether-atmospheres of the earth and the sun could be in relative motion without some evidence of shear forces.

42.2 AN ACOUSTIC ANALOG

If a luminiferous ether exists, the speed of light measured in a reference frame at rest with respect to this ether should be different from the speed determined in a reference frame moving with respect to the ether. Thus, the ether should provide an absolute reference frame to which the motions of all other frames and of all bodies in the universe could be referred. It was clear from the beginning that grave experimental difficulties would accompany any effort to resolve the question of how fast the earth moves relative to the ether or, indeed, to make any meaningful measurements whatever regarding the ether.

In order to see one way in which such measurements might be made, let us consider an experiment to determine whether the air through which sound waves are being propagated is moving relative to the earth. This is directly analogous to the problem of determining the ether "wind" associated with the motion of the earth through the ether. However, the acoustic experiment is vastly simpler than one involving light, because the speed of sound is modest, while the speed of light is great. Imagine two observers A and B a distance d apart. Let A send a sound signal to B when there is a wind blowing from A toward B with a velocity V (Fig. 42.1). The time required for the signal to go from A to B is $d/(c + V)$, where c is the speed of sound relative to air. The time required to transmit a signal from B back to A is $d/(c - V)$, since the propagation speed of the wave relative to B is $c - V$. For example, if $d = 330$ m, $V = 30.0$ m/s, and the speed of sound in air $c = 330$ m/s, the time required for a signal to go from A to B is $330/360 = 0.917$ s, while the time required for the sound to go from B to A is $330/300 = 1.10$ s. The total time for the traversal A to B and return is 2.017 s. Clearly, this is different from the 2.000 s which would be required in still air. While 0.017 s is not a long time, it is easy with modern techniques to measure much shorter times.

Of course, there are simpler ways of telling whether the air is in motion between A and B, but ideally this type of experiment could establish whether A and B are moving or at rest *relative* to the air. By an experiment similar to this, the speed of the earth in its orbital motion about the sun might be measured relative to the ether, which for the moment we assume to be fixed in the solar system. However, a simple calculation shows that the time difference could not be measured directly by ordinary methods. To see this, we observe that the speed of the earth V in its orbital motion around the sun is 30 km/s, roughly one ten-thousandth the speed of light c. If the apparatus were at rest relative to the ether, the time T_r for light to go a distance d from A to B and return to A would be $2d/c$. If the earth were moving with a speed V from A toward B through the ether, the time for the round trip T would be $d/(c + V) + d/(c - V)$. The difference between these times $\Delta T = T - T_r$ is given by

$$\Delta T = T - T_r = \frac{d}{c + V} + \frac{d}{c - V} - \frac{2d}{c} = \frac{2d}{c} \frac{V^2}{c^2 - V^2} \qquad (42.1)$$

Since $V = 10^{-4}c$, we have $\Delta T = 10^{-8}(2d/c) = 10^{-8}T_r$. Even if d were as great as 1,500 m (about a mile), ΔT would be only 10^{-13} s—too short a time to measure directly, as Maxwell pointed out in a letter of 1880. However, in that time green light of $\lambda = 500$ nm travels a distance of 60 wavelengths—a distance readily measurable with an interferometer. To detect the presumed motion of the earth through the ether, Michelson devised an ingenious experiment based on an interferometer (Sec. 27.5), which he invented for this purpose.

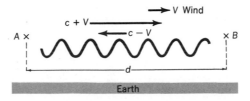

FIGURE 42.1
Transmission of a sound signal from A to B and back in a wind.

42.3 THE MICHELSON–MORLEY EXPERIMENT

The ideas underlying Michelson's inspiration involve the following considerations. In the Michelson interferometer a beam of light is split into two parts, which are sent by different paths and then recombined. Consider first the transit time of a light beam moving perpendicular to the velocity of the apparatus relative to the ether (Fig. 42.2). If the distance between the half-silvered plate P and the mirror is d, the light pulse travels a distance given by $cT_\perp = 2\sqrt{d^2 + (VT_\perp/2)^2}$, where $T_\perp$ is the time required for the light to go from P to M_1' and then to P''. Hence,

$$T_\perp = \frac{2d}{c\sqrt{1 - (V/c)^2}} \tag{42.2}$$

Let us now compare this time with that required for the portion of the wave which moves parallel to the motion, for which

$$T_\parallel = \frac{d}{c + V} + \frac{d}{c - V} = \frac{2cd}{c^2 - V^2} = \frac{2d}{c[1 - (V/c)^2]} \tag{42.3}$$

The round trip of the light takes slightly less time moving perpendicular to V than it does moving parallel to V. The difference $\Delta T = T_\parallel - T_\perp$ is small indeed. To a good approximation

$$\Delta T = T_\parallel - T_\perp = \frac{2d}{c} \frac{V^2}{2c^2} \tag{42.4}$$

This is only one-half the time difference of Eq. (42.2), but now the interferometer offers the possibility of a measurement.

In 1881 Michelson and Morley floated a large Michelson interferometer with $d = 11$ m on a pool of mercury. They watched the interference pattern as the interferometer was rotated, thereby interchanging the parallel and perpendicular paths. If the earth were moving through the ether, the interference fringes should have shifted through about 0.37 of a fringe as the interferometer was rotated. No shift was detected!

Michelson and Morley performed this experiment repeatedly over many months. No effect was observed which could be interpreted as showing a motion of the earth through the ether, in spite of the fact that

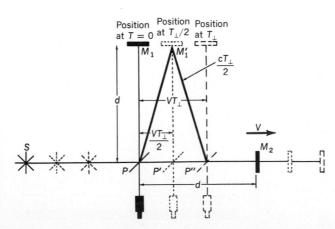

FIGURE 42.2
An electromagnetic signal would take longer to go from P to M_2 and return than from P to M_1 and return if the Michelson interferometer were moving with velocity V relative to an ether. The position of the interferometer is shown dotted for time $\frac{1}{2}T_\perp$, when the light reaches M_1, and dashed for time $T_\perp$, when the light returns to P.

they made measurements throughout the year, when the earth was moving in many different directions relative to the solar system. These results defied satisfactory explanation for 24 years. Michelson first interpreted the experiments as showing that the earth dragged some of the ether along with it, but a subsequent experiment by Lodge showed that a rapidly moving body could not possibly impart to the ether a speed in excess of 0.5 percent of the speed of the body. The most important of the various proposed explanations was that of Lorentz and FitzGerald, who suggested that the arm of the interferometer moving in the direction of the earth's motion shrinks just enough to lead to the null result. As we shall see, this apparently tortured prediction is supported by Einstein's theory of relativity.

42.4 POSTULATES OF THE SPECIAL THEORY OF RELATIVITY

In 1905 Einstein proposed a theory, amazing in its simplicity and genius, which ultimately resolved the Michelson-Morley dilemma. He argued that the reason there is no evidence of relative motion through the ether in any of a number of careful measurements is that any measurement of the speed of light will always yield a value independent of the velocities of both the observer and the source.

Einstein was keenly aware of the successes of classical mechanics, which predicted with great precision the motions of planets, aircraft, and baseballs. It was only when speeds were very great—comparable to that of light—that there was a problem. In classical physics we call a frame of reference in which Newton's laws of motion are valid an *inertial frame*. Any frame of reference which moves with constant velocity relative to an inertial frame is also an inertial frame. Thus, an airplane flying with a constant velocity of 960 km/h northward relative to an inertial frame provides the framework for another inertial system. Einstein proposed that if two postulates are accepted all the strange results which had been so difficult to interpret could be readily explained. These postulates of the special theory of relativity are:

1. *The laws of physics are the same in all inertial frames.*

2. *The speed of light c is the same for every inertial frame and is independent of any motion of the source.*

An immediate implication of Einstein's postulates is that there exists no preferred frame in the universe; there is no ether relative to which everything can be measured. One inertial frame is equivalent to another so far as the speed of light is concerned. This idea requires, however, that the simple relationships used for centuries in going from one inertial frame to a second must be inadequate, although they work well for slowly moving bodies. The correct relationships must reduce to the classical ones for speeds small compared with the speed of light.

42.5 THE LORENTZ TRANSFORMATION

Classically the x, y, z coordinates of a body referred to one inertial frame are related to the x^*, y^*, z^* coordinates of the same body referred to a second frame moving with constant velocity relative to the first by a set of

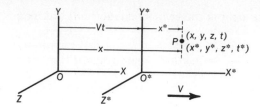

FIGURE 42.3
Reference frame $O^*X^*Y^*Z^*$ moving with constant velocity V in the x direction of reference frame $OXYZ$.

equations known as the *galilean* (or *newtonian*) *transformation*. Suppose we have two inertial frames: $OXYZ$, which we shall regard as at rest, and $O^*X^*Y^*Z^*$, which is moving with a velocity V along the X axis of the rest frame (Fig. 42.3). Let the origins of the two frames coincide at $t = 0$, and assume that time proceeds at the same rate in both frames. If at time t a body is located at a point P specified by (x,y,z) in the rest frame, in the frame $O^*X^*Y^*Z^*$, the body is found at the point (x^*, y^*, z^*), where

$$\begin{aligned} x^* &= x - Vt \\ y^* &= y \\ z^* &= z \end{aligned} \qquad \text{galilean transformation} \qquad (42.5)$$

and $\qquad t^* = t$

If a body has a velocity v in the frame $OXYZ$ with components $v_x, v_y, v_z,$ the corresponding components in the $O^*X^*Y^*Z^*$ frame are

$$v_x^* = v_x - V \qquad v_y^* = v_y \qquad v_z^* = v_z \qquad (42.6)$$

These transformations predict that if a beam of light has a velocity c in the X direction of the rest frame, its velocity in the moving frame is $c - V$. But this is not consistent with the results of the Michelson-Morley experiment. Hence we need a new set of relationships which reduce to the galilean transformation for low velocities but permit us to satisfy Einstein's second postulate for very high velocities. It was Einstein's good fortune that such a transformation had already been derived by Lorentz. His set of equations is

$$\begin{aligned} x^* &= \frac{x - Vt}{\sqrt{1 - V^2/c^2}} \\ y^* &= y \\ z^* &= z \\ t^* &= \frac{t - (V/c^2)x}{\sqrt{1 - V^2/c^2}} \end{aligned} \qquad \text{Lorentz transformation} \qquad (42.7)$$

When V is small compared with c, the Lorentz transformation reduces to the galilean transformation, but as v approaches the speed of light, they become different indeed.

42.6 THE LORENTZ CONTRACTION AND TIME DILATION

Consider a measuring stick of length L^* extending from x_1^* to x_2^* in the moving frame of reference $O^*X^*Y^*Z^*$, so $L^* = x_2^* - x_1^*$ (Fig. 42.4). To an observer in the rest frame the stick has an apparent length $L = x_2 - x_1$, where x_2 and x_1 are measured at the same time t. By Eqs. (42.7), $x_2^* = (x_2 - Vt)/\sqrt{1 - V^2/c^2}$ and $x_1^* = (x_1 - Vt)/\sqrt{1 - V^2/c^2}$. When these are substituted in the expression for the apparent length L in the rest frame, we obtain

$$L^* = x_2^* - x_1^* = \frac{x_2 - Vt}{\sqrt{1 - V^2/c^2}} - \frac{x_1 - Vt}{\sqrt{1 - V^2/c^2}}$$

$$= \frac{x_2 - x_1}{\sqrt{1 - V^2/c^2}} = \frac{L}{\sqrt{1 - V^2/c^2}}$$

or $$L = \sqrt{1 - \frac{V^2}{c^2}} L^* \qquad (42.8)$$

Thus, an object moving relative to an observer appears to be contracted in the direction of motion by the Lorentz factor $\sqrt{1 - V^2/c^2}$.

□ **Example** A meterstick moving parallel to its length at a speed of $0.5c$ passes an observer at rest. What length would the observer find for this meterstick?

By Eq. (42.8),

$$L = \sqrt{1 - \frac{V^2}{c^2}} L^* = \sqrt{1 - 0.25} \ (1.000 \text{ m}) = 0.866 \text{ m} \qquad □$$

Next consider the effect of relative motion on a pendulum clock which is fixed at the origin of the rest frame $OXYZ$ and indicates intervals of time $T = t_2 - t_1$. To an observer in the frame $O^*X^*Y^*Z^*$ moving with velocity V past the clock, the period $T^* = t_2^* - t_1^*$. By Eqs. (42.7), since $x = 0$,

$$T^* = \frac{t_2 - t_1}{\sqrt{1 - V^2/c^2}} = \frac{T}{\sqrt{1 - V^2/c^2}} \qquad (42.9)$$

Thus, when an observer is moving relative to a clock (or the clock relative to the observer), the time interval as seen by this observer is longer than that seen by another observer at rest relative to the clock. To an observer at rest a moving clock runs slow.

□ **Example** Muons are unstable particles which have an average lifetime of 2 μs. If a beam of cosmic-ray muons has a speed of $0.98c$ relative to the earth, find the apparent average life and the distance traversed by the beam during this time.

In the frame in which the muons are at rest, T is 2 μs. Relative to the muons the earth is moving with a speed of $0.98c$. By Eq. (42.9),

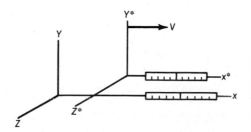

FIGURE 42.4
Frame $O^*X^*Y^*Z^*$, moving with velocity $0.5c$ parallel to the x axis of frame $OXYZ$, has a meterstick lying at rest parallel to the x^* axis; this stick has a length of 0.87 m when measured by an observer at rest in $OXYZ$.

$$T^* = \frac{T}{\sqrt{1 - V^2/c^2}} = \frac{2 \times 10^{-6}\,\text{s}}{\sqrt{1 - 0.96}} = 1 \times 10^{-5}\,\text{s}$$

In this time a muon travels

$$s^* = VT^* = (2.94)\,(10^8\,\text{m/s})\,(1 \times 10^{-5}) = 2{,}940\,\text{m} \qquad \square$$

We now see how observers in fixed and moving frames can find the same speed for light. From the point of view of the observer in the fixed frame, the observer in the moving frame finds the same velocity because metersticks are shortened and clocks run slow in the moving frame.

42.7 VELOCITY TRANSFORMATION

Next consider the values observers in the frames $OXYZ$ and $O^*X^*Y^*Z^*$ would find for the velocity of a body that is moving along the X axis of the rest frame. By the definition of velocity, $v = \Delta x/\Delta t = (x_2 - x_1)/(t_2 - t_1)$. An observer in the moving frame would find $v^* = (x_2^* - x_1^*)/(t_2^* - t_1^*)$, which, by the Lorentz transformation [Eqs. (42.7)], becomes

$$v^* = \frac{(x_2 - Vt_2) - (x_1 - Vt_1)}{(t_2 - Vx_2/c^2) - (t_1 - Vx_1/c^2)} = \frac{(x_2 - x_1)/(t_2 - t_1) - V}{1 - V(x_2 - x_1)/(t_2 - t_1)c^2}$$

or

$$v^* = \frac{v - V}{1 - Vv/c^2} \qquad (42.10)$$

This differs from the galilean velocity transformation by the term vV/c^2, which is negligible for ordinary velocities but becomes highly significant as v approaches c.

□ **Example** A beam of light moving with speed c along the X axis of frame $OXYZ$ (Fig. 42.3) passes an observer in frame $O^*X^*Y^*Z^*$, who determines its speed in his system. Show that he also obtains c, regardless of the value of V, the speed of $O^*X^*Y^*Z^*$ along the X axis of $OXYZ$.

By Eq. (42.10),

$$v^* = \frac{c - V}{1 - Vc/c^2} = c \qquad \square$$

□ **Example** A proton moving eastward with a speed of $0.60c$ in a nuclear-physics laboratory passes an electron moving westward with a speed of $0.90c$. Find the speed of the electron relative to a frame of reference riding with the proton.

If we call eastward the positive direction,

$$V = 0.60c \qquad \text{and} \qquad v = -0.90c$$

By Eq. (42.10),

$$v^* = \frac{-0.90c - 0.60c}{1 - (-0.90)\,(0.60)} = \frac{-1.50c}{1.54} = -0.97c \qquad \square$$

An important consequence of the theory of relativity is that according to its postulates no real body can move and no transfer of energy can occur with a speed which exceeds c, the speed of light in free space.

42.8 RELATIVE MASS

If a constant force F acts on a mass m indefinitely, it is clear that Newton's second law $F = ma$ must break down as the speed of the body approaches the speed of light, since it follows directly from Eq. (42.10) that the speed of light represents an upper limit to velocities involving energy transfer. Thus, a constant force does not cause the body to accelerate forever. We may retain $F = ma$ if, as the speed approaches the speed of light, the mass of the body increases rather than the velocity. Einstein showed by application of the law of conservation of momentum that if m_0 is the mass of a body when it is at rest, its mass m when moving with a velocity v is (Fig. 42.5)

$$m = \frac{m_0}{\sqrt{1 - v^2/c^2}} \tag{42.11}$$

In terms of this relation, the momentum of a body which has rest mass m_0 becomes

$$\text{Momentum} = mv = \frac{m_0 v}{\sqrt{1 - v^2/c^2}}$$

When a constant force acts on a body as its speed approaches the speed of light, the effect of doing work on the body, and thereby supplying it with energy, is to increase its mass rather than its speed. Einstein showed that this can be reconciled with conservation principles if mass is one of the forms of energy and if the additional energy given to the body is $\Delta W = (m - m_0)c^2$. This idea can be extended so that a body of rest mass m_0 has a rest energy $m_0 c^2$; when its velocity is v and its mass m, its total energy W is mc^2. The kinetic energy (energy associated with motion) is

$$\text{Kinetic energy} = mc^2 - m_0 c^2 = W - m_0 c^2 \tag{42.12}$$

$$= m_0 c^2 \left(\frac{1}{\sqrt{1 - v^2/c^2}} - 1 \right) \tag{42.12a}$$

If we expand this square root assuming that v^2/c^2 is small compared with 1, we obtain

$$\text{Kinetic energy} = m_0 c^2 \left[1 + \frac{1}{2} \frac{v^2}{c^2} + \frac{\left(-\frac{1}{2}\right)\left(-\frac{3}{2}\right)}{2!} \frac{v^4}{c^4} + \cdots - 1 \right]$$

$$= \tfrac{1}{2} m_0 v^2 \left(1 + \frac{3}{4} \frac{v^2}{c^2} + \cdots \right)$$

In the limiting case, where v/c is small compared with 1, we obtain the classical formula $\frac{1}{2}mv^2$ for the kinetic energy of a body. As v approaches c, we must turn to the relativistic relation for the kinetic energy as given in Eq. (42.12).

The revolutionary concept that mass is one of the forms of energy has proved to be a cornerstone of modern physics, permitting a new understanding of nuclear physics and leading to the development of nuclear fission as a practical energy source.

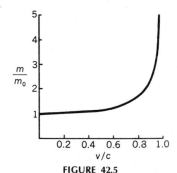

FIGURE 42.5

The ratio of the mass of a body in motion to its rest mass as a function of v/c.

42.9 THE MASS OF THE ELECTRON

The first particle for which the relativistic change of mass with velocity was investigated was the electron, the least massive of charged particles. The experimental results are completely in accord with the predictions of the relativity theory. Since high-energy electrons are usually produced by accelerating them through a large potential difference, it is convenient to measure their energies in electronvolts. *One electronvolt is the energy attained by an electron in falling through a potential difference of one volt* (Sec. 31.8).

Table 42.1 shows how the ratios v/c and m/m_0 vary with the kinetic energy of an electron. In a modern synchrotron, in which electrons are given kinetic energies of over 1 GeV, the mass of the moving electron is greater than the mass of the hydrogen atom at rest!

□ **Example** A Van de Graaff generator accelerates electrons to a kinetic energy of 2 MeV. Find the mass of these electrons. Calculate the speed of the 2-MeV electrons.

$$2 \text{ MeV} = (2 \times 10^6 \text{ eV}) (1.602 \times 10^{-19} \text{ J/eV})$$
$$= 3.2 \times 10^{-13} \text{ J}$$

$$3.2 \times 10^{-13} \text{ J} = (m - m_0)c^2 = (m - m_0) (9 \times 10^{16})$$

$$m - m_0 = 35.6 \times 10^{-31} \text{ kg}$$

$$m = (35.6 + 9.1) \times 10^{-31} \text{ kg}$$
$$= 44.7 \times 10^{-31} \text{ kg} \qquad \text{(almost 5 times the rest mass)}$$

$$44.7 \times 10^{-31} = \frac{9.11 \times 10^{-31}}{\sqrt{1 - v^2/c^2}} \qquad \text{by Eq. (42.11)}$$

Squaring both sides and transposing yields

$$2{,}000 \left(1 - \frac{v^2}{c^2}\right) = 83$$

$$\frac{v^2}{c^2} = \frac{1{,}917}{2{,}000} = 0.959 \text{ m}^2/\text{s}^2$$

$$v = 0.98c = 2.94 \times 10^8 \text{ m/s} \qquad\qquad □$$

TABLE 42.1

Kinetic energy of electron		$\dfrac{v}{c} = \dfrac{\text{speed of electron}}{\text{speed of light}}$	$\dfrac{m}{m_0}$
eV	J		
0	0	0	1
10^2	1.6×10^{-17}	0.0205	1.0002
10^3	1.6×10^{-16}	0.0625	1.002
10^4	1.6×10^{-15}	0.195	1.02
10^5	1.6×10^{-14}	0.548	1.20
10^6	1.6×10^{-13}	0.941	2.96
10^7	1.6×10^{-12}	0.999	20.6
10^8	1.6×10^{-11}	1.000—	197
10^9	1.6×10^{-10}	1.000—	1,960

In our discussion of relativity we have thus far confined our attention to inertial frames of reference. When this restriction applies, we are dealing with what is known as *special relativity*. If we consider frames of reference accelerated relative to an inertial frame, we are led to the *general theory* of relativity, also developed by Einstein. A simple example of an accelerated frame of reference is that of an elevator being accelerated upward. It is a well-known fact that an upward acceleration leads to an increase in the apparent weight of an object in the elevator. However, if an observer were isolated in an elevator on a strange planet, how could he untangle the gravitational contribution to the apparent weight from that associated with the acceleration of the elevator? Einstein concluded that there is no way to distinguish gravitational and acceleration effects in this situation. This conclusion, an important postulate of the general theory of relativity, is known as the *principle of equivalence.*

The special theory leads to most of the relativistic results which are of interest and applicability in modern physics, but the general theory leads to deeper insights and makes predictions which are not derivable from the special theory. We shall mention a few of these predictions.

1. Rays of light passing close to a star should be bent toward it. For example, a ray which just grazes the sun's surface on its way to the earth is deflected through about 1.75 seconds of arc. This observation was one of the first which tested the predictions of general relativity.

2. Physical processes should take place more slowly in regions of low gravitational potential than in regions of high gravitational potential. Thus, atomic vibrations on the sun should appear slowed down relative to their periods on the earth. This leads to the prediction that the spectral lines from atoms of the sun should be shifted very slightly toward the red end of the spectrum when compared with lines from the same elements on the earth.

3. The orbits of the planets should be slightly changed. This effect is important for the planet Mercury, which is closest to the sun. General relativity predicts that, aside from the precession of Mercury's orbit due to the perturbations of other planets, there should be an additional precession of about 43 seconds of arc per century. This precession had been known for many years, but its origin remained a mystery until Einstein developed his general theory. Some questions about the correctness of certain aspects of Einstein's general theory remain.

QUESTIONS

1. The objection has been raised to the theory of relativity that it violates "common sense." How could you refute this objection?
2. Why don't we observe relativistic effects in our everyday lives?
3. What alternatives to the theory of relativity can you propose to explain the results of the Michelson-Morley experiment?
4. For an electron with rest mass m_0 and speed $v = 0.6c$, which is the largest, $\frac{1}{2}mv^2$, $\frac{1}{2}m_0v^2$, or its kinetic energy? Which is smallest?
5. By what factor is the density of a solid body increased when it moves with speed v?
6. Is it conceivable that an electron beam could move *across* the screen of a TV tube or a cathode-ray tube with a speed greater than the speed of light in free space? Why is this not contradictory to special relativity?
7. Photons experience a gravitational attraction characteristic of a body of rest mass $h\nu/c^2$. There

are some regions in the universe where there are such great mass densities over a volume that the energy of the photon is insufficient to permit it to escape. Why are such regions called *black holes?*

8. What is an electronvolt? Why is the electronvolt a convenient unit for measuring energies in atomic and nuclear physics?

PROBLEMS

1. A river 6 km wide has a current of 8 km/h. How much longer will it take a motorboat with a speed relative to the water of 16 km/h to go 6 km upstream and then return than to go straight across the river and return to the same point? *Ans.* 8.04 min

2. A river 3 km wide has a current with a speed of 5 km/h. How much longer will it take a boat with a speed of 8 km/h to go upstream 3 km and return than to go directly across the river and return to the same point?

3. Find the mass of an electron traveling at six-tenths the speed of light. How many times as great as the rest mass is this value?
Ans. 1.14×10^{-30} kg; 1.25

4. Find the speed at which an electron has a mass 20 times its rest mass.

5. How many electronvolts of energy must an electron gain to bring its mass to (a) $1.5m_0$ and (b) $5m_0$? In each case what is the speed of the electron?
Ans. (a) 256 keV; 2.24×10^8 m/s; (b) 2.04 MeV; 2.94×10^8 m/s

6. An electron is given a kinetic energy of 3.066 MeV in a Van de Graaff accelerator. Find the ratio of the electron's mass to its rest mass and calculate the speed of the electron.

7. Find the kinetic energy, total energy, mass, and momentum of an electron moving at a speed of 0.900 times that of light.
Ans. 661 keV; 1.172 MeV; 2.09×10^{-30} kg; 5.64×10^{-22} kg-m/s

8. Find the work which must be done on an electron to increase its speed from $0.5c$ to $0.8c$, where c is the speed of light.

9. Through what potential difference must protons be accelerated to achieve a speed of $0.800c$? What is the momentum of such a proton? The total energy?
Ans. 626 MeV; 6.69×10^{-19} kg-m/s; 1.564 GeV

10. A proton is given a kinetic energy of 8.442 GeV in a proton synchrotron. Find the speed and the mass of the proton at this energy.

11. When an electron leaves the Stanford linear accelerator, its speed is only 0.1 m/s less than the speed of light. Find the mass and the kinetic energy of an electron with that speed. *Hint:* $c^2 - v^2 = (c + v)(c - v) \approx 2c(c - v)$.
Ans. 3.53×10^{-26} kg; 19.8 GeV

12. A proton is given a kinetic energy of 300 GeV at the Fermi National Accelerator Laboratory. Find the ratio of the mass to the rest mass of the proton.

13. An electron is accelerated to a kinetic energy of 51.1 MeV. Find the ratios m/m_0 and v/c. If an electron at rest can be assumed to be spherical, find the ratio of the longitudinal to the transverse diameter for the 51-MeV electron.
Ans. 101; 0.99995; 0.0099

14. An arrow passes an observer with a speed 0.6 times that of light. If the rest length of the arrow is 0.7 m, find its apparent length as it passes the observer.

15. The mean life of a muon at rest is about 2.2 μs. If muons in cosmic rays have an average speed of $0.99c$, find their mean life and the approximate distance they move before decaying.
Ans. 15.6 μs; 4.63 km

16. The mean life of pions at rest is 26 ns. If a beam of pions has a speed of $0.866c$ what will be the observed mean life in the laboratory?

17. Imagine that you are at rest in a frame $O^*X^*Y^*Z^*$ which moves horizontally past an inertial frame $OXYZ$ at a speed of $0.6c$. A boy in that frame drops a ball which, according to your clock, falls for 0.6 s. How long would the ball fall as timed by an observer at rest in the $OXYZ$ frame?
Ans. 0.48 s

18. Derive the following relations between the relativistic kinetic energy K of a particle, its rest mass m_0, and its speed v:

(a) $\quad 1 + K/m_0c^2 = (1 - v^2/c^2)^{-1/2}$

(b) $\quad \dfrac{v}{c} = \left[1 - \dfrac{1}{(1 + K/m_0c^2)^2}\right]^{1/2}$

19. An electron moving to the right with a speed of 2.7×10^8 m/s passes an electron moving to the left with a speed of 2.1×10^8 m/s. Find the speed of one electron relative to the other as predicted by (a) the newtonian-galilean transformation and (b) the relativistic transformation.
Ans. (a) 4.8×10^8 m/s; (b) 2.94×10^8 m/s

20. A particle has a rest energy m_0c^2 and a total energy W. Show that the velocity of this particle is $c\sqrt{1 - (m_0c^2/W)^2}$.

21. An alpha particle moving east with a speed of

0.6c is passed by an electron moving west with a speed of 0.9c. Find the speed of the electron relative to the alpha particle.

Ans. 2.92×10^8 m/s

22. Show that Eqs. (42.7) lead to the transformations

$$x = \frac{x^* + Vt^*}{\sqrt{1 - V^2/c^2}}$$

and

$$t = \frac{t^* + Vx^*/c^2}{\sqrt{1 - V^2/c^2}}$$

23. An observer on earth sees two spaceships approaching each other. Each ship is moving with a velocity of 1.8×10^8 m/s relative to the earth. How fast will an observer in one of the spaceships find that she is approaching the other?

Ans. 2.65×10^8 m/s

24. A free particle moving with speed v has rest energy m_0c^2, momentum $p = mv$, kinetic energy K, total energy W (rest energy plus kinetic energy). Given that $m = m_0/\sqrt{1 - v^2/c^2}$, derive the following relations: (a) $p = Wv/c^2$; (b) $W^2 = p^2c^2 + m_0^2c^4$; (c) $p^2c^2 = K^2 + 2m_0c^2K$.

43

X-RAYS

The discovery of x-rays represents one of the most important developments in physics and remains one of the strong arguments for research in pure science. It stimulated the work which led to the discovery of radioactivity, which in turn laid the ground work for nuclear physics. Both x-rays and the gamma rays from radioactive nuclei are high-energy electromagnetic radiation similar to visible light but characterized by much shorter wavelengths. In this chapter we discuss some of the significant and useful properties of these rays.

43.1 THE DISCOVERY OF X-RAYS

During the autumn of 1895 Röntgen was studying the conduction of electricity through partially evacuated tubes (Sec. 41.3), when he observed a mysterious radiation which was able to penetrate thin layers of material. He called this radiation *x-rays*, with the "x" standing for "unknown." Röntgen promptly performed a broad series of experiments with x-rays and learned many of the properties of this new kind of radiation. Among his findings were the following. X-rays produce fluorescence in all kinds of materials. All substances are somewhat transparent to x-rays, particularly those composed of light elements. Thus paper, wood, and aluminum are almost transparent, while lead and gold are relatively opaque. Photographic emulsions are sensitive to x-rays, although the eye cannot see them. The rays are not deflected by either magnetic or electric fields and hence show no evidence of bearing electric charge. X-rays are produced whenever a beam of high-energy electrons strikes matter. Röntgen was unable to concentrate x-rays with lenses or mirrors and failed to find any evidence of interference or diffraction of x-rays.

We now know that x-rays are electromagnetic radiation similar to light and ultraviolet radiation, except that the wavelengths are much shorter. Although the lines of demarcation between the various kinds of electromagnetic radiation are not sharp, we may think of x-rays as including radiation of wavelengths shorter than about 10 nm.

Among the earliest x-ray photographs were pictures of the human hand and forearm. Soon pictures of this kind were utilized in studying fractures and similar abnormalities. Only 3 months after Röntgen's discovery, x-rays were used in connection with a surgical operation.

The news that rays had been discovered which could take pictures through walls was tremendously exciting. X-rays were front-page news all over the world in 1896. Grossly exaggerated claims and absurd statements made some people fear that privacy was lost forever. A London firm advertised the sale of x-ray-proof underclothing. A bill was introduced in one of the state legislatures "to prohibit the insertion of x-rays, or any device for producing the same into, or their use in connection with, opera glasses, or similar aids to vision."

X-rays have found many important applications. They have given the diagnostic methods of physicians, surgeons, and dentists an exactitude that formerly was impossible. X-rays possess properties valuable in the treatment of certain diseases. Malignant cells are somewhat more readily killed by x-rays than normal cells. Under x-ray treatment inflamed glands shrink in size, and various morbid conditions of the blood and skin clear up. By means of x-rays, the metallurgist determines the effect of heat treatment, tempering, rolling, and aging on metals and alloys. Hidden defects in objects and concealed cracks in metals can be revealed by x-rays, which are a major tool in the nondestructive testing of materials.

43.2 THE PRODUCTION OF X-RAYS

Electromagnetic radiation is emitted when electric charges are accelerated (Sec. 40.10). This radiation appears in the form of photons, the energy of which is given by the product of Planck's constant h and the frequency v. For a typical x-ray photon, λ may be 1 Å and the corresponding energy 12,400 eV.

The most common means of producing x-rays is to accelerate electrons through high potential differences in vacuum tubes and allow them to strike a target of some heavy element. In Röntgen's original experiments he applied a high potential difference between two elements in a partially evacuated tube. The electrons accelerated came from the ionization of the residual gas molecules and from secondary electrons ejected from the cathode by impact of positive ions. Gas tubes (Fig. 43.1) operating on this principle are still used for producing x-rays.

In 1913 Coolidge developed an x-ray tube which was easier to control than the gas type. In the Coolidge tube (Fig. 43.2) the electrons are emitted from a heated filament F placed in a highly evacuated chamber. By varying the temperature of the filament, the number of electrons emitted can be controlled. These electrons are accelerated and strike a target T, from which the x-rays are emitted in every direction, but those which travel into the target are absorbed.

43.3 WAVE PROPERTIES OF X-RAYS

Röntgen's early failure to observe interference and diffraction of x-rays was associated with his apparatus and procedures. The first definite evidence of wave properties for x-rays came out of an experiment devised by Laue, in which a crystal played the role of a diffraction grating. When x-rays fall on a body, the atoms scatter the incident radiation. In a crystalline substance the atoms are arranged in a regular way, and definite phase relationships exist between the scattered rays from neighboring atoms. If a narrow pencil of x-rays falls on a crystal C (Fig. 43.3), the scattered rays reinforce each other in certain directions by constructive interference. When a photographic plate is appropriately exposed, a series of spots is formed (Fig. 43.4) where constructive interference occurs. The theory of this interference was worked out by Laue. His

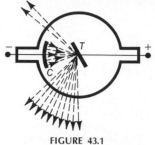

FIGURE 43.1
Electrons accelerated in a gas-type x-ray tube come from the ionization of the residual gas and from secondary emission of electrons when positive ions strike the cathode.

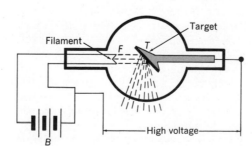

FIGURE 43.2
A hot filament is the source of electrons in a Coolidge x-ray tube.

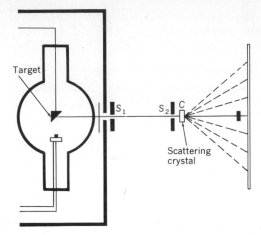

FIGURE 43.3
X-ray diffraction apparatus.

colleagues Friedrich and Knipping successfully exposed a photographic plate to show the Laue spots in 1912. From the positions of these spots and the properties of the scattering crystal Laue computed the range of wavelengths present in their x-ray beam.

Shortly after Laue's work W. L. Bragg proposed a relatively simple way of looking at the process of diffraction by a crystal grating. He pointed out that through any crystal a set of equally separated parallel planes can be drawn which pass through all the atoms of the crystal. If x-rays are incident on this crystal, each atom scatters the radiation. By Huygens' principle the condition for constructive interference of the radiation scattered by every atom lying in a particular plane is that the angle of incidence be equal to the angle of reflection. To form a Laue spot, it is necessary not only to have constructive interference of all scattered wavelets from a given plane of atoms but also to have constructive interference of the radiation scattered from adjacent planes. For the radiation scattered by the second plane of Fig. 43.5 to interfere constructively with that scattered by the first plane, it is necessary that the path

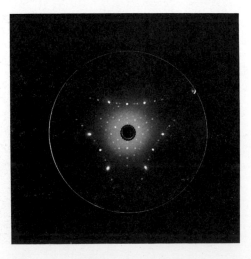

FIGURE 43.4
Laue photograph of a tungsten crystal.
(*RCA Corporation.*)

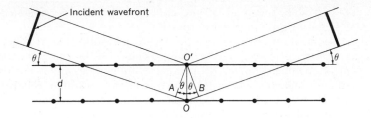

FIGURE 43.5
Constructive interference of the waves scattered from atoms in adjacent planes of a crystal occurs when $n\lambda = 2d \sin \theta$.

difference be an integral number of wavelengths. It can be seen from the figure that the difference in paths is *AOB*. Let the distance between adjacent planes be *d*. Since $AO = OB = d \sin \theta$, it follows immediately that

$$n\lambda = 2d \sin \theta \qquad (43.1)$$

where *n* is an integer, λ is the wavelength of the x-rays, and θ is the glancing angle. This important relation is known as *Bragg's law*.

By use of Bragg's law it is possible to measure the wavelengths found in x-ray beams, knowledge of which has contributed greatly to our understanding of atomic structure. Further, by use of x-rays of known wavelength, one can determine the grating space *d* for crystals of unknown structure. Indeed, by an extension of the ideas of Bragg and Laue, it is possible to infer how atoms are arranged in crystals and in complex molecules, such as proteins.

43.4 THE X-RAY SPECTROMETER

W. H. Bragg, father of W. L. Bragg, developed an x-ray spectrometer (Fig. 43.6) by which the wavelengths of x-rays could be measured. He took an ordinary optical spectrometer, mounted a sodium chloride crystal on the

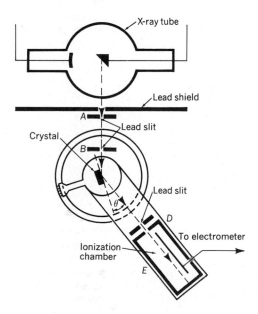

FIGURE 43.6
Spectrometer for measuring the wavelengths of x-rays.

rotating table, and replaced the collimator by an x-ray source and the telescope by an ionization chamber.

An ionization chamber is a device for measuring the number of ion pairs produced in a gas by ionizing radiation or by charged particles. A typical chamber consists of a metal cylinder (Fig. 43.7a) enclosing an insulated metal collector AC, which is charged positive relative to the cylinder. When ionizing radiation enters through a "window" at one end, it ionizes the gas in the chamber; electrons are collected by AC while positive ions go to the chamber walls. The current read by the galvanometer G is a measure of the intensity of the ionizing beam. The current through the conducting gas is not proportional to the applied potential difference; it does not follow Ohm's law. As the potential difference between AC and the cylinder is increased, the current through the galvanometer increases (Fig. 43.7b). If the potential gradient is one or two volts per millimeter, many of the ions recombine before they are collected by the plates. However, an electric field of the order of 2.5 V/mm is ordinarily sufficient to collect virtually all the ions produced, at which point the saturation current is attained. Once saturation is reached, a further moderate increase in field gives no significant change in current. When the potential gradient is increased far above the value required for saturation, the current increases again, at first slowly, then more rapidly as more and more ion pairs are produced by collisions. At a sufficiently high potential gradient, the gas breaks down and a discharge occurs between the chamber and the collector AC.

Using a beam of x-rays from a platinum target, W. H. Bragg measured the ionization current as a function of glancing angle θ and obtained the results shown in Fig. 43.8. When a target of a different element was the source of x-rays, the bumps labeled a, b, and c disappeared and new bumps were found superimposed on the same kind of *continuous* background. Thus the observed x-ray spectrum consisted of bumps or *lines* associated with the particular element of which the target was made and a *continuous spectrum* (Sec. 43.5); the lines, which depend on the target element, constitute the *characteristic spectrum* (Sec. 45.5).

To calculate the wavelengths of the x-ray lines Bragg needed the grating space d for the NaCl crystals (Fig. 43.9). This can be calculated as

(a)

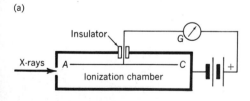

(b)

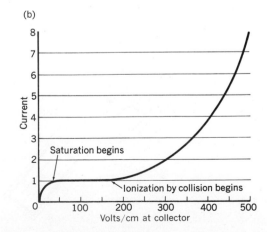

FIGURE 43.7
(a) Ionization chamber for measuring the intensity of beams of x-rays and other ionizing radiations. (b) Current in an ionization chamber as a function of the electric intensity at the collector AC for a particular ionization chamber.

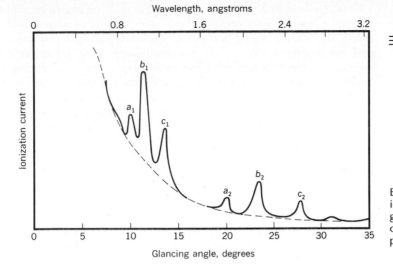

Wavelength, angstroms

Glancing angle, degrees

FIGURE 43.8
Bragg's curve showing the current in his ionization chamber as a function of glancing angle θ; the lines a, b, and c are characteristic lines in the L series of platinum which are superimposed on a continuous spectrum.

follows. If M is the molecular weight of the NaCl, the number of atoms in a mass M is $2N_A$, where N_A is Avogadro's number. The number of atoms per kilogram is $2N_A/M$. To obtain the number of atoms per cubic meter, we multiply by the density ρ to obtain

$$\text{Number of atoms per cubic meter} = \frac{2N_A\rho}{M}$$

The volume per atom is therefore $M/2N_A\rho$. For a cubic crystal we can think of each atom as occupying a cube of length d on a side; thus $d^3 = M/2N_A\rho$, or

$$d = \sqrt[3]{\frac{M}{2N_A\rho}} \qquad \text{for a cubic crystal} \qquad (43.2)$$

The molecular weight of NaCl is 58.45, and the density is 2,160 kg/m³. Since $N_A = 6.02 \times 10^{26}$, $d = 2.82 \times 10^{-10}$ m $= 2.82$ Å.

Bragg measured the wavelengths of many x-ray lines, using sodium chloride crystals. Once the wavelengths of lines are known, it is possible to determine the separations between planes, i.e., the grating spaces, for crystals of more complicated structure.

☐ **Example** With an x-ray spectrometer using a rock-salt crystal, the glancing angle for the first reinforcement ($n = 1$) for the Cu $K\alpha$ line is found to be 15.8°. If the distance between the crystal planes is 2.82 Å, find the wavelength of this line.

$$n\lambda = 2d \sin \theta$$
$$n = 1 \qquad \text{and} \qquad d = 2.82 \text{ Å}$$
$$\sin \theta = 0.273$$
$$\lambda = 2(2.82)(0.273) = 1.54 \text{ Å} = 0.154 \text{ nm} \qquad ☐$$

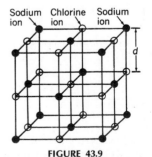

Sodium ion   Chlorine ion   Sodium ion

FIGURE 43.9
Crystal lattice of rock salt (NaCl).

By 1914 it was established that x-rays have wave properties and that they are a form of electromagnetic radiation. Well before this, radioactivity had been discovered, and some radioactive materials were found to emit *gamma* rays, which are of the same fundamental nature as x-rays. Indeed, short-wavelength electromagnetic radiation is called gamma radiation when it comes from a radioactive nucleus and x-radiation when it arises from bombarding a target with high-energy electrons. Since the properties of the rays depend on their wavelengths and not on their origin, we shall discuss radioactivity briefly before we treat the detection and absorption of x-rays and gamma rays in Chap. 44.

Electromagnetic radiation has not only wave characteristics but also particle properties, as pointed out in Sec. 41.6. It comes in *photons* with energy $h\nu$ [Eq. (41.3)], where h is Planck's constant and ν is the frequency. Since $\nu = c/\lambda$, the shorter the wavelength the greater the energy associated with a photon. For x-ray and gamma-ray photons the energy ranges from a few hundred to many millions of electronvolts. Further evidences of the particle properties of electromagnetic radiation are presented in the following three sections.

43.5 THE CONTINUOUS X-RAY SPECTRUM

Two phenomena from the field of x-rays support the blackbody and the photoelectric evidence for the photon hypothesis. They are the short-wavelength limit of the continuous x-ray spectrum and the Compton effect, which are discussed respectively in this and the following section. As the work of Bragg (Sec. 43.4) and others has shown, the wavelengths of the x-rays emitted when a beam of high-energy electrons strike a target vary over a broad range. There is always a *continuous spectrum* upon which certain lines characteristic of the target element can be superimposed. When the intensity of the continuous radiation in a given wavelength interval is plotted as a function of wavelength for different values of the applied potential across the tube, curves similar to those of Fig. 43.10 are produced. Three features of these curves are immediately apparent:

1. The intensity radiated increases at all wavelengths when the potential difference across the tube is raised.

2. The shortest wavelength emitted at a given potential is sharply defined and decreases as the voltage across the tube increases.

3. As the potential difference across the tube is increased, the wavelength at which the maximum energy is radiated shifts toward shorter wavelengths.

We can understand why the spectrum is continuous and predict quantitatively the short-wavelength limit if we invoke the Planck hypothesis that electromagnetic radiation is emitted in photons of energy $h\nu$. X-rays are produced when the kinetic energy of the incident electrons is transformed into electromagnetic radiation through collisions with atoms of the target. Most of the collisions are glancing ones, in which only some moderate fraction of an electron's energy is radiated as a photon. Before being stopped, most electrons have several collisions and produce several photons of widely varying wavelengths. Occasionally an electron has a head-on collision, in which it loses all its energy at once. Such collisions

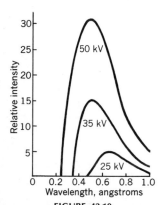

FIGURE 43.10
Wavelength distribution of the continuous x-ray spectrum from an aluminum target.

are relatively rare and are the ones which produce the x-rays at the *short-wavelength limit*. Clearly, the most energy an electron can lose is all it has. This corresponds to Ve, where V is the potential difference across the tube and e the electronic charge. If we equate this energy to that of the most energetic photon, we obtain

$$Ve = h\nu_{max} = \frac{hc}{\lambda_{min}} \qquad (43.3)$$

where c is the speed of light. Solving for λ_{min} yields

$$\lambda_{min} = \frac{hc}{Ve} \qquad (43.3a)$$

This relationship, known as the *Duane-Hunt law* (1915), was used for one of the earliest reliable determinations of Planck's constant h.

In many respects continuous x-ray emission is the inverse of the photoelectric effect. In the production of an x-ray photon the kinetic energy of the incident electron is converted into radiant energy; in the photoelectric effect the radiant energy of a photon is converted, at least in part, into kinetic energy of the electron.

The amount of energy available from each electron is increased as the potential difference across the tube is increased, and we should expect more energy to be radiated per electron. Thus the ordinates in Fig. 43.10 increase as the potential difference across the tube is increased. The area under the curve is proportional to the amount of energy radiated in the form of x-rays. The rest of the energy of the incident electrons appears as heat in the target.

43.6 SCATTERING AND THE COMPTON EFFECT

When a beam of x-rays passes through matter, some energy is scattered out of the beam. According to the classical theory of J. J. Thomson, the atomic electrons are driven to perform simple harmonic motion by the electric intensity E of the incident waves. Since these electrons are accelerated, they radiate a frequency equal to that of the incident waves. Thus the electrons remove energy from the passing wave and reradiate this energy in other directions. Scattering of waves with the frequency of the incident radiation is observed over the entire electromagnetic spectrum; however, it is weak at high frequencies. This unmodified scattering can be treated satisfactorily in terms of classical waves.

In 1923 Compton published the results of careful measurements of the x-ray frequencies scattered by carbon atoms upon which monochromatic x-rays were incident (Fig. 43.11). He found that the scattered beam contains two frequencies, one the same as that of the incident beam, and the second somewhat lower. Figure 43.11 shows the wavelength distribution of the radiation scattered at various angles. At each angle radiation is scattered not only at the wavelength of the incident beam but also at a longer wavelength whose value depends on the scattering angle θ (the angle between the propagation directions of the incident and scattered radiation).

Compton's measurements showed that the change in wavelength $\Delta\lambda$ is independent of the scattering material but depends on the scattering angle θ according to the relation

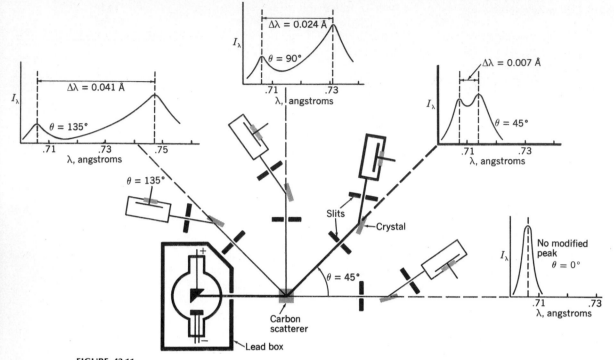

FIGURE 43.11
The scattering of Mo $K\alpha$ x-rays ($\lambda = 0.707$ Å) at an angle θ produces two peaks in the scattered radiation, one at the wavelength of the incident radiation and the second at a wavelength greater by $\Delta\lambda = 0.024(1 - \cos\theta)$ Å.

$$\Delta\lambda = 0.024(1 - \cos\theta) \qquad \text{Å} \qquad (43.4)$$

Compton explained the wavelength shift in the scattering of x-rays by assuming that the incident beam of x-rays consists of a stream of photons of energy $h\nu_0$. These photons possess momentum $h\nu_0/c$ as well as energy. Their collisions with electrons (Fig. 43.12) can be described in terms of the laws of conservation of momentum and conservation of energy. Applying these laws to a collision, we obtain, for the relativistic case in which the electron is initially at rest,

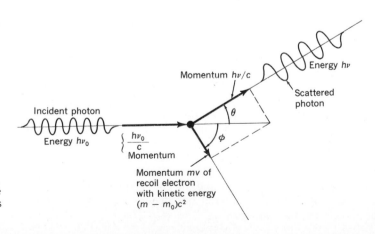

FIGURE 43.12
Elastic collision of a photon with a free electron initially at rest. The photon is scattered, and the electron recoils.

Conservation of energy:

$$h\nu_0 = h\nu + (m - m_0)c^2 \qquad (43.5)$$

Conservation of momentum:

x component: $\quad \dfrac{h\nu_0}{c} = \dfrac{h\nu}{c}\cos\theta + \dfrac{m_0 v}{\sqrt{1 - v^2/c^2}}\cos\phi \qquad (43.6)$

y component: $\quad 0 = \dfrac{h\nu}{c}\sin\theta - \dfrac{m_0 v}{\sqrt{1 - v^2/c^2}}\sin\phi \qquad (43.7)$

where ν is the frequency of the scattered photon, m and v are the mass and speed of the recoil electron, m_0 is its rest mass, θ is the angle through which the photon is scattered, and ϕ is the angle between v and the direction in which the photon was incident. With considerable effort these equations, together with Eq. (42.11), can be solved to find the change in wavelength $\Delta\lambda = c/\nu - c/\nu_0$. The result is that

$$\Delta\lambda = \frac{h}{m_0 c}(1 - \cos\theta) = 0.02426(1 - \cos\theta) \quad \text{Å} \qquad (43.4a)$$

Compton's experiments were a clear indication that electromagnetic radiation has particle as well as wave properties. The Compton effect is convincing evidence that radiation in the x-ray region comes in photons, or quanta, which behave like particles in collisions with electrons.

43.7 PAIR PRODUCTION AND THE POSITRON

Still another phenomenon which requires the photon concept and relativity for explanation was discovered by Anderson in 1932. When photons with energy greater than 1.02 MeV pass through matter, the photon may interact with an atomic nucleus and be transformed into two particles, an electron and a *positron*. A *positron* is a particle identical with an electron except that it bears a positive charge. The higher the energy of a photon, the more probable pair production becomes. It requires a minimum energy of 1.02 MeV to provide the rest mass of the electron and the positron. The positron and electron share whatever energy is left over from the energy of the incident photon. Thus, if pairs are produced by a 4.02-MeV photon, the electron and positron share 3.0 MeV of kinetic energy.

Positrons exist for a very short time. As they move through matter, they lose energy rapidly. As soon as they are stopped, an electron and a positron interact to annihilate each other. Two electron masses vanish, and 1.02 MeV of energy is released. This usually appears in the form of two 0.51-MeV photons moving in opposite directions. The two photons are formed rather than a single 1.02-MeV photon because it is not possible to conserve both energy and momentum for a single photon (unless some additional particle is involved). Both in pair production and in pair annihilation the particle properties of photons are in evidence.

QUESTIONS

1. Why aren't ordinary diffraction gratings with a few hundred lines per millimeter useful for measuring the wavelength of x-rays incident normally in the ordinary way? (They are successfully used at grazing incidence.)
2. The gas between two parallel plates is ionized by a constant x-ray beam. If the potential difference between the plates is increased from zero to a very high value, what happens to the current? Explain carefully. Is Ohm's law obeyed?
3. X-ray photographs are essentially shadowgraphs. What restrictions must be placed on the target area from which the x-rays originate if the x-ray photograph is to have sharp definition?
4. What evidence exists to show the wave properties of x-rays?
5. What do we mean when we say that x-rays are not really reflected by a rock-salt crystal? If the process is not reflection, what is it?
6. How is the short-wavelength limit of the continuous x-ray spectrum related to the photoelectric effect?
7. What factors determine the short-wavelength limit and general shape of the continuous x-ray spectrum as a function of wavelength?
8. Why is the Compton effect readily observed for x-rays but not for visible light? Explain carefully.
9. Is it possible for a Compton recoil electron to have a velocity component opposite that of the incident photon? Is it possible for a Compton-scattered photon to have a velocity component opposite that of the incident photon? Explain.

PROBLEMS

1. Find the angle for first-order Bragg reflection of Mo $K\alpha$ radiation (0.707 Å) from a calcite crystal with a grating space of 3.0356 Å at 18° C. What is the angle for second-order reflection?
   Ans. 6.7°; 13.5°
2. The grating space of calcite is 303.56 pm. Find the wavelength for which the first-order Bragg "reflection" occurs at 25°, that is, when the incident and emergent radiations are inclined 25° with respect to the surface of the calcite. What is the angle for second-order "reflection"?

3. For Ni $K\alpha$ radiation first-order Bragg reflection from NaCl occurs at an angle of 17.1°. Find the wavelength of this radiation. At what angles are the second- and third-order diffraction maxima?
   Ans. 1.658 Å; 36.0°; 61.9°
4. Find the grating space of a KCl crystal which has the same crystal structure as NaCl. The density of KCl is 1,990 kg/m³, and the atomic weights of K and Cl are 39.1 and 35.46, respectively.
5. The grating space of mica is 9.963 Å. At what angle would magnesium $K\alpha$ x-rays with a wavelength of 9.89 Å be reflected? (Such long wavelengths are strongly absorbed in air, so an evacuated spectrometer would be used.)   *Ans.* 29.8°
6. Potassium iodide (KI) crystallizes as a cubic crystal of density 3,130 kg/m³ and structure similar to that of NaCl. The atomic weights of potassium and iodine are 39.1 and 126.9, respectively. Find the grating space of the KI crystal.
7. Calculate the grating space of KBr, which has a molecular weight of 119 and a density of 2,750 kg/m³.   *Ans.* 330 pm
8. Find the grating space of sodium bromide (NaBr) which has a density of 3,203 kg/m³.
9. The shortest wavelength emitted from an x-ray tube arises when all the kinetic energy of the incident electron is converted into radiant energy. Find the shortest wavelength produced in a tube across which a potential difference of 80 kV is maintained.   *Ans.* 15.5 pm
10. A television tube operates at a potential difference of 9 kV. What is the short-wavelength limit of the x-rays emitted?
11. Calculate the short-wavelength limit for the radiation from (a) an x-ray tube operated at 30 kV, (b) a 20-MeV betatron, (c) a 250-MeV synchrotron, and (d) a 50-GeV linear accelerator.
    Ans.   (a) 41.3 pm;   (b) 62 fm;   (c) 4.96 fm; (d) 2.48×10^{-17} m
12. A Coolidge x-ray tube is operating at 180 kV. With what velocity do the electrons strike the anode? (Relativity correction for the change of mass of electrons with speed must be made.) What is the short-wavelength limit of the x-rays produced?
13. Photons of wavelength 4 pm undergo Compton scattering by the electrons of carbon. What is the wavelength of a photon Compton-scattered at 90° with the incident beam? What is the energy in electronvolts of the recoil electrons from these photons?   *Ans.* 6.43 pm; 117 keV
14. X-rays of wavelength 9 pm are scattered by carbon. At what angle will the Compton-scattered photons have a wavelength of 9.5 pm?
15. X-ray photons are scattered through an angle of 60° by a beryllium foil. The scattered photons

have a wavelength of 18 pm. Find the wavelength and the energy of the incident photons.

Ans. 16.8 pm; 73.8 keV

16. A 0.511-MeV photon strikes an electron initially at rest head on and is scattered straight backward. Find the kinetic energy of the recoil electron. *Hint:* 0.511 MeV $= m_0c^2 = h\nu_0 = hc/\lambda_0$.

17. What is the longest wavelength which a photon can have and still transfer one-half its initial energy to a Compton recoil electron?

Ans. 4.85 pm

18. A photon of frequency ν_0 makes a head-on collision with a free electron. Show that the frequency $\nu_{180°}$ of the rebounding photon is given by $\nu_0/(1 + 2h\nu_0/mc^2)$ by applying conservation of momentum and conservation of energy directly to the collision in one dimension.

19. A high-energy photon interacts with a heavy nucleus to produce an electron-positron pair. What was the energy of the photon if $K^+ = 350$ keV and $K^- = 300$ keV, where K^+ stands for the kinetic energy of the positron and K^- the kinetic energy of the electron?

Ans. 1.672 MeV

20. Prove that if a photon is scattered through an angle greater than 60° by a free electron, it can-

not produce an electron-positron pair after the scattering event, regardless of the initial energy of the photon.

21. A high-energy photon interacts with a lead nucleus to create a positron with a kinetic energy of 0.42 MeV and an electron with a kinetic energy of 0.34 MeV. Find the energy of the incident photon.

Ans. 1.78 MeV

22. Show that a free electron at rest cannot absorb a photon unless some other particle (such as a nucleus) is involved in the process. *Hint:* Apply conservation of energy and conservation of momentum.

23. Show that conservation of energy and conservation of momentum require that pair production occur only when a photon interacts with some other particle; i.e., the photon cannot spontaneously transform to a pair without interacting with some particle.

44

RADIOACTIVITY AND THE NUCLEAR ATOM

Only 3 months after Röntgen discovered x-rays, Becquerel discovered radioactivity and a vast new area of physics was opened up. The earliest researches were directed toward finding new radioactive materials and to studying the three types of radiation—alpha, beta, and gamma rays—emitted by radioactive ores. Soon researches with radioactive materials were yielding new insights into the structure of atoms.

44.1 NATURAL RADIOACTIVITY

Röntgen's discovery of x-rays in 1895 stimulated physicists all over the world to generate x-rays and to search for other possible kinds of new rays. In Paris Becquerel was impressed by the apparent association between the production of x-rays and the fluorescence of the glass tube from which the x-rays emerged. Occupying a post previously held by both his father and his grandfather, he had actively continued a research program in fluorescence started by his father. He knew that many substances emit visible fluorescent light when sunlight falls on them, and he reasoned that they might also emit x-rays. In February 1896, to test this possibility he wrapped black paper around a photographic plate, put a familiar fluorescent compound, potassium uranyl sulfate, on top of the wrapped plate, and placed it in the sunlight. When the plate was developed, it was indeed fogged, just as he had hoped.

Becquerel prepared a fresh plate with fluorescent crystals and placed it in a drawer to await another sunny day. A series of cloudy days followed, and eventually Becquerel decided to go ahead with the development of the plate. To his surprise it was strongly fogged. The crystals emitted penetrating radiation without external stimulation! Becquerel found that other uranium compounds also sent out these new rays, which went through matter and ionized air, much like x-rays. Becquerel's discovery marks the starting point of nuclear physics.

It was soon found that the property of emitting penetrating radiations is not confined to uranium and its compounds. Minerals containing thorium and several other elements have this same property. Such substances are said to be *radioactive*, a term introduced by Marie Curie. Two years after Becquerel's discovery of radioactivity Pierre and Marie Curie were able to isolate the highly radioactive element *radium* from the mineral pitchblende. Many other radioactive elements were discovered later.

44.2 ALPHA, BETA, AND GAMMA RAYS

To determine whether the rays from a radioactive ore bore electric charges or not, several experimenters, including Becquerel, passed the rays through a magnetic field. In 1899 Becquerel, and independently Giesel, found that some of the rays bore a negative charge. In 1900 Villard showed that another component of the radiation is uncharged, and in 1903 Rutherford found that still another component carries a positive charge. Thus, three different types of radiaton are emitted by radioactive ores (Fig. 44.1) which contain several kinds of radioactive atoms. Rutherford introduced the names *alpha particles* for the positively charged ones, *beta particles* for the negatively charged ones, and *gamma* rays for the uncharged component.

Alpha particles are the nuclei of helium atoms, as Rutherford was able to show by measuring the ratio of the charge to the mass of the particle. Each alpha particle bears two fundamental units of positive charge. Eventually an alpha particle captures two electrons, thereby becoming a helium atom.

Beta particles are electrons, which Becquerel showed are identical in nature to the cathode rays in discharge tubes and to the particles for which Thomson had measured e/m (Sec. 41.5).

Gamma rays are electromagnetic photons identical in nature and properties to x-ray photons of the same energy. We call the photon a gamma ray if it is emitted by a radioactive nucleus and an x-ray if it arises from the rapid deceleration of an electron when it collides with a target. In the two sections which follow we discuss some of the important characteristics of x-rays and gamma rays.

44.3 THE DETECTION AND MEASUREMENT OF HIGH-ENERGY PHOTONS

There are three properties of high-energy photons which are ordinarily used to detect and measure the intensity of x-rays, gamma rays, and photons of the far ultraviolet:

1. *High-energy photons eject electrons from atoms and thereby produce ionization in a gas.* By collecting the ions in an ionization chamber (Fig. 43.7a) one can determine the energy of radiation which has been absorbed in the chamber.

   Ionization measurements are used to determine exposures of x-rays and gamma rays. The radiation unit, adapted by the International Congress of Radiology, is called the *roentgen* (R), *where the roentgen is that quantity of x- or gamma radiation which produces in dry air a total ionization of 3.33×10^{-10} C per 0.001293 g of dry air.* A roentgen separates 2.08×10^9 ion pairs per cubic centimeter of dry air under standard conditions, which is equivalent to the release of about 88 ergs (8.8 μJ) of energy per gram of air. Since the roentgen refers to a specific result for x-rays and gamma rays in *air*, it does not imply any specific effect in a *biological system,* nor is it applicable to alpha and beta particles or neutrons. For biological purposes a unit called the *rad* (rd) is defined to be the absorbed dose of any radiation which is accompanied by the release of 100 ergs of energy per gram of absorbing material. For photons in the energy range 0.3 to 3 MeV an exposure dose of 1 R *gives rise to about* 1 rd absorbed dose in tissue.

   Although biological effects are produced by all ionizing radiations, the absorbed dose in rads required for a certain effect may be very different for different kinds of radiation. For example, 1 rd of neutrons is far more effective in producing cataracts in the eye than 1 rd of x-rays. This difference in response is taken into account by introducing the *relative biological effectiveness* (RBE), defined for any radiation as the ratio of the absorbed gamma-ray dose in rads to the absorbed dose of the specified radiation which produces the same biological effect. The dose unit for biological effects is called the rem, for *roentgen equivalent man*, where

$$\text{Dose in rem} = \text{RBE} \times \text{dose in rads}$$

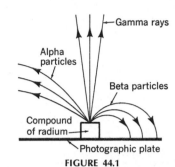

FIGURE 44.1
Alpha and beta particles are deflected in opposite directions by a magnetic field because alpha particles bear positive charges while beta rays are negative; gamma rays are undeflected since they are uncharged. The magnetic field is directed into the paper.

Exposure of the whole body to a single dose in the range of 100 to 200 rem results in illness but is not likely to be fatal. For doses between 200 and 600 rem, hospital treatment is likely to save an exposed person, but a dosage of 1,000 rem is usually fatal. Even small doses may cause serious damage; one should never expose oneself needlessly to substantial radiation doses.

2. *High-energy photons affect photographic emulsions.* Over a fairly wide range the blackening of a photographic plate is proportional to the amount of energy of a given wavelength absorbed in the emulsion.

3. *High-energy photons produce fluorescence.* In many materials this fluorescence is in the visible spectrum. In medical diagnostics the fluorescent screen is a widely used tool; e.g., a fluorescent screen can be used to observe the movement of a mass of bismuth salt as it passes through the digestive tract of a patient. Materials which fluoresce in the ultraviolet when bombarded with high-energy photons are used in scintillation detectors and other devices for studying high-energy photons. When such photons fall on certain crystals, the fluorescence produced is a direct measure of the energy absorbed (Sec. 29.8).

44.4 ABSORPTION OF HIGH-ENERGY PHOTONS

If a sheet of any substance is placed in the path of a beam of high-energy photons such as x-rays, the intensity of the beam is diminished. Let I_0 be the initial intensity of a homogeneous beam and I its intensity after passing through a thickness z of the material. Then

$$I = I_0 e^{-\mu z} \tag{44.1}$$

where μ is the *attenuation coefficient* of the material for the particular wavelength and e is the base of the natural logarithms. The attenuation coefficient varies with the wavelength of the photons; usually increasing as the wavelength increases. It is found experimentally that the absorption depends only on the number and kinds of atoms present in the absorbing layer; it is independent of their physical or chemical state.

A thickness of absorber which is adequate to cut the intensity to one-half is called the *half-value layer.* When a beam of homogeneous x-rays falls on an absorber just thick enough to remove half the beam, a second absorber of the same thickness removes half of what is left, so that one-fourth the initial intensity is transmitted. However, if a nonhomogeneous beam is incident, the second absorber removes less than half of what was transmitted by the first. The half-value layer removes more than half of the less penetrating and less than half of the more penetrating wavelengths of a nonhomogeneous beam. Therefore, the radiation which strikes the second absorber is more penetrating on the average than that which struck the first. Thin layers of copper or aluminum are used to remove the softer, i.e., less penetrating, components of x-rays. If the x-rays were being used to treat a deep tumor, these soft x-rays would only produce burns near the skin.

Photons with energies ranging from a few hundred up to a million electronvolts are absorbed or removed from a beam primarily by photoelectric absorption (Sec. 46.6) and by scattering. Photons with energies above 1.02 MeV may be absorbed by pair production (Sec. 43.7) as well.

We have invoked the quantum hypothesis to achieve a quantitative understanding of these processes. In Compton scattering we found particularly convincing evidence of the particle properties of electromagnetic radiation; another type of scattering is readily explained in terms of wave properties.

44.5 THE MASSES OF ATOMS

Alpha particles from radioactive nuclei played a vital part in unraveling the nature and structure of atoms. Before we discuss the alpha-particle experiments which led Rutherford to propose the nuclear model of the atom, however, we should be familiar with the results of earlier experiments which provided much information about atoms.

When a potential difference is applied between the terminals of a discharge tube, positive ions are accelerated as well as electrons. If a hole is drilled through the cathode (Fig. 44.2), some of the positive ions stream through the hole, giving a fine beam of positive rays. In a series of experiments initiated in 1906, J. J. Thomson used a tube similar to that of Fig. 44.2 to obtain a beam of positive "canal" rays, which he proceeded to deflect by means of electric and magnetic fields. One of his objectives was to determine the ratio of charge to mass for the various positive ions present. In one of his experiments the beam of positive ions passing between the poles of an electromagnet (Fig. 44.3) was deflected horizontally. An electric field parallel to the magnetic flux produced a vertical deflection. As a result, positive ions of different charge-to-mass ratios were deflected to form different parabolic traces on a photographic film. Slow-moving ions of a given kind undergo larger deflections than higher-energy ions of the same kind, primarily because the slower ions spend more time in the deflecting fields.

From the shape and location of such parabolic traces on a photographic plate the ratio of charge to mass can be determined. In this way Thomson showed in 1907 that neon is composed of atoms of two mass numbers, 20 and 22. This was the first proof that atoms of a given chemical element may have different masses. Atoms of the same element which have different masses are called *isotopes*.

Deflecting ions by electric and magnetic fields provides a way of determining the ratio of charge q to mass m for various ions. When it is possible to determine the charge on the ion, a measurement of q/m is sufficient to determine the mass of the ionized atom. However, it is by no means the only tool we have for estimating atomic masses. For example, in Chap. 35 we learned that it takes 9.6487×10^7 C to deposit electrolytically 1 kg equivalent mass of any element. Let us now find

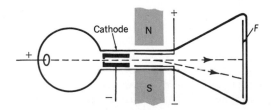

FIGURE 44.2
Positive ions passing through a hole in the cathode are deflected by both electric and magnetic fields.

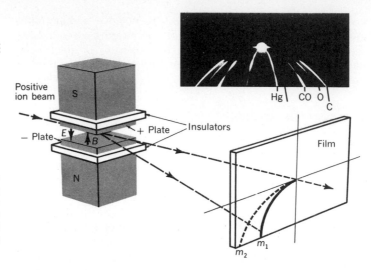

FIGURE 44.3
Parabolic trace on a film is formed by positive ions of the same charge-to-mass ratio but different velocities when deflected by parallel electric and magnetic fields. Reversal of the magnetic field produced the other half of the parabolas in the insert.

Avogadro's number, the number of atoms in a kilogram atomic weight. Since a monovalent atomic ion is a neutral atom with one electron missing, the charge on the ion is 1.602×10^{-19} C. Therefore, Avogadro's number N_A is given by

$$N_A = \frac{9.6487 \times 10^7 \text{ C/kg at wt}}{1.602 \times 10^{-19} \text{ C/atom}}$$

$$= 6.022 \times 10^{26} \text{ atoms/kg at wt}$$

The atomic weight of chlorine is 35.5, and the average mass of chlorine atoms is given by $35.5/N_A = 5.89 \times 10^{-26}$ kg. This average mass does not, of course, necessarily represent the mass of every chlorine atom; chlorine has two stable isotopes, and roughly three-fourths of the atoms have a mass of 5.81×10^{-26} kg, while the other one-fourth have a mass of 6.14×10^{-26} kg. The average mass of atoms of any element can be found by dividing its atomic weight by N_A.

44.6 MASS SPECTROMETERS

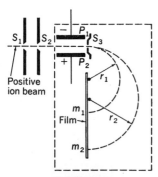

FIGURE 44.4
A Bainbridge-type mass spectrometer for which there is a uniform magnetic field out of the page over the region enclosed by the dashed rectangle.

Several types of instruments have been developed for the accurate determination of the charge-to-mass ratios of isotopes and complex ions. Such instruments are known as *mass spectrometers* (or *mass spectrographs* if they provide photographic recordings). Figure 44.4 is a schematic diagram of a Bainbridge mass spectrograph. Ions from a source are accelerated through a potential difference and pass through slits S_1 and S_2 into a region in which a uniform electric field E is provided by the plates P_1 and P_2; a uniform magnetic field B perpendicular to the plane of the paper is maintained by a magnet not shown. E and B are adjusted so that only ions having a desired velocity v pass through slit S_3. Ions with too low a speed are deflected toward the negative plate, and those with too high a speed toward the positive plate; only ions with $v = E/B$ are undeflected. An arrangement of crossed electric and magnetic fields of this kind is often called a *velocity selector*.

Once the ions have passed S_3, they are still in the magnetic field but no

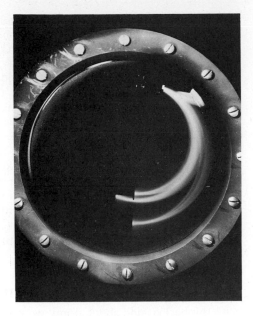

FIGURE 44.5
Paths of atomic beams of three different masses in a mass spectrometer. (*Westinghouse Corporation.*)

longer in an electric field. Consequently the ions move in a circular path with

$$m\frac{v^2}{r} = qvB \qquad (44.2)$$

From measurements of r, B, and E, a straightforward computation gives q/m for the positive ions. Figure 44.5 shows the circular paths of ion beams of three different masses in a mass spectrometer.

It is not necessary for a mass spectrometer to include a velocity selector. If one is satisfied with a somewhat less uniform velocity for the incident ions, one can simply accelerate the ions through a potential difference V and make them incident directly on S_3. The work Vq done on each ion by the accelerating field is equal to the kinetic energy $\frac{1}{2}mv^2$ gained by the ion, or $v = \sqrt{2Vq/m}$. Simpler mass spectrometers of this kind have been widely used to separate isotopes of various elements.

With the mass spectrometer it is possible to determine directly the ratio of the charge of an ion to its mass. If we know the number of atomic charge units borne by the ion, we can calculate its mass. In obtaining a precise set of mass values for the hundreds of isotopes of the known elements, elaborate comparison measurements are made. This is important because, typically, one can determine the mass difference between two ions of nearly the same q/m with far greater precision than that with which one can determine the mass of either ion by itself.

44.7 THOMSON MODEL OF THE ATOM

To the early proponents of the atomic theory of matter, atoms were the fundamental building blocks of nature, each type of atom being indivisible and devoid of structure. However, evidence from the discharge of electricity through gases, from the photoelectric effect, and from thermi-

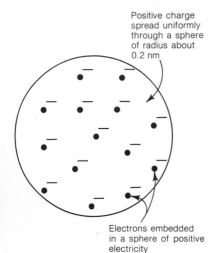

Positive charge spread uniformly through a sphere of radius about 0.2 nm

Electrons embedded in a sphere of positive electricity

FIGURE 44.6
The Thomson model of the atom consisted of a sphere of positive electricity in which electrons are embedded.

onic emission suggests that atoms are not indivisible but instead that the electron is one of the building blocks from which atoms are assembled. The fact that an atom is electrically neutral requires that it have equal amounts of positive and negative charge.

Early in the twentieth century the most widely accepted atom model was that of J. J. Thomson, who suggested that the positive electricity is distributed uniformly over a sphere which has the same radius as the atom (about 10^{-10} m). Inside this sphere he imagined a number of electrons which would correspond roughly to plums in a pudding (Fig. 44.6). Photons, collisions with other atoms, and thermal collisions could eject electrons from such an atom. In 1909 Barkla measured the scattering of x-rays from carbon and obtained evidence that each carbon atom has six electrons. For other elements it appeared that the number of electrons was roughly half the atomic weight. With this information Thomson tried unsuccessfully to calculate how electrons might be arranged in his sphere of positive electricity and what frequencies of spectral lines might be emitted by such an atom.

44.8 THE SCATTERING OF ALPHA PARTICLES

With this background about atoms, we now return to the alpha particles emitted by many radioactive atoms. In 1908 while Rutherford and Geiger were working on an electrical method of counting alpha particles, they had some problems due to the scattering of the particles as they passed through matter. Rutherford suggested that Geiger and Marsden study the scattering of alpha particles by a thin foil (Fig. 44.7). They found that some alpha particles were scattered through an angle greater than 90°, so they came out of the film on the same side at which they entered. In one case Geiger reported that 1 alpha particle out of 8,000 was scattered backward from a platinum foil. Calculations showed that scattering by Thomson-model atoms should be very small indeed.

Geiger and Marsden made scattering measurements for alpha particles of different kinetic energies incident on foils of lead, gold, platinum, tin, silver, copper, iron, and aluminum. In every case their results were completely incompatible with predictions based on the Thomson model.

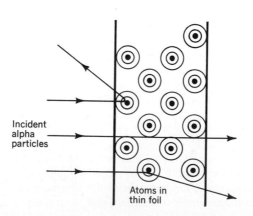

Incident alpha particles

Atoms in thin foil

FIGURE 44.7
Alpha-particle scattering; very large repulsive forces are required to scatter particles backward.

In 1911 Rutherford showed that significant large-angle scattering would be expected if all the positive electricity of the atom were concentrated at the center of the atom while the electrons were distributed over the remaining volume. He was able to calculate the scattering as a function of scattering angle, of the energy of the incident alpha particle, of the thickness of the scattering foil, and of the charge on the nucleus of the atoms which made up the foil. When Rutherford's theoretical values were compared with the experimental ones, they were found to be in good agreement.

In the Rutherford model all the positive electricity is concentrated in a very small volume called the *nucleus*. Surrounding this core are the electrons, held to the nucleus by electrostatic attraction. The radius of even the largest nucleus is only about 10^{-14} m, while the radius of a typical atom is nearer 10^{-10} m. Thus all the positive charge and almost all the mass of the atom are concentrated in about 10^{-12} the volume of the atom.

QUESTIONS

1. How can high-energy photons be detected? Discuss briefly all the ways you can think of.
2. By what processes are high-energy photons removed from a beam. Which process is most effective for long wavelengths? Which for very short wavelengths?
3. Why are the very soft x-rays absorbed out of a beam before the x-rays are used to photograph a broken arm? How can the soft x-rays be removed without eliminating the other wavelengths?
4. When x-rays of the digestive track are made, the patient usually is asked to drink a preparation containing heavy atoms (such as barium). What is the purpose of this preparation?
5. What is the difference between 1 R, 1 rd, and 1 rem? Historically the roentgen was defined and used first. Why were the rad and rem introduced?
6. Why does one obtain parabolic traces of ions in the Thomson apparatus (Fig. 44.3) when an ion beam passes through a region in which there are parallel electric and magnetic fields? Why are the faster ions deflected less than the slower ones by the magnetic field when the force on the ion is proportional to v?
7. What is the experimental evidence that all the positive charge and most of the mass of an atom are concentrated in a nucleus?

PROBLEMS

1. If 50 percent of a homogeneous beam of x-rays is absorbed by an aluminum sheet 4 mm thick, what percentage of the beam will pass through 8 mm of Al? Through 16 mm? Through 24 mm?
 Ans. 25; 6.25; and 1.56%
2. An x-ray beam is studied by means of copper plates. It is found that 600 μm of copper passes 50 percent of the beam, 1.2 mm passes 30 percent, and 2.4 mm passes 10 percent. Explain in detail how this result can be reconciled with the results of Prob. 1.
3. Calculate the attenuation coefficient of aluminum for the beam of Prob. 1. *Ans.* 0.173 mm^{-1}
4. Find the mass of a fluorine atom and a uranium atom if the atomic weight of fluorine is 18.9984 and the uranium atom belongs to the isotope corresponding to an atomic weight of 238.
5. The velocity selector of a mass spectrometer uses crossed electric and magnetic fields. The deflection plates are 12 mm apart and have a potential difference of 9.6 kV between them, and the magnetic induction is 1.1 T. Find the speed of protons which will pass through undeflected. How does this compare with the speed of electrons and of Na$^+$ ions which would pass through undeflected? *Ans.* 7.27×10^5 m/s; same
6. Alpha particles (He^{2+} ions) of the same energy are passing between two parallel charged plates in a field of electric intensity 8×10^5 N/C. A magnetic field perpendicular to the velocity of the alpha particles and to the electric field is varied until the alpha particles are not deflected. At this point $B = 0.2$ T. Find the force exerted on an alpha particle by the magnetic field and the speed of the particles.
7. Alpha particles from radium have a kinetic energy of 4.79 MeV. If these alpha particles strike a gold foil, calculate the distance of closest approach

for an alpha particle making a head-on collision with a gold nucleus held essentially at rest by neighboring gold atoms. *Ans.* 47.5 fm

8. If the magnetic induction of the mass spectrograph of Fig. 44.4 is 0.28 T and the electric intensity between the plates is 70 kN/C, find (a) the speed of ions passed by the velocity selector, (b) the radius of the path of ^{22}Ne$^+$ ions (atomic weight = 22) which reach the photographic plate, and (c) the distance between the line formed by ^{20}Ne$^+$ ions and that formed by ^{22}Ne$^+$ ions for the same velocity selector.

9. A Bainbridge-type mass spectrometer (Fig. 44.4) is to be used to separate the two stable isotopes of silver, ^{107}Ag and ^{109}Ag. The electric field between the plates P_1 and P_2 is 67 kN/C, and the magnetic field is 0.5 T. What is the velocity of the singly ionized silver ions which pass through the velocity selector? Find the distance between the beams of the two isotopes at the collectors which replace the film in Fig. 44.4.

Ans. 1.34 × 10^5 m/s; 11 mm

10. A particle of mass m, charge q, and speed v is moving perpendicular to a magnetic induction B. Show that the momentum of the particle is qBr, where r is the radius of curvature of the path in which the particle moves.

11. Show that if two ions of the same charge and energy but different mass are passing through a uniform magnetic field, the radii of the paths are proportional to the square roots of the masses.

12. An ion of mass m bearing a charge q is accelerated through a potential difference V. It then enters a magnetic field of flux density B, moving perpendicular to the field. Find the velocity (assumed to be small compared to the speed of light) of the ion and the radius of its path in the magnetic field in terms of m, q, V, and B.

Ans. $\sqrt{2Vq/m}$; $\sqrt{2mV/q}/B$

To develop even the simplest atomic model which could predict observed spectral lines required the synthesis of many of the ideas treated in these chapters on modern physics. Needed was the electron, the nucleus, and the concept of the photon—and, indeed, the abandonment of the classical idea that an accelerated charge must emit electromagnetic radiation. With these building blocks Bohr was able to propose a model of the simplest of all atoms, that of hydrogen. In this chapter we consider the optical spectra of hydrogen and the Bohr model.

45

THE HYDROGEN ATOM

45.1 THE BALMER SERIES

Each chemical element emits its own unique spectrum when its electrons are excited in electric discharges or in flames. As we saw in Sec. 29.2, the bright lines characteristic of any particular element offer a means of chemical identification of great simplicity and reliability. It was natural that scientists would endeavor to understand how these lines originate and what determines the particular array of lines associated with a given element.

For an atom with a spectrum as complex as that of iron (Fig. 29.4), the task of explaining quantitatively the origin of all the lines would be forbidding. On the other hand, the visible spectrum of hydrogen (Fig. 45.1) has a challenging regularity and apparent simplicity befitting the simplest and lightest of all atoms. This spectrum consists of an uncountable number of lines which become closer and closer together as the wavelength decreases, until finally at a certain minimum wavelength called the *series limit,* the lines cease. Many physicists had tried in vain to find a magic formula for this series before Balmer showed in 1885 that the wavelengths of these lines are predicted accurately by

$$\lambda = 3.6456 \times 10^{-7} \frac{m^2}{m^2 - 4} \qquad \lambda \text{ in meters}$$

where m is an interger taking the values $3, 4, 5, \ldots$. Since then this group of lines has been known as the *Balmer series.*

The Balmer formula is successful in giving us the wavelengths of the lines, but what is the physics which underlies this purely empirical relation? It was not until 1913 that this question could be answered in terms of an atom model which we now introduce.

45.2 THE BOHR ATOM

In 1912 Niels Bohr spent several months with Rutherford and his group at Manchester, and a year later he made a great forward stride in explaining the structure of atoms. Starting with Rutherford's nuclear model, he assumed that a hydrogen atom is composed of a nucleus called a *proton* and a single electron which revolves about the proton in a circular orbit (Fig. 45.2). The centripetal force required to keep the electron in this orbit is provided by the coulomb electrostatic attraction between the proton and electron:

$$\frac{mv^2}{r} = \frac{1}{4\pi\epsilon_0} \frac{e^2}{r^2} \qquad (45.1)$$

where m, v, and e are the mass, velocity, and charge of the electron and r is the radius of the orbit. Bohr further postulated that the only values

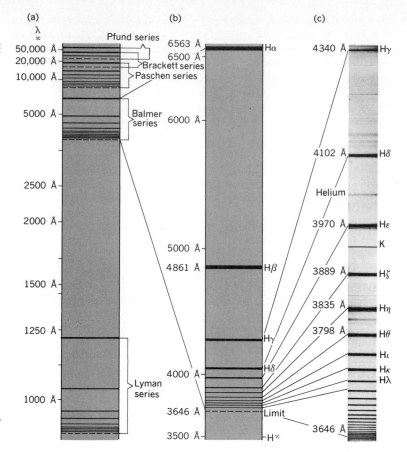

FIGURE 45.1
(a) The spectrum of atomic hydrogen consists of the Lyman series in the ultraviolet, the Balmer series in the visible region, and several series in the infrared region. (b) The Balmer series in greater detail. (c) A portion of the spectrum of the star Zeta Tauri showing more than 20 lines of the Balmer series. (*Photographed by R. H. Curtiss, University of Michigan.*)

which the angular momentum of the electron can have are integral multiples of $h/2\pi$, where h is Planck's constant.

$$mvr = n\frac{h}{2\pi} \tag{45.2}$$

where n is an integer. This condition Bohr adopted from the radiation laws which Planck had developed several years earlier.

Solving Eqs. (45.1) and (45.2) simultaneously for r yields

$$r = 4\pi\epsilon_0 \frac{h^2 n^2}{4\pi^2 m e^2} = \frac{\epsilon_0 h^2}{\pi m e^2} n^2 \tag{45.3}$$

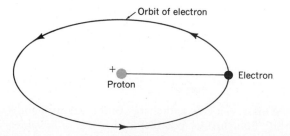

FIGURE 45.2
Bohr model of the hydrogen atom. The sizes of the nucleus and electron are grossly exaggerated. In a more refined model the electron has spin, much as the earth rotates about its axis.

The smallest orbit in which the electron can exist is that given by $n = 1$. If $n = 1$ with the other constants given their proper values, $r_1 = 5.29 \times 10^{-11}$ m. This agrees well with measured values of the radius of the hydrogen atom. The next smallest allowed radius is 4 times as great, the third possible radius is $9r_1$, and so forth. Normally, the hydrogen electron is in the orbit corresponding to $n = 1$, but on occasions it may be "excited" to one of the larger orbits.

The total energy of the hydrogen electron is the sum of the kinetic and potential energies. The kinetic energy $mv^2/2$ is, by Eq. (45.1),

$$\text{Kinetic energy} = \tfrac{1}{2}mv^2 = \frac{e^2}{8\pi\epsilon_0 r}$$

while the potential energy, assumed to be zero when proton and electron are an infinite distance apart, is given by Eq. (31.1):

$$\text{Potential energy} = Ve = -\frac{e^2}{4\pi\epsilon_0 r}$$

from which

$$\text{Total energy} = \text{kinetic} + \text{potential energy} = -\frac{e^2}{8\pi\epsilon_0 r}$$

$$= -\left(\frac{1}{4\pi\epsilon_0}\right)^2 \frac{2\pi^2 me^4}{h^2 n^2} = -\frac{me^4}{8\epsilon_0^2 h^2 n^2} \qquad (45.4)$$

The minus sign means the electron is bound to the nucleus and cannot escape unless it is given enough additional energy.

The hydrogen electron has its lowest possible energy when it is in the orbit characterized by $n = 1$. If the electron is in any other orbit, it may be expected to jump to an orbit of lower energy (Fig. 45.3). When it makes such a jump, a photon of radiation is emitted. The energy of the emitted photon is $h\nu$, where ν is the frequency. It is equal to the energy lost by the electron as it goes from a level A characterized by the integer n_A to another level B characterized by n_B.

$$\text{Energy radiated} = h\nu = \frac{me^4}{8\epsilon_0^2 h^2}\left(\frac{1}{n_B^2} - \frac{1}{n_A^2}\right) \qquad (45.5)$$

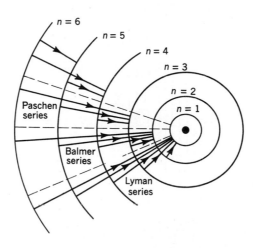

FIGURE 45.3
Bohr orbits and transitions for the excited hydrogen atom (not to scale).

Combining Eq. (45.5) with the relation $c = \nu\lambda$, we obtain

$$\frac{1}{\lambda} = \frac{\nu}{c} = \frac{me^4}{8\epsilon_0^2 h^3 c}\left(\frac{1}{n_B^2} - \frac{1}{n_A^2}\right) = R\left(\frac{1}{n_B^2} - \frac{1}{n_A^2}\right) \qquad (45.6)$$

where R is called the Rydberg constant. It has the value 1.09737×10^7 m^{-1}.

45.3 ENERGY LEVELS OF HYDROGEN

According to Bohr's theory, an electron bound to a proton to form a hydrogen atom can exist only in certain orbits characterized by a total energy given by Eq. (45.4). Modern quantum mechanics does not support the assumption of well-defined electron orbits, but it does predict that the electron must be in one of the energy states (or levels) given by Eq. (45.4). In an electric discharge through hydrogen, electrons are torn free from hydrogen atoms. Eventually the electrons and protons recombine; in this recombination the electron may be captured in one of the many possible energy levels (Fig. 45.4). Atoms with electrons in one of the higher states are said to be *excited*. The electron of an excited hydrogen atom normally makes successive transitions to lower states until it finally reaches the stable level corresponding to $n = 1$. When it jumps from one state to another, radiation is emitted.

The Balmer series arises when electrons from higher levels jump to the level with $n = 2$, as shown schematically in Figs. 45.3 and 45.4. According to Bohr's formula (45.6), the wavelengths of the series of lines emitted

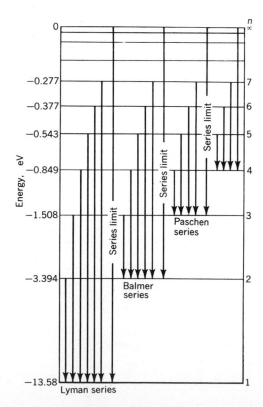

FIGURE 45.4
Energy-level diagram for hydrogen (not to scale).

when electrons jump from higher states described by quantum numbers n_A greater than 2 to the state characterized by $n_B = 2$ are given by

$$\frac{1}{\lambda} = R\left(\frac{1}{4} - \frac{1}{n_A^2}\right)$$

or

$$\lambda = \frac{1}{R}\frac{4n_A^2}{n_A^2 - 4} = 3.65 \times 10^{-7}\frac{n_A^2}{n_A^2 - 4}$$

This agrees with Balmer's formula if we let $n_A = m$.

We expect electrons eventually to make transitions to the state with $n = 1$, since this is the normal state for the electron. For such a transition Eq. (45.5) shows that the emitted frequencies should be in the ultraviolet region. Lyman studied the ultraviolet spectrum of hydrogen in 1906 and found, just as Bohr theory predicts, a series of lines corresponding to electron jumps from excited levels to the level $n = 1$. This series we call the *Lyman* series in honor of its discoverer.

What should happen if electrons in higher excited states make transitions to levels $n = 3$? Equation (45.5) predicts that the frequencies would be in the infrared region of the spectrum. In 1908 this series was discovered by Paschen. Later another series in the far infrared was discovered by Brackett. The wavelengths of the lines in these series are accurately given by Eq. (45.6) when $n_B = 3$ for the Paschen series and $n_B = 4$ for the Brackett series, while n_A takes on all larger integral values.

45.4 OTHER ONE-ELECTRON ATOMS

Consider an ionized helium atom. Here we have a nucleus with a charge of two fundamental units and a single electron. If we apply the Bohr theory to this system, the relations developed for hydrogen are applicable, provided we use $2e$ for the nuclear charge. The energies become 4 times as large in magnitude, and *the predicted frequency of the line emitted for any given transition is 4 times that for hydrogen.* Experiments reveal that this is indeed the case, except for very small shifts which can be explained.

Neutral lithium has three electrons. If two are removed, the remaining *doubly ionized lithium* is found to have a spectrum like that of hydrogen except that all frequencies are greater by a factor of 9. Similarly *triply ionized beryllium* and *quadruply ionized boron* have spectral series like those of hydrogen with frequencies respectively 16 and 25 times greater. Spectral lines for more than a dozen one-electron systems have been observed, and for nuclear charge Ze the frequency is Z^2 times that of the corresponding hydrogen line. In general, if we have a nucleus with positive charge Ze and a single electron, the force between electron and nucleus is $Ze^2/4\pi\epsilon_0 r^2$, and the radius of an allowed orbit is (by the method of Sec. 45.2)

$$r = \frac{\epsilon_0 h^2}{\pi m e^2}\frac{n^2}{Z} \tag{45.3a}$$

while the energy is

$$\text{Total energy} = -\frac{me^4}{8\epsilon_0^2 h^2}\frac{Z^2}{n^2} \tag{45.7}$$

Thus the radius is $1/Z$ times and the energy Z^2 times the corresponding value for hydrogen.

45.5 CHARACTERISTIC X-RAYS AND MOSELEY'S LAW

An early triumph of Bohr's theory was its successful application in explaining the wavelengths of certain characteristic x-ray lines. As we saw in Sec. 43.4, W. H. Bragg found that the x-ray spectrum emitted by targets consists of continuous radiation upon which lines characteristic of the target element are superimposed. The *characteristic lines* of each element are numerous, and their wavelengths depend on the atomic number of the target element in a fairly regular way.

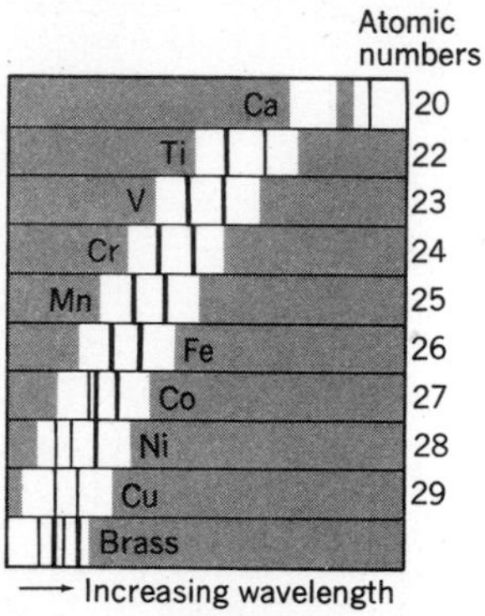

Shortly after Bragg reported his work, the English physicist Moseley measured the wavelengths of characteristic lines from a number of elements (Fig. 45.5) and found that the strongest line observed for each of these elements had a frequency given by the empirical relation

$$\nu = 2.52 \times 10^{15}(Z - 1.13)^2 \qquad \text{Hz} \qquad (45.8)$$

This strong line observed by Moseley is called $K\alpha$, and we now discuss its origin in terms of the Bohr model.

Suppose that a single electron circling a nucleus of charge Ze in the orbit characterized by $n_A = 2$ makes a transition to the $n_B = 1$ orbit. Bohr theory predicts for such a transition that a photon of energy W should be emitted, where, by Eqs. (45.5) and (45.7),

$$W = h\nu = \frac{me^4 Z^2}{8\epsilon_0^2 h^2}\left(\frac{1}{n_B^2} - \frac{1}{n_A^2}\right) = \frac{me^4 Z^2}{8\epsilon_0^2 h^2}\left(\frac{1}{1^2} - \frac{1}{2^2}\right) \qquad (45.9)$$

If this equation is solved for ν with the known values of h, m, and e inserted, we obtain

$$\nu = 2.46 \times 10^{15} Z^2 \qquad \text{Hz} \qquad (45.9a)$$

Subsequent work has shown that this line is indeed associated with the transition of an electron from a level with $n = 2$ to a level with $n = 1$. A plot of the square root of the frequency as a function of atomic number for this line is shown in Fig. 45.6.

Moseley also observed a line called $K\beta$ associated with the transition $n_A = 3$ to $n_B = 1$. This line has a frequency which corresponds well with what one would expect if one placed $n_A = 3$ in Eq. (45.9). The square root of the frequency of $K\beta$ as a function of atomic number is also shown in Fig. 45.6.

Both the slight difference between the constants of Eqs. (45.8) and (45.9a) and the correction to the atomic number in Eq. (45.8) can be

Atomic numbers

Ca	20
Ti	22
V	23
Cr	24
Mn	25
Fe	26
Co	27
Ni	28
Cu	29
Brass	

⟶ Increasing wavelength

FIGURE 45.5

Drawing of Moseley photographs of K x-ray spectra. Brass is an alloy of copper and zinc ($Z = 30$).

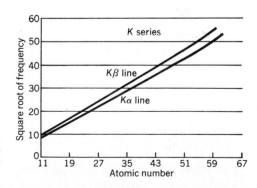

FIGURE 45.6

Moseley plots of the square root of the frequencies (arbitrary units) of K lines as functions of the atomic number of the emitting element.

explained quantitatively as well as qualitatively in terms of effects due to other electrons. A state characterized by $n = 1$ is called a K state and the associated electron a K electron. A K electron moves in a force field which is not very different from that of the nuclear charge Ze alone. The energy, however, is affected by the second K electron in particular and, to a smaller extent, by the other electrons. The net effect of all the other electrons is to make the effective nuclear charge appropriate to Eq. (45.9) about $Z - 1$ rather than Z.

Characteristic x-ray lines involving an electronic transition from a state of higher n to the $n = 1$ level are called K-series lines. Other families of x-ray lines have also been observed. Those corresponding to transitions from higher n to the $n = 2$ level are called L lines (Fig. 45.7), and electrons with $n = 2$ are known as L electrons. Lines corresponding to transitions to levels characterized by $n = 3$ are called M lines, etc. Figure 45.8 shows the wavelengths at which the principal K, L, and M lines are found for various elements.

The x-ray spectra of elements permit us to make an unambiguous assignment of atomic number to every element. At one time there was uncertainty whether potassium or argon had the higher atomic number. Argon has a greater mass, but the chemical properties suggest that potassium has the higher atomic number. X-rays have shown clearly that argon has atomic number 18 while potassium has 19. X-rays resolved a similar paradox in connection with cobalt and nickel.

45.6 ELECTRON SPIN

The $K\alpha$ x-ray lines observed by Moseley actually show up as double lines when examined under higher resolution. In 1925 Uhlenbeck and Goudsmit showed that these and many other features of atomic spectra can be explained by assuming that an electron not only revolves around a nucleus

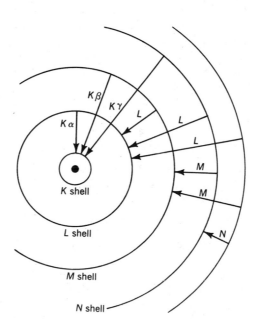

FIGURE 45.7
Transitions involved in the emission of characteristic x-ray lines in terms of the Bohr model, in which the fine structure of the shells is ignored.

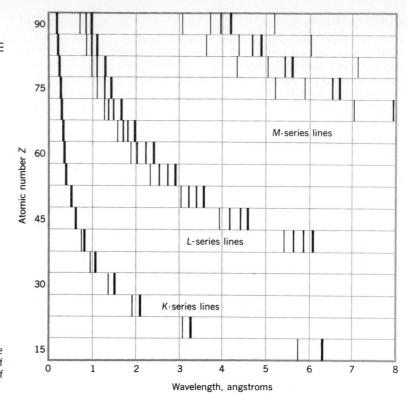

Atomic number Z

M-series lines

L-series lines

K-series lines

Wavelength, angstroms

FIGURE 45.8
Wavelengths of some of the more prominent characteristic x-ray lines of the elements; only a small fraction of the L and M lines is shown.

but has a property, called *spin*, which for our purposes may be visualized as a rotation of the electron about its own axis. A determination of the spin angular momentum about any axis always results in one of two values: either $+h/4\pi$ or $-h/4\pi$, where h is Planck's constant. For a single electron traversing an orbit, the total angular momentum is the vector resultant of the orbital angular momentum and the spin angular momentum. Essentially the spin either adds to or subtracts from the orbital angular momentum an amount $h/4\pi$. The energy is slightly different in the two cases for the following reason. The spinning charge produces a magnetic moment, so the spinning electron resembles a tiny magnet. From the point of view of the electron, the nucleus revolving about the electron produces a magnetic field; the energy of the electron magnet depends on whether its magnetic moment is aligned with this field or opposed to it.

In the optical spectra of sodium and the other alkalies, the energy associated with a given orbit has two values, depending on whether the spin angular momentum of the excited electron is parallel or antiparallel to the orbital angular momentum. Such a splitting of energy levels may give rise to series of double lines. The sodium D lines are examples of such a doublet; they have wavelengths of 5890 and 5896 Å.

1. What is a spectral series? What is a series limit?
2. Is the gravitational attraction between the proton and the electron of a hydrogen atom significant compared with the coulomb attraction? Find the ratio of coulomb to gravitational attraction for an electron in the first Bohr orbit. Does this ratio change from orbit to orbit?
3. Suppose for the moment that all electrons do not have the same charge but that there are three different charges. What effect would this produce in the hydrogen spectrum? Which do you regard as the more convincing evidence that all electrons bear the same charge, the Millikan oil-drop data or the sharpness of lines in the spectra of elements such as hydrogen? What do the spectral results suggest so far as the masses of electrons are concerned?
4. In the Bohr theory what is the significance of the fact that the allowed sharp levels all have negative energy?
5. Can a hydrogen atom in its ground state absorb a photon with an energy of 7.5 eV? One with an energy of 10.2 eV? One with a wavelength equal to the short-wavelength limit of the Balmer series?
6. Can a photon with an energy greater than the binding energy of the hydrogen electron (13.6 eV) be absorbed by a hydrogen atom? Explain quantitatively.
7. What is the maximum number of photons that can be emitted when a hydrogen atom excited to an $n = 4$ level goes to the ground state? What is the minimum number? (Assume there are no nonradiative transitions.)
8. What is the origin of the $K\alpha$ x-ray doublet in copper? How do you explain the observed fact that the $K\alpha$ radiation consists of two lines?

PROBLEMS

1. The ionization potential of a hydrogen atom is 13.58 eV. Use this figure to find the energies of the hydrogen levels with $n = 1$, $n = 2$, and $n = 3$ and to determine the wavelength of the lowest frequency line in the visible (Balmer) spectrum of hydrogen.
   Ans. -13.58, -3.40, and -1.51 eV; 656 nm
2. Use the Balmer formula to calculate the wavelengths of the first four lines of the Balmer series of hydrogen.
3. Compute the wavelengths of the first two lines and of the series limit for the Lyman series of hydrogen. What are the energies of photons with these energies?

Ans. 121.6 nm; 102.7 nm; 91.2 nm; 10.2 eV; 12.1 eV; 13.6 eV

4. Calculate the wavelengths of the first two lines and of the series limit for the series of ionized helium which corresponds to the Balmer series (to $n = 2$) of hydrogen.
5. Find the energies of the photons and the wavelengths of the first three lines of the Paschen series for hydrogen.
   Ans. 0.66 eV, 1.88 μm; 0.966 eV, 1.28 μm; 1.132 eV, 1.096 μm
6. The Brackett series of hydrogen is in the infrared region. Calculate the energies of the photons and the wavelengths of the first two lines and of the series limit for the Brackett series.
7. Calculate the wavelengths of the first two lines and of the series limit of the series for doubly ionized lithium which corresponds to the Lyman series (to $n = 1$) in hydrogen.
   Ans. 13.5 nm; 11.4 nm; 10.1 nm
8. Find the speed of an electron in the first Bohr orbit. How many revolutions does it make each second? Show that the ratio of the speed of an electron in the first Bohr orbit to the speed of light is $\alpha = e^2/2\epsilon_0 hc = \frac{1}{137}$. This ratio is known as the *fine-structure constant.*
9. Calculate the momentum of the photon emitted when a hydrogen atom makes a transition from a state characterized by $n = 3$ to that for which $n = 1$. Find the recoil speed of the hydrogen atom. *Ans.* 6.45×10^{-27} kg-m/s; 3.86 m/s
10. The orbital electron of a singly ionized helium atom is attracted by the nucleus, which has a positive charge twice that of the electron. Find the radius of the first Bohr orbit of ionized helium and the speed of an electron in this orbit.
11. The electron of a Bohr model hydrogen atom revolves around the proton with a speed of 2.19×10^6 m/s at a radius of 5.29×10^{-11} m. To what current does this correspond? Find the magnetic induction at the proton due to the motion of the electron. Compute the magnetic moment due to this orbital motion. (This magnetic moment is known as a *Bohr magneton.*)
    Ans. 0.00105 A; 12.5 T; 9.3×10^{-24} A-m²
12. (*a*) Show that the frequency of revolution of an electron in a circular Bohr orbit of hydrogen is given by $\nu = me^4/4\epsilon_0^2 n^3 h^3$. (*b*) Show that for large n the frequency of revolution is approximately equal to the frequency emitted according to Eq.

(45.5) when the electrons make a transition from state n to state $n - 1$. (This problem is an example of *Bohr's correspondence principle*, which implies that when n is large, quantum physics makes the same predictions as classical physics.)

13. Calculate the angular velocity of a solid sphere of radius 5×10^{-15} m and the mass of an electron if its angular momentum were $h/4\pi$. What would be the tangential velocity at the equator of such a sphere?

Ans. 5.8×10^{24} rad/s; 2.9×10^{10} m/s ($\approx 100c$)!

14. Show that if a neutral hydrogen atom at rest is struck by a moving hydrogen atom, the incident atom must have at least 20.4 eV of energy to raise the atom initially at rest to its first excited state. Why does it take 1.5 times the ionization energy just to excite the atom? *Hint:* Momentum must be conserved.

15. Show that, on the basis of the approximations which led to Eq. (45.9), the slope of the Moseley plot (Fig. 45.7) for $K\beta$ is expected to be $\sqrt{\frac{32}{27}}$ times that for $K\alpha$.

Efforts to understand the structure of more complex atoms and the properties of solids by extending the Bohr postulates met with limited success. Great advances in these areas required first the recognition that the electron has wave properties as well as particle properties and then the development of a wave mechanics capable of predicting the "behavior" of an electron. Although use of this new mechanics involves mathematical skills well beyond those expected here, many of the important ideas can be discussed without carrying through difficult calculations. We now introduce some experimental evidence of wave properties for electrons and explore some of their consequences.

46

ATOMIC STRUCTURE

46.1 DE BROGLIE'S HYPOTHESIS

The brilliant success of the Bohr theory in explaining the spectra of hydrogen and the K-series x-ray lines makes it one of the great triumphs of modern physics. In spite of its success, the Bohr theory is not entirely correct. Indeed, one of its key postulates—that the angular momentum mvr of the electron in its orbit is $nh/2\pi$—is not true. The Bohr theory is a magnificent contribution to physics, not because it is correct in detail but because it opened the way for modern quantum mechanics, in which the electron can no longer be regarded as a tiny particle describing sharp orbits about a nucleus. Instead the electron is found to exhibit wave characteristics which are extremely important in treating its behavior in atoms and in solids.

The first man to suggest that electrons, and indeed other classical particles as well, have wave properties was Prince Louis Victor de Broglie. In 1924 he proposed that since the electromagnetic "waves" of classical physics possessed particle attributes as well as wave attributes, atomic particles might be expected to exhibit wave behavior under proper conditions. His principal argument was based on the fact that nature reveals many symmetries. One possible symmetry was that both electromagnetic radiation and classical particles have both particle and wave properties. If particles have wave properties, how can one predict the wavelengths? Here de Broglie argued that the Compton effect showed that photons had momentum $h\nu/c$, or simply h/λ. Hence one should assign a wavelength $\lambda = h/(\text{momentum})$ to photons, and perhaps to particles as well. For a particle of mass m and velocity v, the resulting *de Broglie wavelength* is

$$\lambda = \frac{h}{mv} \tag{46.1}$$

whether the particle is an airplane or an electron. For macroscopic bodies this wavelength is so tiny that we have no practical means of measuring it, but for electrons and other atomic particles, wavelengths of measurable magnitude are available. Measurements on such particles confirm the de Broglie hypothesis.

□ **Example** What is the de Broglie wavelength of an electron traveling at a speed of 10^7 m/s? Through what potential difference must an electron fall to achieve this speed?

$$\lambda = \frac{h}{mv} = \frac{6.626 \times 10^{-34} \text{ J-s}}{(9.11 \times 10^{-31} \text{ kg})(10^7 \text{ m/s})}$$

$$= 72.7 \text{ nm} = 0.727 \text{ Å}$$

$$Ve = \tfrac{1}{2}mv^2$$
$$V(1.60 \times 10^{-19} \text{ C}) = \tfrac{1}{2}(9.11 \times 10^{-31})(10^{14}) \text{ J}$$
$$V = 284 \text{ J/C} = 284 \text{ V} \qquad \square$$

(a)

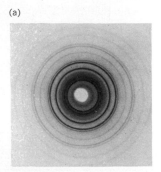

(b)

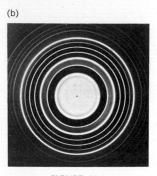

FIGURE 46.1
(a) X-ray diffraction pattern of aluminum. (*Courtesy of Mrs. M. H. Read, Bell Telephone Laboratories, Inc.*) (b) Electron-diffraction pattern of a thin film of cesium iodide. (*Courtesy of L. H. Germer, Bell Telephone Laboratories, Inc.*)

46.2 WAVE PROPERTIES OF ELECTRONS

The first experimental confirmation of de Broglie's hypothesis was made by Davisson and Germer in 1927. They bombarded a single crystal of nickel with a narrow pencil of low-voltage electrons incident perpendicular to the crystal face and determined the intensity of the "reflected" electrons as a function of the angle at which they bounced off the crystal. There were striking maxima and minima in this angular distribution which could be quantitatively explained if the electrons had the wavelengths predicted by Eq. (46.1).

An entirely different experiment by G. P. Thomson also supported the de Broglie hypothesis and emphasized the similarity between electrons and x-rays of the same wavelength. If a pencil of monochromatic x-rays passes through a *powdered* crystalline substance placed at C in Fig. 43.3, the x-rays produce a diffraction pattern on the photographic plate. This pattern consists of a series of concentric circular fringes produced by interference of the scattered x-rays (Fig. 46.1a). If a stream of electrons, all of the same velocity, passes through a thin film of metal, a similar set of rings is produced (Fig. 46.1b). In both cases one can compute the positions of the observed fringes on the assumption that the incident beams consist of waves of appropriate wavelength; for the electrons, the de Broglie wavelength $\lambda = h/mv$.

The resolving power of an optical microscope is limited by the wavelength of the light used. For two neighboring particles to be separable, light of very short wavelength must be used if the particles are very close together. Unfortunately we do not have lenses which can focus x-rays and thereby permit us to build an x-ray microscope. However, when electrons are accelerated through a potential difference in the neighborhood of 10^5 V, their wavelengths are only a few hundredths of an angstrom. With such wavelengths, particles that are as close as 1 nm can be separated and photographed by an electron microscope. The electrons are controlled and focused by electric and magnetic fields.

Electrons are by no means the only classical "particles" for which wave properties have been observed. Other subatomic particles, and even some molecules, have been found to have wavelengths given by Eq. (46.1). Neutrons, which are neutral particles of mass slightly greater than that of the proton, have been studied by use of Bragg diffraction; slow "thermalized" neutrons have wavelengths in the same range as x-rays, and this property makes neutron spectrometers possible and permits physicists to obtain beams of neutrons all of which have virtually the same speed.

46.3 WAVE MECHANICS AND THE UNCERTAINTY PRINCIPLE

In 1926 Schrödinger discovered his famous wave equation; it underlies modern wave mechanics, a great body of theory which takes account of both the wave and particle attributes of electrons. Although the formulation of wave mechanics involves differential equations and relatively advanced mathematical techniques, some of the ideas can be discussed here. When Schrödinger applied his wave equation to the hydrogen problem, he obtained the same values for the energy of the allowed states as Bohr had found [Eq. (45.4)], but his values for the orbital angular momentum did not agree with the values postulated by Bohr. Schrödinger found that the orbital angular momentum mvr is given by $\sqrt{l(l+1)}\,h/2\pi$, where l is zero or an integer less than n, while Bohr had taken the incorrect larger value $nh/2\pi$. Further, for a hydrogen electron in a given n state, the wave mechanics predicts that far from being confined to a circular orbit, the electron might be much closer to or much farther from the nucleus than the appropriate Bohr radius, although the latter is a good measure of the average separation. In wave mechanics the concept of sharp orbits is completely abandoned. Modern wave mechanics, or quantum mechanics, is a fascinating topic. It has led to tremendous advances in physics and in chemistry as well as being a great stimulus to modern philosophy.

One of the startling principles of quantum mechanics, advanced by Heisenberg in 1927, is that there is a basic uncertainty in nature which represents an ultimate limit to the precision with which processes can be described. One form of the Heisenberg *uncertainty principle* can be stated as follows.

In any experiment in which we endeavor to measure both the position and the momentum of a particle, there is an inherent indefiniteness in the results such that if Δx is the limit of uncertainty in the x coordinate of the position and $\Delta(mv_x)$ is the uncertainty in the corresponding component of the momentum, the product of these uncertainties must be at least of the order of magnitude of Planck's constant h divided by 4π.

$$\Delta x\,\Delta(mv_x) \gtrsim \frac{h}{4\pi} \qquad (46.2)$$

A similar indefiniteness accompanies any measurement of an energy. If Δt is the time available for the energy measurement, there is an inherent uncertainty ΔW in the energy such that

$$\Delta W\,\Delta t \gtrsim \frac{h}{4\pi} \qquad (46.3)$$

The constant h is so small that for macroscopic bodies these quantum indefinitenesses do not trouble us, but when we deal with atomic sizes, the uncertainty principle is extremely important.

The uncertainty principle is a fundamental part of wave mechanics. It tells us that we cannot simultaneously know both the precise position and the exact velocity of a particle such as an electron. Indeed, any experiment designed to locate the exact position of a particle is certain to change the state of the system insofar as the momentum is conserved. Conversely, any effort to determine the momentum precisely inevitably introduces uncertainty in the position. As a consequence, an atom model based on sharp orbits for electrons at precisely defined distances from a nucleus can only be a rough approximation. Actually there is a finite possibility of finding the electrons over a wide range of positions. For a hydrogen electron in its lowest energy state, wave mechanics predicts that the most probable distance of the electron from the nucleus is the

Bohr radius, but there is a modest probability that the electron is as much as twice that distance away.

46.4 THE PAULI EXCLUSION PRINCIPLE

When the Schrödinger equation is applied to multielectron atoms, the resulting differential equation is awesome and only approximate solutions are available. However, for some atoms the approximations are extremely good. It is found that one can describe complex atoms reasonably well if one thinks of each electron as existing in a hydrogenlike state and moving in the electric field due to the nucleus and to the average contributions of the other electrons. A still more accurate description includes the effects associated with the magnetic fields due to the motions of the electrons and to their intrinsic magnetic moments. In this very rough model we imagine each electron with its own orbital and spin angular momenta moving in a fuzzy average hydrogenlike "orbit."

When we treat complex atoms from the point of view of hydrogenlike "orbits," we find that we can specify the state of each electron in terms of four numbers,† called *quantum numbers*, each of which has a physical significance. These quantum numbers are:

1. The principal quantum number n, which appears in Eq. (45.7) and is a rough measure of the average distance of the electron from the nucleus. The energy of a state depends strongly on n, which takes on integral values 1, 2, 3, 4, ...

2. The azimuthal quantum number l, which is a measure of the orbital angular momentum in units of $h/2\pi$. It takes on integral values $n - 1$, $n - 2$, . . ., 0. Thus, when $n = 3$, l can be 2, 1, or 0.

3. The quantum number j, which measures the total angular momentum of the electron, spin plus orbital, in units of $h/2\pi$. Its values are limited to the positive half-integers $l - \frac{1}{2}$ and $l + \frac{1}{2}$.

4. The quantum number m_j, which specifies how the total angular momentum is oriented in space relative to some direction prescribed by a magnetic field or some other kind of field. It takes on values $j, j - 1$, $j - 2$, . . ., $-j$.

In 1925 W. Pauli proposed an important rule, known as the *Pauli exclusion principle*, which may be stated for our purposes in the following form: *No two electrons in any one atom can have the same four quantum numbers.* Pauli's great generalization has many applications and forms one of the foundation stones for our understanding of atomic structure.

A neutral atom in its lowest energy state has each electron in the lowest energy level accessible to it. We have seen (Sec. 45.2) that for the ground state of hydrogen the single electron is in the state characterized by $n = 1$. For helium with its two electrons the ground-state configuration has both electrons with $n = 1$. This requires that $l = 0$ and $j = \frac{1}{2}$ for each electron, so one electron has $m_j = \frac{1}{2}$ and the other must have $m_j = -\frac{1}{2}$. It is not possible to have more than two electrons in any one

---

†There are several possible ways of choosing these four numbers; our choice here is based on convenience for describing the inner-electron configurations. A different choice is preferable for discussing the visible spectra of light elements.

atom with $n = 1$, since that would require that two of the three have the same four quantum numbers, a clear violation of the exclusion principle, to which no exceptions are known. Two electrons with $n = 1$ form a closed *shell*, known in x-ray terminology as the *K shell*.

46.5 THE PERIODIC TABLE

The arrangement of atoms in the periodic table (inside back cover) can be explained in terms of the Pauli exclusion principle. In the preceding section we saw that hydrogen and helium constitute the first period (row) in this chart. The next element is *lithium*, atoms of which have three electrons. The third of these electrons cannot have $n = 1$ but goes into the lowest available state for which $n = 2, l = 0$. After lithium comes beryllium ($Z = 4$), having two electrons with $n = 1, l = 0$ and two with $n = 2, l = 0$. For boron ($Z = 5$) the fifth electron has $n = 2, l = 1$. Altogether there can be six electrons in an atom with $n = 2, l = 1$; two with $j = \frac{1}{2}$ and $m_j = \frac{1}{2}$ or $-\frac{1}{2}$ and four with $j = \frac{3}{2}$ and $m_j = \frac{3}{2}, \frac{1}{2}, -\frac{1}{2},$ or $-\frac{3}{2}$. As we fill the $n = 2, l = 1$ subshell we have the elements boron, carbon ($Z = 6$), nitrogen (7), oxygen (8), fluorine (9), and finally neon (10). At this point the $n = 2$ (or L) shell is full.

Sodium has 11 electrons, of which 10 form a neon core; the eleventh is in the state characterized by $n = 3, l = 0$. As we proceed through the table of the elements (see inside back cover), each successive element has one more electron than the preceding and this electron takes always the lowest energy state which is not yet occupied. After all states with $n = 3, l = 0$, and $l = 1$ are filled, the next elements are potassium ($Z = 19$) and calcium ($Z = 20$), for which the last electrons have $n = 4, l = 0$, even though levels with $n = 3, l = 2$ remain empty. However, as we proceed up the periodic table, the $n = 3$ shell is quickly filled, and this gives rise to the first long period in the periodic table. (The period is the principal quantum number n of the outermost electrons.) A detailed explanation of why shells of higher n begin to fill before all the possible levels of lower n are filled is beyond the scope of this text. It is found (with the exception of a few irregularities) that the levels continue to fill in the following order: all levels with ($n = 4, l = 1$), followed by ($n = 5, l = 0$), ($n = 4, l = 2$), ($n = 5, l = 1$), ($n = 6, l = 0$), ($n = 4, l = 3$), ($n = 5, l = 2$), ($n = 6, l = 1$), ($n = 7, l = 0$), ($n = 5, l = 3$). Quantum mechanics, including the Pauli exclusion principle, has given us an understanding of the periodic table and of atomic structure in general, which has led to great advances in physics and chemistry.

The periodic table had its origins in the work of Mendeleev in Russia and (independently) Meyer in Germany, about 1868 to 1870. The arrangement was based on the observation that various chemical and physical properties recur periodically throughout the sequence of the elements in order of their atomic weights. Elements of similar properties were placed in the same column, with the lighter ones above the heavier. We now note a few correlations between chemical and physical properties and the assignment of numbers associated with the Pauli exclusion principle.

The elements lithium, sodium, potassium, rubidium, cesium, and francium all have a single electron alone in a new shell. They are all electropositive elements with similar chemical properties; they react vigorously

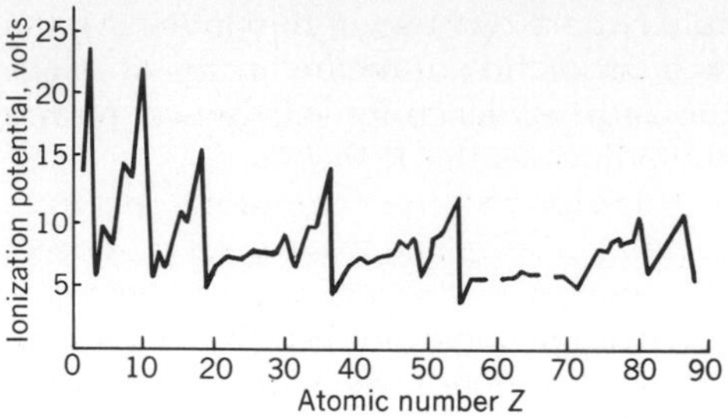

FIGURE 46.2
Ionization potentials of the elements.

with water to release hydrogen. Physically they are all metals with shiny luster and high electric and thermal conductivities. Immediately preceding these *alkali metals* in the periodic table are the *noble gases*, helium, neon, argon, krypton, xenon, and radon, all of which are relatively inert chemically. Except for helium, all have outer electron configurations with eight electrons in their outer shell corresponding to the filling of all levels of this shell with $l = 0$ and $l = 1$. The elements which directly precede the inert gases, fluorine, chlorine, bromine, iodine, and astatine, are called the *halogens;* all are nonmetals, and all are strongly electronegative, forming salts when they combine with the alkali metals. In general, elements which have similar electron configurations in their outer shells have similar chemical and physical properties.

One physical characteristic of the elements which can be correlated with the way in which electrons are added to fill the various shells is the *ionization potential*, or the energy required to remove one of the least tightly bound electrons from an atom, shown graphically in Fig. 46.2. Note that for potassium, sodium, lithium, etc., the ionization potential is exceptionally low compared with that of neighboring atoms. This arises from the fact that in each case there is a single electron in a new outer shell. On the other hand, if there are eight electrons in any given outer shell, a peculiarly stable electronic configuration occurs in which the element is chemically inert, the case for the noble gases. Many other properties show a similar periodicity, among them atomic diameter, compressibility, melting point, and coefficient of thermal expansion. A particularly important example of this periodicity is found in the visible spectra of the elements, which arise for any element when one of the outer electrons of an atom is excited from its normal state to an excited one of higher energy. Atoms with similar electron configurations exhibit similar spectra.

46.6 LOW-LYING ATOMIC ENERGY LEVELS

In the preceding section we have seen how the Pauli exclusion principle can be used to give us insight into the periodic table and the chemical and physical properties associated with the outer-electron configuration of an atom. But can we find any evidence that the four quantum numbers n, l, j, and m_j are useful for describing the inner structure of atoms? Actually the characteristic x-ray spectra (briefly treated in Sec. 45.5) are emitted when an inner electron makes a transition to a lower energy state made vacant by knocking an electron out of this state. A detailed explanation of the many strong characteristic x-ray lines can be made by appropriate use of the quantum numbers n, l, and j on which the energy depends; in the absence of an external field the energy does not depend on m_j. Simpler and more direct evidence comes from the photoelectric absorption of x-rays.

With the aid of an x-ray spectrometer and the continuous spectrum from an x-ray tube, a monoergic beam of x-ray photons can be produced. If a thin foil of some element is placed in the beam, the intensity is reduced from I_0 to I. If t is the thickness of the foil, the absorption coefficient is given by $[\ln (I_0/I)]/t$. Figure 46.3 is a plot of absorption coefficient for a lead foil as a function of wavelength. At 0.14 Å there is a sharp drop, called the *K absorption edge*, due to the fact that photons of wavelength greater than 0.14 Å do not have enough energy to eject one of the K electrons. A K electron is characterized by the quantum numbers $n = 1$, $l = 0$, $j = \frac{1}{2}$.

As the wavelength is increased, the absorption increases, until at 0.78 Å the photons no longer have enough energy to eject electrons with $n = 2$, $l = 0$, $j = \frac{1}{2}$, known in x-ray terminology as L_I electrons. At 0.81 Å comes the L_{II} absorption discontinuity associated with two electrons having $n = 2$, $l = 1$, $j = \frac{1}{2}$ $(m_j = \pm\frac{1}{2})$ while the L_{III} edge at 0.95 Å is due to four electrons with $n = 2$, $l = 1$, $j = \frac{3}{2}$ $(m_j = \frac{3}{2}, \frac{1}{2}, -\frac{1}{2}, -\frac{3}{2})$. At longer wavelengths the absorption increases steadily until one comes to five edges between 3 and 5 Å associated with the $n = 3$ (M) electrons. Thus in a sufficiently heavy atom one can find an absorption discontinuity associated with each possible value of n, l, and j and *no others*. Table 46.1 lists the energy required to eject an electron from each of the states characterized by all possible values of l and j for $n = 1$, 2, and 3.

Further confirmation of these low-lying energy states comes from

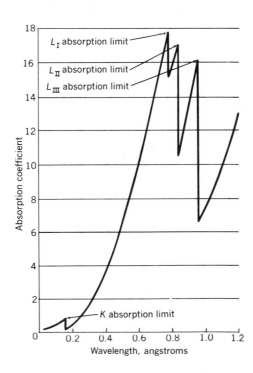

TABLE 46.1
Energies Required to Eject an Electron from X-Ray Levels of Seven Elements, in keV†

n	l	j	X-ray	Al (Z = 13)	Cu (Z = 29)	Mo (Z = 42)	Ag (Z = 47)	W (Z = 74)	Pb (Z = 82)	U (Z = 92)
			Level							
1	0	$\frac{1}{2}$	K	1.562	8.996	20.036	25.556	69.637	88.163	115.06
2	0	$\frac{1}{2}$	L_I	0.1154	1.104	2.872	3.811	12.115	15.892	21.795
2	1	$\frac{1}{2}$	L_{II}	0.00730	0.955	2.633	3.529	11.559	15.231	20.974
2	1	$\frac{3}{2}$	L_{III}	0.00727	0.935	2.528	3.356	10.219	13.061	17.193
3	0	$\frac{1}{2}$	M_I		0.122	0.509	0.718	2.821	3.860	5.556
3	1	$\frac{1}{2}$	M_{II}		0.078	0.413	0.603	2.574	3.566	5.187
3	1	$\frac{3}{2}$	M_{III}		0.076	0.396	0.572	2.279	3.076	4.306
3	2	$\frac{3}{2}$	M_{IV}	0.003	0.234	0.373	1.870	2.591	3.725
3	2	$\frac{5}{2}$	$M_{V'}$		0.231	0.367	1.807	2.482	3.556

†λ in angstrom units is equal to 12.398/(E in keV)

measuring the kinetic energies of photoelectrons ejected from a thin film of some element. If the film is bombarded with photons of energy $h\nu$, the kinetic energy (KE) of a photoelectron ejected from the K shell is given by

$$KE = h\nu - W_K \qquad (46.4)$$

where W_K is the binding energy of the electron in the K shell. If the electron comes from the L_{III} shell, its kinetic energy is $h\nu - W_{L_{III}}$ and so forth. Thus, a knowledge of the energy $h\nu$ of the incident photon and the kinetic energy of an ejected photoelectron permits the determination of the energy required to remove the electron from its bound state.

□ **Example**   X-ray photons of 20.000 keV energy fall on a lead foil and eject photoelectrons, one group of which has a kinetic energy of 6.939 keV. (a) What is the binding energy associated with the level from which these electrons are ejected? (b) What kinetic energy would a photoelectron from the L_I level have?

(a) $$KE = h\nu - W$$
$$6.939 \text{ keV} = 20.000 \text{ keV} - W$$
$$W = 13.061 \text{ keV}$$

which corresponds to the L_{III} level.

(b) $$KE = h\nu - W_{L_I} \qquad \text{(see Table 46.1)}$$
$$= 20.000 \text{ keV} - 15.892 \text{ keV}$$
$$= 4.108 \text{ keV} \qquad \qquad □$$

The characteristic x-ray lines of an element arise from electron transitions from one of the lower levels to a vacancy in a level still lower. From the energy of these photons and the known energy levels of the atom one can determine the initial and final states of the electron making the transition. Data from absorption edges, the kinetic energies of photoelectrons, and the wavelengths of characteristic spectral lines are all mutually compatible with each other and with the requirements of the Pauli exclusion principle.

46.7 THE ZEEMAN AND STARK EFFECTS

The energies of the low-lying energy states are not affected by the application of ordinary electric and magnetic fields, but this is not the case with the energy states of the outer electrons. For example, when a source of light is placed between the poles of a powerful electromagnet, a single spectral line may break up into several components (Fig. 46.4). This separation of a spectral line into components by the action of a magnetic field is known as the *Zeeman effect.* The number of components depends on the particular spectral line and is not the same when the light is viewed in the direction of the magnetic field as when the light is viewed at right angles to this direction. Because of the interaction between the external applied magnetic field and the magnetic fields due to the electrons in the atom, the energy levels in the atom are split up into a number of additional levels. Transitions between these new levels give rise to additional spectral lines. Studies of the Zeeman effect have proved invaluable in resolving problems in spectroscopy and in the development of a detailed understanding of many facets of atomic structure.

An electric field can also split spectral lines. In a sufficiently large electric field, a single spectral line is replaced by a number of components. The greater the electric field, the greater the separation of these components. This separation of spectral line into components by an electric field is known as the *Stark effect* (Fig. 46.5) after its discoverer. The splitting of the spectral line in the figure varies from zero at the top of the figure, where there is no electric field, to its greatest value at the bottom, where the electric field is largest. Data on the amount of separation for different spectral lines give valuable information about the atoms emitting the lines.

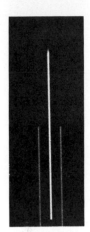

FIGURE 46.4
A magnetic field splits a single spectral line (above) into several components (below). These lines were photographed after the light had passed through a polarizer, which reduced the relative intensities of the outer lines. (*Courtesy of G. H. Dieke.*)

46.8 GASEOUS MOLECULES

That atoms attract each other is clear from the fact that (1) two or more atoms may join together to form a stable gaseous molecule and (2) larger aggregates of atoms are bound together to form liquids and solids. The force between a pair of atoms depends on the outer electron arrangements, and it becomes stronger as their electronic charge clouds begin to overlap. Much progress has been made in understanding interatomic forces since the advent of wave mechanics. We must limit our discussion to molecules consisting of only two atoms. What forces hold simple diatomic molecules together? There is no single answer to this question; there are several kinds of binding, of which we consider two.

Ionic Binding Some molecules are composed of a positive ion bound to a negative ion by coulomb electrostatic attraction. An example is potassium chloride (KCl), in which a potassium atom surrenders its outer (or valence) electron to a chlorine atom. The K^+ ion and the Cl^- ion are held together by *ionic* binding.

Covalent Binding Many gas molecules are composed of two identical atoms; examples are the hydrogen, oxygen, nitrogen, and chlorine molecules. Here we can scarcely invoke the concept of ionic binding, since there is no reason for one of two identical atoms to become positive and

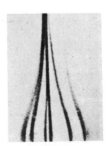

FIGURE 46.5
Stark effect for a line of the helium spectrum, showing the splitting due to an electric field. (*After Foster.*)

the other negative. The search for an alternative mechanism ended with modern wave mechanics. The development of the theory is beyond the scope of this book, but in a qualitative way it is possible to discuss some of the results. In a hydrogen molecule (H_2) the binding comes about from the sharing of two electrons by two hydrogen atoms. This gives rise to an attractive force associated with the electron *exchange*. The occurrence of exchange forces is a strictly wave-mechanical phenomenon for which there is no classical analog. The pair of electrons involved in the exchange must have oppositely directed spins (Sec. 45.7). If the spins of the two electrons are aligned, the atoms repel each other and no molecule is formed. In a similar fashion the binding of oxygen and nitrogen diatomic molecules rises from the sharing of pairs of electrons between the atoms. Binding arising from this type of electron sharing is called *covalent* or *homopolar*.

The extremes of ionic and covalent binding are reasonably well illustrated by potassium chloride and hydrogen molecules, respectively. However, there are many molecules for which the binding is intermediate between covalent and ionic binding. In these molecules the shared electrons are closer to one of the atoms much more of the time than to the other atom; examples are carbon monoxide and water molecules.

An important characteristic of any diatomic molecule is its *dissociation energy*, the energy required to break the molecule into its parts. This, in turn, is equal to the binding energy, the energy released when the two atoms are brought together to form the molecule. Typical binding energies of molecules are of the order of a few electronvolts: for potassium chloride 4.4 eV, for sodium chloride 4.24 eV, and for hydrogen 4.48 eV.

For any pair of atoms there exists an equilibrium interatomic distance at which the energy is minimum (Fig. 46.6). If the two atoms are slightly farther apart, a net restoring force draws them closer. On the other hand, if the atoms approach more closely than the equilibrium distance R_0, strong forces drive them apart.

If the only forces between two atoms were attractive, the atoms would eventually coalesce. Actually, as atoms approach each other, strong repulsive forces arise. Classically, one imagines the atoms to be solid spheres which cannot penetrate each other. However, the atom of modern physics has a tiny nucleus with an electron cloud about it. This picture is used to explain the fact that the atoms occupy a well-defined volume in terms of the Pauli exclusion principle. If we try to bring a second atom so close that its electrons begin to penetrate the occupied shells of the first atom, the Pauli exclusion principle requires that some of

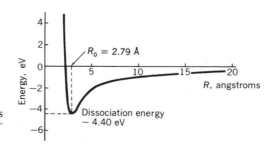

FIGURE 46.6
Energy of a KCl molecule as it varies with the separation of the K+ and Cl− ions.

the electrons go to higher energy states. This requires energy and gives rise to a repulsive force.

46.9 MOLECULAR SPECTRA

In Chap. 45 we considered the spectra emitted by electron transition in individual atoms. Molecules, composed of two or more atoms, also have spectra. Typically, molecular spectra are considerably more complex than atomic spectra, with radiation emitted in three spectral ranges corresponding to different kinds of transitions between molecular energy states. Starting with the longest-wavelength region, we explain these spectra as follows.

Rotational Spectra A simple diatomic molecule can be visualized as a dumbbell (Sec. 15.8). If such a molecule is put into rotation, classically we would expect it to emit radiation of the same frequency as the rotation. However, quantum mechanics limits the rotational energy levels (Fig. 46.7) to those given by the relation.

$$E_J = \frac{h^2}{8\pi^2 I} J(J + 1) \qquad (46.5)$$

where I is the moment of inertia of the molecule about its center of mass, h is Planck's constant, and J is the rotational quantum number, which takes on values 0, 1, 2, 3, ... Radiative transitions occur only when J increases or decreases by 1 ($\Delta J = \pm 1$). The rotational energy levels are spaced in such a way that rotational lines consist of a fundamental frequency and integral multiples of that frequency, all in the far infrared region.

Vibration-Rotation Spectra A diatomic molecule is less analogous to a rigid dumbbell than it is to two masses connected by a strong spring. Such masses can perform vibrations relative to one another under the elastic forces which bind them. The frequencies associated with such vibrations are typically much higher than those associated with pure rotation. Consequently, the vibrational spectra appear in the nearer infrared region. Any time its vibrational spectrum is excited, a molecule has much more than enough energy for the rotational spectra; thus we ordinarily observe rotational motion superimposed on the vibrational.

Electronic Spectra At still higher frequencies we find evidence of transitions of electrons, associated now with the molecule, between allowed quantum states. Since the energies involved are much greater than those associated with either vibration or rotation, electron excitation is generally accompanied by vibrational and rotational excitation. For any single electron state there are many vibrational and rotational states in which the molecule may exist. Consequently, each type of transition between electron energy states produces a large number of closely spaced lines. Such spectra, called *molecular bands* (Fig. 46.8), are often found in the visible and ultraviolet regions.

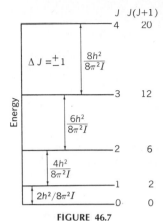

FIGURE 46.7
Lowest radiational levels of a diatomic molecule with a moment of inertia I.

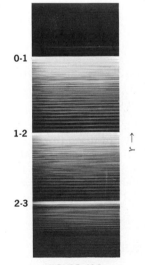

FIGURE 46.8
Bands in the spectrum of nitrogen. (*Courtesy of G. H. Dieke.*)

QUESTIONS

1. What features of the Bohr model of hydrogen are retained in wave mechanics? What features are abandoned?
2. What is the Pauli exclusion principle? How does it help us understand the periodic table?
3. Why don't we observe more evidence of the wave properties of particles in everyday life?
4. What properties of elements show the periodicity of the periodic table?
5. Why do lithium, sodium, potassium, and rubidium have similar chemical properties and similar spectra?
6. What are the evidences that the electrons of atoms have discrete characteristic energies?
7. The atomic weight of argon is greater than that of potassium, and the atomic weight of cobalt is greater than that of nickel. Why do we believe potassium has a greater atomic number than argon and that the atomic number of nickel is greater than that of cobalt?
8. How can the energies of the low-lying atomic levels be determined? What sorts of checks on these levels are available?
9. Physicists sometimes use the rough approximation that the energy needed to remove a K electron from an atom of intermediate atomic number Z is $100\,Z^2/8$ eV. Why is this a better approximation for these elements than $13.6\,Z^2$ eV, which corresponds to Eq. (45.7)?
10. If an electron and a proton have the same de Broglie wavelength, which particle has the greater speed?
11. What is the Zeeman effect? The Stark effect?
12. Why are there three absorption edges for the L shell and five for the M shell? An additional fact you need to know is that the energy of a state depends on n, l, and j but not on m_j (unless a magnetic or electric field is applied).

PROBLEMS

1. Find the speed and the de Broglie wavelength of an electron which has a kinetic energy of 150 eV.
   Ans. 7.26×10^6 m/s; 100 pm
2. Find the de Broglie wavelength of an 8-MeV proton.
3. When ultraviolet photons of a single energy fall on a surface of a metal which has a work function of 4.8 eV, the most energetic photoelectrons have an energy of 2.2 eV. Find the speed and the de Broglie wavelength of the ejected electrons. What is the wavelength of the incident photons?
   Ans. 8.79×10^5 m/s; 8.27 Å; 177 nm
4. Find the momentum and the de Broglie wavelength of an electron in the third Bohr orbit of hydrogen. How does the de Broglie wavelength compare with the circumference of the orbit?
5. Find the de Broglie wavelength associated with neutrons of energies 0.025 and 25 eV.
   Ans. 181 pm; 5.72 pm
6. Find the de Broglie wavelengths associated with electrons of energies 5 meV, 5 eV, and 5 keV.
7. Find the speed and energy of a proton which has a de Broglie wavelength of 50 pm.
   Ans. 7.92 km/s; 5.25×10^{-20} J = 328 meV
8. Find the energy in electronvolts required to strip a calcium atom ($Z = 20$) of its last electron, assuming the other 19 have been removed. How does this compare with the energy required to excite the K x-ray lines of calcium (about 4.04 keV)? Why the difference?
9. If silver $K\alpha_1$ photons (55.8 pm) fall on a molybdenum foil, find the kinetic energy of photoelectrons ejected from the K and L_{II} shells of molybdenum.   *Ans.* 2.19 keV; 19.6 keV
10. If the K, L, and M energy levels of platinum lie at roughly 78, 12, and 3 keV, respectively, compute the approximate wavelengths of the $K\alpha$ and $K\beta$ lines. What minimum potential difference across an x-ray tube is required to excite these lines? At approximately what wavelength is the K absorption edge? The three L edges?
11. Find the wavelengths of the $K\alpha_1$, the $K\alpha_2$, and the $L\alpha_1$ lines of molybdenum ($Z = 42$). $K\alpha_1$ arises from an electron transition to a K vacancy from an L_{III} state; for $K\alpha_2$ the electron goes to a K vacancy from an L_{II} state; $L\alpha_1$ comes from the transition of an electron to an L_{III} vacancy from a M_V state. (See Table 46.1.)
    Ans. 70.7 pm; 71.2 pm; 540 pm
12. Use the data of Table 46.1 to calculate the wavelength of the $K\alpha_1$ (K-L_{III}; electron makes transition to K vacancy from L_{III} level), $K\beta_1$ (K-M_{III}), and $L\alpha_1$ (L_{III}-M_V) lines of tungsten ($Z = 74$).
13. A muonic atom consists of a nucleus of charge Z, a captured negative muon, and $Z - 1$ electrons. The smaller Bohr orbits of the muon are much closer to the nucleus than any of the Bohr orbits of the electrons. The muon has the same charge as an electron but 207 times the mass. Assuming the electrons do not affect the muon significantly, find the radius of the first Bohr orbit for a muon captured by an aluminum nucleus ($Z = 13$). Find the energy of the muon in orbits

characterized by $n = 1$ and $n = 2$ and the wavelength of the photon emitted when the muon makes the transition from the higher of these states to the lower.

Ans. 20 fm; 476 keV; 119 keV; 3.5 pm

14. The moment of inertia of a H_2 molecule in its ground state is 4.55×10^{-48} kg-m². Find the lowest two excited rotational levels for the H_2 molecule and the wavelengths of the first (longest λ) two lines of the rotational spectrum.

15. The moment of inertia of a potassium chloride molecule is 2.38×10^{-45} kg-m². Find the energy of the three lowest excited rotational levels for KCl molecules. What wavelengths are emitted when transitions occur between these levels?

Ans. 4.67, 14.02, and 28.04×10^{-24} J; 0.0213 and 0.0142 m

16. Prove that the de Broglie wavelength of an electron cannot exceed the short-wavelength limit of the x-ray photons which could be produced by this electron. *Hint:* $Ve = (m - m_0)c^2$.

17. The moment of inertia of an HCl molecule is 2.72×10^{-47} kg-m². Find the energies of the first two excited states relative to the ground state and the wavelengths of the first (longest λ) two lines of the rotational spectrum.

Ans. 4.09×10^{-22} J; 1.23×10^{-21} J; 486 μm; 243 μm

SOLID-STATE PHYSICS

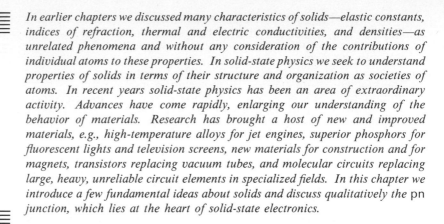

In earlier chapters we discussed many characteristics of solids—elastic constants, indices of refraction, thermal and electric conductivities, and densities—as unrelated phenomena and without any consideration of the contributions of individual atoms to these properties. In solid-state physics we seek to understand properties of solids in terms of their structure and organization as societies of atoms. In recent years solid-state physics has been an area of extraordinary activity. Advances have come rapidly, enlarging our understanding of the behavior of materials. Research has brought a host of new and improved materials, e.g., high-temperature alloys for jet engines, superior phosphors for fluorescent lights and television screens, new materials for construction and for magnets, transistors replacing vacuum tubes, and molecular circuits replacing large, heavy, unreliable circuit elements in specialized fields. In this chapter we introduce a few fundamental ideas about solids and discuss qualitatively the pn *junction, which lies at the heart of solid-state electronics.*

47.1 BINDING IN SOLIDS

As we saw in Sec. 15.1, at a sufficiently low temperature the attractive forces between gas molecules typically are great enough for liquefaction to occur; upon further cooling fluid properties disappear, and eventually there remains a block of substance which maintains both its shape and its volume. Such a mass is commonly known as a *solid*, although some scientists prefer to reserve this term for what we shall call a *crystalline solid*. Of a piece of glass, a bar of gold, and a strip of wood, only the gold is a crystalline solid. The distinction is made on the basis of the kind of order which exists in the material, i.e., on the type of geometrical arrangement in which the atoms fall. By means of x-rays we can often establish the relative positions of various kinds of atoms in a material. Suppose for a moment that we could actually see the individual atoms in a piece of glass. If we locate a silicon atom and examine its neighbors, we find that they are not distributed in a random fashion. There is a definite geometrical order. However, if we look at other silicon atoms we find that their neighbors form a somewhat different pattern. For glass there is no orderly arrangement which persists throughout the material; there is *short-range order* (Fig. 15.1*b*) but not *long-range order*. A material which exhibits short-range order and retains its shape is said to be an *amorphous solid* or sometimes a supercooled liquid.

The geometrical arrangement of atoms in any one region of a crystalline solid is the same as in any other region. The ordering extends in uninterrupted fashion over distances of thousands of atomic diameters. For crystalline solids there are 14 different possible arrays for three-dimensional lattices. The structure of a given crystal, including the locations of specific kinds of atoms within the lattice, can be studied by x-ray diffraction.

There are many examples of the same kinds of atoms arranging themselves in more than one kind of crystal lattice. In these cases the properties of the solid depend on the crystal structure. This is well illustrated by carbon, which forms crystals both in the diamond lattice and in the graphite lattice. Graphite is opaque and soft, rubbing off on paper, while diamond is translucent and hard enough to scratch glass. Carbon crystallizes in a hexagonal lattice as graphite and in a type of cubic lattice as diamond.

The forces of attraction which hold the atoms of a solid in their regular geometrical array determine not only the type of crystal lattice which is formed but also other physical properties of the material such as elastic moduli, optical constants, and conductivities, both electric and thermal. The origin of these forces differs from crystal to crystal; among the many kinds are the following.

Ionic Binding An ionic crystal is composed of positive and negative ions. For example, a crystal of sodium chloride (NaCl) consists of a regular array of Na^+ and Cl^- ions (Figs. 47.1 and 43.9). The binding forces, which come essentially from the coulomb attraction between these charged ions, are relatively large. Once a molecule of NaCl becomes part of the crystal, it is no longer possible to identify a particular chlorine ion as belonging to a particular sodium ion. Instead, each ion has a number of equidistant neighbors. Ordinarily, ionic crystals are relatively transparent to visible and ultraviolet light but show strong absorption in the infrared. At low temperatures they are excellent electric and thermal insulators; at high temperatures they become electric conductors by virtue of the motion of ions through the crystal. A crystal formed from ions exhibits *ionic* or *heteropolar* binding.

FIGURE 47.1
NaCl crystal; the Na^+ ions (dark) have an effective radius of 0.98 Å compared with a radius of 1.81 Å for the Cl^- ions (light).

Covalent Binding In covalent bonds electrons are exchanged between atoms; the average charge density between the atoms may be relatively high. Examples of crystals exhibiting covalent binding are germanium, silicon, and diamond. In a diamond crystal a single carbon atom shares electrons with four other carbon atoms at alternate corners of a cube. At low temperature, covalent crystals are ordinarily hard and often brittle and have high electric resistance. As the temperature is increased, there is a marked decrease in resistance. Practical semiconductors (Sec. 47.7) are based on covalent crystals.

Metallic Binding In metals the valence electrons are not tied to any particular atom or pair of atoms but are free to wander through the crystal. A metallic crystal may be thought of as an array of closely packed positive ions immersed in a sea of uniformly distributed electrons. This type of binding leads to high electric and thermal conductivity and to the optical and mechanical properties characteristic of metals.

Hydrogen Binding Since the hydrogen atom has a single electron, one might expect it to form a bond with only one other atom, but hydrogen frequently forms a strong bond between two atoms. We may think of this bond as primarily ionic in character, since it is formed only with strongly electronegative atoms. The hydrogen atom loses or transfers its electron to one of these atoms, and the proton binds two negative ions, nestling between these very much larger ions. Only hydrogen forms such bonds; no other ion has the very small size of the proton—so small that there are only two nearest neighbors. The hydrogen bond occurs in ice $(H_2O)_n$, solid hydrogen fluoride, and a wide variety of proteins and other organic compounds. Therefore, it is of great interest in biophysics and biochemistry.

Molecular Binding Atoms of argon, helium, and other inert gases are bound in the solid phase by relatively small electric-polarization forces, known as *van der Waals forces.* Because of the weakness of these forces, the binding energies are low. The crystals have low melting points and are highly compressible and mechanically weak.

47.2 THERMAL PROPERTIES OF SOLIDS

Quantum mechanics predicts (and experiment confirms) that at 0 K there still exists in an atom some vibrational energy, known as the *zero-point energy.* At any higher temperature the energy associated with atomic vibrations is greater. The atoms of a solid do not vibrate independently, since they are joined to each other by elastic forces. The motion of any one atom is influenced by, and in turn influences, the motion of its neighbors. Indeed, the vibration of any individual atom can be thought of as a portion of a wave moving through the crystal. In the crystal only those waves exist which satisfy the boundary conditions at the surface of the material. Calculations of the allowed frequencies show that they are great in number for a solid which contains billions of atoms. Nevertheless, it has been possible to confirm by experiment many of the predictions of the theory.

We have seen that electromagnetic waves come in photons of energy $h\nu$. Similarly, vibrational waves in crystals are quantized, with the minimum energy associated with a frequency given by the same relation $h\nu$. The quanta of vibrational waves are called *phonons.* Near 0 K only low-energy phonons are present; as the temperature is raised, phonons of higher frequency come into existence. When we heat a crystal, we excite more phonons. These lattice vibrations make by far the largest contribution to the specific heat, except near 0 K, where the small contribution to the specific heat from the free electrons must be taken into account for a metal.

The transfer of heat by conduction from one point to another in a solid has been discussed phenomenologically in Sec. 16.2. For the solid as a geometrical assembly of atoms, we can calculate the thermal conductivity in terms of lattice vibrations which transfer energy from atom to atom as waves progress through the crystal. When we raise the temperature in one region of the crystal, phonon waves spreading out from this region of increased amplitude augment the amplitudes at points some distance away. This is equivalent to our phenomenological observation that heat is transferred from one region to another. We can compute the thermal conductivities of crystals with considerable accuracy by the phonon description so long as the crystals are not metallic. We do not get valid predictions for metallic conductors because in metals the primary conduction of heat is by free electrons.

47.3 IMPERFECTIONS IN CRYSTALS

A perfect single crystal is one in which a pattern is repeated without variation over the entire volume. Such a crystal is an idealization; all crystals have *imperfections.* Here the term *imperfection* does not imply anything undesirable. For example, we make steel from iron by deliberately introducing imperfections. Many properties of solids—specific heat, thermal conductivity, thermal expansion, and density, for example—are essentially the same for perfect crystals and for those with moderate impurities. On the other hand, some properties of great engineering significance are sensitive to relatively minute imperfections; among these are mechanical strength, resistance to rust, corrosion, and mechanical abrasion, and electric conductivity in semiconductors. Among the common types of crystal imperfections are the following.

Substitutional Impurities No crystal is ever chemically pure. A substitutional impurity is one in which a foreign atom occupies the site of one of the atoms of the host crystal and plays its role. For example, in a *p*-type semiconductor an aluminum atom may replace a germanium atom in the crystal. If the impurity atom has roughly the same size as the host atom, the regularity of the crystal at the lattice site is disturbed only slightly and the mechanical properties of the crystal are unchanged. The change in electrical properties is associated primarily with the fact that the impurity atom brings an electron configuration different from that of a host atom. The impurity atom also introduces additional electron energy levels into the crystal. These may play important roles in changing the color and radiation-absorption properties of the crystal.

Interstitials and Vacancies A defect in which an atom lies in a crystal lattice at a site which is not ordinarily occupied by atoms is known as an *interstitial* (Fig. 47.2) if the atom is simply a misplaced one of the host material or as an *interstitial impurity* if it is an impurity atom. A defect in which a lattice site normally occupied by an atom is unoccupied is called a *vacancy*. Interstitial atoms and vacancies behave much like chemical impurities in that they too bring new energy levels into the crystal. They also scatter electrons moving through the crystal and increase the resistivity.

Dislocations A crystal with no chemical impurities may contain a region, known as a *dislocation*, in which the atoms are not arranged in the perfect lattice structure. Dislocations, which play an important role in determining the strength of ductile materials, are produced during solidification from the liquid phase. Ordinarily the crystal does not grow at a single point as the liquid solidifies. Instead there are many nuclei from which grow a collection of small crystals oriented more or less at random. The resulting solid is said to be *polycrystalline*. At the *grain boundaries* of the small crystals the matching of the crystal pattern is not perfect. The result is a series of dislocations.

47.4 ENERGY BANDS IN SOLIDS

When a group of atoms is brought together to form a solid, the energy levels associated with each atom undergo shifts as a result of the presence of the neighboring atoms, which change the potential energy of an electron at any point. In particular, the potential energies of the outer (or valence) electrons are shifted significantly. Quantum mechanics reveals that as a group of atoms is assembled to form a solid, the energy levels which were sharp for individual atoms become broadened to become *energy bands*. The broadening is particularly great for the higher energy levels (Fig. 47.3).

In a crystal at 0 K the individual atoms are as near to rest as they can ever

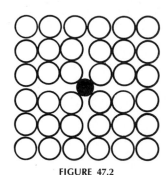

FIGURE 47.2
The imperfection is *interstitial* if it is of the host material and an *interstitial impurity* if it is an atom of some other element.

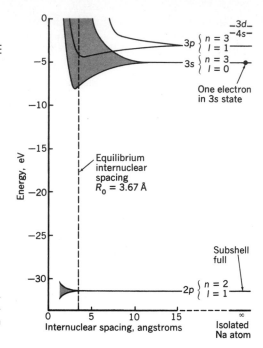

FIGURE 47.3
Energy levels are sharp for gaseous sodium atoms, but the higher energy levels broaden and overlap when sodium atoms crystallize to form a metal.

be, and complications associated with thermal vibrations are minimized. Every electron is in its lowest possible energy state, so the lowest energy levels are filled and no electron is in an excited state. Now we can distinguish between metallic conductors, semiconductors, and insulators by the situations which pertain in the highest energy band as follows.

Metallic Conductors If the energy-level characteristics of the least tightly bound electrons are such that the energy state permits electrons to move freely about the potential hills, the material is a *metallic conductor*. The energy band contains some electrons but is not completely filled (Fig. 47.4a). An electric field can produce a flow of these electrons, and the material is an excellent electric conductor at all temperatures.

FIGURE 47.4
Comparison of energy bands of (a) a metallic conductor, (b) a semiconductor, and (c) an insulator. At 0 K electrons occupy all states indicated in black, while energy states in gray are allowed but unoccupied. Some occupied states (such as K and L levels) lie too low to show in this figure.

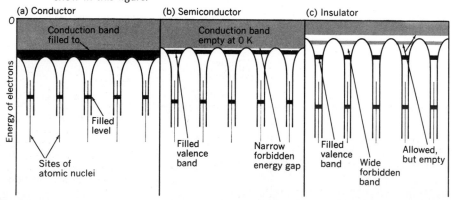

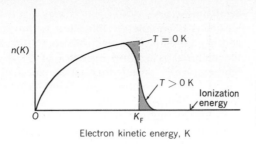

FIGURE 47.5
Fermi distribution, showing the relative number of conduction electrons having a given kinetic energy at 0 K (dashed line) and at room temperature (solid line).

The Pauli exclusion principle requires that free electrons in a metal have a broad range of kinetic energies. Even at 0 K the most energetic electrons have a kinetic energy of a few electronvolts; this maximum is called the *Fermi kinetic energy.* The energy distributions of the conduction electrons in a metal are indicated in Fig. 47.5. In metals (but not in other solids!) the energy E_F of the highest state occupied by electrons at 0 K is called the *Fermi energy* of the metal. When the temperature is increased, a few electrons are raised to slightly higher energies, leaving vacancies in levels just below the Fermi level. Even when the temperature is brought up to 300 K, the average kinetic energy of the electrons is increased very little; only a small fraction of the heat supplied to warm a metal goes into augmenting the kinetic energies of the electrons.

In order to remove one of the least tightly bound electrons from a metal at 0 K, the electron must be supplied enough energy to take it from the Fermi level to the potential just outside the metal (Fig. 47.6). In Sec. 41.6 we found evidence for this statement in the photoelectric effect. We identify the energy W_{min} in Fig. 47.6 as the minimum required by a photon to eject a photoelectron from the Fermi level. Similarly in order to achieve thermionic emission (Sec. 47.5), the temperature must be raised high enough for electrons to escape via thermal excitation.

Insulators If the gap between the highest filled band and the next permitted (but empty) band is large (Fig. 47.4c), modest temperatures are not sufficient to lift electrons to the conduction bands. Electrons in the filled bands are not free, and the material serves as an excellent insulator. Of course, if the temperature is raised sufficiently high, electrons can be excited to conduction levels, so *no material is an insulator at extremely high temperatures.* However, if the energy gap is reasonably large, practically no electrons are lifted to the conduction band at room temperature.

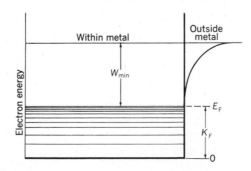

FIGURE 47.6
Approximate energy-level diagram for conduction electrons in a metal. At 0 K all states up to the Fermi level E_F are filled; and energy W_{min} is needed to remove one of the most energetic electrons. The solid curve at the right is the approximate potential energy of an electron just outside the surface of the metal.

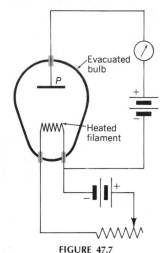

FIGURE 47.7

A hot wire emits electrons, which go to the electrode *P* when it is positive relative to the emitter.

Semiconductors When the energy band associated with the valence electrons is completely filled but within a small energy range there is another band in which electrons could exist and move through the crystal, the material is a good insulator at 0 K (Fig. 47.4*b*). However, as the temperature is raised, thermal agitation gives some electrons enough energy to shift to the higher band. The electrons so excited serve as *conduction electrons.* Further, the *holes* left by the electrons when they move to the higher band also serve as charge carriers. Such an arrangement of energy levels is characteristic of semiconductors, which are insulators at 0 K but become poor conductors at somewhat higher temperature. Semiconductors become steadily better conductors as the temperature is increased because more electrons are transferred to the conduction band. Each electron raised to the conduction band leaves a vacancy, or *hole*, in the valence band.

47.5 THERMIONIC EMISSION

In 1883, when Thomas Edison was developing the incandescent lamp, he made an important discovery. He observed that if he had a third electrode (Fig. 47.7) in one of his lamps, there was a current to this electrode when it was *positive* relative to the incandescent filament, but not when it was *negative. When materials are heated to a high temperature, electrons are emitted.* In Edison's experiments these electrons were attracted to the positive electrode, and a current was registered. When the third electrode was negative, the electrons were repelled. The emission of electrons from heated surfaces is *thermionic emission;* it is sometimes called the *Edison effect.*

The qualitative explanation of thermionic emission is relatively simple. In every material there are some electrons which are less tightly bound than others. In metals, for instance, there are electrons which are free to move about, although they cannot leave the surface of the metal unless they are given additional energy. One way of providing this energy is by raising the temperature. As the temperature goes up, an occasional electron gets enough energy to escape. At higher temperature more electrons leave, some of which have not only enough energy to escape, but enough to have relatively high speeds after they break free of the material. Figure 47.8 shows schematically that both the number and the average speed of escaping electrons increase as the temperature is raised.

Thermionic emission is analogous in many ways to evaporation from the surface of a liquid. In both cases, particles which happen to obtain an unusually large amount of energy when they are at the surface may

FIGURE 47.8

As the temperature of a filament rises, both the number of electrons emitted and their average speed increase.

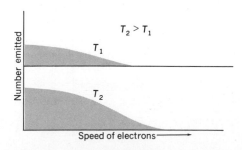

escape. The theoretical work of Richardson showed that the current density J (current I emitted per unit area A) from a heated surface is given by the relation

$$J = \frac{I}{A} = aT^2 e^{-b/T} \qquad (47.1)$$

where a and b are constants, T is the absolute temperature, and e is the base of the natural system of logarithms. The constant a is the same for all pure metals; b varies from metal to metal and has particularly low values for those elements which are good thermionic (and photoelectric) emitters. Indeed, b is the ratio of the work function W_{min} to the Boltzmann constant k (Sec. 14.7; $k = 1.381 \times 10^{-23}$ J/K $= 8.617 \times 10^{-5}$ eV/K).

The number of electrons emitted per unit area per second increases rapidly as the temperature is increased (Fig. 47.9). If we are to be able to measure the total emitted current, the potential difference between the emitting surface and the positive collecting plate must be fairly high. If it is not, there is a cloud of electrons around the emitting surface which repels additional electrons and drives many of them back to the filament. This electron cloud is called a *space charge* (Fig. 47.10). If the collecting plate is made only slightly positive, the current is limited by this space charge and bears no simple relation to the total number of electrons emitted from the filament. Figure 47.11 shows how the current collected by the plate depends on the potential for three temperatures of the filament. In each case the maximum current is determined by the rate at which electrons are emitted. When every electron emitted is collected, the current has reached its *saturation* value. As the temperature of the emitting surface is raised, the saturation current increases. At low plate potentials the current is *space-charge-limited* and does not depend on the temperature so long as the temperature is sufficiently high. Clearly, the current to the collecting plate does not obey Ohm's law.

The oxides of barium and strontium are copious electron emitters at relatively low temperatures. Many radio tubes and other electronic devices which operate at low plate potentials use cathodes coated with such oxides. Where high potentials are required, the bombardment of the negative cathode by positive ions may destroy the oxide surface. If the

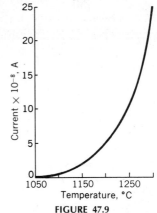

FIGURE 47.9
Thermionic emission increases rapidly with a rise in temperature.

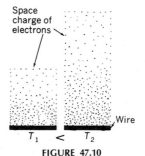

FIGURE 47.10
The space charge of electrons near an emitter increases as the temperature is raised.

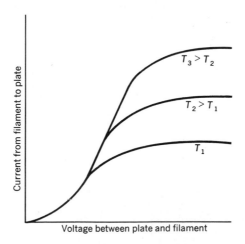

FIGURE 47.11
The electron current to a collecting plate as a function of the plate potential relative to the filament for three filament temperatures.

potentials are not particularly high, a thoriated tungsten filament is commonly used as the emitter. It is produced by adding a small amount of thorium oxide to tungsten. When this material is heated, a layer of thorium is formed on the surface of the tungsten and provides an excellent emitting surface. Where the filament is subject to bombardment by high-energy positive ions, thoriated tungsten fails and pure tungsten emitters are used. It requires far more power to the filament to obtain a given emission with pure tungsten than with thoriated tungsten because the work function for the latter is 2.6 eV compared with 4.5 eV for the pure tungsten. A typical work function for oxide-coated cathodes is 1.0 eV, so obviously the oxide-coated filament provides a given emission current density at a much lower temperature than either pure or thoriated tungsten.

47.6 THE HALL EFFECT

An energy band in which all allowed states are filled by electrons does not give rise to electric conductivity, but when there are not quite enough electrons to fill the band completely, an applied electric field does produce a current. The holes (electron vacancies) move through the material, behaving like positive charges. In 1879, well before the electron was discovered, Hall devised an experimental arrangement by which the sign of the charge carriers in a conductor could be determined. Following Hall, let us consider a thin sheet of conductor (Fig. 47.12) with a pair of contacts C_1 and C_2 attached at opposite sides of the central cross section. For a uniform conductor carrying a current in the $+x$ direction C_1 and C_2 are at the same potential if there is no magnetic field. However, when a magnetic field B_z is applied normal to the sheet, a potential difference develops between C_1 and C_2; C_1 becomes positive relative to C_2 if the charge carriers are negative (electrons) and negative if the carriers are positive (holes).

To see how this potential difference arises consider the case of positive carriers, which drift to the right (Fig. 47.12) if the conventional current is in that direction. The force exerted by the magnetic field on the moving charges deflects them downward (Sec. 36.4), producing a positive layer at the bottom of the strip and a negative one at the top. These charge layers build up until the net upward force on the carriers from the resulting electric field E_y is equal to the average downward force due to the magnetic induction, so that

FIGURE 47.12
The Hall effect tells the sign of the charge carriers in a conductor. With a magnetic field out of the plane of the page and a conventional current in the x direction the charge carriers are deflected downward by the qvB force. If the charge carriers are positive, C_2 becomes positive relative to C_1; for electrons C_2 becomes negative.

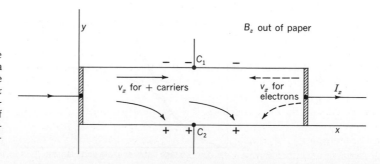

$$qE_y = qv_x B_z \qquad (47.2)$$

where q is the charge and v_x the average drift speed of the carriers.

Now suppose that the carriers are negative, moving to the left for conventional current to the right. Again the force exerted on the carriers by the magnetic field deflects them downward, but now the bottom becomes negative and the top positive. Thus the electric field is now in the $-y$ direction so C_2 is negative relative to C_1, opposite to the case for positive carriers.

We define the *Hall coefficient* R_H for the arrangement of Fig. 47.12 to be

$$R_H = \frac{E_y}{J_x B_z} \qquad (47.3)$$

where J_x is the current density (or ratio of current I to cross-sectional area A of the strip). As we saw in Sec. 36.4, $I = qnAv_x$ so $J_x = qnv_x$, where n is the number of carriers per unit volume. Using this relation and Eq. (47.2) in Eq. (47.3) leads to

$$R_H = \frac{v_x B_z}{qnv_x B_z} = \frac{1}{nq} \qquad (47.4)$$

Thus the experimental determination of the Hall coefficient of a conductor tells us both the sign of the charge carrier and the number of carriers per unit volume since $|q|$ is 1.6×10^{-19} C. The Hall coefficient is negative for electron conduction and positive for holes. For conductors such as sodium, copper, silver, and aluminum, the Hall coefficient is negative and has the right order of magnitude to correspond to one free electron per atom for Na, Cu, and Ag and to three electrons per atom for Al. Elements of valence 2 (such as beryllium, zinc, and cadmium) have positive Hall coefficients, showing that conduction is primarily by holes. Studies of the Hall coefficients have been invaluable in gaining an understanding of semiconductor physics, which is fundamental to transistors and other solid-state devices.

47.7 SEMICONDUCTORS

Crystals of pure germanium are well known as a semiconductor material. Each atom of germanium has four valence electrons, which form four covalent bonds with neighboring atoms. This structure has the semiconductor arrangement of energy states, a completely filled energy band separated from an empty higher-lying conduction level by a small energy gap. Such a crystal is called an *intrinsic semiconductor*.

At any temperature above 0 K a few electrons in the semiconductor are thermally excited to the conduction band. The minimum energy required to move an electron from one of the covalent bonds into the conduction band is called the *energy gap* E_g. For silicon E_g is 1.09 eV; for germanium, 0.72 eV. These are small enough for a significant number of electrons to be excited to the conduction band at room temperature. The resulting holes are available for conduction in the lower band. In this case the Fermi level (Fig. 47.13) is located midway between the top of the valence band and the bottom of the conduction band. Although electrons and holes are separated on an energy diagram, in the crystal itself they exist in the same space. A dynamic equilibrium is achieved between the thermal produc-

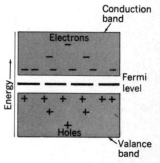

FIGURE 47.13
In an intrinsic semiconductor at room temperature some electrons are excited to the conduction band, leaving vacancies, or holes, in the valence band. Both the conduction electrons and the holes are available for electric conduction. Energies in white are not allowed to the electrons. The Fermi level lies at the middle of the forbidden band.

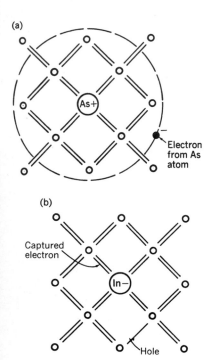

FIGURE 47.14

(a) In a doped germanium crystal an arsenic impurity atom brings five electrons where only four are required to complete the covalent bonds with neighboring germanium atoms; the fifth electron is very loosely bound and readily excited to the conduction band. (b) An indium acceptor atom in a germanium crystal provides only three of the four valence electrons needed to complete the covalent bonds. Much of the time an electron from some other part of the crystal is captured by the indium atom, leaving a *hole* in the crystal.

tion of holes and conduction electrons and the destruction of these carrier pairs by electrons dropping to the lower energy state (accessible because of the holes).

Now consider the effects of adding a small amount of arsenic to a pure germanium crystal. Addition of foreign atoms to an intrinsic semiconductor produces an *impurity semiconductor* (or *extrinsic* or *doped* semiconductor). Most of the important applications of semiconductors call for doped crystals, in which the electrical properties are drastically changed by the impurity atoms. Arsenic impurity atoms occupy sites in the crystal normally filled by germanium atoms (Fig. 47.14a). Each arsenic atom brings a total of five valence electons, of which four are used in forming covalent bonds with the four nearest germanium neighbors. The fifth electron finds no convenient neighbor with which to form a bond and is very loosely tied to the arsenic ion, even at 0 K. Only a small amount of energy is needed to free this electron and permit it to wander about in the lattice. Room temperature is enough to supply this energy. Almost every arsenic atom introduced into the crystal provides one electron which is available for conduction. The arsenic atoms are called *donors*, since each atom donates an electron to the crystal. An intrinsic semiconductor doped with a donor impurity is called an *n*-type semiconductor (*n* stands for negative) because conduction is primarily by electrons provided by the donor atoms. Arsenic is by no means the only donor element for *n*-type semiconductors; other elements from the same group (V_A) of the periodic table (such as phosphorus and antimony) are good donor impurities. Adding a donor impurity to an intrinsic semiconductor raises the position of the Fermi level (Fig. 47.15a) as a consequence of the increase in the number of electrons in the conduction band and the decrease in the number of holes.

Next consider what happens when we add to a germanium crystal atoms of valence 3, such as indium (or gallium, aluminum, boron). Each indium atom (Fig. 47.14b) brings only three electrons to form the four covalent bonds expected from each atom in the germanium crystal, producing a site at which one electron is missing from the four bonds. Such a site readily captures a wandering electron and utilizes it to complete the bonding structure, leaving a vacancy, or *hole*, somewhere in the valence band. Atoms of valence 3 elements are called *acceptor* impurities. They form *p*-type semiconductors, since in a crystal with such impurities there are sites at which an electron is missing and there are positive *holes* available for conduction. Adding an acceptor impurity to an intrinsic semiconductor lowers the Fermi level (Fig. 47.15b), since electrons are captured from both the conduction band and the valence band to fill the acceptor levels. This

FIGURE 47.15

Isolated energy levels are introduced in the forbidden energy band of a semiconductor by doping it with impurity atoms. (a) Donor atoms give levels just below the conduction band and raise the Fermi energy above the middle of the forbidden band. (b) Acceptor atoms introduce levels just above the valence band and lower the Fermi energy below the middle of the forbidden band.

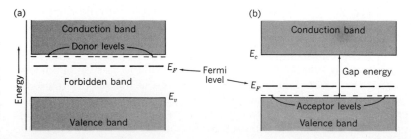

reduces the number of conduction electrons and increases the number of holes. As expected, *p*-type semiconductors exhibit positive Hall coefficients, while *n*-type ones have negative values for R_H.

47.8 *pn* JUNCTIONS

If a region of *n*-type material is in intimate contact with a region of *p*-type material, they form a *pn junction*. Such junctions are of great importance, since they are the building blocks for semiconductor devices. One cannot form a *pn* junction by pressing *n*-type material against *p*-type material, since the act of pressing would bring such a tremendous concentration of impurities and imperfections to the interface that the junction would not have the required properties. One way of making a *pn* junction is to melt an indium pellet which has been placed on top of an *n*-type germanium crystal. Where the germanium has been saturated with molten indium, *p*-type material is produced. A *pn* junction is formed at the interface between the indium-saturated *p* region and the original *n* material.

As we saw in Sec. 30.2, the Fermi level of a material is of great importance. When two isolated materials are placed in contact, electrons move from the one in which the Fermi level is greater to the other. This lowers the Fermi level in the first and raises it in the second. Charges flow until the Fermi levels become equal in the two materials. In an isolated *n*-type semiconductor the Fermi level lies above the center of the forbidden energy region (Fig. 47.15), while for a *p*-type material the Fermi level lies below the center of the energy gap. At a *pn* junction, where *n* and *p* materials are in intimate contact, electrons from the donor atoms near the junction fill the nearby acceptor states just across the junction (Fig. 47.16). This makes the *p* material negative and raises the Fermi level there, while the loss of electrons in the *n* region leaves it positively charged. In the immediate vicinity of the junction there is a scarcity of free carriers of either sign, because the electrons from the *n* side have filled acceptor states and holes on the *p* side. Thus at the junction we have a *depletion layer* perhaps 20 nm ($\approx$50 atom layers) thick which is a relatively poor conductor together with a double layer of bound charges, negative on the *p* side and positive on the *n* side.

When the Fermi level becomes the same on both sides, equilibrium is established. This does not mean that there is no charge transfer across the junction but that the charge transfers in both directions are equal. At the junction in equilibrium there are still many more electrons per unit volume in the conduction band in the *n* material at the left of Fig. 47.16

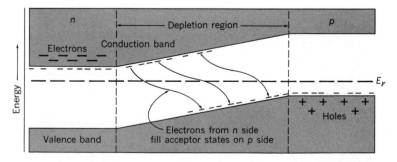

FIGURE 47.16
Energy-level diagram for an unbiased *np* junction. Electrons from donor atoms on the *n* side fill acceptor levels on the *p* side, making the *p* material negatively charged. A scarcity of charge carriers in the transition region increases the resistivity there. At equilibrium electrons diffuse from the *n* side at the same rate as electrons from the *p* side move "downhill" to the *n* side.

than in the p material at the right; consequently, there is an electron current associated with the diffusion of these electrons against the electric field of the double layer. At the same time there is an equal and opposite electron current from electrons which are accelerated "down the hill" by this field. This latter current has a magnitude which depends on the concentration of electrons in the p-type material and on the low rate at which the free electrons appear at the junction. Hence this current has a *saturation* value I_s which is ordinarily small. (The saturation current does not depend on the height of the potential hill any more than the water which goes over a waterfall depends on the height of the fall.) Holes are also exchanged across the boundary; their behavior is directly analogous to that of electrons.

When we apply a potential difference across the pn junction, making the p material positive relative to the n material, the hill down which the electrons slide is reduced. This has essentially no effect on the saturation current I_s due to electrons going from the p to the n side. However, many more electrons from the n material diffuse to the p material, so the electron flow from the n to p side is greatly increased by this bias. The current in what is known as the *forward direction* rises rapidly with increasing forward bias (or potential difference), as shown in Fig. 47.17. On the other hand, if we make the n-type material positive relative to the p type, the hill is increased and electron flow from n to p is substantially reduced while the current I_s remains the same. Consequently, with negative or reverse bias there is a very small current essentially independent of bias. Thus a pn junction acts as a rectifier—not quite a perfect one, since there is some current in the reverse direction for reverse bias, but nevertheless an effective one, since we get a substantial current in the forward direction for a modest potential difference and a very small reverse current for the same (magnitude) reverse bias. The pn junction rectifier performs essentially the same functions as the vacuum-tube diode (Sec. 48.1).

An alternative explanation for the rectifying action of a pn junction can be made in terms of the depletion layer. Reverse bias removes electrons from the n side and adds them to the p side, thus increasing the width of the transition region in which there is a deficiency of carriers. This leads to a small current under reverse bias. On the other hand, forward bias adds electrons to the n side and removes them (leaving holes) in the p region. This reduces the width of the transition region and provides carriers for a substantial forward current.

FIGURE 47.17
How current varies with bias potential difference across a pn junction. Application of bias voltage shifts the Fermi levels in the two sides of the junction as shown.

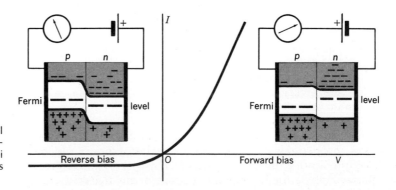

As we have seen, both atoms and molecules have sharply defined allowed energy states. When one of them is in an excited state, it undergoes a series of transitions which bring it eventually to its lowest, or *ground*, state. Many of these transitions, but not necessarily all, involve the emission of a photon. Typically, an isolated excited atom or molecule emits a photon spontaneously and quite independently of its neighbors. The resulting radiation from many such emitters is *incoherent*—a disorganized jumble of waves for which the directions of emission and the phase relationship between the photons are entirely random, even though the frequency may be the same. However, if a photon of a given frequency interacts with an excited atom about to emit this same frequency, the atom may be *stimulated* to radiate energy in the same phase and in the same direction as that of the incident photon. The stimulated radiation from many excited atoms may be *coherent*, with the waves in phase and traveling in the same direction. Thus, if enough identical radiating systems can be excited in some region, a weak beam of the proper frequency can grow to a strong one by stimulating the excited systems to radiate their energy in the same phase and direction, thereby giving a beam of high temporal and spatial coherence.

A device which produces a beam of coherent light by stimulating emission from a large population of excited radiators is called a *laser*, an acronym for *l*ight *a*mplification by *s*timulated *e*mission of *r*adiation. The first stimulated emission amplifiers operated in the microwave region and received the name *masers* with the *m* coming from *microwave*, but *maser* is now a general term for stimulated-emission devices which operate over a broad spectral range, including microwave, infrared, visible, and ultraviolet frequencies. For example, lasers are often called optical masers.

To operate any maser it is necessary (1) to produce an unnaturally large number of excited systems ready to emit the desired radiation and (2) to stimulate the emission in such a way as to obtain a coherent beam in the desired direction. Many techniques have been devised for achieving these objectives. As an example, we shall consider the ruby laser, which was the first to operate (1960).

Ruby is aluminum oxide in which a few aluminum atoms are replaced by chromium atoms; the latter form Cr^{3+} ions in the crystal and are responsible for the ruby color. Pure aluminum oxide is an insulator with a wide forbidden band like that in Fig. 47.4c. Replacing a few aluminum atoms by chromium atoms introduces several allowed energy states in the forbidden band, much as substitutional impurities in silicon and germanium introduce energy states in the forbidden band. However, the new energy states in chromium-doped aluminum oxide are several electronvolts above the top of the valence band and there are many of them, some so close together that they form bands (Fig. 47.18).

In the first laser a cigarette-sized ruby rod was placed on the axis of a powerful xenon flash lamp which provided several thousand joules of light each flash. Part of this light was absorbed by the Cr^{3+} ions, raising many to the bands of levels B_1 and B_2 (Fig. 47.18), from which they quickly decay to the upper laser level E_2 by nonradiative transitions in which the energy goes into heating the ruby. An excited ion remains in this state for a few milliseconds before undergoing a spontaneous transition to the

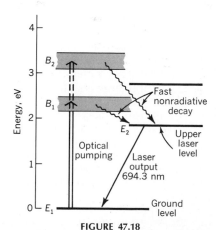

FIGURE 47.18
Simplified energy-level diagram for the Cr^{3+} ion in ruby. Ions in the ground state are pumped by light from a xenon flash bulb to the bands of levels B_1 and B_2, from which they quickly decay by nonradiative transitions to the metastable level E_2. Passing radiation stimulates the transition to the ground level E_1, and coherent radiation is emitted at a wavelength of 694.3 nm.

ground state by the emission of a 694.3-nm photon. However, when a beam of this frequency is passing through the crystal, the excited ions are stimulated to emit in phase with the beam. By polishing both ends of the ruby rod, silvering one end, and half-silvering the other, it is possible to produce particularly large amplification for radiation in the axial direction, some of which escapes through the partially silvered end to form the laser beam.

The process of providing the required number of excited atoms in a laser is known as *pumping*. In the ruby laser there must be enough optical pumping to put more ions in level E_2 than remain in E_1 if laser operation is to be achieved. For 1 kJ of electric energy supplied to the flash lamp there is perhaps 10 J of laser output in a period of 100 μs, about 1 MW of highly directional, monochromatic radiation. In other lasers peak powers over 1 GW have been obtained and power densities in excess of 10^{19} W/m². Such almost incredible optical beams make it possible to detect light reflected from the moon and to burn through a 10-mm steel plate with a single shot. In gas lasers pumping comes from an electric discharge through the gas, and in semiconductor lasers passing a current through a *pn* junction provides the pumping action. Hundreds of different laser and maser systems have been developed to provide intense beams of coherent radiation over a broad spectral range. These devices have found many uses and opened the door to a whole new world of research and application.

QUESTIONS

1. When two materials are rubbed together, what determines which material gains electrons and which loses them?
2. What happens to the Fermi level of a material when electrons are added to it? Why?
3. In what various ways can the "free" electrons be removed from a metal? Discuss.
4. The temperature coefficient of resistance for typical metals is about +0.004 per Celsius degree. Would you expect it to be greater or less in magnitude for a semiconductor? Would it be positive or negative? Why?
5. The energy gap is about 5.3 eV in diamond and 1.09 eV in silicon. Explain why diamond is relatively transparent to visible light while silicon is not.
6. How does the number of conduction electrons compare with the number of holes in (a) an ideal intrinsic semiconductor, (b) a p-type semiconductor, and (c) an n-type semiconductor? Explain.
7. Why do acceptor atoms lower the Fermi level in a crystal?
8. What is the role of donor impurities in a semiconductor? What valence do donor atoms for a germanium crystal usually have?
9. Does an ordinary semiconductor at constant temperature obey Ohm's law? Does a *pn* junction? Explain.
10. Why should the disordered region at an edge dislocation be etched away much faster by an acid than a perfectly ordered region?
11. What is the Hall effect? Why is it of great importance in connection with semiconductors? How could it be used to measure magnetic fields?
12. What is "space charge"? Discuss the importance of space charge in electron tubes.
13. What is the distinction between coherent and incoherent radiation?
14. What are some of the advantages of lasers and masers over the more familiar light sources? What possible uses and applications can you envision for lasers and masers?

PROBLEMS

1. Find the average volume occupied by silver atoms in a silver bar of density 10,500 kg/m³. If each atom were a cube of this volume (which it certainly is not), find the length of one edge.
   Ans. 17.1 × 10⁻³⁰ m³; 257 pm
2. Find the average volume occupied by (a) a copper atom in a copper bar of density 8,930 kg/m³, (b) an osmium atom in a crystal of osmium, the

densest of the metals at 22,500 kg/m³, and (c) a lithium atom in the least dense of metallic crystals at 530 kg/m³.

3. In a NaCl crystal the ions are 2.82 Å from their nearest neighbors. Assuming that each ion is attracted only by the nearest ion in the adjacent layer, calculate the stress (force per unit area) required to pull the crystal apart. Do you expect your answer to be larger or smaller than an experimental determination? Why?
Ans. 3.65×10^{10} N/m²; larger; next nearest neighbors repel

4. In a potassium chloride crystal the ions are 3.14 Å apart. Assuming that each ion is attracted only by the nearest ion in the adjacent layer, find the stress (force per unit area) required to pull the crystal apart. Why is the answer smaller than for NaCl (Prob. 3)?

5. The Fermi kinetic energy for electrons in copper is 7.0 eV. Calculate the speed and the de Broglie wavelength for an electron with this kinetic energy. *Ans.* 1.57×10^6 m/s; 464 pm

6. The Fermi kinetic energy in aluminum is 11.7 eV. Find the speed and the de Broglie wavelength for an electron with this kinetic energy.

7. An experiment on the threshold wavelength for photoelectric emission from zinc yields the value 345 nm. Find the energy of the Fermi surface (relative to the configuration in which the electron is at rest an infinite distance from the metal) in joules and in electronvolts. (This corresponds to the work function W_{min}.) If the Fermi energy of zinc is 9.47 eV, what is the minimum energy an electron must have to escape from zinc referred to the zero of energy in Fig. 47.6?
Ans. 5.76×10^{-19} J; 3.59 eV; 13.1 eV

8. The photocurrent of a cell can be cut to zero by a minimum retarding potential of 2.3 V when monochromatic light of 200 nm is incident. (a) What is W_{min}, which is the energy of the Fermi surface relative to the configuration when the electron is held an infinite distance from the material? (b) What is the maximum kinetic energy of photoelectrons for light of 250 nm wavelength? (c) What is the longest wavelength which can produce photoemission? (d) What would the kinetic energy of an electron be if it were ejected from a conduction state 1.1 eV below the Fermi

level (Fig. 47.6) by a photon of 200 nm wavelength?

9. The Hall coefficient for potassium is -4.2×10^{-10} m³/C. Find the number of free electrons per cubic meter. If the density of potassium is 862 kg/m³, find the average number of free electrons per atom.
Ans. 1.5×10^{28} electrons/m³; 1.1 electrons/atom

10. The Hall coefficient of sodium is -2.5×10^{-10} m³/C, and the density of sodium is 971 kg/m³. From these data estimate the average number of free electrons per cubic meter and per sodium atom.

11. A gold foil 25 μm thick and 15 mm wide bears a current of 10 A in a Hall experiment. There is a magnetic field of 0.75 T normal to the plane of the foil and a potential difference of 21.6 μV is developed across the width of the foil. Find the electric field strength E_y and the Hall coefficient for gold if C_1 is positive in Fig. 47.12.
Ans. -1.44 mN/C; -7.2×10^{-11} m³/C

12. In a Hall experiment on silver a potential difference of 59 μV is developed across a foil 25 mm wide and 0.05 mm thick when the foil bears a current of 28 A in a magnetic field of strength 1.25 T in the direction of the thickness. If C_1 (Fig. 47.12) is positive, calculate the electric field strength E_y and the Hall coefficient of silver.

13. Calculate the ratio of the thermionic emission current from a thoriated tungsten filament to that of a similar pure tungsten filament at 1400 K.
Ans. 6.9×10^6

14. The current in the plate circuit of a vacuum tube is 40 mA. How many electrons are arriving at the plate each second? What area of emitter is required if the cathode is a thoriated tungsten filament at 1500 K and the constant a in the Richardson equation is 1.2×10^6 A/(m²) (K)²?

15. A thoriated-tungsten filament at 1500 K has a total emission current of 1.5 mA. Find the emission current at 1800 K. *Ans.* 62 mA

BASIC ELECTRONICS

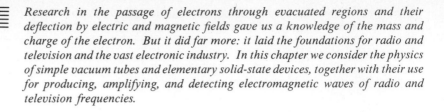

Research in the passage of electrons through evacuated regions and their deflection by electric and magnetic fields gave us a knowledge of the mass and charge of the electron. But it did far more: it laid the foundations for radio and television and the vast electronic industry. In this chapter we consider the physics of simple vacuum tubes and elementary solid-state devices, together with their use for producing, amplifying, and detecting electromagnetic waves of radio and television frequencies.

48.1 RECTIFICATION

Electric energy provided to homes by power companies typically involves alternating currents, which are appropriate for the operation of electric lights, heaters, and ac motors. However, many household appliances require direct-current (dc) sources in their operation; radios, television sets, and battery chargers are examples. A device which provides a dc output from an ac input is called a *rectifier*. There are many types of rectifiers; we shall discuss two, the diode vacuum tube and the *pn* junction.

A two-element vacuum tube containing an electron-emitting cathode and a plate (Sec. 47.5) is called a *diode*. When the cathode is sufficiently hot, such a tube conducts electricity freely when the plate is positive but not when the plate is negative. Thus, when an alternating potential difference is applied between the cathode and the plate, the diode conducts only while the plate is positive.

Suppose we wish to charge a battery but we have available only an alternating potential difference. The direct application of an alternating potential difference to the battery results in no useful charging, since the electric energy stored during one half cycle is returned during the second half cycle. However, if we provide a circuit similar to that of Fig. 48.1*a*, there is a current through the battery only in the proper direction to charge the battery. Figure 48.1*b* shows the alternating potential difference provided for the circuit and the resulting current. In this circuit current exists only during half a cycle. This is therefore called *half-wave rectification.*

A similar half-wave rectification can be obtained by using a suitable *pn* junction in place of the diode (Fig. 48.2*a*). The circuit symbol used for the junction is shown in Fig. 48.2*b*; note that the direction of the conventional current is shown by the orientation of the triangle representing the *p* side of the junction. Major advantages of the junction rectifier are that it requires no power for the cathode and it is much smaller, lighter, and more rugged than the diode. Disadvantages are that the junction rectifier passes a small current in the reverse direction, and it cannot stand as high a reverse voltage.

In the circuits above load current was pulsed, though unidirectional. One way to make the current steadier is to introduce a filter consisting of one or more capacitors in parallel with the load and one or more inductors (called *chokes* when used in this way) in series with the load (Fig. 48.3). By suitable choice of capacitors and inductors it is possible to provide almost any desired degree of constancy in the current. The general action of the capacitors is to store energy in the electrostatic field when the output from the tube is at its maximum and to release this energy to maintain the current as the output from the tube falls to zero. An inductor connected in series with the load stores energy in its magnetic field when the current is maximum and releases this energy to maintain the current as the current decreases.

Thus far we have considered only half-wave rectifiers in that half of each input cycle was not used. The circuits shown in Fig. 48.4 provide *full-wave rectification*. For the vacuum-tube rectifier electrons go from the cathode to the upper plate when the upper terminal of the ac transformer is positive. Conversely, when the lower end of the transformer is positive, electrons go to the lower plate. However, the conventional current through the load is always in the same direction. Similar reasoning applies to the *pn* junction rectifier. Further smoothing of the output of a full-wave rectifier can be achieved by filtering with capacitors, inductors, or a combination of these circuit elements.

48.2 THE TRIODE

In 1907 De Forest had the brilliant idea of introducing a *grid* between the cathode and the plate of a diode to control the current to the plate. This grid consists of an open mesh or helix around the filament, as shown in Fig. 48.5. By keeping the grid at a potential slightly negative relative to the cathode, it is possible to prevent the grid from collecting electrons (Fig. 48.6). Because the grid is close to the electron-emitting cathode, a small change in grid potential makes a significant change in the number of electrons which pass the grid and reach the plate.

If the plate potential of a triode is kept constant and the grid potential is varied, the plate current varies as a function of grid potential as shown in Fig. 48.7. (Note that all potential differences are measured relative to the filament.) If the grid is made sufficiently negative, the current to the plate may be cut off entirely.

Figure 48.8 shows the plate current of a typical triode as a function of plate potential for three different values of the grid potential. Note that at points x and y in this figure the plate current is exactly the same. From x to y the grid potential decreases from -2 to -5 V, while the plate potential increases from 90 to 140 V. Thus, an increase in the plate potential of 50 V has compensated for a decrease in grid potential of 3 V. To put it another way, a reduction in the grid potential of 3 V requires an increase of 50 V in plate potential to keep the plate current constant. Under these conditions of operation, the grid is $\frac{50}{3}$ or 16.7 times as effective in controlling the plate current as the plate. We say that the *amplification factor* is 16.7.

If when the grid potential of a triode is changed by a small amount

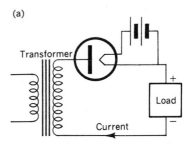

(a)

Transformer

Load

Current

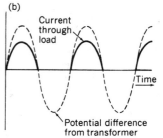

(b)

Current through load

Time

Potential difference from transformer

FIGURE 48.1
(a) Circuit of a diode rectifier used to give a unidirectional current through a load and (b) graph of the resulting current when an alternating potential difference is applied.

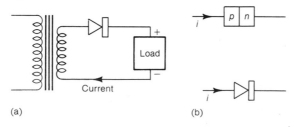

(a)

Load

Current

(b)

i

| p | n |

i

FIGURE 48.2
(a) Circuit using a *pn* junction to give a unidirectional current through a load. The load current in this case is similar to that shown in Fig. 48.1*b*. (b) In the symbol for a *pn* junction the *p* side is represented by a triangle pointing in the direction of conventional current flow for the forward-biased junction (forward bias means that the *p* side is positive).

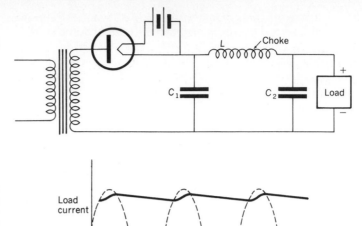

FIGURE 48.3
A filtered rectifier produces an almost
constant current (solid line below)
through the load, in contrast to the current variation in the absence of the filter
(dashed curve).

$-\Delta V_{g'}$, it requires a change in plate potential ΔV_p to keep the plate current constant, the amplification factor μ is defined as

$$\mu = \frac{\Delta V_p}{-\Delta V_g} \qquad I_p \text{ constant} \qquad (48.1)$$

48.3 THE TRIODE AS AN AMPLIFIER

The fact that small changes in the grid potential can produce large changes in the plate current in a triode permits us to obtain large potential variations in the plate circuit when the grid potential is changed by small amounts. Consider the circuit shown in Fig. 48.9. If we apply a small alternating signal to the grid, the plate current changes as indicated in Fig. 48.10. As a result of this change in plate current, the potential difference across the plate load resistor R_L varies. For such a circuit we define the *voltage gain* as

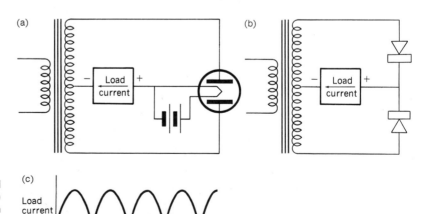

FIGURE 48.4
Full-wave rectification can be achieved
either using a two-plate vacuum tube (*a*)
or a pair of *pn* junctions (*b*). For both
circuits the load current as a function of
time is shown in (*c*).

$$\text{Voltage gain} = \frac{V_{\text{out}}}{V_{\text{in}}} = \frac{R_L \Delta i_p}{\Delta v_g} \qquad (48.2)$$

where R_L is the resistance of the plate resistor and Δi_p and Δv_g are the variations in plate current and grid potential, respectively.

It is possible to apply the changes in potential across the plate (or load) resistor to the grid of a second amplifier tube. In this way signals can be amplified by a factor of many hundreds.

48.4 THE TRIODE OSCILLATOR

In Sec. 40.9 we learned that when a capacitor is discharged through an inductor of low resistance, damped oscillations occur with a frequency ν given by Eq. (40.11): $\nu = 1/(2\pi\sqrt{LC})$. The oscillations are damped because energy is dissipated in heat and sent out as radiation, while no energy is supplied to the circuit. Steady oscillations of constant amplitude can be maintained in such a circuit provided energy is supplied to compensate for the losses. It is possible to use a triode to produce such stable oscillations by providing energy to the oscillating circuit from the plate-potential supply of the triode. One of the ways in which this can be done is indicated in Fig. 48.11. Here the condenser C_1 and the inductor L_1 form the oscillating circuit. The oscillations produce variations in the potential of the grid. This, in turn, results in variations of the plate current. The varying plate current passes through the inductor L_2 and induces, by mutual induction, an emf in the coil L_1, thereby supplying energy to the oscillating grid circuit. The energy is fed back into the grid circuit in such a way as to provide for the losses in this circuit. In this way oscillations of constant amplitude and of frequency determined by the grid-circuit elements are maintained.

If one wishes the oscillation frequency to remain very constant, one uses for C_1 a capacitor which has as dielectric between its plates a carefully cut crystal of quartz. In an oscillating electric field, such a crystal undergoes mechanical vibrations (see Sec. 32.3), the frequency of which depends on the thickness of the quartz. If the oscillating circuit and the crystal are resonant to essentially the same frequency, the oscillator assumes the frequency of the mechanical vibrations of the quartz. In such a case the oscillator is *crystal-controlled*.

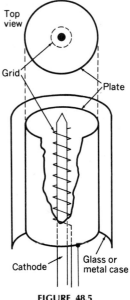

FIGURE 48.5
A simple triode.

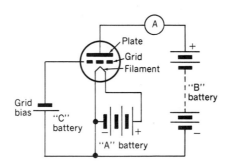

FIGURE 48.6
Circuit for measuring the static characteristics of a triode.

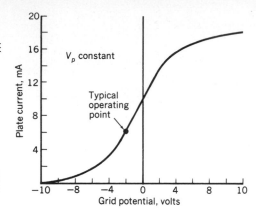

FIGURE 48.7
Relation between plate current and grid potential in a triode when the plate potential is held constant.

48.5 TRANSISTORS

Just as a *pn* junction and a diode vacuum tube can perform the same function in a rectifier circuit, so a solid-state device known as a *transistor* can perform the functions of a triode vacuum tube. Increasingly, solid-state components have replaced vacuum tubes in radio, television sets, and computers. Of the many kinds of solid-state components in common use we shall discuss only the *npn* and the *pnp* transistors, either of which is a possible replacement for a triode in most applications.

An *npn* transistor consists of a thin layer of weakly *p*-type material known as the *base* sandwiched between strongly *n*-type material called the *emitter* and relatively weak *n*-type material which comprises the *collector* (Fig. 48.12). Typically, the base is only of the order of a few micrometers in thickness, and the entire transistor may occupy a volume of only a few thousandths of a cubic millimeter. The base must be sufficiently thin for the probability of an electron's combining with a hole within the base to be small compared with the probability of the electron's passing completely through the base. Essentially all electrons which leave the emitter pass through the base to the collector, so the electron current to the base

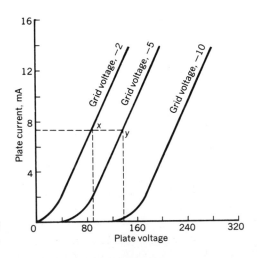

FIGURE 48.8
Plate current in a triode as a function of plate potential for three grid potentials.

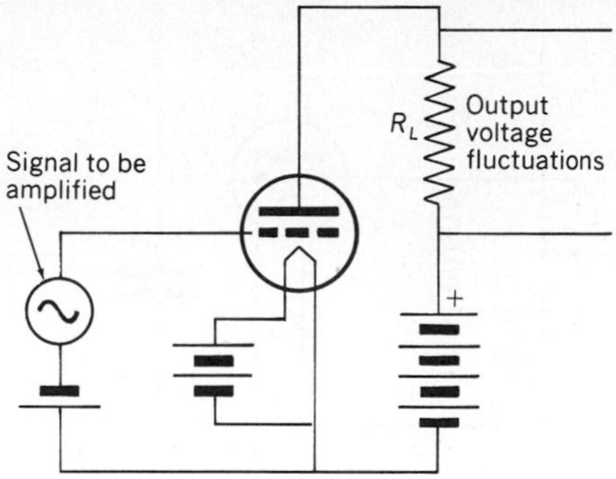

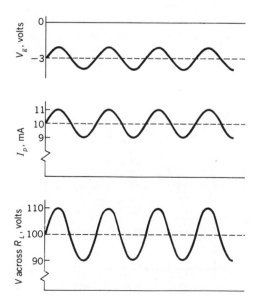

FIGURE 48.9
Circuit using a triode as an amplifier.

is small compared with the electron current to the collector. The emitter is operated at a small forward bias ($\approx$0.5 V) relative to the base (Fig. 48.13), while the collector is reverse-biased by a potential difference of several volts. (Note that in this case the emitter-base junction is an *np* junction and the base-collector junction is a *pn* junction. For any junction *forward bias* means making the *p* side positive relative to the *n* side and *reverse bias* means making the *n* side positive relative to the *p* side.) The applied potential differences appear almost entirely at the junctions themselves, since both the *n* regions are relatively good conductors. In the emitter, thermal agitation provides a generous and continuous supply of conduction electrons, many of which reach the emitter-base interface and diffuse into and through the base. The electrons lost by the emitter are restored through the emitter electrode.

An *npn* transistor in a circuit like that of Fig. 48.13 may function as a voltage amplifier provided the load resistor R_L is sufficiently high. Since the emitter is forward-biased, a small change in emitter potential results in a substantial change in the electron current from emitter to base to

FIGURE 48.10
Variation of grid potential, plate current, and potential difference across the load resistor R_L.

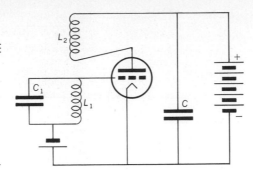

FIGURE 48.11
Simple oscillator circuit using a triode.

(a)

(b)

FIGURE 48.12
(*a*) An *npn* transistor has a very thin layer of weakly doped *p* material sandwiched between a strongly doped *n* emitter and a relatively weakly doped *n* collector. (*b*) Symbol for an *npn* transistor; the arrowhead on the emitter shows the direction of the conventional current.

collector (Fig. 47.17) and hence to a substantial change in the potential difference across the load. A small voltage signal at the emitter leads to a larger change in the potential difference across the load, and the transistor operates as a voltage amplifier with a high impedance load. The emitter-base voltage, the collector current, and the potential difference across the load resistor behave just like the potential, plate current, and the potential difference across the load in Fig. 48.10. Thus, the transistor of Fig. 48.13 and the triode of Fig. 48.9 are both functioning as amplifiers, converting small signal voltages into much larger voltage changes across a load resistor. In general, transistors can perform the same types of functions as triodes and other vacuum tubes.

Although we have discussed only the *npn* transistor so far, *pnp* transistors (Fig. 48.14) are of comparable importance. Their characteristics are similar except that the carriers are holes rather than electrons. Again the emitter-base junction is forward-biased (*p* side positive) and the base-collector junction is reverse-biased. It is a major advantage for circuit designers to be able to choose either a *pnp* or an *npn* transistor; with vacuum tubes there is no comparable choice.

Among the major advantages of the transistor are its extremely small weight and volume, the small potential differences it requires for operation, its relatively low cost, its extremely long life, and its ruggedness. There are disadvantages as well, among them the fact that at sufficiently high temperatures there are so many free electrons in all parts of the transistor that it fails to function properly.

FIGURE 48.13
Circuit for an *npn* transistor used as an amplifier, with an energy-level diagram replacing the standard symbol for the transistor. This is a *common-base* circuit because both the emitter and the collector are biased relative to the base.

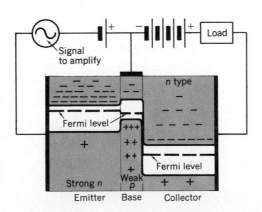

In preceding sections we have seen how simple *pn* junctions, transistors, and vacuum tubes operate. We now consider how these circuit components can be used in simple radio circuits. Remember that for most of these circuits we may use either a vacuum tube or a solid-state device to perform a given function. Since the detailed operation of a radio transmitter or receiver involves many complexities, we emphasize only the basic physics of their operation.

All radio and television stations send out electromagnetic waves. The frequencies involved vary from many thousand to many million hertz. In principle, a simple oscillator such as that of Fig. 48.11 could produce these oscillations; in practice, more sophisticated circuits are used. A primitive type of radio transmitter is shown in Fig. 48.15. It consists of an oscillator, an antenna, and a *modulator*. The oscillator frequency is determined by the resonant grid circuit. For the standard radio band the frequencies range from 535 to 1,605 kHz. The oscillations in the grid circuit produce, through mutual induction, an oscillating current in the inductor L_3 which is part of the antenna circuit. The oscillator thus "drives" the antenna, setting up in it an alternating current of frequency characteristic of the oscillator. The current surging back and forth in the antenna produces the rapidly changing electromagnetic field which we know as radio waves and which can be detected at great distances from the transmitter. In Sec. 40.10 we discussed the electromagnetic waves sent out by an oscillating current (Fig. 40.10) such as the one established in the antenna.

If a simple alternating current from an oscillator is used to drive an antenna, the transmitter sends out a constant-amplitude signal at a high frequency, called the *carrier frequency*. If we wish to convey intelligence by such a signal, we must interrupt it or modify it in some way. The most primitive method of doing this is to turn the oscillator on and off, thus sending a series of dots and dashes. However, it is also possible to send

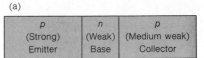

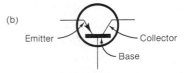

FIGURE 48.14
A *pnp* transistor: (*a*) general structure and (*b*) symbol as a circuit element.

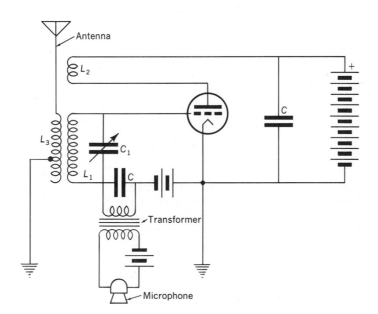

FIGURE 48.15
Simple radio transmitter.

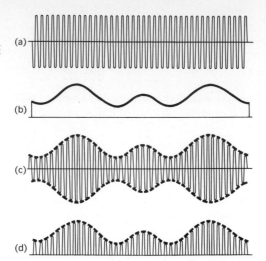

FIGURE 48.16
(a) Unmodulated carrier wave, (b) audio-frequency wave, (c) modulated carrier wave, and (d) rectified wave.

information which leads to accurate reproduction of sound waves. If we speak into the microphone of Fig. 48.15, there are set up in the grid circuit additional small voltage fluctuations which vary the amplitude of the oscillations. When we wish to send a 1,000-Hz note over the transmitter, we "modulate," i.e., vary, the amplitude of the carrier wave at a frequency of 1,000 times a second. Since the carrier frequency is many thousand times as high as the highest sound frequency we wish to transmit, the carrier wave makes many thousands of oscillations in each period of the sound wave. Figure 48.16 shows the general appearance of the unmodulated carrier wave, of the sound frequency with which we modulate this wave, and of the modulated wave which is sent out by the transmitter. This particular method of transmitting intelligence is called *amplitude modulation* (AM). There are, of course, other ways in which the carrier wave can be modified to convey intelligence. One of the possibilities is to introduce fluctuations in the frequency, keeping the amplitude constant. This is *frequency modulation* (FM).

48.7 THE RADIO RECEIVER

The electromagnetic waves sent out by a radio transmitter fall upon the antenna of the receiver shown schematically in Fig. 48.17. These electromagnetic waves set up oscillations in the antenna which induce small potential variations in the grid circuit of the receiver. There are many transmitters from which signals can be received. It is in the grid circuit L_2C_2 that the selection of the signal to be heard is made. By varying the capacitance C_2, the grid circuit L_2C_2 can be made resonant to any desired carrier frequency. Then oscillations in the antenna corresponding to that particular frequency establish changes in potential in the grid circuit, while oscillations of all other frequencies produce negligible effects because the grid circuit is not resonant to them.

The oscillations of potential across C_2 in the resonance circuit are applied to the grid of the triode in Fig. 48.17. If this grid is made sufficiently negative, the plate current increases substantially when the grid

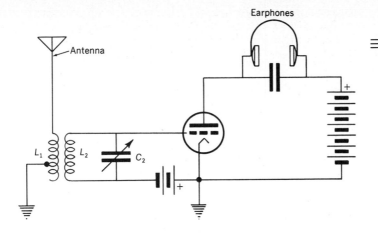

FIGURE 48.17
Simple radio receiver.

potential changes in the positive direction but decreases relatively little (or not at all) when the changes are in the negative direction. In this case the output of the plate circuit is similar to Fig. 48.16d. The variations in plate current at radio frequency are such that the human ear cannot detect them; however, the average plate current over many carrier-wave oscillations fluctuates with the modulation frequency. Thus the current through the earphones oscillates with the modulation frequency, and sound is heard.

To drive a loudspeaker, much more energy must be provided than can be expected from the output of a single triode of the type indicated. In a modern radio receiver several more tubes may be used to provide this additional power. The signal coming into the initial resonant circuit is amplified before the detection process is performed. Then the audio frequency may be amplified several times before it is finally supplied to a *power amplifier* which drives the loudspeaker.

48.8 THE LOUDSPEAKER

A loudspeaker is a device which transforms audio-frequency variations in electric current into sound waves. Figure 48.18 shows schematically the principal parts of a dynamic loudspeaker. A light thin coil of wire is attached to the apex of a light cone. The coil is placed in a magnetic field. Current from the power amplifier is fed to this coil; the fluctuating current produces fluctuating forces upon the coil, causing it to vibrate. The cone vibrates with the coil and produces waves in the surrounding air.

48.9 THE KENNELLY-HEAVISIDE LAYERS

Radio reception from distant stations is usually superior at night. Sometimes reception is excellent at distances of several thousand miles when a receiving station only a few hundred miles away may not be able to detect the signal. In order to explain these phenomena it was suggested by Kennelly and by Heaviside that there are layers of ionized air in the upper atmosphere which act as refractors and reflectors of radio waves. Ions are produced in the outer layers by a variety of processes, chiefly by ultraviolet

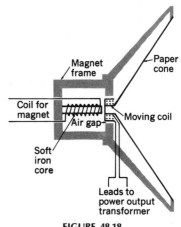

FIGURE 48.18
Dynamic loudspeaker.

light from the sun. The height of the ionized layers changes from time to time; as a result, the transmission of electric waves varies with time. In general, the ionizing layers form at different altitudes at different times of day, being higher and more uniform at night.

48.10   THE CATHODE-RAY OSCILLOSCOPE

One of the most useful of all types of vacuum tubes is the so-called *cathode-ray tube*. A heated cathode (Fig. 48.19) emits electrons. These electrons are accelerated to a cylindrical anode, which has a hole in its center through which the cathode rays pass. The electron beam may be controlled by the grid. Beyond the anode is a beam of rapidly moving electrons of well-defined energy.

This beam passes between a pair of flat plates P_1 whose plane is horizontal. If a potential difference is applied betwen these plates, an electric field is created which can move the beam of electrons up or down. Next the beam passes between a pair of plates P_2 whose plane is vertical. If a potential difference is applied between these plates, the beam can be moved to the right or left. The beam of electrons strikes a screen coated with a fluorescent material that emits light where the electrons strike. The position of the point on the screen at which electrons strike can be controlled by the potentials applied to the deflecting plates.

In one common application of the oscilloscope the horizontal deflectors P_2 have applied to them a sawtooth potential variation which sweeps the beam at a constant rate from left to right and then suddenly switches the beam back to the left edge. If a sinusoidal potential difference is applied to the deflecting plates P_1, the beam traces a sine curve on the face of the oscilloscope. If some other type of potential variation is applied to the plates P_1, the picture on the oscilloscope screen reproduces the fluctuations in potential across the deflecting plates.

When a cathode-ray tube is combined with electronic circuits for amplifying signals and applying potentials to the horizontal and vertical deflecting plates, the unit is called a *cathode-ray oscilloscope*. The cathode-ray oscilloscope is one of the most convenient, versatile, and powerful tools of modern physics.

48.11   TELEVISION

Television receivers use a cathode-ray tube in which horizontal and vertical deflections are usually produced by magnetic rather than electric fields. The magnetic deflection is obtained through a pair of coils, one on

FIGURE 48.19
Cathode-ray tube.

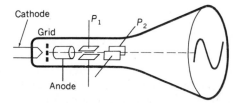

each side of the picture tube, which produce a horizontal magnetic field, and another pair, one above and one below the neck of the tube, which produce a vertical field. By sending high-frequency alternating currents through these coils it is possible to make the cathode-ray beam move back and forth across the tube, producing the familiar lines of the television picture.

If one moved the cathode-ray beam across the face of the oscilloscope, then down a little and across the screen again, etc., only a number of white lines would be drawn on the face of the screen. However, by controlling the intensity of the beam by means of the grid in the tube, it is possible to draw lines which vary in their lightness, depending on the potential applied to the grid at the particular instant the spot is drawing a given segment of the line. Thus the cathode-ray beam draws lines of varying brightness. There are 525 horizontal lines drawn to make a complete picture. In order to transmit pictures of moving objects, it is necessary that the picture be "redrawn" many times a second. In current television practice the even-numbered lines are drawn in $\frac{1}{60}$ s and the odd-numbered lines in the next $\frac{1}{60}$ s. Thus, 30 complete pictures or "frames" are drawn each second.

At the sending station the information transmitted is obtained in roughly the following way. An image of the picture to be transmitted is produced by lenses on the photosensitive screen of a tube known as an *image orthicon tube.* Each element of the sensitive surface emits photoelectrons, the number of which depends upon the intensity of the incident light. By a complex process the number of photoelectrons emitted from each of the tiny sensitive elements is determined and is converted into information which can be used to modulate the carrier wave. At the receiver this information is used to control the grid potential at the instant the corresponding element of the picture is to be reproduced on the receiving screen.

48.12 RADAR

The wavelengths of ordinary radio waves are of the order of hundreds of meters. Such waves bend readily around obstacles such as buildings or airplanes. If radio waves of much higher frequency, and therefore shorter wavelength, are produced, the waves are readily reflected by aircraft and other objects of comparable size. Further, these waves can be focused into sharp beams by a reasonably small antenna. If a sharply focused beam of high-frequency waves is used to scan the sky, it is reflected by airplanes or other objects in the sky. By measuring the time lapse between the transmission of a pulse and the reception of the reflected pulse, one can determine the distance of the object from the transmitter since radio waves are transmitted with the speed of light.

It is possible to scan a relatively large area by means of a radar transmitter and show on an oscilloscope screen the areas from which there is substantial reflection and those from which the reflection is much smaller. In this way one can make a radar map of an area which is not visible because of clouds, fog, or darkness.

One of the earliest and simplest examples of the use of radio waves to measure distance is found in the radio altimeter. If a radio wave is sent down from an airplane, it is reflected by the earth. If this radiation is sent in pulses, and if it is possible to measure the time it takes the wave to travel to the earth and return, the altitude of the plane above the terrain can be calculated automatically and displayed for the pilot.

QUESTIONS

1. How can a simple diode be used as a rectifier to charge a battery? Draw an appropriate circuit.
2. A simple diode rectifier provides a varying unidirectional current. How can one obtain an almost constant direct current from such a rectifier by adding suitable components to the circuit?
3. Draw a diagram of a full-wave solid-state rectifier using a transformer as a power source to charge a battery.
4. How does a triode act as an oscillator? From what source does the energy come to maintain the oscillations?
5. How can a triode be used to amplify a small signal? Explain with the aid of suitable curves. Repeat for a transistor amplifier.
6. When either a transistor or a triode is used as an amplifier, it is desirable to operate in a region of the characteristic curve which is linear (nearly a straight line). Why is this desirable? What happens if a nonlinear region is used?
7. Why do transistors malfunction or fail if the temperature becomes too great?
8. How can a triode be used as a detector? Why is the grid of a triode that is used as a detector biased more negatively than the same grid would be if the triode were used as an amplifier?
9. Draw a *pnp* transistor circuit analogous to the *npn* transistor circuit of Fig. 48.13. Explain the operation of the *pnp* transistor, for which the bias voltages are opposite those of the *npn* transistor.
10. How does one change stations in an ordinary radio? In a television receiver? Why the difference?
11. How does a cathode-ray oscilloscope work? What are some of its uses?
12. How is the picture drawn on the screen of a television tube?
13. Why do solid-state radios and television receivers use much less power than those using only vacuum tubes?
14. What advantages in addition to smaller power consumption do solid-state circuit elements have over vacuum tubes? Under what circumstances are vacuum tubes better than solid-state circuit elements?

PROBLEMS

1. For a certain triode the plate current changes $60 \mu A$ for a 5-V change in plate potential. If it requires a 0.4-V change in grid potential to restore the plate current to its original value, calculate the amplification factor. *Ans.* 12.5
2. A vacuum tube has an amplification factor of 15. If the grid voltage is decreased 0.26 V, how much must the plate potential be raised for the plate current to remain constant?
3. For a 12AU7-A vacuum tube the current is 15 mA for a 200-V plate potential and a −4-V grid potential. It is also 15 mA at a 276-V plate potential and a −8-V grid potential. What is the amplification factor? *Ans.* 19
4. The plate current of a vacuum tube is 5 mA when the grid potential is −3.1 V and the plate potential is 150 V. It is also 5 mA when the grid potential is −2.9 V and the plate potential is 142 V. What is the amplification factor?
5. The signal available from an antenna has an amplitude of $5 \mu V$. If this signal is passed through two stages of amplification each with a voltage gain of 20, what is the amplitude of the potential fluctuations in the output circuit? Through how many more stages of the same voltage gain would the signal have to be passed to obtain an output voltage in excess of 750 mV?
Ans. 2 mV; 2
6. How many stages of voltage amplification factor 15 are required to have an output signal of 3 V amplitude from an input signal of $20 \mu V$ amplitude?
7. A triode is used as an amplifier. If the grid is biased at −3 V and a plate resistor R_L of 8 kΩ is used, the plate current is 20 mA. If the grid potential is varied from −3.2 to −2.8 V, the plate current varies from 19.5 to 20.5 mA. Find the maximum and minimum potential difference across the plate resistor and the voltage gain.
Ans. 164 V; 156 V; 20
8. When an ac signal of 25 mV amplitude is impressed on the grid of a triode which is biased at −2.5 V, the plate current varies from 6 to 6.2 mA. If the resistance R_L in the plate circuit is 9 kΩ, find the voltage amplification of the triode.

9. A triode amplifier uses a plate resistor of 15 kΩ. When no signal is applied to the grid, the plate current is 6 mA. If the voltage gain is 50, find the amplitude of the potential fluctuations across the plate resistor when a signal of amplitude 4 mV is applied to the grid. Between what limits does the plate current fluctuate?

Ans. 0.2 V; 5.987 and 6.013 mA

10. Electrons in a television tube are accelerated through a potential difference of 10 kV. Find the speed of the electrons using the appropriate relativistic equations. How much smaller is this than the erroneous value calculated by using $\frac{1}{2}mv^2$ for the kinetic energy of the electrons? What is the short-wavelength limit of the x-rays produced inside the picture tube?

11. An electron is projected along the axis of a cathode-ray tube midway between two parallel plates with a velocity of 1.5×10^7 m/s. The plates are 12 mm apart and 25 mm long and have a potential difference of 180 V between them. Find (a) the angle the electron makes with the axis as it leaves the plates, (b) the distance of the electron from the axis as it leaves the plates, and (c) the distance from the axis at which the electron strikes a fluorescent screen 0.24 m beyond the plates.

Ans. (a) 0.28 rad; (b) 3.7 mm; (c) 73 mm

12. A cathode-ray tube has electrostatic deflection plates 15 mm square with a separation of 5 mm located 0.25 m from the fluorescent screen of the tube. The electrons are accelerated through a potential difference of 1,500 V between emitter and anode. Calculate the approximate deflection of the beam on the screen for a 50-V potential applied to the plates.

13. If the cathode-ray tube of Prob. 12 uses magnetic rather than electrostatic deflection, find the deflection produced by a magnetic field of 0.003 T which is assumed uniform over a length of 15 mm and zero elsewhere.

Ans. 95 mm

49
NUCLEI AND NUCLEAR ENERGY

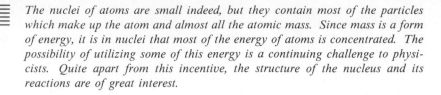

The nuclei of atoms are small indeed, but they contain most of the particles which make up the atom and almost all the atomic mass. Since mass is a form of energy, it is in nuclei that most of the energy of atoms is concentrated. The possibility of utilizing some of this energy is a continuing challenge to physicists. Quite apart from this incentive, the structure of the nucleus and its reactions are of great interest.

49.1 BUILDING BLOCKS OF NUCLEI

In our discussion of atoms thus far, we have been concerned primarily with the energy states of the electrons which are bound to the nuclei by coulomb attraction. We have attributed to each nucleus a charge Z and a mass number A, which is the integer closest to the atomic weight of the nuclear species to which the atom belongs. In this chapter we consider the structure and properties of the nuclei.

The simplest of all nuclei is that of ordinary hydrogen, an atom of which we designate by $_1^1$H. Note that we shall indicate the atomic number Z of an atomic species with a subscript at the lower left and the mass number A with a superscript at the upper left.† The nucleus of the hydrogen atom, called the *proton* (p), bears a positive charge equal to the negative charge on the electron. It has a mass of 1.67265×10^{-27} kg (1,836 times the mass of the electron) or 1.007276 atomic mass units (u). *One unified atomic mass unit is one-twelfth the mass of a carbon 12 ($_6^{12}$C) atom.*

The proton has an intrinsic spin angular momentum such that the component in the direction of an applied external field is either $\frac{1}{2}h/2\pi$ or $-\frac{1}{2}h/2\pi$. Associated with this spin is a small magnetic moment (Sec. 36.5) of the order of magnitude one would expect from a spinning positive charge the size of the proton, which has a radius of about 10^{-15} m.

Since all nuclei have charges which are integral multiples of the protonic charge, it is reasonable and fruitful to assume that protons are fundamental building blocks of all nuclei. A nucleus of atomic number Z contains Z protons. However, for a typical nucleus only about one-half the mass can be attributed to the protons. It was suggested by Rutherford that there might be a second building block of roughly the same mass as the proton but having no charge. This particle, the *neutron*, was discovered in 1932 (Sec. 50.2). The neutron mass is 1.008665 u, slightly greater than that of the proton. It has the same intrinsic angular momentum, and, although chargeless, it does have a magnetic moment about seven-tenths that of the proton and of opposite sign (as though negative charge were rotating). There is evidence that the neutron has a positive core with enough negative charge in its outer reaches to make it electrically neutral. A spinning neutral system of this kind would have a negative magnetic moment.

Protons and neutrons are the building blocks with which we can construct any nucleus. The number of protons in a nucleus determines the atomic number Z, which is the positive charge on the nucleus in elementary units and the number of electrons associated with the neutral atom. The sum of the number of protons and the number of neutrons is the *mass number A*. Protons and neutrons in a nucleus are collectively

† We follow the recommendation of the International Union of Pure and Applied Physics. Many books and papers indicate the mass number A by a superscript at the right ($_1$H^1, $_6$C^{12}), a form which requires less composition time.

referred to as *nucleons*. The mass number A is then the number of nucleons in the nucleus. In the quantum-mechanical treatment of nuclear structure, the strong binding of nuclei can be explained in terms of the exchange of charge between nucleons. A nucleon may be a proton at one instant and become a neutron the next instant by transferring its charge to another nucleon, which transmutes from neutron to proton. In addition to mass, each nucleon contributes angular momentum and magnetic moment to the nucleus. However, these are vector quantities, and their resultant is usually not large—indeed, it is typically zero for nuclei with even numbers of both protons and neutrons. This can be understood by applying the Pauli exclusion principle to nuclei.

The chemical behavior of any atom is determined by its electronic configuration and hence by its atomic number Z. All atoms with $Z = 1$ are hydrogen, all with $Z = 8$ are oxygen, and all with $Z = 92$ are uranium. However, atoms with the same number of protons may have different numbers of neutrons. Atoms with the same atomic number but different mass number are *isotopes* (Sec. 44.5). Most elements have at least two stable isotopes; tin has ten, the largest number for any element. In addition to stable isotopes, one or more radioactive isotopes are known for every element.

49.2 THE SIZE OF NUCLEI

If a typical atom were expanded to a radius of 10 m, its nucleus would have a radius of less than 1 mm. The volume of a nucleus is only about 10^{-12} of the volume of the atom. To put it another way, a nucleus viewed from an outer electron subtends an angle of 1 percent of the angle subtended by the sun at the earth ($0.5°$). In a sense an atom is a very open structure largely comprising empty space through which electrons are passing.

Although nuclei are exceedingly small, several types of measurements have been devised to determine nuclear radii. While there are minor differences in the empirical values, all methods agree on the order of magnitude. It is not surprising that different experiments yield slightly different radii. For one thing not all nuclei are exactly spherical; we know that some are *prolate*, i.e., lengthened along the axis of rotation, and others *oblate*, i.e., shorter along the axis than in the equatorial plane, as the earth is. Further, the nucleus does not have a sharp surface like a billiard ball. Nevertheless, it is reasonable to think of a nucleus as having a roughly spherical shape.

The volumes of all nuclei are roughly proportional to the number of nucleons A. If we assume a roughly spherical shape, we find that the radius (in meters) of a nucleus of mass number A is given approximately by

$$r = 1.2 \times 10^{-15} \sqrt[3]{A} \qquad (49.1)$$

The heaviest of all common nuclei is that of uranium 238, for which the nuclear radius is somewhat less than 10^{-14} m, as compared with the typical atomic radius of about 10^{-10} m.

The densities of nuclei of all kinds are roughly equal, about 2×10^{17} kg/m³. A liter of packed nucleons would have a mass of 2×10^{14} kg.

49.3 THE DEUTERON

In 1932 Urey and his collaborators discovered a stable isotope of hydrogen with mass number 2. Such a hydrogen atom is known as *deuterium* or *heavy hydrogen*. Its nucleus, the *deuteron*, is composed of one proton and one neutron and has a mass of 2.01355 u. When a proton and a neutron combine to form a deuteron, a mass of 0.00239 u (1.007276 + 1.008665 − 2.01355) disappears and is transformed into a gamma photon.

According to Einstein's theory of relativity, the mass-energy equivalence is expressed (Sec. 42.8) by

$$\text{Energy} = mc^2 \tag{49.2}$$

By Eq. (49.2),

$$(1 \text{ u})c^2 = (1.66 \times 10^{-27} \text{ kg})(3 \times 10^8 \text{ m/s})^2 = 1.49 \times 10^{-10} \text{ J}$$

Since $1 \text{ eV} = 1.60 \times 10^{-19} \text{ J}$,

$$1 \text{ u} = \frac{931.5 \text{ MeV}}{c^2} \tag{49.2a}$$

The mass which disappears when a deuteron is formed corresponds to 2.225 MeV. In order to break a deuteron into a proton and a neutron, we must supply this energy. Indeed, one of our best methods of measuring the neutron mass involves finding the lowest-energy photon which can split the deuteron into a proton and a neutron, a process called *photodisintegration*.

The fact that energy is released in the reaction $^1_1\text{H} + ^1_0 n \longrightarrow ^2_1\text{H}$ and is required to break up the deuteron is analogous to the situation in chemical reactions. When carbon and oxygen unite to form carbon dioxide, energy is set free. If we wish to decompose carbon dioxide into carbon and oxygen, we must add this amount of energy to the system. The total energy which would be released if we could perform a series of operations to build a nucleus from neutrons and protons is called the *binding energy* of the nucleus. Alternatively, the binding energy is the energy required to tear the nucleus apart into protons and neutrons.

A proton and a neutron are bound together only if their spins are essentially aligned in the same direction. As a result the spin angular momentum of the deuteron is one unit of $h/2\pi$, and the magnetic moment of the deuteron is only about one-third that of the proton, because the neutron contributes negative magnetic moment to the system.

There is a third isotopic form of hydrogen, called *tritium*, which has mass number 3. Its nucleus, the *triton*, is composed of one proton and two neutrons. It is unstable, half a given sample transforming to helium 3 in 12 years.

49.4 NUCLEAR FORCES AND BINDING ENERGIES

The forces which hold nuclear particles together are not like any of the other forces we have studied. Physicists today distinguish four basic forces which may exist between particles in nature; in order of decreasing strength they are (1) the strong (nuclear) force, (2) the electromagnetic force, (3) the weak interaction, and (4) the gravitational interaction. The strong interaction at a distance of 10^{-15} m is about 100 times the electric force between two protons. It is the electromagnetic force which is responsible for atomic structure and interatomic binding. The weak interaction governs the decay of beta emitters and many subatomic particles; it is roughly 10^{-14} times the strength of the strong interaction. The weakest of the forces is the gravitational attraction; for two protons the gravitational attraction is only about 10^{-36} times the electrostatic repulsion. Not every particle participates in all four forces, although the proton does. The electron is involved in all except the strong force. Particles which interact through the strong force are called *hadrons;* more than 100 kinds of hadrons have been identified, the proton and the neutron being the best known.

Nuclear forces are so great and the amount of energy involved when nucleons are brought together to form a new nucleus is so tremendous that we can observe the difference in masses before and after. The difference between the mass of the component nucleons from which a nucleus is composed and the actual mass of the nucleus is known as the nuclear *binding energy in mass units*. To find the binding energy in joules we multiply the mass difference in kilograms by c^2, the square of the speed of light (Sec. 42.8). To obtain the binding energy in megaelectronvolts we make use of the fact that 1 u corresponds to 931.5 MeV [Eq. (49.2a)]. In practice, mass spectrometers determine the masses of heavy atoms (or ions) rather than the masses of nuclei directly. It is convenient to think of any neutral atom of mass number A and atomic number Z as being composed of Z hydrogen atoms and $A - Z$ neutrons. The hydrogen atoms bring the protons for the nucleus and the electrons for the shells. Then the binding energy is given by the difference between the mass of Z hydrogen atoms plus $A - Z$ neutrons and the mass of the neutral atom:

$$\text{Binding energy} = [Zm_\text{H} + (A - Z)m_n - M]c^2 \qquad (49.3)$$

where m_H, m_n, and M are the masses of the hydrogen atom, the neutron, and the neutral atom, respectively. Table 49.1 lists the masses of a number of common atomic species.

To obtain the mass of the nucleus from the known atomic mass, one subtracts the mass of the Z electrons.

$$\text{One electron mass} = 0.0005486 \text{ u} = \frac{0.511 \text{ MeV}}{c^2}$$

The binding energy increases with the mass number A. For purposes of comparison it is convenient to divide the total binding energy by the number of nucleons to obtain the *binding energy per nucleon*. A plot of the binding energy per nucleon for many stable nuclei is shown in Fig. 49.1.

□ **Example** Find the binding energy and the binding energy per nucleon for ^{7}Li, which is composed of three hydrogen atoms and four neutrons.

$$3 \times 1.007824 = 3.023472$$
$$4 \times 1.008665 = \underline{4.034660}$$
$$7.058132$$
$$^7\text{Li} \qquad \underline{7.01600}$$
$$\text{Binding energy} = 0.0421 \text{ u} \times c^2$$
$$0.0421 \times 931.5 = 39.3 \text{ MeV binding energy}$$
$$\text{Binding energy per particle} = \frac{39.3}{7} = 5.6 \text{ MeV/nucleon} \qquad □$$

TABLE 49.1
Masses of Atoms
(Based on ^{12}C as 12.00000)

Element	Z	A	Mass, u
(Neutron)	0	1	1.008665
Hydrogen	1	1	1.007824
(Deuterium)		2	2.01410
(Tritium)		3	3.01605
Helium	2	3	3.01603
		4	4.002603
Lithium	3	6	6.01513
		7	7.01600
Carbon	6	12	12.000000
Nitrogen	7	14	14.00307
Oxygen	8	16	15.99491
Sodium	11	23	22.98977
Sulfur	16	32	31.97207
Nickel	28	58	57.93534
Copper	29	63	62.92959
Tin	50	120	119.90220
Lead	82	208	207.97665
Uranium	92	238	238.05077

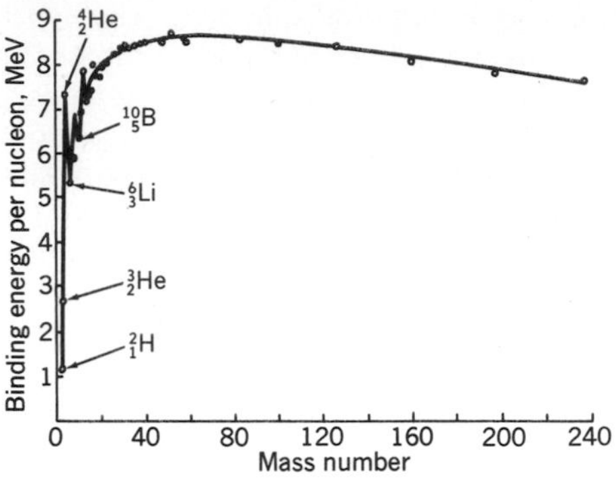

FIGURE 49.1
Binding energy per nucleon as a function of mass number for stable nuclei.

49.5 STABLE NUCLEI

We do not know the exact nature of the forces which hold nucleons together, but we do know what stable nuclei exist in nature and some of the conditions which must be satisfied if a nucleus is to be stable. Among the light elements we never find stable nuclei which have radically different numbers of protons and neutrons. For example, oxygen has three stable isotopes with masses 16, 17, and 18. The 8 protons of the oxygen nucleus form stable configurations with 8, 9, or 10 neutrons, but not with 4 neutrons or with 20 neutrons. As we go up the periodic table, the number of neutrons increases more rapidly than the number of protons, until in ^{238}U there are almost 1.6 neutrons for every proton.

The increase in binding energy per nucleon for light elements (Fig. 49.1) is associated in part with the fact that on the average each nucleon has more neighbors to which it is bound. However, the rise is not steady; for example, the binding energy per nucleon is greater for $^4_2$He and $^{16}_8$O than for nuclei with one or two more nucleons. This can be explained in terms of a *shell model* of nuclei, which predicts that 2, 8, 20, 50, 82, and 126 nucleons of either type form particularly stable arrays. Throughout the realm of nuclei there is strong evidence of the pairing of two protons of opposite spin and of two neutrons of opposite spin to form very stable nuclei. Except for hydrogen, elements of even atomic number are several times as abundant as those of odd atomic number. Elements of odd Z have at most two stable isotopes, both of which ordinarily have an even number of neutrons (prominent exceptions: $^2_1$H, $^6_3$Li, $^{10}_5$B, and $^{14}_7$N). On the other hand, elements of even Z often have many isotopes, most of which have also an even number of neutrons.

For the heavier elements the mutual electrostatic repulsion of the protons increases rapidly with Z, and consequently the net binding energy per nucleon falls off. To this coulomb repulsion we attribute the fact that the number of neutrons exceeds the number of protons for stable nuclei heavier than $^{40}_{20}$Ca. It is also the reason for the instability of the very heavy nuclei. Whenever the arrangement of the nucleons is unstable, the nucleus is radioactive and undergoes one of several kinds of transition. It was in the observation of such transitions that nuclear physics had its birth.

49.6 ALPHA PARTICLES

In Sec. 44.2 we learned that naturally radioactive nuclei emit three types of radiation, named *alpha*, *beta*, and *gamma* rays. A few nuclear species, for example, $^{227}_{80}$Ac and $^{218}_{84}$Po, emit all three, but most emit gamma rays and either alpha particles or beta particles but not both.

The alpha particle is the nucleus of the helium atom. This fact was established by separating an alpha emitter from an evacuated region by means of a thin wall through which alpha particles could pass. After a few hours enough gas collected in the evacuated region so that its spectrum could be excited. This showed that the gas was helium. Alpha particles have been found to bear a double elementary positive charge, substantiating the fact that they are helium nuclei.

Alpha particles are easily absorbed by metal foils or by a few centimeters of air. They affect a photographic plate, cause many materials to fluoresce brilliantly, and ionize the air through which they pass. When alpha particles strike screens of fluorescent material, they produce tiny flashes of light called *scintillations*. Many of the early researches on alpha particles were carried out in darkened rooms by patient observers who sat for hours counting these flashes. If an alpha emitter is placed at R in Fig. 49.2b, a fluorescent screen S receives many scintillations each second when it is close to the source. As the screen is moved away from the alpha emitter, the number of scintillations per second remains roughly constant over a considerable distance. Then the number drops off sharply (Fig. 49.2a). As the particles move through the gas, they lose energy by ionizing the gas. In passing through the first few centimeters of air, the alpha particles are slowed down significantly, but few are lost. This type of range curve is characteristic of particles which lose energy gradually as they pass through matter. A curve like that of Fig. 49.2a suggests that all the alpha particles from the source had the same energy. For any given alpha emitter we find that either all the alpha particles have the same energy or that there are a few groups of alpha particles emitted, each group with its own discrete energy.

In general, a nucleus is unstable against alpha decay if its mass exceeds that of an alpha particle plus that of the residual nucleus. This is equivalent to saying that a neutral atom is unstable against alpha decay if its mass exceeds that of a helium atom plus that of the remaining atom. In the reaction

$$^{226}_{88}\text{Ra} \longrightarrow {}^{222}_{86}\text{Rn} + {}^{4}_{2}\text{He} + \text{energy}$$

the energy released is the difference between the mass of the radium (226.02536 u) and the sum of the masses of the radon (222.01753) and the helium (4.00260). This is 0.00523 u, or 4.87 MeV.

49.7 BETA EMISSION

Beta particles are electrons ejected from nuclei with speeds which may exceed $0.9c$, where c is the speed of light. These electrons have a penetrating power far greater than that of alpha particles.

The energies of the beta rays from a given radioactive nucleus vary continuously from very low energy to a maximum (Fig. 49.3). There is a *continuous energy spectrum of the beta rays*. This is in sharp contrast with the energy spectrum of the alpha particles, which has at most a few discrete energies. If beta rays were the only particles coming out of a given kind of nucleus, one would expect the energies of the beta rays to be discrete. However, if some other particle were emitted simultaneously with the electron, this particle would share the total energy availa-

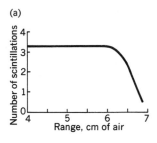

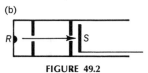

FIGURE 49.2
(a) Graph of range of 7.68-MeV alpha particles from $^{214}_{84}$Po in air, as measured in apparatus (b).

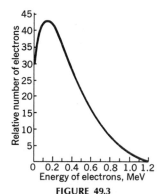

FIGURE 49.3
Beta-ray spectrum of $^{210}_{82}$Bi showing the relative number of electrons emitted at various energies.

ble and the electron energy spectrum would be continuous. There is convincing evidence that another particle is indeed ejected from beta emitters along with the electron. It has no charge or rest mass but has a spin equal in magnitude to that of the electron. This particle was originally called the *neutrino*, meaning "little neutral one," but for reasons suggested in Sec. 50.9 it seems preferable to call it the *antineutrino* (represented by $\bar{\nu}$).

When a nucleus contains too many neutrons for the number of protons present, the unbalance can be corrected by the transformation of a neutron into a proton with the emission of a beta ray and an antineutrino:

$$\,_0^1 n \longrightarrow \,_1^1 H + \,_{-1}^0 \beta + \bar{\nu}$$

An isolated neutron is not a stable particle but decays by the reaction above with a half-life of about 12 min, i.e., in 12 min one-half of any sample of neutrons not bound in a nucleus decays.

The beta-decay process results in a product nucleus of the same mass number A but atomic number Z higher by 1. Beta decay is expected whenever the mass of the product nucleus is less than that of the original nucleus. As an example, the mass of $\,_6^{14} C$ exceeds that of $\,_7^{14} N$, and

$$\,_6^{14} C \longrightarrow \,_7^{14} N + \,_{-1}^0 \beta + \bar{\nu}$$

When an element of low to moderate mass has too many protons for the number of neutrons, a transmutation occurs which results in the changing of a proton into a neutron. This may happen in one of two ways: (1) a positron may be emitted, or (2) an orbital electron may be captured. In either case a neutrino is also involved. An example of a positron emitter is carbon 11, which decays by the reaction

$$\,_6^{11} C \longrightarrow \,_5^{11} B + \,_1^0 \beta + \nu$$

where $\,_1^0 \beta$ represents a positron and ν a neutrino.

49.8 GAMMA RAYS

High-energy electromagnetic rays are called gamma rays when they are emitted by radioactive nuclei. The properties of gamma rays are identical with those of x-rays of the same wavelength.

When a radioactive nucleus emits an alpha or beta particle, it is likely that the residual nucleus will be left not in its most stable arrangement but in an excited state. Ordinarily, when the nucleus goes to a more stable configuration, the energy released is radiated as a gamma ray. Like the alpha particles, the gamma rays from a given radioactive species have discrete energies.

49.9 THE URANIUM-RADIUM SERIES

The most abundant radioactive nucleus in nature is that of the uranium isotope of mass 238 u, which we abbreviate ^{238}U. This material decays by the emission of an alpha particle with a half-life of 4.51×10^9 years. *The half-life is the time required for one-half the nuclei in a sample to disintegrate.* For example, the half-life of radon is 3.85 days; one-half of a sample of radon disappears in 3.85 days. At the end of 7.7 days three-quarters of it has disappeared, and one-fourth remains. After 15.4 days, only one-sixteenth of the original radon is present.

The reaction in which ^{238}U emits an alpha particle can be written

$$^{238}_{92}U \longrightarrow {}^{4}_{2}He + {}^{234}_{90}Th$$

Note that the emission of an alpha particle reduces the atomic number (left subscript) by 2 and the mass number (left superscript) by 4. The ^{234}Th decays by beta emission to protactinium 234, which in turn emits a beta ray and becomes ^{234}U:

$$^{234}_{90}Th \longrightarrow {}^{0}_{-1}\beta + {}^{234}_{91}Pa$$
$$^{234}_{91}Pa \longrightarrow {}^{0}_{-1}\beta + {}^{234}_{92}U$$

^{234}U is an alpha emitter with the daughter ^{230}Th. Ultimately the decay series ends with lead 206. The detailed list of reactions and half-lives for the uranium-radium series is presented in Table 49.2.

In addition to the uranium-radium series, three other natural radioactive series exist. The longest-lived member of the *thorium series* is ^{232}Th, and the final stable nucleus formed is again lead, but this time ^{208}Pb. The *neptunium series* has ^{237}Np as its longest-lived member, and the radioactive chain ends with bismuth 209, while the *actinium series* has ^{235}U as its longest-lived member and ^{207}Pb as its final stable nucleus.

49.10 RADIOACTIVE DATING

Since each alpha particle is a helium nucleus, helium is in the process of formation in all minerals containing alpha-emitting radioactive substances. The number of alpha particles given out by 1 g of uranium in equilibrium with its radioactive products has been found to be 9.7×10^4

TABLE 49.2
The Uranium-Radium Series

Nuclide	Element	Early name	Half-life	Energy of rays, MeV		
				Alpha	Beta	Gamma
$^{238}_{92}U$	Uranium	Uranium I	4.5×10^9 yr	4.18		
$^{234}_{90}Th$	Thorium	Uranium X$_1$	24.5 days		0.103	0.09
$^{234}_{91}Pa$	Protactinium	Uranium X$_2$	1.14 min		2.32	0.8
$^{234}_{92}U$	Uranium	Uranium II	2.33×10^5 yr	4.76		
$^{230}_{90}Th$	Thorium	Ionium	8.3×10^4 yr	4.66		
$^{226}_{88}Ra$	Radium	Radium	1,620 yr	4.79		0.19
$^{222}_{86}Rn$	Radon	Ra emanation	3.8 days	5.49		
$^{218}_{84}Po$	Polonium	Radium A	3.05 min	6.00		
$^{214}_{82}Pb$	Lead	Radium B	26.8 min		0.65	0.29
$^{214}_{83}Bi$	Bismuth	Radium C	19.7 min	5.5	3.15	1.8
$^{214}_{84}Po$	Polonium	Radium C'	0.15 ms	7.68		
or						
$^{210}_{81}Tl$	Thallium	Radium C''	1.32 min		1.80	
$^{210}_{82}Pb$	Lead	Radium D	22 yr		0.026	0.047
$^{210}_{83}Bi$	Bismuth	Radium E	5.0 days		1.17	
$^{210}_{84}Po$	Polonium	Radium F	138 days	5.3		0.8
$^{206}_{82}Pb$	Lead	Radium G	Stable			

per second. If this helium were all occluded and retained by the mineral, the ratio of the amount of helium to the amount of uranium would give an estimate of the age of the mineral. Without doubt some of the helium escapes, and this estimate of the age of the mineral would be too low. A determination of the uranium and helium in different kinds of rocks has shown that the ratio of the amount of helium to the amount of uranium is largest in those formations which, from geological considerations, are known to be the oldest. Studies of many kinds of radioactive decay point to an age for the earth of the order of 5×10^9 years.

For determining the ages of artifacts of archaeological interest, radioactive carbon 14 has proved to be a highly useful tool. Radiocarbon decays by beta emission with a half-life of 5,570 years. The $^{14}_{6}C$, produced by the bombardment of atmospheric nitrogen with neutrons, is in equilibrium in the earth's atmosphere and is present in all living matter in equilibrium concentration through exchange with the atmosphere. However, once the matter ceases to live, exchange stops, and the $^{14}_{6}C$ concentration decreases steadily. Libby has established the dates when life ceased in a wide variety of organic materials. This development is an important tool for geology and archaeology.

49.11 FUSION

The alpha particle is composed of four nucleons, two protons, and two neutrons. The mass of a ^4He atom is 4.002603 u, while the sum of the masses of two hydrogen atoms and two neutrons is 4.033095 u. The binding energy is 0.03049 u, or 28.4 MeV.

If we could build helium nuclei from protons and neutrons, we would have a source of tremendous energy. Actually, the probability of getting two protons and two neutrons simultaneously in so small a volume that they would interact to form an alpha particle is small. However, we might well carry out the process by first having a proton and a neutron combine to form a deuteron. Then two deuterons might be brought close enough together to interact. When we do this, so much energy is available that the four particles do not ordinarily stick together. Instead, either a high-energy proton is emitted, leaving us a triton, or a neutron is emitted, leaving us with a ^3He nucleus. The reactions are

$$^2_1H + {}^2_1H \begin{cases} \nearrow {}^3_2He + {}^1_0n + 3.26 \text{ MeV energy} \\ \\ \searrow {}^3_1H + {}^1_1H + 4.03 \text{ MeV energy} \end{cases}$$

The energy released in a nuclear reaction like this is calculated by subtracting the mass of the reaction products from the mass of the interacting particles. One of the many ways in which we can get an alpha particle from these products is to bring a deuteron and a triton together. The resulting reaction is

$$^2_1H + {}^3_1H \longrightarrow {}^4_2He + {}^1_0n + 17.6 \text{ MeV energy}$$

The process of building helium from hydrogen is called *fusion*. As we have seen, 28.4 MeV is released for each helium nucleus built from the building blocks. If we produce fusion reactions on a large scale, we release a colossal amount of energy in a very short time and thus create a great explosion. The hydrogen bomb is an example of a fusion device. If deuterons and tritons are to come in contact with each other in spite of the repulsion between the positive charges, they must have high speeds. In a hydrogen bomb these speeds can be achieved by bringing the particles to an exceedingly high temperature, thereby producing *thermonuclear reactions*. One way to produce temperatures of this magnitude is by a nuclear explosion utilizing fission. An alternative is to pour vast amounts of energy into a plasma of ionized deuterium and tritium. In achieving the temperature needed (tens of millions of kelvins) in a laboratory fusion experiment, the hot plasma is confined by magnetic fields, since any type of material container would be vaporized by contact with the plasma.

The release of energy by means of the combination of nucleons into heavier particles (fusion) is directly analogous to the release of chemical energy by the combination of atoms. In the nuclear case one can release energy by combining a proton and a neutron to form a deuteron. In the analogous chemical situation one can combine an atom of sodium with an atom of chlorine to form sodium chloride. One great difference between these two reactions lies in the amount of energy released per interacting particle. In the case of proton and neutron the energy resulting from the combination is 2.2 MeV per proton, while in the case of the sodium chlorine reaction the energy is about 4 eV per sodium atom.

49.12 FISSION

One can obtain energy from chemical reactions not only by combining atoms of various substances to form molecules but also by breaking very large and complex molecules, such as trinitrotoluene molecules, into smaller and more tightly bound units. Similarly, heavy and complex nuclei can be made to break up with the release of energy.

When any heavy nucleus is highly excited, it may break up into two roughly equal parts. This process is called *fission*. Some heavy nuclei fission readily; others do not. The excitation can be achieved in several ways. For example, gamma rays may induce fission. The most familiar method is by introducing a neutron into a nucleus of uranium 235. When a slow neutron collides with a ^{235}U nucleus, it may be captured. The resulting ^{236}U nucleus breaks up into two roughly equal parts, plus a few additional neutrons. A typical reaction is

$$^{235}_{92}U + ^{1}_{0}n \longrightarrow ^{236}_{92}U \longrightarrow ^{145}_{56}Ba + ^{88}_{36}Kr + 3\,^{1}_{0}n + \text{energy}$$

The mass of the products is less than that of the fissioning nucleus; thus energy is released, on the average about 200 MeV per fission. The excited ^{236}U nucleus may break up in many different ways. Figure 49.4 shows the relative fission yields of the many mass numbers which can be created as fission fragments.

When a heavy nucleus fissions, the product nuclei usually contain several too many neutrons for stability. For example, one common fission product is strontium 95. The heaviest stable isotope of strontium is ^{88}Sr. The radioactive ^{95}Sr nucleus undergoes three successive beta decays and finally reaches stability as niobium 95. Some fission products transmute through beta decays of long half-life and produce serious contamination after a nuclear explosion.

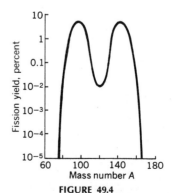

FIGURE 49.4
Fission yield of ^{236}U as it varies with the mass numbers of the fragments.

49.13 CHAIN REACTIONS AND THE REACTOR

Since two to three neutrons are released in a typical fission, two or three other atoms may be fissioned if these neutrons are captured by ^{235}U nuclei. The neutrons released in these fissions can produce four or five more fissions, and a *chain reaction* is set off if a sufficient supply of ^{235}U atoms is present in a small volume. Such a chain reaction does not occur in ordinary uranium because a large fraction of the neutrons are captured by ^{238}U nuclei, which are 140 times as abundant in natural uranium as ^{235}U nuclei. Few of the ^{238}U nuclei fission.

To produce the first nuclear weapons, ^{235}U was separated from ^{238}U so a large fraction of the neutrons emitted by fissioning nuclei would be captured by ^{235}U nuclei and thereby produce additional fissions. However, even pure ^{235}U does not support a chain reaction unless an amount known as a *critical mass* is present, because too large a fraction of the neutrons escapes from the surface of a subcritical mass. If two almost critical masses are brought together, the system becomes critical, and an explosion occurs.

The neutrons released in fission are *fast*, with energies of several mega-electronvolts. They are quickly slowed down by collisions with nuclei. ^{235}U does not capture fast neutrons readily, but it captures slow neutrons (energy less than 1 eV) very effectively. On the other hand, ^{238}U captures neutrons of intermediate energies more readily than slow ones. If the neutrons from fissions in natural uranium are slowed down in some other material so the ^{238}U does not capture them, a sustained reaction using natural uranium can be maintained in a nuclear *reactor* (or *pile*). Figure 49.5 shows a reactor in which carbon is used to slow down or *moderate* the neutrons. (Heavy water, containing deuterium, is another material which is an excellent *moderator*.) Fissions produce fast neutrons which enter the carbon, where they have many collisions and slow down to thermal velocities before they reach another uranium slug. The probability that a slow neutron will be captured by a ^{235}U nucleus is much greater

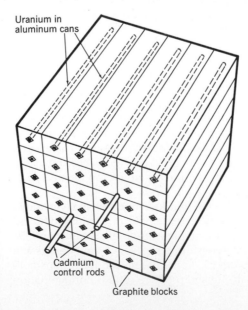

FIGURE 49.5
Uranium-graphite reactor, in which uranium slugs are "canned" in aluminum; cadmium rods are used to control the power level at which the reactor operates. In newer reactors, uranium oxide cylinders are packed in zirconium tubes.

Uranium in aluminum cans

Cadmium control rods

Graphite blocks

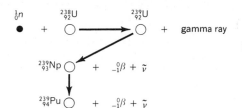

FIGURE 49.6
Production of plutonium from ^{238}U.

than the probability it will be captured by a ^{238}U nucleus, even though there are 140 times as many of the latter. Cadmium control rods are moved into or out of the reactor to regulate the power level of the pile and keep it from blowing up. Cadmium 113 is an excellent capturer of slow neutrons. A small fraction (<1 percent) of the neutrons are not emitted until several seconds have elapsed after a fission has occurred. These *delayed neutrons* make the orderly control of a reactor feasible.

Each fission in a reactor releases about 200 MeV of energy, and the total energy transformed from mass to thermal energy in a large reactor is great. Nuclear energy is being used for generating electric power, for heating, for propelling submarines, and for nuclear weapons. As oil, natural gas, and coal reserves are consumed, it is possible that more and more of our total energy requirements may be supplied from nuclear sources. However, the problem of disposing of the highly radioactive wastes from nuclear plants is a serious and difficult one which may limit our dependence on nuclear energy.

49.14 PLUTONIUM

When neutrons are captured in ^{238}U, fission sometimes occurs, but usually the ^{239}U formed does not split up. However, it does emit a beta particle (half-life 23 min) to become neptunium 239 (Fig. 49.6), a beta emitter with a half-life of 2.3 days which becomes plutonium 239. ^{239}Pu is an alpha emitter with a half-life of 24,400 years. This plutonium fissions readily. Practically no plutonium exists in natural materials. Substantial quantities have been produced for use in nuclear weapons. It is produced in reactors which use natural uranium as fuel. As a result, any country with nuclear reactors and suitable plants for separating the plutonium from the reactor waste has the potential of producing nuclear weapons. The plutonium can also be used as a reactor fuel.

QUESTIONS

1. The two stable isotopes of neon have mass numbers of 20 and 22. How many neutrons does each isotope have?
2. How many protons are associated with the nucleus $^{226}_{88}Ra$? How many neutrons are associated with the nucleus?
3. How do atomic number and atomic weight change when a nucleus emits (*a*) an alpha particle, (*b*) a beta particle, (*c*) a gamma ray, and (*d*) a positron?

4. Why is ^3_2He stable rather than ^3_1H? Why is $^{17}_8O$ stable rather than $^{17}_9F$? What factors determine the numbers of protons and neutrons in a stable nucleus of mass number A?
5. Why do the atomic weights of many elements differ greatly from integer values? Are the binding energies of the nucleons sufficient to explain this fact?
6. What is meant by the binding energy of a nucleus? How do the binding energies of nuclei vary with atomic number? The binding energy per particle?

7. What are the properties of alpha, beta, and gamma rays?
8. Why did early observations of beta decay seem to be in conflict with one of the conservation laws? How has this conflict been resolved?
9. A neutron is not charged, but it does have a magnetic dipole moment associated with its spin. How can this be explained on the basis of classical electromagnetism?
10. Which of the following properties of neutrinos and antineutrinos are not zero? (a) Kinetic energy; (b) rest mass; (c) linear momentum; (d) angular momentum.
11. Why are fission products ordinarily radioactive while fusion products are not?
12. (a) A nucleus of thorium 232 ($^{232}_{90}$Th) eventually becomes lead 208 ($^{208}_{82}$Pb). In the series of alpha and beta decays by which this occurs, how many alpha and beta particles are emitted? (b) After a series of alpha and beta decays plutonium 239 ($^{239}_{94}$Pu) becomes lead 207 ($^{207}_{82}$Pb). How many alpha and beta particles are emitted in the complete decay scheme?
13. The half-life of radium 226 is 1,622 years. There is considerable evidence that the age of the earth may be 3 million times as great. If this is true, how can you explain the natural occurrence of radium in moderate abundance?
14. In a typical fission reaction a neutron is captured by a $^{235}_{92}$U nucleus, and three neutrons and two fission fragments are formed. If one of the fragments is a $^{92}_{38}$Sr nucleus, what must the other fragment be?
15. What is the function of graphite in a typical reactor?
16. In a certain reactor there are three free neutrons released per fission. If half the neutrons absorbed in the reactor are captured by nonfissioning nuclei, what is the smallest fraction of the neutrons which can escape and still have the reactor critical?

PROBLEMS

1. An oxygen 16 atom is assembled from eight hydrogen atoms and eight neutrons. Find the energy released and the average binding energy per nucleon. *Ans.* 127.6 MeV; 7.98 MeV/nucleon

2. How much energy would be released if a carbon 12 atom were assembled from six hydrogen atoms and six neutrons? What is the average binding energy per nucleon?
3. Find the binding energy and the binding energy per nucleon for $^{40}_{20}$Ca, which has a mass of 39.9626 u. *Ans.* 342 MeV; 8.55 MeV/nucleon
4. Find the binding energy and the binding energy per nucleon for tin 120.
5. Find the minimum energy required for the photodisintegration of an alpha particle into (a) ^{3}He + $^1_0$n, (b) ^{3}H + ^{1}H, and (c) ^{2}H + ^{2}H. *Ans.* (a) 20.6 MeV; (b) 19.8 MeV; (c) 23.8 MeV
6. In Sec. 49.11 three fusion reactions involving interactions of deuterium with deuterium and tritium are written. Check the energy released in each case by finding the mass converted to energy in the reaction.
7. Find the maximum energy of the beta particles from the reaction $^1_0n \longrightarrow {}^1_1p + {}^0_{-1}\beta + \bar{\nu}$. If the half-life of free neutrons is 636 s, what fraction of a sample of free neutrons remains after 42.4 min? *Ans.* 783 keV; one-sixteenth
8. Find the binding energy and the binding energy per nucleon for sulfur 32.
9. Find the binding energy and the binding energy per nucleon for tritium. The maximum energy of beta rays emitted in the decay of tritium to $^3_2$He is 18.6 keV. Find the mass difference between $^3_1$H and $^3_2$He. *Ans.* 8.48 MeV; 2.83 MeV/nucleon; 20 μu
10. (a) A radioactive nucleus has a half-life of 20 min. What fraction of an initial sample has decayed after 2 h? (b) At the end of 12 days one-eighth of a sample of radioactive material remains. What is the half-life?
11. Carbon 14 decays by beta emission to nitrogen 14 by beta decay with a half-life of 5,570 years. A gram of carbon from a living tree undergoes 720 disintegrations per hour. If 1 g of carbon from a wooden relic undergoes 180 disintegrations per hour, approximately what is the age of the relic? (Assume that when the tree was cut, its radioactivity corresponded to that of a living tree.) *Ans.* 11,000 to 12,000 years
12. When a ^{238}U nucleus captures a slow neutron, uranium 239, with a half-life of 23.5 min, is produced. If a sample of radioactive material from a nuclear reactor contains 10 g of ^{239}U what mass of this isotope remains at the end of 47 min? 94 min?
13. Strontium 90 is one of the radioactive nuclei arising in the decay of fission products. It is a beta emitter with a half-life of 28 years. It has been found in milk from cows grazing on grass on which fission products have been precipitated. How many years must one wait before

natural decay reduces this radioactive species to less than 1 percent of its initial value?

Ans. About 186 years

14. (*a*) The half-life of sodium 24 is 15 h. What fraction of a sample of ^{24}Na remains after 30 h? After 75 h? (*b*) The half-life of iodine 131 is 8.1 days. What fraction of a given sample of ^{131}I remains after 24.3 days? After 40.5 days?

15. Polonium 212 emits 8.78-MeV alpha particles. Find the speed of the alpha particles and the mass of the $^{212}_{84}$Po atom. (The mass of $^{208}_{82}$Pb is given in Table 49.1.)

Ans. 2.06×10^7 m/s; 211.98867 u

16. How close can an 8.78-MeV alpha particle approach a uranium nucleus assuming the latter remains at rest in its crystal?

17. It is estimated that the energy released in the nuclear bomb explosion at Hiroshima was about 7.6×10^{13} J, equivalent to 20,000 tons of TNT. If an average of 200 MeV was released per fission, and if all fissions were by neutron capture in ^{235}U, find the number of ^{235}U atoms fissioned and the mass of ^{235}U consumed. If 20 percent of the ^{235}U atoms fissioned, what mass of ^{235}U was needed for the bomb? *Ans.* 2.37×10^{24}; 0.93 kg; 4.63 kg

18. When a slow neutron is captured by a ^{235}U nucleus, a fission releasing 200 MeV results. How much mass disappears? What fraction of the mass of the reacting particles is converted into energy?

19. The heat of combustion of coal (assume pure carbon) is about 3×10^7 J/kg. (*a*) When 1 kg of coal burns, what mass is converted to energy? (*b*) How many electronvolts are released per atom of carbon burned? (*c*) If one-tenth the mass of carbon were annihilated, how many joules could be obtained per kilogram?

Ans. (*a*) 3.3×10^{-10} kg; (*b*) 3.7 eV/atom; (*c*) 9×10^{15} J

20. A nuclear reactor converts 10 mg of mass to energy each day. Calculate the electric power output if 9 percent of the nuclear power is converted into electric power. Compute the power output in megawatts. If the average fission yields 200 MeV, how many nuclei fission each second?

21. How many fissions per day are required for a reactor to develop electric power at the rate of 20 MW if 12 percent of the energy released in fission is converted into electric energy?

Ans. 4.5×10^{23}

22. How close can a 4.18-MeV alpha particle from the decay of ^{238}U approach a lead nucleus if the latter remains at rest?

23. Uranium 238 decays to thorium 234 by the emission of an alpha particle of energy 4.18 MeV. Assuming that essentially all the energy released in this decay goes to the alpha particle, calculate the mass of thorium 234. What is the speed of the alpha particle?

Ans. 234.04368 u; 1.42×10^7 m/s

24. A ^{222}Rn nucleus emits an alpha particle with an energy of 5.49 MeV. Find the speed of the alpha particle and the recoil speed of the ^{218}Po nucleus.

25. A fission fragment of 150 u mass has an energy of 80 MeV. Find its speed and the temperature at which molecules of this mass would have an average energy of 80 MeV.

Ans. 1.01×10^7 m/s; 6.19×10^{11} K

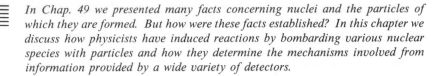

50

NUCLEAR REACTIONS

In Chap. 49 we presented many facts concerning nuclei and the particles of which they are formed. But how were these facts established? In this chapter we discuss how physicists have induced reactions by bombarding various nuclear species with particles and how they determine the mechanisms involved from information provided by a wide variety of detectors.

50.1 THE FIRST ARTIFICIAL TRANSMUTATION

The discovery of radioactivity in 1896 stimulated research activity which led to an understanding of both the decay schemes of the heavy elements such as uranium and the properties of the alpha, beta, and gamma rays arising from these natural transmutations. In 1919 Rutherford found that when alpha particles passed through air, a penetrating charged particle was produced which had a range as great as 400 mm, although the range of the incident alpha particles was only 70 mm. The more penetrating particles were not produced when the alpha particles passed through oxygen or carbon dioxide, but they were observed whenever nitrogen was present. Rutherford showed that the penetrating particles were high-energy protons, produced by the capture of alpha particles by nitrogen nuclei (Fig. 50.1).

When an alpha particle is captured by a nitrogen nucleus, a highly excited fluorine 18 nucleus is produced. This excited nucleus has too much energy to remain together; it breaks up by ejecting one of the protons, leaving a residual $^{17}_{8}O$ nucleus. We write the reaction as

$$^4_2He + {}^{14}_7N \longrightarrow {}^{18}_9F^* \longrightarrow {}^{17}_8O + {}^1_1H + energy$$

Here the asteisk indicates that the ^{18}F nucleus is excited.

Rutherford's discovery of artificial transmutation suggested that physicists could change one element into another by shooting high-speed particles into nuclei. In Rutherford's day alpha particles from radioactive materials were the only convenient high-energy projectiles available. Only high-energy charged particles can enter nuclei, because positive charges repel each other; these repulsive forces become exceedingly great before the two particles are close enough for the attractive nuclear forces to become dominant. Rutherford's discovery was a great stimulus toward the development of accelerators to produce high-energy charged particles. Until such high-voltage machines could be developed, there was always the possibility of using alpha particles from radioactive materials as bombarding projectiles. Here again Rutherford's discovery provided a major stimulus.

50.2 THE DISCOVERY OF THE NEUTRON

In 1930 Bothe and Becker observed that several of the lighter elements emit a very penetrating radiation when bombarded by alpha particles from polonium. They thought this radiation consisted of gamma rays, since it was not affected by magnetic fields and was therefore uncharged. The radiation was so penetrating that when alpha particles bombarded beryllium, it required 20 mm of lead to cut the intensity in half.

Two years later the Joliots found that these penetrating rays were fairly readily absorbed in paraffin, water, or cellophane. They concluded that the penetrating radiation was absorbed through some reaction which resulted in the ejection of protons from hydrogenous materials.

Later that same year Chadwick repeated the Joliot experiment and measured the energy of the ejected protons. He concluded that this energy was much too high to have been received in any sort of gamma-ray reaction. Twelve years earlier Rutherford had suggested that there might be a neutral particle with approximately the mass of the proton. If the penetrating radiation consisted of high-speed particles of this kind, they should produce not only recoil protons but also recoils in other nuclei. It turned out that Feather had already published data on the recoil velocities of nitrogen nuclei when they were bombarded by the mysterious particles. By applying the laws of conservation of energy and of momentum to what he assumed were elastic collisions between these particles and target nuclei, Chadwick was able to show that both his own data on proton recoils and those of Feather were consistent with the idea that the penetrating radiation consisted of neutral particles of mass slightly greater than the mass of the proton. These particles are called *neutrons*. The reaction by which neutrons are produced when beryllium is bombarded with alpha particles (Fig. 50.2) is

$$\,^{9}_{4}\text{Be} + \,^{4}_{2}\text{He} \longrightarrow \,^{13}_{6}\text{O}^{*} \longrightarrow \,^{12}_{6}\text{C} + \,^{1}_{0}n + Q$$

where Q represents the energy released through the conversion of mass.

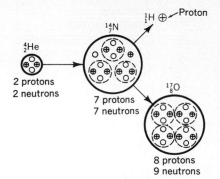

FIGURE 50.1
A nitrogen 14 nucleus captures a bombarding alpha particle to yield a proton and an oxygen 17 nucleus.

50.3 REACTIONS INDUCED BY ACCELERATED PARTICLES

Soon after Rutherford's discovery of artificial transmutation, physicists in several countries began to build accelerators to produce high-energy protons. The first device which was successful in accelerating charged particles artificially to produce nuclear reactions was put into operation by Cockcroft and Walton in England in 1930. Their accelerator produced a beam of protons of 0.5 MeV energy by utilizing high-voltage transformers, rectifiers, and capacitors.

Cockcroft and Walton found that when lithium was bombarded with protons of 150 keV energy, alpha particles were emitted with energies of about 8.6 MeV (Fig. 50.3). The following reaction occurred:

$$\,^{7}_{3}\text{Li} + \,^{1}_{1}\text{H} \longrightarrow \,^{8}_{4}\text{Be}^{*} \longrightarrow \,^{4}_{2}\text{He} + \,^{4}_{2}\text{He} + 17.3 \text{ MeV energy}$$

The energy released is shared equally by the alpha particles, since this is the only way momentum can be conserved.

The source of the energy released in this and other nuclear reactions is the conversion of mass into energy. The masses of the lithium and hydrogen atoms add to 8.02382 u (Table 49.1), while the two helium atoms have a total mass of 8.00521 u. The decrease in mass is 0.01861 u, which corresponds to 17.3 MeV.

Shortly after Cockcroft and Walton made their discovery, other types of accelerators were put into operation, some of which are described in Secs. 31.8 and 50.5.

When a beam of high-energy particles (electrons, protons, deuterons, alpha particles, or other ions) bombards nuclei, a wide variety of reactions take place in which the bombarding particle is captured to produce a radioactive "compound" nucleus, which then emits one or more particles before it reaches a stable state. So long as the bombarding particles

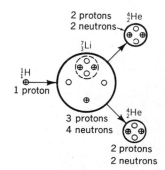

FIGURE 50.2
A neutron may be ejected when an alpha particle is captured by a beryllium 9 nucleus, leaving a carbon 12 nucleus.

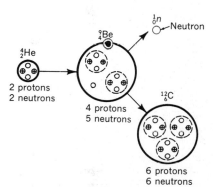

FIGURE 50.3
A lithium 7 nucleus captures a proton, to produce a beryllium 8 nucleus, which promptly breaks into two alpha particles.

have energies less than about 100 MeV, the reactions can be described in terms of protons, neutrons, electrons, positrons, neutrinos, and antineutrinos, but when the bombarding particles have energies of hundreds and thousands of megaelectronvolts, new and strange particles are created—a whole new science of particle physics is born. Before introducing these particles, we consider some of the reactions induced by nuclear projectiles of moderately high energy and describe some of the detection devices which physicists use to study reactions.

50.4 ARTIFICIAL TRANSMUTATIONS

The development of high-voltage accelerators made it possible to bombard atomic nuclei with a variety of swiftly moving projectiles. When an accelerated particle is captured by a nucleus, the resulting system of nucleons is initially in a state of high excitation. It may get rid of its excess energy by emitting one or more particles. Hundreds of artificial transmutations have been studied. Table 50.1 lists some typical reactions. Figures 50.4 and 50.5 show two reactions in schematic form. A reaction in which a proton is captured and a neutron is emitted is called a (p, n) reaction, etc. In this shorthand the entering particle appears first, and the emitted particle after the comma.

If a target is bombarded by protons (or some other particles), it is common to find that protons of one energy are strongly captured while protons of slightly higher or lower energy are rejected. We say that such capture reactions show *resonances*. A proton is likely to be captured if it brings in just the right energy and angular momentum to form a compound nucleus in one of its allowed energy states. A proton with more or less energy than this required amount is not captured.

Many nuclei created by transmutations are unstable and eventually decay. In some cases these unstable nuclei have relatively long half-lives. When a proton is captured by carbon 12, nitrogen 13 is formed. This nitrogen nucleus is unstable, with a half-life of 10.1 min. It decays to ^{13}C by the emission of a positron. When tellurium 130 is bombarded with neutrons, ^{131}Te is formed but decays by electron emission to iodine 131. ^{131}I is also a beta emitter; its half-life is 8 days. This radioactive isotope is used in treating hyperthyroid conditions.

TABLE 50.1
Some Typical Nuclear Transmutations

Kind	Example
(α,p)	$^{27}_{13}\text{Al} + ^{4}_{2}\text{He} \longrightarrow ^{1}_{1}\text{H} + ^{30}_{14}\text{Si}$
(α,n)	$^{27}_{13}\text{Al} + ^{4}_{2}\text{He} \longrightarrow ^{1}_{0}n + ^{30}_{15}\text{P}$
(p,α)	$^{19}_{9}\text{F} + ^{1}_{1}\text{H} \longrightarrow ^{4}_{2}\text{He} + ^{16}_{8}\text{O}$
(p,n)	$^{11}_{5}\text{B} + ^{1}_{1}\text{H} \longrightarrow ^{1}_{0}n + ^{11}_{6}\text{C}$
(p,γ)	$^{31}_{15}\text{P} + ^{1}_{1}\text{H} \longrightarrow ^{32}_{16}\text{S} + \gamma$
(d,α)	$^{27}_{13}\text{Al} + ^{2}_{1}\text{H} \longrightarrow ^{4}_{2}\text{He} + ^{25}_{12}\text{Mg}$
(d,p)	$^{14}_{7}\text{N} + ^{2}_{1}\text{H} \longrightarrow ^{1}_{1}\text{H} + ^{15}_{7}\text{N}$
(d,n)	$^{25}_{12}\text{Mg} + ^{2}_{1}\text{H} \longrightarrow ^{1}_{0}n + ^{26}_{13}\text{Al}$
$(d,2n)$	$^{3}_{1}\text{H} + ^{2}_{1}\text{H} \longrightarrow 2^{1}_{0}n + ^{3}_{2}\text{He}$
(n,α)	$^{27}_{13}\text{Al} + ^{1}_{0}n \longrightarrow ^{4}_{2}\text{He} + ^{24}_{11}\text{Na}$
(n,p)	$^{106}_{48}\text{Cd} + ^{1}_{0}n \longrightarrow ^{1}_{1}\text{H} + ^{106}_{47}\text{Ag}$
$(n,2n)$	$^{107}_{47}\text{Ag} + ^{1}_{0}n \longrightarrow 2^{1}_{0}n + ^{106}_{47}\text{Ag}$
(n,γ)	$^{113}_{48}\text{Cd} + ^{1}_{0}n \longrightarrow ^{114}_{48}\text{Cd} + \gamma$
(γ,n)	$^{9}_{4}\text{Be} + \gamma \longrightarrow ^{1}_{0}n + ^{8}_{4}\text{Be}$
(γ,p)	$^{27}_{13}\text{Al} + \gamma \longrightarrow ^{1}_{1}\text{H} + ^{26}_{12}\text{Mg}$

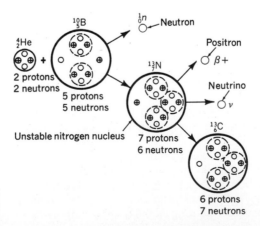

FIGURE 50.4
A boron 10 nucleus captures an alpha particle, producing a neutron and a nitrogen 13 nucleus; the $^{13}_{7}\text{N}$ nucleus is unstable and decays to carbon 13 by the emission of a positron and a neutrino.

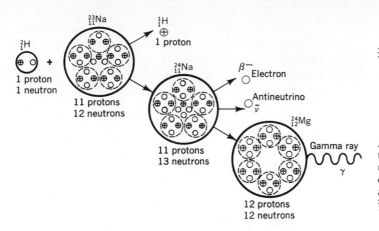

FIGURE 50.5
A sodium 23 nucleus captures a deuteron, emitting a proton and leaving an unstable sodium 24 nucleus, which decays to magnesium 24 by the emission of an electron and an antineutrino; the $^{24}_{12}$Mg nucleus is left in an excited state and emits a gamma photon.

50.5 CHARGED-PARTICLE ACCELERATORS

In both nuclear and atomic physics we use beams of high-energy ions (in particular, protons and deuterons) for studying the reactions induced in various targets. A variety of particle accelerators have been developed, many of which depend on magnetic fields to constrain the particle to move in a circular orbit while electric fields produce the accelerations. Although particle accelerators are basic tools of nuclear physics, the underlying principles are those which govern the interactions of ions with electric and magnetic fields.

The Cyclotron The first major accelerator based on circular orbits in a magnetic field was the cyclotron, developed in 1932 by Lawrence and Livingston. The key parts of the cyclotron are a large electromagnet, which provides a constant magnetic field over a large area, and an oscillating electric field between two semicircular hollow elements, called *dees* because they have the shape of a D. Ions are produced at the center of the magnetic field between the two dees, as shown in Fig. 50.6. A charged particle which is accelerated toward the lower dee gains energy as it crosses the gap between the dees. Once in the lower dee its path is circular, since

$$qvB = \frac{mv^2}{r} \tag{50.1}$$

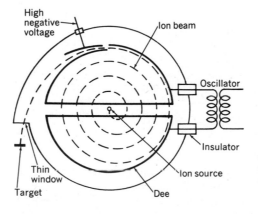

FIGURE 50.6
Cyclotron, showing the path of a positively charged particle in the vacuum chamber. There is a magnetic field into the paper.

Soon the particle finds itself in the region between the two dees once more. If now the upper dee is negative, and the lower dee positive, the particle is accelerated across the gap and enters the upper dee with a higher velocity, where it traverses a semicircular arc of greater radius. When it again reaches the gap between the dees, the electric field has reversed, and the particle is accelerated downward. At each passing of the gap it gains energy, and the radius of its path in the magnetic field becomes greater, until eventually the particle has a very high energy and its orbit approaches the circumference of the magnetic poles. It is then removed from the cyclotron and directed against a target. The time $T_{1/2}$ required for the particle to make the half circle in one of the dees is

$$T_{1/2} = \frac{\pi r}{v} \qquad (50.2)$$

In this time the accelerating field should be reversed; therefore this is the time for one-half cycle of the alternating potential difference applied to the dees. The number of complete revolutions made per second is

$$\nu = \frac{v}{2\pi r} \qquad (50.3)$$

Substituting the ratio v/r from Eq. (50.1), we obtain

$$\nu = \frac{Bq}{2\pi m} \qquad (50.4)$$

If we apply the constant frequency given by Eq. (50.4) to the dees, ions are accelerated each time they pass the gap.

The time required for a charged particle to make half a revolution in the cyclotron is independent of radius only so long as the mass of the particle is constant. As we saw in Sec. 42.8, the mass of any particle varies rapidly with velocity when v approaches c, the speed of light. If the cyclotron is used to accelerate a particle to a velocity which is more than a few percent of the speed of light, the mass of the particle increases with time. By Eq. (50.4), the accelerating field can be kept in step with the particles only by decreasing the frequency as the mass increases. In very large cyclotrons the speed of the particles becomes great enough so that relativistic effects are important. Then the frequency of the cyclotron must be varied during an acceleration cycle. A cyclotron operated in such a way is said to be *frequency-modulated* and is sometimes called a *synchrocyclotron*.

The Betatron   The betatron is used to produce high-speed electrons. It consists essentially of a large electromagnet between the poles of which is placed an evacuated tube in the form of a doughnut (Fig. 50.7). Alternating current is applied to the coils of the electromagnet, and the accelerating potential difference is induced by the changing magnetic field, which acts on the electrons in the vacuum tube as if the tube were the secondary of a transformer. The electrons previously injected into the tube are whirled around in circular paths. As the electrons gain energy, the magnetic field is increased at such a rate that it is just strong enough to keep the electrons rotating in the same circular path. While the magnetic field builds up to its maximum value, the electrons travel around the vacuum tube many thousand times and acquire a velocity which closely approaches the speed of light.

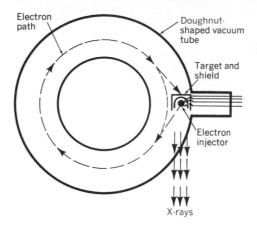

FIGURE 50.7
Doughnut-shaped vacuum tube of a betatron with magnetic field into the plane of the page.

The Synchrotron In a synchrotron particles are accelerated at essentially constant radius in a doughnut-shaped evacuated tube. The particles are confined to the tube by a magnetic field and are accelerated by a radio-frequency electric field applied across one or more gaps, as in a cyclotron. (Figure 50.8 shows the accelerator tube and a portion of the bending magnets in the tunnel of the Brookhaven National Laboratory 33-GeV alternating-gradient synchrotron which began operation in 1961.) As the particle gains energy, the magnetic field must be increased since qvB must be equal to mv^2/r, the centripetal acceleration. As the particle gains speed, the time to traverse the circumference of the orbit decreases, so the frequency of the accelerating electric field must be increased. In

FIGURE 50.8
Tunnel of the 33-GeV alternating-gradient synchrotron where it joins the 50-MeV linear accelerator which injects protons into the "doughnut." At the right is a small portion of the magnet ring providing the guide field for the 880-m-circumference ring. (*Brookhaven National Laboratory*.)

order to keep this frequency change and the change in the magnetic field in reasonable bounds, it is customary to inject the particles with an energy sufficient to make the speed v only slightly less than the speed of light c. Under these circumstances the acceleration results primarily in an increase of the mass of the accelerated particle, while the speed approaches only slightly closer to c. Synchrotons have ordinarily been designed to accelerate either electrons or protons; several successful synchrotons are in operation at laboratories around the world.

At the Fermi National Laboratory (Fig. 50.9) a proton synchrotron with an orbit radius of 1 km was placed in service in 1972; it is capable of producing 500-GeV protons. A linear accelerator is used to inject 8-GeV protons into the synchrotron, which accelerates them for up to 4 s; each second the protons go around the 6.5-km circumference almost 50,000 times and a pulse of about 2×10^{13} protons is produced every 10 s. (The reason the circumference of the ring exceeds $2\pi r$ is that there are six circular sectors separated by 50-m straight sections.) There is an internal research area at one of the straight sections where targets can be bombarded by the proton beam. Alternatively, the protons may emerge from the ring near the Central Laboratory (which appears at the lower left side of the picture and consists of twin 16-floor towers) and be delivered by evacuated tubes to one of the three experimental areas: one for studying direct proton-induced reactions, a second for neutrino experiments, and a third for research with mesons. It is nearly 3 km from the main ring to the neutrino facility.

A second ring using superconducting magnets is being constructed in the same tunnel occupied by the main ring. Because these magnets may have a field as great as 4.5 T (double that of the conventional magnets), the new ring is expected to produce a beam of protons with an energy of about 1 teraelectronvolt (10^{12} eV).

50.6 THE DETECTION OF NUCLEAR REACTIONS

In nuclear reactions the target nuclei, the bombarding particle, and the resulting products are far too small to see and too light to weigh on the

FIGURE 50.9

An aerial view of the Fermi National Accelerator Laboratory, Batavia, Illinois. The large circle is the main ring; it has a radius of 1 km. Protons extracted tangentially from the main ring at a point near the 16-story twin-towered Central Laboratory can be delivered to the three experimental areas at the left. (*Fermi National Accelerator Laboratory.*)

Meson area

Neutrino area

Proton area

Central laboratory

most sensitive balance. The problem of studying a reaction is a difficult one, since all judgments about what has happened must be based on indirect evidence. However, scientists have developed a number of ingenious methods for studying nuclear reactions and detecting the products of the reactions.

Nuclear Emulsions The first detector of nuclear reactions was the photographic plate. Gamma rays striking such a plate produce blackening, just as visible light and x-rays do. By using an exceedingly fine-grained emulsion, the path of a high-energy proton, alpha particle, or electron can be studied under a microscope.

The Cloud Chamber The Wilson cloud chamber consists of a cylinder with a glass top and a movable piston at the bottom (Fig. 50.10). The volume of the chamber is filled with air and saturated water vapor or alcohol. When the piston is moved downward quickly, the mixture expands adiabatically, the temperature falls, and the formerly saturated vapor becomes supersaturated. Vapor molecules condense on any tiny particle which is present to serve as nucleus for a droplet. In particular, droplets form readily on charged ions. If a charged particle is moving through the chamber when the expansion occurs, droplets form on the ions produced by the passage of this charged particle, and a *track* is produced. Figure 50.11 shows cloud-chamber pictures of the breakup of nuclei struck by high-energy neutrons. The ionized fragments leave visible tracks.

When a strong magnetic field is impressed on a cloud chamber, charged particles moving across the field are deflected, producing curved paths. From the radius of curvature of a path, its length, and the number of ion pairs formed per unit path length, it is often possible to determine the mass and charge of the particle responsible for the track. It is not practical to build cloud chambers large enough to show the entire path

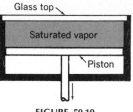

FIGURE 50.10
Wilson cloud chamber.

FIGURE 50.11
Cloud-chamber photographs of (*a*) carbon and oxygen atoms breaking apart when struck by high-energy neutrons and (*b*) oxygen stars produced by 90-MeV neutrons in a 1.4-T magnetic field. (*The University of California.*)

(a)

(b)

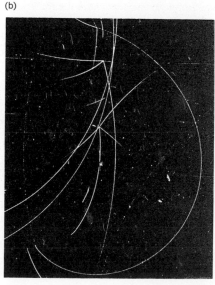

of an extremely high-energy particle. To observe the paths and the reactions induced by such particles, photographic emulsions or bubble chambers are used.

The Bubble Chamber In 1952 Glaser invented the bubble chamber, in which a charged particle moves through a superheated liquid rather than a supersaturated gas. The liquid is maintained at a temperature just under its boiling point at an elevated pressure. When the path of a particle is desired, the pressure is suddenly reduced. Ions produced by the passing particles offer nuclei on which bubbles form preferentially before general boiling begins. The bubble tracks are photographed (Fig. 50.12). Bubble chambers containing liquid hydrogen have proved to be peerless detectors of many high-energy reactions for which the particle tracks are too long to be shown completely in a cloud chamber of reasonable size. A charged particle loses its energy in a much shorter distance in a liquid than in a gas because it interacts with many more atoms per unit path length.

The Geiger-Müller Counter A common detector of beta and gamma rays is the Geiger-Müller counter. An elementary counter can be made by stretching an insulated wire along the axis of a conducting metal cylinder (Fig. 50.13). This chamber is evacuated and then filled to a pressure of about 100 mmHg with a mixture of argon and methyl alcohol. The central wire is made about 1 kV positive relative to the metal cylinder. When an ionizing particle passes through this counter, electrons are ejected from atoms. The electrons move to the central wire, while the positive ions move outward. The electric field near the wire is sufficiently great for the gas to break down, and an avalanche results, during which a large instantaneous current is drawn. As the current rises, the potential difference across the Geiger counter falls, while that across the resistance R (Fig. 50.13) rises. Soon the potential difference across the Geiger counter becomes too low to maintain the discharge. The current drops to zero, and the potential difference across the counter again rises to 1 kV, so that the counter is once more ready to detect an incident particle. Each time the gas breaks down in the Geiger counter, the potential difference across the resistor R rises sharply; this rise can be made to trigger a scaling circuit, which automatically counts the pulses.

(a)

(b)

FIGURE 50.12
(a) Bubble-chamber photogram showing the path of a π^+ meson entering at the lower left and curving to top center, where it decays into a muon (and a mu neutrino, which leaves no track). The μ^+ moves a short distance to the right and down before decaying to a positron, a neutrino, and a mu antineutrino. The positron leaves an almost circular path, while the other particles leave no tracks. There is a magnetic field B perpendicular to the plane of the paper. (b) Explanatory sketch. (*Lawrence Radiation Laboratory.*)

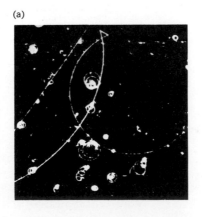

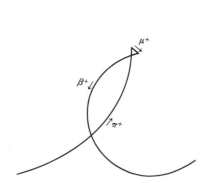

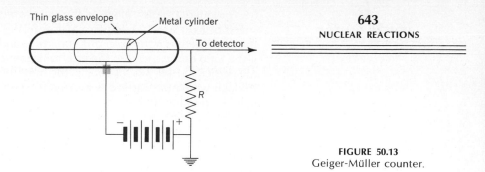

FIGURE 50.13
Geiger-Müller counter.

The Scintillation Detector A detector which is the modern descendent of the scintillation screen of Sec. 49.6 is the scintillation counter. When gamma rays fall on a crystal of sodium iodide with a little thallium impurity, absorption of the gamma rays results in the fluorescent emission of ultraviolet and visible light. In the scintillation detector shown in Fig. 50.14, the gamma-ray energy is absorbed in the crystal, producing light which falls on the photosensitive surface of the cathode of an *electron multiplier tube*. Electrons are ejected by the photoelectric effect and accelerated by a potential difference of the order of 100 V to electrode 1, which is made of a material which has the property of copious secondary emission. When they strike this electrode, several secondary electrons are ejected for each incident electron. The electrons from electrode 1 are accelerated to electrode 2, where a further multiplication is produced, and so forth. At the final collector an electron pulse is collected which may be 1 million times as great as that from the photocathode. Each time a gamma ray strikes and is absorbed in the crystal, a charge pulse is delivered to the anode. The charge collected is a measure of the energy of the gamma ray. The scintillation counter has the great advantage over the Geiger counter of not only telling when a gamma ray has been absorbed in the crystal but also yielding a charge pulse which is a measure of the energy.

Cerenkov Detectors When a charged particle passes through a medium with a speed greater than the speed of light in the medium, an electromagnetic "shock wave" of light is produced which is closely analogous to acoustic shock waves (Sec. 22.10). For example, electrons with a speed of

FIGURE 50.14
Scintillation detector with photomultiplier tube.

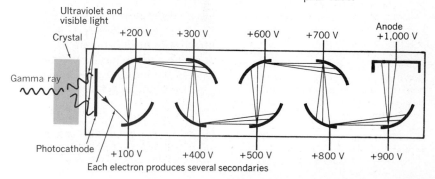

0.9c produce a bluish glow as they pass through water, in which the speed of light is $c/1.33 = 0.75c$. Just as in the acoustic case, the sine of the angle between the path of the particle and the wavefront generated is the ratio of the speed of the electromagnetic wave c/n to the speed of the particle. (Here n is, of course, the index of refraction of the material through which the particle is moving.) Measurement of this angle permits the physicist to determine the speed of the particle generating the radiation, called *Cerenkov radiation* in honor of the physicist who made an early investigation of the phenomenon.

50.7 COSMIC RAYS

There are many accelerators which give charged particles (chiefly electrons and protons) energies in the range from 300 MeV to hundreds of gigaelectronvolts (GeV; billion electronvolts is sometime used in the United States). In addition to these accelerators, there is another source of high-energy particles which is of great importance in studying nuclear reactions. This source is *cosmic radiation*. The primary cosmic rays consist of charged particles, mostly (85 percent) protons, with energies as great as 10^{20} eV. Also present are alpha particles, as well as some primary (and many secondary) electrons and positive nuclei with atomic numbers from 3 to 26. Few primary cosmic rays reach the surface of the earth. As they pass through the upper atmosphere, they collide with atmospheric nuclei and blast them apart. At the earth's surface the progeny of the primaries are high-energy photons, electrons, positrons, and other particles (Sec. 50.8). We call these particles *secondary cosmic rays*. A single proton may initiate a "shower" containing thousands of these secondaries.

The cosmic radiation which reaches the earth changes drastically in time, both in intensity and in energy spectrum. The major variations, which occur chiefly in particles with energies below 50 GeV, are closely related to solar magnetic activity. The particles are accelerated by magnetic fields in interstellar space and by accelerating fields associated with disturbances such as solar flares on the sun and other stars.

50.8 PIONS AND MUONS

When cosmic-ray particles collide with nuclei in the upper atmosphere, particles called *pions* (or π mesons) are often produced. Pions, which can be produced equally well by beams from accelerators, may be positive, negative, or neutral. A charged pion has a mass 273 times that of an electron, while the neutral pion is less massive, about 264 electron masses. Neutral π mesons disappear with the emission of two photons, which share the 135 MeV of energy released. Negative pions are strongly attracted to nuclei and are captured quickly. Positive pions are repelled by nuclei and are less likely to be captured. A π^+ decays (Fig. 50.12) in a very short time to a positive *muon* (μ^+) and a *mu neutrino* (ν_μ), while a negative pion decays according to the reaction

$$\pi^- \longrightarrow \mu^- + \bar{\nu}_\mu$$

where the bar above the symbol signifies an antiparticle. The mu neutrino is similar to, but distinguishable from, the neutrino of beta decay.

Both the μ^+ and μ^- particles have a mass of 207 electron masses. They are short-lived, decaying through the weak interaction by the reactions

$$\mu^- \longrightarrow \beta^- + \bar{\nu} + \nu_\mu \quad \text{and} \quad \mu^+ \longrightarrow \beta^+ + \nu + \bar{\nu}_\mu$$

Pions are intimately linked to the attractive force between nucleons, each of which is sometimes imagined to be surrounded by a virtual pion cloud. In terms of this picture, the binding forces in nuclei are attributed to the exchange of pions much as we attribute covalent binding between hydrogen atoms to electron-exchange forces.

50.9 PARTICLES AND ANTIPARTICLES

In 1955 a particle identical with the proton except for a negative charge was discovered by Segré, Chamberlain, and their collaborators at the University of California at Berkeley. The negative proton, or *antiproton*, was created by bombarding protons in a target with 6-GeV protons, thereby inducing the reaction

$$p + p + \text{energy} \longrightarrow p + p + \bar{p} + p$$

where $\bar{p}$ indicates an antiproton. The energy of the bombarding proton is converted into a proton-antiproton pair plus the kinetic energy of the four residual particles. As soon as the antiproton is slowed down, it is annihilated by a proton; in a typical annihilation reaction the rest mass of the annihilating pair may appear as five pions and their kinetic energy (Fig. 50.15):

$$p + \bar{p} \longrightarrow \pi^+ + \pi^- + \pi^+ + \pi^- + \pi^0$$

One year later the antineutron $\bar{n}$ was discovered at the same laboratory. Since the neutron bears no charge, the antineutron is also neutral. It is quickly annihilated, either by a proton or a neutron, usually with the production of several pions. If an antineutron is not annihilated by a nucleon, it decays by the reaction $\bar{n} \longrightarrow \bar{p} + \beta^+ + \nu$.

We note a remarkable similarity here to the behavior of the electron and positron, ordinarily created together and disappearing by the annihilation reaction $\beta^- + \beta^+ \longrightarrow 2\gamma$. We therefore identify the positron as the *antielectron*. Of course, any antiparticle has a charge opposite to that of the corresponding particle, or else charge would not be conserved in annihilations. Many elementary particles decay in such a short time that they rarely find an antiparticle to annihilate. For example, the π^+ is antiparticle to the π^-, and vice versa. However, as we saw in Sec. 50.8 the most common fates for a π^- are (1) capture by a nucleus and (2) decay to a μ^-. If we know a decay mode of any particle, we can often write a corresponding possible decay mode for its antiparticle by changing each particle in the reaction to its antiparticle (just as was done for the muon decays in Sec. 50.8).

There is every reason to believe that a positron and a negative proton could form an atom of *antihydrogen*, which would have a spectrum similar to that of ordinary hydrogen. Indeed, from a collection of antiprotons, antineutrons, and positrons a world of *antimatter* could be constructed which would be indistinguishable from our world so long as everything were made of antiparticles. However, if some of this *contraterrene* matter

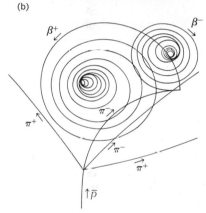

(a)

(b)

FIGURE 50.15

Bubble-chamber photograph showing the path of an antiproton entering from below. Near the bottom of the picture the antiproton annihilated a proton, forming two π^+, two π^-, and a π^0, which leaves no track. In the upper quadrant of the picture a π^- interacted with a proton. Among the resulting products were a positron, which formed the counterclockwise spiral at top center, and an electron, which produced the smaller clockwise spiral. (*b*) Explanatory sketch. (*Lawrence Radiation Laboratory.*)

were to come in contact with terrene matter, particle-antiparticle annihilation would occur, and almost the entire mass of the annihilating particles would be converted to energy of other forms. Contrast this to a fission reaction in which only about 0.1 percent of the mass is transformed.

50.10 PARTICLES OF MODERN PHYSICS

When a beam of protons with very high energies is incident on a target, many kinds of particles may be produced (Fig. 50.16). Other types of experiments also produce new kinds of particles; e.g., two very high energy beams of protons may collide head-on or a beam of electrons with energies of several GeV may collide head-on with a beam of positrons of the same energy. Many of the more important particles of modern physics are listed together with their antiparticles in Table 50.2. In addition to the particles listed the existence of many others has been established and there is less conclusive evidence for many more.

Most, but not all, of the particles observed in nature can be classified as either *leptons* or *hadrons*. Leptons (light particles) do not participate in the strong reaction (Sec. 49.4), while the hadrons do. The known leptons consist of just four particles plus their antiparticles; they are the electron, the negative muon, the electron neutrino, and the muon neutrino. The μ^- and μ^+ are known to decay (Sec. 50.8), but the other leptons are considered elementary in that (so far, at least) they cannot be broken into smaller entities. They have no measurable size and have shown no signs of internal structure. All leptons have a spin quantum number of $\frac{1}{2}$ (in units of $h/2\pi$). When we assign a lepton number of $+1$ to the electron and the μ^-, ν, and ν_μ and a lepton number of -1 to the positron, μ^+, $\bar{\nu}$, and $\bar{\nu}_\mu$, we find that there is a conservation of lepton number in all observed reactions. When a neutron decays according to $n \longrightarrow p + \beta^- + \bar{\nu}$, we

FIGURE 50.16
A 300-GeV proton entering a 30-in. hydrogen bubble chamber from the right produces 26 charged particles, each of which leaves a visible track. Uncharged particles are also produced. (*Fermi National Accelerator Laboratory.*)

TABLE 50.2
Selected Particles of Modern Physics

Name	Symbol and charge	Anti-particle	Rest mass $\times c^2$, MeV	Spin in $h/2\pi$	Quark content	Mean life, s	Dominant decay modes
Photon	γ		0	1			
Leptons:							
Electron neutrino	ν	$\bar{\nu}$	0	$\frac{1}{2}$			Stable
Muon neutrino	ν_μ	$\bar{\nu}_\mu$	0	$\frac{1}{2}$			Stable
Electron	β^- or e^-	β^+ or e^+	0.5110	$\frac{1}{2}$			Stable
Muon	μ^-	μ^+	105.66	$\frac{1}{2}$		2.2×10^{-6}	$\beta^- + \nu_\mu + \nu$
Hadrons:							
Mesons:							
Pion	π^0		134.96	0	$u\bar{u} + d\bar{d}$	8.28×10^{-17}	Two photons ($\gamma\gamma$)
	π^+	π^-	139.57	0	$u\bar{d}$	2.60×10^{-8}	$\mu^+ + \nu_\mu$
Kaon	K^+	K^-	493.7	0	$u\bar{s}$		$\mu^+ + \nu_\mu; \pi^+ + \pi^0$
	K_S^0		497.7	0	$d\bar{s}$	8.93×10^{-11}	$\pi^+ + \pi^-; 2\pi^0$
	K_L^0		497.7	0		5.18×10^{-8}	$3\pi^0; \pi\beta\nu; \pi^+\pi^-\pi^0$
Eta	$\eta^{(0)}$		548.6	0			$\gamma\gamma; 3\pi^0; \pi^+\pi^-\pi^0$
Phi	$\phi^{(0)}$		1,020	1	$s\bar{s}$	10^{-22}	$K^+K^-; K_L K_S$
Psi or J	ψ or J		3,098	1	$c\bar{c}$	10^{-20}	Hadrons
Baryons:							
Proton	p^+	$\bar{p}$ or p^-	938.27	$\frac{1}{2}$	uud		Stable
Neutron	$n^{(0)}$	$\bar{n}$	939.57	$\frac{1}{2}$	udd	918	
Lambda†	$\Lambda^{(0)}$	$\bar{\Lambda}$	1,115.6	$\frac{1}{2}$	uds	2.58×10^{-10}	$p + \beta^- + \bar{\nu}$
Sigma†	Σ^+		1,189.4	$\frac{1}{2}$	uss	8.0×10^{-11}	$p + \pi^0; n + \pi^+$
	Σ^0		1,192.5	$\frac{1}{2}$	uds	$(<10^{-14})$	$\Lambda + \gamma$
	Σ^-		1,193.4	$\frac{1}{2}$	dds	1.48×10^{-10}	$n + \pi^-$
Xi†	Ξ^0		1,314.9	$\frac{1}{2}$	uss	3.0×10^{-10}	$\Lambda + \pi^0$
	Ξ^-		1,321.3	$\frac{1}{2}$	dss	1.65×10^{-10}	$\Lambda + \pi^-$
Omega	Ω^-		1,672.2	$\frac{3}{2}$	sss	1.3×10^{-10}	$\Xi^0\pi^-; \Xi^-\pi^0; \Lambda K^-$

†Hyperon.

note that one lepton (β^-) and one antilepton ($\bar{\nu}$) appear, for a net of zero leptons. In the μ^- decay we find one electron, one μ neutrino, and one antineutrino, one lepton before the decay and a net one lepton after the decay. The principle of *conservation of leptons* is a key conservation law in modern particle physics in the sense that any time a lepton appears or disappears there is the corresponding creation or annihilation of an antilepton.

The most familiar of the hadrons are the proton and the neutron, but many other kinds of hadrons have been discovered. They are complex particles, and there is strong evidence that they have internal structure. All have masses greater than the heaviest lepton, the muon. Originally the hadrons were divided into *mesons* and *baryons*, with baryon designating *heavy* particles and meson an intermediate-mass particle. However, this distinction is not truly appropriate because some mesons have masses greater than most baryons; a true distinguishing characteristic is that mesons all have zero or integer spin in units of $h/2\pi$, while baryons have half-integer ($\frac{1}{2}$, $\frac{3}{2}$, . . .) spin. Of all the hadrons only the proton appears to be stable; all the others (except bound neutrons in stable nuclei) undergo one or more types of decay.

If to each ordinary baryon we assign the baryon number +1 and to the corresponding antibaryon −1, we find again that in all observed reactions the net baryon number does not change. Thus it appears that *the conservation of baryons is also a valid law of particle physics.* On the other hand, *there is no conservation of mesons,* just as there is no conservation of photons.

50.11 THE QUARK HYPOTHESIS

The task of organizing the observed particles in physics in some understandable way is herculean, but it can also be highly rewarding. For one example, the logical development of a hypothesis may suggest that some specific undiscovered particle should exist. Several times in the past particles have been detected in experiments after their existence was predicted by theoretical considerations. Particle physics is currently one of the vigorously active areas of physics, both in theory and in experiment. It is perhaps fitting that we conclude this book, which has been filled with well-established facts and ideas, with a look at some current speculative ideas about particles. Here we are dealing with physics emerging from a sea of ignorance. As facts about particle interactions and decays are established by experiments, physicists are advancing hypotheses compatible with the observations and then exploring the predictions of these hypotheses. We must remember that there are probably many more particles to be found and that new data may force changes in our way of thinking about them; thus we are looking now at physics in the process of development.

In 1963 Gell-Mann, and independently Zweig, proposed that all hadrons are composed of three types of elementary particles called *quarks*† (and their antiparticles); later in 1974 a fourth quark was added. The four types of quarks are named *up* (*u*), *down* (*d*), *strange* (*s*), and *charmed* (*c*); their antiparticles are $\bar{u}, \bar{d}, \bar{s}$, and $\bar{c}$. All have spin quantum number $\frac{1}{2}$. The *u* and *c* quarks have charge $\frac{2}{3}e$, while the *d* and *s* quarks have charge $-\frac{1}{3}e$ (Table 50.3); antiquarks have the opposite charges. In addition to these properties two new characteristics named strangeness and charm were introduced; each is described by a quantum number which can take on only zero or integer (±) values.

Table 50.2 contains the presumed quark composition of some of the hadrons listed. Starting with quarks as building blocks, particle physicists assume that all mesons are composed of one quark and one antiquark (or

---

†The name comes from a line in James Joyce's *Finnegans Wake*, "three quarks for Mr. Marks."

TABLE 50.3
Properties of Quarks‡

Name	Symbol	Charge, atomic units	Spin in $h/2\pi$	Strangeness	Charm	Mass $\times c^2$, MeV
Up	*u*	$+\frac{2}{3}$	$\frac{1}{2}$	0	0	336
Down	*d*	$-\frac{1}{3}$	$\frac{1}{2}$	0	0	338
Strange	*s*	$-\frac{1}{3}$	$\frac{1}{2}$	−1	0	540
Charmed	*c*	$+\frac{2}{3}$	$\frac{1}{2}$	0	1	1,500

‡For each quark there is a corresponding antiquark. In some theories each quark and antiquark comes in three different "colors." Additional quarks have been postulated.

at least equal numbers of each, since to get 0 or integer spin, we need to assemble an even number of particles with half-integer spin). A π^+ meson is composed of a u quark and a d antiquark; clearly the charges add up to $+e$ and the spin to 0 or 1 (0 is observed). The K^+ meson, composed of a u quark and an s antiquark, has charge $+e$, spin 0, and strangeness 1.

Baryons are presumably composed of three quarks. For example, the proton is assumed to have the composition uud and the antiproton $\bar{u}\bar{u}\bar{d}$. With this scheme it has been possible to make assignments for all particles discovered before 1976 in terms of the four quarks and their antiparticles (although some of the assignments are complicated). The quark model offers an explanation of why the observed particles exist and many of their measured properties. However, a word of caution is needed. Before 1974 it appeared that three quarks (u, d, and s) were enough to construct all known hadrons; in 1974 it became necessary to add the charmed quark. In 1976 Lederman and his colleagues found evidence of a particle named *upsilon* which has a mass corresponding to 9.5 GeV, 9 times that of the proton. The four-quark hypothesis does not appear to cope with the upsilon, and a pair of new quarks has been proposed. It is ironic that the quark hypothesis was introduced to simplify the observed spectrum of hadrons, which was growing with no limit in sight. Now it seems that it is the quarks which are proliferating.

Perhaps an even more serious objection to the quark hypothesis has been the fact that, in spite of intensive searches for an isolated quark, the existence of individual quarks has not yet (1978) been established. This has led some physicists to the view that quarks may be "mathematical entities" only. This makes the impressive successes of the quark theory most mysterious.

Clearly there is much more physics to be done. How many particles remain undiscovered? Will important uses be found for the new particles? Are there regions in the universe where stars and planets are composed of antiparticles? All these questions have been speculated upon; definitive answers await further research.

QUESTIONS

1. How can nuclear events be detected? Describe as many methods as you can.
2. How are charged particles accelerated to high energies to produce nuclear reactions when they collide with a nucleus? (There are many more kinds of particle accelerators than are discussed in this book.)
3. Why are low-energy neutrons much more effective in producing nuclear reactions than low-energy protons or alpha particles?
4. Why is it more difficult to produce atomic disintegration of heavy nuclei than of light nuclei by proton bombardment from a 20-MeV cyclotron?
5. How does a cyclotron give protons millions of electronvolts of energy even though the highest potential difference involved is only a few kilovolts?
6. Relativity theory imposes a practical limit to the maximum energy which a proton can acquire from a simple cyclotron. Why does this limit arise? How is this limitation overcome in the synchrocyclotron and the synchrotron?
7. How are tracks produced in a Wilson cloud chamber? In a bubble chamber?
8. In Fig. 50.14 why are the electron and positron tracks spirals?
9. Neutrons incident on the upper atmosphere collide with and are captured by ^{14}N nuclei with the subsequent emission in many cases of protons or alpha particles. What nuclei remain in each case?
10. What are cosmic rays? What is the distinction between primary and secondary cosmic rays?
11. What is a meson? A lepton? A baryon? A hadron?

12. Roughly in what direction do you suppose the π^0 meson produced in the p-$\bar{p}$ annihilation of Fig. 50.15 went? Why?

13. When a π^+ meson at rest decays to a μ^+ lepton, the muon always has the same energy, but when a μ^+ lepton decays to a positron, the positron may have any of a broad range of energies. Why?

14. In what sense are leptons and hadrons conserved? Are mesons conserved? In addition to the conservation laws, in what other respects do leptons and hadrons share common properties not associated with mesons and photons?

PROBLEMS

1. One way of obtaining tritium ^3_1H is to bombard ^6_3Li with neutrons. Find the Q (energy released) in the ^6_3Li (n, α) ^3_1H reaction (see Table 49.1). *Ans.* 4.79 MeV

2. Find the Q (energy released) in the ^9_4Be (α, n) $^{12}_6C$ reaction if the mass of ^9_4Be is 9.012186 u.

3. The first observed artificial transmutation was reported by Rutherford in 1910. He bombarded $^{14}_7N$ with alpha particles, and protons came out of the reaction. Find the Q (energy released) in the reaction if the mass of $^{17}_8O$ is 16.99913 u. (In this case Q is negative, meaning that energy must be supplied to the system.) *Ans.* −1.2 MeV

4. Find the Q (energy released) in the reaction $^{19}_9F$ (p, α) $^{16}_8O$ if the mass of $^{19}_9F$ is 18.99841 u.

5. From the masses of ^6Li and ^7Li compute the binding energy of a neutron captured by a ^6Li nucleus. What is the wavelength of the gamma ray emitted in the ^6Li (n, γ) ^7Li reaction? *Ans.* 7.26 MeV; 171 fm

6. Complete each of the following reactions:

^{27}Al (n,α)? $\qquad$ ^{31}P (p,α)? $\qquad$ ^3H (γ,n)?
^{44}Ca (p,n)? $\quad$ $^{58}Cu \longrightarrow$ $^{58}Ni + ? + ?$ $\quad$ ^{16}O (α,p)?
^7Li (γ,α)? $\qquad$ ^{109}Ag (n,γ)? $\qquad$ ^{24}Mg (d,α)?

7. Find the energy released in the decay of a pion to a muon. When a pion at rest decays, the muon receives about 4.1 MeV of kinetic energy. Find the momentum carried off by the neutrino. *Ans.* 33.9 MeV; 1.58×10^{-20} kg-m/s

8. Complete the following reactions:

$^{23}Na + ? \longrightarrow$ $^{24}Mg + {}^1n$
$^{35}Cl + ? \longrightarrow$ $^{32}S + {}^4He$
$^{10}B + ? \longrightarrow$ $^7Li + {}^4He$
$? + {}^2H \longrightarrow$ $^{24}Na + {}^1H$

9. Find the energy released in the decay of a positive muon to a positron (plus neutrino and antineutrino). *Ans.* 105.1 MeV

10. A π^0 meson at rest decays into two photons, which must have equal and opposite moments. Find the energy and wavelength of one of the photons.

11. A proton of rest mass 938 MeV (1.67×10^{-27} kg) is given a kinetic energy of 18.76 GeV in a proton synchrotron. Find the mass of the proton at this speed. What must the radius of the orbit be if the vertical magnetic field is 1.2 T? *Ans.* 3.51×10^{-26} kg; 54.6 m

12. A neutron of mass m and initial speed v_i collides head on with a nucleus of mass M at rest. Show that if the collision is elastic, the energy W transferred to the nucleus is $2Mm^2v_i{}^2/(M + m)^2$. Find the fraction of the energy of the neutron transferred to (a) a proton of mass m, (b) an alpha particle of mass $4m$, and (c) a uranium atom of mass $238m$.

13. A 4.19-MeV alpha particle (mass 4 u) collides elastically with an oxygen atom (mass 16 u) at rest. Calculate the maximum kinetic energy which can be transferred to the oxygen atom. (Use nonrelativistic equations.) *Ans.* 2.68 MeV

14. Show that in a cyclotron the maximum kinetic energy which can be given an ion is $q^2B^2r^2/2m$, where B is the magnetic intensity and r the maximum allowable radius.

15. An alpha particle with a speed of 5×10^7 m/s is moving perpendicular to a uniform magnetic field of 1.25 T in a cyclotron. Find the force on the alpha particle and the radius of the circular path which it follows. What frequency must be applied to the dees to accelerate this alpha particle? If the radius of the cyclotron is 1.5 m, what is the maximum energy an alpha particle can be given in this cyclotron? *Ans.* 2.00×10^{-11} N; 0.829 m; 9.6 MHz; 169 MeV

16. A synchrocyclotron has a magnetic field of 1.25 T. Find the appropriate frequency for accelerating protons when the speed of the protons is (a) $0.0100c$ and (b) $0.800c$.

17. Calculate the maximum energy of protons obtainable from a cyclotron with a dee diameter of 1.2 m and a magnetic induction of 1.1 T. At what frequency must the cyclotron be operated? If the average energy gain per dee passage is 30 keV, how many revolutions do the protons make? *Ans.* 20.9 MeV; 16.8 MHz, 348 r

18. Calculate the maximum energy which the cyclotron of Prob. 17 can give alpha particles (He^{2+} ions). At what frequency must the cyclotron be operated?

19. A cyclotron has pole faces with a diameter of 1.4 m and operates at a frequency of 15 MHz.

Find the time required for a proton to traverse a semicircle in one dee, the magnetic field required to accelerate protons, and the maximum kinetic energy which protons can receive under these conditions. *Ans.* 33.3 ns; 0.984 T; 22.7 MeV

20. A negative muon may be captured by a nucleus and occupy a "Bohr orbit" until it decays or interacts with a proton. Find the radius of the first Bohr orbit of a muon captured by gold and show that this orbit lies inside the nucleus.

21. A nucleus of charge Z may temporarily capture a pion, which may occupy a hydrogenlike level of very low energy. If a negative pion with a mass of 273 electron masses is captured by a chromium atom, find (*a*) the Bohr radius and binding energy for a pion in the ground ($n = 1$) state and (*b*) the energy of the photon emitted when the pion makes a transition from the $n = 2$ to the $n = 1$ state. *Ans.* (*a*) 8.1 fm; 2.14 MeV; (*b*) 1.6 MeV

22. The main ring accelerator at the Fermi National Laboratory (Fig. 50.9) is a synchrotron which receives 8-GeV protons from its injector and accelerates them to an energy of 500 GeV in a roughly circular doughnut-shaped vacuum tube with a radius of 1 km. Find the magnetic intensity B required at the orbit to keep the protons in a circular path of 1 km radius (*a*) when they are injected at 8 GeV energy and (*b*) when they reach

the final 500 GeV energy. How long does it take a 500-GeV proton to cover the 6.5-km circumference of the vacuum tube (which has six straight sections between arcs of the circle with 1 km radius). The protons gain 125 GeV/s during the acceleration phase. How much energy do they receive per cycle from the accelerating cavities located in one of the straight sections? At approximately what frequency must the accelerating cavities operate?

23. By using superconducting magnets to provide a higher magnetic field, Fermilab plans to raise the peak energy to 1 TeV (one teraelectronvolt = 10^{12} eV). (*a*) What value of B is required to keep 1-TeV protons in a circular orbit of 1 km radius? (*b*) At some places in the accelerator conductors must carry 4,600 A in fields which vary up to 4.5 T. Calculate the force required per meter to hold the conductor at rest where it is carrying 4,600 A perpendicular to the field of 4.5 T. *Ans.* (*a*) 3.34 T; (*b*) 20,700 N/m

APPENDIX

MATHEMATICAL FORMULAS

A.1 TRIGONOMETRIC RELATIONS

Consider angle θ in the right triangle ABC (Fig. A.1) which has sides of lengths a, b, and c. By definition, the *sine of* θ is the *ratio of the side opposite* θ *to the hypotenuse*, the *cosine of* θ is the *ratio of the side adjacent to* θ *to the hypotenuse*, and the *tangent of* θ is the *ratio of the side opposite* θ *to the side adjacent to* θ. The values of these functions are listed in Table A.1. The cotangent of θ is the reciprocal of tan θ.

$$\sin \theta = \frac{a}{c} \qquad \cos \theta = \frac{b}{c} \qquad \tan \theta = \frac{a}{b} \qquad \cot \theta = \frac{b}{a}$$

Angle ϕ is the complement of θ. Clearly,

$$\sin \phi = \frac{b}{c} = \cos \theta = \cos (90° - \phi)$$

and

$$\cos \phi = \frac{a}{c} = \sin \theta = \sin (90° - \phi)$$

By the pythagorean theorem, $c^2 = b^2 + a^2$. From this it follows immediately that $\sin^2 \theta + \cos^2 \theta = 1$. Other trigonometric identities which are useful are

$$\sin (\theta \pm \phi) = \sin \theta \cos \phi \pm \cos \theta \sin \phi$$
$$\cos (\theta \pm \phi) = \cos \theta \cos \phi \mp \sin \theta \sin \phi$$

In any plane triangle having angles α, β, and γ with opposite sides a, b, and c, respectively (Fig. A.2),

$$\frac{\sin \alpha}{a} = \frac{\sin \beta}{b} = \frac{\sin \gamma}{c} \qquad \text{law of sines}$$

and

$$\cos \alpha = \frac{b^2 + c^2 - a^2}{2bc} \qquad \text{law of cosines}$$

If an angle exceeds 90°,

$$\sin \beta = \sin (180° - \beta) \qquad \text{and} \qquad \cos \beta = -\cos (180° - \beta)$$

If an angle θ is expressed in radians,

$$\sin \theta = \theta - \frac{\theta^3}{3!} + \frac{\theta^5}{5!} - \frac{\theta^7}{7!} + \cdots$$

$$\cos \theta = 1 - \frac{\theta^2}{2!} + \frac{\theta^4}{4!} - \frac{\theta^6}{6!} + \cdots$$

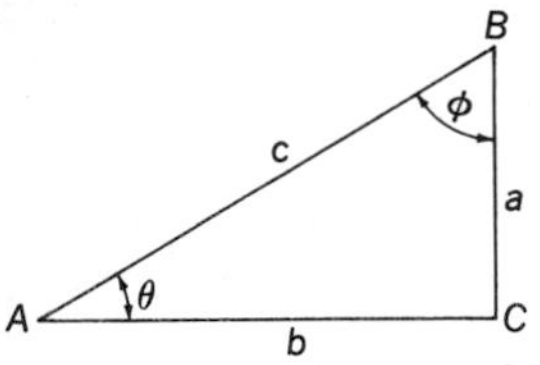

FIGURE A.1

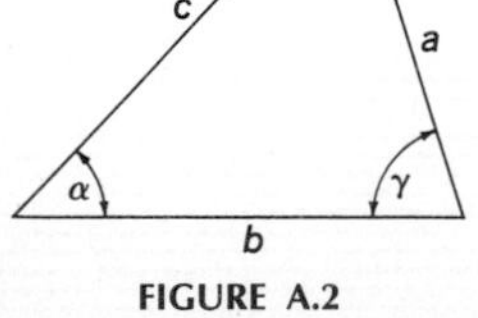

FIGURE A.2

A.2 QUADRATIC EQUATIONS

Any quadratic equation can be put in the form $ax^2 + bx + c = 0$, for which the solutions are

$$x = \frac{-b \pm \sqrt{b^2 - 4ac}}{2a}$$

A.3 MENSURATION FORMULAS

The circumference of a circle of radius r is $2\pi r$ ($\pi = 3.1416$).
The area of a circle of radius r is πr^2.
The area of an ellipse with semiaxes a and b is πab.
The surface area of a sphere of radius r is $4\pi r^2$, and its volume is $\frac{4}{3}\pi r^3$.

Angle	Sine	Cosine	Tangent	Angle	Sine	Cosine	Tangent
1°	.01745	.9998	.01746	46°	.7193	.6947	1.0355
2°	.03490	.9994	.03492	47°	.7314	.6820	1.0724
3°	.05234	.9986	.05241	48°	.7431	.6691	1.1106
4°	.06976	.9976	.06993	49°	.7547	.6561	1.1504
5°	.08716	.9962	.08749	50°	.7660	.6428	1.1918
6°	.1045	.9945	.1051	51°	.7771	.6293	1.2349
7°	.1219	.9925	.1228	52°	.7880	.6157	1.2799
8°	.1392	.9903	.1405	53°	.7986	.6018	1.3270
9°	.1564	.9877	.1584	54°	.8090	.5878	1.3764
10°	.1736	.9848	.1763	55°	.8192	.5736	1.4281
11°	.1908	.9816	.1944	56°	.8290	.5592	1.4826
12°	.2079	.9781	.2126	57°	.8387	.5446	1.5399
13°	.2250	.9744	.2309	58°	.8480	.5299	1.6003
14°	.2419	.9703	.2493	59°	.8572	.5150	1.6643
15°	.2588	.9659	.2679	60°	.8660	.5000	1.7321
16°	.2756	.9613	.2867	61°	.8746	.4848	1.8040
17°	.2924	.9563	.3057	62°	.8829	.4695	1.8807
18°	.3090	.9511	.3249	63°	.8910	.4540	1.9626
19°	.3256	.9455	.3443	64°	.8988	.4384	2.0503
20°	.3420	.9397	.3640	65°	.9063	.4226	2.1445
21°	.3584	.9336	.3839	66°	.9135	.4067	2.2460
22°	.3746	.9272	.4040	67°	.9205	.3907	2.3559
23°	.3907	.9205	.4245	68°	.9272	.3746	2.4751
24°	.4067	.9135	.4452	69°	.9336	.3584	2.6051
25°	.4226	.9063	.4663	70°	.9397	.3420	2.7475
26°	.4384	.8988	.4877	71°	.9455	.3256	2.9042
27°	.4540	.8910	.5095	72°	.9511	.3090	3.0777
28°	.4695	.8829	.5317	73°	.9563	.2924	3.2709
29°	.4848	.8746	.5543	74°	.9613	.2756	3.4874
30°	.5000	.8660	.5774	75°	.9659	.2588	3.7321
31°	.5150	.8572	.6009	76°	.9703	.2419	4.0108
32°	.5299	.8480	.6249	77°	.9744	.2250	4.3315
33°	.5446	.8387	.6494	78°	.9781	.2079	4.7046
34°	.5592	.8290	.6745	79°	.9816	.1908	5.1446
35°	.5736	.8192	.7002	80°	.9848	.1736	5.6713
36°	.5878	.8090	.7265	81°	.9877	.1564	6.3138
37°	.6018	.7986	.7536	82°	.9903	.1392	7.1154
38°	.6157	.7880	.7813	83°	.9925	.1219	8.1443
39°	.6293	.7771	.8098	84°	.9945	.1045	9.5144
40°	.6428	.7660	.8391	85°	.9962	.08716	11.4301
41°	.6561	.7547	.8693	86°	.9976	.06976	14.3007
42°	.6691	.7431	.9004	87°	.9986	.05234	19.0811
43°	.6820	.7314	.9325	88°	.9994	.03490	28.6363
44°	.6947	.7193	.9657	89°	.9998	.01745	57.2900
45°	.7071	.7071	1.0000	90°	1.0000	.00000	∞

TABLES OF DATA

TABLE B.1
Heat Constants of Solids

Substance	Melting point, °C	Coefficient of linear expansion per Celsius degree	Specific heat, kcal/(kg)(C°)	Heat of fusion	
				kcal/kg	Btu/lb
Aluminum	660	0.0000255	0.22	76.8	140
Bismuth	271	0.000013	0.030	12.6	22.7
Brass		0.0000193	0.090		
Copper	1084	0.0000167	0.093	43	77
Glass		0.0000083	0.20		
Gold	1064	0.0000142	0.031		
Ice	0	0.000051	0.50	79.8	144
Iron	1535	0.000012	0.11	30	54
Lead	327	0.000029	0.031	5.4	9.7
Mercury	−38.9		0.033	2.8	5.4
Nickel	1452	0.000013	0.106	4.6	8.3
Platinum	1772	0.000009	0.032	27	48.6
Silver	962	0.000019	0.057	22	39
Steel		0.000012	0.11		
Tungsten	3410	0.0000045	0.032		
Zinc	420	0.000032	0.092	28.1	50.6

TABLE B.2
Heat Constants of Liquids

Substance	Boiling point, °C	Volume expansion per Celsius degree	Specific heat, kcal/(kg)(C°)	Heat of vaporization	
				kcal/kg	Btu/lb
Alcohol (ethyl)	78.1	0.0011	0.55	205	369
Ammonia	−34			294	529
Aniline	184		0.514	110	198
Benzene	80.3	0.00124	0.34	94.4	170
Chloroform	61	0.00126	0.232	58	106
Ether (ethyl)	34.5	0.00163	0.56	88.4	159
Gasoline	70–90	0.0012		71–81	128–146
Glycerin	290	0.00053	0.58		
Mercury	358	0.000182	0.0332	68	122
Turpentine	159	0.00094	0.42	70	126
Water	100	0.00030	1.00	540	970

Note: Densities are tabulated in Sec. 12.8.

TABLE B.3
Boiling Point of Water [Boiling points of water at pressures near standard atmospheric pressure; the pressures are given in millimeters of mercury at 0°C and in kilonewtons per square meter (kilopascals).]

Pressure		Tempera-	Pressure		Tempera-	Pressure		Tempera-
kPa	mmHg	ture, °C	kPa	mmHg	ture, °C	kPa	mmHg	ture, °C
97.7	733	98.99	99.8	749	99.59	102.0	765	100.18
98.0	735	99.07	100.1	751	99.67	102.2	767	100.26
98.2	737	99.14	100.4	753	99.74	102.5	769	100.33
98.5	739	99.22	100.6	755	99.82	102.8	771	100.40
98.8	741	99.29	100.9	757	99.89	103.0	773	100.47
99.0	743	99.37	101.2	759	99.96	103.3	775	100.55
99.3	745	99.44	101.4	761	100.04	103.6	777	100.62
99.6	747	99.52	101.7	763	100.11	103.8	779	100.69

TABLE B.4
Properties of Saturated Water Vapor

Tem- pera- ture, °C	Pressure		Volume		Heat units per unit mass, kcal/kg		
	Pa	lb/in.^{2}	m^3/kg	ft^3/lb	Of water	Latent heat	Total heat of vapor
0	6.2×10^2	0.089	204.970	3,283.00	0	594.7	594.7
10	1.23×10^3	0.178	106.620	1,707.60	10	589.4	599.4
20	2.31×10^3	0.336	58.150	931.48	20	584.1	604.1
30	4.20×10^3	0.61	33.132	530.72	30	578.8	608.8
40	7.32×10^3	1.06	19.670	314.77	40.1	573.4	613.5
50	1.23×10^4	1.78	12.091	193.68	50.1	567.9	618
60	1.98×10^4	2.88	7.695	123.26	60.1	562.4	622.6
70	3.11×10^4	4.51	5.050	80.89	70.2	556.8	627
80	4.72×10^4	6.86	3.4085	54.60	80.3	551	631.5
90	7.00×10^4	10.16	2.3592	37.79	90.4	545.2	635.6
100	1.013×10^5	14.70	1.6702	26.754	100.5	539.6	639.7
110	1.433×10^5	20.79	1.2073	19.339	110.7	532.9	643.6
120	1.986×10^5	28.83	0.8894	14.247	120.9	526.6	647.4
130	2.705×10^5	39.26	0.6664	10.675	131.1	520	651
140	3.621×10^5	52.56	0.5071	8.123	141.3	513.2	654.5
150	4.771×10^5	69.24	0.3917	6.274	151.6	506.2	657.8
160	6.197×10^5	89.93	0.3065	4.91	161.9	498.9	660.8
170	7.942×10^5	115.27	0.2429	3.891	172.2	491.4	663.7
180	1.005×10^6	145.90	0.1945	3.116	182.6	483.7	666.3
190	1.258×10^6	182.56	0.1575	2.523	193.1	475.7	668.8
200	1.557×10^6	226.00	0.1288	2.063	203.6	467.5	671.1
210	1.910×10^6	277.20	0.1063	1,703	214.1	459.1	673.2

DERIVATIONS

C.1 ENERGY STORED IN AN INDUCTIVE CIRCUIT

The energy stored in the magnetic field associated with a circuit for which the coefficient of self-induction is L is equal to the work required to establish the current in the circuit, or the energy which will be released when the current is eliminated. The back emf due to the self-inductance in the circuit is

$$e = -L\frac{di}{dt}$$

The work done against this back emf in the time dt is

$$ei\,dt = Li\frac{di}{dt}\,dt = Li\,di$$

The whole work in building up a current I in a time T is

$$W = \int_0^T ei\,dt = \int_0^I Li\,di = \tfrac{1}{2}LI^2$$

C.2 EFFECTIVE VALUE OF AN ALTERNATING CURRENT

By definition the effective value of an alternating current is the steady current which dissipates heat in a resistor at the same average rate as the alternating current. Let the alternating current be

$$i = I_{\text{max}} \sin 2\pi\nu t$$

The power developed instantaneously in a resistor R is

$$i^2R = I_{\text{max}}^2 R \sin^2 2\pi\nu t$$

The heat developed during one cycle is

$$I_{\text{eff}}^2 RT = \int_0^T I_{\text{max}}^2 R \sin^2 2\pi\nu t\,dt$$

where T is the period ($T = 1/\nu$). Therefore,

$$I_{\text{eff}}^2 T = I_{\text{max}}^2 \int_0^T \sin^2 \frac{2\pi}{T} t\,dt = \frac{I_{\text{max}}^2 T}{2}$$

from which

$$I_{\text{eff}} = \frac{I_{\text{max}}}{\sqrt{2}} = 0.707 I_{\text{max}}$$

C.3 AC SERIES CIRCUITS

Application of Kirchhoff's second law to the circuit of Fig. 40.3 yields

$$v = \mathcal{E}_{\text{max}} \sin 2\pi\nu t = L\frac{di}{dt}$$

from which

$$i = \int \frac{\mathcal{E}_{\text{max}} \sin 2\pi\nu t\,dt}{L} = -\frac{\mathcal{E}_{\text{max}} \cos 2\pi\nu t}{2\pi\nu L}$$

and

$$I_{max} = \frac{\mathcal{E}_{max}}{2\pi\nu L}$$

Similarly, when an alternating potential difference $\mathcal{E}_{max} \sin 2\pi\nu t$ is applied across a capacitor,

$$\mathcal{E}_{max} \sin 2\pi\nu t = \frac{q}{C}$$

Since $i = dq/dt$,

$$\frac{i}{C} = \frac{d(\mathcal{E}_{max} \sin 2\pi\nu t)}{dt} = 2\pi\nu\mathcal{E}_{max} \cos 2\pi\nu t$$

Therefore,

$$i = \frac{\mathcal{E}_{max} \cos 2\pi\nu t}{1/2\pi\nu C} \qquad \text{and} \qquad I_{max} = \frac{\mathcal{E}_{max}}{1/2\pi\nu C}$$

For an ac series circuit containing a source of emf $e = \mathcal{E}_{max} \sin 2\pi\nu t$ and resistance, inductance, and capacitance (Fig. 40.7), application of Kirchhoff's second law at any instant leads to

$$e = \mathcal{E}_{max} \sin 2\pi\nu t = Ri + L\frac{di}{dt} + \frac{q}{C}$$

If we differentiate with respect to time,

$$2\pi\nu\mathcal{E}_{max} \cos 2\pi\nu t = R\frac{di}{dt} + L\frac{d^2i}{dt^2} + \frac{i}{C}$$

The solution of this differential equation is

$$i = \frac{\mathcal{E}_{max} \sin(2\pi\nu t - \theta)}{\sqrt{R^2 + [2\pi\nu L - (1/2\pi\nu C)]^2}}$$

where

$$\tan \theta = \frac{2\pi\nu L - 1/2\pi\nu C}{R}$$

and

$$I = \frac{\mathcal{E}}{\sqrt{R^2 + [2\pi\nu L - (1/2\pi\nu C)]^2}}$$

If either the inductor or the capacitor is removed from the circuit and the circuit is completed once more, the appropriate term vanishes from the equations above. This leads to Eq. (40.4) if the capacitor is removed and to Eq. (40.6) if there is no inductance.

C.4 POWER IN AN AC CIRCUIT

The instantaneous power supplied to an ac circuit is given by $p = ei$ [Eq. (40.9)]. The average power over a complete cycle is

$$P = \frac{1}{T}\int_0^T p \, dt = \frac{1}{T}\int_0^T ei \, dt$$

$$= \frac{1}{T}\int_0^T (\mathcal{E}_{max} \sin 2\pi\nu t) I_{max} \sin(2\pi\nu t - \theta) \, dt$$

where T is the period ($T = 1/\nu$). Since

$$\sin(2\pi\nu t - \theta) = \sin 2\pi\nu t \cos \theta - \cos 2\pi\nu t \sin \theta$$

P becomes

$$P = \frac{\mathcal{E}_{max}I_{max}}{T} \int_0^T \left(\sin^2 \frac{2\pi t}{T} \cos \theta + \sin \frac{2\pi t}{T} \cos \frac{2\pi t}{T} \sin \theta \right) dt$$

$$= \frac{\mathcal{E}_{max}I_{max} \cos \theta}{2} = \mathcal{E}I \cos \theta$$

which is Eq. (40.10).

INDEX

PERIODIC TABLE OF THE ELEMENTS

(Based on carbon 12 as 12.00000)

s

$d(l = 2)$

$p(l = 1)$

$s(l = 0)$

Inert gases

n		I_A	II_A	III_B	IV_B	V_B	VI_B	VII_B		VIII		I_B	II_B	III_A	IV_A	V_A	VI_A	VII_A	2 He 4.0026
1		1 H 1.0079	2 He 4.0026																2 He 4.0026
2		3 Li 6.94	4 Be 9.0122											5 B 10.81	6 C 12.011	7 N 14.0067	8 O 15.9994	9 F 18.9984	10 Ne 20.18
3		11 Na 22.9898	12 Mg 24.305											13 Al 26.9815	14 Si 28.086	15 P 30.9738	16 S 32.06	17 Cl 35.453	18 Ar 39.948
4		19 K 39.098	20 Ca 40.08	21 Sc 44.956	22 Ti 47.90	23 V 50.941	24 Cr 51.996	25 Mn 54.9380	26 Fe 55.847	27 Co 58.9332	28 Ni 58.71	29 Cu 63.546	30 Zn 65.38	31 Ga 69.72	32 Ge 72.59	33 As 74.9216	34 Se 78.96	35 Br 79.904	36 Kr 83.80
5		37 Rb 85.47	38 Sr 87.62	39 Y 88.9059	40 Zr 91.22	41 Nb 92.906	42 Mo 95.94	43 Tc 98.91	44 Ru 101.07	45 Rh 102.905	46 Pd 106.4	47 Ag 107.868	48 Cd 112.41	49 In 114.82	50 Sn 118.69	51 Sb 121.75	52 Te 127.60	53 I 126.9045	54 Xe 131.30
6		55 Cs 132.905	56 Ba 137.33	†	72 Hf 178.49	73 Ta 180.948	74 W 183.85	75 Re 186.2	76 Os 190.2	77 Ir 192.22	78 Pt 195.09	79 Au 196.967	80 Hg 200.59	81 Tl 204.37	82 Pb 207.2	83 Bi 208.9808	84 Po (209)	85 At (210)	86 Rn (222)
7		87 Fr (223)	88 Ra 226.025	‡															

Transition elements

$f(l = 3)$

d

	57 La 138.91	58 Ce 140.12	59 Pr 140.9077	60 Nd 144.24	61 Pm (145)	62 Sm 150.4	63 Eu 151.96	64 Gd 157.25	65 Tb 159.925	66 Dy 162.50	67 Ho 164.930	68 Er 167.26	69 Tm 168.934	70 Yb 173.04	71 Lu 174.97
† Lanthanide series (*rare earths*)	57 La 138.91	58 Ce 140.12	59 Pr 140.9077	60 Nd 144.24	61 Pm (145)	62 Sm 150.4	63 Eu 151.96	64 Gd 157.25	65 Tb 159.925	66 Dy 162.50	67 Ho 164.930	68 Er 167.26	69 Tm 168.934	70 Yb 173.04	71 Lu 174.97
‡ Actinide series	89 Ac (227)	90 Th 232.038	91 Pa 231.033	92 U 238.029	93 Np 237.048	94 Pu (244)	95 Am (243)	96 Cm (247)	97 Bk (247)	98 Cf (251)	99 Es (254)	100 Fm (257)	101 Md (258)	102 No (259)	103 Lw (260)